T0142353

Communications
in Computer and Information Science 1524

Editorial Board Members

Joaquim Filipe ⓘ
  Polytechnic Institute of Setúbal, Setúbal, Portugal

Ashish Ghosh
  Indian Statistical Institute, Kolkata, India

Raquel Oliveira Prates ⓘ
  Federal University of Minas Gerais (UFMG), Belo Horizonte, Brazil

Lizhu Zhou
  Tsinghua University, Beijing, China

More information about this series at https://link.springer.com/bookseries/7899

Michael Kamp · Irena Koprinska · Adrien Bibal ·
Tassadit Bouadi · Benoît Frénay · Luis Galárraga ·
José Oramas · Linara Adilova et al. (Eds.)

Machine Learning and Principles and Practice of Knowledge Discovery in Databases

International Workshops of ECML PKDD 2021
Virtual Event, September 13–17, 2021
Proceedings, Part I

 Springer

For the full list of editors *see next page*

ISSN 1865-0929 ISSN 1865-0937 (electronic)
Communications in Computer and Information Science
ISBN 978-3-030-93735-5 ISBN 978-3-030-93736-2 (eBook)
https://doi.org/10.1007/978-3-030-93736-2

© Springer Nature Switzerland AG 2021
This work is subject to copyright. All rights are reserved by the Publisher, whether the whole or part of the material is concerned, specifically the rights of translation, reprinting, reuse of illustrations, recitation, broadcasting, reproduction on microfilms or in any other physical way, and transmission or information storage and retrieval, electronic adaptation, computer software, or by similar or dissimilar methodology now known or hereafter developed.
The use of general descriptive names, registered names, trademarks, service marks, etc. in this publication does not imply, even in the absence of a specific statement, that such names are exempt from the relevant protective laws and regulations and therefore free for general use.
The publisher, the authors and the editors are safe to assume that the advice and information in this book are believed to be true and accurate at the date of publication. Neither the publisher nor the authors or the editors give a warranty, expressed or implied, with respect to the material contained herein or for any errors or omissions that may have been made. The publisher remains neutral with regard to jurisdictional claims in published maps and institutional affiliations.

This Springer imprint is published by the registered company Springer Nature Switzerland AG
The registered company address is: Gewerbestrasse 11, 6330 Cham, Switzerland

Editors

Michael Kamp (iD)
IKIM, Ruhr-University Bochum
Bochum, Germany

Monash University
Melbourne, Australia

Adrien Bibal (iD)
University of Namur
Namur, Belgium

Benoît Frénay (iD)
University of Namur
Namur, Belgium

José Oramas (iD)
University of Antwerp
Antwerp, Belgium

Yamuna Krishnamurthy (iD)
Royal Holloway University of London
Egham, UK

Christine Largeron
Université Jean Monnet
Saint-Etienne cedex 2, France

Tiphaine Viard
Telecom Paris
Paris, France

Massimiliano Ruocco
Norwegian Univesity of Science
and Technology
Trondheim, Norway

Claudio Gallicchio
University of Pisa
Pisa, Italy

Franz Pernkopf (iD)
Graz University of Technology
Graz, Austria

Holger Fröning (iD)
Heidelberg University
Heidelberg, Germany

Riccardo Guidotti (iD)
University of Pisa
Pisa, Italy

Irena Koprinska (iD)
University of Sydney
Sydney, NSW, Australia

Tassadit Bouadi (iD)
University of Rennes 1
Rennes, France

Luis Galárraga (iD)
Inria
Rennes, France

Linara Adilova
Ruhr University Bochum
Bochum, Germany

Bo Kang (iD)
Ghent University
Ghent, Belgium

Jefrey Lijffijt (iD)
Ghent University
Gent, Belgium

Pascal Welke (iD)
University of Bonn
Bonn, Germany

Erlend Aune
BI Norwegian Business School
Oslo, Norway

Gregor Schiele (iD)
University of Duisburg-Essen
Essen, Germany

Michaela Blott (iD)
Xilinx Research
Dublin, Ireland

Günther Schindler
Heidelberg University
Heidelberg, Germany

Anna Monreale (iD)
University of Pisa
Pisa, Italy

Salvatore Rinzivillo (iD)
ISTI-CNR
Pisa, Italy

Eirini Ntoutsi (iD)
Freie Universität Berlin
Berlin, Germany

Bodo Rosenhahn (iD)
Leibniz University Hannover
Hannover, Germany

Daniela Cialfi (iD)
University of Chieti-Pescara
Chieti, Italy

Maxwell Ramstead (iD)
McGill University
Montreal, Canada

Pedro M. Ferreira (iD)
University of Lisbon
Lisboa, Portugal

Donato Malerba (iD)
Universita di Bari Aldo Moro
Bari, Italy

Philippe Fournier-Viger (iD)
Shenzhen University
Shenzhen, China

Sebastian Ventura (iD)
University of Córdoba
Córdoba, Spain

Min Zhou (iD)
Noah's Ark Lab, Huawei
Beijing, China

Ilaria Bordino
UniCredit
Rome, Italy

Francesco Gullo (iD)
Unicredit
Rome, Italy

Lorenzo Severini
Unicredit
Rome, Italy

Przemyslaw Biecek (iD)
Warsaw University of Technology
Warsaw, Poland

Mykola Pechenizkiy (iD)
Eindhoven University of Technology
Eindhoven, The Netherlands

Christopher Buckley (iD)
University of Sussex
Brighton, UK

Pablo Lanillos (iD)
Radboud University Nijmegen
Nijmegen, The Netherlands

Tim Verbelen (iD)
Ghent University
Ghent, Belgium

Giuseppina Andresini (iD)
University of Bari Aldo Moro
Bari, Italy

Ibéria Medeiros (iD)
University of Lisbon
Lisbon, Portugal

M. Saqib Nawaz (iD)
Harbin Institute of Technology
Harbin, China

Meng Sun (iD)
Peking University
Beijing, China

Valerio Bitetta
UniCredit
Milan, Italy

Andrea Ferretti
UniCredit
Milan, Italy

Giovanni Ponti
ENEA Headquarters
Portici, Italy

Rita Ribeiro (iD)
University of Porto
Porto, Portugal

João Gama (iD)
University of Porto
Porto, Portugal

Lee Cooper (iD)
Northwestern University
Chicago, IL, USA

Jonas Richiardi (iD)
University of Lausanne
Lausanne, Switzerland

Diego Saldana Miranda (iD)
F. Hoffmann–La Roche Ltd
Basel, Switzerland

Guilherme Graça (iD)
University of Lisbon
Lisbon, Portugal

Ricard Gavaldà (iD)
UPC BarcelonaTech
Barcelona, Spain

Naghmeh Ghazaleh (iD)
PD Personalised Healthcare
Basel, Switzerland

Damian Roqueiro (iD)
ETH Zurich
Basel, Switzerland

Konstantinos Sechidis (iD)
Novartis Pharma AG
Basel, Switzerland

Preface

The European Conference on Machine Learning and Principles and Practice of Knowledge Discovery in Databases (ECML PKDD) is the premier European conference on machine learning and data mining. In 2021, ECML PKDD was held virtually during September 13–17 due to the COVID-19 pandemic.

The program included workshops on specialized topics held during the first and last day of the conference. This two-volume set includes the proceedings of the following workshops:

1. Workshop on Advances in Interpretable Machine Learning and Artificial Intelligence (AIMLAI 2021)
2. Workshop on Parallel, Distributed, and Federated Learning (PDFL 2021)
3. Workshop on Graph Embedding and Mining (GEM 2021)
4. Workshop on Machine Learning for Irregular Time Series (ML4ITS 2021)
5. Workshop on IoT, Edge, and Mobile for Embedded Machine Learning (ITEM 2021)
6. Workshop on eXplainable Knowledge Discovery in Data Mining (XKDD 2021)
7. Workshop on Bias and Fairness in Artificial Intelligence (BIAS 2021)
8. Workshop on Workshop on Active Inference (IWAI 2021)
9. Workshop on Machine Learning for Cybersecurity (MLCS 2021)
10. Workshop on Machine Learning in Software Engineering (MLiSE 2021)
11. Workshop on Mining Data for Financial Applications (MIDAS 2021)
12. Workshop on Data Science for Social Good (SoGood 2021)
13. Workshop on Machine Learning for Pharma and Healthcare Applications (PharML 2021)
14. Workshop on Machine Learning for Buildings Energy Management (MLBEM 2021)

Each workshop section contains the papers from the workshop and a preface from the organizers.

We would like to thank all participants and invited speakers, the Program Committees and reviewers, and the ECML PKDD conference and workshop chairs – thank you for making the workshops a successful event. We are also grateful to Springer for their help in publishing this volume.

October 2021

Michael Kamp
on behalf of the volume editors

Organization

Workshop Chairs

AIMLAI 2021

Adrien Bibal	University of Namur, Belgium
Tassadit Bouadi	University of Rennes, France
Benoît Frénay	University of Namur, Belgium
Luis Galárraga	Inria, France
José Oramas	University of Antwerp, Belgium

PDFL 2021

Michael Kamp	IKIM, Ruhr-University Bochum, Germany/Monash University, Australia
Linara Adilova	Ruhr-University Bochum and Fraunhofer IAIS, Germany
Yamuna Krishnamurthy	Royal Holloway, University of London, UK

GEM 2021

Bo Kang	Ghent University, Belgium
Christine Largeron	Université Jean Monnet, France
Jefrey Lijffijt	Ghent University, Belgium
Tiphaine Viard	Telecom Paris, France
Pascal Welke	University of Bonn, Germany

ML4ITS 2021

Massimiliano Ruocco	Norwegian University of Science and Technology, Norway
Erlend Aune	BI Norwegian Business School, Norway
Claudio Gallicchio	University of Pisa, Italy

ITEM 2021

Gregor Schiele	University of Duisburg-Essen, Germany
Franz Pernkopf	Graz University of Technology, Austria
Michaela Blott	Xilinx Research, Ireland
Holger Fröning	Heidelberg University, Germany
Günther Schindler	Heidelberg University, Germany

XKDD 2021

Riccardo Guidotti   University of Pisa, Italy
Anna Monreale   University of Pisa, Italy
Salvatore Rinzivillo   ISTI-CNR, Italy
Przemyslaw Biecek   Warsaw University of Technology, Poland

BIAS 2021

Eirini Ntoutsi   Freie Universität Berlin, Germany
Mykola Pechenizkiy   Eindhoven University of Technology,
                                   The Netherlands
Bodo Rosenhahn   Leibniz Universität Hannover, Germany

IWAI 2021

Christopher Buckley   University of Sussex, UK
Daniela Cialfi   University of Chieti-Pescara, Italy
Pablo Lanillos   Donders Institute, The Netherlands
Maxwell Ramstead   McGill University, Canada
Tim Verbelen   Ghent University, Belgium

MLCS 2021

Pedro Ferreira   University of Lisbon, Portugal
Michael Kamp   IKIM, Ruhr-University Bochum,
                                   Germany/Monash University, Australia
Giuseppina Andresini   University of Bari Aldo Moro, Italy
Donato Malerba   University of Bari Aldo Moro, Italy
Ibéria Medeiros   University of Lisbon, Portugal

MLiSE 2021

Philippe Fournier-Viger   Shenzhen University, China
M. Saqib Nawaz   Harbin Institute of Technology, China
Sebastian Ventura   University of Cordoba, Spain
Meng Sun   Peking University, China
Min Zhou   Huawei, China

MIDAS 2021

Valerio Bitetta   Unicredit, Italy
Ilaria Bordino   Unicredit, Italy
Andrea Ferretti   Unicredit, Italy
Francesco Gullo   Unicredit, Italy
Giovanni Ponti   ENEA, Italy
Lorenzo Severini   Unicredit, Italy

SoGood 2021

Ricard Gavaldà	UPC BarcelonaTech, Spain
Irena Koprinska	University of Sydney, Australia
João Gama	University of Porto, Portugal
Rita P. Ribeiro	University of Porto, Portugal

PharML 2021

Lee Cooper	Northwestern University, USA
Naghmeh Ghazaleh	F. Hoffmann–La Roche Ltd., Switzerland
Jonas Richiardi	Lausanne University Hospital and University of Lausanne, Switzerland
Damian Roqueiro	ETH Zurich and D-BSSE, Switzerland
Diego Saldana	F. Hoffmann–La Roche Ltd., Switzerland
Konstantinos Sechidis	Novartis Pharma AG, Switzerland

MLBEM 2021

Pedro Ferreira	University of Lisbon, Portugal
Guilherme Graça	University of Lisbon, Portugal

Contents – Part I

Parallel, Distributed, and Federated Learning

Graph Embedding and Mining

Machine Learning for Irregular Time Series

IoT, Edge, and Mobile for Embedded Machine Learning

eXplainable Knowledge Discovery in Data Mining

Bias and Fairness in AI

International Workshop on Active Inference

Contents – Part II

MIning DAta for financial applicationS

Sixth Workshop on Data Science for Social Good (SoGood 2021)

Machine Learning for Pharma and Healthcare Applications

Machine Learning for Buildings Energy Management

Advances in Interpretable Machine Learning and Artificial Intelligence

Workshop on Advances in Interpretable Machine Learning and Artificial Intelligence (AIMLAI 2021)

Recent technological advances rely on accurate decision support systems that can be perceived as black boxes due to their overwhelming complexity. Given the pervasiveness of automated decision-making nowadays, such a lack of transparency can incur in technical, ethical, trust, and even legal issues. For example, if the control module of a self-driving car failed at detecting a pedestrian, it becomes crucial to know why the system erred. In some other cases, the decision system may reflect unacceptable biases that can generate distrust. The General Data Protection Regulation (GDPR), approved by the European Parliament in 2018, suggests that individuals should be able to obtain explanations of the decisions made from their data by automated processing, and to challenge those decisions. All these reasons have given rise to the domain of interpretable AI/ML: the study of approaches to make the rationale behind algorithmic decisions understandable to humans.

The AIMLAI workshop aims at gathering researchers, experts, and professionals interested in the topic of interpretable AI/ML. The workshop encourages interdisciplinary collaborations, with particular emphasis on knowledge management, infovis, human computer interaction, and psychology. It also welcomes applied research for use cases where interpretability matters.

Acknowledgements

This workshop was partially supported by the following:

- HyAIAI project (https://project.inria.fr/hyaiai/)
- The TAILOR network - Foundations of Trustworthy AI – Integrating Reasoning, Learning and Optimization (https://tailor-network.eu)
- FNRS EOS O.0055.18F "Verifying Learning Artificial Intelligence Systems".
- FWO Fundamental Project (G0A4720N) "Design and Interpret: A New Framework for Explainable Artificial Intelligence"
- BOF DOCPRO4-NZ project (ID 41612) "Multimodal Relational Interpretation for Deep Models"

Organization

AIMLAI 2021 Organizers

Adrien Bibal	University of Namur, Belgium
Tassadit Bouadi	IRISA/University of Rennes, France
Benoît Frénay	University of Namur, Belgium
Luis Galárraga	IRISA/Inria, France
José Oramas	University of Antwerp/imec-IDLab, Belgium

Program Committee

Michaël Aupetit	QCRI, Qatar
Laure Berti-Equille	IRD, France
Pedro Bizarro	Feedzai, Portugal
Chaofan Chen	University of Maine, USA
Miguel Couceiro	University of Lorraine, France
Françoise Fessant	Orange Labs, France
Alex Freitas	University of Kent, UK
Johannes Fürnkranz	Johannes Kepler University, Austria
Romaric Gaudel	ENSAI, France
Sébastien Gambs	University of Quebec, Canada
Tias Guns	KU Leuven, Belgium
Thomas Guyet	Agrocampus Rennes, France
Haiyang Huang	Duke University, USA
Hamed Behzadi Khormouji	University of Antwerp, Belgium
Christine Largouët	Agrocampus Ouest, France
Paulo Lisboa	Liverpool John Moores University, UK
Véronique Masson	University of Rennes I, France
Amedeo Napoli	CNRS, France
Philippe Preux	University of Lille, France
Yanou Ramon	University of Antwerp, Belgium
Laurence Rozé	INSA, France
Lesia Semenova	Duke University, USA
Cagatay Turkay	University of Warwick, UK
Kaili Wang	KU Leuven, Belgium
Tong Wang	University of Iowa, USA
Jean-Daniel Zucker	IRD, France

TS-MULE: Local Interpretable Model-Agnostic Explanations for Time Series Forecast Models

Udo Schlegel[(✉)], Duy Lam Vo, Daniel A. Keim, and Daniel Seebacher

University of Konstanz, Konstanz, Germany
u.schlegel@uni-konstanz.de

Abstract. Time series forecasting is a demanding task ranging from weather to failure forecasting with black-box models achieving state-of-the-art performances. However, understanding and debugging are not guaranteed. We propose *TS-MULE*, a local surrogate model explanation method specialized for time series extending the LIME approach. Our extended LIME works with various ways to segment and perturb the time series data. In our extension, we present six sampling segmentation approaches for time series to improve the quality of surrogate attributions and demonstrate their performances on three deep learning model architectures and three common multivariate time series datasets.

Keywords: Explainable AI · LIME · Time series

1 Introduction

Time series forecasting is an essential task with applications in a broad range of domains, such as industrial process control, finance, and risk management, since predicting future trends and events is a critical input into many types of planning and decision-making processes [1]. Recently, deep learning methods have increasingly found their way into the field of time series forecasting as a result of their successful application in other domains such as natural language processing [2] and object detection [3]. A major drawback of such models is that, due to their non-linear, multi-layered structure, they are black box models that suffer from a lack of explainability. Such a lack of explainability prevents deep learning from being used in production in sensitive domains, such as healthcare [4], as opposed to statistical methods [5], or is complicated by laws, such as the EU General Data Protection Regulation [6], which enforces a right for explanations. Thus, agencies such as DARPA introduced the explainable AI (XAI) initiative [7] to promote the research around interpretable Machine Learning (ML).

Gaining the necessary understanding of these complex models to provide explanations globally for the whole input space is often infeasible, leading to the development of methods that provide only local explanations of the underlying prediction function, such as LIME [8]. LIME is an XAI technique that can explain

© Springer Nature Switzerland AG 2021
M. Kamp et al. (Eds.): ECML PKDD 2021 Workshops, CCIS 1524, pp. 5–14, 2021.
https://doi.org/10.1007/978-3-030-93736-2_1

the predictions of any classifier by learning and providing an interpretable surrogate model around the classification. An advantage of LIME in terms of interpretability is that it perturbs the input by changing components that make sense to humans (e.g., words or parts of an image), even if the model is using much more complicated components as features (e.g., word embeddings) [8].

For images, such interpretable components can be superpixels, which are a perceptual grouping of pixels, or for texts, it can be individual words or sentences. However, finding such semantically meaningful components for univariate or even multivariate time series data is not trivial. Segmenting the time series into fixed-width windows might miss meaningful elements between windows by weighting them equally or are larger or smaller than the chosen window size. Thus, such a fixed segmentation can potentially miss important subsequences in the time series by splitting them. One possible approach could identify motifs in the time series. Such motifs are subsequences of the time series very similar to each other. However, even optimized algorithms can have a worst-case complexity of $\mathcal{O}(n^2)$ [9] and are, thus, not suitable to identify potential patterns beforehand.

To tackle such issues, we propose *TS-MULE*, an extension to LIME by improving the segmentation, for local explanations of univariate and multivariate time series. We provide five novel algorithm approaches to provide a meaningful segmentation of time series to enable local interpretable model-agnostic explanations of time series forecasting models. To provide such meaningful segmentation, we incorporate the matrix profile [10] as well as the SAX transformation [11] and extend the results of these algorithms with binning or top-k approaches to incorporate the findings of these techniques. We evaluate these segmentation algorithms against each other and the baseline of a uniform segmentation on three standard forecasting datasets with three different black-box models.[1]

2   Related Work

An important distinction when selecting methods for explaining complex machine learning models is for which user group these XAI methods must be accessible. Most of the proposed XAI methods used, especially for time series deep learning models, are usually only accessible to model developers. For instance, by examining the activation of latent layers [12], or via relevance back-propagation [13]. However, especially for other groups, particularly model users (see Spinner et al. for an overview of user groups [14]), such approaches are less practical since explanations need to be provided at a higher level of abstraction. Available approaches with a higher level of abstraction currently come primarily from the computer vision domain for explaining image classifications [15].

There are already first works that apply these concepts in time series classification and prediction. For example, the approach of Suresh et al. [16] replaces each time series observation with uniform noise to study the impact on model performances and thus determine feature importance. Since replacing features

---

[1] Source code and evaluation results are available at: https://github.com/dbvis-ukon/ts-mule.

with out-of-domain noise can lead to arbitrary changes in model output, Tonek-aboni et al. use data distributions to produce reliable counterfactuals [17]. Both previous approaches rely on observation-level replacement and thus, cannot identify important larger patterns in time series. Two recent approaches tackle this issue by using longer time segments as input for the perturbation and replacing it with, for instance, linear interpolations, constant values or segments from other time series [18], or with zeros, local or global mean values, or local or global noise [19]. However, both of these approaches rely on fixed window sizes. Thus they are incapable of modeling, e.g., semantically meaningful patterns in the time series, which can have variable lengths. Additionally, they might miss important patterns if the predefined window size is smaller or longer than the pattern or if patterns lie between the fixed time segments.

Hence, we provide an extension of the LIME approach to identify superpixels-like patterns, i.e., semantically related data regions, in time series data. This paper presents a set of suitable segmentation algorithms and evaluates their suitability for providing explanations under various data characteristics.

3   Post-hoc Local Explanations with LIME

Creating explanations for decisions of black-box models has various alternatives. One of these possibilities is the post-hoc approach LIME by Ribciro et al. [8]. Local Interpretable Model-Agnostic Explanations, shortly LIME, uses an interpretable surrogate model to create explanations for black-box models. In the first step, a chosen sample to explain and a model to be explained are given as input to the approach. The sample is then segmented by a previously chosen segmentation algorithm, e.g., a superpixel segmentation for images [8]. LIME then creates masks for the sample deactivating segments or replacing them with non-informative values. In many cases, this step is called perturbation and is something different than the perturbation mentioned later. These newly generated (perturbed) samples are predicted with the input model to get new predictions. LIME collects these predictions and trains a new interpretable classifier, often a linear model, on the masks with the predictions as the target. In the case of a linear model, the coefficients are used to weigh the different input segments and to explain the model for the given sample. Figure 1 demonstrate the described approach on time series with a uniform segmentation.

LIME is generally applicable for any data type, but there are some challenges due to the necessity of segmentation. Valuable segmentation makes sense to humans as it incorporates their domain knowledge. For instance, superpixel segmentation identifies perceptual groups in images, which in most cases correspond to a human interpretable object. As time series are generally hard to segment without domain knowledge, a general approach is rather difficult, even with domain knowledge not applicable. A forecasting black box model often just uses a window as input to predict the target value in many cases. Such a window is fixed beforehand and slides over the data, thus having no strict segmentation in itself. Finding such segmentation is a significant challenge for time series as it needs to be generally applicable.

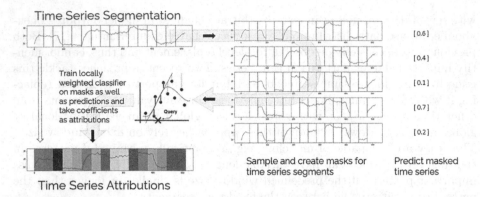

Fig. 1. The LIME approach applied on time series starting with the uniform segmentation on a time series sample. Next, doing the masking, perturbing, and predicting step of LIME to generate more local samples. Afterward, a linear interpretable model is trained on the masks and predictions using local weighting. At last, extracting the coefficients of the model leads to the wanted attributions for the initial sample.

4   Finding Suitable Segmentation Mappings

We propose *TS-MULE*, extending the LIME [8] approach for time series with novel segmentation algorithms. Our approach presents five segmentation techniques created for time series and three different replacement strategies.

4.1   Using Static Windows

Uniform segmentation is the most basic method to segment a time series into windows. In this approach, we split the time series $ts = \{t_0, t_1, t_2, ..., t_n\}$ into equally and non-overlapping m-sized windows $ws = \{w_0, w_1, w_2, ..., w_d\}$ with $d = \lceil n/m \rceil$. If n is not a multiple of m, the final windows may have more or less time points. We expand the uniform segmentation to exponential windows, which ignores the size m and has longer windows at the end. A time series ts in *exponential segmentation* is split into $d = log(n)$ windows and its length increases with $[e^0], [e^1], [e^2], ..., [e^d]$. To cover all the points of the time series, in the final window, we adjust its length by $n - \sum([e^0], [e^1], [e^2], ..., [e^{d-1}])$. A benefit of such segmentation is that we put more weight on the latest points with longer windows.

4.2   Using the Matrix Profile

A matrix profile is a vector that stores the z-normalized Euclidean distance between any subsequence within a time series and its nearest neighbor [10]. Such a matrix profile can be used to identify motifs as well as outlier subsequences in large time series [10]. We introduce the *slope* and *bin* segmentation based on the matrix profile on time series to incorporate local trends and patterns.

The *slope segmentation* has the parameters window size m as input for the matrix profile and k for the number of partitions for the segmentation. The basic idea behind this segmentation approach consists in the opportunity to find patterns in the time series using the matrix profile. By further focusing on the slope of the matrix profiles distances, we can identify drastic changes in the nearest neighbors to find not only possible patterns but also uncommon changes in the time series itself. Such a uncommon changes can be used as plausible splits for the segmentation as the pattern are still included in the segments. We calculate the matrix profile $mp = \{d_1, d_2, ..., d_j\}$ with our previously adjusted window size m so that $j = n - m + 1$ to find interesting distances, e.g., to identify motifs. Afterward, we either calculate the gradient on the resulting matrix profile ∇mp and take the absolute value $|\nabla mp|$ to identify peaks as steep slopes. Or, depending on the configuration, we sort the resulting matrix profile vector ascending and compute the slope to identify jumps in distances to find significant changes in the time series. We sort the resulting vector in both cases and take the k-largest values to find segment borders. The time series indices of these k values segment our time series and describe drastic changes in the time series.

We further present *bin segmentation* based on the matrix profile with the same parameters m and k as above. Again the idea behind this approach enables finding patterns in the time series by not using the gradient to find drastic changes in the nearest neighbor but using bins to combine similar distances in the matrix profile to segments. We calculate and sort the matrix profile again. However, we further split the min-max range of the matrix profile into k-bins. Afterward, we label the k-bins numerically so that lower numbers have a low and higher a high matrix profile. We convert our matrix profile to the corresponding bin number and assign our base value to the max or min bin. Next, we slide over the resulting profile with a window length m. Due to the sliding window approach, a time point can be either in the segment seg_i or seg_j. For our *bins-min segmentation*, we assign the time point t_t to seg_i if bin_i is smaller than bin_j. Our *bins-max segmentation*, oppositely, uses the seg_i if bin_i is larger than bin_j.

4.3   Using the SAX Transformation

SAX segmentation introduces a segmentation based on horizontal binning of a time series with k partitions as the parameter. The basic idea behind this segmentation approach includes the changes in the range of the values by splitting the overall distribution of possible values into bins. The SAX transformation [11] converts a time series ts into a sequence of symbols $sa = \{s_0, s_1, s_2, ..., s_n\}$ with $s_i \in \{a, b, c, ...\}$ based on a continuous binning of intervals in the vertical direction. We incorporate a base number of bins $b = 3$ for the SAX algorithm and use repeating symbols as segments, e.g., $sa = \{a_1, a_2, a_3, b_1, b_2, a_1, c_1, c_2, c_3\}$ involves four segments leading to $\{a_1, a_2, a_3\}, \{b_1, b_2\}, \{a_1\}, \{c_1, c_2, c_3\}$. At each iteration, the amount of bins is increased $b+ = 1$ to finally achieve a previously selected k partitions as more bins generally convert to more partitions. For some cases, the exact partition size is not possible, and we allow a difference of ten percent to the selected partition size to mitigate such edge cases.

4.4   Comparing the Segmentation Algorithms

Existing and proposed segmentation algorithms lead to different segments representing potentially suitable techniques for various data sets. Figure 2 presents these algorithms on two differently scaled time series features. Especially, comparing the uniform segmentation with the others demonstrates the advantages of the other approaches. Depending on the algorithm, different segments are visible and present some more focused parts of the time series samples. Choosing from a broader range of techniques can lead to improved explanations for humans.

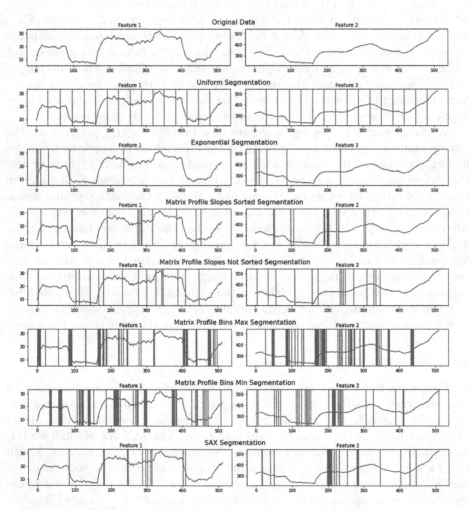

Fig. 2. Comparison of the different segmentation variants. Red stripes show segment splits. Some segmentation algorithms proposed end up with more as well as very short segments than the uniform segmentation with default parameters. (Color figure online)

5   Evaluating *TS-MULE* on Time Series Forecasting

The evaluation of our proposed segmentation and perturbation approaches is based on the perturbation analysis for fidelity by Schlegel et al. [15,20] adapted to forecasting tasks using the mean squared error. As datasets for our evaluation, we use the Beijing Air Quality 2.5, Beijing Multi-Site Air Quality, and the Metro Interstate Traffic data to show the results on divers multivariate time series. For the air quality datasets, we use a fixed input size of 24. The metro traffic forecasting has an input length of 72. We use three different basic implementations of black-box models: a basic one-dimensional convolutional neural network, a deep neural network, and a recurrent neural network (LSTMs [21]).

The perturbation analysis by Schlegel et al. [20] consists of three steps: explanation generation, data perturbation based on explanations, and perturbation evaluation. At first, a selected dataset, e.g., the test data, is evaluated with a quality metric (e.g., accuracy), and explanations are generated for every sample. Next, every sample of the selected dataset is perturbed such that time points with high relevances for the explanation are replaced with non-information holding values. As non-information holding values for time series are challenging to find, we focus on the proposed ones (zero, inverse, mean) by Schlegel et al. [20]. Often the high relevance attributions are identified by using a threshold. Lastly, the perturbed data gets evaluated, and the quality metric change is calculated. The assumption is that a value change of the predicted data at highly relevant input positions decreases the quality metric performance of the model as the data loses valuable information. Such an assumption then leads to the conclusion that a working XAI technique decreases the performance more than a random change.

Table 1. Evaluation results of the perturbation analysis for every segmentation technique for three datasets and three models. We calculate the perturbation analysis results based on the percentage change to the original prediction and the randomized change. A larger value shows a better explanation.

Zero	Beijing Air Quality 2.5	CNN	DNN	RNN	Beijing AQ Multi Site	CNN	DNN	RNN	Metro Interstate Traffic	CNN	DNN	RNN
Uniform		**2.31**	**4.24**	2.32		1.50	9.00	7.67		2.43	0.22	6.55
Exponential		0.56	1.12	1.41		0.62	0.16	**11.52**		0.55	0.01	0.62
Slopes		1.31	2.11	1.95		1.3	6.76	3.97		**3.39**	0.18	**9.29**
Bins Min		0.35	3.43	**3.6**		0.41	**10.46**	5.71		1.25	0.4	7.38
Bins Max		1.69	1.22	2.38		**1.52**	1.68	2.67		1.44	0.44	2.68
SAX		1.24	2.58	2.23		1.10	8.00	4.15		1.55	**1.16**	7.34

We extend the assumptions to calculate a score for improved comparability of the results by focusing on the percentage increase in relation to a random change of the time series. Schlegel et al. [20] propose to take the 90th percentile value of the attribution values of the sample as a threshold. However, we have to scale our *TS-MULE* values because we observed that depending on the segment count, the distribution of the attribution changes. Such a distribution change leads to either more or less highly relevant time points for the perturbation as, e.g., there are more attribution values above the threshold value. Thus, we take the initial prediction scores *orig*, the perturbed prediction scores *pert*, and the random position change prediction score *rand* and calculate the increase of the perturbed: $pert_c = \frac{pert-orig}{orig}$ and random: $rand_c = \frac{rand-orig}{orig}$. We set these in relation to get our final score: $score = \frac{|pert_c|}{|rand_c|}$. A score below one depicts a worse performance than random guessing. Scores larger than one illustrate plausible explanations better than guessing. Through this scaling, the segmentation algorithms can be compared. Larger results demonstrate better segmentation. Table 1 presents such a perturbation analysis on fidelity with our proposed segmentation approaches.

Our preliminary results for a zero perturbation, see Table 1, show that *uniform* is working well for short time series windows (Beijing Air Quality with 24) while *slopes* generate better performances on long windows (Metro Interstate Traffic with 72). However, also our proposed *bins-min*, *bins-max* and *SAX* illustrate promising results for short windows and can be further tuned by adding more parameters. Also, by further adding a minimum length for segments, these algorithms can be improved. The DNN for the Metro Interstate Traffic dataset is interesting as non of the proposed segmentation strategies seem to work. However, such an effect can be caused as the model's performance is way worse than the other two models. In general, the *uniform* segmentation works well as a starting point, but exchanging it with our proposed algorithms enables more diverse and improved attributions.

6   Conclusion

We present *TS-MULE*, a local interpretable model-agnostic explanation extraction technique for time series. For *TS-MULE*, we extend the LIME approach with novel time series segmentation techniques and replacement methods to enforce a better non-informed values exchange. Thus, we contribute five novel time series segmentation algorithms and the *TS-MULE* framework for time series forecasting. We show on three forecasting datasets that *TS-MULE* performs better than randomly perturbing data and thus reveals relevant input values for the prediction of a model. Further, we demonstrate that our proposed segmentation algorithms lead to improved attributions in most cases. As future work, we want to compare the performance of *TS-MULE* against other XAI techniques applied to time series in the framework of Schlegel et al. [20]. We also want to identify shapelets to generate segments with more in-depth domain knowledge and to investigate into similar attribution techniques like SHAP [22].

Acknowledgements. This work has received funding from the European Union's Horizon 2020 research and innovation programme under grant agreement No. 826494.

References

1. Montgomery, D.C., Jennings, C.L., Kulahci, M.: Introduction to Time Series Analysis and Forecasting. Wiley (2015)
2. Vaswani, A., et al.: Attention is all you need. In: Advances in Neural Information Processing Systems, vol. 30 (2017)
3. Zhao, Z.-Q., Zheng, P., Shou-tao, X., Xindong, W.: Object detection with deep learning: a review. IEEE Trans. Neural Netw. Learn. Syst. **30**(11), 3212–3232 (2019)
4. Rudin, C.: Stop explaining black box machine learning models for high stakes decisions and use interpretable models instead. Nat. Mach. Intell. **1**(5), 206–215 (2019)
5. Chuah, M.C., Fu, F.: ECG anomaly detection via time series analysis. In: Thulasiraman, P., He, X., Xu, T.L., Denko, M.K., Thulasiram, R.K., Yang, L.T. (eds.) ISPA 2007. LNCS, vol. 4743, pp. 123–135. Springer, Heidelberg (2007). https://doi.org/10.1007/978-3-540-74767-3_14
6. European Union: European General Data Protection Regulation. Technical report (2018)
7. Gunning, D.: Explainable Artificial Intelligence (XAI) DARPA-BAA-16-53. Technical report (2016)
8. Ribeiro, M., Singh, S., Guestrin, C.: "Why should i trust you?": explaining the predictions of any classifier. In: Proceedings of the 22nd ACM SIGKDD International Conference on Knowledge Discovery and Data Mining, San Francisco, CA, USA, 13–17 August 2016, pp. 1135–1144. ACM (2016)
9. Mueen, A., Keogh, E.J., Zhu, Q., Cash, S., Brandon Westover, M.: Exact discovery of time series Motifs. In: SIAM International Conference on Data Mining SDM (2009)
10. Yeh, C.-C.M., et al.: Matrix profile i: all pairs similarity joins for time series: a unifying view that includes motifs, discords and shapelets. In: IEEE International Conference on Data Mining (2016)
11. Lin, J., Keogh, E., Lonardi, S., Chiu, B.: A symbolic representation of time series, with implications for streaming algorithms. In: ACM SIGMOD Workshop on Research Issues in Data Mining and Knowledge Discovery (2003)
12. Siddiqui, S.A., Mercier, D., Munir, M., Dengel, A., Ahmed, S.: TSViz: demystification of deep learning models for time-series analysis. IEEE Access **7**, 67027–67040 (2019)
13. Bach, S., Binder, A., Montavon, G., Klauschen, F., Müller, K.-R., Samek, W.: On pixel-wise explanations for non-linear classifier decisions by layer-wise relevance propagation. PLOS ONE **10**, e0130140 (2015)
14. Spinner, T., Schlegel, U., Schäfer, H., El-Assady, M.: explAIner: a visual analytics framework for interactive and explainable machine learning. IEEE Trans. Vis. Comput. Graph. **26**, 1064–1074 (2019)
15. Schlegel, U., Arnout, H., El-Assady, M., Oelke, D., Keim, D.A.: Towards a rigorous evaluation of XAI methods on time series. In: ICCV Workshop on Interpreting and Explaining Visual Artificial Intelligence Models (2019)
16. Suresh, H., Hunt, N., Johnson, A., Celi, L.A., Szolovits, P., Ghassemi, M.: Clinical intervention prediction and understanding using deep networks. arXiv preprint arXiv:1705.08498 (2017)
17. Tonekaboni, S., Joshi, S., Duvenaud, D., Goldenberg, A.: Explaining time series by counterfactuals (2020)

18. Guillemé, M., Masson, V., Rozé, L., Termier, A.: Agnostic local explanation for time series classification. In: 2019 IEEE 31st International Conference on Tools with Artificial Intelligence (ICTAI), pp. 432–439. IEEE (2019)
19. Mujkanovic, F., Doskoč, V., Schirneck, M., Schäfer, P., Friedrich, T.: timeXplain – a framework for explaining the predictions of time series classifiers. arXiv preprint arXiv:2007.07606 (2020)
20. Schlegel, U., Oelke, D., Keim, D.A., El-Assady, M.: An empirical study of explainable AI techniques on deep learning models for time series tasks. In: Pre-registration Workshop NeurIPS (2020)
21. Hochreiter, S., Schmidhuber, J.: Long short-term memory. Neural Comput. 9(8), 1735–1780 (1997)
22. Lundberg, S., Lee, S.-I.: A unified approach to interpreting model predictions. In: Advances in Neural Information Processing Systems (2017)

Interpretable Models via Pairwise Permutations Algorithm

Troy Maasland[2], João Pereira[1,2], Diogo Bastos[1,2], Marcus de Goffau[1],
Max Nieuwdorp[1], Aeilko H. Zwinderman[3], and Evgeni Levin[1,2(✉)]

[1] Department of Vascular Medicine, Academic Medical Center,
University of Amsterdam, 1105 AZ Amsterdam, The Netherlands
e.levin@amsterdamumc.nl
[2] Horizon B.V., Delft, The Netherlands
[3] Department of Clinical Epidemiology, Biostatistics and Bioinformatics,
University Medical Center, Amsterdam, The Netherlands

Abstract. One of the most common pitfalls often found in high dimensional biological data sets are correlations between the features. This may lead to statistical and machine learning methodologies overvaluing or undervaluing these correlated predictors, while the truly relevant ones are ignored. In this paper, we will define a new method called *pairwise permutation algorithm* (PPA) with the aim of mitigating the correlation bias in feature importance values. Firstly, we provide a theoretical foundation, which builds upon previous work on permutation importance. PPA is then applied to a toy data set, where we demonstrate its ability to correct the correlation effect. We further test PPA on a microbiome shotgun dataset, to show that the PPA is already able to obtain biological relevant biomarkers.

Keywords: Permutation · Importance · Correlation · PPA · Diabetes

1 Introduction

Measuring feature importance has often been plagued by high feature correlations. One important drawback is the lack of a theoretical definition for variable importance, in case variables are correlated [12,14], even in linear models [12]. From a clinical perspective, correlated biomarkers are of high interest because they both may play a role in a shared biological pathway identified by the model and yet exhibit different behaviour in other circumstances. The method proposed in this paper, which will be referred to as *pairwise permutation algorithm* (PPA), allows us to calculate the importance of features without having to rely on the previously mentioned selection approaches. Highly correlated features, which have a similar relation with the output value, should have close importance ranks since they explain the same variability in the data. The *pairwise*

T. Maasland and J. Pereira—Equal contribution to this work.

© Springer Nature Switzerland AG 2021
M. Kamp et al. (Eds.): ECML PKDD 2021 Workshops, CCIS 1524, pp. 15–25, 2021.
https://doi.org/10.1007/978-3-030-93736-2_2

permutation algorithm aims to provide feature importance values while avoiding the use of aggressive pre-selection techniques, since these techniques might remove relevant information from the data. It also manages to retain model interpretability by generating an importance value per feature, even when applied to black box models. Moreover, when working with highly dimensional biological data sets, it is simply not feasible to try and address each of the correlations in the data individually.

2   Related Work

In this work, we focus on model-agnostic procedures which can be divided into local and global methods. Local-based methods such as LIME and its variants [6,41] attempt to explain predictions on single data points by perturbing it and building a simple, yet interpretable model on the perturbed predictions. Similarly, SHAP [5], offers a local explanation based on the additional prediction value each feature has when adding it to all possible feature subsets. Unlike local-based methods, global methods are concerned with determining the overall model behaviour and what features it values for its prediction. For example, in clinical research, the goal is to determine biomarkers that can identify a condition in the general population, or potential targets for novel drug development. Therefore, in this setting, we are mainly concerned with a more holistic view of feature importance i.e. global. A notable example is that of permutation importance which was first introduced by Breiman [2] in random forests as a way to understand the interaction of variables that is providing the predictive accuracy. Suppose that for a certain feature i in data-set $\mathbf{X}$, we randomly permute the instances' values, and denote the resultant data-set by $\mathbf{X}_i^\pi$. Permutation importance is defined as the difference in the expected model loss on the original dataset and the original one:

$$PI_{\{i\}}(f) := E\left[f(\mathbf{X}_i^\pi)\right] - E\left[f(\mathbf{X})\right] \tag{1}$$

For random forests, there is already available work that analyzes the behaviour of this permutation importance, including the cases when high correlations are present. Gregorutti et al. [14] provided a theoretical description of the effect of correlations on the permutation importance, a phenomenon already observed by Toloşi and Lengauer [13,14]. Furthermore, a feature selection procedure was introduced, which was more efficient in selecting important, highly correlated variables [14]. Strobl et al. showed that the larger feature importance values for correlated predictors in random forests were due to the preference for such predictors in the early splits of the trees. A new conditional permutation-based feature importance calculation was suggested, in order to circumvent this inflation, as well as the depreciation for its correlated predictor [4]. Furthermore, Hooker and Mentch proposed the 'permute and relearn' approach [11]. Based on this approach we define the relearned permutation importance as

$$PI_j^{\pi L} = E\left[f^{\pi j}(\mathbf{X_t})\right] - E\left[f(\mathbf{X_t})\right] \tag{2}$$

In which $f^{\pi j}$ is the model trained on the train dataset $\mathbf{X}^{\pi \mathbf{j}}$, in which feature j is permuted, f the model trained on the original train dataset $\mathbf{X}$ and $\mathbf{X_t}$ the test dataset. One drawback of this approach was also mentioned in the context of correlated features, as this resulted in the compensation effect, in which the importance of the correlated features was reduced [11]. Local based methods, such as the ones introduced earlier, are focused on the contribution of each feature towards individual predictions, whereas permutation importance gives us a more broad estimation, since it is based on the overall accuracy of the model. While the former approach provides a higher degree of interpretability, the latter is usually more appropriate in a research environment, in which the aim would be to discover new leads which could help researchers to investigate the underlying biological mechanisms.

3 Pairwise Permutations Algorithm

3.1 Notation

We will refer to a single instance of the data-set as instance or point interchangeably throughout the paper. We denote matrices, 1-dimensional arrays and scalars with capital bold and regular text, respectively (e.g. $\mathbf{X}$, $\mathbf{x}$, α). Matrices' columns and rows will be denoted by $\mathbf{X}[:, i]$ and $\mathbf{X}[j, :]$, respectively. The expected loss of a function given by: $\frac{1}{N} \sum_{i=1}^{N} l\left[y, f(\mathbf{x}_i)\right]$ will be denoted by $E\left[f(\mathbf{X})\right]$.

3.2 Intuition

Features that are equally important for the output value should have similar feature importance ranks, and these should not be affected by feature correlation. In an attempt to prevent the compensation effect for correlated features mentioned by Hooker and Mentch, we have chosen to permute all the feature pairs and calculate the corresponding permutation importance of the pair. A key assumption in our method is that the higher the correlations, the larger should be the correction to that feature individual importance.

3.3 Definition

In this section, we define the pairwise permutation importance (PPI) as the weighted average of the permutation importance values, computed using the 'permute and relearn' approach defined in Eq. 2. The correlations between the feature pairs will act as the weights. Let $\mathbf{R}$ be the correlations matrix between all the features and $\mathbf{R}_{i,j}$ the correlation value between features i and j. Let $PI_{i,j}$ define the relearn permutation importance (see Eq. 2) when both the feature i and j have been permuted together, and $PI_{i,i}$ the relearn permutation importance, when only feature i is permuted.

$$PPI_i = \underbrace{\frac{1}{\sum\limits_{j=1}^{M} |\mathbf{R}_{i,j}|}}_{q} \underbrace{\left(PI_{i,i} + \sum\limits_{\substack{j=1 \\ j \neq i}}^{M} |\mathbf{R}_{i,j}| \cdot PI_{i,j} \right)}_{p} \tag{3}$$

Note that when a feature has no correlations in the data, according to the previous equation, the PPI will actually follow the relearn permutation importance, which in our terminology is the single permutation importance (SPI). Since for complex data sets with thousands of features the computational time can become infeasible ($\mathcal{O}(N^2)$), one possible simplification is to set a threshold and consider only the permutation pairs with a correlation above it. We define this procedure in Algorithm 1.

3.4 Expected Difference

It might be tempting to compute the expected loss of the model, perform the permutation analysis and then compute the difference of the expected losses. This is actually how Fisher et al. [3] defined the permutation importance. However, we note that this procedure is sub-par as we show in the following theorem:

Theorem 1. *For a given function $f : \mathbb{R}^M \to \mathbb{R}$, let $\mathbf{X}$ and $\mathbf{x}$ be a sample and an instance from the domain of f, respectively, $\mathbf{X}_i^\epsilon$ be $\mathbf{X}$ with permuted values for the r.v. X_i and $\tilde{\mathbf{x}}$ an instance from $\mathbf{X}_i^\epsilon$. Then, for any loss function $l\,[y, f(\mathbf{x})]$ and norm function $||\cdot|| : \mathbb{R}^M \to \mathbb{R}$ it holds that:*
$$\mathbf{E}\left[||l\,[y, f(\mathbf{x})] - l\,[y, f(\tilde{\mathbf{x}})]||\right] \geq ||\mathbf{E}\left[l\,[y, f(\mathbf{x})]\right] - \mathbf{E}\left[l\,[y, f(\tilde{\mathbf{x}})]\right]||$$

Proof. Consider the following convex function $\varphi(x) = ||x||$ for $x = l\,[y, f(\mathbf{x})] - l\,[y, f(\tilde{\mathbf{x}})]$. Then, by Jensen's inequality:

$$\mathbf{E}\left[\varphi(x)\right] \geq \varphi\left(\mathbf{E}\left[x\right]\right) \Leftrightarrow \mathbf{E}\left[||l\,[y, f(\mathbf{x})] - l\,[y, f(\tilde{\mathbf{x}})]||\right] \geq ||\mathbf{E}\left[l\,[y, f(\mathbf{x})] - l\,[y, f(\tilde{\mathbf{x}})]\right]||$$
$$= ||\mathbf{E}\left[l\,[y, f(\mathbf{x})]\right] - \mathbf{E}\left[l\,[y, f(\tilde{\mathbf{x}})]\right]||$$

$\square$

This means that computing the expected value of the normed difference of individual loss values is more robust to non-linear relationships between the input variables then computing the difference of the normed expected loss values.

4   Simulations with Toy Dataset

To see how our new PPA would behave for correlated features, we generated a toy dataset, based on the one used by Hooker and Mentch [11]. The data was created by assuming a linear regression model:

$$y_i = x_{i1} + x_{i2} + x_{i3} + x_{i4} + x_{i5} + 0x_{i6} + 0.5x_{i7} + 0.8x_{i8} + 1.2x_{i9} + 1.5x_{i10}. \tag{4}$$

Algorithm 1: Pairwise permutations algorithm

Input: X, $\mathbf{X_t}$, $\mathbf{y_{test}}$, $E\left[f(\mathbf{X_t})\right]$, R, α
Output: v

for *feature i in* **X** do
    $p \leftarrow 0$, $q \leftarrow 0$ (Eq. 3);
    for *feature j in* **X** do
        if $|\mathbf{R}_{i,j}| > \alpha$ then
            Permute the feature pair (i, j) together in **X**;
            Retrain the model with the permuted input data $\mathbf{X}_{i,j}^{\tau}$;
            Calculate the model's error, $E\left[f^{\pi,i,j}(\mathbf{X_t})\right]$, on the test data;
            Calculate $PI_{i,j}$ through the relearn formula
              $E\left[f^{\pi,i,j}(\mathbf{X_t})\right] - E\left[f(\mathbf{X_t})\right]$;
            if $i=j$ then
                $p \leftarrow p + PI_{i,i}$;
            else
                $p \leftarrow p + |\mathbf{R}_{i,j}| \cdot PI_{i,j}$;
            end
            $q \leftarrow q + |\mathbf{R}_{i,j}|$;
        end
    end
    $PPI_i \leftarrow p/q$;
    v.append(PPI_i);
end

This was then turned into a classification model, by generating the binary outcome y with the classification rule in Eq. 5 with $\varepsilon \sim N(0, 0.1)$.

$$y_i = \begin{cases} 1, & \text{for } y_i + \varepsilon_i \geq \overline{y} \\ 0, & \text{otherwise} \end{cases}, \tag{5}$$

All features were generated from a multivariate normal distribution $N(0, \Sigma)$ with Σ equal to the identity matrix, except that $\Sigma_{12} = \Sigma_{21} = \rho = 0.9$. All features were then transformed into a uniform distribution, mimicking how Hooker and Mentch [11] generated their data. In total, 1000 samples were generated.

In case the features are in the same scale, the coefficients in the linear model can be seen as the conditional importance of the feature on all other variables [4,11]. Therefore, based on the magnitude of the coefficients, we can rank the features on their importances, where features with the same coefficients should be equally important, while a feature with a higher coefficient should get higher importance than a feature with a lower coefficient. The order of the features should not be affected by any correlations between the features.

Using XGBoost with the logistic loss function as the classification algorithm [17,18], we performed 50 stratified shuffle splits (70%train/30%test) and measured the ROC AUC after adding a noise feature to the dataset and standard scaling it. We found the optimal hyperparameters using a 5-fold cross validation grid search. To compute the PPIs, a correlation threshold of 0.3 was used. Also, the SPIs were obtained.

4.1   Results

The classification model obtained an average AUC of 0.97 ± 0.01. As shown by the average feature ranks in Fig. 1(a), our new PPA is able to retrieve the right order of feature importances, in which x_{10} is clearly the most important one, followed by x_9. As expected, x_6 and the random variable are identified as the least important features. The results for the SPI are shown in Fig. 1(b). It is clearly shown that the PPA outperforms this approach, as the SPI decreased the importance of the correlated features x_1 and x_2 and was not able to retrieve the right order of feature importances. This was also observed for the random forest algorithm by Hooker and Mentch [11]. The toy dataset showed that in case two features have the same coefficient in the linear model and are correlated, the PPA is able to retrieve the right order for the feature importances. We also analysed the effect of a correlation of $\rho = 0.9$ between x_1 and x_6, by changing the covariance matrix Σ to $\Sigma_{16} = \Sigma_{61} = 0.9$ and setting the correlation between x_1 and x_2 to 0. This represents a case in which an important feature is correlated to an irrelevant feature. However, we saw in this case that the importance of x_1 was decreased by x_6, while the importance of x_6 was increased by x_1, as shown in Fig. 1(c). This could be expected as the grouped importance is shared equally between both features, while in the case of features with different importances, this might not be the right assumption. In this case, the PPA may not be the appropriate choice

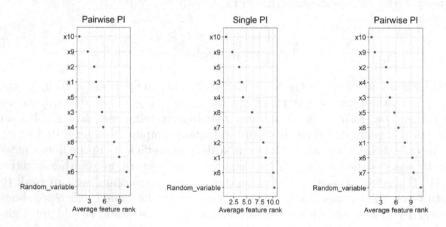

(a) x_1 and x_2 with $\rho = 0.9$ (b) x_1 and x_2 with $\rho = 0.9$ (c) x_1 and x_6 with $\rho = 0.9$

Fig. 1. Average rank $\pm$ standard error for each feature based on the pairwise permutation importance algorithm for (a) and (c) and the single permutation importance algorithm for (b).

5   Microbial Biomarkers for Type 2 Diabetes Mellitus

In this section we test the PPA on a real-world dataset, specifically microbiome data. The goal is to obtain biologically relevant markers. Therefore, we downloaded the Qin 2012 microbiome dataset from MLRepo [20,21]. This curated classification dataset contained shotgun data for 124 samples, representing Chinese healthy controls (n = 59) and Type 2 Diabetes Mellitus (T2D) patients (n = 65). For full details of the preprocessing of the raw sequence reads for datasets in MLRepo, see [20]. We used the same procedure as in the previous section with some additional preprocessing. First, the read counts were rarefied to 28 358 reads per sample, which was the lowest observed number of reads in a sample. After that, features with less than 6 reads per sample on average, representing a relative abundance of 0.02%, were removed. The final dataset consisted then of 124 samples with 377 microbial OTUs.

5.1   Results

The classification model was able to achieve an average ROC AUC score of 0.92 ± 0.05, as depicted in Fig. 2(a). Figure 2(b) represents the top 15 most predictive OTUs in the classification model. Analyzing these OTUs (and several more beyond the top 15) primarily highlights 2 main patterns. The strongest pattern observed in the data, most likely represents an effect that T2D has on the dietary behavior of these Chinese T2D patients. *Lactobacillus acidophilus*, *Acidaminococcus intestini* and *Anaerostipes caccae* are strongly associated with T2D and with each other in this dataset. A regular dose of *L. acidophilus* is commonly recommended in Chinese Medicine [22]. Fermented soybean products are popular in China (i.a.) and various of these products commonly contain *L. acidophilus* [23–25]. Indeed, there is evidence that supports the beneficial claims regarding these fermented products and T2D [26,27]. Trans-aconitic acid in the urine is a biomarker for the consumption of soy products [28] and *A. intestini* is known to be able to oxidise trans-aconitate [29] converting it to acetate. *A. caccae*, is an acetate and lactate consuming butyrate producer. Cross-feeding interactions between *L. acidophilus* and *A. caccae* have been analyzed in detail in vitro [30]. Other butyrate producing species, like *Roseburia intestinalis*, can have similar cross-feeding interactions [31] but were not part of this specific pattern, but with the 2nd main pattern (see below), suggesting that *A. caccae* was part of same fermented soybean product popular with, or given to, these Chinese T2D patients that likely also contained *L. acidophilus* and *A. intestini*.

The second pattern involves several butyrate producers (the *Roseburia, Faecalibacterium, Coprococcus* genera, several *Eubacterium* species and *Anaerostipes hadrus*) in a cross-feeding relationship with various acetate producing dietary fibre degrading species (*Blautia* and *Ruminococcus* representatives). This cluster of species is generally found to be negatively associated with T2D, not just in this study throughout the diabetes microbiome field [21,32–35]. Insufficient butyrate production has been associated with both T1D and T2D development both in rats, mice and in humans [36–39]. Besides being used by colonocytes as

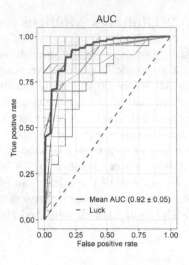 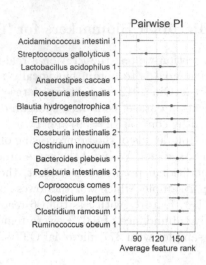

(a) The ROC-AUC plot.   (b) Top 15 feature importance ranks.

Fig. 2. (a) Individual ROC AUC curves for each shuffle and average ROC AUC plot for all shuffles. (b) Average rank ± standard error for the top 15 ranked features based on the pairwise permutation importance algorithm.

a primary energy source [40] butyrate is a powerful inhibitor of histone deacetylase, which has emerged as a target in the control of insulin resistance [42–44]. Animal and in vitro studies have generally found a beneficial effect of butyrate and acetate on glucose homeostasis and insulin sensitivity [45].

6   Conclusions

In this paper, we have set a first step in correcting the compensation effect, observed for 'permute and relearn' permutation importances in case correlated features are present. Our new PPA is able to obtain the right ranking of features, when two features are highly correlated and have the same importance, stated by the magnitude of their coefficient, in linear models. Furthermore, while not yet optimal for correlations between more than 2 features or correlated features with unequal importance related to the output variable, our PPA is already able to obtain relevant biological insights in a Chinese Type 2 Diabetes microbiome dataset.

Acknowledgments. We would like to thank Manon Balvers for helping with computational experiments.

References

1. Kiers, H., Smilde, A.: A comparison of various methods for multivariate regression with highly collinear variables. Stat. Meth. Appl. **16**, 193 (2007)

2. Breiman, L.: Random forests. Mach. Learn. **45**(1), 5–32 (2001)
3. Fisher, A., Rudin, C., Dominici, F.: Model class reliance: variable importance measures for any machine learning model class, from the "Rashomon" perspective (2018)
4. Strobl, C., Boulesteix, A.L., Kneib, T., Augustin, T., Zeileis, A.: Conditional variable importance for random forests. BMC Bioinform. **9**, 307 (2008). https://doi. org/10.1186/1471-2105-9-307
5. Lundberg, S.M., Lee, S.-I.: A unified approach to interpreting model predictions. In: Advances in Neural Information Processing Systems (2017)
6. Ribeiro, M., Singh, S., Guestrin, C.: Why should i trust you?: explaining the predictions of any classifier. eprint arXiv:1602.04938 (2016)
7. Pereira, J., Groen, A.K., Stroes, E.S.G., Levin, E.: Graph space embedding. In: Proceedings of the 28th International Joint Conference on Artificial Intelligence Main Track, pp. 3253–3259 (2019). https://doi.org/10.24963/ijcai.2019/451
8. Shrikumar, A., Greenside, P., Kundaje, A.: Learning important features through propagating activation differences. arXiv preprint arXiv:1704.02685 (2017)
9. Kootte, R.S., et al.: Improvement of insulin sensitivity after lean donor feces in metabolic syndrome is driven by baseline intestinal microbiota composition. Cell Metab. **26**, 611–619 (2017)
10. Ojala, M., Garriga, G.C.: Permutation tests for studying classifier performance. J. Mach. Learn. Res. **11**, 1833–1863 (2010)
11. Hooker, G., Mentch, L.: Please stop permuting features an explanation and alternatives. arXiv preprint arXiv:1905.03151v1 (2019)
12. Grömping, U.: Variable importance assessment in regression: linear regression versus Random Forest. Am. Stat. **63**(4), 308–319 (2009). https://doi.org/10.1198/tast.2009.08199
13. Tolosi, L., Lengauer, T.: Classification with correlated features: unreliability of feature ranking and solutions. Bioinformatics **27**(14), 1986–1994 (2011)
14. Gregorutti, B., Michel, B., Saint-Pierre, P.: Correlation and variable importance in random forests. Stat. Comput. **27**(3), 659–678 (2016). https://doi.org/10.1007/s11222-016-9646-1
15. Imangaliyev, S., Keijser, B., Crielaard, W., Tsivtsivadze, E.: Personalized microbial network inference via co-regularized spectral clustering. Methods **83**, 28–35 (2015)
16. Ishwaran, H., et al.: Variable importance in binary regression trees and forests. Electron. J. Stat. **1**, 519–537 (2007)
17. Caruana, R., Niculescu-Mizil, A., Crew, G., et al.: Ensemble selection from libraries of models. In: 21st International Conference on Machine Learning, ICML 2004, vol. 18. ACM (2004)
18. Chen, T., Guestrin, C.: XGBoost: a scalable tree boosting system. In: Proceedings of the 22nd ACM SIGKDD International Conference on Knowledge Discovery and Data Mining, pp. 785–794. ACM (2016)
19. Meinshausen, N., Bühlmann, P.: Stability selection. J. R. Stat. Soc. Ser. B (Stat. Methodol.) **72**, 417–473 (2010)
20. Vangay, P., Hillmann, B.M., Knights, D.: Microbiome Learning Repo (ML Repo): a public repository of microbiome regression and classification tasks. GigaScience **8**, 1–12 (2019)
21. Qin, J., Li, Y., Cai, Z., et al.: A metagenome-wide association study of gut microbiota in type 2 diabetes. Nature **490**, 55–60 (2012)
22. Cohen, M.R.: The New Chinese Medicine Handbook: An Innovative Guide to Integrating Eastern Wisdom with Western Practice for Modern Healing, Fair Winds Press (2015)

23. Chang, S.Y., Kim, D.-H., Han, M.J.: Physicochemical and sensory characteristics of soy yogurt fermented with Bifidobacterium breve K-110, Streptococcus thermophilus 3781, or Lactobacillus acidophilus Q509011. Food Sci. Biotechnol. **19**, 107–113 (2010). https://doi.org/10.1007/s10068-010-0015-0

24. Bedani, R., Rossi, E.A., Isay Saad, S.M.: Impact of inulin and okara on Lactobacillus acidophilus La-5 and Bifidobacterium animalis Bb-12 viability in a fermented soy product and probiotic survival under in vitro simulated gastrointestinal conditions. Food Microbiol. **34**(2), 382–389 (2013)

25. Kanda, H., Wang, H.L., Hesseltine, C.W., et al.: Yoghurt production by Lactobacillus fermentation of soybean milk. Process Biochem. **11**(4), 23 (1976)

26. Kwon, D.Y., Daily, J.W., III., Kim, H.J.: Antidiabetic effects of fermented soybean products on type 2 diabetes. Nutr. Res. **30**(1), 1–13 (2010)

27. Mueller, N.T., Odegaard, A.O., Gross, M.D., et al.: Soy intake and risk of type 2 diabetes mellitus in Chinese Singaporeans. Eur. J. Nutr. **51**, 1033–1040 (2012)

28. Münger, L.H., Trimigno, A., Picone, G., et al.: Identification of urinary food intake biomarkers for milk, cheese, and soy-based drink by untargeted GC-MS and NMR in healthy humans. J. Proetome Res. **16**(9), 3321–3335 (2017)

29. Cook, G.M., Wells, J.E., Russell, J.B.: Ability of Acidaminococcus Fermentans to oxidize trans-aconitate and decrease the accumulation of tricarballylate, a toxic end product of ruminal fermentation. Appl. Environ. Microbiol. **60**(7), 2533–2537 (1994)

30. Moens, F., Verce, M., De Vuyst, L.: Lactate- and acetate-based cross-feeding interactions betweeen selected strains of Lactobacilli. Bifidobacteria and colon bacteria in the presence of inulin-type fructans. Int. J. Food Microbiol. **241**, 225–236 (2017)

31. Saulnier, D.M.A., Spinler, J.K., Gibson, G.R., et al.: Mechanisms of Probiosis and Prebiosis: considerations for enhanced functional foods. Curr. Opin. Biotechnol. **20**(2), 135–141 (2009)

32. de Goffau, M.C., Luopajärvi, K., Knip, M., et al.: Fecal microbiota composition differs between children with beta-cell autoimmunity and those without. Diabetes **62**(4), 1238–1244 (2013)

33. Hur, K.Y., Lee, M.-S.: Gut microbiota and metabolic disorders. Diabetes Metab. J. **39**(3), 198–203 (2015)

34. Hartstra, A.V., Bouter, K.E.C., Bäckhed, F., et al.: Insights into the role of the microbiome in obesity and type 2 diabetes. Diabetes Care **38**(1), 159–165 (2015)

35. Murri, M., Leiva, I., Gomez-Zumaquero, J.M., et al.: Gut microbiota in children with type 1 diabetes differs from that in healthy children: a case-control study. BMC Med. **11**, 46 (2013)

36. Noureldein, M.H., Bitar, S., Youssef, N.: Butyrate modulates Diabetes-linked gut dysbiosis: epigenetic and mechanistic modifications. J. Mol. Endocrinol. **64**(1), 29–42 (2020)

37. Endesfelder, D., Engel, M., Davis-Richardson, A.G., et al.: Towards a functional hypothesis relating anti-islet cell autoimmunity to the dietary impact on microbial communities and butyrate production. Microbiome **4**, 17 (2016)

38. Jia, L., Li, D., Feng, N., et al.: Anti-diabetic effects of clostridium butyricum CGMCC0313.1 through promoting the growth of gut butyrate-producing bacteria in Type 2 Diabetic Mice. Sci. Rep. **7**(1), 7046 (2017)

39. Khan, S., Jena, G.: Sodium butyrate reduces insulin-resistance, fat accumulation and dyslipidemia in Type-2 Diabetic rat: a comparative study with metformin. Chem. Biol. Interact. **254**, 124–134 (2016)

40. Donohoe, D.R., Garge, N., Zhang, X., et al.: The microbiome and butyrate regulate energy metabolism and autophagy in the mammalian colon. Cell Metab. **13**(5), 517–526 (2012)
41. Pereira, J., Groen, A.K., Stroes, E.S.G., Levin, E.: Graph space embedding. In: Proceedings of the 28th International Joint Conference on Artificial Intelligence, IJCAI 2019 (2019)
42. Sharma, S., Taliyan, R.: Histone deacetylase inhibitors: future therapeutics for insulin resistance and type 2 diabetes. Pharmacol. Res. **113**(Pt A), 320–326 (2016)
43. Dirice, E., Ng, R.W.S., Martinez, R., et al.: Isoform-selective inhibitor of histone deacetylase 3 (HDAC3) limits pancreatic islet infiltration and protects female nonobese diabetic mice from diabetes. J. Biol. Chem. **292**(43), 17598–17608 (2017)
44. Khan, S., Jena, G.: The role of butyrate, a histone deacetylase inhibitor in diabetes mellitus: experimental evidence for therapeutic intervention. Epigenomics **7**(4), 669–680 (2015)
45. Canfora, E.E., Jocken, J.W., Blaak, E.E.: Short-chain fatty acids in control of body weight and insulin sensitivity. Nat. Rev. Endocrinol. **11**(10), 577–591 (2015)

A Classification of Anomaly Explanation Methods

Véronne Yepmo Tchaghe, Grégory Smits[(✉)], and Olivier Pivert

Univ Rennes, IRISA - UMR 6704, 22305 Lannion, France
{veronne.yepmo-tchaghe,gregory.smits,olivier.pivert}@irisa.fr

Abstract. The usage of algorithms in real-world situations is strongly desired. But, in order to achieve that, final users need to be reassured that they can trust the outputs of algorithms. Building this trust requires algorithms not only to produce accurate results, but also to explain why they got those results. From this last problematic a new field has emerged: eXplainable Artificial Intelligence (XAI). Deep learning has greatly benefited from that field, especially for classification tasks. The considerable amount of works and surveys devoted to deep explanation methods can attest that. Other machine learning tasks, like anomaly detection, have received less attention when it comes to explaining the algorithms outputs. In this paper, we focus on anomaly explanation. Our contribution is a categorization of anomaly explanation methods and an analysis of the different forms anomaly explanations may take.

Keywords: Anomaly explanation · Outlier interpretation · XAI

1 Introduction

An outlier/anomaly/irregularity is *an observation which deviates so much from other observations as to arouse suspicions that it was generated by a different mechanism* [8]. Finding those deviating observations constitutes anomaly detection. Anomaly detection has many applications, ranging from spam detection in mail servers to the identification of cancerous cells in MRI photographs. It can be dealt with a binary classification task in which there is high imbalance between the classes (because anomalies are few in comparison to regular data points). But this requires knowing in advance the types of anomalies that can be found (even in real-world situations), which is not a suitable assumption because new anomalies, different from the ones learned, can appear after the model was developed. A more realistic design of anomaly detection uses density: low density regions are more likely to contain outliers (e.g.: Local Outlier Factor (LOF)). Another outlier identification setting is to build a model for the regular instances and consider as anomalies instances which do not fit the model (e.g.: clustering, isolation forest, one-class Support Vector Machines (one-class SVMs)).

In regular classification, when explaining an instance, the interest is on what makes the instance similar to the other instances of the same class; which common properties are shared by the instances of the same class. In contrast, in

© Springer Nature Switzerland AG 2021
M. Kamp et al. (Eds.): ECML PKDD 2021 Workshops, CCIS 1524, pp. 26–33, 2021.
https://doi.org/10.1007/978-3-030-93736-2_3

anomaly explanation, the knowledge sought is about how the anomaly differs from the other instances. That is why anomaly explanation cannot be entirely managed like classification explanation and deserves a particular attention. This paper provides a taxonomy of anomaly explanation methods. After a review of the existing works on the topic in Sect. 2, the proposed categorization will be detailed in Sect. 3. For each category, its advantages and limits, the purpose of the explanations generated, along with some examples will be given.

2   Existing Comparison Criteria of Anomaly Explanations

Although there is no work entirely dedicated to a review of the anomaly explanation field, existing approaches to anomaly explanation are sometimes compared according to their properties and to the nature of the generated output. In [4,23], two types of anomaly explanation methods are mentioned: model-agnostic explanations and model-specific explanations. A model-specific method is a method developed for a particular machine learning algorithm, while a model-agnostic method can be used with any algorithm. In [4], another categorization is introduced in addition to the previous one: local vs global anomaly explanation methods. A local method explains one outlier at a time and a global method provides explanations for all the anomalies of the dataset at once. In [7], feature-based explanations, semantic explanations, visualisation techniques, metrics, model-specific methods and model-agnostic methods are used as categories of anomaly explanation methods. In [15], several categories are identified: anomaly detectors with explanations, outlier explanations for groups of outliers, outlying aspects mining (which identifies which subset of features makes a data point different from the rest of the dataset), outlying property detection (which finds the feature that makes a data point different from the data points that are the most similar to it), pictorial explanations, decision rules and sequential feature explanations. Although the work in [18] is mainly a survey on anomaly detection algorithms, the problem of anomaly explanation is also discussed. A distinction is made between model-agnostic explanation methods and neuralization which is the conversion of machine learning models into neural networks in order to use explanation methods developed for neural networks, whatever they are used for. In the other works related to the topic, anomaly explanation methods are listed without a particular classification. The coarsest taxonomies (local vs global and model-agnostic vs model-specific) do not take into account the specificities of the anomaly detection topic. The other taxonomies provide categories which are not really well-distinguishable: in [7] for instance, the feature-based explanations can be model-specific or model-agnostic; semantic explanations and visualisation techniques can be used for feature-based explanations.

3   Taxonomy of Anomaly Explanation Methods

We will consider the following example: in Table 1, we have a list of products along with their model, unit weight (W) and unit price (P). We want to identify the anomalous products, using the information in the last two columns of the table which represent the true weight (TW) of each product and its true price range (TP) observed on online merchants.

Table 1. List of products and their true characteristics

ID	Model	W (g)	P (USD)	TW (g)	TP (USD)
1	iPhone X	174	550	174	[500–600]
2	iPhone 11	194	600	194	[800–1000]
3	iPhone 12	300	500	164	[1100–1500]
4	Galaxy S20	163	850	163	[800–900]
5	Galaxy S21	169	900	169	[900–1200]
6	Galaxy Note 20	250	900	192	[550–700]
7	MI 11	100	500	196	[450–600]
8	MI 10S	208	300	208	[100–350]
9	POCO F2 Pro	260	800	210	[200–300]

From the table, it can be seen that the anomalies are: the product *2* because of its low price, the product *3* because of its high weight and its low price, the products *6* and *9* because of their high weight and their high price, and the product *7* because of its low weight.

According to the reasons why a product is an anomaly, it is shown hereafter that four types of explanations may be envisaged: **explanation by feature importance**, **explanation by feature values**, **explanation by data points comparisons** and **data structure aware explanation**.

3.1   Anomaly Explanation by Feature Importance

For an algorithm which aims at recognizing in a set of images which ones are cat images and which ones are dog images, the most natural way to tell users why the algorithm tagged a picture as a cat instead of a dog is to return the group of pixels that helped the algorithm to make the difference. This group of pixels can represent the whiskers of the cat on each image for example. In this way, the user will notice that the whiskers are an attribute that the cat possesses, and not the dog, and will therefore understand why the algorithm decided that it is a cat picture. In general, identifying the features/attributes which contributed the most to the decision of an algorithm is a good start and a classical method to provide explanations. Anomaly detection is also concerned. In Fig. 1a below, to mark the square data point as anomalous, we can look only at the feature f_1 for all the instances: in comparison to the regular data points in blue for which the values of the attribute f_1 vary between -1 and 8, it takes the value 12. The same cannot be told for the feature f_2 since the square instance has a value of 2.5 for that attribute, which is normal when compared to the values of f_2 for the regular instances. Consequently, to explain that anomaly to the user, it can be said that attribute f_1 contributed to the abnormality of the square data point. This first category of anomaly explanation is **feature importance**. The contribution of each feature can be weighted or not. With weighted feature importance, the feature *unitprice* will receive a higher weight than the feature *unitweight* for outlier *9*. Both attributes contribute to making the instance anomalous, but the feature *unitprice* contributes the most because it is further away from the regular values than *unitweight* is, for that instance. This type of explanations is the most explored one. The works in [1,4,9,16] for example provide weighted feature importance explanations. Those in [5,14,20] are non-weighted feature importance anomaly explanation methods. They just return the most important features without quantifying their priority with respect to anomaly identification.

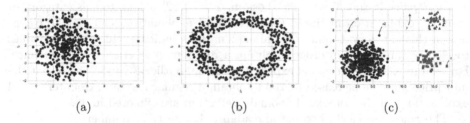

(a)   (b)   (c)

Fig. 1. The necessity of anomaly explanation by feature importance (a) by feature values (b) and by structure analysis (c) (Color figure online)

Anomaly explanation methods based on feature importance do not only provide information about why a specific data point is anomalous, but they can also give a global understanding of the anomalies by identifying the features that explain a set of anomalies or all the anomalies. Furthermore, feature importance can help identify different groups of anomalies, like in [16] where the authors propose a clustering of the anomalies based on the features gradients to identify the types of anomalies present in the data set. But if the original features are transformed prior to the anomaly detection, feature importance scores will not be meaningful to the final users as they will not recognize the features presented by the explanation system. In addition to that, just telling which features are important is sometimes not enough. In Fig. 1b, when trying to explain the abnormality of the square data point using feature importance, we will observe that both features have equal importance, because one attribute does not help the algorithm to identify the anomaly more than the other: the isolated instance has a regular value for each of the features taken independently. It is the combination of the values for both attributes which makes the data point irregular. In this case, explanation by feature importance will return the attribute pair $\langle f_1, f_2 \rangle$, and that is little information since the complete attribute space is returned. In two dimensions it is easy for the user to plot and observe. But, if we are in higher dimension, which is almost always the case, displaying a list of features with more than two having the same importance is not really helping the user. In these situations, it would have been more helpful to say, for instance, that the data point in Fig. 1b is anomalous because it has a value for the feature f_1 around 7.5 and a value for the feature f_2 around 9. This second category of explanation is called **anomaly explanation by feature values**.

3.2 Anomaly Explanation by Feature Values

All the explanations coming from decision-tree-based anomaly detection algorithms lie in this category. Explanations are in the Disjunctive Normal Form (DNF), and each literal of the DNF is a conjunction of predicates. Each predicate is a condition on the value of a feature which has the form $f\,s\,v$ where f is a feature, s is one of the signs $<, \leq, =, >, \geq$ and v is a feature value. As an illustration, an explanation by feature values of outlier 9 can be: $unitweight \geq 210$ and $unitprice \geq 300$. Works like those in [2,3,10,22] belong to this category of anomaly explanation methods. Counterfactual explanations can also be classified in this category. Counterfactual explanations indicate which feature values to change (and how) in order to obtain a different prediction for an instance. For example, a counterfactual explanation of the outlier 2 will indicate that the unit price must be increased by 200 to obtain a regular instance. Counterfactual explanations in the context of anomaly detection are explored in [6].

The rules can easily become unreadable due to their number. As a result, some authors choose to return a short list of rules, each rule having a limited number of predicates. This can be sub-optimal because some less important (but still important) information about why an instance is anomalous may be ignored. Another flaw of this type of explanations is that, unlike feature importance, it is a bit complicated to explain anomalies globally. In addition to that, extracting and consolidating rules is more complex in terms of time processing. However, rules remain the most natural way of explaining anomalies, and translating rules to a pseudo natural language is relatively easy.

With the two previous categories of explanations, we just have information about the anomaly. We do not know concretely what is the difference between anomalies and regular data points. With the example in Fig. 1b, after discovering that the instance is anomalous because $f_1 = 7.5$ and $f_2 = 9$, the user can ask if a data point with $f_1 = 8$ and $f_2 = 7$ would be anomalous (without plotting the data set of course). Explanations by feature importance and by feature values do not provide an answer to this question. An answer would be provided if the anomaly was explained by directly comparing it to regular data points. This third category of explanations will be called **anomaly explanation by data points comparison**.

3.3 Anomaly Explanation by Data Points Comparisons

Angle-Based Outlier Detection (ABOD) [11] is an unsupervised anomaly detection method providing explanations. To give explanations on why an instance is outlying, ABOD finds its closest instance in the nearest cluster, then computes and returns the difference vector between the two data points. Works in [13,17,21] also belong to this category.

Displaying similar instances and showing the differences between the anomalous instance and similar instances allow the user to concretely and easily perceive why a data point is irregular. But these explanations are very limited by the choice of a distance/similarity metric and require distances computation to find similar instances. Plus, this kind of explanation is not very informative when used alone. When used in combination with the two first categories of explanations, it can provide richer explanations to anomalies.

But if there are different clusters of regular data points in the data set, and each of these clusters has some anomalies as shown in Fig. 1c where there are 3 clusters and 4 anomalies (x_1, x_2, y and z), the most complete explanation that can be provided is telling that x_1 and x_2 are anomalies for the cluster of round instances and why it is the case, that y is an anomaly for the triangles and why, and finally that z is an anomaly for the squares and why.

To provide this kind of detailed explanations, an analysis of the intrinsic structure of the data set is required, followed by a comparison of the anomaly(ies) with this intrinsic structure. This last category of explanations will be called **explanation by structure analysis**. It starts at the anomaly detection level by identifying groups of anomalies and individual anomalies with respect to different groups of regular data points.

3.4 Anomaly Explanation by Structure Analysis

Analyzing the structure means discovering in the dataset groups of regular data points, groups of irregular data points, instances which deviate from each group and instances that are in groups where they are not supposed to be. In the example from Table 1, products can be grouped according to the model in order to identify and explain the anomalies of each model. For example, outlier 2 is an outlier for the model *iPhone 12* because its price is lower than usual, for products of this model. An explanation by structure analysis should provide this information. Besides that, regular products can be grouped according to the true price range, in order to obtain different ranges of products. For example in Table 1, high-end products can be those which true prices range in the interval [800–1500], low-end products those which true prices range from 100 to 400 and, an intermediate range of products can contain those for which *unitprice* ∈ [450–700]. With this breakdown, an explanation by structure analysis for the outlier 2 is that according to its unit price it is a mid-range product, but it is not normal because products of this model are supposed to be high-end products. This kind of explanations can be provided by analyzing in details (possibly manually) the detected anomalies, but the goal is to simplify the process as much as possible, for humans and for the computer. Identifying the anomalies and giving directly this type of detailed explanations could be very useful. Two works have been identified along these lines [12,19], but this type of explanation is sorely lacking references.

Anomaly explanation by structure analysis provides the most detailed information about why instances are anomalous and it is certainly the kind of explanation the most expected in various applicative contexts. But it has not been deeply explored yet. The works identified as belonging to this category are a sequence of steps (anomaly detection -> clustering -> analysis of the clusters). No method in the literature has provides so far a unified algorithm going directly from the detection to the detailed explanations. Also, the two methods identified in the literature explain anomalies in groups. But structure analysis should also be able to explain why a specific data point is anomalous, and not only why a set of instances are anomalous.

4   Conclusion

This work aimed at providing a categorization of anomaly explanation methods and at opening directions for future works on that crucial and topical field. Four categories were defined in order to provide a taxonomy which takes into account the particularities of anomaly detection and which is more refined than the taxonomies existing in the literature: **feature importance**, **feature values**, **data points comparisons** and **structure analysis**. Anomaly explanation by feature importance has been widely explored, in contrast to structure analysis which provides the most detailed explanations. For this last category of explanations, the integration of human experts can be investigated: a human expert can help describe the structure of the regularities or irregularities so as to facilitate the identification and the explanation of anomalies. In conclusion, a lot can still be done in relation to anomaly explanation. Although the field can leverage the methods developed for other tasks (like classification) or for neural networks, there is a need for explanation methods specifically built for anomalies.

References

1. Antwarg, L., Shapira, B., Rokach, L.: Explaining anomalies detected by autoencoders using SHAP. arXiv preprint arXiv:1903.02407 (2019)
2. Barbado, A., Corcho, Ó., Benjamins, R.: Rule extraction in unsupervised anomaly detection for model explainability: application to oneclass SVM. arXiv preprint arXiv:1911.09315 (2019)
3. Baseman, E., Blanchard, S., DeBardeleben, N., Bonnie, A., Morrow, A.: Interpretable anomaly detection for monitoring of high performance computing systems. In: Outlier Definition, Detection, and Description on Demand Workshop at ACM SIGKDD, San Francisco, August 2016 (2016)
4. Carletti, M., Terzi, M., Susto, G.A.: Interpretable anomaly detection with DIFFI: depth-based feature importance for the isolation forest. arXiv preprint arXiv:2007.11117 (2020)
5. Gupta, N., Eswaran, D., Shah, N., Akoglu, L., Faloutsos, C.: Beyond outlier detection: LOOKOUT for pictorial explanation. In: Berlingerio, M., Bonchi, F., Gärtner, T., Hurley, N., Ifrim, G. (eds.) ECML PKDD 2018. LNCS (LNAI), vol. 11051, pp. 122–138. Springer, Cham (2019). https://doi.org/10.1007/978-3-030-10925-7_8

6. Haldar, S., John, P.G., Saha, D.: Reliable counterfactual explanations for autoencoder based anomalies. In: 8th ACM IKDD CODS and 26th COMAD, pp. 83–91 (2021)
7. Hamelers, L.: Detecting and explaining potential financial fraud cases in invoice data with Machine Learning. Master's thesis, University of Twente (2021)
8. Hawkins, D.M.: Identification of Outliers, vol. 11. Springer, Dordrecht (1980). https://doi.org/10.1007/978-94-015-3994-4
9. Kauffmann, J., Müller, K.R., Montavon, G.: Towards explaining anomalies: a deep Taylor decomposition of one-class models. Pattern Recogn. **101**, 107198 (2020)
10. Kopp, M., Pevný, T., Holeňa, M.: Anomaly explanation with random forests. Exp. Syst. Appl. **149**, 113187 (2020)
11. Kriegel, H.P., Schubert, M., Zimek, A.: Angle-based outlier detection in high-dimensional data. In: Proceedings of the 14th ACM SIGKDD International Conference on Knowledge Discovery and Data Mining, pp. 444–452 (2008)
12. Macha, M., Akoglu, L.: Explaining anomalies in groups with characterizing subspace rules. Data Min. Knowl. Disc. **32**(5), 1444–1480 (2018). https://doi.org/10.1007/s10618-018-0585-7
13. Mejia-Lavalle, M.: Outlier detection with innovative explanation facility over a very large financial database. In: 2010 IEEE Electronics, Robotics and Automotive Mechanics Conference, pp. 23–27. IEEE (2010)
14. Micenková, B., Ng, R.T., Dang, X.H., Assent, I.: Explaining outliers by subspace separability. In: 2013 IEEE 13th International Conference on Data Mining, pp. 518–527. IEEE (2013)
15. Mokoena, T.: Why is this an anomaly? Explaining anomalies using sequential explanations. Ph.D. thesis (2019)
16. Nguyen, Q.P., Lim, K.W., Divakaran, D.M., Low, K.H., Chan, M.C.: GEE: a gradient-based explainable variational autoencoder for network anomaly detection. In: 2019 IEEE Conference on Communications and Network Security (CNS), pp. 91–99. IEEE (2019)
17. Rieck, K., Laskov, P.: Visualization and explanation of payload-based anomaly detection. In: 2009 European Conference on Computer Network Defense, pp. 29–36. IEEE (2009)
18. Ruff, L., et al.: A unifying review of deep and shallow anomaly detection. In: Proceedings of the IEEE (2021)
19. Shukla, A.K., Smits, G., Pivert, O., Lesot, M.J.: Explaining data regularities and anomalies. In: 2020 IEEE International Conference on Fuzzy Systems (FUZZ-IEEE), pp. 1–8. IEEE (2020)
20. Siddiqui, M.A., Fern, A., Dietterich, T.G., Wong, W.K.: Sequential feature explanations for anomaly detection. ACM Trans. Knowl. Disc. Data (TKDD) **13**(1), 1–22 (2019)
21. Smith-Renner, A., Rua, R., Colony, M.: Towards an explainable threat detection tool. In: IUI Workshops (2019)
22. Song, F., Diao, Y., Read, J., Stiegler, A., Bifet, A.: EXAD: a system for explainable anomaly detection on big data traces. In: 2018 IEEE International Conference on Data Mining Workshops (ICDMW), pp. 1435–1440. IEEE (2018)
23. Zhang, X., Marwah, M., Lee, I.T., Arlitt, M., Goldwasser, D.: ACE - an anomaly contribution explainer for cyber-security applications. In: 2019 IEEE International Conference on Big Data (Big Data), pp. 1991–2000. IEEE (2019)

Bringing a Ruler Into the Black Box: Uncovering Feature Impact from Individual Conditional Expectation Plots

Andrew Yeh[1,2(✉)] and Anhthy Ngo[1,3]

[1] New York University, New York, NY 10011, USA
{ay1626,an3056}@nyu.edu
[2] The Wharton School of the University of Pennsylvania, Philadelphia, PA 19104, USA
ayeh21@upenn.edu
[3] The MITRE Corporation, McLean, VA 22102, USA
ango@mitre.org

Abstract. As machine learning systems become more ubiquitous, methods for understanding and interpreting these models become increasingly important. In particular, practitioners are often interested both in what features the model relies on and how the model relies on them – the feature's impact on model predictions. Prior work on feature impact including partial dependence plots (PDPs) and Individual Conditional Expectation (ICE) plots has focused on a visual interpretation of feature impact. We propose a natural extension to ICE plots with ICE feature impact, a model-agnostic, performance-agnostic feature impact metric drawn out from ICE plots that can be interpreted as a close analogy to linear regression coefficients. Additionally, we introduce an in-distribution variant of ICE feature impact to vary the influence of out-of-distribution points as well as heterogeneity and non-linearity measures to characterize feature impact. Lastly, we demonstrate ICE feature impact's utility in several tasks using real-world data.

1 Introduction

As machine learning (ML) systems become more ubiquitous in human decision making, transparency and interpretability have grown significantly in importance [14]. Some models may not require user trust due to a low-risk nature, e.g. movie recommendation systems. Other problems don't require top performance and safely rely on highly interpretable models that may not perform as well as black box models. However, when a problem space combines a high risk nature with demands for superior performance, earning the user's trust in the model is essential.

We distinguish three phases to "trusting" a model: strong performance, model understanding, and prediction understanding (See Fig. 1). To distinguish a feature's contribution to model performance from its contribution to model predictions, we call the former "feature importance" and the latter "feature impact" [11].

We thank David Rosenberg for his insightful feedback, engagement, and advice in shaping this project from proposal to paper as well for introducing us to the original ICE paper. We thank Anu-Ujin Gerelt-Od without whom this project would not be possible. Last but not least, we thank Lee Kho for her valuable input, ideas, and support.

© Springer Nature Switzerland AG 2021
M. Kamp et al. (Eds.): ECML PKDD 2021 Workshops, CCIS 1524, pp. 34–48, 2021.
https://doi.org/10.1007/978-3-030-93736-2_4

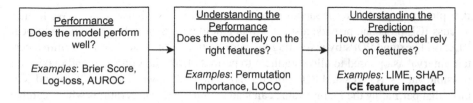

Fig. 1. Three stages in model trust

There exist several visual methods to display feature impact, the relationship between features and predictions, most notably partial dependence plots (PDPs) [6] and individual conditional expectation (ICE) plots [7]. PDPs aggregate the effects of a feature while ICE plots disaggregate divergent effects by plotting individual observations.

Visual tools are highly intuitive and can convey a lot of information in a single plot. However, they have some weaknesses as well. Firstly, visual interpretation is imprecise which makes comparison between features difficult. Secondly, ICE plots in particular can only plot a subset of the observations in the dataset to avoid overcrowding, which can hide outlier observations or overfit extrapolations from view. Thirdly, the cost of visual inspection does not scale well to the number of features–visually inspecting the plots for millions of features, for example, is infeasible.

In this paper, we address these issues and extend ICE plots by extracting feature impact metrics from them ("ICE feature impact"). ICE feature impact is model- and performance-agnostic, meaning it measures the impact of each feature on the prediction only, without regarding the accuracy of that prediction. ICE feature impact also addresses the issues with the visual approach discussed above: it is a precise metric, allowing comparisons between different ICE plots; it takes into account every observation, including outliers, instead of only a subset; and it can be ranked to prioritize inspection of ICE plots to only the most impactful features, allowing the usefulness of ICE plots to scale with the number of features.

We also introduce an in-distribution version of feature impact with a hyperparameter to reduce the influence of out-of-distribution points, and we supplement ICE feature impact with measures of heterogeneity and non-linearity to add depth. Together, these metrics provide a quantitative perspective for understanding feature impact complementary to the qualitative nature of inspecting ICE plots.

2   Related Work

First introduced by Friedman [6], partial dependence plots (PDPs) are a model and performance agnostic method of illustrating the relationships between one or more input variables and the predictions of a black-box model. PDPs estimate the partial dependency by marginalizing over all other features – essentially permuting the at-issue features to specific values across the observed range and then averaging the resulting predictions across training observations.

Individual Conditional Expectation (ICE) [7] plots disaggregate the average feature impact curve of PDPs into its component, individual observation-curves. This allows

ICE plots to capture heterogeneous relationships that PDPs otherwise miss. We further discuss ICE plots and provide a specific methodology in Sect. 3.1. Accumulated Local Effects [1] extend PDPs by restricting the permutation of at-issue features within a certain interval as opposed to allowing them to permute from the minimum and maximum possible values as PDPs and ICE plots do. This addresses a weakness in PDPs and ICE plots that permuting the feature value can lead to unrealistic observations when features are correlated and motivates in-distribution ICE feature impact.

Parr et al. [11] distinguishes the idea of "feature impact" from standard feature importance metrics as follows: while feature importance metrics measure how important a feature is to the model's performance, feature impact metrics measure how variations in feature values impact the prediction, irrespective of performance.

LIME [12] uses an interpretable surrogate model to approximate the feature impact on a local scale around the prediction. Parr et al. [11] proposes a non-parametric feature impact methodology that does not interrogate a fitted model. Instead, they extend the concept of PDPs by calculating the empirical partial dependence of the prediction on the at-issue feature based on the data and then approximating the area under the resulting partial dependence curve with a Riemann's Sum.

Shapley values [13] detail how to fairly determine the total contribution of each feature to the overall prediction–making it a feature impact metric–by taking into account both a feature's individual contribution and collaborative contribution together with all possible subsets of features. Shapley values themselves are highly computationally expensive to calculate precisely, though they can be approximated with a Monte Carlo approach [15], Kernel SHAP [10], or Tree SHAP [9]. Tree SHAP differs from other approaches as it relies solely on the training data without interventionist means like permuting the value of features.

3   Methodology

An implementation of ICE feature impact as described below is available in Github.[1]

3.1   ICE Plot Replication

We establish terminology and notation for the remainder of the paper by detailing the ICE replication methodology we use. To replicate ICE plots, we create "phantom observations" from each "real observation" where all not "at-issue features(s)" are constant, but we permute the "at-issue feature(s)". We then use the phantom observations to interrogate the model.

The exact algorithm is as follows: for at-issue feature(s) x_S, fitted model $\hat{f}$, and feature matrix $\mathbf{X} \in \mathbb{R}^{N \times p}$, let there be $n_{\mathbf{x}_S}$ unique values of x_S found in the data.

1. For each observation $x^{(i)}$, create $\mathbf{n_{x_S}}$ observations with all features the same as in $x^{(i)}$, except for $\mathbf{x_S}$. Replace $\mathbf{x_S}$ with the $n_{\mathbf{x}_S}$ unique values of feature p found above. This results in $n_{\mathbf{x}_S}$ new observations for each $x^{(i)}$.

---

[1] https://github.com/mixerupper/mltools-fi_cate.

2. We call the resulting observations "phantom observations", denoted $x^{(i)}[k]$ which is the kth phantom observation for $x^{(i)}$ with $k = 1, \ldots, n_{\mathbf{x_S}}$. For each observation $x^{(i)}$, one of its phantom observations is exactly identical to $x^{(i)}$, and the others are identical except for a permuted $\mathbf{x_S}$. Combine all $n \cdot n_{\mathbf{x_S}}$ phantom observations into a new feature matrix.
3. Use fitted model $\hat{f}$ to predict $\hat{y}$ for all phantom observations.
4. For each original observation, plot a line composed of the corresponding phantom points with the at-issue feature on the x-axis and $\hat{y}$ on the y-axis. This results in n lines, with each line composed of $n_{\mathbf{x_S}}$ phantom points.

Additionally, if n is large, we sample uniformly from each quantile of $\mathbf{x_S}$ if $\mathbf{x_S}$ is continuous and each value of $\mathbf{x_S}$ if $\mathbf{x_S}$ is categorical to capture the whole distribution.

3.2 ICE Feature Impact

While ICE plots allow visual inspection of feature impact, it does not output any quantitative metrics for comparability. We elicit a numeric feature impact metric from ICE plots in the form of ICE feature impact.

For the sequence of points that make up each observation-curve, we calculate the absolute change in prediction divided by the change in feature ($|\frac{dy}{dx}|$) for each consecutive point. This uses rise over run to quantify the impact of the feature on the prediction value. Then, ICE feature impact is the mean of all the $|\frac{dy}{dx}|$ terms over all phantom points that make up an observation and all observations. To account for features of different scales, we multiply by the standard deviation of that feature. We will see that ICE feature impact has an analogous interpretation to coefficients in a linear model.

The exact algorithm is as follows: for feature $\mathbf{x}_S$, let $\sigma_{\mathbf{x}_S}$ denote the standard deviation of $\mathbf{x}_S$, let n be the number of observations, $n_{\mathbf{x}_S}$ be the number of unique values of $\mathbf{x}_S$, $x^{(i)}$ be the ith observation, $x^{(i)}[k]$ be the kth phantom observation corresponding to $x^{(i)}$, $\mathbf{x}_S^{(i)}$ be the value of $\mathbf{x}_S$ in observation $x^{(i)}$, $\mathbf{x}_S^{(i)}[k]$ be the value of $\mathbf{x}_S$ in the kth phantom observation corresponding to $x^{(i)}$, and $\hat{y}$ be the fitted model. Thus, the **ICE feature impact** is:

$$
\begin{aligned}
\mathbf{FI}(\mathbf{x}_S) &= \frac{\sigma_{\mathbf{x}_S}}{n \cdot (n_{\mathbf{x}_S} - 1)} \sum_{i=1}^{n} \sum_{k=2}^{n_{\mathbf{x}_S}} \left| \frac{d\hat{y}(x^{(i)}[k])}{dx_S^{(i)}[k]} \right| \\
&\approx \frac{\sigma_{\mathbf{x}_S}}{n \cdot (n_{\mathbf{x}_S} - 1)} \sum_{i=1}^{n} \sum_{k=2}^{n_{\mathbf{x}_S}} \left| \frac{\hat{y}(x^{(i)}[k]) - \hat{y}(x^{(i)}[k-1])}{x_S^{(i)}[k] - x_S^{(i)}[k-1]} \right|
\end{aligned}
\tag{1}
$$

The ICE feature impact of $\mathbf{x}_S$ can be interpreted as the absolute change in the predicted value of $\hat{y}$ for each one-unit change in $\mathbf{x}_S$ if $\mathbf{x}_S$ was standardized to a standard deviation of 1 and all other features remained constant. Note that ICE feature impact gives the magnitude of impact, not the direction. Average direction of feature impact can be determined by comparing the ICE feature impact with the value of Eq. 1 without an absolute value on the inner summation term.

3.3 In-Distribution ICE Feature Impact

One of the drawbacks of ICE feature impact as introduced in Sect. 3.2 is that it weights evenly across all phantom points, no matter their likelihood of occurrence in the true feature distribution. This may be concerning if features are highly correlated, and permuting the at-issue feature $\mathbf{x}_S$ takes us out of the feature distribution, e.g., taking the health data from a 9 year old and changing the age to 70 while leaving the other features untouched would give us a phantom observation that has a low likelihood of occurring in reality.

This is a missing data problem with the missing value being the likelihood of the observation. The likelihood is 1 for all true observations and missing for all phantom observations. Let us denote the likelihood of phantom observation $x^{(i)}[k]$ for at-issue feature $\mathbf{x}_S$ with $L_{\mathbf{x}_S}(x^{(i)}[k])$. Then, given this likelihood, the in-distribution ICE feature impact of $\mathbf{x}_S$ is:

$$
\mathbf{IDFI}(\mathbf{x}_S) \approx \frac{\sigma_{\mathbf{x}_S}}{\sum_{i=1}^{n}\sum_{k=2}^{n_{\mathbf{x}_S}} L_{\mathbf{x}_S}} \sum_{i=1}^{n}\sum_{k=2}^{n_{\mathbf{x}_S}} L_{\mathbf{x}_S}(x^{(i)}[k]) \left| \frac{\hat{y}(x^{(i)}[k]) - \hat{y}(x^{(i)}[k-1])}{\mathbf{x}_S^{(i)}[k] - x_S^{(i)}[k-1]} \right|
\tag{2}
$$

To estimate $L_{\mathbf{x}_S}(x^{(i)}[k])$, we model likelihood as exponentially decaying with respect to the absolute distance of the at-issue feature's permutation divided by the feature's standard deviation:

$$
L_{\mathbf{x}_S}(x^{(i)}[k]) = \lambda^{\frac{|\mathbf{x}_S^{(i)}[k] - x_S^{(i)}|}{\sigma_{\mathbf{x}_S}}}
\tag{3}
$$

where $0 < \lambda \leq 1$ is a hyperparameter that measures how quickly the weight decays as the phantom feature value differs from the real feature value. Note that $\lambda = 1$ gets us back to ICE feature impact without out-of-distribution considerations.

We can estimate $\sigma_{\mathbf{x}_S}$ as the sample standard deviation of $\mathbf{x}_S$ in the data or as an arbitrarily sophisticated estimate of the standard deviation for the at-issue feature based on the value of all other features for the observation. For example, [5] proposes estimating the conditional distribution of a feature based on all other features using a pseudo-maximum likelihood problem estimated via a single self-attention architecture.

The in-distribution ICE feature impact weights phantom observations closer to the real observation more heavily when measuring feature impact, giving us a perspective on feature impact that is more "true to the data" [4].

4 Real Data

To examine ICE feature impact, we use UC Irvine's cervical cancer risk factors dataset.[2] The dataset contains medical information for 858 patients from *Hospital Universitario de Caracas*. There are 32 features including age, number of pregnancies, and use of IUD. The target variable is Biopsy, which is binary.

---

[2] Cervical Cancer (Risk Factors) Data Set contains a detailed description of the dataset.

4.1 Complementary to Feature Importance

First, we show that ICE feature impact presents an additional dimension to understanding models beyond feature importance.

We train a random forest classifier [2] on the dataset.[3] We then calculate the following metrics for each feature: ICE feature impact, Tree SHAP [9], Random Forest feature importance [3], and permutation feature importance [2] and normalize them to be positive and sum to 100. We take the correlation between ICE feature impact and alternative metrics and find that the correlation is low (See Table 1). This indicates that ICE feature impact differs substantially from alternatives instead of fulfilling the same function.

Table 1. Pearson correlation of feature importance and impact metrics with ICE feature impact. All metrics were first normalized to sum to 100. Tree SHAP values were additionally first made positive to remove direction before normalizing to sum to 100.

Metric	Correlation w/ICE FI
In-distribution ICE FI ($\lambda = 0.75$)	0.99
Random Forest feature importance	0.36
Permutation feature importance	0.35
Tree SHAP values	0.17

Table 2 shows the features with the two most positive differences and the features with the two most negative differences between their random forest feature importance and ICE feature impact values.[4] While Age and Number of Sexual Partners are highly predictive features and are helpful in reducing impurity of classification, they do not have a strong impact on the model's predictions itself. On the opposite end of the spectrum, STDs:molluscum contagiosum and STDs:pelvic inflammatory disease have highly imbalanced feature distributions with the majority of values equal to 0 and therefore are not as helpful for reducing impurity. However, when these factors are present – specifically, when the value is missing and the mean is imputed – they contribute strongly to the model prediction, explaining the higher feature impact.

4.2 Interpretability: Analogous to Linear Regression Coefficients

In the base case of analyzing a linear regression model, ICE feature impact values are exactly the absolute value of the linear regression coefficients. We also calculated ICE feature impact for the pseudo-linear models of Logistic Regression and linear SVMs. Table 3 shows that the resulting model coefficients are strongly correlated with the corresponding ICE feature impact values.

---

[3] We use the sklearn package with parameters of 500 trees, a random state seed of 20, and the default values for the remaining parameters. As this exercise is about model interpretability, we did not tune the model to improve performance.

[4] See Appendix C for the full feature impact table, Appendix A for the ICE plots for all features, and Appendix B for the centered ICE plots (c-ICE) [7] for all features.

Table 2. Feature impact table for features in cervical cancer dataset with two largest and most negative difference between Random Forest feature importance and ICE Feature Impact.

Feature	ICE FI	Native feature importance	Difference
STDs:molluscum contagiosum	9.8	0.1	9.6
STDs:pelvic inflammatory disease	8.9	0.1	8.8
Number of sexual partners	1.0	9.9	−8.9
Age	3.4	17.8	−14.3

Table 3. Pearson correlation of ICE feature impact values with absolute value of coefficients of linear and pseudo-linear models.

Model	ICE feature impact	
	Base	In-Dist
Linear regression	1	1
Logistic	0.73	0.8
SVM	0.9	0.98

These results show that ICE feature impact can be interpreted analogously to linear regression coefficients with features standardized to a unit standard deviation.

4.3 Quantifying Heterogeneity and Non-linearity

In linear models, knowing feature impact means knowing exactly where predictions come from. In non-linear models, however, the relationship between features and the model prediction can be more complex: in particular, the relationship can be heterogeneous – different across observations – or non-linear – different across the feature's support. We propose measures of heterogeneity and non-linearity to allow the practitioner a more nuanced understanding of ICE feature impact.

Let heterogeneity be the degree to which the pattern of ICE curves varies across observations, i.e. the feature impact is heterogeneous when its impact is higher on some observations and lower on others. Then, following the notation described in Sect. 3.1, the heterogeneity of feature $\mathbf{x}_S$ is:

$$\mathbf{HE}(\mathbf{x}_S) = \frac{\sigma_{\mathbf{x}_S}}{n_{\mathbf{x}_S}} \sum_{k=1}^{n_{\mathbf{x}_S}} SD_{i \in \{1,\dots,n\}} \left(\frac{\hat{y}(x^{(i)}[k]) - \hat{y}(x^{(i)}[k-1])}{x_S^{(i)}[k] - x_S^{(i)}[k-1]} \right) \quad (4)$$

where the standard deviation is taken for fixed k across all real observations. The lower the heterogeneity metric, the more similar the shape of observation-curves are at each point. For linear regressions and additive models like GAM [8], the heterogeneity metric is zero since the effect of a feature on the prediction is the same across all observations.

Let non-linearity be the degree to which features have a non-linear relationship with the model's predictions, i.e. how much the effect of a feature varies across the support for a given observation. For features with low non-linearity, the corresponding ICE feature impact can be interpreted as close to a linear regression coefficient, even if the underlying model is non-linear. We quantify non-linearity as follows:

$$\mathbf{NL}(\mathbf{x}_S) = \frac{\sigma_{\mathbf{x}_S}}{n} \sum_{i=1}^{n} SD_{k \in \{1, \ldots, n_{\mathbf{x}_S}\}} \left(\frac{\hat{y}(x^{(i)}[k]) - \hat{y}(x^{(i)}[k-1])}{x_S^{(i)}[k] - x_S^{(i)}[k-1]} \right) \quad (5)$$

where the standard deviation is taken for fixed i across all corresponding phantom observations. For linear regressions, the non-linearity is equal to 0 as desired since the effect of a feature is constant across the feature's support.

Table 4 shows the heterogeneity and non-linearity of the features listed in Table 2.[5] Note that the features with the largest positive differences between feature impact and feature importance have higher heterogeneity but similar non-linearity compared to the features with the largest negative differences. This is because ICE feature impact captures heterogeneity through taking the absolute value of the feature impact $\frac{dy}{dx}$ units but does not discriminate between non-linear or linear relationships.

Table 4. ICE feature impact, heterogeneity, and non-linearity for features in cervical cancer dataset with the two most positive and most negative differences between Random Forest feature importance and ICE Feature Impact.

Feature	Feature impact	Heterogeneity	Non-linearity
STDs:molluscum contagiosum	9.8	0.27	0.19
STDs:pelvic inflammatory disease	8.9	0.23	0.17
Number of sexual partners	1.0	0.05	0.04
Age	3.4	0.11	0.18

5 Discussion

Building upon efforts to interpret machine learning models, we extend ICE plots by drawing out ICE feature impact, a measure of the relationship between features and model predictions. ICE feature impact is uncorrelated with alternative feature importance metrics, highlighting features that are impactful to predictions but do not contribute as strongly to model performance. It has a highly interpretable form and is analogous to linear regression coefficients.

We also propose in-distribution ICE feature impact to downweight out-of-distribution observations and the heterogeneity and non-linearity measures that add dimensionality to our characterization of ICE feature impact.

Altogether, ICE feature impact provides a different perspective from traditional feature importance methods, complements ICE plots, and serves as an alternative to SHAP values in understanding where a model's predictions come from.

---

[5] See Appendix D for heterogeneity and non-linearity for all features.

Appendix A   ICE Plots for Cervical Cancer Data

(See Fig.2)

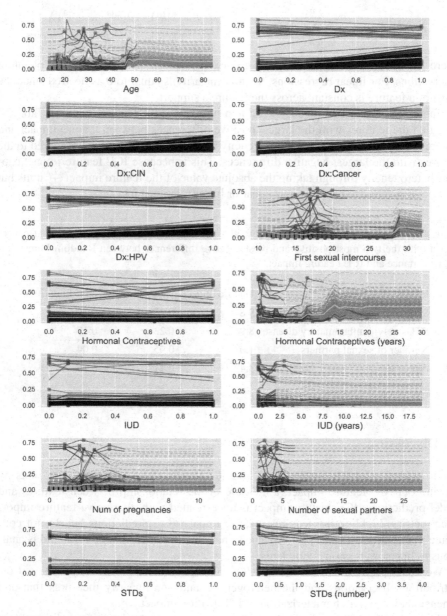

Fig. 2. ICE plots for all features in cervical cancer dataset, following the methodology described in Sect. 3.1. Each green dot represents a different observation, and the corresponding line shows how varying the observation's at-issue feature value affects the model's prediction. Observation-lines are solid within $\frac{1}{2}$ a standard deviation (of the at-issue feature) and dotted outside that range.

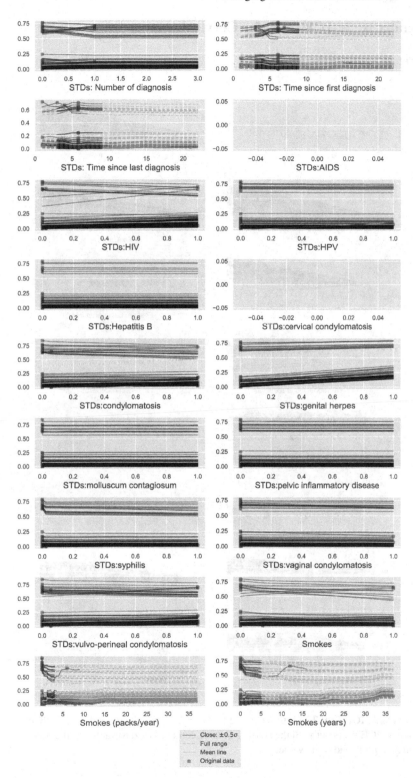

Appendix B   c-ICE Plots for Cervical Cancer Data

(See Fig.3)

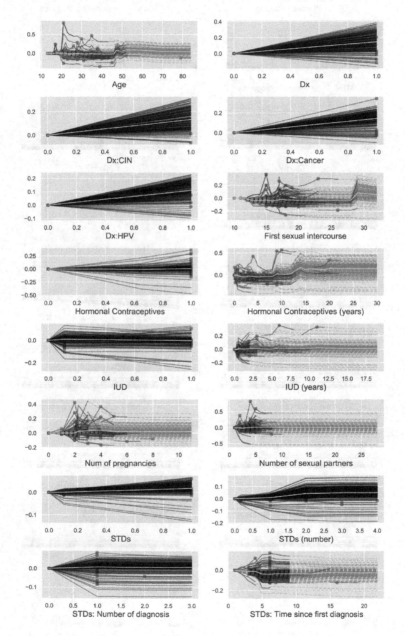

Fig. 3. Centered ICE (c-ICE) plots [7] for all features in cervical cancer dataset. c-ICE plots are equivalent to ICE plots but with the starting $\hat{y}$ value centered to zero such that the lines represent the change in $\hat{y}$ instead of its value.

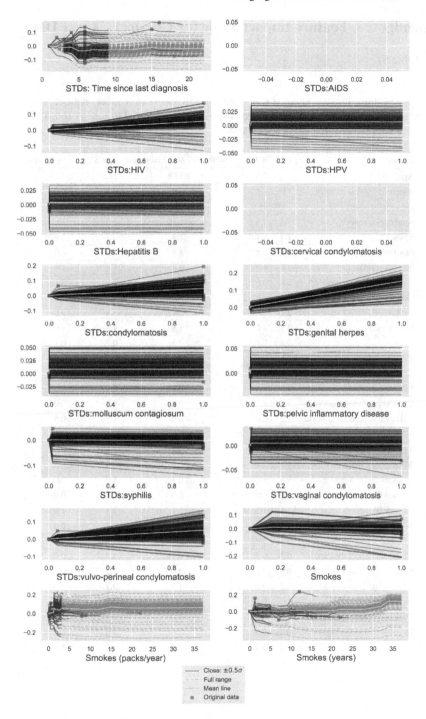

Appendix C   Feature Impact Table for Cervical Cancer Data

(See Table 5)

Table 5. Feature impact table for all features in cervical cancer dataset. All feature impact/importance metrics have been made positive and normalized to sum to 100. The ordering of features is as ordered in the original dataset.

Feature	ICE	ICE ID ($\lambda = 0.75$)	Random Forest	Tree SHAP	Permutation
Age	3.4	2.6	17.8	11.3	13.3
Number of sexual partners	1.0	1.1	9.9	6.3	9.9
First sexual intercourse	16.9	16.9	12.3	7.2	12.1
Num of pregnancies	1.3	1.4	10.0	0.0	13.1
Smokes	1.3	1.1	1.4	0.2	1.9
Smokes (years)	1.8	1.8	3.9	6.8	4.6
Smokes (packs/year)	7.0	6.6	3.7	1.9	4.4
Hormonal Contraceptives	0.8	0.6	2.9	10.8	5.1
Hormonal Contraceptives (ye	9.0	7.8	15.6	7.8	15.9
IUD	2.2	2.0	2.2	3.4	3.0
IUD (years)	3.7	4.0	3.6	4.6	4.6
STDs	0.6	0.4	0.5	0.7	0.1
STDs (number)	0.6	0.6	1.1	1.2	1.0
STDs:condylomatosis	1.2	1.0	0.6	1.8	0.5
STDs:cervical condylomatosis	0.0	0.0	0.0	0.0	0.0
STDs:vaginal condylomatosis	3.8	4.1	0.3	0.8	0.0
STDs:vulvo-perineal condylo	1.1	0.9	0.6	1.5	0.5
STDs:syphilis	2.1	2.2	0.4	1.1	0.0
STDs:pelvic inflammatory di	8.9	9.9	0.1	1.3	0.0
STDs:genital herpes	6.8	7.4	0.9	2.4	0.5
STDs:molluscum contagiosum	9.8	10.8	0.1	1.0	0.0
STDs:AIDS	0.0	0.0	0.0	0.0	0.0
STDs:HIV	1.2	1.2	1.1	1.8	1.9
STDs:Hepatitis B	5.7	6.3	0.2	1.5	0.0
STDs:HPV	5.4	5.9	0.1	1.0	0.0
STDs: Number of diagnosis	0.2	0.2	0.7	4.7	0.0
STDs: Time since first diag	0.3	0.3	2.0	1.2	1.5
STDs: Time since last diagn	0.3	0.3	1.7	2.7	0.0
Dx:Cancer	1.0	0.7	1.7	3.5	2.1
Dx:CIN	0.6	0.5	1.2	3.9	1.1
Dx:HPV	0.8	0.6	1.6	4.0	1.6
Dx	1.2	0.9	1.7	3.7	1.1

Appendix D   Heterogeneity and Non-linearity of ICE Feature Impact for Cervical Cancer Data

(See Table 6)

Table 6. Heterogeneity and non-linearity dimensions for all features in cervical cancer dataset. The raw ICE feature impact is presented without normalization to sum to 100.

Feature	ICE feature impact	ICE heterogeneity	ICE non-linearity
Age	0.08	0.11	0.18
Number of sexual partners	0.02	0.05	0.04
First sexual intercourse	0.38	0.59	1.63
Num of pregnancies	0.03	0.05	0.07
Smokes	0.03	0.04	0.02
Smokes (years)	0.04	0.06	0.12
Smokes (packs/year)	0.16	0.32	0.60
Hormonal Contraceptives	0.02	0.04	0.01
Hormonal Contraceptives (years)	0.20	0.34	0.47
IUD	0.05	0.07	0.04
IUD (years)	0.08	0.12	0.28
STDs	0.01	0.02	0.01
STDs (number)	0.01	0.02	0.01
STDs:condylomatosis	0.03	0.03	0.01
STDs:cervical condylomatosis	0.00	0.00	0.00
STDs:vaginal condylomatosis	0.08	0.11	0.07
STDs:vulvo-perineal condylo	0.02	0.03	0.01
STDs:syphilis	0.05	0.07	0.04
STDs:pelvic inflammatory di	0.20	0.23	0.17
STDs:genital herpes	0.15	0.18	0.13
STDs:molluscum contagiosum	0.22	0.27	0.19
STDs:AIDS	0.00	0.00	0.00
STDs:HIV	0.03	0.03	0.02
STDs:Hepatitis B	0.13	0.17	0.11
STDs:HPV	0.12	0.14	0.10
STDs: Number of diagnosis	0.00	0.00	0.00
STDs: Time since first diag	0.01	0.01	0.02
STDs: Time since last diagn	0.01	0.01	0.02
Dx:Cancer	0.02	0.01	0.00
Dx:CIN	0.01	0.01	0.00
Dx:HPV	0.02	0.01	0.00
Dx	0.03	0.02	0.00

References

1. Apley, D.W., Zhu, J.: Visualizing the effects of predictor variables in black box supervised learning models. J. Roy. Stat. Soc. Ser. B (Stat. Methodol.) **82**(4), 1059–1086 (2020)

2. Breiman, L.: Random Forests. Mach. Learn. **45**(1), 5–32 (2001)
3. Breiman, L.: Manual on setting up, using, and understanding random forests v3.1, pp. 1–58. Statistics Department, University of California Berkeley (2002)
4. Chen, H., Janizek, J.D., Lundberg, S., Lee, S.: True to the model or true to the data? arXiv preprint arXiv:2006.16234 (2020)
5. Fakoor, R., Mueller, J., Erickson, N., Chaudhari, P., Smola, A.J.: Fast, accurate, and simple models for tabular data via augmented distillation. In: Larochelle, H., Ranzato, M., Hadsell, R., Balcan, M., Lin, H. (eds.) Advances in Neural Information Processing Systems 33. NeurIPS (2020)
6. Friedman, J.H.: Greedy function approximation: a gradient boosting machine. Ann. Stat. **29**(5), 1189–1232 (2001)
7. Goldstein, A., Kapelner, A., Bleich, J., Pitkin, E.: Peeking inside the black box: visualizing statistical learning with plots of individual conditional expectation. J. Comput. Graph. Stat. **24**, 44–65 (2013)
8. Hastie, T., Tibshirani, R.: Generalized additive models. Stat. Sci. **1**(3), 297–310 (1986)
9. Lundberg, S.M., Erion, G.G., Lee, S.: Consistent individualized feature attribution for tree ensembles. arXiv preprint arXiv:1802.03888 (2018)
10. Lundberg, S.M., Lee, S.I.: A unified approach to interpreting model predictions. In: Guyon, I., et al. (eds.) Advances in Neural Information Processing Systems 30. Curran Associates, Inc. (2017)
11. Parr, T., Wilson, J.D., Hamrick, J.: Nonparametric feature impact and importance. arXiv preprint arXiv:2006.04750 (2020)
12. Ribeiro, M.T., Singh, S., Guestrin, C.: "Why should I trust you?": explaining the predictions of any classifier. In: KDD, pp. 1135–1144. ACM (2016)
13. Shapley, L.: A value for n-person games. In: Kuhn, H., Tucker, A. (eds.) Contributions to the Theory of Games, vol. 2. Princeton Press, Princeton (1953)
14. Varshney, K.R.: Engineering safety in machine learning. In: ITA, pp. 1–5. IEEE (2016)
15. Štrumbelj, E., Kononenko, I.: Explaining prediction models and individual predictions with feature contributions. Knowl. Inf. Syst. **41**, 647–665 (2013)

Reject and Cascade Classifier with Subgroup Discovery for Interpretable Metagenomic Signatures

Maxence Queyrel[1,2]([✉]), Alexandre Templier[1], and Jean-Daniel Zucker[2,3]

[1] Quinten France, 8 rue Vernier, 75017 Paris, France
[2] Sorbonne University, IRD, UMMISCO, 93143 Bondy, France
[3] Sorbonne University, INSERM, NUTRIOMICS, 75013 Paris, France

Abstract. Over the past decade, technological advances have made high-speed, high-resolution sequencing of genetic material possible at ever lower cost (from millions to one hundred dollars). In this context, the human microbiome has demonstrated its ability to support the stratification and the classification of various human diseases. Thus, the gut microbiota is set to play a key role in precision medicine as a "super-integrator" of patient status. Identifying metagenomic signatures is becoming increasingly important in precision medicine. To address the interpretability/accuracy trade off, we propose a hybrid approach based on a cascade classifier combining a first step of Subgroup Discovery (for interpretability) and then a classifier model (for accuracy). With this approach, different interpretable signatures stratify the maximum possible number of patients while those remaining are defined by a default non-interpretable signature. Several datasets from the NCBI public repository on different diseases (colorectal cancer, cirrhosis, diabetes, obesity) have been used to evaluate the interest of our approach to build both accurate and interpretable metagenomic diseases signatures. The results show that the approach reaches comparable or superior performances to the state-of-the-art approaches while offering better interpretability than black box.

Keywords: Interpretable learning · Cascade model · Subgroup discovery · Metagenomics

1 Introduction

Our vision of the human being seen as an entity in itself has evolved to leave room for a more global vision where the ecosystems that we shelter are taken into account. The term "holobiont" is used to define this concept and characterizes a host (a person, an animal, a plant...) with all the microorganisms living inside. The microorganisms inside an ecosystem (e.g., the gut) can form inter-member organizations, sometimes referred to as "guilds", and create localized interactions that influence their ecosystem (*Wu et al.* 2021). It is thus essential to analyze the composition and the interactions of microbiota members when we study them. Metagenomics manages these objectives and refers to

© Springer Nature Switzerland AG 2021
M. Kamp et al. (Eds.): ECML PKDD 2021 Workshops, CCIS 1524, pp. 49–66, 2021.
https://doi.org/10.1007/978-3-030-93736-2_5

the set of methods supporting microbiota by sequencing and analyzing DNA of several individuals of different taxa[1] in the same environment.

In modern medicine, targeting certain patient populations that would benefit from a particular treatment is becoming an important goal (*Loh et al.* 2019; *Korepanova* 2021). This medicine, called precision medicine, contrasts with the "one-size-fit-all" vision where all patients receive similar treatments, improving the health of only 4–25% of the population (*Petrosino* 2018). Precision medicine is defined as the tailoring of medical treatments to the characteristics of individuals, classifying them into sub-populations that differ in their reaction to a disease or their response to a treatment. Practitioners are increasingly using diagnostics based on new accessible "big data" such as clinical, environmental, "omics" (genomics, transcriptomics, radiomics, etc.), including so-called metagenomic data, which is the quantifying of the metagenome. Thus, it aims to propose more effective therapeutic solutions by acting specifically on the potential causes of disease.

Metagenomic data collection is performed using Next-generation high-throughput DNA sequencing technologies (NGS) allowing to sequence the DNA of any organism providing precise biological information. The outputs of NGS are small sequences of nucleotides, called reads, representing small DNA fragments. Complex bioinformatic workflows, called quantitative metagenomics, have been designed to manipulate metagenomic data generated by NGS and yield quantitative measurements of biological objects such taxa (*Kunin et al.* 2008; *Nayfach and Pollard* 2016; *Wen et al.* 2017; *Quince* et al. 2017). This creates taxa abundance tables generally reported as relative abundance corresponding to a ratio to the total number of taxa present in the microbiota (see Appendix A). The rows and the columns represent N samples (patients, environments, ...) and D features (species, genus, ...) respectively. Based on these data, the goal of this paper is to define an interpretable and personalized stratification method, called *Q-Classifier*, to enhance the interest of precision medicine in metagenomics. Indeed, the approach calculates different interpretable metagenomic signatures that gather certain individuals allowing to define personalized treatments.

2   Related Work

Many studies have focused on supervised learning on abundance tables by comparing several state-of-the-art (SOTA) algorithms to assess the strength of microbiome-phenotype associations by evaluating the generalization of disease-predictive models across cohorts (*Edoardo Pasolli* et al. 2015; *Pasolli* et al. 2016; *Oh and Zhang* 2020). *Thomas* et al. (2019) also analyze, through several studies gathering 969 fecal metagenomes, the reproducibility of metagenomic biomarkers potentially linked to colorectal cancer with random effect models on the microbial richness and diversity. *Harris* et al. (2019) were interested in the results obtained by classifiers on quantitative metagenomic data constructed from the different approaches cited above namely read-based

---

[1] Taxon: (plural taxa), it is an entity grouping all living organisms having in common certain well-defined characteristics. The term taxon is used in phylogenetic classification to group (from the most general to the most specific: domain, kingdom, phylum, class, order, family, genus, species) living beings according to various criteria.

taxonomy profiling and assembly-based method. They found that there was not a large difference between the two approaches, the random forest model achieving the best results in both cases. Moreover, the development of deep learning in metagenomics has led some studies to create an end-to-end pipeline to make prediction from raw metagenomics data based on DNA embeddings learnt with neural network models (*Queyrel* et al. 2021).

Other studies developed statistical analyses as biomarker discovery, species interaction clustering or patient stratification (*Quince* et al. 2017). Biomarker discovery is an important challenge in medicine because it has the goal to yield meaningful biological information. In the case of metagenomics, it solves the problem of finding which microorganisms are likely to explain the difference between certain samples. *Segata* et al. (2011) propose a method named *LefSe* (for linear discriminant analysis (LDA) effect size) making high-dimensional biomarker discovery to identify genomic features that distinguish sample classes. Some recent methods based on microbial ecosystem interactions address this problem. *Prifti* et al. (2020) propose *Predomics*, a family of classification algorithms which explore biomarkers and are a simplification of linear models to be even more interpretable. Another approach, called *GutBalance*, (*Yang* et al. 2021) uses discriminative balance analysis (DBA) method in order to select distal balances of pairs and trios of bacteria. Other

3   Applications of Subgroup Discovery to Metagenomics for Phenotype Status Prediction

3.1   Subgroup Discovery

Subgroup Discovery (SD) defines a category of models providing interpretable predictions that can be interpreted via simple formulas unlike well-known SOTA ML algorithms (e.g., *SVM*, *Random Forest* or also *Neural Networks*) returning black box patterns (*Imparato* 2012). SD algorithms are supervised learning creating individual rules which are subgroups of the dataset describing the properties of individual groups specific to a target class. The rule creation is done in two steps: the first is "rule generation" which defines subgroups of individuals from the dataset and the second is "rule selection" which filters the subgroups based on specific metrics. For precision medicine in metagenomics, SD could help decision making by providing results defining metagenomic profiles related to certain phenotypes, i.e., the impact of the bacterial balance regulating the microbiota on the phenotype.

3.2   Overview and Concepts of the *Q-Classifier*

The *Q-Classifier* is inspired by the *Q-Finder*, a SD algorithm (*Esnault* et al. 2020). *Q-Finder* has demonstrated its advantages in terms of interpretability and statistical robustness on the International Diabetes Management Practices Study (IDMPS), an observational study problem. SD does not try to cover the entire database regarding prediction, which prevents this approach from being used as a classification algorithm. Therefore, personalized prediction can only be done on patients who are in subgroups.

It is necessary to reformulate the algorithm so that it can be used to classify metagenomic data based on taxa abundance. We want to combine the individual stratification aspect to the explicability of the model predictions. *Q-Classifier* is built to be a multiclass supervised learning algorithm that generates statistically credible subgroups to discriminate examples in different classes. It is a hybrid model between subgroup discovery and classifier because some examples have an interpretable prediction when they can be predicted by rules, while the more complex ones are predicted by a SOTA supervised algorithm (*SVM, Random Forest*, etc. ...). Furthermore, the compositional nature of the data is taken into account with a preprocessing step based on centered log-ratio transformation (see Appendix B for more details about the method). *Q-Classifier* is in line with precision medicine because it may create different interpretations adapted for each individual as they may respond to a given pathology based on different feature values. Moreover, no example is excluded from the prediction as it is the case for other SD algorithms. This makes it possible to have total coverage and not to reduce the classification scores. The rule generation is done with the *Q-Finder* algorithm, while the rule selection still relies on credibility metrics and subgroup ranking but with some modifications more adapted for classification tasks and are described in the next section (Sect. 3.3). The *Q-Classifier* algorithm is summarized in Fig. 2 in the Appendix C while the data processing step is detailed in Appendix D.

3.3   Statistical Metrics and Optimal Union

Statistical Metrics. They are inspired by the work of *Esnault* et al. (2020), defining an order and thresholds that are experimentally adapted for our study. The metrics are defined by five statistical criteria to evaluate the classification scores of the subgroups to allow them to be filtered and ranked in the model *Q-Classifier*.:

1. **Coverage criterion:** It is defined by the ratio between the subgroup's size and the dataset's size. It prevents from considering too small subgroups and from being clinically relevant. The default minimum threshold is 20%.
2. **F1-score criterion:** A more relevant classification score than accuracy on unbalanced data that combines precision and recall metrics so that they have equal relative contributions. The formula is defined by $\frac{2 \times TP}{2 \times TP + FN + FP}$ with TP as true positive, FN as false negative and FP as false positive. The default threshold for this criterion is set to be greater than 0.5.
3. **Basic pattern[2] contribution criterion:** It is specified to ensure that each new basic pattern in the rule has a sufficiently large contribution to the F1-score of the subgroup. The contribution is calculated over the subgroup's score by removing or not the basic pattern (*absolute contribution*) and by comparing the ratio between the minimum and the maximum of the basic patterns' score (contribution ratio) (*Esnault* et al. 2020). Default thresholds are arbitrarily set to be greater than 0.03 for the *absolute contribution* and lower than 1.05 for the *contribution ratio*.
4. **Effect significance criterion:** The hypergeometric p-value which is computed as the sum of the mass probability functions of the hypergeometric law in the interval

---

[2] A basic pattern corresponds to an elementary unit of a rule characterized by a variable, a comparator and a value (e.g. age > 10).

$[TP, TP+FP]$. This discrete law is chosen because it is well adapted to the analysis of the confusion matrix from a prediction model. We set the default threshold value to at most 5%

5. **Effect adjusted significance criterion:** The Bonferroni correction is applied to control the type 1 errors. The hypergeometric p-value is multiplicated by the number of generated subgroups and the threshold is 5%.

Rules Aggregation. It is composed of different ranks discriminating the subgroups according to the previous criteria. The larger the rank number, the more statistically robust the rule is because it satisfies more criteria:

- Rank 1: subgroups that satisfy the coverage criterion
- Rank 2: subgroups of rank 1 that also satisfy the F1-score criterion
- Rank 3: subgroups of rank 2 that also satisfy the basic patterns contribution criterion
- Rank 4: subgroups of rank 3 that also satisfy the significance criterion
- Rank 5: subgroups of rank 4 that also satisfy the effect adjusted significance criterion.

Optimal Union. This approach, whose pseudo-code is written in appendix through Algorithm 1, consists in creating a set of the best subgroups, from the set of all subgroups having passed the rank 5, which maximizes a predefined metric (in our case the F1 score). The algorithm is initialized with the subgroup that has the best score according to the chosen metric. It then performs two steps called "forward" and "backward". The "forward" phase searches for the rules which, once added, improve the score of the union the most. If no improvement can be made by the "forward" phase, then the algorithm ends. Otherwise, the "backward" phase is executed after the addition of the new rule to check if, on the contrary, removing a rule (except the one with the best score) would improve the score of the union. Any rule that has been added to the optimal union is no longer considered afterwards even if it has been removed by a backward phase. We iterate the process from the forward phase until the algorithm ends (i.e., no rule improves the optimal union).

3.4   Rejection and Delegation Concepts to Adapt SD for Prediction

To combine SD and supervised learning, the *Q-Classifier* does not make a majority vote between models or an average of the probability prediction, it therefore forms a cascading prediction that follows these principles:

a. All samples present in at least one subgroup or in several subgroups with the same class are stratified by the SD approach with an interpretable prediction.
b. All samples not assigned to any subgroups or in several subgroups with different classes are rejected.

In the training step, the SOTA classifier uses the entire training set. However, the algorithm adds a higher weight to the samples rejected in the SD stage (case **b**) in order to reinforce its learning on the most difficult examples to classify. The weights are arbitrary set to 3 times higher compared to accepted samples (case **a**), however an elaborate optimization of this hyper parameter could be investigated. This allows the classifier to focus more on complex data and improve its performance. In the classification step, samples in case **a** are only predicted by the SD approach and are excluded from the classifier prediction, while samples in case **b** are first attempted to be predicted by the SD approach and then rejected for delegation to the SOTA classifier. The training (resp. classification) step is summarized in Fig. 3 (resp. Fig. 4) in Appendix C and its pseudo code is provided in appendix in Algorithm 3 (resp. Algorithm 2).

4   Experiments

4.1   Survey of Existing Metagenomics Datasets

Existing metagenomic datasets generated by Illumina shotgun sequencers are manipulated in several studies (*Edoardo Pasolli* et al. 2015; *Thomas* et al. 2019; *Pasolli* et al. 2016; *Prifti* et al. 2020). These datasets represent a basis for comparing our different methods with the state of the art. They are accessible from NCBI website[3] which repertories open-source studies. The Sect. 4.1 resumes the four public datasets manipulated during our experiments:

Table 1. Information about the metagenomic datasets

Dataset name	Disease	Control subjects	Case subjects	Control-to-Case	Reference
Colorectal	Colorectal Cancer	73[a]	48	60.3%	(*Zeller* et al. 2014)
Cirrhosis	Liver Cirrhosis	114	118	49.1%	(*N. Qin* et al. 2014)
Obesity	Obesity	89	164	35.2%	(*MetaHIT consortium* et al. 2013)
T2D	Type 2 diabetes	174	170	50.6%	(*J. Qin* 2012)

[a] 15 patients had an adenoma; this is a benign tumor, so they have been labeled as control cases as in *Pasolli* et al. 2016.

4.2   Benchmark on Real-World Metagenomic Data

Each result is calculated by 10-fold cross validation with 80% in train and 20% in test. The length of the rules (complexity) of the *Q-Classifier* is set to 2 and the default parameters are used during the rule selection part (see Sect. 3.3), while a random search is performed to tune the parameters of the SOTA classifiers (SVM, Random Forst, Gradient Boosting and Ada Boost) and the one with the best scores is selected. Feature selection is performed by recursive feature elimination trained by 3-fold cross-validation with a minimum of 40 features kept. *MetaML* (*Pasolli* et al. 2016) and *Predomics* (*Prifti* et al. 2020) are the reference methods using species-level relative abundances. Four approaches are experimented on all datasets, *Q-Classifier* is trained either on the *MetaPhlAn2* species abundance data from the *Prifti* et al. 2016 study or on those from the *Metagenome2Vec* (*Queyrel* et al. 2021) pipeline using *FastDNA* (*Menegaux and Vert* 2019) and the data preprocessing is performed or not with CLR transformation. The *FastDNA* algorithm is trained with specific simulated data for this study (see

---

[3] www.ncbi.nlm.nih.gov.

Appendix F for more details about the simulation). Results are summarized in Table 2. The initial rejected rate (IRR) written in tables defines the percentage of rejected and delegated samples (case **b** in Sect. 3.4) by the *Q-Classifier*. Accuracy, F1-score, Precision and Recall are the computed metrics. The standard deviation is calculated for the accuracy score and written with the symbol ±.

Classification Performances: The best approach is the *Q-Classifier* on *MetaPhlAn2* data with CLR transformation reaching higher scores for almost all metrics in real-world datasets. The CLR transformation systematically improves the scores either from *MetaPhlAn2* or from *FastDNA* abundances. During our experiments, the *Q-Classifier* algorithm was tested by providing as input the abundances generated by the deep learning algorithm *FastDNA* and obtained promising results although slightly worse than those obtained by the software *MetaPhlAn2*. This means that the *Q-Classifier* method could be used in an end-to-end way without having to use gene catalogs. The IRR value varies from one method and dataset to another ranging from 17% to 87%. We note that

Table 2. Classification results on four real-world benchmark datasets (Table 1). Results are reported for two reference methods (*MetaML* and *Predomics*). *Q-Classifier* on *MetaPhlAn2* or *FastDNA* species abundance with CLR transformation of not are our methods tested in this experiment.

Method	Metrics	Colorectal	Cirrhosis	Obesity	T2D
MetaML	Accuracy	0.81 ± 0.068	0.88 ± 0.043	0.64 ± 0.028	0.66 ± 0.052
	precision	0.82	0.89	0.54	0.67
	recall	0.81	0.88	0.64	0.66
	F1-score	0.79	0.88	0.54	0.66
Predomics	Accuracy	–	0.84 ± 0.035	0.66 ± 0.035	0.68 ± 0.030
Q-Classifier MetaPhlAn2	Accuracy	0.57 ± 0.074	0.88 ± 0.046	0.62 ± 0.010	0.65 ± 0.062
	Precision	0.61	0.95	0.66	0.64
	Recall	0.8	0.81	0.84	0.67
	F1-score	0.69	0.87	0.74	0.65
	IRR	0.62	**0.17**	0.87	0.76
Q-Classifier MetaPhlAn2 CLR	Accuracy	0.85 ± 0.047	0.94 ± 0.049	0.79 ± 0.033	0.81 ± 0.053
	Precision	**0.84**	**0.97**	**0.81**	**0.80**
	Recall	**0.93**	**0.91**	**0.89**	0.80
	F1-score	**0.88**	**0.94**	**0.85**	**0.80**
	IRR	**0.54**	0.20	**0.76**	0.77
Q-Classifier FastDNA	Accuracy	0.70 ± 0.049	0.81 ± 0.036	0.64 ± 0.048	0.68 ± 0.084
	Precision	0.61	0.79	0.67	0.64
	Recall	0.56	0.88	0.87	0.74
	F1-score	0.58	0.83	0.76	0.68
	IRR	0.63	0.73	0.8	0.70
Q-Classifier FastDNA CLR	Accuracy	0.75 ± 0.075	0.86 ± 0.023	0.71 ± 0.076	0.81 ± 0.023
	Precision	0.66	0.88	0.76	0.79
	Recall	0.67	0.88	0.84	**0.82**
	F1-score	0.66	0.88	0.79	**0.80**
	IRR	0.61	0.28	0.8	**0.54**

the IRR remains quite high and thus the samples predicted by a SOTA classifier represent more than half of the colorectal dataset and more than three quarters of the obesity and T2D datasets. This represents a weakness of the algorithm because a majority of samples do not have interpretable predictions. To strengthen the understanding of the predictions made by the SOTA classifiers, one method (not tested in our experiments) could be to use an interpretability tool such as the Shapley values (*Lundberg and Lee* 2017) which give the contribution of features to the prediction of a sample relative to the average prediction of the dataset.

Metagenomics Signatures as Rules. In the *Cirrhosis* dataset using a CLR transformation, *Q-Classifier* generated 1287 rules on the training set (80% of the 232 subjects) which passed the 5 criteria, 1130 are of complexity 1 and 157 are of complexity 2. The optimal union reduces this set of rules by taking only 5 rules of complexity 1. These rules are listed below (RCO (resp. RCA) refers to the rules of the control (resp. case) samples) and their metrics are computed on the validation set (47 subjects: 23 controls and 24 cases) and adjusted for multiplicity (FDR) :

RCA1 If *Veillonella Unclassified* $\geq$ 4.42, then class is **Case** with *coverage* = 53%, $F1\text{-}score$ = 0.94 and $p\text{-}value$ = 1.23e−9

RCA2 If *Streptococcus Parasanguinis* $\geq$ 4.92, then class is **Case** with *coverage* = 26%, $F1\text{-}score$ = 0.67 and $p\text{-}value$ = 5.18e−5

RCA3 If *Streptococcus Anginosus* $\geq$ 3.13, then class is **Case** with *coverage* = 26%, $F1\text{-}score$ = 0.67 and $p\text{-}value$ = 5.18e−5

RCO1 If *Veillonella Unclassified* $\leq$ 5.40, then class is **Control** with *coverage* = 55%, $F1\text{-}score$ = 0.90 and $p\text{-}value$ = 3.93e−8

RCO2 If *Veillonella Dispar* $\leq$ 1.67, then class is **Control** with *coverage* = 38%, $F1\text{-}score$ = 0.78 and $p\text{-}value$ = 1.53e−5

Each of these rules represent a metagenomic signature of either case or control samples. Compared to the interpretable model provided by *predomics* from the study of *Prifti* et al. (2020), there are 3 species in common: *Veillonella Unclassified*, *Streptococcus Anginosus* and *Veillonella Dispar*. Although *predomics* models are formula (sums, difference or ratio) of abundance, the fact that both *predomics* and *Q-Classifier* share critical bacterial species proves a level of consistency as some of these bacterial species are known to be related to the disease (here cirrhosis). Appendix H provides a Venn diagram visualization of the disjunction and union of the subgroups described above.

5   Conclusion

To address the interpretability/Accuracy trade off in classification model from metagenomics data, we propose to learn a hybrid model returning, for a set of samples, an interpretable prediction with SD approach and classifying the remaining samples with a SOTA classifier. We adapted the subgroup discovery *Q-Finder* algorithm into such a classifier model that we called the *Q-Classifier*. It creates metagenomic signatures for personalized and interpretable stratification of patients' diseases. It takes as inputs the taxa abundance tables, created either by bioinformatic methods (e.g., *MetaPhlAn2*) or by deep learning methods (e.g., *FastDNA*). This algorithm is thus a cascading classifier of a state-of-the-art subgroup discovery and a state-of-the-art classifier model. We also took into account the compositional characteristic of the data by applying a CLR transformation that supports normalizing the data while preserving the structure of the data. This is necessary to guarantee the interpretability of the rules. This preprocessing step effectively improved the results obtained in the experiments. On the 4 real-world datasets, the *Q-Classifier* reached comparable or superior performances to the state-of-the-art while bringing more interpretability with simple formula (rules).

Appendix

A   Example of Metagenomic Abundance Table

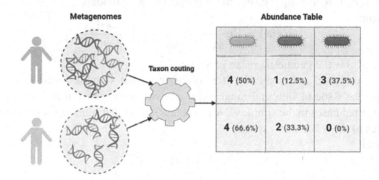

Fig. 1. An example of an abundance table where two metagenomes have different numbers of taxa. For yellow DNA, both have an absolute abundance equal to four, but the relative abundances in percentage are different: 50% for the former versus 66.6% for the latter. The relative abundance thus allows us to obtain the proportion of one taxon in relation to the others.

B   Compositional Data and Log-Ratio Transformations

One of the major difficulties often underestimated is the composition of the quantitative metagenomic data. Indeed, the number of sequences generated by NGS is not the same and varies from one sample or study to another. When the biological objects included in the samples are counted, it should not be restricted to an absolute count because it would not be representative of the real composition. A normalization step must therefore be applied and consist of dividing each abundance by the total number of taxonomic units, resulting in a table of relative abundance (see Fig. 1). They are characterized as compositional data that are defined by covering all vectors representing parts of a whole that carry only relative information. This is the case of relative abundance tables because they represent relative information, namely the percentage of total abundance, which restricts them to a sample space with the constraints of having the sum of each characteristic always equal to 1 and having their values included in the interval $[0, 1]$. These constraints require specific mathematical transformations to avoid misinterpretations or irreproducible analyses (*Yang* et al. 2021). The data processing often used are log-ratio transformations and refer to Additive Log-ratio (ALR), Centered Log-Ratio (CLR) and Isometric Log-Ratio (ILR). The choice of the method is defined by the desired interpretation as described below:

- ALR: Isomorphic[4] and not isometric[5]. Transforms the original D features to $D - 1$ features space.
  Formula:
  $$alr(x) = [ln\frac{x_1}{x_D}, ln\frac{x_2}{x_D}, ..., ln\frac{x_{D-1}}{x_D}]$$
- CLR: Both isometric and isomorphic. It removes the value-range restriction, but it does not remove the sum constraint. It does not change the dimension of the basis as the ALR or ILR making it easier to train interpretable models
  Formula:
  $$clr(x) = [ln\frac{x_1}{g(x)}, ln\frac{x_2}{g(x)}, ..., ln\frac{x_{D-1}}{g(x)}$$

  Where g(x) is the geometric mean of x.
- ILR: Isomorphic and isometric. It is often the most suitable transformation that manage the issue of sum and range value constraints because it is associated with orthonormal bases in the simplex. Nevertheless, as ALR it transforms the original D features to $D - 1$ features space.
  Formula:
  $$ilr(x) = clr(x) \cdot \Psi'$$

  $$\Psi\Psi' = I_{D-1}$$

  Where Ψ is a $(D - 1, D)$-matrix whose rows are $clr(e_i)$ and $e_1, e_2, ..., e_{D-1}$ is a generic orthonormal basis of the simplex S^D.

As ILR transformations are difficult to interpret, recent studies have defined a method called balance (*Quinn and Erb* 2020; *Yang* et al. 2021), which is the log-ratio of the geometric means of two non-overlapping groups of features defined by a sequential binary partition (SBP). In that way, balances are more interpretable than common log-ratio transformations. Metagenomic compositionality is also managed by *Friedman and Alm* 2012 to create a clustering graph network interaction of species. They proposed a robust approximation method called *SparCC* to derive the correlation matrix based on a rough estimate of the variance of the ratio-log of species.

---

[4] Isomorphic: meaning that the mapping between the simplex and the new basis is preserved.
[5] Isometric: meaning that the distances in the simplex are equivalent to the distances of the new transformed values.

C *Q-Classifier* Schemes and Overview

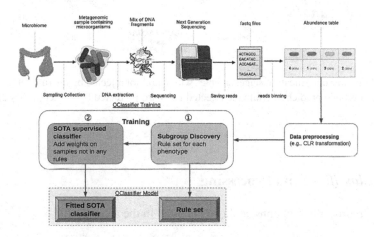

Fig. 2. *Q-Classifier* overview: The algorithm takes as input the calculated metagenomic abundance data and starts by preprocessing them to the selected parameters (such as CLR transformation). The training phase is composed by one step of statistically credible rule generation followed by state-of-the-art classifier training. At the end, the algorithm consists of a set of rules and a state-of-the-art classifier cascaded during the classification step.

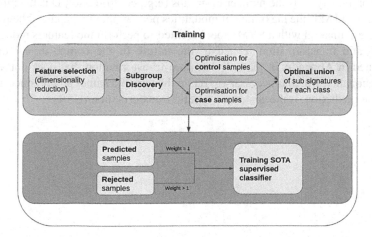

Fig. 3. *Q-Classifier* training stage: An optional feature selection is first processed, then statistically credible subgroups on all classes (control and case) are generated. Optimal unions of metagenomic signatures for each class are computed and gathered. Finally, a SOTA classifier is trained by adding more weight to the data that has been rejected.

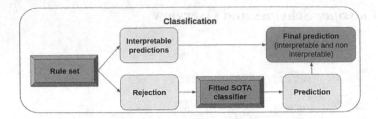

Fig. 4. *Q-Classifier* classification stage: samples which are not rejected by the rule set have therefore an interpretable prediction while the rejected ones are predicted by a fitted SOTA classifier.

D   *Q-Classifier* Data Processing

Two data preprocessing options could be included in the algorithm:

- A dimensionality reduction operation to limit the number of rules generated by the *Q-Finder*, which reduces runtime and improves statistical power when adjusting the p-value with the Bonferroni correction. Indeed, one of the weaknesses of the *Q-Finder*, and thus of the *Q-Classifier*, is its high complexity which is equal to $O(G \times (F \times (M \times D)^C)$. Where C is the rule complexity, F is the aggregation rules complexity, G is the number of groups (e.g., control/case), D is the number of variables and M is the maximum of modalities per variable. In practice, the recursive feature elimination with a SVM model is fitted to perform the features reduction.
- A log-ratio transformation related to the nature of the compositional data. As explained in Appendix B, the only log-ratio transformation that does not alter the dimensionality, which is necessary to retain the variable names and produce an interpretable prediction, is the centered log-ratio transformation (CLR).

E   Pseudo Code of Optimal Union

F   Simulated Dataset Used to Train Taxa Classifier

To learn a good representation of the sequences and improve the classification power of the models, it is necessary to ensure that the simulated dataset has close abundances for each genome according to the taxonomy level. We used the *CAMISIM* (*Fritz* et al. 2019) framework which takes a file of genome abundances as configuration. *CAMISIM* takes in consideration the size of the genome in addition to its abundance, so to generate an abundance equally proportionate between genomes we have the following formula:

$$A_g = \frac{1}{L_g}$$

$$AG_g = \frac{A_g}{\sum_i^{|G|} A_i}, g \in G$$

Algorithm 1 *optimal_union*: From the subgroups that pass the rank 5 of the credibility criteria, takes those that maximize a specific metric (*metric_to_maximize*)

input	:	G_{metric}: Set of subgroups with computed metrics
		metric_to_maximize: Metric maximized in optimal union
		nb_rules_max: (Default=10) Number maximum of rules in the optimal union
		threshold. (Default=0.001): Minimum gain when adding a rule in the optimal union
output	:	G_{opti}: Set, optimal union of subgroups in G_{metric}

1 $G_{opti} = \{\}$ score_prev_forward = -inf forward_done = False **while** *not forward_done* **do**
     // Case where G_{opti} is higher than k we stop **if** $G_{opti} \geq nb\_rules\_max$ **then**
2 | break
3 **end**
     /* Forward: Compute scores for each rule in the set if they are added in the optimal union G_{opti}) */ scores = [ ] **for** *rule in G_{metric}* **do**
4 | scores.append(metric_to_maximize(G_{opti} U {rule}))
5 **end**
     // best_score return the best score in scores   score_forward = best_score(scores)   **if** *score_forward - score_prev_forward < threshold* **then**
6 | forward_done = True
7 **else**
     // best_rule return the rule with the best score in scores G_{opti}.append(best_rule(scores)) score_prev_forward = score_forward   /* Backward: Compute scores for each rule in the optimal union (G_{opti}) if they are removed from it   */ backward_done = False   **while** *not backward_done* **do**
8 | | backward_done = True // There is at least 3 rules **if** *length(G_{opti}) > 2* **then**
         /* Do not look at the first rule (index 0) to ensure not removing the best rule   */ scores = [ ] **for** *rule in $G_{opti}[1:]$* **do**
9 | | | scores.append(metric_to_maximize(G_{opti} \ {rule}))
10 | | **end**
         score_backward = best_score(scores) **if** *score_backward > score_forward* **then**
11 | | G_{opti}.remove(best_rule(scores)) backward_done = False
12 | | **end**
13 | **end**
14 | **end**
15   **end**
16 **end**
17 **end**

Where G is the ensemble of genomes, L_g is the base pair length of the genome g and AG_g stands for the equally balanced abundance between genomes for the genome g. However, if a certain taxonomic level is defined, such as species, the abundance formula should be modified accordingly. This will avoid that a species appears too often or not often enough, which could be the case when several genomes have the same species. The formulas are defined by:

$$L_s = \sum_i^{|G_s|} L_i \; ; \; L\_norm_s = \frac{L_s}{\sum_i^{|S|} L_i}, \; s \in S$$

$$A_g = \frac{1}{L\_norm_s \times (L_g/L_s)} \; ; \; A_g = \frac{A_g}{\sum_i^{|G_s|} A_i}$$

$$AS_g = \frac{A_g}{\sum_i^{|G|} A_i}$$

Where S is the ensemble of species, G_s is the ensemble of genomes for the species s, L_s is the base pair length of the species s and AS_g stands for the abundance equally balanced between genomes at the species level for the genome g.

3.5M and 1.5M Illumina reads with an average of 150 base pairs have been simulated for the train and validation set respectively, corresponding to a depth of coverage (mapping depth) of 27% over all the genomes. The initial abundance given was calculated with the formulas above to be equally proportionate at the species level. Indeed, the model is trained at the species level rather than the genome level because genomes of the same species are close enough to have the same prediction and it is easier to train the model with a smaller number of classes. An almost equal number of sequences for each species is not representative of real metagenomes where abundance follows exponential distributions. Nevertheless, in the case of read classification modeling, this prevents the classifier from focusing and predicting the predominant classes while learning a more robust embedding. These simulated data have the advantage over real NGS data of providing information on the origin of a sequence, allowing the training of a supervised algorithm.

G   Pseudo Code of *Q-Classifier*'s Training and Classification Stage

---

Algorithm 2 *Q-Classifier* classification stage

---

input   : **X**: Matrix, test data processed in the same way as train data
   G_{opti}: Dictionary, optimal union of subgroups for each class
   SOTA: Algo, a fitted SOTA classifier
output : X_{pred}: Matrix, the X matrix with a prediction for each sample
// $X_{subgroup\_pred}$ contains samples with interpretable prediction
// X_{reject} contains samples that have been rejected $X_{subgroup\_pred}$,
X_{reject} = predict(X, G_{opti})   // Use only the rejected samples for the
prediction X_{sota_pred} = SOTA.predict(X_{reject}) // The final prediction is
the concatenation of both prediction from rules and SOTA X_{pred} =
concatenate($X_{subgroup\_pred}$, X_{sota_pred})

Algorithm 3 *Q-Classifier* training stage

---

input : **X**: Matrix, train data
       y: Vector, train label
       algo_subgroup: Algo, to generate subgroups
       algo_metrics: Algo, to compute some metrics over subgroups
       SOTA: Algo, a SOTA classifier
       metric_to_maximize: Metric maximized in optimal union
       feature_reduction: Algo, to keep only best features
       transformation: Algo, to process raw data (e.g. CLR)
       weight_reject: Float, weight on rejected to train SOTA
output : G_{opti}: Dictionary, optimal union of subgroups for each class
       SOTA: Algo, a fitted SOTA classifier

18 **if** *feature_reduction* **then**
19 | X = recursive_feature_elimination(X)
20 **end**
21 **if** *transformation* **then**
22 | X = tranformation(X, min_feature)
23 **end**
24 $G = algo\_subgroup(X)$ // Set of generated subgroups $G_{opti} = dict()$
    // Compute subgroup metrics for each class   **for** *class in unique(y)* **do**
25 | G_{metric} = algo_metrics(X, class, G) $G_{opti}[class]$ = optimal_union(G_{metric}, metric_to_maximize)
26 **end**
    // predict: function that assigns subgroups to samples
    $X_{subgroup\_pred}$, X_{reject} = predict(X, G_{opti})   // Add more weights on rejected samples   weights = [] **for** *x in X* **do**
27 | **if** *x in $X_{subgroup\_pred}$* **then**
28 | | weights.append(1)
29 | **else**
30 | | weights.append(weight_reject)
31 | **end**
32 **end**
33 SOTA.fit(X, y, weigths)

H   Rules Coverage Analysis on the *Cirrhosis* Dataset

(See Fig. 5)

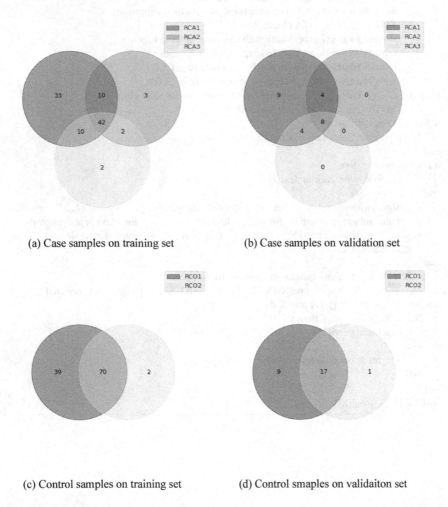

(a) Case samples on training set   (b) Case samples on validation set

(c) Control samples on training set   (d) Control smaples on validaiton set

Fig. 5. Venn diagram of the subgroups in the optimal union of the *Cirrhosis* dataset. Each circle corresponds to a subgroup characterized by a rule (with a name referring to the one in Sect. 4.2). The values inside the circles correspond to the number of samples in the subgroups. When a value lies between several circles, it represents the number of samples shared by the corresponding subgroups.

The best rule for the control class and the case class are respectively "RCO1" and "RCA1". We notice that the subgroups associated with these rules have a high intersection with the other subgroups in the optimal union. This is even more important on the validation set where sometimes one subgroup is completely included in another. As the optimal union of the rules is computed on the training set, it is possible in validation to obtain these results. This visualization allows us to determine the disjunction or union of the subgroups' samples.

References

Pasolli, E., Truong, D.T., Malik, F.: Machine learning meta-analysis of large metagenomic datasets: tools and biological insights (2015)

Esnault, C., Gadonna, M.-L., Queyrel, M., Templier, A., Zucker, J.-D.: Q-Finder: an algorithm for credible subgroup discovery in clinical data analysis - an application to the international diabetes management practice study. Front. Artif. Intell. **3**, 559927 (2020)

Friedman, J., Alm, E.J.: Inferring correlation networks from genomic survey data. PLoS Comput. Biol. **8**(9), e1002687 (2012)

Fritz, A., Hofmann, P., Majda, S., et al.: CAMISIM: simulating metagenomes and microbial communities. Microbiome **7**(1), 17 (2019)

Harris, Z.N., Dhungel, E., Mosior, M., Ahn, T.-H.: Massive metagenomic data analysis using abundance-based machine learning. Biol. Direct **14**(1), 12 (2019)

Imparato, A.: Interactive Subgroup Discovery, p. 134 (2012)

Korepanova, N.: Subgroup discovery for treatment optimization. In: Workshop on Data Analysis in Medicine, WDAM 2017, pp. 48–41 (2017)

Kunin, V., Copeland, A., Lapidus, A., Mavromatis, K., Hugenholtz, P.: A bioinformatician's guide to metagenomics. Microbiol. Mol. Biol. Rev. **72**(4), 557–578 (2008)

Loh, W.-Y., Cao, L., Zhou, P.: Subgroup identification for precision medicine: a comparative review of 13 methods. Wiley Interdisc. Rev. Data Min. Knowl. Disc. **9**(5), 604–621 (2019)

Lundberg, S.M., Lee, S.-I.: A unified approach to interpreting model predictions, p. 10 (2017)

Menegaux, R., Vert, J.-P.: Continuous embeddings of DNA sequencing reads and application to metagenomics. J. Comput. Biol. **26**(6), 509–518 (2019)

Le Chatelier, E., Nielsen, T., et al.: Richness of human gut microbiome correlates with metabolic markers. Nature **500**(7464), 541–546 (2013)

Nayfach, S., Pollard, K.S.: Toward accurate and quantitative comparative metagenomics. Cell **166**(5), 1103–1116 (2016)

Oh, M., Zhang, L.: DeepMicro: deep representation learning for disease prediction based on microbiome data. Sci. Rep. **10**(1), 6026 (2020)

Pasolli, E., Truong, D.T., Malik, F., Waldron, L., Segata, N.: Machine learning meta-analysis of large metagenomic datasets: tools and biological insights. PLOS Comput. Biol. **12**(7), e1004977 (2016)

Petrosino, J.F.: The microbiome in precision medicine: the way forward. Genome Med. **10**(1), 12 (2018)

Prifti, E., Chevaleyre, Y., Hanczar, B., et al.: Interpretable and accurate prediction models for metagenomics data. GigaScience **9**(3), giaa010 (2020)

Qin, J., et al.: A metagenome-wide association study of gut microbiota in type 2 diabetes, p. 6 (2012)

Qin, N., Yang, F., Li, A., et al.: Alterations of the human gut microbiome in liver cirrhosis. Nature **513**(7516), 59 64 (2014)

Queyrel, M., Prifti, E., Templier, A., Zucker, J.-D.: Towards end-to-end disease prediction from raw metagenomic data. Int. J. Biomed. Biol. Eng. **15**(6), 234–246 (2021)

Quince, C., Walker, A.W., Simpson, J.T., Loman, N.J., Segata, N.: Shotgun metagenomics, from sampling to sequencing and analysis, p. 27 (2017)

Quinn, T.P., Erb, I.: Interpretable log contrasts for the classification of health biomarkers: a new approach to balance selection. mSystems **5**(2), e00230-19 (2020)

Segata, N., Izard, J., Waldron, L., et al.: Metagenomic biomarker discovery and explanation. Genome Biol. **12**(6), R60 (2011)

Thomas, A.M., Manghi, P., Asnicar, F., et al.: Metagenomic analysis of colorectal cancer datasets identifies cross-cohort microbial diagnostic signatures and a link with choline degradation. Nat. Med. **25**(4), 667–678 (2019)

Wen, C., Zheng, Z., Shao, T., et al.: Quantitative metagenomics reveals unique gut microbiome biomarkers in ankylosing spondylitis. Genome Biol. **18**(1), 142 (2017)

Wu, G., Zhao, N., Zhang, C., Lam, Y.Y., Zhao, L.: Guild-based analysis for understanding gut microbiome in human health and diseases. Genome Med. **13**(1), 22 (2021)

Yang, F., Zou, Q., Gao, B.: GutBalance: a server for the human gut microbiome-based disease prediction and biomarker discovery with compositionality addressed. Brief. Bioinf. **22**(5), bbaa436 (2021)

Zeller, G., Tap, J., Voigt, A.Y., et al.: Potential of fecal microbiota for early stage detection of colorectal cancer. Mol. Syst. Biol. **10**(11), 766 (2014)

Demystifying Graph Neural Network Explanations

Anna Himmelhuber[1,2]($\boxtimes$), Mitchell Joblin[1], Martin Ringsquandl[1],
and Thomas Runkler[1,2]

[1] Siemens AG, Munich, Germany
{anna.himmelhuber,mitchell.joblin,martin.ringsquandl,
thomas.runkler}@siemens.com
[2] Technical University of Munich, Munich, Germany

Abstract. Graph neural networks (GNNs) are quickly becoming the
standard approach for learning on graph structured data across several
domains, but they lack transparency in their decision-making. Several
perturbation-based approaches have been developed to provide insights
into the decision making process of GNNs. As this is an early research
area, the methods and data used to evaluate the generated explanations
lack maturity. We explore these existing approaches and identify common pitfalls in three main areas: (1) synthetic data generation process,
(2) evaluation metrics, and (3) the final presentation of the explanation.
For this purpose, we perform an empirical study to explore these pitfalls along with their unintended consequences and propose remedies to
mitigate their effects.

1 Introduction

Many important real-world data sets are graphs or networks. These include
social networks, knowledge graphs, protein-protein interaction networks, the
World Wide Web and many more. Graph neural networks leverage link structure to encode information as well as incorporate node feature information [12]
and currently achieve state-of-the-art on many prediction tasks [9]. Similarly to
other connectionist models, GNNs lack transparency in their decision-making.
Explaining GNNs is currently in the early stages of research, but since graphs
are particularly expressive by encoding contexts, they are a promising candidate
when it comes to producing rich explanations [4]. The most popular type of
GNN explainer methods is perturbation-based, where the output variations are
studied with respect to different input perturbations [14].

When developing any explainable method, it is important to evaluate the performance of the method with respect to valid procedures and metrics [1]. With
this in mind, we explore the evaluation methods employed by perturbation-
based explainer methods in the GNN domain. Validating explanations is generally a challenging task because a ground-truth explanation is not always available. Even for synthetically generated datasets with ground-truth explanations,

© Springer Nature Switzerland AG 2021
M. Kamp et al. (Eds.): ECML PKDD 2021 Workshops, CCIS 1524, pp. 67–75, 2021.
https://doi.org/10.1007/978-3-030-93736-2_6

this approach can be error prone. While validation schemes do exist, they lack maturity.

Many explainer methods come with differing evaluation protocols, which makes their comparison difficult. However, some of these protocols overlap or have been adopted by others, e.g. with several papers [5,13,15] using state-of-the-art explainer method [10] and its evaluation protocol as benchmark. As GNN explainer methods become more and more popular, its vital to avoid evaluation pitfalls and, in the next step, introduce a standard evaluation approach. That's why we perform an empirical study on perturbation-based explanations for GNNs, with focus on [10]. Our contributions include identifying pitfalls in three main areas: (1) synthetic data generation process, (2) evaluation metrics, and (3) the final presentation of the explanation. For each pitfall we propose a remedy to increase the validity of the evaluation.

2   Terminology and Concepts

For perturbation-based explainer methods for GNNs, the output consists of masks, indicating important input features, including node masks, edge masks or node feature masks depending on the explanation task. We can observe three different types of masks that have been proposed, including soft masks (GNNExplainer [10], CF-GNNExplainer [5]), discrete masks (ZORRO [13]) and approximated discrete masks (PGExplainer [15])[1]. These mask are then applied to the input graph(s) and fed into the trained GNNs to carry out predictions, which is targeted by the objective function to be similar to the original prediction. The currently overarching established explanation evaluation scheme for perturbation-based explainer methods consists of a data generation, GNN training, and mask generation step as is shown in Fig. 1. The generated synthetic data is comprised of a base graph and a specific motif (1), which are connected randomly and additionally perturbed by noise. A GNN is applied to the graph execute a prediction task, e.g. node classification (2). In the next step, the explainer method generates masks of the receptive field (3). If the explainer method outputs soft masks, thresholding is needed to arrive at the final explainer subgraph (4). Please refer to the Appendix A for detailed background on GNNs and perturbation-based explainer methods.

2.1   Terminology

– **Ground-truth explanation:** The ground-truth explanation is a particular motif that is used during the synthetic data set generation, e.g. the "house motif" shown in Fig. 1 (1). The motif is embedded into a larger graph and perturbed with noise.
– **Ground-truth label:** The ground-truth label is the respective class a node (or graph) is assigned to.

---

[1] Please refer to the Appendix A for more details on GNNs and explainer methods.

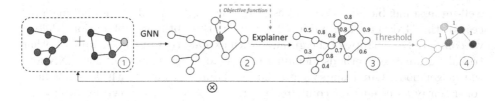

Fig. 1. Data generation, training and explanation process (Color figure online)

- **Explainer subgraph** with importance scores: The explainer method assigns importance scores to the edges, indicating their influence in the prediction by the GNN, as shown in Fig. 1 (3).
- **Threshold application:** In order to arrive at a compact subgraph, a threshold is applied to reduce the explainer subgraph to the most important edges by removing all edges that fall below the threshold.
- **Final (explainer) subgraph:** A reduced final explainer subgraph remains, as shown in Fig. 1 (4).
- **Label flip:** If the input to the GNN is changed, e.g. using the reduced final explainer subgraph instead of the original receptive field, a label flip can occur. This means that a different class is predicted than in the original prediction.

2.2 Synthetic Data

BA-Shapes: Node classification dataset with a base graph of 300 nodes and a set of 80 five-node "house"-structured network motifs, which are attached to randomly selected nodes of the base graph and function as ground-truth explanations. The resulting graph is further perturbed by adding $0.1N$ random edges. Nodes are assigned to 4 classes based on their structural roles. In a house-structured motif as can be seen in Fig. 1 (1), there are 3 types of roles: the top (yellow), shoulder (orange) and bottom (blue) node of the house and nodes that do not belong to a house (grey).

Tree-Cycles: Node classification dataset with two different labels, that consists of a base 8-level balanced binary tree and 80 six-node cycle motifs, which are attached to random nodes of the base graph and function as ground-truth explanations.

3 Pitfalls of Evaluation and Possible Remedies

3.1 Pitfall 1: Data Generation

For all 4 introduced explainer methods [5, 10, 13, 15], the synthetic datasets BA-Shapes and Tree-Cycles are used for evaluation. Their advantage is their intuitive motifs and labelling, which is understandable by humans. However, the defined ground-truth explanation e.g. the "house-motif", while being intuitively well-understandable, does not necessarily align with the decision-making process of the GNN and hence doesn't represent the optimal explanation. Below, we compare the entropy of the ground-truth explanation to the entropy of other trivial explanation methods including the entire receptive field of the GNN and the target node. Our results[2] show the proclaimed ground-truth does not have consistently lower entropy compared to trivial baselines as shown in Table 1.

Table 1. Average baseline entropies for BA-Shapes

Method	Top nodes	Shoulder nodes	Bottom nodes
Ground-truth	1.21	0.96	0.95
Receptive field	1.31	0.93	1.16
Target node	1.25	1.24	1.24

For each house node type we see a need for a different type of ground-truth explanation, given the differences in entropy and accuracy performance (see Sect. 3.2). Considering Occam's razor, which suggests the simplest explanation is best, we can see for several types of nodes, that the house motif is not the optimal ground-truth explanation. Figure 2 (left) shows a number of different possible ground-truth explanations, including the top triangle, the bottom square, the target node "left shoulder" and right shoulder node, a bottom node, or the top node. It can be seen that while the assigned ground-truth explanation "house motif" does lead to the correct prediction in nearly all cases, so do the more compact motifs square and triangle with a similarly low entropy and would be the more compact ground-truth explanation.

Remedy: We propose a new ground-truth explanation generation in order to better evaluate and compare the explanations. Properties of the dataset should include the lowest possible entropy matching with the identified ground-truth explanation, e.g. a specific motif. One way to achieve this, is to do a motif search: The entropy and prediction accuracy of several different potential motifs around the target node are calculated, similarly to our approach in Fig. 2. The best result is then chosen to be the ground-truth explanation, which ensures maximal compactness and therefore comprehensibility of the explanation. This motif search step implies some additional work and resources, but it is a worthwhile trade-off as it ensures that the ground-truth explanation for a specific class prediction from a specific GNN is known and therefore confirms the validity of the evaluation outcome. It also ensures, that only essential parts of the graph are in the ground-truth explanation.

3.2   Pitfall 2: Evaluation Metrics

For evaluating the accuracy of an explainer method, the ground-truth explanation has to be known. For synthetic datasets, graph motifs can be used as

---

[2] A 3-layer vanilla Graph Convolutional Network is used carry out experiments.

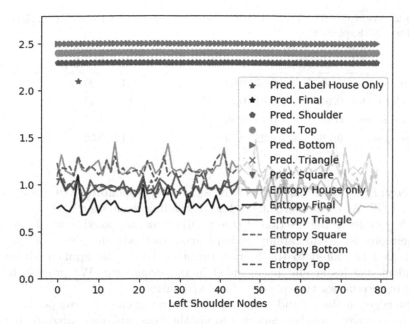

Fig. 2. Entropies for different motifs and markers (stars) indicating incorrect predictions for the node classification of the left shoulder nodes in the "house motif". The markers represent whether the respective motif leads to the correct prediction. A marker is equivalent to the prediction of the incorrect label.

an approximation ground-truth, even though the GNN might not make predictions as intuitively expected as discussed in Sect. 3.1. When choosing a metric, many papers use the general term "accuracy" with wildly differing definitions. In [13], accuracy is used, defined as the matching rate for important edges in explanations compared with those in the ground truths, whereas in [5], it is the proportion of explanations that are "correct". In [10] and [15], an accuracy is formalized according to a binary classification task, where edges in the ground-truth explanation are treated as labels and the importance scores are viewed as prediction scores. The accuracy is equivalent to the calculation of the area under the curve (AUC) of the receiver operating characteristic curve (ROC), which is calculated on the prediction score and not on the predicted classes. ROC AUC has limitations in its capacity to evaluate the explanations, as it only gives us the probability that a randomly chosen positive instance (edges in the ground-truth explanation) is ranked higher than a randomly chosen negative instance. However, when evaluating an explanation, we do care about the actual probability of the evaluation being correct, instead of just the ranking. Even more problematic is the ROC AUC's tendency to be misleading in situations with high class imbalance [8], as is the case here. Due to the large number of true negatives, edges that are neither in the ground-truth explanation nor in the final explainer subgraph, the false positive rate is pulled down substantially, which leads to an overly optimistic result.

Table 2. Avg. ROC AUC, Avg. PR AUC and average recall with standard deviations (SD) for GNNExplainer

Class	ROC AUC	SD	PR AUC (proposed)	SD	Recall (proposed)	SD
Top nodes	0.98	0.07	0.69	0.19	0.65	0.18
Shoulder nodes	**0.98**	0.91	**0.51**	0.19	0.51	0.13
Bottom nodes	0.93	0.18	0.56	0.22	0.57	0.21
Cycle nodes	0.71	0.22	0.55	0.16	0.52	0.14

Remedy: Precision-Recall curves (PR) and the corresponding AUC, as opposed to the ROC AUC, can provide a less misleading evaluation due to the fact that they evaluate the fraction of true positives among positive predictions [8]. Furthermore, for comparability to hard mask methods and since for the final explanation presented to the user, a threshold has to be applied, threshold-dependent metrics should be included in the evaluation. We propose to use recall, to account for the sparsity of an explanation.

The edges in the ground-truth explanation represent the true positives and the false negatives are the edges that are in the final explainer subgraph but not in the ground-truth explanation as it provides information about the compactness and therefore comprehensibility of the explainer subgraph. Table 2 shows the difference between avg. ROC AUC and avg. PR AUC. As we expect, PR AUC does not achieve the same level as the ROC AUC with a difference of up to 47% points, providing a more comprehensive picture of the explanation quality, similarly to the average recall.

3.3 Pitfall 3: Threshold Application

Reducing the size of the original subgraph is a post-processing step, executed after training the GNN. It is possible that the originally predicted label can flip, resulting in fidelity of the explanation not being ensured. Fidelity refers to an explanation being faithful to the model it aims to explain. For example for [10] a label flip occurs for 66% of the top node explainer subgraphs with $T = 6$. In this case, the final explainer subgraph would lead to a different prediction than the original subgraph and therefore defeating its purpose of explaining the original prediction. Overall, for the BA-Shape dataset, in [10] 19% of labels flip, in [5] 39% of labels flip and in [15] 18% of labels flip.

Additionally, for soft mask explainer methods [5,10], the size of the final explainer subgraph is parameterized. In the established evaluation scheme a dedicated hyperparameter T controls the size of the final explainer subgraph. For synthetic datasets, this hyperparameter is set according to knowledge about the ground-truth motif. Using this approach to configure the threshold leads to leaking information and unfairly biases the result. Therefore, the resulting evaluation is flawed, since it can be assumed that the ground-truth size of the explainer subgraph is not typically known. Choosing an appropriate threshold is not trivial, as the resulting recall can differ substantially for different thresholds as can be seen in Table 3, showing a necessary trade-off between the compactness and completeness of the final explainer subgraph.

Table 3. Average recall and precision for different thresholds for GNNExplainer

Class	Recall th = 6	Recall th = 20
Top nodes	0.65	0.98
Shoulder nodes	0.51	0.82
Bottom nodes	0.57	0.75
Cycle nodes	**0.52**	**0.97**

Remedy: The integration of an additional test to ensure that no label flip occurred for the final explainer subgraph is highly recommended to avoid an explanation that leads to an different decision - in other words, ensuring explanation fidelity. A simple if-then mechanism, that moves on to the next optimal explanation, in case a label flip occurs, would be sufficient.

Furthermore, to ensure that no knowledge about the ground truth leaks into the evaluation of the explanation and biases the result for soft mask approaches, we recommend to configure the size of the final explainer subgraph not to a fixed number of edges, but to carry out a grid-search on a test set to choose the optimal threshold.

4 Conclusion

The expressive nature of graphs makes them a promising candidate for producing rich explanations for GNN decision-making. But since a mature standardized approach to evaluating explanations for GNN explainer methods is missing, a valid comparison of different methods can be challenging. For this reason, we find it important to examine existing evaluation methods closely to uncover potential pitfalls. In this paper, we show the implications of three identified evaluation pitfalls in the context of GNNs and propose remedies to avoid them.

A   Background on GNNs and Perturbation-Based Explainer Methods

For a GNN, the goal is to learn a function of features on a graph $G = (V, E)$ with edges E and nodes V. The input is comprised of a feature vector x_i for every node i, summarized in a feature matrix $X \in \mathbb{R}^{n \times d_{in}}$ and a representative description of the link structure in the form of an adjacency matrix A. The output of the convolutional layer is a node-level latent representation matrix

$Z \in \mathbb{R}^{n \times d_{out}}$, where d_{out} is the number of output latent dimensions per node. Therefore, every convolutional layer can be written as a non-linear function:

$$H^{(l+1)} = f(H^{(l)}, A),$$

with $H^{(0)} = X$ and $H^{(L)} = Z$, L being the number of stacked layers. The vanilla GNN model employed here, uses the propagation rule [3]:

$$f(H^{(l)}, A) = \sigma(\hat{D}^{-\frac{1}{2}} \hat{A} \hat{D}^{-\frac{1}{2}} H^{(l)} W^{(l)}),$$

with $\hat{A} = A + I$, I being the identity matrix. $\hat{D}$ is the diagonal node degree matrix of $\hat{A}$, $W^{(l)}$ is a weight matrix for the $l-th$ neural network layer and σ is a non-linear activation function. Taking the latent node representations Z of the last layer we define the logits of node v_i for a node classification task as follows:

$$\hat{y}_i = \mathrm{softmax}(z_i W_c^{\top}),$$

where $W_c \in \mathbb{R}^{d_{out} \times k}$ projects the node representations into the k dimensional classification space.

GNNExplainer: The GNNExplainer takes a trained GNN and its prediction(s), and it returns an explanation in the form of a small subgraph of the input graph together with a small subset of node features that are most influential for the prediction. For their selection, the mutual information between the GNN prediction and the distribution of possible subgraph structures is maximized through optimizing the conditional entropy.

CF-GNNExplainer: The CF-GNNEXPLAINER works by perturbing input data at the instance-level. The instances are nodes in the graph since it is focused on node classification. The method iteratively removes edges from the original adjacency matrix based on matrix sparsification techniques, keeping track of the perturbations that lead to a change in prediction, and returning the perturbation with the smallest change w.r.t. the number of edges, after adding different edges to the subgraph.

ZORRO: ZORRO employs discrete masks to identify important input nodes and node features through a greedy algorithm, where nodes or node features are selected step by step. The goodness of the explanation is measured by the expected deviation from the prediction of the underlying model. A subgraph of the node's computational graph and its set of features are relevant for a classification decision if the expected classifier score remains nearly the same when randomizing the remaining features.

PGExplainer: The PGExplainer learns approximated discrete masks for edges to explain the predictions. Given an input graph, it first obtains the embeddings for each edge by concatenating node embeddings. Then the predictor uses the edge embeddings to predict the probability of each edge being selected, similarly to an importance score. The approximated discrete masks are then sampled via the reparameterization trick. Finally, the objective function maximises the mutual information between the original predictions and new predictions.

References

1. Arrieta, A.B., et al.: Explainable Artificial Intelligence (XAI): concepts, taxonomies, opportunities and challenges toward responsible AI. Inf. Fus. **58**, 82–115 (2020)
2. Huang, Q., et al.: GraphLIME: local interpretable model explanations for graph neural networks. arXiv preprint arXiv:2001.06216 (2020)
3. Kipf, T.N., Welling, M.: Semi-supervised classification with graph convolutional networks. In: ICLR (2017)
4. Lecue, F.: On the role of knowledge graphs in explainable AI. Sema. Web **11**(1), 41–51 (2020)
5. Lucic, A., et al.: CF-GNNExplainer: counterfactual Explanations for Graph Neural Networks. arXiv preprint arXiv:2102.03322 (2021)
6. Molnar, C., Casalicchio, G., Bischl, B.: Interpretable machine learning – a brief history, state-of-the-art and challenges. In: Koprinska, I., et al. (eds.) ECML PKDD 2020. CCIS, vol. 1323, pp. 417–431. Springer, Cham (2020). https://doi.org/10.1007/978-3-030-65965-3_28
7. Robnik-Šikonja, M., Bohanec, M.: Perturbation-based explanations of prediction models. In: Zhou, J., Chen, F. (eds.) Human and Machine Learning. HIS, pp. 159–175. Springer, Cham (2018). https://doi.org/10.1007/978-3-319-90403-0_9
8. Saito, T., Rehmsmeier, M.: The precision-recall plot is more informative than the ROC plot when evaluating binary classifiers on imbalanced datasets. PLOS ONE **10**(3), e0118432 (2015)
9. Xu, K., et al.: How powerful are graph neural networks? In: ICLR (2018)
10. Ying, R., et al.: GNNExplainer: generating explanations for graph neural networks. In: Advances in Neural Information Processing Systems 32, p. 9240 (2019)
11. Yuan, H., et al.: XGNN: towards model-level explanations of graph neural networks. In: Proceedings of the 26th ACM SIGKDD International Conference on Knowledge Discovery & Data Mining (2020)
12. Zhou, J., et al.: Graph neural networks: a review of methods and applications. AI Open **1**, 57–81 (2020)
13. Funke, T., Khosla, M., Anand, A.: Hard masking for explaining graph neural networks (2020)
14. Yuan, H., et al.: Explainability in graph neural networks: a taxonomic survey. arXiv preprint arXiv:2012.15445 (2020)
15. Luo, D., et al.: Parameterized explainer for graph neural network. In: Advances in Neural Information Processing Systems (2020)
16. Anonymous: Causal screening to interpret graph neural networks. Submitted to International Conference on Learning Representations (2021, under review). https://openreview.net/forum?id=nzKv5vxZfge
17. Yuan, H., Yu, H., Wang, J., Li, K., Ji, S.: On explainability of graph neural networks via subgraph explorations. arXiv preprint arXiv:2102.05152 (2021)

On the Transferability of Neural Models of Morphological Analogies

Safa Alsaidi[1], Amandine Decker[1], Puthineath Lay[1],
Esteban Marquer[1(✉)], Pierre-Alexandre Murena[2], and Miguel Couceiro[1]

[1] Université de Lorraine, CNRS, LORIA, 54506 Vandoeuvre-lès-Nancy, France
{esteban.marquer,miguel.couceiro}@loria.fr
[2] HIIT, Aalto University, Helsinki, Finland
pierre-alexandre.murena@aalto.fi

Abstract. Analogical proportions are statements expressed in the form
"A is to B as C is to D" and are used for several reasoning and classification tasks in artificial intelligence and natural language processing
(NLP). In this paper, we focus on morphological tasks and we propose a
deep learning approach to detect morphological analogies. We present an
empirical study to see how our framework transfers across languages, and
that highlights interesting similarities and differences between these languages. In view of these results, we also discuss the possibility of building
a multilingual morphological model.

Keywords: Morphological analogy · Deep learning · Transferability ·
Analogy classification

1 Introduction

An analogy, or analogical proportion, is a relation between two word pairs or
four elements A, B, C, and D, meaning "A is to B as C is to D", often written
as $A : B :: C : D$. An analogical proportion becomes an equation if one of its
four objects is unknown [9].

Analogies have been extensively studied in Natural Language Processing,
which resulted in different formalizations with noteworthy applications in various
domains such as derivational morphology [10]. Analogies on words can refer
exclusively to their morphology as in the following example: "*apple* is to *tree*
as *apples* is to *trees*". It is based on morphological variations of two words:
"apple" and "tree". The question of the correctness of an analogy $A : B :: C : D$
is a difficult task; however, it has been tackled both formally and empirically
[6,8,10]. Recent empirical works propose data-oriented strategies to learn the
correctness of analogies from past observations. These strategies are based on
machine learning approaches.

This research was partially supported by TAILOR, a project funded by EU Horizon
2020 research and innovation programme under GA No. 952215, and the Inria Project
Lab "Hybrid Approaches for Interpretable AI" (HyAIAI).

© Springer Nature Switzerland AG 2021
M. Kamp et al. (Eds.): ECML PKDD 2021 Workshops, CCIS 1524, pp. 76–89, 2021.
https://doi.org/10.1007/978-3-030-93736-2_7

Analogical classification and inferences are two main problems worth addressing when dealing with the axiomatic settings of analogies. Multiple attempts have been conducted in terms of formulating and manipulating analogies [6,7]. Most algorithmic approaches to solving morphological analogies rely on the formal characterization of proportional analogies (see Sect. 3).

However, other approaches to detecting analogies have been used as well. Some of which include Fam and Lepage (2018) [2] and the Alea algorithm by Langlais *et al.* (2009) [5]. Fam and Lepage's approach relies on feature vectors to detect analogies within a list of words and create analogical grids. As a result, analogies between more than 4 words can be generated. The axioms of Lepage [7] have been also reformulated by Yvon [13] to give a closed form solution, which is computed by Alea in a Monte-Carlo setting, where character strings A, B and C are randomly sliced and merged to obtain potential solutions to $A : B :: C : x$.

An approach by Murena *et al.* (2020) [10] outperformed both Alea and Fam and Lepage's approaches. In this approach, analogical equations $A : B :: C : x$ are solved by finding the x that minimizes the Kolmogorov complexity, which is evaluated by using a simple description language for character strings and an associated binary code.

Another approach by Lim *et al.* [8] propose to learn an analogy operator directly from analogies. They used neural networks to learn semantic analogies, and proposed different models for analogical classification and inferences (regression). They also used pretrained GloVe embeddings [11], which achieved competitive results on analogy classification and completion. Moreover, they relied on the properties of formal analogies to increase the amount of training data as introduced in Sect. 3.1.

In this paper, we adapt the approach developed by Lim *et al.* [8] for semantic word analogies and apply it on morphological analogies. For this approach, we had to develop and train various morphological word embedding models. Furthermore, we explore the potential of transferring our neural analogy model across 11 languages, which is the core of this paper. Our method achieves competitive results on each language and promising ones in the transferability settings. This allowed us to explore building a single multilingual model that could work with several languages.

This paper is organized as follows. In this introduction, we recall some of the related works particularly the one by Lim *et al.* [8] from which our framework was inspired. In Sect. 2, we explain how we adapt this approach to morphological analogy and the models we developed and trained. In Sect. 3, we introduce our datasets and describe how the models trained on each language perform when transferred to the other languages. We further investigate different transferability settings in Sect. 4. We work with 4 language models that reveal interesting results.

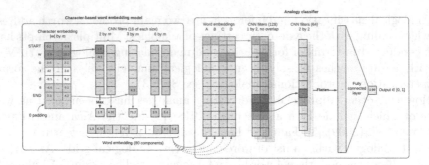

Fig. 1. The character-level word embedding model and the CNN classifier.

2   Proposed Approach

As aforementioned, we adapt the approach by Lim *et al.* [8] in terms of using the same architecture for our classification model. However, we have developed our own custom embedding models, which is trained along with the classifier.

2.1   Classification Model

The architecture of the classification model is a Convolutional Neural Network (CNN), which takes as input the embeddings of size n of four elements A, B, C and D stacked into a n by 4 matrix. The CNN has 3 layers as follows (see Fig. 1). A first convolutional layer with filter of 1 by 2 is applied on the embeddings. This layer respects the boundaries between the two pairs, where it analyses each pair and extracts the differences and similarities between $A : B$ and $C : D$. We use 128 of such filters with a Regularized Linear Unit (ReLU) as activation function.

A second convolutional layer with 64 filters of 2 by 2 is applied on the resulting 128 by n by 2 tensor, after which the result is flattened into a $64 \times (n - 1)$ unidimensional vector. ReLU is also used as an activation function. This layer should compare the results of the analysis of both pairs: if A and B are different in the same way as C and D, then $A : B :: C : D$ is a valid analogy. In the third layer, the results are flattened and used as input of a dense layer. We use a sigmoid activation to get a result between 0 and 1 as we work with a binary classifier.

2.2   Embedding Model

In [8], Lim *et al.* achieved good results on semantic analogies by using pre-trained GloVe embeddings that are supposed to encode the semantics of words. But when we tried to apply it to morphological analogies, we got poor performance where after investigation it turned out that the model could not learn but rather produced tensors full of zeros. Therefore, since we work with morphological analogies and with languages with no embedding models already available, we had to develop and train our own models. We developed a character-level word embedding model able to capture the morphological aspect of words. To our knowledge no such embedding model is available for the languages we manipulate, so we decided to train an embedding model together with our classifier. We use a very simple model architecture based on Convolutional Neural Network (CNN) as described in [12]. This architecture is designed to embed individual words for morphological tasks. In our case it is composed of two layers as follows.

First, we have a character embedding layer to encode each character of the word with a vector of size m. Characters never encountered during the training phase are embedded with vectors full of 0 by default. At the beginning and at the end of the word, we add special vectors to signify the boundaries of the word. For a word of $|w|$ characters, we obtain a $(|w| + 2)$ by m matrix.

Then we use a convolutional layer with 16 filters of each of the sizes 2 by m, 3 by m, ..., 6 by m, for a total of $16 \times 5 = 80$ filters. The filters are used on the embeddings so that they overlap along the character dimension. We expect those filters to capture morphemes of up to 6 characters.

A max pooling layer is then applied to the output where we keep only the greatest number produced by each of the 80 filters so that only the most important patterns appear in the final embedding. We finally concatenate the results to produce an embedding of size 80.

3   Experiments

We worked with 11 languages in our experiments: 10 from SIGMORPHON 2016 [1] and the other from JAPANESE BIGGER ANALOGY TEST SET [3].

3.1   Datasets and Augmentation

The SIGMORPHON 2016 [1] datasets contains training, development and test data. Data is available for 10 languages: Spanish, German, Finnish, Russian, Turkish, Georgian, Navajo, Arabic (Romanized), Hungarian and Maltese (see Table 1 in Appendix). It is separated in 3 subtasks: inflection, reinflection and unlabeled reinflection. In our experiments, we focused on the data from the inflection task, which is made up of triples $\langle A, F, B \rangle$ of a source lemma A (ex: "cat"), a set of features F (ex: pos = N, num = PL) and the corresponding inflected form B (ex: "cats"). On the other hand, Japanese dataset contains several files, each of them containing pairs of linguistically related words. We were interested in inflectional and derivational morphology relations for which the dataset contains respectively 515 and 502 pairs of words. For each two pairs with the same relation, we produced an analogy. However, Japanese produced poor results when solving analogies which may be related to the small size of the dataset.

Deep learning approaches require large amounts of data. Therefore we took advantage of some properties of analogies to produce more data based on our datasets, this process is called *data augmentation*. Given a valid analogy A :

$B :: C : D$, we can generate 7 more valid analogies, namely, $A : C :: B : D$, $D : B :: C : A$, $C : A :: D : B$, $C : D :: A : B$, $B : A :: D : C$, $D : C :: B : A$, $B : D :: A : C$, and 3 invalid analogies, namely, $B : A :: C : D$, $C : B :: A : D$, and $A : A :: C : D$.[1]

3.2 Transferability Results

In this section, we will explain the results we obtained once we applied the different models trained on each language to the other languages of the dataset. The objective is to explore the generalization capabilities of the CNN model, and to test its dependence on the training language. To evaluate their generalization capacity, we ran the evaluation using each model on all the languages. In next subsections we introduce our results for full and partial transfer. In full transfer, we describe the results when transferring both the embedding and classification models. Then in partial transfer, we transfer only the classification model (the data is embedded with the "right" embedding model). We test them on 3 settings: base, negative, and positive analogies. Positive and negative analogies stand for the valid and invalid forms of analogies, which are presented in the previous section. Base is the initial analogy form $A : B :: C : D$, we do not use the data augmentation for this evaluation.

Full Transfer. The results for full transfer (*i.e.,* using both the embeddings and the classifier of a language on another language) on positive and negative data are presented in Fig. 4 in Appendix. The results for positive data are most of the time above 90% except for Arabic and Navajo words, compared to the results of negative data, which were more heterogeneous. This is probably due to differences in alphabet between the source and target language that cause a large portion of the target alphabet to be unrecognized. As a result, characters unknown to the model are embedded as zeros and tend to be ignored by the model. The very low performance on negative classification can be explained in that the classifier answers "valid analogy" by default when encountering unrecognized characters.

Overall, most languages transfer reasonably well, with an accuracy between 50% to 80%. However, models have an accuracy close to 0% when transferred across languages that don't share most of the alphabets. Interestingly, models trained on Turkish and Hungarian perform slightly better when compared to those trained on other languages. Further analysis and experiments are introduced in Sect. 4.

---

[1] Lim *et al.* [8] generate the 3 invalid analogies for each of the 8 valid analogies, resulting in 24 invalid analogies. We experimented with 24 invalid analogies but the performance of the model was worse than when using only 3 invalid analogies, despite the increase in the amount of training data. The decrease in performance is likely due to the stronger imbalance between invalid and valid examples.

Partial Transfer. To solve the issues we had with character dictionaries for full transfer, we tried to transfer only the classifier. The results for partial transfer (*i.e.,* using the classifier model of a language on another language using the target language's embedding model) are presented in Fig. 5 in Appendix.

The results are similar to those of full transfer, where the model transfers well for positive data but results remain heterogeneous for negative data. As we used the embedding model corresponding to the language, we have less combinations producing an accuracy of 0% for negative data. To elaborate, the performance is above 10% for all but three cases (from Georgian and Spanish to Japanese and from Spanish to Arabic), and above 25% in most cases, which is a clear improvement from full transfer.

The main difference between both experiments is that for partial transfer, the alphabet gap is no longer an issue which results in a significant improvement in the performance. But since the embedding model was not trained together with the classifier, this may result in a mismatch in the representation. We tried to address this potential issue with our experiments with multilingual models.

3.3 Discussion

The quality of the transfer is not symmetric, which means that a model that performs well when transferred from a language A to a language B may perform badly when trained on language B and transferred to language A. The Hungarian model transfers really well to other languages while most of the other languages are not that efficient with Hungarian. This could be explained due to the fact that Hungarian has "particularly rich morphology" [4] using inflection, derivation and compounding.

The Arabic model is efficient for full transfer but not really for partial transfer. It could be possible that the Arabic embedding model encodes the sub-words differently from the other models, as Arabic words are formed by roots and word patterns. This would mean that the embedding and classification model would be strongly related. As a result, it performs poorly once applied to other languages in case of partial transfer.

For Japanese, the accuracy is not always of 0% for negative data when the models transfer to Japanese. None of the Japanese characters are present in the dictionaries of the other models so we could expect all the analogies to look like $\varepsilon : \varepsilon :: \varepsilon : \varepsilon$ and thus all the said invalid ones to be classified as valid. The reason why we do not have 0% in those cases is still unknown.

Eventually Hungarian seem to be very efficient models in terms of transfer learning. In average, Hungarian is one of the languages closest to all the others in the family hierarchy, which may explain why it transfers well. As a result, we decided to run more experiments to further evaluate and improve our models in terms of transferability. Further details are introduced in the next chapter.

4   Toward a Multilingual Model

Following our observations, we decided to work on building a multilingual model. Our idea was that using several languages during the training phase would increase the generalization capacity of our model. We thus trained different models with different languages as training data in order to see which setting produces the best results in terms of transferability. We then evaluated the transferability of the models across the 11 languages to determine if we obtain better performance than with only one language during the training phase.

4.1   Models

We explored two different settings regarding the languages we used: on the one hand we trained models with two languages as training data and on the other hand we trained models with all our eleven languages or 10 without Japanese.

Models with Two Languages. For our bilingual models, we worked with two pairs of languages: Hungarian-Finnish and Hungarian-Turkish. We chose Hungarian and Finnish because they are close to all the other languages in the family hierarchy. As for Hungarian and Turkish, they are the languages that produced the best results in terms of transferability.

Models with All Languages. There are two types of multilingual models. In one case, we trained a *single embedding model* for all the languages. We thus use data of all the languages. Regarding the training set data, it is shuffled so that the model is not trained on one language after the other. In the other case, we train one embedding model per language (called *multi-embedding model*); therefore, there will be 10 or 11 embedding models for 10 or 11 languages model. For this model, the training sets of data are concatenated one after another.

4.2   Results

Models with Two Languages. As we could expect, for full transfer, the results (see Fig. 2 in Appendix) turned out to be very good when transferred along the same languages (*i.e.*, Hungarian and Finnish when applied to Hungarian or Finnish produces good results and same for the other model). The results were very bad for Russian and Georgian (*i.e.*, the accuracy is around 2%) and for Japanese (*i.e.*, the accuracy is 25%) when the Hungarian and Finnish model is applied. Navajo and Arabic is slightly worse in full transfer compared to other languages for both models.

For partial transfer, applying the Hungarian and Finnish model to the Hungarian or Finnish gave slightly worse results as each of the individually trained embedding model are not being perfectly adapted to the classifier. However, the results for the Hungarian and Turkish were very good when applied to Hungarian or Turkish. For positive (valid) examples, the results of the Hungarian and Finnish model was very good when applied to the rest of the languages. It was slightly lower for negative examples (*i.e.,* the accuracy is still above 30% for invalid), except for Japanese. In comparison, the results for the Hungarian and Turkish model were very strange when applied to the rest of the languages. Lower performance can be observed on negative examples for Georgian and much lower performance on Japanese between base and positive examples.

Models with All Languages. In the experiment, we train on both 10 languages and 11 languages (with and without Japanese). So, we train 2 models of single embedding model and others 2 of multi-embedding model (see Sect. 4.1). The accuracy results (see Fig. 3 in Appendix) of the four models are comparable, except for Maltese, where the performance is lower than those of other languages. Both models with Japanese obtain better results, especially on invalid examples. The single embedding model with Japanese provides the best result, especially the result with 91% of Japanese invalid accuracy.

4.3   Discussion

Overall, the bilingual models achieve better results than the multilingual ones. However, the bilingual models produce low results for a few languages while the multilingual models are more constant. Moreover, we used less than 5000 analogies per language to train the multilingual models, while we use 25000 analogies per language for the bilingual models. We could expect the performance of our multilingual model to increase with more training data.

During these experiments, we also tested another training setting. Instead of using 8 valid analogies for 3 invalid ones, we applied the properties on the 3 invalid forms. In the end we had 8 valid analogies for 24 invalid ones. However this setting produced worse results than the one with 3 invalid analogies. This could be related to the stronger imbalance between the positive and negative data. To solve this issue, it would be interesting to train models on balanced datasets: we would work with the 8 valid forms and 8 invalid forms randomly selected among the 24.

Eventually we tried to perform an analysis of the similarities between our 11 languages by building dendrograms based on the first transfer results. However none of the dendrograms produced were similar to the family hierarchy, which may indicate that either the differences we observe in our results match some other linguistic properties than the language families or that the differences are due to empirical differences in the data.

5   Conclusion

We successfully adapted Lim's *et al.* semantic analogies approach [8], applied it to morphological analogies, and achieved competitive results. Compared to the model of Lim *et al.*, our CNN model is more flexible in many aspects: *(i)* it is able to model any words even those never encountered in the training phase; *(ii)* it is able to carry over domain and language specificities from the training process; *(iii)* it has strong potential to carry over models of analogy when using an adapted embedding model as shown with our transfer experiments; *(iv)* it has strong potential to learn under different sets of axioms to adapt to other analogical settings. Our early experiments on transferability highlighted the potential to transfer and reuse our neural approach across domains. Therefore, we developed two bilingual models and a multilingual model that we applied across the 11 languages and tested different transferability settings.

Appendix

Table 1. Number of analogies for each language before data augmentation.

Language	Train	Dev	Test
Arabic	373240	7671	555312
Finnish	1342639	22837	4691453
Georgian	3553763	67457	8368323
German	994740	17222	1480256
Hungarian	3280891	70565	66195
Maltese	104883	3775	3707
Navajo	502637	33976	4843
Russian	1965533	32214	6421514
Spanish	1425838	25590	4794504
Turkish	606873	11518	11360

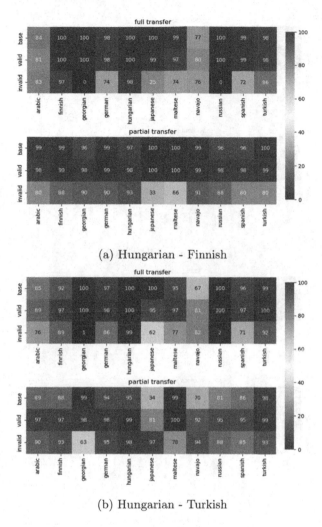

(a) Hungarian - Finnish

(b) Hungarian - Turkish

Fig. 2. Accuracy (in %) of fully transferred bilingual models on SIGMORPHON 2016 and JAPANESE BIGGER ANALOGY TEST SET.

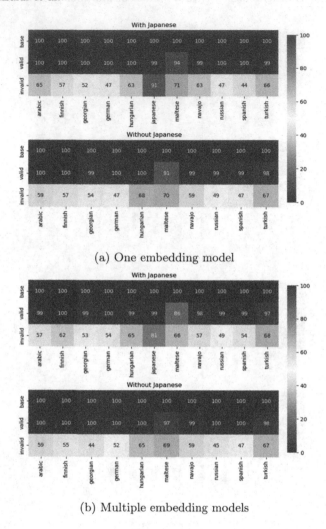

(a) One embedding model

(b) Multiple embedding models

Fig. 3. Accuracy (in %) of fully transferred multilingual models on SIGMORPHON 2016 and JAPANESE BIGGER ANALOGY TEST SET.

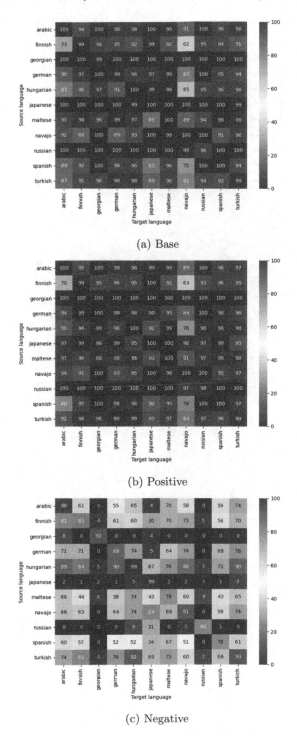

(a) Base

(b) Positive

(c) Negative

Fig. 4. Accuracy (in %) of fully transferred models on Sigmorphon 2016 and Japanese Bigger Analogy Test Set.

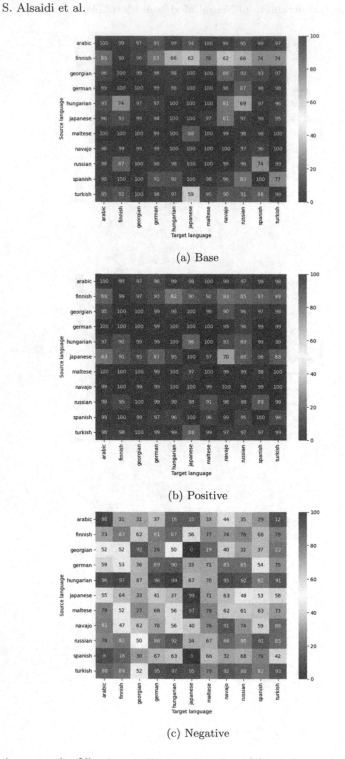

(a) Base

(b) Positive

(c) Negative

Fig. 5. Accuracy (in %) of partially transferred models on SIGMORPHON 2016 and JAPANESE BIGGER ANALOGY TEST SET.

References

1. Cotterell, R., et al.: The SIGMORPHON 2016 shared task–morphological reinflection. In: the Proceedings of the ACL 2016 Meeting of SIGMORPHON, pp. 10–22 (2016)
2. Fam, R., Lepage, Y.: Tools for the production of analogical grids and a resource of n-gram analogical grids in 11 languages. In: 11th International Conference on Language Resources and Evaluation, LREC 2018. European Language Resources Association (ELRA) (2018)
3. Karpinska, M., et al.: Subcharacter information in Japanese embeddings: when is it worth it? In: Proceedings of the ACL Workshop on the Relevance of Linguistic Structure in Neural Architectures for NLP, pp. 28–37 (2018)
4. Kiefer, F.: Hungarian. Revue belge de philologie et d'histoire, tome 88, fasc. 3, Langues et littératures modernes (2010)
5. Langlais, P., Yvon, F., Zweigenbaum, P.: Improvements in analogical learning: application to translating multi-terms of the medical domain. In: 12th Conference of the European Chapter of the Association for Computational Linguistics, EACL 2009, pp. 487–495 (2009)
6. Lepage, Y.: De l'analogie rendant compte de la commutation en linguistique. Habilitation à diriger des recherches (2003). https://tel.archives-ouvertes.fr/tel-00004372
7. Lepage, Y.: Analogy and formal languages. In: Joint meeting of the 6th Conference on Formal Grammar and the 7th Conference on Mathematics of Language, vol. 53, pp. 180 191 (2004)
8. Lim, S., Prade, H., Richard, G.: Solving word analogies: a machine learning perspective. In: Kern-Isberner, G., Ognjanović, Z. (eds.) ECSQARU 2019. LNCS (LNAI), vol. 11726, pp. 238 250. Springer, Cham (2019). https://doi.org/10.1007/978-3-030-29765-7_20
9. Miclet, L., Bayoudh, S., Delhay, A.: Analogical dissimilarity: definition, algorithms and two experiments in machine learning. J. Artif. Intell. Res. **32**, 793–824 (2008)
10. Murena, P.-A., et al.: Solving analogies on words based on minimal complexity transformation. In: The Proceedings of the 29th International Joint Conference on Artificial Intelligence, IJCAI 2020, pp. 1848–1854 (2020)
11. Pennington, J., Socher, R., Manning, C.D.: GloVe: global vectors for word representation. In: Empirical Methods in Natural Language Processing (EMNLP), pp. 1532–1543 (2014)
12. Vania, C.: On understanding character-level models for representing morphology. Ph.D. thesis, University of Edinburgh (2020)
13. Yvon, F.: Finite-state transducers solving analogies on words. Rapport GET/ENST&LTCI (2003)

Behavior of k-NN as an Instance-Based Explanation Method

Chhavi Yadav[(✉)] and Kamalika Chaudhuri

University of California, San Diego, USA
cyadav@ucsd.edu, kamalika@cs.ucsd.edu

Abstract. Adoption of DL models in critical areas has led to an escalating demand for sound explanation methods. Instance-based explanation methods are a popular type that return selective instances from the training set to explain the predictions for a test sample. One way to connect these explanations with prediction is to ask the following counterfactual question - how does the loss and prediction for a test sample change when explanations are removed from the training set? Our paper answers this question for k-NNs which are natural contenders for an instance-based explanation method. We first demonstrate empirically that the representation space induced by last layer of a neural network is the best to perform k-NN in. Using this layer, we conduct our experiments and compare them to influence functions (IFs) [6] which try to answer a similar question. Our evaluations do indicate change in loss and predictions when explanations are removed but we do not find a trend between k and loss or prediction change. We find significant stability in the predictions and loss of MNIST vs. CIFAR-10. Surprisingly, we do not observe much difference in the behavior of k-NNs vs. IFs on this question. We attribute this to training set subsampling for IFs.

Keywords: k-NN · Explainability · Instance-based · Interpretability · Representations · Influence functions

1 Introduction

Deep Learning (DL) models have shown a great deal of promise in providing high accuracy predictions; however, these accurate predictions come at the cost of high opacity. As they are being adopted in critical applications, it is becoming increasingly important to explain their predictions and interpret them.

A popular class of explanation methods is *instance-based explanations*. Here the goal is to select instances from the training set to explain why an input is assigned a particular label. k-Nearest Neighbors (k-NN), which provide the top k most similar instances to a test sample, are natural contenders for instance-based explanations. The idea of using k-NNs as an explanation method dates back to the 1990s [2]. It has recently made a come back like in [7]. However, the utility of k-NNs as an instance-explanation method has not been investigated thoroughly.

© Springer Nature Switzerland AG 2021
M. Kamp et al. (Eds.): ECML PKDD 2021 Workshops, CCIS 1524, pp. 90–96, 2021.
https://doi.org/10.1007/978-3-030-93736-2_8

One way to establish the connection between explanations and prediction is to see how the prediction and loss at a test sample change when explanations are removed from the training set. We call this the *'counterfactual question'*. In this work, we take a closer look at k-NNs as instance-based explanations in the context of the above counterfactual question. We compare our k-NN results with removing the most influential training points according to influence function (IF) score from [6].

To answer this, we first find what is a good representation space in which to choose the nearest neighbors? In the case of a neural network f, representation spaces are induced by different layers of f. We find that the last layer is the best – in the sense that k-NN using representations of the last layer approximates the decision boundary of f the best amongst all layers.

Section 3 discusses our experimental setups and results in detail. Overall, we do see a change in loss and predictions sans the k nearest neighbors. We observe a significant stability in the predictions and loss of MNIST compared to CIFAR-10. However, we do not observe an increasing trend between the value of k and loss or prediction change. In terms of k-NN vs. IFs, we do not observe much difference in their behaviors. We attribute this to subsampling the training set in the IF case and call for a deeper investigation.

2 Related Work

Instance-based explanations are a popular interpretability tool. Several such methods exist including Counterfactual explanations [3,9], Prototypes and Criticisms [4,8], k-NNs [2,7] and Influence Functions [1,5,6]. Each of them has a different definition and hence a different way of selecting the instance explanations.

Counterfactual explanation method starts with any training instance & modifies it to reduce the counterfactual loss. The output explanation instance hence need not be from the training set. Prototype based explanation methods first find a representative set of points to describe the data using maximum mean discrepancy and provide the nearest-prototype as explanations for a given test instance. Influence functions estimate the influence of a training sample or set of training samples on a test sample by estimating the change in loss without those training samples. However, they tend to return outliers as noted in [1]. Given a test sample, k-NNs simply return the k nearest neighbors in a specified representation space as explanations. In this work, we dive deeper into the properties of k-NN as an explanation method.

3 Experiments and Results

Our goal in this section is to address the following questions:

- Which representation space induced by the different layers of a neural network f is best for finding k-NNs?

– How does removing the top k nearest neighbors (offered as explanations) from the training data affect the prediction on a test input?

These questions are considered in the context of two standard datasets – MNIST and CIFAR-10. For MNIST we use a custom model as shown in Table 1, while for CIFAR-10 we used a Wide-ResNet [10] with depth and width 10.

3.1   The Choice of Representation

Since k-NN is a distance based method, the distance function and representation space in which these distances are found, are two significant aspects that affect its output. In a neural network f, each layer induces its own representation space ϕ_{layer}, giving k-NN an option to choose amongst these.

Label Agreement: We say that label agreement happens for a test point when the label predicted by nearest neighbor and the neural network are the same. We find the fraction of test points for which label agreement happens in different layers of f. As this fraction approaches 1, gap between the decision boundaries of k-NN and f reduces, becoming approximately the same, when tested with large number of samples. If this occurs, then k-NN is a global approximator of the neural network.

Setup: We take 1000 samples from the test set of each dataset, with equal contributions from all classes. We find the 1-NNs from the training set using representations from various layers of the neural network. We compare labels from 1-NN to those predicted by the neural network. We use cosine distance function.

Result: From Tables 1 and 2 it is clear that as we go from first layer to the last, the fraction of points with label agreement increases. For MNIST, the fraction approaches 1 at the last layer, while for CIFAR-10, it approaches 0.9. Since k-NN in the last layer approximates decisions of f the best, we use ϕ_{last} as our k-NN representation space.

Table 1. Fraction of points for which the label predicted by 1-NN and neural network are same on MNIST

Layer	Fraction
conv_relu	0.962
conv_relu_maxpool	0.977
conv_relu	0.979
conv_relu_maxpool	0.983
fc	0.995
relu_drop_fc	0.997
relu_fc	0.998

Table 2. Fraction of points for which the label predicted by 1-NN and neural network are same on CIFAR-10

Layer	Fraction
conv	0.328
block1_block2_block3	0.756
batchnorm_relu_avgpool	0.877
fc	0.881

3.2 Effect of Removing Explanatory Instances

Given a test sample x, our k-NN explainer (in ϕ_{last}) would return the top k nearest neighbors of x from the training set as explanations. In this section, we want to understand the effect of removing these k samples from the training set on the prediction and loss at x. In [6], the authors define highly influence points to be those training points that when removed from the training set, lead to the largest change in loss at x. Therefore, we compare our results to influence functions.

Setup: We take 500 random samples from the test set and find their top k nearest neighbors from the training set using ϕ_{last} and cosine distance function. $k \in [1, 5, 10, 15, 20]$. For each test sample x, we remove these k nearest neighbors from the training set and retrain a model from scratch using the new training set, keeping other parameters fixed. We note the loss change (LC) and predicted labels for the test samples before and after retraining. A Positive LC implies that the loss at x increased due to removal of the k nearest neighbors from the training set. Similarly, a change in predicted label before and after retraining, which we call a label flip, implies some effect was caused by removal of the k nearest neighbors.

We compare our results to Influence Functions (IFs). For each test sample x, we find the top k influential points using the definition in [6] and follow the same procedure as above. Calculating influence is an expensive task since for each test sample, it takes $O(np + rtp)$ time, where $rt = O(n)$, n = size of dataset, r = repeats, t = iterations & p = model parameters, as mentioned in [6]. Hence, due to computational constraints, we use some commonplace tricks. We test using 100 random test samples. When searching for influential samples in the training set we look in a subsampled training dataset of size 1000. However, we retrain with the complete training set minus the top k influential samples. We freeze all but last layer weights.

Results: As is evident from Table 3 and 5, we do see a loss and prediction change when top k nearest neighbors are removed from training set. However, there is no particular increasing trend between the value of k and loss change or label flips. This is counter-intuitive since one would expect removing more samples would lead to potentially worse learning and hence higher loss change. A plausible explanation can be that there is enough information in the rest of the training samples to learn. Hence, removing a few does not matter much, unless the test sample is noisy or an outlier.

We also note that predictions on MNIST are more stable than CIFAR-10, in the sense that the loss change and percentage of label flips from Table 5 for MNIST are way lower than that for CIFAR-10. This reiterates the belief that MNIST is an easier dataset and has less intraclass diversity in the samples.

Contrasting the results of k-NN and Influence functions, we do not observe much difference in behavior. Removing top k nearest neighbors vs. influential points has almost similar effects. This maybe due to subsampling the training set for influence functions for computational reasons & calls for further investigation.

Table 3. k-NN: For various k, loss change (LC) at test samples between retrained and original models. Average (Avg.) LC is the average across all test samples, Avg. +ve & −ve LC are the average positive & negative loss change across respective test samples, Max. & Min. LC note the maximum & minimum loss change across all test samples, +ve LC % notes the percentage of test samples with a ≥ 0 LC.

Dataset	k	Avg. LC	Avg. +ve LC	Avg. −ve LC	Max. LC	Min. LC	+ve LC %
MNIST	1	0.009 ± 0.177	0.015 ± 0.180	-0.040 ± 0.146	2.622	-0.752	89.6
	5	0.019 ± 0.499	0.026 ± 0.525	-0.040 ± 0.148	11.107	-0.786	89.2
	10	0.029 ± 0.452	0.036 ± 0.473	-0.036 ± 0.133	9.392	-0.624	90.4
	15	0.024 ± 0.327	0.031 ± 0.340	-0.040 ± 0.136	5.375	-0.553	91.0
	20	0.034 ± 0.535	0.043 ± 0.560	-0.050 ± 0.163	8.354	-0.747	90.4
CIFAR-10	1	-0.407 ± 2.018	0.751 ± 1.597	-0.908 ± 1.975	8.514	-19.012	30.2
	5	-0.446 ± 1.866	0.647 ± 1.238	-0.928 ± 1.893	6.169	-15.23	30.6
	10	-0.342 ± 2.643	1.009 ± 3.227	-0.954 ± 2.057	35.885	-19.441	31.2
	15	-0.075 ± 3.134	1.516 ± 4.383	-0.851 ± 1.844	44.61	-19.101	32.8
	20	-0.526 ± 1.954	0.483 ± 1.084	-0.965 ± 2.082	6.984	-18.344	30.3

Table 4. Influence Functions: For various k, loss change (LC) at test samples between retrained and original models. Average (Avg.) LC is the average across all test samples, Avg. +ve & −ve LC are the average positive & negative loss change across respective test samples, Max. & Min. LC note the maximum & minimum loss change across all test samples, +ve LC % notes the percentage of test samples with a ≥ 0 LC.

Dataset	k	Avg. LC	Avg. +ve LC	Avg. −ve LC	Max. LC	Min. LC	+ve LC %
MNIST	1	-0.00019 ± 0.002	$1.3e-05 \pm 8e-05$	-0.00285 ± 0.007	0.0006	-0.02	93
	5	$7e-05 \pm 0.003$	0.00029 ± 0.002	-0.00346 ± 0.008	0.0185	-0.02	94
	10	-0.00021 ± 0.002	$2e-06 \pm 2e-05$	-0.00232 ± 0.006	0.0001	-0.02	91
	15	-0.00018 ± 0.002	$4e-06 \pm 3e-05$	-0.00229 ± 0.006	0.0003	-0.018	92
	20	-0.00018 ± 0.002	$2e-05 \pm 1e-04$	-0.00343 ± 0.008	0.0007	-0.02	94
CIFAR-10	1	-0.43 ± 1.544	0.464 ± 1.039	-0.87 ± 1.562	5.373	-8.501	33
	5	-0.349 ± 1.334	0.666 ± 1.187	-0.688 ± 1.201	4.416	-4.998	25
	10	-0.447 ± 1.637	0.494 ± 1.037	-0.89 ± 1.68	4.319	-8.891	32
	15	-0.539 ± 1.623	0.378 ± 0.948	-1.077 ± 1.694	4.323	-6.95	37
	20	-0.475 ± 1.351	0.314 ± 0.699	-0.797 ± 1.418	3.398	-7.776	29

Table 5. Percentage of label flips over test samples, between the retrained and original models when the top k nearest neighbors & the top k influential samples are removed from the training set.

k	MNIST k-NN	MNIST IF	CIFAR k-NN	CIFAR IF
1	0.8	0	21.2	23
5	0.6	0	23.2	19
10	1	0	22.8	22
15	0.6	0	23.4	19
20	0.8	0	19.48	21

4 Conclusion

We present an investigation into the behavior of k-NN as an instance-based explanation method. First, we demonstrate that representation space induced by the last layer of a neural network is the best to find neighbors in. Following this, we try to connect explanations and predictions by probing the counterfactual question - what happens when the k nearest neighbors of a test sample are removed from the training set? How are the loss and prediction affected?

From our experiments we do see a change in loss and predictions. However, we do not find signs of an increasing relationship between the value of k and loss or prediction change. We also find evidence for stable predictions on MNIST than CIFAR-10.

We do not observe much difference between the behavior of k-NN vs. IFs. We believe that this could be due to subsampling the training set for computational constraints. A future direction can be to investigate this more thoroughly with the complete training set, though it is a difficult task in reality.

We conclude by saying that when choosing an explanation method, it's of utmost importance to think carefully about what the application is and if the properties of the explanation method match the requirements. In line of this, we call for a deeper investigation into the properties of explanation methods.

References

1. Barshan, E., Brunet, M.E., Dziugaite, G.K.: RelatIF: identifying explanatory training samples via relative influence. In: International Conference on Artificial Intelligence and Statistics, pp. 1899–1909. PMLR (2020)
2. Caruana, R., Kangarloo, H., Dionisio, J.D., Sinha, U., Johnson, D.: Case-based explanation of non-case-based learning methods. In: Proceedings of the AMIA Symposium, p. 212. American Medical Informatics Association (1999)
3. Karimi, A.H., Barthe, G., Balle, B., Valera, I.: Model-agnostic counterfactual explanations for consequential decisions. In: International Conference on Artificial Intelligence and Statistics, pp. 895–905. PMLR (2020)
4. Kim, B., Koyejo, O., Khanna, R., et al.: Examples are not enough, learn to criticize! criticism for interpretability. In: NIPS, pp. 2280–2288 (2016)

5. Koh, P.W., Ang, K.S., Teo, H.H., Liang, P.: On the accuracy of influence functions for measuring group effects. arXiv preprint arXiv:1905.13289 (2019)
6. Koh, P.W., Liang, P.: Understanding black-box predictions via influence functions. In: International Conference on Machine Learning, pp. 1885–1894. PMLR (2017)
7. Rajani, N.F., Krause, B., Yin, W., Niu, T., Socher, R., Xiong, C.: Explaining and improving model behavior with k nearest neighbor representations. arXiv preprint arXiv:2010.09030 (2020)
8. Rousseeuw, L., Kaufman, P.: Clustering by means of medoids (1987)
9. Wachter, S., Mittelstadt, B., Russell, C.: Counterfactual explanations without opening the black box: automated decisions and the GDPR. Harv. JL & Tech. **31**, 841 (2017)
10. Zagoruyko, S., Komodakis, N.: Wide residual networks. arXiv preprint arXiv:1605.07146 (2016)

Enhancing Performance of Occlusion-Based Explanation Methods by a Hierarchical Search Method on Input Images

Hamed Behzadi-Khormouji[1,2](✉)(iD) and Habib Rostami[2](iD)

[1] imec-IDLab, University of Antwerp, Antwerpen, Belgium
Hamed.behzadikhormouji@uantwerpen.be
[2] Computer Engineering Department, Persian Gulf University, 75168 Bushehr, Iran
Habib.Rostami@pgu.ac.ir

Abstract. In this work, we address some drawbacks of back-propagation-based and perturbation-based visualization methods by proposing an explanation method called Fast Multi-resolution Occlusion (FMO). FMO, opposite to the back-propagation-based methods that cannot be applied on all types of Convolutional Neural Networks (CNNs), can highlight the important input features independent of the architecture. Also, FMO introduces a novel fast occlusion strategy called multi-resolution occlusion which not only efficiently addresses the time-consumption issue of the traditional Occlusion Test method but also outperforms the well-known perturbation-based methods. We assess the methods on CNNs DenseNet121, InceptionV3, InceptionResnetV2, MobileNet, and ResNet50 using three datasets ILSVRC2012, PASCAL VOC07, and COCO14.

Keywords: Perturbation-based visualization · Explaining deep models

1 Introduction

Convolutional Neural Networks (CNNs) have complex architectures such that it is not apparent how they make decisions and achieve such promising results. So, it is highly needed to dissect the reasoning process in these networks by explanation and interpretation methods to highlight the important input features, especially in real-world applications such as medicine and self-driving cars [1–3].

Explanation methods that aim to recognize pixels contributing to the model's decisions can be categorized as backpropagation-based visualization [4–8] and perturbation-based visualization methods [9–13]. One of the massive advantages of the backpropagation methods is their computational speed which makes them popular. However, producing some undesirable artifacts as a result of a gradient in discontinuities, and the inability to apply on all architectures can be considered as some drawbacks of these methods. In contrast, perturbation-based

© Springer Nature Switzerland AG 2021
M. Kamp et al. (Eds.): ECML PKDD 2021 Workshops, CCIS 1524, pp. 97–104, 2021.
https://doi.org/10.1007/978-3-030-93736-2_9

methods come with the advantage of being totally independent of the architecture. However, from a computational point of view, these methods are extremely expensive, which makes them unsuitable for many real-world applications.

According to the mentioned challenges, we aim to propose an extended abstract of an explanation method called Fast Multi-resolution Occlusion (FMO) [14], which is independent of the architecture opposite to the backpropagation-based methods and is computationally more efficient than the perturbation-based methods in terms of time consumption and visualization accuracy. FMO provides multi-resolution heatmaps, which is a novel attribute among explanation methods. According to the experiments, FMO outperforms well-known perturbation-based methods that show it is highly applicable to real-world problems. In the following sections, the proposed method is put forward in Sect. 2. Experimental results and conclusions are shown in Sects. 3 and 4.

2   Proposed Method

FMO gets a CNN model, an input image X, its predicted class Y, the corresponding probability Z pertaining to class Y as determined by the model, and the number of visualization resolution levels N. The output of FMO is N probability matrixes $P_{n_i * n_i}^{R_i}$ to be fused to the input image to highlight the regions of interest. The matrix dimensions of each level are calculated according to Eq. 1.

$$n_i = (ceil(\frac{k_i - \omega^{R_i}}{\tau^{R_i}}) + 1) * n_{i-1} \;\; subject\; to \begin{cases} k_0 = size\; of\; input\; image & i = 0 \\ k_i = \omega^{R_{i-1}} & i = 1...N \\ n_0 = ceil(\frac{k_0 - \omega^{R_0}}{\tau^{R_0}}) + 1 \end{cases}$$

(1)

where ω^{R_i} and τ^{R_i} are the size and stride of the occlusion window A at the resolution level R_i, respectively. In each resolution level, occlusion window slides on the input image to perturb an area. Then, the occluded image is fed to the model to obtain output probability. The variables are obtained as follows.

$$\omega^{R_i} = \frac{\omega^{R_{i-1}}}{\alpha} \quad \tau^{R_i} = \frac{\tau^{R_{i-1}}}{\beta} \;\; such\; that \begin{cases} i & 1,...,N \\ \alpha, \beta & reduction\; coefficients \end{cases}$$

(2)

To produce probability matrixes $P_{n_i * n_i}^{R_i}$ from the inputs, the occluded input image is fed to the model and predicts the output index $\hat{Y}$ and its corresponding probability $\hat{Z}$, whereby the target class Y is obtained. If $Y = \hat{Y}$ then the value of the row h_i^j and the column w_i^j of the probability matrix $P_{n_i * n_i}^{R_i}$ is filled with the normalized change of the probability (Eq. (3)).

$$P_{n_i * n_i}^{R_i}[h_i^j . w_i^j] = \frac{|Z - \hat{Z}|}{Z} \quad j = 0, ..., n_i - 1, index\; of\; elements$$

(3)

Each cell of this matrix corresponds to the output probability of the model fed by an image occluded with an area including some pixels.

If $Y \neq \hat{Y}$ then it is supposed that the occlusion window A^{R_i} covers a region of interest. Consequently, FMO drills down into the region covered by A^{R_i} and searches it with a higher resolution R_{i+1} to discover more fine-grained regions of interest. After searching all the resolutions, the final probability matrix P is calculated by a combination of all weighted matrixes $P^{R_i}_{n_i * n_i}$ (Eq. 4) and super-imposed on the input image to produce a visualization.

$$P = \sum_{i=1}^{N} \gamma^{R_i} * P^{R_i}_{n_i * n_i} \quad \textit{probability matrix weight } \gamma^R_i \in [0,1] \qquad (4)$$

3   Experimental Results

The performance of the proposed method has been compared with well-known methods such as Occlusion Test [9], LIME [10], RISE [11], Meaningful Perturbation [12], and Extremal Perturbation [13] in terms of time consumption and visualization accuracy. The datasets used in this research work are ILSVRC2012 [15], PASCAL VOC07 [16], and COCO14 [17]. According to the time-consumption experiments, FMO was averagely 2.32 times faster than LIME, 24.84 times faster than Occlusion Test, 11.87 times faster than RISE, 8.72 times faster than Meaningful Perturbation and 10.03 times faster than Extremal Perturbation on CNNs DenseNet121 [18], InceptionV3 [19], InceptionResnetV2 [20], MobileNet [21], and ResNet50 [22]. Also, FMO was compared with the methods quantitatively based on Pointing Game metric [23] on two CNNs. In this experiment, the visualization accuracy of FMO on DenseNet121 and Resnet50 were 86.0% and 66.2%, respectively, which were higher than those of other compared methods. FMO, opposite to the compared methods, can produce multi-resolution heatmaps. According to the Fig. 1, the image A, illustrates the visualization of the lowest resolution, whereas the image B shows the visualization of the highest resolution. The image C illustrates the combination of the heatmaps of the two resolutions. We provide more experimental details in Appendix B and C.

A   B   C

Fig. 1. A and B show heatmaps in low and high resolutions, and C: combined resolutions.

4   Conclusion

We propose an explanation method which is more efficient than perturbation-based methods in terms of time-consumption and visualization accuracy.

Appendix A: More Details of the Equations

The index j in Eq. (3) indicates the index of elements in the probability matrix $P_{n_i * n_i}^{R_i}$. Also Z is the probability of the original unoccluded image I belonged to the class index Y, and $\hat{Z}$ also shows the probability belonged to the class index Y, but when the occluded image I has been passed through the model. Therefore, to record the changes in the output probability, opposite to the Occlusion Test method that just records the output probability of occluded image, we record the normalized change of probability. As a result, each cell $[h_i^j . w_i^j]$ in the probability matrix $P_{n_i * n_i}^{R_i}$ indicates the normalized change of probability pertaining to a region of original image. The value in this cell shows the importance of that region in the form of normalized change of probability.

γ_i^R in Eq. (4) indicates the probability matrix weight in the resolution R_i. In order to see the heatmap in each resolution R_i, the weight of the resolution R_i is set to 1 and the weight of the others is set to 0. In this equation, before performing the weighted sum, all probability matrix $P_{n_i * n_i}^{R_i}$ are resized to the shape of the original image.

Appendix B: Details of Time Consumption

Table 1 shows the average time consumption of the FMO, RISE, LIME, Extremal Perturbation and Meaningful Perturbation methods on the models DenseNet121, InceptionV2, Inception V3, MobileNet, ResNet50. As can be seen, the proposed method, FMO, had the lowest time consumption over all models in comparison to the other methods, whereas Occlusion Test method had the highest time consumption. For example, FMO takes 1.90 s, 4.86 s, 2.71 s, 0.59 s and 2.70 s. On models DenseNet121, InceptionResNetV2, InceptionV3, MobileNet and ResNet50, respectively, which are far less than those of the Occlusion Test, RISE, LIME, Extremal Perturbation and Meaningful Perturbation methods in all of the five models.

Table 1. Average time consumption of the FMO, RISE, LIME, Extremal Perturbation and Meaningful Perturbation methods

Method	Models				
	DenseNet121	Inception V2	Inception V3	MobileNet	ResNet50
Occlusion Test	54.35	142.10	63.46	16.63	40.29
RISE	22.55	65.05	34.97	9.60	19.24
LIME	4.30	10.60	6.75	2.72	5.32
Extremal Perturbation	30.90	49.69	22.74	11.35	13.29
Meaningful Perturbation	30.10	43.10	15.60	9.35	13.13
FMO	**1.90**	**4.86**	**2.71**	**0.59**	**2.70**

Appendix C: Details of Visual Accuracy

Table 2 shows the localization accuracy of the methods on DenseNet121 and ResNet50. As can be seen, FMO outperforms other methods in terms of localization accuracy and time consumption on two datasets VOC07 and COCO14.

Table 2. Localized accuracy of each method on two hard datasets.

Method	VOC07 Test Set	COCO14 Val Set
	DenseNet121 (percent)/Time (second)	ResNet50 (percent)/Time (second)
Meaningful Perturbation	76.2/31.18	51.8/13.42
Extremal Perturbation	80.4/31.85	56.5/14.19
RISE	84.1/22.72	54.7/19.28
Occlusion Test	45.6/58.87	30.7/39.26
FMO	**86.0/6.50**	**66.2/1.48**

Figure 2 and 3 illustrate the visualization results on two datasets VOC07 and COCO14. According to these figures, FMO and Meaningful Perturbation methods can highlight properly the regions of interest in comparison to other methods.

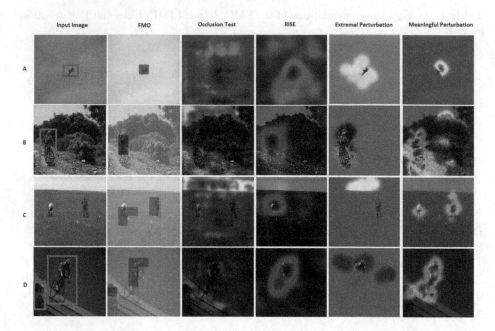

Fig. 2. The visualization output of the methods with the PASCAL VOC dataset on model DenseNet121.

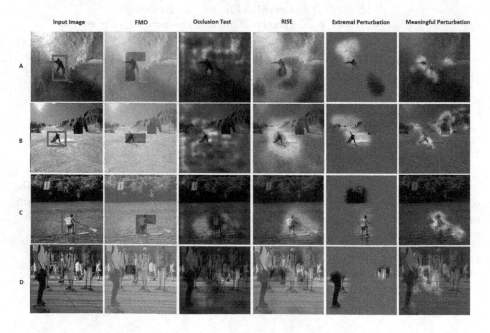

Fig. 3. The visualization output of the methods with COCO dataset on the ResNet50 model.

References

1. Behzadi-khormouji, H., et al.: Deep learning, reusable and problem-based architectures for detection of consolidation on chest X-ray images. Comput. Meth. Program. Biomed. **185**, 105162 (2020). ISSN 0169-2607. https://doi.org/10.1016/j.cmpb.2019.105162
2. Gupta, A., Anpalagan, A., Guan, L., Khwaja, A.S.: Deep learning for object detection and scene perception in self-driving cars: survey, challenges, and open issues. Array **10**, 100057 (2021). ISSN 2590-0056. https://doi.org/10.1016/j.array.2021.100057
3. Xiao, D., Yang, X., Li, J., Islam, M.: Attention deep neural network for lane marking detection. Knowl. Based Syst. **194**, 105584 (2020). https://doi.org/10.1016/j.knosys.2020.105584
4. Zhou, B., Khosla, A., Lapedriza, A., Oliva, A., Torralba, A.: Learning deep features for discriminative localization. arXiv arXiv:1512.04150 (2015)
5. Simonyan, K., Vedaldi, A., Zisserman, A.: Deep inside convolutional networks: visualising image classification models and saliency maps. In: 2nd International Conference on Learning Representations, ICLR 2014 (Workshop Track Proceedings) (2014)
6. José Oramas, M., Wang, K., Tuytelaars, T.: Visual explanation by interpretation: improving visual feedback capabilities of deep neural networks. In: 7th International Conference on Learning Representations, ICLR 2019 (2019)
7. Selvaraju, R.R., Cogswell, M., Das, A., Vedantam, R., Parikh, D., Batra, D.: Grad-CAM: visual explanations from deep networks via gradient-based localization. arXiv arXiv:1610.02391 (2017)
8. Shrikumar, A., Greenside, P., Kundaje, A.: Learning important features through propagating activation differences. arXiv arXiv:1704.02685 (2017)
9. Zeiler, M.D., Fergus, R.: Visualizing and understanding convolutional networks. arXiv arXiv:1311.2901 (2014)
10. Ribeiro, M.T., Singh, S., Guestrin, C.: "Why should i trust you?" Explaining the predictions of any classifier. arXiv arXiv:1602.04938 (2016)
11. Petsiuk, V., Das, A., Saenko, K.: RISE: randomized input sampling for explanation of black-box models, v1 (2018). http://arxiv.org/abs/1806.07421
12. Fong, R.C., Vedaldi, A.: Interpretable explanations of black boxes by meaningful perturbation. arXiv arXiv:1704.03296 (2018)
13. Fong, R., Patrick, M., Vedaldi, A.: Understanding deep networks via extremal perturbations and smooth masks. In: 2019 Proceedings of the IEEE International Conference on Computer Vision, pp. 2950–2958 (2019). https://doi.org/10.1109/ICCV.2019.00304
14. Behzadi-Khormouji, H., Rostami, H.: Fast multi-resolution occlusion: a method for explaining and understanding deep neural networks. Appl. Intell. **51**(4), 2431–2455 (2020). https://doi.org/10.1007/s10489-020-01946-3
15. Deng, J., Dong, W., Socher, R., Li, L.-J., Li, K., Fei-Fei, L.: ImageNet: a large-scale hierarchical image database. In: 2009 IEEE Conference on Computer Vision and Pattern Recognition, pp. 2–9 (2009). https://doi.org/10.1109/CVPR.2009.5206848
16. Everingham, M., Ali Eslami, S.M., Van Gool, L., Williams, C.K.I., Winn, J.M., Zisserman, A.: The Pascal visual object classes challenge - a retrospective. Int. J. Comput. Vis. **111**, 98–136 (2014)
17. Lin, T.-Y., et al.: Microsoft COCO: common objects in context. In: Fleet, D., Pajdla, T., Schiele, B., Tuytelaars, T. (eds.) ECCV 2014. LNCS, vol. 8693, pp. 740–755. Springer, Cham (2014). https://doi.org/10.1007/978-3-319-10602-1_48

18. Shakeel, M.S., Lam, K.M.: Deep-feature encoding-based discriminative model for age-invariant face recognition. Pattern Recogn. **93**, 442–457 (2019). https://doi.org/10.1016/j.patcog.2019.04.028
19. Szegedy, C., Vanhoucke, V., Shlens, J., Wojna, Z.: Rethinking the inception architecture for computer vision. arXiv arXiv:1512.00567 (2015)
20. Szegedy, C., Ioffe, S., Vanhoucke, V., Alemi, A.: Inception-v4, Inception-ResNet and the impact of residual connections on learning. arXiv arXiv:1602.07261 (2016)
21. Sandler, M., Howard, A., Zhu, M., Zhmoginov, A., Chen, L.C.: MobileNetV2: inverted residuals and linear bottlenecks. In: 2018 Proceedings of the IEEE Computer Society Conference on Computer Vision and Pattern Recognition, pp. 4510–4520 (2018). https://doi.org/10.1109/CVPR.2018.00474
22. He, K., Zhang, X., Ren, S., Sun, J.: Deep residual learning for image recognition (2015). https://doi.org/10.1109/CVPR.2016.90
23. Zhang, J., Bargal, S.A., Lin, Z., Brandt, J., Shen, X., Sclaroff, S.: Top-down neural attention by excitation backprop. Int. J. Comput. Vis. **126**(10), 1084–1102 (2017). https://doi.org/10.1007/s11263-017-1059-x

Post-hoc Counterfactual Generation
with Supervised Autoencoder

Victor Guyomard[1,2]([envelope]), Françoise Fessant[1], Tassadit Bouadi[2],
and Thomas Guyet[3][ID]

[1] Orange Labs, Lannion, France
victor.guyomard@orange.com
[2] Inria, CNRS, IRISA, University of Rennes, Rennes, France
[3] Institut Agro/IRISA UMR6074, Rennes, France

Abstract. Nowadays, AI is increasingly being used in many fields to
automate decisions that largely affect the daily lives of humans. The
inherent complexity of these systems makes them so-called black-box
models. Explainable Artificial Intelligence (XAI) aims at solving this
issue by providing methods to overcome this lack of transparency. Counterfactual explanation is a common and well-known class of explanations
that produces actionable and understandable explanations for end-users.
However, generating realistic and useful counterfactuals remains a challenge. In this work, we investigate the problem of generating counterfactuals that are both close to the data distribution, and to the distribution
of the target class. Our objective is to obtain counterfactuals with likely
values (*i.e.* realistic). We propose a model agnostic method for generating realistic counterfactuals by using class prototypes. The novelty of
this approach is that these class prototypes are obtained using a supervised auto-encoder. Then, we performed an empirical evaluation across
several interpretability metrics, that shows competitive results with the
state-of-the-art method.

Keywords: Counterfactual explanation · Interpretability · Prototypes

1 Introduction

The black-box nature of most of the current machine learning systems makes it
difficult to build trust in algorithmic decision making, especially in sensitive and
safety critical areas where humans are directly affected. Post-hoc explanation
methods aims at reinstalling this trust by providing additional information to
support the interpretability of decisions. A class of explanations is through counterfactual whose aim is to disturb samples to change model's original decision.
Thus, it illustrates the impact of some features of the example. The generated
samples can act as a recommendation for end-users to achieve their desired outputs. Most counterfactual explanation methods perturb the original instance
under various constraints until the desired model prediction is achieved thanks

© Springer Nature Switzerland AG 2021
M. Kamp et al. (Eds.): ECML PKDD 2021 Workshops, CCIS 1524, pp. 105–114, 2021.
https://doi.org/10.1007/978-3-030-93736-2_10

to the optimization of a loss function [10]. According to the expected properties of the explanation some extra terms are added to this loss function. Thus the counterfactual is expected to be both actionable and realistic. Proximity and sparsity can serve as intuitive proxies for actionability since they represent the magnitude and number of perturbations suggested by a counterfactual and it has been argued that people find it easier to understand shorter explanations [8]. Realism relates to how well the counterfactual resembles or fits-in with the known data distribution. A way to increase realism is to incorporate an auto-encoder (AE) reconstruction loss in the counterfactual search process [1,6,7]. An AE uses a neural network to project an instance onto a latent space and then tries to reconstruct this instance [2]. The error represents how successful the instance is reconstructed and how similar it is to training data. A higher error represents a data point farther from the data manifold. However, adhering to training data distribution is not sufficient for the counterfactual explanation. The counterfactual explanation should also belong to the feature distribution of its target class. To investigate this issue, we propose in this paper to extend the work of [6] by replacing the AE with a supervised AE whose latent space is organized thanks to class label information. The purpose of this is to generate counterfactual explanations that are faithful to the whole distribution of the data but also to the distribution of the data of the target class. In next sections we present our approach and empirically evaluate it with various metrics.

2   Post-hoc Counterfactual Generation

Let $\mathcal{X} \subseteq \mathbb{R}^p$ represents the p-dimensional feature space. Let $D = \{(x_i, y_i)\}_{i=1}^n$ be the training dataset with $x_i \in \mathcal{X}$ and $y_i \in \{1, \ldots, l\}$ for each $i \in \{1, \ldots, n\}$.

In a post-hoc explanation approach, we are willing to explain the outputs of a given probabilistic decision model denoted $f_{\text{pred}} : \mathcal{X} \mapsto \{1, \ldots, l\}$. This decision model predicts which class an example belongs to among l classes. $[f_{\text{pred}}(x)]_y$ gives the probability of an example x to belong to class y.

Then, in this context, the generation of counterfactuals is usually seen as an optimization problem [10]. Let x_0 be a new example (*i.e.* the instance to explain). Intuitively, a counterfactual is an example $x_{cf} = x_0 + \delta$ that is closed to x_0 but classified by f_{pred} in a different class. Then, generating a counterfactual may consist in finding the *closest example* (with the smallest δ) belonging to a different predicted class than $f_{pred}(x_0)$.

Such an approach may generate poorly actionable or realistic examples. Indeed, the closest example is not necessarily the most meaningful for the user. Firstly, the notion of closeness is basically measured in the example space, but it can be more interesting to embed the example in a latent space to capture better the example semantics [5]. In this case, we can consider an autoencoder AE_D trained on D. This autoencoder is composed of an encoder (ENC_D) and a decoder (DEC_D) such as $\text{AE}_D(x) = \text{DEC}_D(\text{ENC}_D(x))$ for all $x \in \mathcal{X}$. Its latent space will then be used to measure the distance between examples. Secondly, additional constraints on the counterfactual search process may improve the realism of the counterfactual.

Van Looveren *et al.* [6] suggest such a post-hoc strategy to construct counterfactuals through an optimization problem with several penalization terms in their objective function. More precisely, a counterfactual instance $x_{cf} = x_0 + \delta \in \mathcal{X}$ is found by optimizing the following objective function:

$$\min_{\delta} \left(c \cdot f_{\kappa}(x_0, \delta) + \|\delta\| + \gamma \cdot L_{AE} + \theta \cdot L_{\text{proto}} \right) \qquad (1)$$

$f_{\kappa}(x_0, \delta)$ encourages the predicted class y_i of the counterfactual x_{cf} to be different from the predicted class y_0 of the original instance x_0:

$$f_{\kappa}(x_0, \delta) = \max \left([f_{\text{pred}}(x_0 + \delta)]_{y_0} - \max_{y_i \neq y_0} [f_{\text{pred}}(x_0 + \delta)]_{y_i}, -\kappa \right)$$

where $\kappa \geq 0$ caps the divergence between $[f_{\text{pred}}(x_0 + \delta)]_{y_0}$ and $[f_{\text{pred}}(x_0 + \delta)]_{y_i}$.
$\|\delta\| = \beta \cdot \|\delta\|_1 + \|\delta\|_2^2$ encourages to minimize the distance between x_0 and x_{cf} with the aim of generating sparse perturbations.

$L_{AE} = \|x_0 + \delta - \text{AE}_D(x_0 + \delta)\|_2^2$ represents the reconstruction error of x_{cf} evaluated by the autoencoder AE. Intuitively, it penalizes counterfactuals that would be far from examples in $\mathcal{X}$.

Finally, the last term L_{proto} is defined as follows:

$$L_{\text{proto}} = \|\text{ENC}_D(x_0 + \delta) - \text{proto}_{y_j}\|_2^2,$$

where proto_{y_j} is defined as the mean of the K nearest examples of x_0 in latent space that are labeled y_j by f_{pred}:

$$\text{proto}_{y_j} = \frac{1}{K} \sum_{k=1}^{K} \text{ENC}_D \left(x_k^j \right) \qquad (2)$$

where x_k^j is the k-th example in latent space labeled y_j and $y_j = \text{argmin}_{y_i \neq y_0} \|\text{ENC}(x_0) - \text{proto}_{y_i}\|_2$.

L_{proto} fosters a counterfactual instance that would be near the prototype proto_{y_j} of the nearest class $y_j \neq y_0$ of x_0. The idea is to guide the counterfactual search around proto_{y_j}. It is worth noting that proto_{y_j} is an *abstract* prototype as it lies in the latent space and not in the example space. Nonetheless, a concrete interpretable prototype may be regenerated by $DEC_D(\text{proto}_{y_j})$.

K is an hyperparameter of the method that represents the number of examples used to compute prototypes. The other hyper-parameters $(c, \beta, \gamma, \theta)$, and κ are the weights of the loss function terms. For a detailed description of these parameters and their impact on the counterfactual search, we refer the readers to [6].

3   Counterfactuals Based on Latent Prototypes

Van Looveren *et al.* [6] suggest a method to obtain counterfactuals that lie closer to the data distribution of a given class by guiding them with prototypes in a latent space. Nonetheless, we observe that it is not always the case in practice. Indeed it can sometimes lead to counterfactuals that are far from the data distribution of their corresponding class. Figure 1 gives an intuition of this phenomenon, with an example from the MNIST dataset[1]. The example to be explained is a digit predicted as a 5 by a classifier, and the corresponding counterfactual is predicted as a 3. What can be observed is that the counterfactual lies closer to the data distribution of class 5 than the data distribution of its predicted class (class 3). As latent prototypes are representative examples for each class in a latent space, it seems relevant to us to take into account the class in the construction of the latent space.

We propose to learn such a latent space with a supervised autoencoder. Supervised autoencoders are types of autoencoders that learn a supervised task jointly with the reconstruction task [4]. This type of network has two outputs: one for the encoded/decoded example and the other for the classification decision. The global loss function used to train the autoencoder is defined as a weighted sum of a classification loss and a reconstruction loss:

$$L((f_{\text{pred}}, \text{AE}), D) = \underbrace{E(f_{\text{pred}}, D)}_{\text{Classification loss}} + \lambda \underbrace{R(\text{AE}, D)}_{\text{Reconstruction loss}} \tag{3}$$

where

- The reconstruction loss is defined as a mean of L_2 distances over the training set of n samples:

$$R(\text{AE}, D) = \frac{1}{n} \sum_{i=1}^{n} \| \text{AE}(x_i) - x_i \|_2^2$$

- The classification loss is defined as cross-entropy on the training set (D):

$$E(f_{\text{pred}}, D) = \frac{1}{n} \sum_{i=1}^{n} \sum_{k=1}^{l} -\mathbb{1}_{[y_i=k]} \log \left([f_{\text{pred}}(x_i)]_k \right)$$

λ is an hyperparameter to adjust the trade-off between the two terms.

The intuition behind the supervised autoencoder is that the latent space will benefit from the weighted loss function (Eq. 3), by producing an organized latent space according to classes. Our idea is to use this supervised autoencoder to generate counterfactuals with the same principle of Sect. 2. The difference lies in the way the auto-encoder is trained. Since prototypes are mean examples computed in the latent space, they are expected to be more representative of a given class, and hence allow for the construction of more accurate counterfactuals. Moreover as shown in [4], supervised autoencoder never harms classification performances and can even improve generalization.

---

[1] http://yann.lecun.com/exdb/mnist/.

Example,5   Counterfactual example,3

Fig. 1. MNIST original example (on the left) and its corresponding counterfactual generated with Van Looveren *et al.* [6] method (on the right).

4 Experiments and Results

The goal of our experiments is to compare the quality of the counterfactuals generated with the two methods: with and without supervision of the autoencoder (baseline). This quality is evaluated using a set of various metrics. The experimental protocol can be summarized by:

1. (a) Train a supervised autoencoder that combines classification layers and autoencoder by minimizing a weighted loss that penalizes reconstruction and classification (see Sect. 3).
   (b) Extract trained autoencoder and classifier, then construct counterfactuals with the method of Sect. 2.
2. (a) Train the same autoencoder architecture with only a reconstruction loss (baseline autoencoder).
   (b) Construct counterfactuals with this autoencoder and classifier of step 1a.
3. Compute metrics on both counterfactual sets.

4.1 Metrics

To evaluate counterfactual quality we used prediction gain, realism and actionability from Nemirovsky *et al.* [9].

- **Prediction gain:** The prediction gain is given by the difference between the predicted probability of the counterfactual and the predicted probability of the example, according to the counterfactual class.

$$\text{Gain} = [f_{\text{pred}}(x_{cf})]_{y_i} - [f_{\text{pred}}(x_0)]_{y_i} \tag{4}$$

where y_i denotes the predicted class for the counterfactual.
The range for the prediction gain is $[0, 1]$ and a higher value means being more confident in the class change of the counterfactual.
- **Realism:** The realism corresponds to the reconstruction error of the counterfactual evaluated by an autoencoder.

$$\text{Realism} = \|\text{AE}_{\text{evaluate}}(x_{cf}) - x_{cf}\|_2^2 \tag{5}$$

where $AE_{evaluate}$ is another autoencoder used for evaluation (more details in Sect. 6). A low value indicates that the counterfactual lies closer to the data manifold.

Table 1. Counterfactual metrics comparison. The arrows indicate whether larger ↑ or lower ↓ values are better, and the best results are in bold.

Metrics	Baseline	Supervised autoencoder
↑ Prediction gain	0.552 ± 0.106	$\mathbf{0.839 \pm 0.160}$
↓ Realism	0.253 ± 0.010	$\mathbf{0.249 \pm 0.012}$
↓ Actionability	$\mathbf{26.174 \pm 13.762}$	38.360 ± 18.465

– **Actionability:** The actionability is the L_1 distance between an example and its corresponding counterfactual.

$$\text{Actionability} = \|x_{cf} - x_0\|_1 = \|\delta\|_1 \qquad (6)$$

A low value indicates a sparser perturbation, which corresponds to fewer changes in the counterfactual.

We also evaluated on the two metrics (IM1, IM2) based on the reconstruction errors of autoencoders proposed in [6], but the results are not reported or discussed here because we did the same observations as in [3]. The authors pointed out the lack of robustness of the metrics and show that it is possible to obtain a better IM1 values with unrealistic examples.

4.2 Results

We evaluate our protocol on the MNIST dataset, based on the metrics defined in Sect. 4.1. We have chosen a random sample of 5,000 test examples for the generation of the corresponding counterfactuals, and used the hyperparameters and architectures detailed in Supplementary Material (Sect. 6) for the training of the autoencoders. Performances relative to training are also provided in Supplementary material.

Metric Evaluation: Table 1 provides the results for baseline and supervised autoencoder. We compute the mean and standard deviation of each metric for the 5000 randomly selected test examples. We observe that our method achieves a better prediction gain (the mean value is 0.287 points higher), with an equivalent realism. A better prediction gain is interesting, since it means more confidence in the predicted class of counterfactuals. This gain is obtained by modifying more pixels in the original space in order to obtain a digit of a different predicted class. As a consequence, counterfactuals are less sparse (and indeed we can observe that the actionability is higher).

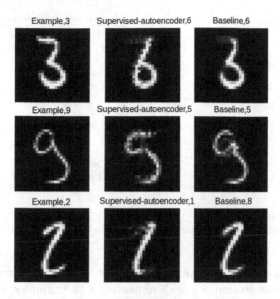

Fig. 2. MNIST Examples (on the left column) and corresponding counterfactuals generated for supervised AE (in the middle) and baseline AE (on the right).

Discussion. The left column shows the original examples to explain, the second and third columns represent respectively the counterfactuals obtained with supervised autoencoder (in the middle) and the baseline (on the right). We observe more pixel changes in the case of supervised autoencoder. In the first row, an example predicted as a 3 is converted into a 6 in each case. Nonetheless, in the case of supervised autoencoder, pixels are added in order to create the typical loop of a 6. Besides, the counterfactual achieves a greater predicted probability for class 6 (0.99 vs 0.41), leading to a better prediction gain. In the second row, the example is predicted as 9 and counterfactuals as 5. We observe the same behaviour, in the case of the supervised autoencoder, pixels are removed in the 9 loop in order to make a 5. Fewer pixels are removed in the case of baseline, leading to less visual changes (the counterfactual is predicted as a 5 but visually looks like a 9). In the last row, the example is predicted as a 2 and produced counterfactual is predicted as 1 in the case of supervised autoencoder, and 8 in the case of baseline. With baseline, the perturbation produced a counterfactual that appears to be visually similar to the example. Nonetheless, the class prediction change to 8, even if it looks like a 2 (the predicted probability for class 8 is 0.43). With the supervised autoencoder, the counterfactual is converted into a 1 by deleting pixels from the extremities, the predicted probability for class 1 is 0.99 (Fig. 2).

Visualization of the Latent Space. In order to better understand the impact of the supervision of the autoencoder on the latent space, we train a supervised autoencoder on MNIST and also the same autoencoder without the supervision

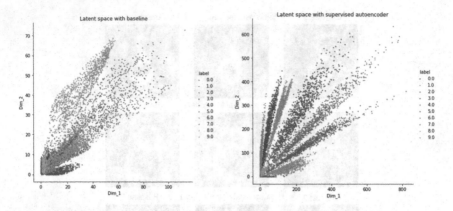

Fig. 3. Latent space visualization on a multiclass classification problem

(baseline autoencoder). For visualization purpose, we fixed the latent space dimension of size 2 in this section. Figure 3 represents test examples with their corresponding predicted classes in the latent space, with the baseline and supervised autoencoder. This experiment supports our intuition of Sect. 3. Indeed, in the case of supervised autoencoder, the latent space is organized according to classes, meaning that classes are organized by separate clusters. This is not the case with baseline autoencoder, where the representation does not take classes into account, examples of different classes are mixed in the latent space.

5   Conclusion and Future Work

In this work we have proposed a post-hoc counterfactuals generation method with a supervised autoencoder. Our method is an adaptation of Van Looveren *et al.* [6] work to improve the faithfulness of counterfactuals to the data distribution of the target class by using a supervised autoencoder. In this way, prototypes are computed from a latent space organized with the class labels, leading to more representative counterfactuals. This work has been evaluated on the MNIST dataset on various metrics. Our method increases the prediction gain with an equivalent realism but at the cost of a higher actionability, meaning that counterfactuals are less sparse. We plan now to confirm these results on other types of data, in particular tabular data. Using tabular data is challenging because we lose the visualisation capabilities we relied on here to show that an example is closer to a given class.

6   Supplementary Material

This section describes used architectures and hyperparameters.

Supervised Autoencoder: This architecture is composed of an auto-encoder part, and two dense layers (classification layers) next to encoder. The final output

Table 2. Accuracy and reconstruction error for each λ on test set

λ	↑ Accuracy	↓ Reconstruction error
0.1	0.984	0.0069
1	0.984	0.0028
10	0.981	0.0012
100	0.979	0.0011

is the concatenation of the last dense layer and the decoder output. Auto-encoder part is the same as those used by Van Looveren and Klaise [6]. It consists of an encoder with two convolution layers, the first two contains 16 filters of size 3×3 and ReLU activations and are followed by a 2×2 max-pooling layers and finally a convolution layer with a filter of size 3×3 and linear activation. The decoder part takes encoded instances as input and pass them to a convolutional layer with 16 filters of size 3×3 and RelU activation, then to a 2×2 upsampling layer and, a convolutional layer with 16 filters of size 3×3 and RelU activation again and finally a convolution layer with a filter of size 3×3 and linear activation. The classification part is composed of 2 dense layers stacked next to the encoder. It takes flatten encoded instances from the encoder and pass them to a dense layer of size 128 with RelU activation and L_1 regularization. This dense layer is followed by a *softmax* layer with 10 units. Loss is defined as summation of reconstruction loss and classification loss as shown in Sect. 3. To fix λ of Eq. 3, we train our model for different lambda values on the training set, and choose the best trade-off between accuracy and reconstruction error on the test set. These results are shown in Table 2, and best performances are obtained for $\lambda = 10$.

As labels are one-hot, classification loss will be a categorical cross-entropy, and reconstruction loss will be defined as a mean squared error loss. Training is achieve with a batch size of 128, and for 25 epochs, optimizer is set as Adam optimizer.

Baseline Autoencoder: This architecture is the same as autoencoder part of supervised autoencoder. Loss is defined as mean squared error, and training is achieved with a batch size of 128, and for 18 epochs, optimizer is set as Adam optimizer. We achieve to train our model, on the training set for a reconstruction error of 0.0016 on test set.

Counterfactual Search Hyperparameters: Hyperparameters values are fixed to those used by Van Looveren *et al.* [6] ($\gamma = 100$, $\kappa = 0$, $c = 1$, $\beta = 0.1$, $\theta = 100$, $K = 5$).

Autoencoder for Evaluation: We used the same autoencoder architecture as Van Looveren *et al.* [6]. The training set is the same as the one used for supervised and baseline autoencoder.

References

1. Dhurandhar, A., et al.: Explanations based on the missing: towards contrastive explanations with pertinent negatives. In: Proceedings of the International Conference on Neural Information Processing Systems (NIPS), pp. 590–601 (2018)
2. Kramer, M.A.: Nonlinear principal component analysis using autoassociative neural networks. AIChE J. **37**(2), 233–243 (1991)
3. Labaien, J., Zugasti, E., Carlos, X.D.: DA-DGCEx: ensuring validity of deep guided counterfactual explanations with distribution-aware autoencoder loss. arXiv arXiv:2104.09062 (2021)
4. Le, L., Patterson, A., White, M.: Supervised autoencoders: Improving generalization performance with unsupervised regularizers. In: Proceedings of the International Conference on Neural Information Processing Systems (NIPS) (2018)
5. Li, O., Liu, H., Chen, C., Rudin, C.: Deep learning for case-based reasoning through prototypes: a neural network that explains its predictions (2017)
6. Looveren, A.V., Klaise, J.: Interpretable counterfactual explanations guided by prototypes. arXiv arXiv:1907.02584 (2020)
7. Mahajan, D., Tan, C., Sharma, A.: Preserving causal constraints in counterfactual explanations for machine learning classifiers. In: Proceedings of the Workshop Microsoft at NIPS - "CausalML: Machine Learning and Causal Inference for Improved Decision Making" (2019)
8. Miller, T.: Explanation in artificial intelligence: insights from the social sciences. arXiv arXiv:1706.07269 (2017)
9. Nemirovsky, D., Thiebaut, N., Xu, Y., Gupta, A.: CounteRGAN: generating realistic counterfactuals with Residual Generative Adversarial Nets. arXiv arXiv:2009.05199 (2020)
10. Wachter, S., Mittelstadt, B.D., Russell, C.: Counterfactual explanations without opening the black box: automated decisions and the GDPR. arXiv arXiv:1711.00399 (2017)

Parallel, Distributed, and Federated Learning

Workshop on Parallel, Distributed, and Federated Learning (PDFL 2021)

In the past few years, as large data volumes are being generated on edge devices, such as mobile phones or autonomous vehicles, more and more machine learning systems are moving towards processing data in-situ, that is, on (or close to) the data generating devices. By learning models directly at the data sources—which often have computational power of their own such as mobile phones, smart sensors, and tablets—network communication is reduced by orders of magnitude. Moreover, it facilitates obtaining a global model without centralizing privacy-sensitive data, thereby contributing to the development of trustworthy AI systems. This form of parallel, distributed, and federated machine learning has gained substantial interest in recent years, both from researchers and practitioners, and may allow for disruptive changes in areas such as smart assistants, machine learning on medical or industrial data, and autonomous driving.

This workshop was the fourth edition in the series of successful PDFL (previously DMLE) workshops at ECML PKDD, following the events held in 2018, 2019, and 2020. It was held virtually due to the COVID-19 pandemic. The workshop included two invited talks, by Bharat Rao (CEO of Carenostics) and Blaise Agüera y Arcas (Google AI), and technical presentations. It was attended by around 30 people.

The accepted papers presented interesting novel aspects of decentralized machine learning, especially in the context of trustworthy AI. We want to thank the authors for their valuable contributions, great presentations, and lively and fruitful discussions. We would also like to thank the PDFL 2021 Program Committee, whose members made the workshop possible with their rigorous and timely reviews. Finally, we would like to thank ECML PKDD for hosting the workshop and its workshop chairs, Alipio Jorge and Yun Sing Koh, for their valuable support.

Michael Kamp
Yamuna Krishnamurth
Linara Adilova

Organization

PDFL 2021 Chairs

Michael Kamp	Monash University, Australia
Yamuna Krishnamurthy	Royal Holloway, University of London, UK
Linara Adilova	Fraunhofer IAIS, Germany

Program Committee

Wray Buntine	Monash University, Australia
Janis Keuper	Fraunhofer ITWM, Germany
Mark Jelasity	University of Szeged, Hungary
Henning Petzka	Lund University, Sweden
Michael Mock	Fraunhofer IAIS, Germany
Tim Wirtz	Fraunhofer IAIS, Germany
Pascal Welke	University of Bonn, Germany
Jochen Garcke	Fraunhofer SCAI, Germany
Christian Bauckhage	Fraunhofer IAIS, Germany
Sven Giesselbach	Fraunhofer IAIS, Germany
Dorina Weichert	Fraunhofer IAIS, Germany

Differentially Private Learning from Label Proportions

Timon Sachweh$^{(\boxtimes)}$(iD), Daniel Boiar(iD), and Thomas Liebig(iD)

TU Dortmund University, Dortmund, Germany
{timon.sachweh,daniel.boiar,thomas.liebig}@tu-dortmund.de
http://www-ai.cs.tu-dormtund.de

Abstract. Due to IoT and Industry 4.0, more and more data is collected by sensor nodes, which send their data to a central data lake. This approach results in high data traffic and privacy risk, which we want to address in this paper. Therefore we use an existing Learning from Label Proportions (LLP) algorithm, to use the decentralized properties and extend this approach by applying Differential Privacy to the transferred data. This yields to reduced data transfer and increased privacy.

Keywords: Differential Privacy · Learning from label proportions · Distributed learning · Spatio-temporal · Traffic · IoT · Industry 4.0

1 Introduction

Over the last few years decentralized data collection has become more and more popular. This trend is driven by IoT devices and its measured sensor values. In order to create added value for companies, this data is usually collected in a centralized manner, which results in high data transfer and data protection risks. Especially data protection risks are important to address by organizations, due to the introduction of GDPR in all EU countries in 2018 [5]. Organizations want to use this data in order to gain more information or predict future sensor states, e.g. "Will the traffic flow stay the same in the next 15–30 min?" and have to be compliant according to GDPR at the same time.

Therefore we extend the decentralized learning approach from [12,13] by applying *Differential Privacy* to label proportions sent between the different decentralised IoT devices to result in a privacy preserving algorithm. Before we present our approach, we will first formally define the term *Differential Privacy*.

This research has been funded by the Federal Ministry of Education and Research of Germany as part of the competence center for machine learning ML2R (01IS18038A) and by the German Research Foundation DFG under grant SFB 876 "Providing Information by Resource-Constrained Data Analysis" project B4 "Analysis and Communication for Dynamic Traffic Prognosis".

© Springer Nature Switzerland AG 2021
M. Kamp et al. (Eds.): ECML PKDD 2021 Workshops, CCIS 1524, pp. 119–127, 2021.
https://doi.org/10.1007/978-3-030-93736-2_11

Differential Privacy. Assumed, there is an algorithm $M : D \to R$ with domain D and range R. So D can be any set of input data and R the set of all possible outputs of M. Furthermore $D', D'' \in D$ have characteristic $||D' - D''||_1 \leq 1$, so that they do not differ in more then one element. This algorithm is ϵ-differentially private, if for all $S \subseteq R$ and for all D' and D'' the following formula applies [3]:

$$Pr[M(D') \in S] \leq e^{\epsilon} Pr[M(D'') \in S] \tag{1}$$

In more detail this definition formalizes, that the probability, that an algorithm M outputs the same results S by using different inputs D' and D'' differs in maximum by e^{ϵ}. Therefore by modifying the parameter ϵ, the degree of privacy, that will be applied, can be specified.

With $\epsilon = 0$, the probabilities of both outputs can be the same. This means that a single data point no longer has any influence on the output, making the algorithm completely privacy compliant. This is because the output does not allow any inference of a removed datum. However, you lose any information of the data, which makes the dataset useless.

Therefore $\epsilon = 0.1$ is usually chosen to increase the degree of privacy and still not reduce the information level of the data too much. The concrete implementation, how the algorithm becomes compliant with respect to equation, is described in Sect. 3.

In this paper, we analyze the influence of adding noise to the label proportions. Therefore we will use the original *Learning from Label Proportions (LLP) algorithm* [12] and a simple centralized *k Nearest Neighbor (kNN)* [2,10] classifier, to set benchmarks in aspects of accuracy and transferred data. We compare our results of the *modified LLP algorithm (p-LLP)*, that gains privacy aspects, with those benchmark results.

Firstly, we will give a brief introduction into other approaches, that can preserve privacy in machine learning (see Sect. 2). Afterwards, we will describe the *LLP* algorithm and our modifications (see Sect. 3), as well as analyze the performance on a city traffic dataset (see Sect. 4). Finally, we will give a short conclusion.

2   Related Work

As already mentioned, data protection has become more relevant in the application of data-driven learning methods due to the GDPR. It has to be considered that decentralized data collection is the basis of the approach described here, as it allows to prevent the extraction of user data. There are various methods for increasing privacy preservation in case of decentralized data collection, such as data aggregation, data masking, or encryption. The current state of research for these methods is outlined in the following.

The basic idea of **data aggregation** is to hide individual data in the data of many. The data is collected over a longer period of time or from different data sources and aggregated using functions, which are usually additive. The resulting single characteristic per class contains thereby only one comparison value, which can be compared with the other aggregated class values. Current methods that rely on aggregation for privacy protection include termSlice-Mix-AggRegaTe (SMART) [8], an improved version of this [16], the PriSense algorithm [11], or even the approach described in [6]. The latter does not aggregate the data, but the trained machine learning procedure. However, this method is only applicable if a decentrally trained model is used that can subsequently be updated globally. [8] and [11] partially send unique data to neighboring nodes, which reduces the security to be guaranteed, while [16] requires very high performance at the decentral nodes.

Data masking is another option for processing data that prevents private data from being read. For this purpose, the original data is enriched with certain values - so-called camouflage values -, or random values, so that the exact distribution, as well as the absolute data values, no longer correspond to the original. A simple approach for the aggregation functions minimum and maximum is described in [7]. The problem here is that Minimum and Maximum are sufficient just for very few use cases. The Cluster-based Private Data Aggregation (CPDA) approach [8] uses polynomial computations to aggregate within a cluster, but the reconstruction for the principal node of a cluster is very expensive. In contrast to this, the Federated Learning approach with Secure Aggregation described in [1] forwards updates of partial models to the server only in aggregate, using encrypted communication. Due to the partial encryption of the data and multiple data transfer it has high performance requirements as well, which often cannot be implemented in real life use cases. These high performance requirements are resulting from the partial encryption of the data and a multiple data transfer.

To increase data security through **encryption**, data is first encrypted, aggregated with other encrypted data, and finally decrypted again (secure aggregation [15]), whereas the type of encryption may vary. For example, the LVPDA approach [14] uses the Pallier Homomorphic Cryptosystem for encryption. Also [4] is based on this type of encryption and additionally uses a blockchain. As with all encryption-based approaches, both the amount of data to be transferred and the computing power requirements for the decentralized nodes are very high for these approaches.

In order to address the problems described above like insecurity or high performance requirements *LLP* approach was designed. This is achieved, by storing the measured data in a decentralized way. Each node stores its own collected data and only *Label Proportions*, are sent to neighboring nodes. This approach takes advantage of the fact that there is a correlation between distance of different measuring points and the predictive relevance. However, this aggregation does not guarantee that information about individuals is not disclosed. Therefore we have developed the following approach, which extends the *LLP* approach by differential privacy.

3   Algorithm

In general the *LLP* algorithm stays the same as proposed in [12]. Because it is important to know the structure and flow of the algorithm, we will briefly

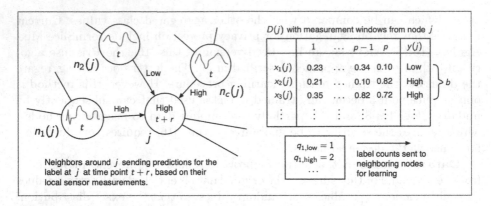

Fig. 1. Distributed Learning of Local Models, cited in verbatim from [12]

discuss it by describing Fig. 1. There are m wireless sensor nodes $(n_1, n_2, \ldots n_m)$, which store their measurements in $D(i) \forall i \in 1..m$. During explanation of the general algorithm and node setup, D denotes the measured data points ordered by window size and combined with the corresponding label as shown in Fig. 1. Each row in $D(i)$ consists of $[t-w, t]$ measurements, where t denotes a timestamp and w is the *windows size* of the last w measurements. Each row is assigned a label, which is taken from a measured value from a future timestamp $t + r$. In the first place those measurements are split in batches $B_1, \ldots, B_h$ where $h = \lceil |D(i)|/b \rceil$ and b denotes the size of the batches, in which $D(i)$ will be divided. The *batch size* is also shown along the rows of the table in Fig. 1. The *batches* are then used to calculate *label proportions* for each batch. The generated *label proportions* are sent to the closest c neighbors. Each node uses the received *label proportions* to train $c+1$ models $f_j(k)$, where $k \in 1, \ldots, c+1$ and j is the current node. The models $f_j(k)$ are learned by doing a k-means clustering on own measured node data $D(j)$. Initially each cluster, a random label is assigned. These cluster labels are exchanged by a more efficient local search with multi start strategy, introduced by [12]. In this phase, the label proportions of the neighboring node come in place for calculating the loss between the sent label proportions and the label proportions of the predicted labels, calculated by $f_j(k)$. Whenever the loss gets better, the exchanged cluster label assignment will be persisted and therefore result in a better model. The final prediction is done by doing a majority voting of the $c+1$ trained models.

This approach has the advantage, that we make use of more than only local measured datapoints, but keeping the bandwidth of transferred data low, because only *aggregated data* is sent between the nodes. However privacy cannot be guaranteed by this approach. Assuming, we have traffic flow measurement values, with labels $0, 1, 2, 3, 4$ and over a time frame of size b is only label 4 present. Then, from the label proportion, it can be inferred that everyone drove that fast during the period.

We solve this issue, by applying the *Differential Privacy* definition (see Sect. 1) to the label proportions. Firstly, we have to calculate the l_1-sensitivity function to know, how much influence a single datapoint can make to the output of a function $f : D \to R$:

$$\triangle f = \max_{\substack{D',D'' \in D, \\ ||D'-D''||_1=1}} ||f(D') - f(D'')||_1 \tag{2}$$

For this scenario, D is the current batch B_i and R is the resulting *Label Count*. Considering that we have a simple counting query, a single datapoint can have a maximum influence of 1 (see [3] example 3.1). Finally we can use the *Laplace Distribution* to generate noise, which can be added to the *Label Counts*, to be privacy compliant under ϵ-Differential Privacy [3]:

$$lap(x, \sigma, \mu) = \frac{1}{2\sigma} e^{-\frac{|x-\mu|}{\sigma}} \tag{3}$$

$$lap(x, \frac{\triangle f}{\epsilon}, 0) = \frac{\epsilon}{2 \triangle f} e^{-\frac{|x-0|\epsilon}{\triangle f}} \tag{4}$$

In the formula above the *position* parameter μ is set to 0 and the *scale* parameter is set to $\frac{\triangle f}{\epsilon}$. This parameters have to be set like this, to be compliant with the *Differential Privacy* definition (proof can be found in [3] theorem 3.6).

The modified algorithm for calculating label counts can be seen in below. As mentioned before, the batches B_i are already generated and possible labels Y are also known. The output $Q(j)$ contains differentially private label proportions of all batches.

Require: $B_1, \ldots, B_h, Y$
Ensure: $Q(j)$
 1: $Q(j) \leftarrow matrix(h, |Y|)$
 2: **for** i in $1..h$ **do**
 3:      **for** j in $1..|Y|$ **do**
 4:          $Q(j)_{i,j} \leftarrow sum(B_i == Y_j)$
 5:      **end for**
 6:      // adding noise to label counts
 7:      $m \leftarrow sum(Q(j)_i)$
 8:      **for** j in $1..|Y|$ **do**
 9:          $Q(j)_{i,j} \leftarrow Q(j)_{i,j} + lap(e = 0, s = 1/\epsilon)$
10:         clip $Q(j)_{i,j}$ to bounds $[0.001, m]$
11:         normalize $Q(j)_i$
12:      **end for**
13: **end for**

Initially $Q(j)$ is created with dimensions count batches (h) and count possible labels $(|Y|)$. Afterwards the label proportions are calculated iterative for each batch as follows. Firstly, the label counts (see lines 3–5) and the total sum (see line 7) are calculated. Then the Laplace noise, which is calculated by the sensitivity and ϵ is applied. Afterwards the new value is clipped to the maximum bounds, to prevent to large, or negative values. Finally, the label counts with noise are normalized. The resulting proportion is stored in $Q(j)$.

4   Experimental Evaluation

For evaluation the traffic flow data at junctions in the city of Dublin, recorded by the Sydney Co-ordinated Adaptive Traffic System (SCATS) [9] is used. The dataset contains average traffic flow values for every 15 minutes in January 2013. To compute the needed sensor data $D(j)$ for the LLP algorithm, a sliding window with $w = 5$ is used. Corresponding labels $y(j)$ are the discretized traffic flow values at horizon $r = 1$. The discrete ranges are 0–5, 5–30, 30–60, 60–150, 150–260. The following test is based on this data of 4 sensor nodes and their 3 closest neighbors based on euclidean distance of geo-coordinates. All models are trained with 10-fold cross validation.

kNN with $k = 16$ is used, to give a reference, what accuracy results a centralised supervised learning approach can achieve. LLP is evaluated by varying the batch size $b \in \{8, 16, 32, 64, 128\}$, as well as varying the cluster count $k \in \{8, 16, 32, 64\}$ of integrated k-means algorithm. While modifying b the cluster count is fixed to $k = 16$, whereas when cluster count is modified, $b = 50$ is set. Both the cluster variations and the batch variations are performed with $(p\text{-}LLP)$ and without (LLP) privacy applied. When privacy is applied $\epsilon = 0.1$ is set.

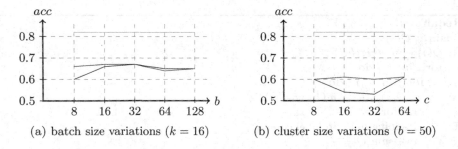

(a) batch size variations ($k = 16$)   (b) cluster size variations ($b = 50$)

Fig. 2. Average accuracy results by applying different batch sizes (a) or different cluster sizes (b). Blue plot are results of LLP, the black ones are of $p\text{-}LLP$ ($\epsilon = 0.1$). The green plot is kNN with $k = 16$ as benchmark (Color figure online)

As shown in Fig. 2 the average accuracy of LLP is in the range between 65% and 67%. The best results are achieved with a batch size of $b = 16$. By comparing the accuracy of $p\text{-}LLP$ one can see, that results mostly differ when $b = 8$ is set. For batch sizes over 32, the accuracy varies by maximum of 0.5%, which can happen due to tolerances. The results are understandable, since with small batch size the influence of a data point is proportionally large and therefore much noise is calculated on the label proportion, which yields in less information.

However the results in Fig. 2(b) are not so understandable. The cluster size is used to train the k-means algorithm, that is trained with measurement data from the own node. Therefore applying *Differential Privacy* to the variation of cluster sizes should not affect the average accuracy to drop by 6%.

As last evaluation step, we chose the best batch size $b = 32$ and best cluster size $k = 16$ from the previous evaluation, to analyze the influence of ϵ on *p-LLP*. As shown in Fig. 3 the accuracy of 67% of *LLP* cannot be reached by *p-LLP*. However, it can be seen that with a very small epsilon ($\epsilon = 0.01$ or $\epsilon = 0.05$) the dataset becomes unusable, which is evident from the accuracy of 32%, respectively 48%. From an $\epsilon = 0.1$ the prediction accuracy of 63% more and more reaches the accuracy of *LLP*. At the same time, sufficient privacy is still guaranteed with $\epsilon = 0.1$.

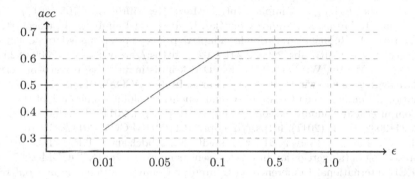

Fig. 3. Accuracy of *LLP* with $b = 32$ and $k = 16$ (blue). Accuracy of *p-LLP* with $b = 32$, $k = 16$ and varying ϵ (black) (Color figure online)

5 Conclusion

In this paper we extended the *LLP* algorithm by applying *Differential Privacy* to the label proportions, that are sent between nodes. Moreover we showed in the evaluation, that it is possible to achieve nearly the same accuracy as *LLP* when setting $\epsilon = 0.1$ for *p-LLP*. Therefore the *p-LLP* algorithm can be seen as superior, due to a more privacy preserving mechanism, which is important regarding GDPR. Next *p-LLP* has to be evaluated based on a larger node network, as well as more neighboring nodes. Transferability to other subject areas must also be examined.

References

1. Bonawitz, K., et al.: Practical secure aggregation for privacy-preserving machine learning. In: Thuraisingham, B.M., Evans, D., Malkin, T., Xu, D. (eds.) Proceedings of the 2017 ACM SIGSAC Conference on Computer and Communications Security, CCS 2017, Dallas, TX, USA, October 30–03 November, 2017, pp. 1175–1191. ACM (2017). https://doi.org/10.1145/3133956.3133982
2. Buitinck, L., et al.: API design for machine learning software: experiences from the scikit-learn project. In: ECML PKDD Workshop: Languages for Data Mining and Machine Learning, pp. 108–122 (2013)
3. Dwork, C., Roth, A.: The algorithmic foundations of differential privacy. Found. Trends Theor. Comput. Sci. **9**(3-4), 211–407 (2014). https://doi.org/10.1561/0400000042
4. Fan, H., Liu, Y., Zeng, Z.: Decentralized privacy-preserving data aggregation scheme for smart grid based on blockchain. Sensors **20**(18), 5282 (2020). https://doi.org/10.3390/s20185282
5. Goddard, M.: The EU general data protection regulation (GDPR): European regulation that has a global impact. Int. J. Mark. Res. **59**(6), 703–705 (2017)
6. Grama, M., Musat, M., Muñoz-González, L., Passerat-Palmbach, J., Rueckert, D., Alansary, A.: Robust aggregation for adaptive privacy preserving federated learning in healthcare. CoRR abs/2009.08294 (2020). https://arxiv.org/abs/2009.08294
7. Groat, M.M., He, W., Forrest, S.: KIPDA: k-indistinguishable privacy-preserving data aggregation in wireless sensor networks. In: INFOCOM 2011. 30th IEEE International Conference on Computer Communications, Joint Conference of the IEEE Computer and Communications Societies, 10–15 April 2011, Shanghai, China, pp. 2024–2032. IEEE (2011). https://doi.org/10.1109/INFCOM.2011.5935010
8. He, W., Liu, X., Nguyen, H., Nahrstedt, K., Abdelzaher, T.F.: PDA: privacy-preserving data aggregation in wireless sensor networks. In: INFOCOM 2007. 26th IEEE International Conference on Computer Communications, Joint Conference of the IEEE Computer and Communications Societies, 6–12 May 2007, Anchorage, Alaska, USA, pp. 2045–2053. IEEE (2007). https://doi.org/10.1109/INFCOM.2007.237
9. McCann, B.: A review of scats operation and deployment in Dublin. In: Proceedings of the 19th JCT Traffic Signal Symposium & Exhibition (2014)
10. Pedregosa, F., et al.: Scikit-learn: machine learning in Python. J. Mach. Learn. Res. **12**, 2825–2830 (2011)
11. Shi, J., Zhang, R., Liu, Y., Zhang, Y.: Prisense: privacy-preserving data aggregation in people-centric urban sensing systems. In: INFOCOM 2010. 29th IEEE International Conference on Computer Communications, Joint Conference of the IEEE Computer and Communications Societies, 15–19 March 2010, San Diego, CA, USA, pp. 758–766. IEEE (2010). https://doi.org/10.1109/INFCOM.2010.5462147
12. Stolpe, M., Liebig, T., Morik, K.: Communication-efficient learning of traffic flow in a network of wireless presence sensors. In: Proceedings of the Workshop on Parallel and Distributed Computing for Knowledge Discovery in Data Bases (PDCKDD 2015) (2015)
13. Stolpe, M., Morik, K.: Learning from label proportions by optimizing cluster model selection. In: Gunopulos, D., Hofmann, T., Malerba, D., Vazirgiannis, M. (eds.) ECML PKDD 2011. LNCS (LNAI), vol. 6913, pp. 349–364. Springer, Heidelberg (2011). https://doi.org/10.1007/978-3-642-23808-6_23

14. Zhang, J., Zhao, Y., Wu, J., Chen, B.: LVPDA: a lightweight and verifiable privacy-preserving data aggregation scheme for edge-enabled IoT. IEEE Internet Things J. **7**(5), 4016–4027 (2020). https://doi.org/10.1109/JIOT.2020.2978286
15. Zhang, W.: Secure data aggregation. In: van Tilborg, H.C.A., Jajodia, S. (eds.) Encyclopedia of Cryptography and Security, 2nd edn., pp. 1104–1105. Springer (2011). https://doi.org/10.1007/978-1-4419-5906-5_639
16. Zhang, X., Liu, X., Yu, J., Dang, N., Qi, X., Zhang, Q.: Energy-efficient privacy preserving data aggregation protocols based on slicing. In: 2019 International Conference on Internet of Things (iThings) and IEEE Green Computing and Communications (GreenCom) and IEEE Cyber, Physical and Social Computing (CPSCom) and IEEE Smart Data (SmartData), iThings/GreenCom/CPSCom/SmartData 2019, Atlanta, GA, USA, July 14–17, 2019, pp. 546–551. IEEE (2019). https://doi.org/10.1109/iThings/GreenCom/CPSCom/SmartData.2019.00109

Approaches to Uncertainty Quantification in Federated Deep Learning

Florian Linsner[1]([✉]), Linara Adilova[1], Sina Däubener[1], Michael Kamp[2], and Asja Fischer[1]

[1] Faculty of Mathematics, University of Bochum, Bochum, Germany
{florian.linsner,linara.adilova,sina.daubener,asja.fischer}@rub.de
[2] Department of Data Science and AI, Faculty of IT, Monash University, Melbourne, Australia
michael.kamp@monash.edu

Abstract. Trustworthy machine learning allows data privacy and a robust assessment of the uncertainty of predictions. Methods for quantifying uncertainty in deep learning have recently gained attention, while federated deep learning allows to utilize distributed data sources in a privacy-preserving manner. In this paper, we integrate several approaches for uncertainty quantification in federated deep learning. In particular, we show that prominent approaches such as MC-dropout and stochastic weight averaging Gaussian (SWAG) can be extended efficiently to federated setup. Moreover, we demonstrate that deep ensembles allow for natural integration in the federated learning framework. Our empirical evaluation confirms that a trustworthy uncertainty quantification on out-of-distribution data is possible in federated learning with little (SWAG) to no (MC-dropout, ensembles) additional communication. While all methods perform well in our empirical analysis and should serve as baselines in future developments in this field, deep ensembles and MC-dropout allow for better uncertainty based identification of out-of-distribution data and wrong classified data.

Keywords: Federated deep learning · Uncertainty · OOD detection

1 Introduction

Deep learning is used in many critical application areas, such as healthcare [44, 47] or autonomous driving [19,29].

However, predictions of deep learning models can be mistaken, especially on unseen data. When machine learning is applied as a tool, not minding the possible flaws of the models can be very costly, e.g., for medical diagnosis or nuclear power plants control decisions. In such critical environments it is important that a model can quantify the certainty of its predictions. While for classification an uncertainty score can be derived from the widely used softmax output, it is often uncalibrated and over-confident or misleading [13].

© Springer Nature Switzerland AG 2021
M. Kamp et al. (Eds.): ECML PKDD 2021 Workshops, CCIS 1524, pp. 128–145, 2021.
https://doi.org/10.1007/978-3-030-93736-2_12

To obtain reliable uncertainty estimates, multiple methods for uncertainty quantification for deep learning models have been proposed [8,10,23,26,28,32, 34]. Bayesian neural networks dating back to Neal [36] offer a natural way of quantifying uncertainty by marginalizing over the parameter posterior distribution. Moreover, the induced stochasticity can be used for evaluating the uncertainty of the prediction in terms of prediction variance [9]. While Neal [36] introduced the Hamiltonian Monte Carlo method for deriving the posterior, simultaneously MacKay [31] analyzed methods based on the Laplace approximation. Both approaches have been further extended for scalability [41,46], however challenges like the convergence of the Markov chain in the first, and calculating the Hessian in the second remain. Another line of research uses approximate variational inference [12,15], where a simpler parametrized distribution is fit to approximate the true posterior by maximizing the variational lower bound of the log likelihood, also referred to as evidence lower bound (ELBO). For better scalability practical approximate Bayesian implementations like MC-dropout [10] or SWAG [32] have been developed, as well as more empirical approaches such as deep ensembles [28].

With increasing amounts of data, the overhead of quantifying uncertainty can be challenging when training a neural network. For the efficient training of neural networks on large-scale distributed datasets, various parallelization methods have been proposed. On-device, edge, or in-situ processing, has been well studied in the context of stream processing [11,42] and monitoring functions over sensor networks [5,7,24]; averaging models have been used in online learning from distributed data streams [20,22]. For deep learning, training models in-situ and averaging their parameters on a coordinator node was termed *federated learning* [25,33]. Because of its inherent communication-efficiency [19,33] and its preservation of the privacy of sensitive local data [1,43], it has gained substantial interest in the community [35,48,51], including studies on the convergence of the distributed system and the quality of the resulting model [2,30,39,50]. While research in federated deep learning is usually focused on achieving a lower amount of communication, preserving the error rate, and/or preserving privacy [25,48], quantifying uncertainty in the federated setup is not fully understood, yet.

In their work, Boughorbel et al. [3] propose using uncertainty, measured as generalization ability of the model, for weighted aggregation in the global model. Nevertheless, they do not concentrate on the uncertainty quantification techniques for federated learning.

In this work we extend the use of ensembles in federated deep learning as well as include two other popular approaches for uncertainty quantification: MC-dropout and stochastic weight averaging Gaussians (SWAG).

2   Preliminaries

In this section we give a brief overview about approaches to uncertainty quantification, as well as federated deep learning.

2.1   Uncertainty Quantification in Deep Learning

Let $D := \{(x_i, y_i)\}_{i \in [n]}$ be a dataset consisting out of n independent input-output tuples sampled from the same data generating distribution. A neural network with parameters θ trained on D forms a model $p(Y|x^*, \theta)$ of the conditional probability distribution of the output Y given the input x^*. For making a prediction one is usually interested in finding the output y with the highest probability, i.e. $argmax_y p(Y = y|x^*, \theta)$[1]. One way to assess the uncertainty of such a prediction is to calculate the *Shannon entropy* of the predictive distribution. Let the output variable Y take values in a discrete set with K states, then the Shannon entropy is given by

$$H\left(p\left(Y|x^*, \theta\right)\right) = -\sum_{k=1}^{K} p\left(y_k|x^*, \theta\right) \cdot \log p\left(y_k|x^*, \theta\right). \tag{1}$$

The entropy reaches a minimal value of 0 iff one output value has probability 1 and the others probability 0, i.e. when the model has maximal certainty.

Bayesian models incorporate model uncertainty by taking the posterior distribution $p(\theta|D)$ of the parameters θ given the dataset D into account and estimating the expected prediction as

$$p(y|x^*, D) = \int p(y|x^*, \theta)p(\theta|D)d\theta. \tag{2}$$

Since the posterior distribution for Bayesian neural networks is intractable, different approximation techniques have been proposed. In the following we will briefly introduce the approaches to uncertainty quantification that we have incorporated into a distributed framework in this paper.

Deep Ensembles. The most straightforward and simple approach for estimating uncertainty in neural networks is based on ensembles. The use of ensemble techniques in machine learning has been intensively studied. It is known as a way to improve the performance of weak classifiers in practice, not only for neural networks but also for other models, e.g. random forests [4]. Neural network ensembles were further leveraged by Lakshminarayanan et al. [28] to quantify the uncertainty of neural network predictions. They suggest to train a neural network S-times with different random initializations leading to S different models with parameter sets $\theta_1, \ldots, \theta_S$. For the final prediction all the predictions of every single model are averaged:

$$p(y|x^*) := \frac{1}{S} \sum_{s=1}^{S} p(y|x^*, \theta_s). \tag{3}$$

---

[1] For simplicity we will write $p(y|x^*, \theta)$ for $p(Y = y|x^*, \theta)$ in the following.

Next to assessing the uncertainty based on the entropy of $p(y|x^*)$ one can now also assess uncertainty by calculating the variance between network predictions, also referred to as *predictive variance*:

$$\sigma^2 = \frac{1}{S} \sum_{s=1}^{S} p\left(y|x^*, \theta_s\right)^2 - p\left(y|x^*\right)^2. \tag{4}$$

While surprisingly simple, uncertainty estimates from deep ensembles have shown to be competitive or even superior to the mathematically grounded uncertainty from Bayesian methods [28, 38]. Note, that while the implementation is straightforward, the computational burden is S-times as much as for training a single model.

Monte Carlo Dropout. Dropout is a regularization method first proposed by Hinton et al. [16] for reducing overfitting in deep learning by preventing complex co-adaptions of neurons [45]. The term *dropout* refers to the random deactivation of neurons of a neural network with probability α, called the droprate, during training time. When dropout is used for regularization it is only applied during training.

Gal and Ghahramani [10] showed, that if the network is trained with L2-regularization in addition and nodes are dropped also during inference (with the same probability as during training) the procedure will become an approximate Bayesian method. For deriving the final prediction, the probability of an output given a certain input is estimated multiple times for different sub-nets resulting from randomly dropping neurons and averaged output probabilities as in Eq. (3) and the predictive variance can be estimated as in Eq. (4), where θ_s now represents the parameters of the s-th sub-network.

Stochastic Weight Averaging Gaussian. Maddox et al. [32] propose another approximate Bayesian method, that exploits the trajectory of stochastic gradient decent (SGD). Their method is inspired by *Stochastic Weight Averaging* (SWA) [17], where starting from a pretrained solution, averaging the network parameters along the trajectory of SGD improves generalization. In SWAG normal distributions are placed over the parameters, where the mean of each parameter is calculated during training as in SWA by:

$$\theta_{\text{SWA}} = \frac{1}{T} \sum_{i=1}^{T} \theta_i, \tag{5}$$

where T is the number of SWA epochs. To efficiently calculate the covariance of the parameters, SWAG computes a running average of the second uncentered moment for each weight:

$$\overline{\theta^2} = \frac{1}{T} \sum_{i=1}^{T} \theta_i^2. \tag{6}$$

After the last training step $\overline{\theta^2}$ and θ_{SWA} are combined to form a (in our case diagonal) covariance matrix over the parameters by

$$\Sigma_{diag} = diag(\overline{\theta^2} - \theta_{\text{SWA}}^2). \tag{7}$$

During inference, the derived parameter distribution $\mathcal{N}(\theta_{\text{SWA}}, \Sigma_{diag})$ is treated as an approximate Bayesian posterior, equivalent to $p(\theta|D)$ in Eq. (2). The integral is approximated by a Monte Carlo estimate, i.e. by averaging over samples from the approximate posterior. In their original work Maddox et al. [32] experimentally showed that SWAG approximates the shape of the true posterior while being much less computationally expensive than traditional Bayesian methods.

2.2   Federated Deep Learning

In federated deep learning, the goal is to train a global neural network model f_{global} - which is a mapping from the input space to the output space - on m workers, each worker f_i with $i \in [m]$ holding a local dataset D_i drawn iid from the same data distribution. For that, each worker trains a local model with the same network structure as the global model and shares its model parameters θ_i with a coordinator. This coordinator averages the model parameters of local models and redistributes the averaged parameters

$$\overline{\theta} = \frac{1}{m} \sum_{i=1}^{m} \theta_i \tag{8}$$

so that the local workers continue training from $\overline{\theta}$. This process is iterated until a suitable stopping criterion is met.

The vanilla variant of federated learning averages all local models after a fixed number of training steps. The amount of communication spent on achieving a good performing global model is a critical characteristic of federated learning, since there exists a correlation: Investing more communication is expensive, but it leads to a better model. To reduce communication, random subsets of models can be averaged [33] or communication intervals can be adjusted dynamically [19]. To improve model quality, averaging can be replaced by other aggregation techniques, such as the geometric median [40] or the Radon point [21].

In our evaluation vanilla variant of federated learning serves as a baseline for the other approaches, since none of the special uncertainty quantification techniques are integrated there. We will refer to this approach as *global model*— as uncertainty is measured for the final global model at coordinator.

3   Leveraging Uncertainty in Federated Deep Learning

Given the preliminaries described in Sect. 2, we now leverage the uncertainty methods into the federated learning setting.

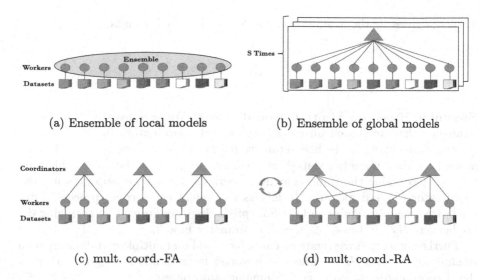

(a) Ensemble of local models   (b) Ensemble of global models

(c) mult. coord.-FA   (d) mult. coord.-RA

Fig. 1. Schema for different ways to build ensembles in a federated setup.

3.1 Ensembles in Federated Deep Learning

Different strategies to form an ensemble in a federated training scenario can be introduced, as illustrated in Fig. 1 and described in the following.

Ensemble of Local Models. A naive way to incorporate deep ensemble based uncertainty quantification into the federated setup is to consider the workers' local models as members of an ensemble. In order not to have the same model at every worker, one does not perform any communication with the coordinator, which leads to m separately trained models. These trained models f_i are used for the final prediction derived by averaging $f_i(x^*)$ and by replacing $p(y|x^*, \theta_s)$ in Eq. (1) and Eq. (4) by $f_i(x^*)$ for deriving the uncertainty measures. Note, that the idea of federated learning, where the local models benefit from others without seeing their data, is lost here.

Ensemble of Global Models. Here we describe another straightforward approach where the benefits of federated learning are kept, however, at the cost of a massive computational overhead. In *ensemble of global models* each worker trains S neural networks $f_{i,s}$, by using different random initialization to start with at each local model. For each of the S models we conduct the same procedure like in federated learning, increasing the computational effort by S times. However, because each run is independent of the others, this approach can easily be parallelized. It is still not practical in a real world setting with multiple computationally weak devices due to the increased computational effort and storage restrictions. The prediction for a new input x^* during the evaluation in each worker is given by the average over the S models, which results in

$$\hat{y} = \frac{1}{S} \sum_{s=1}^{S} f_{i,s}(x^*).$$ (9)

Ensemble Based on Multiple Coordinators. We also investigate ensemble strategies that are based on employing several coordinators in the federated training framework. In the first setup, we refer to as *fixed assignment (FA)* , the m workers are randomly grouped into A small subgroups, where each subgroup $(a \in [A])$ contains the same amount of workers. Each subgroup gets its own coordinator C^a and individually follows the federated learning process. After the last communication period the final predictions are derived by averaging the predictions $f(x^*; \bar{\theta}^a)$, $a \in [A]$ of each coordinator model.

Furthermore, we investigate an ensemble based on multiple coordinators with *random association (RA)*, where each worker is randomly reassigned to one of the A coordinators C^a after each communication phase.

3.2   MC-Dropout in Federated Deep Learning

We straightforwardly apply *federated MC-dropout* by transforming each local network f_i into a network where dropout with a droprate of $\alpha_i \in (0,1)$ is applied during training and prediction, c.f. Fig. 2. The chosen droprate is the same for all workers, i.e. $\alpha_i = \lambda$, $\forall i \in [m]$. Note that this does not result in a change of the communication needed compared to the baseline. During inference each worker samples multiple subnets by randomly dropping neurons and the final prediction is derived by averaging the predictions of subnets in analogy to the centralized setting.

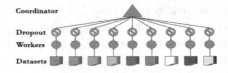

Fig. 2. Using dropout in a federated setup.

3.3   SWAG in Federated Deep Learning

We conclude this section by leveraging SWAG to the federated setting as depicted in Fig. 3. *Federated SWAG* is implemented by first conducting vanilla federated training, but with one communication period less. This is followed by one period where each worker calculates the SWAG estimates of mean θ_{SWA} and variance Σ_{diag} as described in Sect. 2. In the last communication period, each worker sends those two vectors (with a dimension each equal to the size of the network parameters θ) to the coordinator, which averages all mean values and all variance values and sends the averages back to the workers. Since both, the aggregated θ_{SWA} and aggregated Σ_{diag}, are distributed back, the amount of communication in this period is doubled. For inference each worker conducts predictions based on draws from the estimated approximate posterior distribution $\mathcal{N}(\theta_{\mathrm{SWA}}, \Sigma_{diag})$.

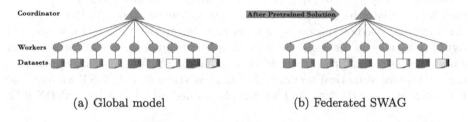

(a) Global model   (b) Federated SWAG

Fig. 3. Baseline and federated SWAG.

4   Empirical Evaluation

We start with a description of the experimental setup. We conducted the experiments on two datasets, MNIST [49], which consists out of black and white images of size 28×28 of handwritten digits from zero to nine, and CIFAR-10 [27], which consists out of colored 32×32 images of ten different classes of common things like cars, cats, or airplanes. We used the standard training and test data split. For the experiments on MNIST we implemented a simulated distributed environment[2], where code was run on a single GPU, and did not take into account issues of real distributed systems, like race conditions during communication. The experiments on CIFAR-10 are conducted using the already existing DL-Platform [18], which enables federated deep learning on a large scale. For both datasets the hyper-parameters, especially the learning rate, were tuned to maximize validation set accuracy. The tuning was done separately for each experiment. An overview of the used experimental settings is given in Appendix A. For the experiments on MNIST we used a simple fully connected network with two hidden layers with 128 neurons each and stochastic gradient decent as optimizer. Each experiment was run ten times with random initializations of the network parameters. For CIFAR-10 we used a ResNet18 architecture, and ran the experiment five times with different random initializations and using dropout for regularization[3]. For this, as well as for MC-dropout we applied a dropout-rate of $\lambda = 0.5$ for

---

[2] Sourcecode available at: https://github.com/FloLins/Approaches-to-Uncertainty-Quantification-in-Federated-Deep-Learning.

[3] Using dropout for regularization was necessary to reach a reasonable accuracy on CIFAR in our experiments. For comparison, results for the same setting without regularization can be found in the appendix in Table 6.

both datasets. For the federated setup we use 20 local learners and 1 coordinator. In the case of FA and RA we introduce 4 coordinators, analogously for the ensemble of global models we train 4 models at each local worker.

To investigate the ability of the different approaches to quantify uncertainty we measure the ability of the models to distinguish between (i) correctly classified and missclassified test examples and (ii) out of distribution data (OOD data) and the original in distribution data. To measure these abilities we first estimate the Shannon entropy given in Eq. (1) and the predictive variance given in Eq. (4) and then we calculate the *area under the receiver operating curve* (AUROC) [14] for classification based on this quantities. For calculating the AUROC for wrong and right predictions on the test dataset we sampled an equal amount of correctly and wrongly classified samples (to avoid class imbalances). We repeat this procedure three times for statistical accuracy. For models trained on MNIST we used the KMNIST [6] as OOD data, and for models trained on CIFAR-10— SVHN [37].

Table 1. Performance of ensemble approaches in comparison to vanilla federated training (denoted as global model).

Approach	Accuracy	Ent. AUROC	Var. AUROC
MNIST (10 runs) with KMNIST as out-of-distribution data			
Global model	97.70 ± 0.001	0.909 ± 0.004	–
Ensemble of local models	94.13 ± 0.001	0.899 ± 0.001	0.893 ± 0.001
Ensemble of global models	**97.88** ± 0.001	**0.927** ± 0.002	**0.931** ± 0.002
Ensemble w. mult. coord.-FA	97.03 ± 0.001	0.916 ± 0.002	0.914 ± 0.003
Ensemble w. mult. coord.-RA	97.36 ± 0.004	0.906 ± 0.002	0.878 ± 0.006
MNIST (10 runs) uncertainty for wrongly classified data			
Global Model	97.70 ± 0.001	0.968 ± 0.008	–
Ensemble of local models	94.13 ± 0.001	0.934 ± 0.005	0.930 ± 0.006
Ensemble of global models	**97.88** ± 0.001	**0.969** ± 0.006	**0.966** ± 0.006
Ensemble w. mult. coord.-FA	97.03 ± 0.001	0.959 ± 0.007	0.950 ± 0.007
Ensemble w. mult. coord.-RA	97.36 ± 0.004	0.962 ± 0.006	0.945 ± 0.011
CIFAR-10 on ResNet18 (5 runs) with dropout and SVHN as out-of-distribution data			
Global model	86.58 ± 0.289	0.924 ± 0.011	–
Ensemble of local models	72.65 ± 0.251	0.679 ± 0.023	0.689 ±0.033
Ensemble of global models	**89.00** ± 0.136	**0.937** ± 0.009	**0.804** ± 0.012
Ensemble w. mult. coord.-FA	83.43 ± 0.449	0.860 ±0.014	0.740 ± 0.012
Ensemble w. mult. coord.-RA	86.72± 0.511	0.920 ± 0.004	0.750 ± 0.026
CIFAR-10 on ResNet18 (5 runs) with dropout, uncertainty for wrongly classified data			
Global model with dropout	86.58 ± 0.289	0.888 ± 0.006	–
Ensemble of local models	72.65 ± 0.251	0.788 ±0.006	0.687 ± 0.007
Ensemble of global models	**89.00** ± 0.136	**0.891** ± 0.008	**0.868** ± 0.006
Ensemble w. mult. coord.-FA	83.43 ± 0.449	0.858 ± 0.005	0.815 ± 0.007
Ensemble w. mult. coord.-RA	86.72± 0.511	0.887 ± 0.005	0.861 ± 0.009

Because each setup differs in the amount of averaged predictions and computational complexity, we only compare them separately to the global model found by standard federated deep learning which serves as a baseline.

Ensembles. We report the accuracy and the results for uncertainty based OOD detection and detection of wrongly classified data for the investigated ensemble methods in Table 1. The *ensemble of local models* has a lower accuracy than the baseline on both datasets, which can be attributed to the fact that each local model only sees a limited amount of data. Even on the simpler MNIST dataset no satisfactory results are achieved.

Like in a centralized setting the *ensemble of global models*, where we used four global models, increases both accuracy and uncertainty quantification quality, but at the cost of an increased computational payload. This overhead could decrease its practical usability in the case that the training is performed on multiple computationally weak devices. In contrast, just increasing the amount of coordinators from one to four does not increase the computational complexity of the approach. A fixed association of workers to coordinators lead to less accurate results than a random association (compare Ensemble w. mult. coord.-FA to Ensemble w. mult. coord.-RA in Table 1), which could be explained by the fact that each coordinator sees more data. We also experimented with randomly choosing the size of the random subgroups (results not shown), which however did not lead to significant differences in the results.

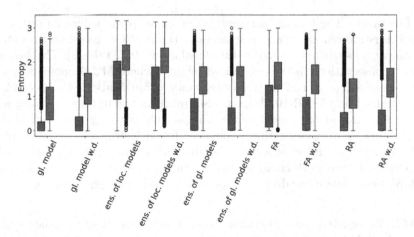

Fig. 4. The spread of entropy values for right (green) and wrong (red) predictions of CIFAR-10. (Color figure online)

In Figs. 4 and 5 we show the uncertainty for wrong and right predictions for the in distribution test dataset of CIFAR-10. One can see that generally the highest entropy values are produced by the ensemble of local models but entropy is high for correct and incorrect predictions. The other models, like

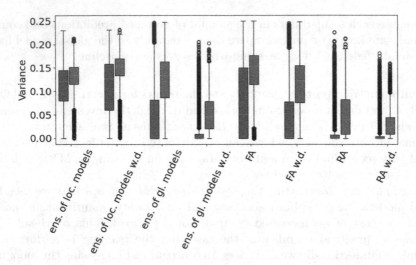

Fig. 5. The spread of variance values for right (green) and wrong (red) predictions of CIFAR-10. (Color figure online)

ensemble of global models, show a significantly larger difference in entropy values for missclassified and correctly classified examples, which is reflected by the higher AUROC values.

MC-Dropout. The findings for applying dropout in the federated setting are twofold: First, in our experiments it increases the AUROC based on the entropy for both datasets when compared to the global model (c.f. Table 2). This behavior relates to observations in the centralized setting, where MC-dropout is used as a simple and effective method for uncertainty quantification [10]. Second, for models trained on CIFAR-10, dropout also improves the test set accuracy when compared to the global model and is comparable with the accuracy derived from dropout as a regularization (global model with dropout). Because of the significant accuracy increase when using dropout during training, one possible direction for future work could be to investigate if the inherent prevention of co-adaptation while using dropout is beneficial during weight averaging.

SWAG. We conclude our experimental discussion with analyzing the effects of federated SWAG on test set accuracy and uncertainty. The results presented in Table 3 demonstrate that SWAG produces a slightly higher test set accuracy compared to the global model on both datasets. Further, the AUROC as well as the box-plot in Fig. 6 shows that the ability to indicate erroneous predictions based on the entropy is also increased. Because SWAG can be applied to existing (already trained) models by using the current state as pre-trained solution the found improvements can easily be archived by applying one communication period of the SWAG algorithm, i.e. without a lot of computational overhead compared to vanialla ferderated learning. Variance based OOD detection as well as wrong and right prediction distinction (Fig. 7) is always inferior to using the entropy.

Table 2. Comparison of federated MC-dropout to global model approach.

Approach	Accuracy	Ent. AUROC	Var. AUROC
MNIST (10 runs) with KMNIST as out-of-distribution data			
Global model	**97.70** ± 0.001	0.909 ± 0.004	–
Federated MC-dropout	96.77 ± 0.001	**0.920** ± 0.002	**0.871** ± 0.003
MNIST (10 runs) uncertainty for wrong classified data			
Global Model	**97.70** ± 0.001	**0.968** ± 0.008	–
Federated MC-dropout	96.77 ± 0.001	0.948 ± 0.009	**0.914** ± 0.013
CIFAR-10 on ResNet18 (5 runs) with SVHN as out-of-distribution data			
Global model	77.88 ± 0.279	0.742 ± 0.019	–
Global model with dropout	**86.58** ± 0.289	**0.924** ± 0.011	–
Federated MC-dropout	86.32 ± 0.253	0.913 ± 0.010	**0.714** ± 0.007
CIFAR-10 on ResNet18 (5 runs) uncertainty for wrong classified data			
Global Model	77.88 ± 0.279	0.838 +0.006	–
Global model with dropout	**86.58** ± 0.289	**0.888** ± 0.006	–
Federated MC-dropout	86.32 ± 0.253	0.880 ± 0.007	**0.849** ± 0.008

Table 3. Comparison of federated SWAG to global model approach.

Approach	Accuracy	Ent. AUROC	Var. AUROC
MNIST (10 runs) with KMNIST as out-of-distribution data			
Global model	97.70 ± 0.001	0.909 ± 0.004	–
Federated SWAG	**98.16** ± 0.001	**0.918** ± 0.004	**0.893** ± 0.005
MNIST (10 runs) uncertainty for wrong classified data			
Global Model	97.70 ± 0.001	0.968 ± 0.008	–
Federated SWAG	**98.16** ± 0.001	**0.974** ± 0.006	**0.969** ± 0.009
CIFAR-10 on ResNet18 (5 runs) with dropout and SVHN as out-of-distribution data			
Global model	86.58 ± 0.289	0.924 ± 0.011	–
Fed. SWAG	**87.14** ± 0.170	**0.924** ± 0.009	**0.807** ± 0.025
CIFAR-10 on ResNet18 (5 runs) with dropout, uncertainty for wrong classified data			
Global model	86.58 ± 0.289	0.888 ± 0.006	–
Fed. SWAG	**87.14** ± 0.170	**0.892** ± 0.006	**0.856** ± 0.007

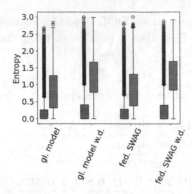
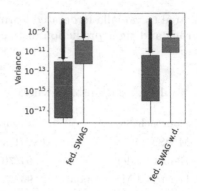

Fig. 6. The spread of entropy values for right (green) and wrong (red) predictions of CIFAR-10. (Color figure online)

Fig. 7. The spread of variance values for right (green) and wrong (red) predictions of CIFAR-10. (Color figure online)

5   Discussion and Conclusion

Due to the crucial relevance of uncertainty in trustworthy real world applications we investigated how approaches for uncertainty quantification can be applied in federated deep learning. More precisely we investigated ways to incorporate deep ensembles, SWAG and MC-dropout into the federated learning setup by changing network structure, communication protocols and training procedures. We believe, that the distributed setup provides further opportunities, as well as challenges for quantification of uncertainty and position our work as a baseline for possible future approaches.

The empirical results suggest that *ensembles of global models* and *federated SWAG* retain the model quality of standard federated learning while at the same time improve upon the standard setting in terms of out-of-distribution (OOD) detection and detection of missclassified test data. The disadvantage of this approach is that it requires additional computation and storage for each global model that needs to be computed. One needs to evaluate this requirements for a real world application, since in applications with computational weak devices this approach could be infeasible.

Federated SWAG is less computational demanding, while still improving the prediction accuracy. In terms of prediction accuracy and uncertainty estimation abilities we found that the *federated SWAG* reaches values close to the *ensemble of global models*. Therefore we recommend the usage of *federated SWAG* due to the precise results and the high usability.

Further, the flexibility of the federated learning protocols allows multiple applications of uncertainty quantification in practice. *Ensembles based on multiple coordinators* could be beneficial for federated learning in moving objects, like mobiles or cars, because our empirical analysis shows that adding additional coordinators and assigning workers randomly to subgroups does not decrease both accuracy and capabilities for uncertainty quantification. Further no modification of the training algorithm nor network structure are needed to apply this method. Even adding uncertainty quantification methods to a running system is possible by using *federated SWAG*, since the actual state of model can be used as pre-trained solution.

Concluding, in this work we have empirically shown, that federated deep learning can benefit from approaches for uncertainty quantification in terms of accuracy and certainty of prediction while gaining additional design flexibility.

Acknowledgments. Funded by the Deutsche Forschungsgemeinschaft (DFG, German Research Foundation) under Germany's Excellence Strategy - EXC 2092 CASA - 390781972 and the BMBF Project *Intrusion Detection in der Industrie 4.0 durch Fusion physikalischer Sensordaten mittels KI - mINDFUL*.

A   Appendix

The configuration for each experimental setup is summarized in Table 4. Furthermore the optimized learning rate for each individual setup is presented in Table 5.

Table 4. Hyperparameters used during the experiments.

Parameter	MNIST	CIFAR-10
Epochs per Communication Period	10	0.064
Batch Size	100	16
#Coordinators	4	4
#Workers	20	20
#Communication Periods	10	3125
OOD-Dataset	KMNIST	SVHN
Weight Decay	1e−5	1e−4
Repetitions for Ensemble of Global Models	4	4
Optimizer	SGD	SGD
Loss	Cross Entropy Loss	Cross Entropy Loss

Table 5. Learning rates used during the experiments.

Approach	MNIST	CIFAR-10
Global Model	0.1	0.01
Global Model w. dropout	–	0.1
Ensemble of local models	0.1	0.01
Ensemble of local models w. dropout	–	0.1
Ensemble of global models	0.1	0.01
Ensemble of global models w. dropout	–	0.1
Ensemble w. mult. coord.-FA	0.1	0.01
Ensemble w. mult. coord.-FA w. dropout	–	0.1
Ensemble w. mult. coord.-RA	0.1	0.01
Ensemble w. mult. coord.-RA w. dropout	–	0.1
Federated MC-dropout	0.05	0.1
Federated SWAG	0.1	0.01
Fed. SWAG w. dropout	–	0.1

Table 6 shows the results of CIFAR-10 when dropout was not applied as regularization method.

Table 6. Performance of CIFAR-10 without dropout as regularisation .

CIFAR-10 on ResNet18 (5 runs) with SVHN as out-of-distribution data			
Approach	Accuracy	Ent. AUROC	Var. AUROC
Global model	77.88 ± 0.279	0.742 ± 0.019	-
Ensemble of local models	66.77 ± 0.564	0.468 ± 0.019	0.648 ± 0.031
Ensemble of global models	84.02 ± 0.218	0.776 ± 0.02	0.719 ± 0.015
Ensemble w. mult. coord.-FA	73.24 ± 0.791	0.584 ± 0.023	0.631 ± 0.019
Ensemble w. mult. coord.-RA	78.65 ± 0.291	0.760 ± 0.028	0.704 ± 0.024
Federated SWAG	78.72 ± 0.57	0.749 ± 0.03	0.708 ± 0.033
CIFAR-10 on ResNet18 (5 runs)uncertainty for wrong classified data			
Global Model	77.88 ± 0.279	0.838 ± 0.006	-
Ensemble of local models	66.77 ± 0.564	0.760 ± 0.004	0.630 ± 0.006
Ensemble of global models	84.02 ± 0.218	0.862 ± 0.004	0.823 ± 0.008
Ensemble w. mult. coord.-FA	73.24 ± 0.791	0.808 ± 0.004	0.742 ± 0.012
Ensemble w. mult. coord.-RA	78.65 ± 0.291	0.847 ± 0.006	0.826 ± 0.006
Federated SWAG	78.72 ± 0.57	0.845 ± 0.007	0.825 ± 0.007

References

1. Abadi, M., et al.: Deep learning with differential privacy. In: Proceedings of the 2016 ACM SIGSAC Conference on Computer and Communications Security, pp. 308–318. ACM (2016)
2. Adilova, L., Rosenzweig, J., Kamp, M.: Information-theoretic perspective of federated learning. In: Workshop on Information Theory and Machine Learning 2019. Association for Information Systems (2019)
3. Boughorbel, S., Jarray, F., Venugopal, N., Moosa, S., Elhadi, H., Makhlouf, M.: Federated uncertainty-aware learning for distributed hospital ehr data. In: Machine Learning for Health (ML4H) Workshop at NeurIPS (2019)
4. Breiman, L.: Random forests. Mach. Learn. **45**(1), 5–32 (2001). https://doi.org/10.1023/A:1010933404324
5. Burdakis, S., Deligiannakis, A.: Detecting outliers in sensor networks using the geometric approach. In: 2012 IEEE 28th International Conference on Data Engineering (ICDE), pp. 1108–1119. IEEE (2012)
6. Clanuwat, T., Bober-Irizar, M., Kitamoto, A., Lamb, A., Yamamoto, K., Ha, D.: Deep learning for classical Japanese literature. ArXiv abs/1812.01718 (2018)
7. Deligiannakis, A., Kotidis, Y., Roussopoulos, N.: Processing approximate aggregate queries in wireless sensor networks. Inf. Syst. **31**(8), 770–792 (2006)
8. Däubener, S., Fischer, A.: Investigating maximum likelihood based training of infinite mixtures for uncertainty quantification. In: Workshop on Uncertainty in Machine Learning ECML/PKDD (2020)
9. Feinman, R., Curtin, R.R., Shintre, S., Gardner, A.B.: Detecting Adversarial Samples from Artifacts. arXiv preprint arXiv:1703.00410 (2017)
10. Gal, Y., Ghahramani, Z.: Dropout as a bayesian approximation: Representing model uncertainty in deep learning. In: International Conference on Machine Learning, pp. 1050–1059 (2016)
11. Giatrakos, N., Deligiannakis, A., Garofalakis, M., Sharfman, I., Schuster, A.: Prediction-based geometric monitoring over distributed data streams. In: Proceedings of the 2012 ACM SIGMOD International Conference on Management of Data, pp. 265–276. ACM (2012)
12. Graves, A.: Practical variational inference for neural networks. In: Advances in Neural Information Processing Systems, vol. 24. Curran Associates, Inc. (2011). https://proceedings.neurips.cc/paper/2011/file/7eb3c8be3d411e8ebfab08eba5f49632-Paper.pdf
13. Guo, C., Pleiss, G., Sun, Y., Weinberger, K.Q.: On calibration of modern neural networks. In: Proceedings of the 34th International Conference on Machine Learning, pp. 1321–1330. Proceedings of Machine Learning Research, PMLR, 06–11 August 2017
14. Hanley, J., Mcneil, B.: The meaning and use of the area under a receiver operating characteristic (roc) curve. Radiology **143**(1), 29–36 (1982)
15. Hinton, G.E., van Camp, D.: Keeping the neural networks simple by minimizing the description length of the weights. In: Proceedings of the Sixth Annual Conference on Computational Learning Theory, COLT 1993, pp. 5–13. Association for Computing Machinery, New York (1993). https://doi.org/10.1145/168304.168306
16. Hinton, G.E., Srivastava, N., Krizhevsky, A., Sutskever, I., Salakhutdinov, R.R.: Improving neural networks by preventing co-adaptation of feature detectors (2012)
17. Izmailov, P., Podoprikhin, D., Garipov, T., Vetrov, D., Wilson, A.G.: Averaging weights leads to wider optima and better generalization. arXiv preprint arXiv:1803.05407 (2018)

18. Kamp, M., Adilova, L.: Distributed Learning Platform (2020). https://github.com/fraunhofer-iais/dlplatform
19. Kamp, M., et al.: Efficient decentralized deep learning by dynamic model averaging. In: Berlingerio, M., Bonchi, F., Gärtner, T., Hurley, N., Ifrim, G. (eds.) ECML PKDD 2018. LNCS (LNAI), vol. 11051, pp. 393–409. Springer, Cham (2019). https://doi.org/10.1007/978-3-030-10925-7_24
20. Kamp, M., Boley, M., Keren, D., Schuster, A., Sharfman, I.: Communication-efficient distributed online prediction by dynamic model synchronization. In: Calders, T., Esposito, F., Hüllermeier, E., Meo, R. (eds.) ECML PKDD 2014. LNCS (LNAI), vol. 8724, pp. 623–639. Springer, Heidelberg (2014). https://doi.org/10.1007/978-3-662-44848-9_40
21. Kamp, M., Boley, M., Missura, O., Gärtner, T.: Effective parallelisation for machine learning. In: Thirty-first Conference on Neural Information Processing Systems, pp. 6477–6488. Curran Associates (2017)
22. Kamp, M., Boley, M., Keren, D., Schuster, A., Sharfman, I.: Communication-efficient distributed online prediction by dynamic model synchronization. In: Calders, T., Esposito, F., Hüllermeier, E., Meo, R. (eds.) ECML PKDD 2014. LNCS (LNAI), vol. 8724, pp. 623–639. Springer, Heidelberg (2014). https://doi.org/10.1007/978-3-662-44848-9_40
23. Kendall, A., Gal, Y.: What uncertainties do we need in bayesian deep learning for computer vision? (2017)
24. Keren, D., Sharfman, I., Schuster, A., Livne, A.: Shape sensitive geometric monitoring. IEEE Trans. Knowl. Data Eng. 24(8), 1520–1535 (2012)
25. Konečný, J., McMahan, H.B., Yu, F.X., Richtarik, P., Suresh, A.T., Bacon, D.: Federated learning: Strategies for improving communication efficiency. In: NIPS Workshop on Private Multi-Party Machine Learning (2016)
26. Kong, L., Sun, J., Zhang, C.: Sde-net: equipping deep neural networks with uncertainty estimates (2020)
27. Krizhevsky, A., Nair, V., Hinton, G.: Cifar-10 (canadian institute for advanced research). Learning Multiple Layers of Features from Tiny Images (2009). http://www.cs.toronto.edu/~kriz/cifar.html
28. Lakshminarayanan, B., Pritzel, A., Blundell, C.: Simple and scalable predictive uncertainty estimation using deep ensembles (2016)
29. Lechner, M., Hasani, R., Amini, A., Henzinger, T., Rus, D., Grosu, R.: Neural circuit policies enabling auditable autonomy. Nature Mach. Intell. 2, 642–652 (2020)
30. Li, D., Wang, J.: Fedmd: heterogenous federated learning via model distillation. arXiv preprint arXiv:1910.03581 (2019)
31. MacKay, D.J.C.: A practical bayesian framework for backpropagation networks. Neural Comput. 4(3), 448–472 (1992). https://doi.org/10.1162/neco.1992.4.3.448
32. Maddox, W.J., Izmailov, P., Garipov, T., Vetrov, D.P., Wilson, A.G.: A simple baseline for bayesian uncertainty in deep learning. In: Advances in Neural Information Processing Systems, pp. 13132–13143 (2019)
33. McMahan, B., Moore, E., Ramage, D., Hampson, S., Arcas, B.A.: Communication-efficient learning of deep networks from decentralized data. In: Artificial Intelligence and Statistics, pp. 1273–1282 (2017)
34. Meijerink, L., Cinà, G., Tonutti, M.: Uncertainty estimation for classification and risk prediction on medical tabular data (2020)
35. Mohri, M., Sivek, G., Suresh, A.T.: Agnostic federated learning. arXiv preprint arXiv:1902.00146 (2019)
36. Neal, R.M.: Bayesian Learning for Neural Networks. Ph.D. thesis, University of Toronto, CAN (1995)

37. Netzer, Y., Wang, T., Coates, A., Bissacco, A., Wu, B., Ng, A.Y.: Reading digits in natural images with unsupervised feature learning. In: NIPS Workshop on Deep Learning and Unsupervised Feature Learning 2011 (2011)
38. Ovadia, Y., et al.: Can you trust your model's uncertainty? evaluating predictive uncertainty under dataset shift. In: Advances in Neural Information Processing Systems. vol. 32. Curran Associates, Inc. (2019). https://proceedings.neurips.cc/paper/2019/file/f1ea154c843f7cf3677db7ce922a2d17-Paper.pdf
39. Peterson, D., Kanani, P., Marathe, V.J.: Private federated learning with domain adaptation. arXiv preprint arXiv:1912.06733 (2019)
40. Pillutla, K., Kakade, S.M., Harchaoui, Z.: Robust aggregation for federated learning. arXiv preprint arXiv:1912.13445 (2019)
41. Ritter, H., Botev, A., Barber, D.: A scalable laplace approximation for neural networks. In: International Conference on Learning Representations (2018). https://openreview.net/forum?id=Skdvd2xAZ
42. Sharfman, I., Schuster, A., Keren, D.: A geometric approach to monitoring threshold functions over distributed data streams. ACM Trans. Database Syst. (TODS) **32**(4), 23 (2007)
43. Sharma, M., Hutchinson, M., Swaroop, S., Honkela, A., Turner, R.E.: Differentially private federated variational inference. arXiv preprint arXiv:1911.10563 (2019)
44. Silva, S., Gutman, B.A., Romero, E., Thompson, P.M., Altmann, A., Lorenzi, M.: Federated learning in distributed medical databases: Meta-analysis of large-scale subcortical brain data. In: 2019 IEEE 16th international symposium on biomedical imaging (ISBI 2019), pp. 270–274. IEEE (2019)
45. Srivastava, N., Hinton, G., Krizhevsky, A., Sutskever, I., Salakhutdinov, R.: Dropout: a simple way to prevent neural networks from overfitting. J. Mach. Learn. Res. **15**(56), 1929–1958 (2014). http://jmlr.org/papers/v15/srivastava14a.html
46. Welling, M., Teh, Y.W.: Bayesian learning via stochastic gradient langevin dynamics. In: Proceedings of the 28th International Conference on International Conference on Machine Learning, pp. 681–688. ICML (2011)
47. Xu, J., Glicksberg, B.S., Su, C., Walker, P., Bian, J., Wang, F.: Federated learning for healthcare informatics. J. Healthcare Inf. Res. **5**(1), 1–19 (2021)
48. Yang, Q., Liu, Y., Chen, T., Tong, Y.: Federated machine learning: Concept and applications (2019)
49. LeCun, Y., Cortes, C., Christopher, J.C.: The mnist database (visited on 2021-04-02). http://yann.lecun.com/exdb/mnist/
50. Yao, X., Huang, T., Wu, C., Zhang, R., Sun, L.: Towards faster and better federated learning: a feature fusion approach. In: 2019 IEEE International Conference on Image Processing (ICIP), pp. 175–179. IEEE (2019)
51. Zhao, Y., Li, M., Lai, L., Suda, N., Civin, D., Chandra, V.: Federated learning with non-iid data. CoRR (2018)

Optimized Federated Learning on Class-Biased Distributed Data Sources

Yongli Mou[1]([✉]), Jiahui Geng[2], Sascha Welten[1], Chunming Rong[2],
Stefan Decker[1,3], and Oya Beyan[3,4]

[1] Chair of Computer Science 5, Information Systems and Databases,
RWTH Aachen University, Aachen, Germany
{mou,welten,decker}@dbis.rwth-aachen.de
[2] Department of Electrical Engineering and Computer Science,
University of Stavanger, Stavanger, Norway
{jiahui.geng,chunming.rong}@uis.no
[3] Fraunhofer Institute for Applied Information Technology (FIT),
Sankt Augustin, Germany
[4] Faculty of Medicine and University Hospital Cologne, Institute for Medical
Informatics, University of Cologne, Cologne, Germany
oya.beyan@uni-koeln.de

Abstract. Due to privacy protection, the conventional machine learning
approaches, which upload all data to a central location, has become less
feasible. Federated learning, a privacy-preserving distributed machine
learning paradigm, has been proposed as a solution to comply with privacy requirements. By enabling multiple clients collaboratively to learn a
shared global model, model parameters instead of local private data will
be exchanged under privacy restrictions. However, compared with centralized approaches, federated learning suffers from performance degradation when trained on non-independently and identically distributed
(non-i.i.d.) data across the participants. Meanwhile, the class imbalance
problem is always encountered in machine learning in practice and causes
bad prediction on minority classes. In this work, We propose FedBGVS
to alleviate the class bias severity by employing a balanced global validation set. The model aggregation algorithm is refined by using the
Balanced Global Validation Score (BGVS). We evaluate our methods
by experiments conducted on both the classical benchmark datasets
MNIST, SVHN and CIFAR-10 and a public clinical dataset ISIC-2019.
The empirical results demonstrate that our proposed methods outperform the state-of-the-art federated learning algorithms in label distribution skew and class imbalance settings.

Keywords: Federated learning · Statistical heterogeneity · Class
imbalance · Image classification

1 Introduction

The ever-increasing computational power and amounts of data collected provide
the opportunity for the explosive growth of deep learning. In recent years, we

---

Y. Mou and J. Geng—The authors contributed equally to this work.

© Springer Nature Switzerland AG 2021
M. Kamp et al. (Eds.): ECML PKDD 2021 Workshops, CCIS 1524, pp. 146–158, 2021.
https://doi.org/10.1007/978-3-030-93736-2_13

are experiencing a prosperous era of deep learning and have witnessed great success in many fields, such as computer vision and natural language processing. However, scandals like Facebook-Cambridge Analytica [5] have raised concerns about privacy and ethics. Conventional approaches that require data centralization have become less feasible due to the potential risks of privacy leakage while collecting and sharing sensitive data. The promulgation of data protection laws such as the General Data Protection Regulation (GDPR) accelerates the study of privacy-preserving techniques. The concept shift of bringing algorithms to data instead of vice versa is considered as a new solution. In disciplines such as medicine and healthcare, infrastructures such as Personal Health Train (PHT) [2,20] and DataSHIELD [23] are implemented to support distributed approaches that allow the analysis of distributed sensitive individual data while circumventing data privacy restrictions. Federated learning [10,11] is a new parallel paradigm for privacy-preserving distributed machine learning that has attracted a great deal of attention in recent years. It aims to collaboratively train the global machine learning models while keeping data decentralized.

As a decentralized approach, federated learning faces several challenges due to its decentralized architecture, such as expensive communication, systems heterogeneity, and statistical heterogeneity [14]. Studies show that statistical heterogeneity causes non-trivial performance degradation and slow convergence, sometimes even divergence, making it the critical challenge in federated learning. For example, medical radiology images in different hospitals are acquired from different devices using different standards, which causes feature shift. Like conventional machine learning, federated learning suffers from data bias in real datasets, such as class imbalance problems. Models trained on the dataset with class imbalance problems always have performance degradation on minority classes since losses like cross-entropy optimize based on overall accuracy.

In this work, we focus on the challenges of statistical heterogeneity and class imbalance in federated learning. We propose a novel approach FedBGVS that uses a balanced global validation dataset on the server to evaluate the client models and calculate the client weights in model aggregation. Experiments on benchmark datasets MNIST, SVHN, and CIFAR-10 and clinical dataset ISIC-2019 demonstrate that the proposed approach outperforms several state-of-the-art federated learning algorithms in label distribution skew of non-iid and class imbalance settings.

2   Preliminaries

To formulate the problem, we consider supervised learning tasks in horizontal data partitioning scenario. We assume that a global dataset $\mathcal{D} = \{\mathcal{D}_1, \cdots, \mathcal{D}_K\}$ is distributed on K clients, where $\mathcal{D}_k = \{(\mathbf{x}_k, y_k) | \mathbf{x}_k \in \mathcal{X}, y_k \in \mathcal{Y}\}$ and $|\mathcal{D}_k| = N_k$. The data in D_k are drawn from the distribution $(\mathbf{x}_k, y_k) \sim \mathcal{P}_k(X, Y)$.

2.1   Federated Learning

Federed learning is originally proposed by Google [10,11] and enables mobile devices to collaboratively train a shared model while keeping all the training data on device. We formalize the federated learning as the optimization problem in the form of Eq. 1 where F_k is the local loss functions with data draw from different distributions and $\lambda_k > 0$, $\sum_k \lambda_k = 1$.

$$\min_w \sum_{k=1}^{K} \lambda_k F_k(\omega) = \sum_{k=1}^{K} \mathbb{E}_{(\mathbf{x}_k, y_k) \sim \mathcal{P}_k}[F_k(\omega|x_k, y_k)] \qquad (1)$$

In a federated learning system, a server orchestrates the training process by repeating the following five steps (as shown in Algorithm 1): client selection (Line 3), global model broadcast (Line 6), client computation (Line 7), model aggregation and model update (Line 9) [7]. FedAvg [18,19] has been the standard approach for federated learning as shown in Algorithm 1. Compared to traditional FedSGD, the clients in FedAvg update their local model with multiple local epochs to decrease the number of communication rounds, which is much more communication-efficient.

Algorithm 1: FedAvg Algorithm

Input: D_k: dataset on client k, K: number of clients, η: learning rate, T: number of communication rounds, E: number of local epochs, L: loss function

Output: ω^T: final model

Server executes:

1 Initialize ω^0

2 **for** $t = 0, 1, \cdots, T$ **do**

3 | Sample a subset of clients S_t

4 | $n \leftarrow \sum_{i \in S_t} |D_i|$

5 | **for** $i \in S_t$ in parallel **do**

6 | | Send the global model ω^t to client i

7 | | $\omega_i^t \leftarrow$ **LocalTraining**(i, ω^t)

8 | **end**

9 | $\omega^{t+1} \leftarrow \sum_{i \in S_t} \lambda_i \omega_i^t$, where $\lambda_i = \frac{|D_i|}{n}$

10 **end**

11 Return ω^T

Client executes:

12 **LocalTraining**(i, ω^t)

13 $\omega_i^t \leftarrow \omega^t$

14 **for** epoch $e = 1, 2, \cdots, E$ **do**

15 | **for** each batch b of D_i **do**

16 | | $\omega_i^t \leftarrow \omega_i^t - \eta \sum_{(\mathbf{x}, y) \in b} \Delta L(\mathbf{x}, y, \omega_i^t)$

17 | **end**

18 **end**

19 Return ω_i^t

2.2 Statistical Heterogeneity

Statistical heterogeneity refers to non-i.i.d. problem, i.e., data distributions are usually non-independent and identically distributed across the clients.

$$\mathcal{P}_i(X, Y) \neq \mathcal{P}_j(X, Y), \forall i \neq j$$

Based on Bayes' law, the joint distribution equals the conditional distribution times marginal distribution, i.e., $\mathcal{P}_k(X, Y) = \mathcal{P}_k(Y|X)\mathcal{P}_k(X) = \mathcal{P}_k(X|Y)\mathcal{P}_k(Y)$. According to the characteristics of the marginal distribution and conditional distribution, statistical heterogeneity is classified into the five categories [7]: feature distribution skew, label distribution skew, same label different features, same feature different labels and quantity skew. In this work, we focus on the label distribution skew, where $\mathcal{P}_k(Y)$ vary across clients, even if $\mathcal{P}_k(Y|X)$ are same.

3 Related Work

Many studies have shown that FedAvg suffers from performance degradation (low accuracy), slow convergence and even divergence on non-i.i.d. data [16,25]. Multiple research works have investigated various techniques for non-i.i.d. challenges by optimizing federated learning training process. Zhao et al. [25] firstly propose to solve this problem by utilizing a small globally shared training sub-dataset. Although this approach has been shown to be effective in improving model performance in non-i.i.d scenarios, it violates the principle that federal learning does not share data. In addition, federated learning lacks an estimation of the distribution of the dataset and does not guarantee a fair and unbiased training set. Another idea is to make all model updates trained on separated local data converge to the globally shared model. Li et al. [14] proposed FedProx, adding a proximal term in the local training stage as a regularization term, which effectively suppressed the divergence of model updates brought by heterogeneity. They also provided a mathematical convergence guarantee for the framework. Acar et al. further considered the performance degradation problem as the inconsistency of the local empirical loss and the global empirical loss and proposed FedDyn [1] with a dynamic regularizer for each client, aligning the global and local loss. SCAFFOLD [8] considered the non-i.i.d as the variance introduced from the parties and applied the variance reduction technique to mitigate the impact. Specifically, control variates are designed to estimate the update direction of the server and client models; then local updates are corrected by adding the drift in the local training. Yaganeh et al. proposed Inverse Distance Aggregation (IDA) [24] to reduce the inconsistency among the updated local parameters. Wang et al. proposed FedMA [22] that shares the global model in a layer-wise manner. From the perspective of feature space shift, Li et al. [17] claimed that the local normalization of features helps with the robustness and the convergence of the optimization process. They proposed FedBN that preserves Batch Normalization (BN) on each local client and excludes BN parameters from model aggregation. Some works make efforts on strategies of selecting clients for each round of model aggregation. Kopparapu et al. [12] made use of accuracy tested on the server to guide the client selection and exclude poorly performed clients from joining the training. Wang et al. introduced Reinforcement Learning (RL) in the client selection stage. They proposed FAVOR [21] that learns to select a subset of clients in each communication round based on deep Q-learning to maximize a reward encouraging the increase of validation accuracy and penalizes the number of communications.

4   Methods

4.1   Federated Learning Algorithms

Simply averaging the model weights in model aggregation could mislead to slower convergence and lousy performance because the contribution of each client differs due to statistical heterogeneity in federated learning. We introduce two federated learning algorithms based on FedAvg and use them in comparison experiments. FedProx [15] introduces an additional proximal term (as shown in Eq. 2 during local training (Line 16 in Algorithm 1) to limit the $L2$-distance between the local model and the global model.

$$L(\mathbf{x}, y, \omega_i^t) = L(\mathbf{x}, y, \omega_i^t) + \frac{\mu}{2}\|\omega^t - \omega_i^t\| \qquad (2)$$

IDA [24] modifies the client weights (Line 9 in Algorithm 1) in model aggregation by measuring the distance between client models and the average of all client models as shown in Eq. 3, where Z is normalization factor.

$$\lambda_i = \frac{1}{Z}\left\|\omega_{avg}^t - \omega_i^t\right\|^{-1}, \text{ where } Z = \sum_{i \in S_t}\left\|\omega_{avg}^t - \omega_i^t\right\|^{-1} \qquad (3)$$

4.2   Proposed Approach

Inspired by IDA, we proposed FedBGVS (Balanced Global Validation Score), a novel model aggregation strategy. To alleviate the impact of label distribution skewness, we introduce a balanced global validation dataset to score the performance of client models, which is very straightforward. A separated balanced global validation dataset is created, which is only available on the server-side in model aggregation and cannot be accessed by any clients without violating data privacy protection.

Now, we present how to calculate the balanced global validation score BVGS. After receiving models from selected clients, we obtain the validation scores s_i by evaluating the model's performance from client i on the balanced global validation dataset. Then we normalize the validation score as shown in Eq. 4, where S_t is the set of selected clients in communication round t. Finally, the client weights λ_i is calculated by Eq. 5, where $\frac{|D_i|}{n}$ is the ratio of number of data in dataset D_i to the global dataset. The factor λ controls how much the balanced global validation score will influence the model aggregation. When $\lambda = 0$, the algorithm FedBGVS is identical to FedAvg.

$$\tilde{s}_i = \frac{s_i}{\sum_{i \in S_t} s_i} \tag{4}$$

$$\lambda_i = \gamma(\tilde{s}_i - \frac{|D_i|}{n}) + \frac{|D_i|}{n} \tag{5}$$

We tried different metrics as the validation score, such as accuracy, IoU, and cross-entropy loss. For accuracy and IoU, we set $s_i = Acc_i$ and $s_i = IoU_i$, respectively, while for cross-entropy loss we use $s_i = \frac{1}{CE_i}$.

5 Experiments

We conduct comprehensive experiments on classical machine learning benchmark datasets MNIST, SVHN, and CIFAR-10 for proof-of-concept, as well as a public clinical dataset from the International Skin Imaging Collaboration 2019 Challenge (ISIC-2019).

5.1 Dataset and Data Distribution

MNIST and SVHN contain digit images from handwriting and stress view respectively (see in Fig. 1 Row 1 and 2). The CIFAR-10 contains images with 10 common objects in real life (see in Fig. 1 Row 3). The ISIC-2019 dataset contains 25,331 dermoscopic images among eight different diagnostic categories (see in Fig. 2), namely melanoma (MEL), melanocytic nevus (NV), basal cell carcinoma (BCC), actinic keratosis (AK), benign keratosis (BKL). dermatofibroma (DF), vascular lesion (VASC), and squamous cell carcinoma (SCC). Unlike MNIST, SVHN and CIFAR-10, classes in ISIC-2019 are naturally unbalanced. As shown in Table 1, more than half of the data is classified as NV, while four classes, AK, DF, VASC, and SCC, only occur less than 10% in total.

Table 1. Class distribution of ISIC-2019 dataset

MEL	NV	BCC	AK	BKL	DF	VASC	SCC
4522	12875	3323	867	2624	239	253	628

We split each dataset into three parts (as shown in Table 2: on clients for training, on the server for global validation, and a hold-out test set. Since we want to build a balanced global validation dataset, the size of the global validation dataset of ISIC-2019 is limited due to the number of samples in the minority class, so we only have 144 images in the global validation dataset. To simulate the label distribution skewness of non-i.i.d. cases, we use distribution-based label imbalance [13]. Each client is allocated a proportion of the samples of each label

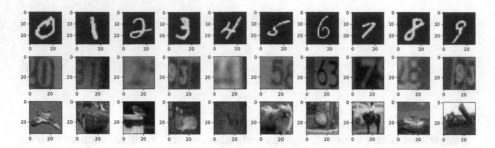

Fig. 1. Row 1: MNIST, Row 2: SVHN, Row 3: CIFAR-10

(a)   (b)   (c)   (d)   (e)   (f)   (g)   (h)

Fig. 2. (a) MEL, (b) NV, (c) BCC, (d) AK, (e) BKL, (f) DF, (g) VASC and (h) SCC

according to Dirichlet distribution $\mathcal{P}_k(Y) \sim Dir(\beta)$ with a random control vector β, i.e., we allocate a $\mathcal{P}_k(Y = y)$ proportion of the instances of class y to client k. Figure 3 shows the label distributions of ISIC-2019. Finally, 20% of data on each client is used for local validation.

Table 2. Data split

Dataset	Clients	Server	Test
MNIST	60,000	5,000	5,000
SVHN	73,257	13,016	13,016
CIFAR-10	50,000	5,000	5,000
ISIC-2019	20,265	144	4,922

5.2 Experimental Setup

In our experiments, we employ ResNet-18 [6] for the image classification that built with PyTorch framework and pretrained on ImageNet [4]. Adam optimizers [9] with an initial learning rate of 0.0001 and a weight decay of 0.0005 are used to minimize cross-entropy loss with the batch size of 16. We apply center cropping on ISIC-2019 and resize the cropped images into the size of 256×256. For comparison, we choose Cyclic Institutional Incremental Learning (CIIL) [3], FedAvg, FedProx and IDA as our baselines. For CIIL, we train the model for 50 rounds (cycles), visit four clients in each round and run 1 epoch on each client.

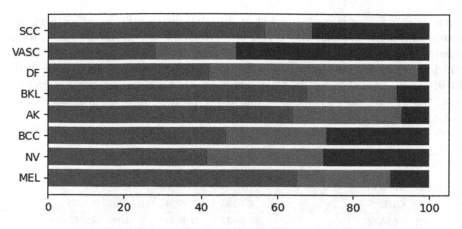

Fig. 3. Label distributions of ISIC-2019 on each client

For federated learning, we train the model for 50 rounds, run 1 epoch on each client to update the local model and aggregate the model updates after each round.

6   Results

For quantitatively measuring the performance of our proposed approach, we report four metrics: accuracy, mean precision, mean recall and mean intersection over union on MNIST, SVHN, CIFAR-10 and ISIC-2019 in Tables 3, 4, 5 and 6 respectively.

The results have shown strong correlation among the four metrics, i.e., models performed well in accuracy has also higher mean recall, mean precision and mean IoU. In general, CIIL performs the worst compared with other federated learning approaches in class-biased settings. As the feature complexity of dataset increased, the performance gaps between incremental learning and federated

Table 3. Results on MNIST dataset

Methods	Accuracy	Precision	Recall	IoU
CIIL	0.9928	0.9927	0.9928	0.9856
FedAvg	0.9932	0.9930	0.9930	0.9861
IDA	0.9926	0.9925	0.9926	0.9852
FedProx	0.9948	0.9947	0.9948	0.9896
FedBGVS (acc, $\gamma = 1.0$)	0.9934	0.9932	0.9932	0.9866
FedBGVS (FedProx, $\gamma = 0.25$)	**0.9956**	**0.9955**	**0.9956**	**0.9912**
FedBGVS (ce, $\gamma = 0.5$)	0.9948	0.9945	0.9947	0.9893
FedBGVS (iou, $\gamma = 0.5$)	0.9942	0.9941	0.9941	0.9883

learning approaches become more clear. In label distribution skew settings, federated learning approachs with BVGS outperforms FedAvg, IDA and FedProx in general. However, the mean precision and recall performances of FedBVGS are still limited in the naturally extremely unbalanced situation, such as on ISIC-2019. We observe that FedBGVS combined with FedProx performs overall the best. On MNIST and SVHN, it obtained the highest accuracy and was second to FedBGVS on CIFAR-10, 0.18% behind.

Table 4. Results on SVHN dataset

Methods	Accuracy	Precision	Recall	IoU
CIIL	0.9386	0.9299	0.9412	0.8770
FedAvg	0.9637	0.9630	0.9598	0.9255
IDA	0.9644	0.9633	0.9623	0.9275
FedProx	0.9620	0.9600	0.9606	0.9237
FedBGVS (acc, $\gamma = 0.75$)	0.9628	0.9608	0.9598	0.9230
FedBGVS (FedProx, $\gamma = 1.0$)	**0.9654**	**0.9636**	**0.9624**	**0.9287**
FedBGVS (ce, $\gamma = 0.5$)	0.9632	0.9611	0.9599	0.9241
FedBGVS (iou, $\gamma = 0.5$)	0.9630	0.9601	0.9614	0.9243

Table 5. Results on CIFAR-10 dataset

Methods	Accuracy	Precision	Recall	IoU
CIIL	0.9020	0.9013	0.9130	0.8272
FedAvg	0.9374	0.9377	0.9373	0.8829
IDA	0.9462	0.9466	0.9462	0.8987
FedProx	0.9398	0.9397	0.9404	0.8874
FedBGVS (acc, $\gamma = 1.0$)	**0.9562**	**0.9562**	**0.9562**	**0.9167**
FedBGVS (FedProx, $\gamma = 0.5$)	0.9544	0.9541	0.9542	0.9130
FedBGVS (ce, $\gamma = 0.5$)	0.9556	0.9554	0.9558	0.9156
FedBGVS (iou, $\gamma = 0.5$)	0.9550	0.9553	0.9559	0.9152

Table 6. Results on ISIC-2019 dataset

Methods	Accuracy	Precision	Recall	IoU
CIIL	0.6952	0.5302	**0.6346**	0.3726
FedAvg	0.7133	0.5474	0.5915	0.3880
IDA	0.7070	**0.5610**	0.6129	0.3923
FedProx	0.7156	0.5474	0.5900	0.3742
FedBGVS (acc, $\gamma = 0.25$)	**0.7249**	0.5515	0.6034	**0.3955**
FedBGVS (FedProx, $\gamma = 0.25$)	0.7152	0.5351	0.6195	0.3788
FedBGVS (ce, $\gamma = 0.5$)	0.7229	0.5284	0.5578	0.3899
FedBGVS (iou, $\gamma = 0.5$)	0.7110	0.5342	0.5075	0.3643

In each round, we also validate the local models on their local validation dataset as well as on the validation datasets of the other clients. Figure 4 and Fig. 5 (a), (b), (c), (d) and (e) show the accuracy and mean IoU of models trained on client 1, 2, 3 and 4 and aggregated global model in each communication round, respectively. The plots show that model trained on client i has higher performance on the corresponding local validation dataset but not always converges on

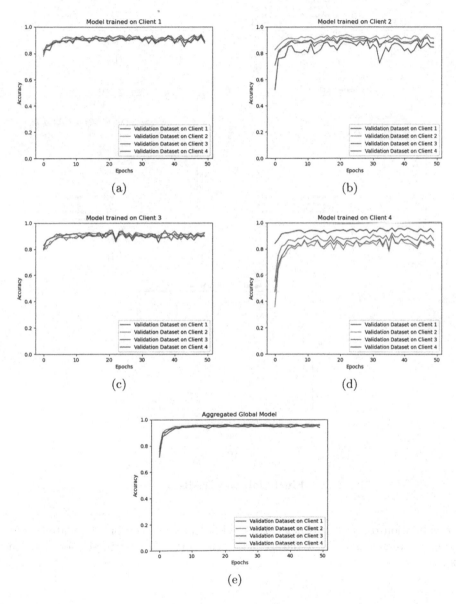

Fig. 4. Accuracy on CIFAR-10

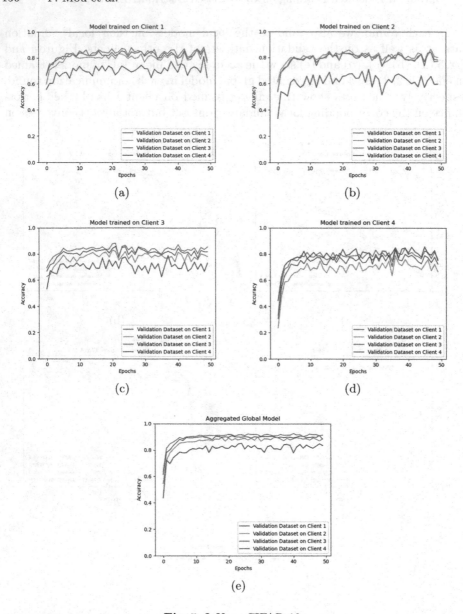

Fig. 5. IoU on CIFAR-10

the validation dataset of other clients, while the accuracy of aggregated global model converges on all local datasets, i.e., the curves are getting closer along the axis.

7   Conclusion

In this paper, we proposed FedBGVS that employs an additional balanced global validation dataset on the server used for scoring and adjusting client weights in model aggregation to alleviate the class bias severity. With comprehensive comparison experiments with the state-of-the-art methods, e.g., FedProx and IDA, we demonstrated that a globally balanced guidance in the model aggregation does contribute to better performance in the class-biased settings. This work also shows the feasibility of federated learning and the universality of our proposed methods in image classification.

For further work, we plan to study the impact of the global validation set size on our method. We will also explore other metrics like the similarity between the models or the performance of the model on the data set to improve the performance of the model aggregation, and try to give a theoretical proof of the model and its impact on communication rounds.

References

1. Acar, D.A.E., Zhao, Y., Matas, R., Mattina, M., Whatmough, P., Saligrama, V.: Federated learning based on dynamic regularization. In: International Conference on Learning Representations (2020)
2. Beyan, O., et al.: Distributed analytics on sensitive medical data: The personal health train. Data Intell. **2**(1–2), 96–107 (2020)
3. Chang, K., et al.: Distributed deep learning networks among institutions for medical imaging. J. Am. Med. Inform. Assoc. **25**(8), 945–954 (2018)
4. Deng, J., Dong, W., Socher, R., Li, L.J., Li, K., Fei-Fei, L.: Imagenet: a large-scale hierarchical image database. In: 2009 IEEE Conference on Computer Vision and Pattern Recognition, pp. 248–255. IEEE (2009)
5. González, F., Yu, Y., Figueroa, A., López, C., Aragon, C.: Global reactions to the cambridge analytica scandal: a cross-language social media study. In: Companion Proceedings of the 2019 World Wide Web Conference, pp. 799–806 (2019)
6. He, K., Zhang, X., Ren, S., Sun, J.: Deep residual learning for image recognition. In: Proceedings of the IEEE Conference on Computer Vision and Pattern Recognition, pp. 770–778 (2016)
7. Kairouz, P., et al.: Advances and open problems in federated learning. arXiv preprint arXiv:1912.04977 (2019)
8. Karimireddy, S.P., Kale, S., Mohri, M., Reddi, S., Stich, S., Suresh, A.T.: Scaffold: stochastic controlled averaging for federated learning. In: International Conference on Machine Learning, pp. 5132–5143. PMLR (2020)
9. Kingma, D.P., Ba, J.: Adam: A method for stochastic optimization. arXiv preprint arXiv:1412.6980 (2014)
10. Konečný, J., McMahan, B., Ramage, D.: Federated optimization: distributed optimization beyond the datacenter. arXiv preprint arXiv:1511.03575 (2015)
11. Konečný, J., McMahan, H.B., Ramage, D., Richtárik, P.: Federated optimization: distributed machine learning for on-device intelligence. arXiv preprint arXiv:1610.02527 (2016)
12. Kopparapu, K., Lin, E., Zhao, J.: FEDCD: improving performance in non-iid federated learning. arXiv preprint arXiv:2006.09637 (2020)

13. Li, Q., Diao, Y., Chen, Q., He, B.: Federated learning on non-iid data silos: an experimental study. arXiv preprint arXiv:2102.02079 (2021)
14. Li, T., Sahu, A.K., Talwalkar, A., Smith, V.: Federated learning: challenges, methods, and future directions. IEEE Signal Process. Mag. **37**(3), 50–60 (2020)
15. Li, T., Sahu, A.K., Zaheer, M., Sanjabi, M., Talwalkar, A., Smith, V.: Federated optimization in heterogeneous networks. arXiv preprint arXiv:1812.06127 (2018)
16. Li, X., Huang, K., Yang, W., Wang, S., Zhang, Z.: On the convergence of fedavg on non-iid data. arXiv preprint arXiv:1907.02189 (2019)
17. Li, X., Jiang, M., Zhang, X., Kamp, M., Dou, Q.: Fedbn: federated learning on non-iid features via local batch normalization. Training **1**, 1–2 (2021)
18. McMahan, B., Moore, E., Ramage, D., Hampson, S., Arcas, B.A.: Communication-efficient learning of deep networks from decentralized data. In: Artificial Intelligence and Statistics, pp. 1273–1282. PMLR (2017)
19. McMahan, H.B., Moore, E., Ramage, D., y Arcas, B.A.: Federated learning of deep networks using model averaging. arXiv preprint arXiv:1602.05629 (2016)
20. Mou, Y., Welten, S., Jaberansary, M., Ucer Yediel, Y., Kirsten, T., Decker, S., Beyan, O.: Distributed skin lesion analysis across decentralised data sources. In: Public Health and Informatics, pp. 352–356. IOS Press (2021)
21. Wang, H., Kaplan, Z., Niu, D., Li, B.: Optimizing federated learning on non-iid data with reinforcement learning. In: IEEE INFOCOM 2020-IEEE Conference on Computer Communications, pp. 1698–1707. IEEE (2020)
22. Wang, H., Yurochkin, M., Sun, Y., Papailiopoulos, D., Khazaeni, Y.: Federated learning with matched averaging. arXiv preprint arXiv:2002.06440 (2020)
23. Wilson, R.C., et al.: Datashield - new directions and dimensions. Data Sci. J. **16**, 21 (2017). https://doi.org/10.5334/dsj-2017-021
24. Yeganeh, Y., Farshad, A., Navab, N., Albarqouni, S.: Inverse distance aggregation for federated learning with Non-IID data. In: Albarqouni, S., et al. (eds.) DART/DCL –2020. LNCS, vol. 12444, pp. 150–159. Springer, Cham (2020). https://doi.org/10.1007/978-3-030-60548-3_15
25. Zhao, Y., Li, M., Lai, L., Suda, N., Civin, D., Chandra, V.: Federated learning with non-iid data. arXiv preprint arXiv:1806.00582 (2018)

Splitting Algorithms for Federated Learning

Saber Malekmohammadi[1,2(✉)], Kiarash Shaloudegi[2], Zeou Hu[1,2],
and Yaoliang Yu[1]

[1] School of Computer Science, University of Waterloo, Waterloo, Canada
{saber.malekmohammadi,zeou.hu,yaoliang.yu}@uwaterloo.ca
[2] Noah's Ark Lab, Huawei Technologies, Montreal, Canada
kiarash.shaloudegi@huawei.com

Abstract. Over the past few years, the federated learning (FL) community has witnessed a proliferation of new FL algorithms. However, our understating of the theory of FL is still fragmented, and a thorough, formal comparison of these algorithms remains elusive. Motivated by this gap, we show that many of the existing FL algorithms can be understood from an operator splitting point of view. This unification allows us to compare different algorithms with ease, refine previous convergence results and uncover new algorithmic variants. In particular, our analysis reveals the vital role played by the step size in FL algorithms. We perform numerical experiments on both convex and nonconvex models to validate our findings.

1 Introduction

The accompanying (moderate) computational power of modern smart devices such as phones, watches, home appliances, cars, etc., and the enormous data accumulated from their interconnected ecosystem have fostered new opportunities and challenges to train/tailor modern deep models. Federated learning (FL), as a result, has recently emerged as a massively distributed framework that enables training a shared or personalized model without infringing user privacy. Tremendous progress has been made since the seminal work of [31], including algorithmic innovations [23,35,38,47], convergence analysis [19,20,25,29], personalization [12,13,30,48], privacy protection [1,33], model robustness [3,6], fairness [17,24,32], standardization [9,16], applications [37,41], just to name a few. We refer to the excellent surveys [18,22,44] and the references therein for the current state of affairs in FL.

The main goal of this work is to take a closer examination of some of the most popular FL algorithms, including FedAvg [31], FedProx [23], and FedSplit [35], by connecting them with the well-established theory of operator splitting in optimization. In particular, we show that FedAvg [31] corresponds to forward-backward splitting and we demonstrate a trade-off in its step size and number of local epochs. FedProx [23], on the other hand, belongs to backward-backward

© Springer Nature Switzerland AG 2021
M. Kamp et al. (Eds.): ECML PKDD 2021 Workshops, CCIS 1524, pp. 159–176, 2021.
https://doi.org/10.1007/978-3-030-93736-2_14

splitting, or equivalently, forward-backward splitting on a related but regularized problem, which has been somewhat overlooked in the literature. Interestingly, our results reveal that the recent personalized model in [13] is exactly the problem that `FedProx` aims to solve. We show that when the step size in `FedProx` diminishes (sublinearly fast), then it actually solves the original problem, which is contrary to the observation in [35] where the step size is fixed and confirms again the importance of step size in `FL`. These results largely clarify the pros and cons of `FedProx`. Furthermore, `FedSplit` [35] corresponds to Peaceman-Rachford splitting [28,36] and we show that its convergence (in theory) heavily hinges on the strong convexity of the objective and hence might be less stable for nonconvex problems.

Inspecting `FL` through the lens of operator splitting allows us to immediately uncover new `FL` algorithms, by adapting existing splitting algorithms. Indeed, we show that Douglas-Rachford splitting [14,28] (more precisely the method of partial inverse [43]) yields a (slightly) slower but more stable variant of `FedSplit`. Our holistic view of `FL` algorithms suggests a very natural unification, building on which we show that the aforementioned algorithms reduce to different parameter settings in a grand scheme. We believe this is an important step towards standardizing `FL` from an algorithmic point of view, in addition to standard datasets, models and evaluation protocols as already articulated in [9,16]. Our unification also allows one to compare and implement different `FL` algorithms with ease. We perform experiments on both convex and nonconvex models to validate our findings, and compare the above-mentioned `FL` algorithms on an equal footing.

We proceed in Sect. 2 with some background introduction. Our main contributions are presented in Sect. 3, where we connect `FL` with operator splitting, shed new insights on existing algorithms, suggest new algorithmic variants and refine convergence analysis, and in Sect. 4, where we present a unification of current algorithms. We conclude in Sect. 5 with some future directions.

2   Background

In this section we recall the federated learning (`FL`) framework of [31]. We consider m users (edge devices), where the i-th user aims at minimizing a function $f_i : \mathbb{R}^d \to \mathbb{R}, i = 1, \ldots, m$, defined on a shared model parameter $\mathbf{w} \in \mathbb{R}^d$. Typically, each user function f_i depends on the respective user's local (private) data $\mathcal{D}_i$.

Following [31], many existing `FL` algorithms fall into the natural formulation that optimizes the (arithmetic) average performance:

$$\min_{\mathbf{w} \in \mathbb{R}^d} f(\mathbf{w}), \quad \text{where} \quad f(\mathbf{w}) := \sum_{i=1}^m \lambda_i f_i(\mathbf{w}). \tag{1}$$

The weights λ_i here are nonnegative and w.l.o.g. sum to 1. Below, we assume they are specified *beforehand* and fixed throughout.

To facilitate our discussion, we recall the Moreau envelope and proximal map of a function[1] $f : \mathbb{R}^d \to \mathbb{R} \cup \{\infty\}$, defined respectively as:

$$\mathsf{M}_f^\eta(\mathbf{w}) = \min_{\mathbf{x}} \; \tfrac{1}{2\eta} \|\mathbf{x} - \mathbf{w}\|_2^2 + f(\mathbf{x}), \qquad \mathsf{P}_f^\eta(\mathbf{w}) = \arg\min_{\mathbf{x}} \; \tfrac{1}{2\eta} \|\mathbf{x} - \mathbf{w}\|_2^2 + f(\mathbf{x}), \tag{2}$$

where $\eta > 0$ is a parameter acting similarly as the step size. We also define the reflector

$$\mathsf{R}_f^\eta(\mathbf{w}) = 2\mathsf{P}_f^\eta(\mathbf{w}) - \mathbf{w}. \tag{3}$$

Clearly, when $f = \iota_C$ is the indicator function of a (closed) set $C \subseteq \mathbb{R}^d$, $\mathsf{P}_f^\eta(\mathbf{w}) \equiv \mathsf{P}_C(\mathbf{w})$ is the usual (Euclidean) projection onto the set C while $\mathsf{R}_f^\eta(\mathbf{w}) \equiv \mathsf{R}_C(\mathbf{w})$ is the reflection of $\mathbf{w}$ w.r.t. C.

We are now ready to reformulate our main problem of interest (1) in a product space:

$$\min_{\mathbf{w} \in H} \; \mathsf{f}(\mathbf{w}) = \langle \mathbf{1}, \mathbf{f}(\mathbf{w}) \rangle, \quad \text{where} \;\; \mathbf{f}(\mathbf{w}) := \big(f_1(\mathbf{w}_1), \dots, f_m(\mathbf{w}_m)\big), \tag{4}$$

$\mathbf{1}$ is the vector of all 1s, $H := \{\mathbf{w} = (\mathbf{w}_1, \dots, \mathbf{w}_m) \in \mathbb{R}^{dm} : \mathbf{w}_1 = \cdots = \mathbf{w}_m\}$, and we equip with the inner product $\langle \mathbf{w}, \mathbf{z} \rangle := \sum_i \lambda_i \mathbf{w}_i^\top \mathbf{z}_i$. Note that the complement $H^\perp = \{\mathbf{w} : \sum_i \lambda_i \mathbf{w}_i = \mathbf{0}\}$,

$$\mathsf{P}_H(\mathbf{w}) = (\bar{\mathbf{w}}, \dots, \bar{\mathbf{w}}), \quad \text{where} \;\; \bar{\mathbf{w}} := \sum_i \lambda_i \mathbf{w}_i, \;\; \text{and} \;\; \mathsf{R}_H(\mathbf{w}) = (2\bar{\mathbf{w}} - \mathbf{w}_1, \dots, 2\bar{\mathbf{w}} - \mathbf{w}_m). \tag{5}$$

We define the (forward) gradient update map w.r.t. to a (sub)differentiable function f:

$$\mathsf{G}_f^\eta := \mathrm{id} - \eta \cdot \partial f, \qquad \mathbf{w} \mapsto \mathbf{w} - \eta \cdot \partial f(\mathbf{w}). \tag{6}$$

When f is differentiable and convex, we note that $\mathsf{R}_f^\eta = \mathsf{G}_f^\eta \circ \mathsf{P}_f^\eta$ and $\nabla \mathsf{M}_f^\eta = \frac{\mathrm{id} - \mathsf{P}_f^\eta}{\eta}$.

3   FL as Operator Splitting

Following [35], in this section we interpret existing FL algorithms (FedAvg, FedProx, FedSplit) from the operator splitting point of view. We reveal the importance of the step size, obtain new convergence guarantees, and uncover some new and accelerated algorithmic variants.

---

[1] We allow the function f to take ∞ when the input is out of our domain of interest.

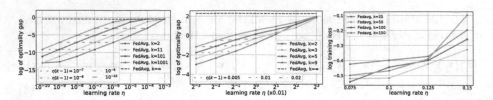

Fig. 1. Optimality gap $\{f(\mathbf{w}^*_{\text{FedAvg}}) - f^*\}$ or training loss $\{f(\mathbf{w}^*_{\text{FedAvg}})\}$ of (approximate) fixed-point solutions of FedAvg for different learning rates η and local epochs k. Different colored lines are for different numbers of local epochs, and dashed lines for different product values $\eta(k-1)$. Left: least squares (closed-form solution); Middle: logistic regression (6000 communication rounds); Right: non-convex CNN on the MNIST dataset (200 communication rounds).

3.1 FedAvg as Forward-Backward Splitting

The FedAvg algorithm of [31] is essentially a k-step version of the forward-backward splitting of [8]:

$$\mathbf{w}_{t+1} \leftarrow \mathsf{P}_H \mathsf{G}^{\eta_t}_{\mathsf{f},k} \mathbf{w}_t, \quad \text{where} \quad \mathsf{G}^{\eta_t}_{\mathsf{f},k} := \underbrace{\mathsf{G}^{\eta_t}_{\mathsf{f}} \circ \mathsf{G}^{\eta_t}_{\mathsf{f}} \circ \cdots \circ \mathsf{G}^{\eta_t}_{\mathsf{f}}}_{k \text{ times}}, \tag{7}$$

i.e. we perform k steps of (forward) gradient updates w.r.t. f followed by 1 step of (backward) proximal update w.r.t. H. The forward step $\mathsf{G}^{\eta_t}_{\mathsf{f},k}$ is performed in parallel by the users while the backward step P_H is performed by the server. Of course, in practice we employ stochastic (i.e. minibatch) approximations of the gradient and sample the users in each communication round.

The number of local steps k turns out to be a key factor: Setting $k = 1$ reduces to the usual (stochastic) forward-backward algorithm and enjoys the well-known convergence guarantees. On the other hand, setting $k = \infty$ (and assuming convergence on each local function f_i) amounts to (repeatedly) averaging the chosen minimizers of f_i's, and eventually converges to the average of minimizers of all m user functions. In general, the performance of the fixed point solution of FedAvg appears to depend on the number of local steps k and the step size η. Let us illustrate this with quadratic functions $f_i(\mathbf{w}) = \frac{1}{2}\|A_i\mathbf{w} - \mathbf{b}_i\|_2^2$ and non-adaptive step size $\eta_t \equiv \eta$, where we obtain the fixed-point $\mathbf{w}^*_{\text{FedAvg}}(k)$ of FedAvg in closed-form [35]:

$$\mathbf{w}^*_{\text{FedAvg}}(k) = \left(\sum_{i=1}^m A_i^\top A_i \cdot \tfrac{1}{k}\sum_{j=0}^{k-1}(I - \eta A_i^\top A_i)^j\right)^{-1}\left(\sum_{i=1}^m \tfrac{1}{k}\sum_{j=0}^{k-1}(I - \eta A_i^\top A_i)^j \cdot A_i^\top \mathbf{b}_i\right). \tag{8}$$

For small η we may apply Taylor expansion and ignore the higher order terms:

$$\tfrac{1}{k}\sum_{j=0}^{k-1}(I - \eta A_i^\top A_i)^j \approx \tfrac{1}{k}\sum_{j=0}^{k-1}(I - j\eta A_i^\top A_i) = I - \tfrac{\eta(k-1)}{2}A_i^\top A_i. \tag{9}$$

Thus, we observe that the fixed point $\mathbf{w}^*_{\text{FedAvg}}(k)$ of FedAvg depends on $\eta(k-1)$, up to higher order terms. In particular, when $k = 1$, the fixed point does not

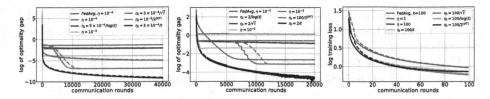

Fig. 2. Effect of step size η and averaging on `FedProx`. Left: least squares; Middle: logistic regression; Right: CNN on MNIST. The dashed and solid lines with the same color show the results obtained with and without the ergodic averaging step in Theorem 1, respectively. For exponentially decaying η_t, we use period T equal to 500 for both least squares and logistic regression experiments, and 10 for CNN experiment.

depend on η (as long as it is small enough to guarantee convergence), but for $k > 1$, the solution that `FedAvg` converges to does depend on η. Moreover, when η is small, the final performance is almost completely determined by $\eta(k-1)$ in this quadratic setting, as we verify in experiments, see Fig. 1 left, where details on generating A_i and $\mathbf{b}_i$ can be found in Appendix A.1.

In Fig. 1, we run `FedAvg` in (7) on least squares, logistic regression and a non-convex CNN on the MNIST dataset [21]. The experiments[2] are run with pre-determined numbers of communication rounds and different configurations of local epochs k and learning rate η. By examining the dependencies of `FedAvg`'s final performance on k and η we conclude that, in general, smaller local epochs and learning rates gives better final solutions (assuming sufficient communication rounds to ensure convergence). Moreover, for least squares and logistic regression, the product $\eta(k-1)$ we derived in (9) almost completely determines the final performance, when k and η are within a proper range. For the nonconvex CNN (Fig. 1, right), especially with limited communications, the approximation in (9) is too crude to be indicative. We note that similar results were also reported in e.g. [10,29].

3.2 `FedProx` as Backward-Backward Splitting

The recent `FedProx` algorithm [23] replaces the gradient update in `FedAvg` with a proximal update:

$$\mathsf{w}_{t+1} = \mathsf{P}_H \mathsf{P}_f^{\eta_t} \mathsf{w}_t, \tag{10}$$

where as before we may use a minibatch to approximate P_f^{η} or select a subset of users to approximate P_H. Written in this way, it is clear that `FedProx` is an instantiation of the backward-backward splitting algorithm of [27,34]. In fact, this algorithm traces back to the early works of *e.g.*[2,11,26], sometimes under the name of the Barycenter method. It was also rediscovered in [45,46] in the ML community under a somewhat different motivation. The recent work [35] pointed

---

[2] Complete details of our experimental setup are given in Appendix A while all non-convex experimental results are averaged over 4 runs with different random seeds.

out that `FedProx` does not solve the original problem (1). While technically correct, this conclusion did not take many other subtleties into account, which we explain next.

Following [5], we first note that, with a constant step size $\eta_t \equiv \eta$, `FedProx` is actually equivalent as `FedAvg` but applied to a "regularized" problem:

$$\min_{\mathbf{w} \in H} \tilde{f}(\mathbf{w}), \quad \text{where} \quad \tilde{f}(\mathbf{w}) := \langle \mathbf{1}, \mathsf{M}_{\mathbf{f}}^{\eta}(\mathbf{w}) \rangle, \quad \text{and} \quad \mathsf{M}_{\mathbf{f}}^{\eta}(\mathbf{w}) = (\mathsf{M}_{f_1}^{\eta}(\mathbf{w}_1), \ldots, \mathsf{M}_{f_m}^{\eta}(\mathbf{w}_m)). \quad (11)$$

Interestingly, [13] proposed exactly (11) for the purpose of personalization, which we now realize is automatically achieved if we apply `FedProx` to the original formulation (1). Indeed, $\nabla \mathsf{M}_{\mathbf{f}}^{\eta}(\mathbf{w}) = [\mathbf{w} - \mathsf{P}_{\mathbf{f}}^{\eta}(\mathbf{w})]/\eta$ hence $\mathsf{G}_{\tilde{f}}^{\eta}(\mathbf{w}) = \mathbf{w} - \eta \nabla \tilde{f}(\mathbf{w}) = \mathsf{P}_{\mathbf{f}}^{\eta}(\mathbf{w}) = (\mathsf{P}_{f_1}^{\eta}(\mathbf{w}_1), \ldots, \mathsf{P}_{f_m}^{\eta}(\mathbf{w}_m))$. This simple observation turns out to be crucial in understanding `FedProx`.

Another important observation is that[3] as $\eta \to 0$, $\mathsf{M}_f^{\eta} \to f$ (pointwise or uniformly if f is Lipschitz) while $\mathsf{M}_f^{\eta} \to \min f$ as $\eta \to \infty$. Therefore as $\eta \to 0$, `FedProx` tends to solve the original problem (1), which can indeed be formalized as follows:

Theorem 1. *Assuming each user participates in every round, the step size η_t is diminishing and non-summable (i.e. $\eta_t \to 0$ and $\sum_t \eta_t = \infty$) and the functions $\{f_i\}$ are convex, then the averaged iterates $\bar{\mathbf{w}}_t := \frac{\sum_{s=1}^{t} \eta_s \mathbf{w}_s}{\sum_{s=1}^{t} \eta_s}$ of FedProx converge to a correct solution of the original problem (1).*

The proof is essentially due to [27,34], which we merely adapt to our `FL` setting. In Appendix B, we give examples to show that this step size condition is both sufficient and necessary. We emphasize that $\bar{\mathbf{w}}_t$ converges to a solution of the original problem (1), not the regularized problem (11). The subtlety is that we must let the step size η_t approach 0 reasonably slowly, a possibility that was not discussed in [35] where they always fixed the step size η_t to a constant η. Theorem 1 also makes intuitive sense, as $\eta_t \to 0$ slowly, we are effectively tracking the solution of the regularized problem (11) which itself tends to the original problem (1): recall that $\mathsf{M}_f^{\eta} \to f$ as $\eta \to 0$.

Even the ergodic averaging step in Theorem 1 can be omitted in some cases:

Theorem 2 ([34]). *Under the same assumptions as in Theorem 1, if $\mathbf{f}$ is strongly convex or the solution set of (1) has nonempty interior, then the vanilla iterates $\mathbf{w}_t$ of FedProx also converge.*

We remark that convergence is in fact linear for the second case. Nevertheless, the above two conditions are perhaps not easy to satisfy or verify in most applications. Thus, in practice, we recommend the ergodic averaging in Theorem 1 since it does not create additional overhead (if implemented incrementally) and in our experiments it does not slow down the algorithm noticeably.

---

[3] These results are classic and well-known, see *e.g.* [40].

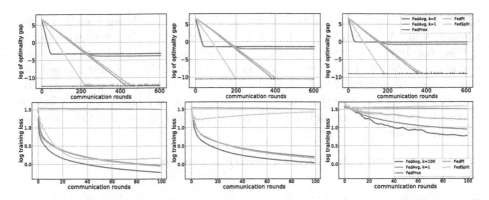

Fig. 3. The effect of data heterogeneity on the performance of different splitting methods; top row: least squares problem. Bottom row: nonconvex CNN model. Top-Left: small data heterogeneity with $H \approx 119 \times 10^3$; Top-Middle: moderately data heterogeneity with $H \approx 7.61 \times 10^6$; Top-Right: large data heterogeneity with $H \approx 190.3 \times 10^6$. Bottom-Left: i.i.d. data distribution; Bottom-Middle: non-i.i.d. data distribution with maximum 6 classes per user; Bottom-Right: non-i.i.d. data distribution with maximum 2 classes per user.

In Fig. 2, we show the effect of step size η on the convergence of FedProx, and compare the results with that of FedAvg on both convex and non-convex models. We run FedProx with both fixed and diminishing step sizes. In the experiments with diminishing step sizes, we have set the initial value of η (i.e. η_0) to larger values - compared to the constant η values - to ensure that η does not get very small after the first few rounds. From the convex experiments (Fig. 2, left and middle), one can see that FedProx with a fixed learning rate converges fast (in a few hundred rounds) to a suboptimal solution. In contrast, FedProx with diminishing η converges slower, but to better quality solutions. It is interesting to note that only when the conditions of Theorem 1 are satisfied (i.e. η diminishes neither too fast nor too slow), FedProx converges to a correct solution of the original problem (1), e.g. see the results for $\eta_t \propto 1/t$ which satisfies both conditions in Theorem 1. Surprisingly, for the nonconvex setting (Fig. 2, right), the best results are achieved with larger learning rates. A similar observation about FedAvg in nonconvex settings was reported in [31, Fig. 5 & 6].[4] Moreover, from the results on both convex and nonconvex models, one can see that ergodic averaging does not affect the convergence rate of FedProx.

3.3 FedSplit as Peaceman-Rachford Splitting

Pathak and Wainwright [35] introduced the FedSplit algorithm recently:

$$w_{t+1} = R_H R_f^{\eta_t} w_t, \tag{12}$$

---

[4] Note that the results of FedAvg and FedProx for $\eta = 100$ and $100/\log(t)$ overlap with each other.

which is essentially an instantiation of the Peaceman-Rachford splitting algorithm [28,36]. As shown by [28], FedSplit converges to a correct solution of (1) if **f** is strictly convex, and the convergence rate is linear if **f** is strongly convex and smooth (and η is small). Pathak and Wainwright [35] also studied the convergence behaviour of FedSplit when the reflector R_f^{η} is computed approximately. However, we note that convergence behaviour of FedSplit is not known or widely studied for nonconvex problems. In particular, we have the following surprising result:

Theorem 3. *If the reflector R_f^{η} is a (strict) contraction, then f must be strongly convex.*

The converse is true if f is also smooth and η is small [15,28]. Therefore, for non-strongly convex or nonconvex problems, we cannot expect FedSplit to converge linearly (if it converges at all).

3.4   FedPi as Douglas-Rachford Splitting

A popular alternative to the Peaceman-Rachford splitting is the Douglas-Rachford splitting [14,28], which, to our best knowledge, has not been applied to the FL setting. The resulting update, which we call FedPi, can be written succinctly as:

$$\mathsf{w}_{t+1} = \frac{\mathsf{w}_t + R_H R_f^{\eta t} \mathsf{w}_t}{2},\tag{13}$$

i.e. we simply average the current iterate and that of FedSplit evenly. Strictly speaking, the above algorithm is a special case of the Douglas-Rachford splitting and was rediscovered by [42] under the name of partial inverse (hence our name FedPi). The moderate averaging step in (13) makes FedPi much more stable:

Theorem 4 ([28,42]). *Assuming each user participates in every round, the step size $\eta_t \equiv \eta$ is constant, and the functions $\{f_i\}$ are convex, then the vanilla iterates w_t of FedPi converge to a correct solution of the original problem (1).*

Compared to FedSplit, FedPi imposes less stringent condition on f_i. However, when f_i is indeed strongly convex and smooth, as already noted by [28], FedPi will be slower than FedSplit by a factor close to $\sqrt{2}$ (assuming the constant step size is set appropriately for both). More importantly, it may be easier to analyze FedPi on nonconvex functions, as recently demonstrated by [39].

We remark that in FedProx we need the step size η_t to diminish in order to converge to a solution of the original problem (1) whereas FedPi achieves the same with a constant step size η, although at the cost of doubling the memory cost at the server side.

In Fig. 3, we compare the performance of different splitting algorithms and how they respond to different degrees of user heterogeneity. We use least squares and a convolutional neural network (CNN) model on MNIST for the convex and nonconvex experiments, respectively. The details of the experimental setup

Table 1. A unifying framework (14)–(16) for FL. Note that (a) FedAvg replaces the proximal update P_f^η with a gradient update $\mathsf{G}_{f,k}^\eta$; (b) ? indicates properties that remain to be studied; (c) "sampling" refers to selecting a subset of users while "stochastic" refers to updating with stochastic gradient.

Algorithm	α	β	γ	$\eta_t \equiv \eta$	$\eta_t \to 0, \sum_t \eta_t = \infty$	nonconvex	sampling	stochastic
FedAvg	1	1	1	Eq. 1	Eq. 1	✓	✓	✓
FedProx	1	1	1	Eq. 11	Eq. 1	✓	✓	✓
FedSplit	2	2	1	Eq. 1	–	?	?	?
FedPi	2	2	$\frac{1}{2}$	Eq. 1	–	✓	?	?

are described in Appendix A.1. For the convex setting, as expected, FedSplit, FedPi, and FedAvg with $k = 1$ achieve the smallest optimality gaps. The performance of all the algorithms deteriorates as users' data become more heterogeneous (see Appendix A.1). For the nonconvex setting, the best results can be achieved by FedProx, FedPi and FedAvg with a big k. It is noteworthy that in the non-convex setting, the performance of FedAvg with $k = 1$ is significantly worse than that with $k = 100$.

4   Unification

The operator splitting view of FL not only allows us to compare and understand the many existing FL algorithms but also opens the door for unifying them. Indeed, we now introduce a grand scheme that unifies all aforementioned FL algorithms:

$$\mathsf{z}_{t+1} = (1 - \alpha_t)\mathsf{u}_t + \alpha_t \mathsf{P}_f^{\eta_t}(\mathsf{u}_t) \tag{14}$$

$$\mathsf{w}_{t+1} = (1 - \beta_t)\mathsf{z}_{t+1} + \beta_t \mathsf{P}_H(\mathsf{z}_{t+1}) \tag{15}$$

$$\mathsf{u}_{t+1} = (1 - \gamma_t)\mathsf{u}_t + \gamma_t \mathsf{w}_{t+1}. \tag{16}$$

Table 1 confirms that the FL algorithms discussed in Sect. 3 are all special cases of this unifying scheme, which not only provides new (adaptive) variants but also clearly reveals the similarities and differences between seemingly different algorithms. In appendix C, we study the effect of α, β and γ and found that γ mostly affects the convergence speed: the closer γ is to 1, the faster the convergence is, while α and β mostly determine the final optimality gap: the closer they are *both* to 2 (as in FedSplit and FedPi), the considerably smaller the final optimality gap is. However, setting only one of them close to 2 only has a minor effect on optimality gap or convergence speed.

5   Conclusions

We have connected FL with the established theory of operator splitting, revealed new insights on existing algorithms and suggested new algorithmic variants and analysis. Our unified view makes it easy to understand, compare and implement different FL algorithms in a streamlined and standardized fashion. Our experiments demonstrate some interesting differences in the convex and nonconvex settings, and in the early and late communication rounds. In the future we plan to study the effect of stochasticity and extend our analysis to nonconvex functions.

Appendix for *Splitting Algorithms for Federated Learning*

A   Experimental Setup

In this section we provide more experimental details that are deferred from the main paper.

A.1   Experimental setup: Least Squares and Logistic Regression

For simulating an instance of the aforementioned least squares and logistic regression problems, we follow the experimental setup of [35].

Least Squares Regression: We consider a set of m devices with local loss functions $f_i(\mathbf{w}) := \frac{1}{2}\|A_i\mathbf{w} - \mathbf{b}_i\|_2^2$, and the main goal is to solve the following minimization problem:

$$\min_{\mathbf{w}\in\mathbb{R}^d} F(\mathbf{w}) := \sum_{i=1}^{m} f_i(\mathbf{w}) = \frac{1}{2}\sum_{i=1}^{m}\|A_i\mathbf{w} - \mathbf{b}_i\|_2^2,$$

where $\mathbf{w} \in \mathbb{R}^d$ is the optimization variable. For each user i, the response vector $\mathbf{b}_i \in \mathbb{R}^{n_i}$ is related to the design matrix $A_i \in \mathbb{R}^{n_i \times d}$ via the linear model

$$\mathbf{b}_i = A_i\mathbf{w}_\star + \epsilon_i,$$

where $\epsilon_i \sim N(0, \sigma^2 I_{n_i})$ for some $\sigma > 0$ is the noise vector. The design matrix $A_i \in \mathbb{R}^{n_i \times d}$ is generated by sampling its elements from a standard normal, $A_i^{k,l} \sim N(0,1)$. For the least squares experiments, we instantiated the problem with the following set of parameters:

$$m = 25, \quad d = 100, \quad n_i = 5000, \quad \sigma^2 = 0.25.$$

Binary Logistic Regression: There are m users and the design matrices $A_i \in \mathbb{R}^{n_i \times d}$ for $i = 1, \ldots, m$ are generated as described before. Each user i has a label vector $\mathbf{b}_i \in \{-1, 1\}^{n_i}$. The conditional probability of observing $\mathbf{b}_{ij} = 1$ (the j-th label in $\mathbf{b}_i$) is

$$\mathbf{P}\{\mathbf{b}_{ij} = 1\} = \frac{e^{\mathbf{a}_{ij}^\top \mathbf{w}_0}}{1 + e^{\mathbf{a}_{ij}^\top \mathbf{w}_0}}, \quad j = 1, \ldots, n_i,$$

where $\mathbf{a}_{ij}$ is the j-th row of A_i. Also, $\mathbf{w}_0 \in \mathbb{R}^d$ is fixed and sampled from $N(0,1)$. Having generated the design matrices A_i and sampled the labels $\mathbf{b}_i$ for all users, we find the maximum likelihood estimate of $\mathbf{w}_0$ by solving the following convex program, which has a solution $\mathbf{w}_*$:

$$\min_{\mathbf{w} \in \mathbb{R}^d} F(\mathbf{w}) := \sum_{i=1}^{m} f_i(\mathbf{w}) = \sum_{i=1}^{m} \sum_{j=1}^{n_i} \log(1 + e^{-b_{ij}\mathbf{a}_{ij}^{\top}\mathbf{w}}) + \frac{\|\mathbf{w}\|_2^2}{2mn_i}.$$

Following [35], we set

$$m = 10, \quad d = 100, \quad n_i = 1000.$$

Data Heterogeneity Measure: We adopt the data heterogeneity measure of [20] to quantify the amount of heterogeneity in users' data for the least squares and logistic regression problems by

$$H := \frac{1}{m} \sum_{i=1}^{m} \| \nabla f_i(\mathbf{w}_*) \|_2^2,$$

where $\mathbf{w}_*$ is a minimizer of the original problem (1). When users' data is homogeneous, all the local functions f_i have the same minimizer of $\mathbf{w}_*$ and $H = 0$. In general, the more heterogeneous the users' data is, the larger H becomes.

Other Parameters: Through the experiments, we used local learning rate $\eta = 10^{-5}$ for least squares, and $\eta = 10^{-2}$ for logistic regression, unless otherwise specified. In addition, FedAvg is run with $k = 5$ for both least squares and logistic regression, unless otherwise specified.

A.2   Experimental setup: MNIST datasets

We consider a distributed setting with 20 users. In order to create a non-i.i.d. dataset, we follow a similar procedure as in [31]: first we split the data from each class into several shards. Then, each user is randomly assigned a number of shards of data. For example, in Fig. 3 to guarantees that no user receives data from more than 6 classes, we split each class of MNIST into 12 shards (i.e., a total of 120 shards for the whole dataset), and each user is randomly assigned 6 shards of data. By considering 20 users, this procedure guarantees that no user receives data from more than 6 classes and the data distribution of each user is different from each other. The local datasets are balanced–all users have the same amount of training samples. The local data is split into train, validation, and test sets with percentage of 80%, 10%, and 10%, respectively. In this way, each user has 2400 data points for training, 300 for test, and 300 for validation. We use a simple 2-layer CNN model with ReLU activation, the detail of which can be found in Table 2. To update the local models at each user using its local data, unless otherwise is stated, we apply gradient descent with $\eta = 0.01$ for FedAvg, and gradient descent with $k = 100$ and $\eta = 0.01$ for proximal update in the splitting algorithms.

Table 2. Details of the CNN model on MNIST.

Layer	Output Shape	# of Trainable Parameters	Activation	Hyper-parameters
Input	$(1, 28, 28)$	0		
Conv2d	$(10, 24, 24)$	260	ReLU	kernel size $= 5$; strides $= (1, 1)$
MaxPool2d	$(10, 12, 12)$	0		pool size $= (2, 2)$
Conv2d	$(20, 8, 8)$	5020	ReLU	kernel size $= 5$; strides $= (1, 1)$
MaxPool2d	$(20, 4, 4)$	0		pool size $= (2, 2)$
Flatten	320	0		
Dense	20	6420	ReLU	
Dense	10	210	softmax	
Total		11910		

B   Proofs

We first recall the following convenient result:

Lemma 1 ([7],[4, Theorem 5.36]). *Let $\{\mathbf{w}_t\}$ and $\{\mathbf{z}_t\}$ be two sequences in $\mathbb{R}^d$, $\emptyset \neq \mathsf{F} \subseteq \mathbb{R}^d$, and $W_k := \mathrm{cl}\,\mathrm{conv}\,(\mathbf{w}_t : t \geq k)$. Suppose that*

1. for all $\mathbf{w} \in \mathsf{F}$, $\|\mathbf{w}_t - \mathbf{w}\|_2^2 \to p(\mathbf{w}) < \infty$;
2. for all k, $\mathrm{dist}(\mathbf{z}_t, W_k) \to 0$ as $t \to \infty$.

Then, the sequence $\{\mathbf{z}_t\}$ has at most one limit point in F. Therefore, if additionally

3. all limit points of $\{\mathbf{z}_t\}$ lie in F,

then the whole sequence $\{\mathbf{z}_t\}$ converges to a point in F.

Below, we use well-known properties about firm nonexpansions and Fejér monotone sequences, see the excellent book [4] for background.

Theorem 1. *Assuming each user participates in every round, the step size η_t is diminishing and non-summable (i.e. $\eta_t \to 0$ and $\sum_t \eta_t = \infty$) and the functions $\{f_i\}$ are convex, then the averaged iterates $\bar{\mathbf{w}}_t := \frac{\sum_{s=1}^t \eta_s \mathbf{w}_s}{\sum_{s=1}^t \eta_s}$ of FedProx converge to a correct solution of the original problem (1).*

Proof. We simply verify Lemma 1.

Let $\mathbf{w} \in \mathrm{dom}\,\mathsf{f} \cap H$, $\mathbf{a}^* \in \partial f(\mathbf{w})$ and $\mathbf{b}^* \in H^\perp$. Applying the firm nonexpansiveness of $\mathsf{P}_f^{\eta_t}$:

$$\|\mathsf{P}_f^{\eta_t}\mathbf{w}_t - \mathbf{w}\|_2^2 = \|\mathsf{P}_f^{\eta_t}\mathbf{w}_t - \mathsf{P}_f^{\eta_t}(\mathbf{w} + \eta_t \mathbf{a}^*)\|_2^2 \tag{17}$$

$$\leq \|\mathbf{w}_t - \mathbf{w} - \eta_t \mathbf{a}^*\|_2^2 - \|\mathbf{w}_t - \mathsf{P}_f^{\eta_t}\mathbf{w}_t - \eta_t \mathbf{a}^*\|_2^2 \tag{18}$$

$$= \|\mathbf{w}_t - \mathbf{w}\|_2^2 - \|\mathbf{w}_t - \mathsf{P}_f^{\eta_t}\mathbf{w}_t\|_2^2 + 2\eta_t \langle \mathbf{w} - \mathsf{P}_f^{\eta_t}\mathbf{w}_t; \mathbf{a}^* \rangle, \tag{19}$$

$$\|\mathsf{P}_H \mathsf{P}_f^{\eta_t}\mathbf{w}_t - \mathbf{w}\|_2^2 \leq \|\mathsf{P}_f^{\eta_t}\mathbf{w}_t - \mathbf{w}\|_2^2 - \|\mathsf{P}_f^{\eta_t}\mathbf{w}_t - \mathsf{P}_H \mathsf{P}_f^{\eta_t}\mathbf{w}_t\|_2^2 + 2\eta_t \langle \mathbf{w} - \mathsf{P}_H \mathsf{P}_f^{\eta_t}\mathbf{w}_t; \mathbf{b}^* \rangle. \tag{20}$$

Summing the above two inequalities and applying the inequality $-\|\mathbf{x}\|_2^2 + 2\langle\mathbf{x};\mathbf{y}\rangle \leq \|\mathbf{y}\|_2^2$ repeatedly:

$$\|\mathsf{P}_H \mathsf{P}_f^{\eta_t} \mathsf{w}_t - \mathsf{w}\|_2^2 \leq \|\mathsf{w}_t - \mathsf{w}\|_2^2 + 2\eta_t \langle\mathsf{w} - \mathsf{w}_t; \mathbf{a}^* + \mathbf{b}^*\rangle + \eta_t^2 [\|\mathbf{a}^* + \mathbf{b}^*\|_2^2 + \|\mathbf{a}^*\|_2^2]. \qquad (21)$$

Summing over t and rearranging we obtain for any $\mathsf{w} \in \operatorname{dom} f \cap H, \mathsf{w}^* = \mathbf{a}^* + \mathbf{b}^*$:

$$2\langle\mathsf{w} - \bar{\mathsf{w}}_t; \mathsf{w}^*\rangle + [\|\mathbf{a}^*\|_2^2 + \|\mathsf{w}^*\|_2^2]\sum_{k=0}^{t}\eta_k^2/\Lambda_t \geq (\|\mathsf{w}_{t+1} - \mathsf{w}\|_2^2 - \|\mathsf{w}_1 - \mathsf{w}\|_2^2)/\Lambda_t, \tag{22}$$

where $\Lambda_t := \sum_{k=1}^{t}\eta_k$. Using the assumptions on η_t we thus know

$$\liminf_{t\to\infty}\ \langle\mathsf{w} - \bar{\mathsf{w}}_t; \mathsf{w}^*\rangle \geq 0. \tag{23}$$

Since w is arbitrary and w^* is chosen to be its subdifferential, it follows that any limit point of $\{\bar{\mathsf{w}}_t\}$ is a solution of the original problem (1), i.e. condition 3 of Lemma 1 holds. If $\{\bar{\mathsf{w}}_t\}$ is bounded, then $\mathsf{F} \neq \emptyset$, which we assume now. Let $\mathsf{w} \in \mathsf{F}$ and set $\mathsf{w}^* = \mathbf{0}$ we know from (21) that $\{\mathsf{w}_t\}$ is quasi-Fejér monotone w.r.t. F (i.e. condition 1 in Lemma 1 holds). Lastly, let $\bar{\eta}_{t,k} := \eta_k/\Lambda_t$ and we verify condition (II) in Lemma 1:

$$\operatorname{dist}(\bar{\mathsf{w}}_t, W_k) \leq \left\|\bar{\mathsf{w}}_t - \sum_{s=k}^{t}\bar{\eta}_{t,s}\mathsf{w}_s / \sum_{\kappa=k}^{t}\bar{\eta}_{t,\kappa}\right\|_2 \tag{24}$$

$$\leq \sum_{\kappa=0}^{k-1}\bar{\eta}_{t,\kappa}\left[\|\mathsf{w}_\kappa\|_2 + \left\|\sum_{s=k}^{t}\bar{\eta}_{t,s}\mathsf{w}_s\right\|_2 / \sum_{\kappa=k}^{t}\bar{\eta}_{t,\kappa}\right] \tag{25}$$

$$\xrightarrow{t\to\infty} 0, \tag{26}$$

since w_t is bounded and for any k, $\bar{\eta}_{t,k} \to 0$ as $t \to \infty$. All three conditions in Lemma 1 are now verified.

We remark that the step size condition is tight, as shown by the following simple example:

Example 1. Let $f_{\pm}(w) = \frac{1}{2}(w \pm 1)^2$. Simple calculation verifies that

$$\mathsf{P}_{f_{\pm}}^{\eta}(w) = \frac{w \mp \eta}{1 + \eta}. \tag{27}$$

Therefore, the iterates of **FedProx** for the two functions f_+ and f_- are:

$$w_{t+1} = \frac{w_t}{1 + \eta_t} = \prod_{\tau=0}^{t}\frac{1}{1 + \eta_\tau}w_0, \tag{28}$$

which tends to the true minimizer $w_\star = 0$ for any w_0 iff

$$\prod_{\tau=0}^{t}\frac{1}{1 + \eta_\tau} \to 0 \iff \sum_{t}\eta_t \to \infty. \tag{29}$$

Table 3. Explanation for the parameter tuning in Fig. 4. For each adaptive tuning experiment, we change the adaptive parameter values from round 0 to round T. After round T, all parameters are fixed (to the values as in `FedSplit`).

Algorithm	α	β	γ	Adaptive parameters setting
α adaptive	adaptive	2	1	$\alpha = \min(1 + \frac{t}{T}, 2)$
β adaptive	2	adaptive	1	$\beta = \min(1 + \frac{t}{T}, 2)$
γ adaptive	2	2	adaptive	$\gamma = \min(\frac{1}{2} + \frac{t}{2T}, 1)$
α, β adaptive	adaptive	adaptive	1	$\alpha = \beta = \min(1 + \frac{t}{T}, 2)$
α, β, γ adaptive	adaptive	adaptive	adaptive	$\alpha = \beta = \min(1 + \frac{t}{T}, 2)$ and $\gamma = \frac{\alpha}{2}$

If we consider instead f_+ and $2f_-$, then

$$2w_{t+1} = \frac{w_t - \eta_t}{1 + \eta_t} + \frac{w_t + 2\eta_t}{1 + 2\eta_t}. \tag{30}$$

Passing to a subsequence if necessary, let $\eta_t \to \eta \neq 0$ and suppose $w_t \to w_\star = \frac{1}{3}$, then passing to the limit we obtain

$$2 = \frac{1 - 3\eta}{1 + \eta} + \frac{1 + 6\eta}{1 + 2\eta} \iff 2(1 + \eta)(1 + 2\eta) = (1 - 3\eta)(1 + 2\eta) + (1 + 6\eta)(1 + \eta) \tag{31}$$

$$\iff \eta = 0, \tag{32}$$

contradiction. Therefore, it is necessary to have $\eta_t \to 0$ in order for `FedProx` to converge to the true solution $w_\star = \frac{1}{3}$ on this example.

Theorem 2. *If the reflector R_f^η is a (strict) contraction, then f must be strongly convex.*

Proof. Since $R_f^\eta = R_{\eta f}$ and f is strongly convex iff ηf is so, w.l.o.g. we may take $\eta = 1$. Suppose R_f is γ-contractive for some $\gamma \in (0, 1)$, i.e. for all $\mathbf{w}$ and $\mathbf{z}$:

$$\|R_f \mathbf{w} - R_f \mathbf{z}\|_2 \leq \gamma \cdot \|\mathbf{w} - \mathbf{z}\|_2. \tag{33}$$

It then follows that the proximal map $P_f = \frac{\text{id} + R_f}{2}$ is $\frac{1+\gamma}{2}$-contractive. Moreover, $\frac{2}{1+\gamma} P_f$, being nonexpansive, is the gradient of the convex function $\frac{2}{1+\gamma} M_{f^*}$. Thus, $\frac{2}{1+\gamma} P_f$ is actually firmly nonexpansive [4, Corollary 18.17]. But,

$$\frac{2}{1+\gamma} P_f = \frac{2}{1+\gamma}(\text{id} + \partial f)^{-1} = [\text{id} + (\partial f - \frac{1-\gamma}{1+\gamma}\text{id}) \circ \frac{1+\gamma}{2}\text{id}]^{-1}, \tag{34}$$

and hence $(\partial f - \frac{1-\gamma}{1+\gamma}\text{id}) \circ \frac{1+\gamma}{2}\text{id}$ is maximal monotone [4, Proposition 23.8], i.e. f is $\frac{1-\gamma}{1+\gamma}$-strongly convex.

C   Over-relaxation

The parameters α_t, β_t and γ_t are fixed to various constants in existing FL algorithms (see Table 1), but of course in practice we may adaptively adjust them. In general, we keep the following principle in mind: choosing these parameters in $[0, 1]$ leads to more stable and potentially slower variants (known as under-relaxation in numerical analysis) while choosing them in $[1, 2]$ leads to potentially faster but less stable variants (known as over-relaxation). To see why over-relaxation may accelerate convergence, let us use (14) as an example. Reversing time, a (backward descending) proximal update becomes a (forward ascending) gradient update:

$$\tilde{u}_t := P_f^{\eta_t}(u_t) \iff u_t = \tilde{u}_t + \eta_t \cdot \partial f(\tilde{u}_t), \tag{35}$$

whence follows from (14) that

$$z_{t+1} = \tilde{u}_t - (\alpha_t - 1)\eta_t \cdot \partial f(\tilde{u}_t). \tag{36}$$

Thus, if $\alpha_t \geq 1$, through over-relaxation we obtain a free gradient *descent* update on top of the proximally updated $\tilde{u}_t$, which could further decrease the function value of f. When f is L-smooth and σ-strongly convex, a suitable choice is to set

$$(\alpha_t - 1)\eta_t = \frac{2}{\sigma+L} \implies \alpha_t = 1 + \frac{2}{\eta_t(\sigma+L)}. \tag{37}$$

In contrast, when $\alpha_t \in [0, 1]$, through under-relaxation we obtain a free gradient *ascent* step which is likely to increase the function value of f. Nevertheless, averaging does make the update more stable. To summarize, in practice we recommend starting with relaxation parameters (i.e. $\alpha_t, \beta_t, \gamma_t$) less than 1 and gradually increasing them to 2 as we approach the solution.

Effect of Over-Relaxation. As can be seen from Fig. 4, setting γ to values close to 1 enhances the speed of convergence to the final solution. However, it does not affect the final optimality gap (final solution quality). On the other hand,

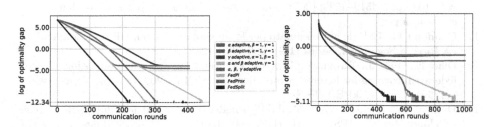

Fig. 4. The effect of parameters α, β and γ on the final optimality gap and the convergence speed of splitting algorithms. Left: least squares; Right: logistic regression. In each experiment, we change the value of adaptive parameters as in Table 3. T is equal to 200 and 600, for least squares and logistic regression, respectively.

α and β mostly determine the final optimality gap in the sense that setting *both* α and β to values close to 2 reduces the final optimality gap considerably (the difference between `FedSplit` and `FedPi` algorithms compared to `FedProx`). However, setting only one of them to values close to 2 has a minor effect on the final optimality gap or the convergence speed.

References

1. Augenstein, S., et al.: Generative models for effective ml on private, decentralized datasets. In: ICLR (2020). https://openreview.net/forum?id=SJgaRA4FPH
2. Auslender, A.: Méthodes Numériques pour la Résolution des Problèmes d'Optimisation avec Contraintes. Ph.D. thesis, Faculté des Sciences, Grenoble, France (1969)
3. Bagdasaryan, E., Veit, A., Hua, Y., Estrin, D., Shmatikov, V.: How to backdoor federated learning. In: AISTATS. Proceedings of Machine Learning Research, vol. 108, pp. 2938–2948 (2020). http://proceedings.mlr.press/v108/bagdasaryan20a.html
4. Bauschke, H.H., Combettes, P.L.: Convex Analysis and Monotone Operator Theory in Hilbert Spaces. Springer, 2nd edn. (2017). https://link.springer.com/book/10.1007/978-3-319-48311-5
5. Bauschke, H.H., Combettes, P.L., Reich, S.: The asymptotic behavior of the composition of two resolvents. Nonlinear Anal. Theory Methods Appl. **60**(2), 283–301 (2005). https://doi.org/10.1016/j.na.2004.07.054
6. Bhagoji, A.N., Chakraborty, S., Mittal, P., Calo, S.: Analyzing federated learning through an adversarial lens. In: ICML, vol. 97, pp. 634–643 (2019). http://proceedings.mlr.press/v97/bhagoji19a.html
7. Brézis, H., Browder, F.E.: Nonlinear ergodic theorems. Bull. Am. Math. Soc. **82**(6), 959–961 (1976). https://projecteuclid.org:443/euclid.bams/1183538367
8. Bruck, R.E.: On the weak convergence of an ergodic iteration for the solution of variational inequalities for monotone operators in hilbert space. J. Math. Anal. Appl. **61**(1), 159–164 (1977). https://doi.org/10.1016/0022-247X(77)90152-4
9. Caldas, S., Wu, P., Li, T., Konecny, J., McMahan, H.B., Smith, V., Talwalkar, A.: Leaf: a benchmark for federated settings (2018). https://arxiv.org/abs/1812.01097,arXiv:1812.01097
10. Charles, Z., Konečný, J.: Convergence and accuracy trade-offs in federated learning and meta-learning. In: AISTATS. Proceedings of Machine Learning Research, vol. 108, pp. 4519–4529 (2021). http://proceedings.mlr.press/v130/charles21a/charles21a.pdf
11. Cimmino, G.: Calcolo approssimato per le soluzioni dei sistemi di equazioni lineari. La Ricerca Scientifica, Series **II**(9), 326–333 (1938)
12. Diao, E., Ding, J., Tarokh, V.: Heterofl: Computation and communication efficient federated learning for heterogeneous clients. In: ICLR (2021). https://openreview.net/forum?id=TNkPBBYFkXg
13. Dinh, C.T., Tran, N., Nguyen, J.: Personalized federated learning with moreau envelopes. In: NeurIPS, pp. 21394–21405 (2020). https://proceedings.neurips.cc/paper/2020/file/f4f1f13c8289ac1b1ee0ff176b56fc60-Paper.pdf
14. Douglas, J., Jr., Rachford, H.H., Jr.: On the numerical solution of heat conduction problems in two and three space variables. Trans. Am. Math. Soc. **82**(2), 421–439 (1956). https://doi.org/10.2307/1993056

15. Gabay, D.: Applications of the method of multipliers to variational inequalities **15**(9), 299–331 (1983). https://doi.org/10.1016/S0168-2024(08)70034-1
16. He, C., et al.: FedML: a research library and benchmark for federated machine learning (2020). arXiv:2007.13518
17. Hu, Z., Shaloudegi, K., Zhang, G., Yu, Y.: Fedmgda+: federated learning meets multi-objective optimization (2020). arXiv:2006.11489
18. Kairouz, P., McMahan, H.B., Avent, B.: Advances and open problems in federated learning (2019). arXiv:1912.04977
19. Khaled, A., Mishchenko, K., Richtárik, P.: First analysis of local gd on heterogeneous data (2020). arXiv:1909.04715
20. Khaled, A., Mishchenko, K., Richtarik, P.: Tighter theory for local SGD on identical and heterogeneous data. In: AISTATS. Proceedings of Machine Learning Research, vol. 108, pp. 4519–4529 (2020). http://proceedings.mlr.press/v108/bayoumi20a.html
21. LeCun, Y., Cortes, C., Burges, C.: Mnist handwritten digit database (2010). http://yann.lecun.com/exdb/mnist, available Under the Terms of the Creative Commons Attribution-Share Alike 3.0 License
22. Li, T., Sahu, A.K., Talwalkar, A., Smith, V.: Federated learning: Challenges, methods, and future directions (2019). arXiv:1908.07873
23. Li, T., Sahu, A.K., Zaheer, M., Sanjabi, M., Talwalkar, A., Smith, V.: Federated optimization in heterogeneous networks. In: Proceedings of Machine Learning and Systems, vol. 2, pp. 429–450 (2020). https://proceedings.mlsys.org/paper/2020/file/38af86134b65d0f10fe33d30dd76442e-Paper.pdf
24. Li, T., Sanjabi, M., Beirami, A., Smith, V.: Fair resource allocation in federated learning. In: ICLR (2020). https://openreview.net/forum?id=ByexElSYDr
25. Li, X., Huang, K., Yang, W., Wang, S., Zhang, Z.: On the convergence of fedavg on non-iid data. In: ICLR (2020). https://openreview.net/forum?id=HJxNAnVtDS
26. Lions, J.L., Temam, R.: Une méthode d'éclatement pes opérateurs et des contraintes en calcul des variations. Comptes rendus mathématiques de l'Académie des Sciences, Paris 263, 563–565 (1966). https://gallica.bnf.fr/ark:/12148/bpt6k6426017v/f241
27. Lions, P.L.: Une methode iterative de resolution d'une inequation variationnelle. Israel J. Math. **31**(2), 204–208 (1978). https://doi.org/10.1007/BF02760552
28. Lions, P.L., Mercier, B.: Splitting algorithms for the sum of two nonlinear operators. SIAM J. Numer. Anal. **16**(6), 964–979 (1979). https://doi.org/10.1137/0716071
29. Malinovskiy, G., Kovalev, D., Gasanov, E., Condat, L., Richtarik, P.: From local SGD to local fixed-point methods for federated learning. In: ICML, vol. 119, pp. 6692–6701 (2020). http://proceedings.mlr.press/v119/malinovskiy20a.html
30. Mansour, Y., Mohri, M., Ro, J., Suresh, A.T.: Three approaches for personalization with applications to federated learning (2020). https://arxiv.org/abs/2002.10619, arXiv:2002.10619
31. McMahan, B., Moore, E., Ramage, D., Hampson, S., y Arcas, B.A.: Communication-efficient learning of deep networks from decentralized data. In: AISTATS, vol. 54, pp. 1273–1282 (2017). http://proceedings.mlr.press/v54/mcmahan17a/mcmahan17a.pdf
32. Mohri, M., Sivek, G., Suresh, A.T.: Agnostic federated learning. In: ICML, vol. 97, pp. 4615–4625 (2019). http://proceedings.mlr.press/v97/mohri19a.html

33. Nasr, M., Shokri, R., Houmansadr, A.: Comprehensive privacy analysis of deep learning: Passive and active white-box inference attacks against centralized and federated learning. In: IEEE Symposium on Security and Privacy (SP), pp. 739–753 (2019). https://doi.org/10.1109/SP.2019.00065
34. Passty, G.B.: Ergodic convergence to a zero of the sum of monotone operators in hilbert space. J. Math. Anal. Appl. **72**(2), 383–390 (1979). https://doi.org/10.1016/0022-247X(79)90234-8
35. Pathak, R., Wainwright, M.J.: Fedsplit: an algorithmic framework for fast federated optimization. In: NeurIPS (2020). https://proceedings.neurips.cc//paper/2020/hash/4ebd440d99504722d80de606ea8507da-Abstract.html
36. Peaceman, D.W., Rachford, Jr., H.H.: The numerical solution of parabolic and elliptic differential equations. J. Soc. Ind. Appl. Math. **3**(1), 28–41 (1955). https://www.jstor.org/stable/2098834
37. Qiang, Y.: Federated recommendation systems. In: IEEE International Conference on Big Data, pp. 1–1 (2019). https://doi.org/10.1109/BigData47090.2019.9005952
38. Reddi, S., et al.: Adaptive Federated Optimization (2020). arXiv:2003.00295
39. Rockafellar, R.T.: Progressive decoupling of linkages in optimization and variational inequalities with elicitable convexity or monotonicity. Set-Valued Variational Anal. **27**, 863–893 (2019). https://doi.org/10.1007/s11228-018-0496-1
40. Rockafellar, R.T., Wets, R.J.B.: Variational Analysis. Springer (1998). https://doi.org/10.1007/978-3-642-02431-3
41. Smith, V., Chiang, C.K., Sanjabi, M., Talwalkar, A.S.: Federated multitask learning. In: NeurIPS (2017). https://papers.nips.cc/paper/2017/hash/6211080fa89981f66b1a0c9d55c61d0f-Abstract.html
42. Spingarn, J.E.: Partial inverse of a monotone operator. Appl. Math. Optim. **10**, 247–265 (1983). https://doi.org/10.1007/BF01448388
43. Spingarn, J.E.: Applications of the method of partial inverses to convex programming: decomposition. Math. Program. **32**, 199–223 (1985). https://doi.org/10.1007/BF01586091
44. Yang, Q., Liu, Y., Chen, T., Tong, Y.: Federated machine learning: concept and applications. ACM Trans. Intell. Syst. Technol. **10**(2) (2019). https://doi.org/10.1145/3298981
45. Yu, Y.: Better approximation and faster algorithm using the proximal average. In: NeurIPS (2013). https://proceedings.neurips.cc/paper/2013/file/49182f81e6a13cf5eaa496d51fea6406-Paper.pdf
46. Yu, Y., Zheng, X., Marchetti-Bowick, M., Xing, E.P.: Minimizing nonconvex non-separable functions. In: AISTATS, vol. 38, pp. 1107–1115 (2015). http://proceedings.mlr.press/v38/yu15.html
47. Yurochkin, M., Agarwal, M., Ghosh, S., Greenewald, K., Hoang, N., Khazaeni, Y.: Bayesian nonparametric federated learning of neural networks. In: ICML, vol. 97, pp. 7252–7261 (2019). http://proceedings.mlr.press/v97/yurochkin19a.html
48. Zhang, M., Sapra, K., Fidler, S., Yeung, S., Alvarez, J.M.: Personalized federated learning with first order model optimization. In: ICLR (2021). https://openreview.net/forum?id=ehJqJQk9cw

Migrating Models: A Decentralized View on Federated Learning

Péter Kiss[1]([✉])(iD) and Tomáš Horváth[1,2](iD)

[1] Department of Data Science and Engineering, ELTE – Eötvös Loránd University, Faculty of Informatics, Pázmány Péter sétány 1/C., Budapest 1117, Hungary
peter.kiss@inf.elte.hu
[2] Pavol Jozef Šafárik University, Faculty of Science, Institute of Computer Science, Jesenná 5, 040 01 Košice, Slovakia

Abstract. Federated learning (FL) researches attempt to alleviate the increasing difficulty of training machine learning models, when the training data is generated in a massively distributed way. The key idea behind these methods is moving the training to locations of data generation, and periodically collecting and redistributing the model updates. We present our approach for transforming the general training algorithm of FL into a peer-to-peer-like process. Our experiments on baseline image classification datasets show that omitting central coordination in FL is feasible.

Keywords: Federated learning · Peer-to-peer · Neural networks

1 Introduction

The goal of supervised machine learning (ML) is to predict some missing values or attributes (labels) of data points given observed values (input features) using some model of the data distribution.

Given this model, the learning process usually corresponds to empirical risk minimization, that aims at finding its parameters $\mathbf{w} \in \mathbb{R}^d$, that are able to minimize a loss function l over the training data, which tells us how the learned model distribution differs from the data distribution.

Nowadays, without doubt, artificial neural networks (NNs) are the most popular ML models due to their applicability for a wide range of tasks and to the end-to-end nature of their learning process. In this paper we focus on federated training of this class of models (specifically for image classification). However, as we will indicate, the proposed approach can be applied, with more or less restrictions, to a broader family of ML models as well.

In case of a general loss surface, as that one of NNs, training happens using versions of mini-batch gradient descent, where the weights $\mathbf{w}$ are iteratively

This work was partially supported by the project "Application Domain Specific Highly Reliable IT Solutions" financed by the National Research, Development and Innovation Fund of Hungary (TKP2020-NKA-06).

© Springer Nature Switzerland AG 2021
M. Kamp et al. (Eds.): ECML PKDD 2021 Workshops, CCIS 1524, pp. 177–191, 2021.
https://doi.org/10.1007/978-3-030-93736-2_15

changed (moved) in the negative direction of gradients computed from the loss on a random subset (*mini-batch* $\mathcal{B}$) of training data:

$$\mathbf{w}_{t+1} = \mathbf{w}_t - \eta_t \nabla l^{\mathcal{B}}(\mathbf{w}_t). \tag{1}$$

This training method requires huge amount of training data and computational resources. In fact, without the loss on generality, we can say that the more the data and training time the better the model becomes.

As the amount of data to be processed has been growing at a higher rate than the computation and storage power of machines (where the training takes place), parallelization of learning gained more and more importance. *Data parallel* distributed training methods of NNs [5, 7, 23] focused, at first place, on data center based learning and the exploitation of multi-core or multi-GPU architectures but later gave a framework for distributed learning in data centers.

Formally, the setup of distributed ML can be described as follows: given a set of nodes $\mathcal{V} = \{v^1, v^2, \ldots, v^K\}$ with $K = |\mathcal{V}|$ and n data points allocated into sets $\mathcal{D}^k$ of indices of data points stored at nodes v^k ($1 \leq k \leq K$) with $n^k = |\mathcal{D}^k|$ being the number of data points at the node v^k. Without the loss of generality we usually assume that $\mathcal{D}^k \cap \mathcal{D}^l = \emptyset$ whenever $l \neq k$, thus $n = \sum_{k=1}^{K} n^k$. The task at hand, defining the local loss for node v^k as $l^k(\mathbf{w}) = \frac{1}{n^k} \sum_{i \in \mathcal{D}^k} l_i(\mathbf{w})$, is to solve the following optimization problem:

$$\min_{\mathbf{w} \in \mathbb{R}^d} l(\mathbf{w}) = \sum_{k=1}^{K} \frac{n^k}{n} l^k(\mathbf{w}) \tag{2}$$

where $l_i(\mathbf{w})$ denotes the loss on ith data point $(\mathbf{x}_i, y_i)$ given the parametrization $\mathbf{w}$, with $y_i \in \mathcal{C}$, where $\mathcal{C}$ stands for the range of the missing values, in our case the set of classes of images.

To solve the problem in Eq. 2 for non convex losses, as that of in cases of NNs, the most widely used methods are different versions of distributed mini-batch gradient descent, where a coordinator collects and aggregates the gradients (Eq. 1) coming from the k different nodes:

$$\mathbf{w}_{t+1} = \mathbf{w}_t - \eta_t \frac{1}{K} \sum_{k=1}^{K} \nabla l^{\mathcal{B}^k}(\mathbf{w}_t). \tag{3}$$

After the aggregation is performed, the new model with parameters $\mathbf{w}_{t+1}$ will be redistributed to the worker nodes, and this loop continues until a reasonably good global performance is reached.

The idea of FL [19] was that we could exploit the computation power of the huge amount of user devices for the training, where the data is generated anyway. This way we can better protect users' privacy and, in the same time, save the data centers from the not negligible burden of storing and processing such huge amount of data.

The setup of FL, thus, differs from the traditional distributed mini-batch gradient descent (MBGD) based training in the following characteristics: (i) the number of nodes can be much larger than the average number of training instances stored on a single node, (ii) the data on each node can be drawn from a different distribution, and, (iii) the number of data instances on different nodes may vary by orders of magnitude.

Since here the process is distributed over geographically widespread network, instead of a data center, the cost of the communication becomes a real problem. Thus, nodes execute multiple updates on their local models before they send their "new" models (or, equivalently, the differences to the old models) to the coordinator:

$$\mathbf{w}_{t+1} = \mathbf{w}_t - \frac{n}{n^k} \sum_{k=1}^{K} \Delta^{(k)}, \text{ with } \Delta^{(k)} = \sum_{i=0}^{r} \nabla l^{\mathcal{B}_{t_i}^k}(w_{t_i}^k), \tag{4}$$

where $\mathbf{w}_{t_{i+1}}^{\mathbf{k}} = \mathbf{w}_{t_i}^{\mathbf{k}} - \eta \nabla l^{\mathcal{B}_{t_i}^k}(\mathbf{w}_{t_i}^k)$, $\mathbf{w}_{t_0}^k = \mathbf{w}_t$ and $r = \beta \cdot n^k/|\mathcal{B}|$ the number of local updates, for an epoch number β, and batch size $|\mathcal{B}|$.

In Federated Averaging (FedAvg) [23], that is most probably the most widely used method of FL, the global model is distributed across only a random subset of nodes, each being chosen with probability γ. In [23] it has been empirically shown that with a value $\gamma = 0.1$ this method reaches, if not outperforms, the convergence rate of full aggregation, despite of the very significant communication and computation savings.

1.1 Related Work

Since, according to the problem statement of FL, the number of nodes K is extremely large, the management of these nodes even with the sub-setting that is introduced in FedAvg might be a challenging task. Consequently one of the most apparent practical issue of FedAvg is delays that stem from the centralized synchronous nature of the algorithm. Delays are introduced by *overcrowded channels* around the coordinator and by so-called *struggling nodes*, both leading to a slow-down in the training process. To solve this problem a variety of techniques have been proposed providing strategies to orchestrate updates, such that, applying *traditional scheduling techniques* [34]; using *federated client selection* [24] that prefers nodes with the best communication and computation capabilities; dynamically adapting of training scheme to available resources [31]; or allowing *asynchronous communication patterns* [6]. Another viable way for reducing communication burden seeks to decrease the size of data to be communicated through some kind of *quantization* [1,2,9,25,27]. Methods for compressing updates also include techniques that build on characteristics of NN training, such that pruning updates [10] through variational dropout [18] or evaluating the importance of parameter layers in NNs [6].

Assuming the convexity of the loss function, a range of innovative methods has been proposed, mostly built on the "communication-efficient distributed dual coordinate ascent" (CoCoA) framework using dual optimization [16], with the main goal to minimize the number of communication rounds during the learning.

A different perspective to deal with communication difficulties is to *decentralize the training* across multiple parameter servers as it has been already proposed in DistBelief [7]. Going further, for convex loss functions, a number of completely peer-to-peer gossip-based [8] asynchronous algorithms has been proposed such as [11], to mention only one example. Dual optimization has been used in peer-to-peer setup as well, for example the alternating direction method of multipliers [4,32] in [3,28].

A second important issue in FL is the degradation of performance, which might be caused by *weight divergence* [36,38], resulting from averaging the updates. An interesting approach to tackle the problem is the so-called neuron matching [36], to mention only one work. Another cause of the dropped performance might be the fact, that updates applied on the common model of FL can be viewed as a MBGD with updates being computed over extremely large "mini batches", that can cause serious generalization gaps [22].

In general, higher number of participating nodes, bigger local batch sizes and stronger divergence in the local data distributions lead to degraded performance and, especially for more complex tasks and models, they often prevent the system to learn any interpretable consensus model.

1.2 Our Contribution

The two migrating model (MM) approaches we present in this paper can be derived from FedAvg in the following way: (i) splitting the coordinator into multiple smaller processes, that is, multiple "coordinators" collect the updates from a smaller amount of nodes (per coordinator) and (ii) instead of having a fixed node for the coordinator, the models to be trained are passed from node to node, each adding its update when it "owns" the model. Finally, (iii) to simulate the model averaging step, the updates are written in a buffer and their average will be applied on the model with a predefined frequency. (iv) We will also discuss a special case of the MM approach, in which we omit the buffer and apply the updates without delay.

The contributions of the presented MM approach are:

1. reduction of the complications of centralized FL through evening the communication burden over the whole network by simulating FedAvg in a peer-to-peer environment;
2. by reducing the number of updates used for model averaging (or excluding averaging in the special case) and, thus, using smaller effective batch sizes (at the price of more biased updates) it is possible to dramatically reduce the general communication and computation cost in exchange for, according to our empirical results, only a slightly worst training efficiency;
3. using a simple mechanism to involve those nodes in the training process, that promise the most performance gain, incentivating training on as diverse data as possible (closer to iid. wrt. whole distribution);
4. any node can initiate an optimization of new models, so the network optimizes various models in parallel, completely asynchronously.

For inference, the nodes can use some of the previously seen models, potentially in an ensemble fashion to obtain a kind of "global model", with competitive performance to FedAvg, at least according to our experiments. For performance evaluation at first place we wanted to compare our method to FedAvg over at least as many nodes, as in our methods, thus we used bagging ensembles, along with measurements for the performance of a single model.

In a real-world scenario, a *tracker* can be deployed to provide information about the network, the various models and their migration within the network.

2   Migrating Models (MM)

Let us assume that nodes $v^1, v^2, \ldots, v^{K'} \in V$ participating in the training are organized into a graph $G(V, E)$ and the models with parameters $\mathbf{w}^1, \ldots, \mathbf{w}^{K'}$ ($K' \leq K$) are travelling along the edges $e \in E$ of the graph.

The initial phase of the learning algorithm starts with random initialization of K' model parameter sets $\mathbf{w}^1, \ldots, \mathbf{w}^{K'}$ at a subset of nodes (in our simulation, K' is a hyper-parameter, in the real world it might change dynamically). These initial weights need not be aligned across the nodes. Different initializations may even help to explore a bigger portion of the parameter space. As long as the input and output dimensions are aligned across the system, any kind of models can be used in the presented MM approach, whose training is done by MBGD (i.e. not only NNs).

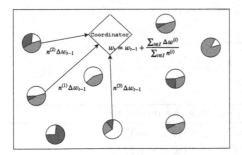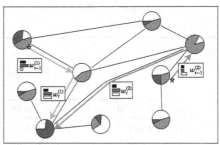

Fig. 1. Visualisation of FedAvg (left) and MM (right) approaches. Pie charts represent data distributions at nodes with colors corresponding to classes. FedAvg randomly picks some nodes, refines the common model on them, then adds the averaged update to it. In MM, a node initiates a model (indicated by stars referring to two different models), trains it on the local data and computes the belief vector about its performance (rectangle with confidence per class as the colored bars). In the following step the model will be moved to the next available node with the largest expected gain.

Updating the models, as visualized in the Fig. 1, happens in a completely asynchronous manner: When a model has been trained on a partial set of data, it looks for an appropriate next location to move, preferably a node, where the data is likely to help improving the model.

Choosing Next Location. To decide to which node should a model $\mathbf{w}^k$ be moved for further training, each model maintains a *belief vector* $\mathbf{p}^k \in \mathbb{R}^C$, that corresponds to a guess for its (per class) performance on the global data for the C classes.

All the entries of $\mathbf{p}^k$ start with 0 and, after training on a dataset, the belief vector is updated according to the Algorithm 1. The idea is to set believes to performance of the updated model on an i.i.d. test subset of local data, if the class is present, and discount the values of the absent ones, since we expect this performance to decrease. After obtaining the new belief vector, a new location for the model $\mathbf{w}^k$ is chosen, based on data sets residing at the neighboring nodes $N_G(v^k)$ from its current host node v^k. Selection of the next node to move the model to, is based on an expected performance gain, that is described in the Algorithm 2, where I is the indicator function returning 1 if its parameter is true. In this step the node broadcasts its belief vector to the neighbours, who will push up the expected performance on classes that they possess (we set it to 1 for simplicity), and send the sum of the entries back. The intuition is that the highest sum promises the most performance gain.

Algorithm 1. Updating the belief vector $\mathbf{p}^k$ of the model $\mathbf{w}^k$ based on F1-scores $\mathbf{t}^k$ on a test set of the given node k, with a discount rate ξ

1: **procedure** BELIEFUPDATE($\mathbf{p}^k = (p_1^k, p_2^k, \ldots, p_C^k), \mathbf{t^k} = (t_1^k, t_2^k, \ldots, t_C^k)$)
2:     $\mathbf{p^k} \leftarrow (1 - \xi)\mathbf{p^k}$
3:     **for** $c \leftarrow 1$ to C **do**
4:         $p_c^k \leftarrow \max\{p_c^k, t_c^k\}$
5:     **return** $\mathbf{p}^k$

Algorithm 2. Finding the best node to migrate the model to

---

procedure GETMAXBENEFITNODE($v^k, \mathbf{p}^k = (p_1^k, p_2^k, \ldots, p_C^k)$)

$$j^* \leftarrow \arg\max_{j|v^j \in N_G(v^k)} \underbrace{\sum_{c \in C} \max\{I(\exists i \in \mathcal{D}^j(y_i = c)), p_c^k\}}_{\text{benefit at the node } v^j \text{ (computed at the node } v^j)}$$

    return v^{j^*}

Update Buffer. To simulate the aggregation step, for each model $\mathbf{w}^k$ we specify a buffer size σ^k that defines that, in a given "training round" r, how many nodes the model should visit before the aggregated (weighted averaged) update will be applied on its parameters $\mathbf{w}_{r-1}^k$. By updating the believes before each model relocation, we hope to get an aggregated update more similar to one that would have been resulted from an i.i.d. training run (wrt. general distribution).

Algorithm 3. Migrating models – Experimental algorithm K' – number of models β, η – number of epochs and the learning rate at local training

1: **procedure** TRAINING(K', β, η)
2:    initialize graph $G(V, E)$
3:    $v^1, \ldots, v^{K'} \leftarrow$ Pick K' nodes randomly from G
4:    initialize belief vectors $\mathbf{p}^1, \ldots, \mathbf{p}^{K'} \leftarrow \mathbf{0}$
5:    initialize buffer sizes $\sigma^1, \ldots, \sigma^{K'} \leftarrow 1$
6:    initialize $\mathbf{w}_0^1, \ldots, \mathbf{w}_0^{K'}$ randomly
7:    $r \leftarrow 0$
8:    **repeat**
9:       **for all** $k \in \{1, \ldots, K'\}$ **do** in parallel
10:         $\delta \leftarrow 0$
11:         count $\leftarrow 0$
12:         $l_r^k \leftarrow$ accuracy($\mathbf{w}_r^k, \mathcal{D}^k$)   ▷ Test accuracy at current node's test set
13:         $\sigma^k \leftarrow$ UpdateBufferSize(l_r^k, σ^k)   ▷ Extend the buffer size if needed
14:         **for** $s \leftarrow 1$ to σ^k **do**   ▷ Collecting the σ^k updates
15:            $\delta \leftarrow \delta +$ ClientUpdate($k, \mathbf{w}_r^k, \beta, \eta$) $* n^k$
16:            count $\leftarrow$ count $+ |D^k|$
17:            $\mathbf{t}^k \leftarrow$ Test($\mathcal{D}^k, \mathbf{w}_r^k$)   ▷ Test the F1-score of $\mathbf{w}$ on $\mathcal{D}^k$
18:            $\mathbf{p}^k \leftarrow$ BeliefUpdate($\mathbf{p}^k, \mathbf{t}^k$)
19:            $v^k \leftarrow$ GetMaxBenefitNode($v^k, \mathbf{p}^k$)
20:         $\mathbf{w}_{r+1}^k \leftarrow \frac{\delta}{\text{count}}$
21:      $r \leftarrow r + 1$
22:    **until** stop

Algorithm 4. Migrating models – Client update

1: **procedure** CLIENTUPDATE($k, \mathbf{w}, \beta, \eta$)
2:    $\mathbf{w}' \leftarrow \mathbf{w}$
3:    $B \leftarrow$ split$\mathcal{D}^k$ to set of batches
4:    **for** each local epoch i from 1 to β **do**
5:       **for all** $\mathcal{B} \in B$ **do**
6:          $\mathbf{w}' \leftarrow \mathbf{w}' - \eta \nabla^{\mathcal{B}} l(\mathbf{w})$
7:    **return** $\mathbf{w}'$

To keep the computation and communication costs low, we start from $\sigma^k = 1$ for all $k = 1, \ldots, K'$. Then, during the training, we monitor the improvement of the loss functions l^k and, based on some heuristics, we extend the buffer if we experience that performance is not improving at the expected rate[1]. The MM method is described in the Algorithms 3 and 4.

Simple Migrating Models (sMM). As a specific case of MM, we also experimented with the simplest possible setup, denoted here as sMM, where $\sigma = 1$. In this case,

---

[1] A simple heuristic we have used here was to extend the buffer size if the exponential moving average of the model performance on new datasets (on the new nodes) shows no improvement after θ steps, where θ is a hyper-parameter.

at the price of highly biased updates, we can further reduce the communication and training costs of MM for a single update. Besides, there are two factors which are believed to decrease the performance of FedAVG such that (i) using large batch sizes and (ii) performing model averaging. In sMM, which for models with MBGD training is equivalent to a single node MBGD, both of these factors are excluded. It might be worth to note, that since we are not averaging over the updates of the nodes, sMM can be applied for different training methods, consequently different models as well (for example decision trees).

3 Experiments

Data Sets and Models. Experiments were run with three image classification task such that MNIST, Fashion-MNIST and CIFAR-10. The number of nodes for the two simplest cases, MNIST and Fashion-MNIST, was set to 200. For CIFAR-10, due to heavy computational load and the danger of stagnation in learning process, the number of nodes was set to 50. Similarly to the experimental settings in [23], 90% of each data set have been divided into equally sized one or two class chunks. 10% of this data, that has been assigned to each node, was assigned to a local test set (with the same distribution as in the local train set). The remaining 10% of the data was retained for evaluating the overall performance of the algorithms.

For the above mentioned three image classification tasks two types of NNs were utilized: For MNIST and Fashion-MNIST data sets we used a fully connected single hidden layer network (FCN)[2] For the CIFAR-10 data set we used a convolutional neural network (CNN)[3]. Experiments with FedAvg and the proposed MM and sMM approaches were performed.

Hyper-parameters, such that the size $|\mathcal{B}|$ of batches b, the number of epochs β and the learning rate η of the optimization process were tuned using grid search with the following values: $\beta \in \{1, 3, 5, 10\}$, $|\mathcal{B}| \in \{10, 32, 64\}$, $\eta \in \{1, 0.1, 0.01\}$ when we used FCNs and $\eta \in \{0.1, 0.001, 0.0001\}$ for the case of CNNs. Besides these, we run tests with different maximal buffer sizes (number of gradients to be collected) $\sigma \in \{1, 3, 5\}$ and also tested the general performance of bagging ensembles of a given number of models $K' \in \{1, 4\}$. Here, bagging refers to predictions resulting from averaging unnormalized per-class activations of a selected

---

[2] Following the Keras reference model for MNIST (not available anymore):
input: 784 dimension vector ($=28 \times 28$) $\rightarrow$ dropout $\rightarrow$ dense with 128 units $\rightarrow$ sigmoid activation $\rightarrow$ dropout $\rightarrow$ dense with 10 units $\rightarrow$ softmax.

[3] Following the Keras reference model for CIFAR-10 (not available anymore):
input: $32 \times 32 \times 3$ image$\rightarrow$
2d convolution with 32 3×3 filter and same padding and ReLU$\rightarrow$ 2d convolution with 32 3×3 filter and same padding and ReLU$\rightarrow$ 2×2 maxpooling $\rightarrow$ Dropout$\rightarrow$ 2d convolution with 64 3×3 filter and same padding and ReLU$\rightarrow$ 2d convolution with 64 3×3 filter and same padding and ReLU $\rightarrow$ 2×2 maxpooling $\rightarrow$ Dropout $\rightarrow$ dense with 512 units and ReLU $\rightarrow$ Dropout $\rightarrow$ dense with 10 units $\rightarrow$ softmax.

subset of the models $\mathbf{w}^1, \mathbf{w}^2, \ldots, \mathbf{w}^{K'}$, that are trained in our network. These hyper-parameters were chosen in a way, that the communication and computation costs should be upper bounded by that one of FedAvg. We defined the number of maximum collected node updates such that the number of communication rounds will be always upper bounded by that of FedAvg[4].

Dropout. Due to the strongly skewed nature of the local data sets, after each training round the sMM models tend to overfit on the local data. Thus, a strong regularization is necessary. Therefore, we applied dropout, as described in [13], with a high probability: 0.25, 0.25, 0.5 for the layers of the used CNN, and 0.5 for the single hidden layer of the used FCN. For the control experiments, we did not use any dropout since we found out that it has a bad impact on FedAvg. Thus, for the buffered learning we decreased the dropout rate π of the model $\mathbf{w}^k$ by a factor σ^k, i.e. $\pi^k = \frac{\pi}{\sigma^k}$.

Graph. We used a full graph topology, where for simulating a more realistic network, we randomly picked 5 nodes and chose the most promising relocation targets among those. For the discount rate at the belief update step we used $\xi = 0.05$.

3.1 Results

Results are summarized in Table 1 and visualized in Figs. 2, 3 and 4. The data series of these figures show the average performance of the 10 best results for the four versions of federated learning, namely, (i) MM with bagging ensembles (denoted as MM), (ii) the FedAvg baseline (FedAvg), (iii) sMM with bagging ensembles (sMM), and, (iv) MM with only one single model (Single MM). On the left side of these figures the performance comparison in terms of accuracy is shown, the middle charts depict the communication costs (i.e. how many times the weights of the model had to be forwarded) while on the right side the corresponding computational costs of these algorithms are showed.

The computation costs has been calculated for a training round r by multiplying N_r, the number of nodes participating in the training in the round (that is $N_r = K'$, with K' being our initial choice in sMM, $N_r = \sum_{k=1}^{K'} \sigma_r^k$ in MM, while $N_r = K' = \gamma K$ for FedAvg), and the number of epochs β. Thus, the accumulated computation costs at round r are computed by $\text{cost}_r^{\text{comp}} = \text{cost}_{r-1}^{\text{comp}} + N_r * \beta$ The communication cost values were calculated in a similar way, such that $\text{cost}_r^{\text{comm}} = \text{cost}_{r-1}^{\text{comm}} + \omega_r$, where ω is the number of transmissions of the model weights, that is necessary for updating all trained models. For sMM, $\omega_r = N_r$ and, for MM, $\omega_r = \sum_{k=1}^{K'} \sigma_r^k$. That equals to the number of participating nodes in a training round, because collecting the gradients from σ_r^k nodes means the same number of transmissions. The value of ω for FedAvg is, on the other hand,

---

[4] For the buffer size extension we set a threshold for improvement to $\theta = 15$, that is, if the accuracy did not improve in the last 15 relocations then we extend the number of models to aggregate.

$2 * N_r$ since each active node has to first acquire the recent model and then to send back the updates. At these values, we calculated the number of transmissions necessary only for the training and did not include the costs of broadcasting the common models for inference to each node.

The performance of the algorithms is measured via accuracy, since for the MM and sMM methods averaging ensembles were used for making predictions.

Results have shown that the sMM algorithm is very simple and viable method to train NNs at a very low computational and communication cost. Its performance, however, stays under the performance of the FedAvg baseline, even with ensembles.

The MM approach closed this gap in a trade-off for increasing costs. We believe, however, that producing a similar performance to FedAvg with a decentralized algorithm, such that the proposed MM, is promising[5, 6].

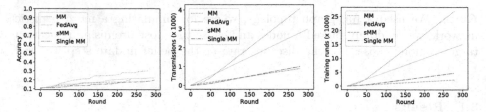

Fig. 2. Accuracy, Communication cost and Training cost on `CIFAR-10`

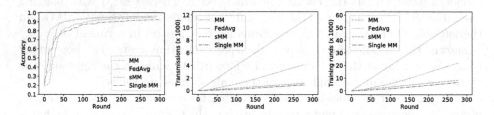

Fig. 3. Accuracy, Communication cost and Training cost on `MNIST`

---

[5] The accuracy, along with the communication and computation costs, of ensemble of MMs exceeds that one of FedAvg. The reason for that is that, due to resource intensity of `CIFAR-10` training, the number of participating nodes have been reduced to 50, which means fewer models have been used in FedAvg, while the hyper-parameters (K' and maximum σ) of sMM and MM have been kept unchanged.

[6] Hyper-parameter search was not performed for the γ parameter of FedAvg, $\gamma = 1/10$ was used according to [23]. The main reason was our limited computation possibilities. However, since the settings for γ also affect sMM and MM (e.g. the buffer size or the ensemble count), it is possible that the gap between the communication and computation costs might be different in case of a large-scale hyper-parameter search.

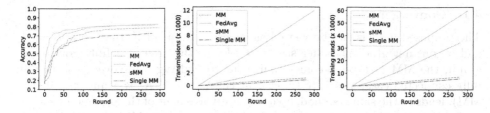

Fig. 4. Accuracy, Communication cost and Training cost on `Fashion-MNIST`

Table 1. The averages of best 10 results in terms of accuracy of each methods, with the corresponding communication and computations costs. Bold fonts denote the best values i.e. best accuracy, lowest communication and computation cost.

	Accuracy				Parameter transmission ($\times 10^4$)				Training epochs($\times 10^4$)			
	Fed Avg	MM	sMM	Single MM	Fed Avg	MM	sMM	Single MM	Fed Avg	MM	sMM	Single MM
C10	0.22	**0.30**	0.2	0.19	14.95	26.90	**2.24**	4.67	2.99	4.07	**0.89**	1.00
MN	0.95	**0.96**	0.94	0.92	59.8	22.27	8.54	**6.74**	11.96	4.14	1.17	**0.92**
F-M	0.81	**0.82**	0.78	0.72	59.8	34.14	7.08	**5.59**	11.96	4.03	1.23	**0.93**

4   Discussion

Besides the promising results of the experiments, a few issues have to be mentioned concerning the presented methods, though. These are the following:

Ensembles. A disadvantage of the proposed methods is the increased resource demand of the ensembles in inference time at the nodes, if one decides to boost performance this way. This could be alleviated trough distillation [12], however, at the price of additional computation load. Another question is, how a node acquires the necessary number of high quality models for inference. In our experiment, we trained the exact number of models that constitute the predictor and used each model in the tested ensemble.

Stagnation. The potentially advantageous effect of omitting model averaging is probably balanced out with the loss of its regularization effect. Despite of the strong regularization, the local models overfit on the most recent data sets. This leads, in a significant number of cases, to a stagnating ensemble accuracy. However, we experienced a similar phenomenon in the case of FedAvg as well. An analysis for the poor performance of FedAvg in strongly skewed data sets can be found for example in [15] or in [29].

Real World Usage. In our experiment, we have been working with a fully connected graph, that is naturally unrealistic in real world scenarios. The distribution of data over the nodes, along with the number of nodes, is pretty far from the characteristics given by [19]. However, using similar set-ups is a common practice as we have seen in the literature.

4.1   Convergence

The proposed models can be viewed as generalizations of FedAvg: For $K' = 1$ and a constant gradient buffer size $\sigma = \gamma K$, in the case of a fully connected graph, the MM method becomes equivalent to FedAvg. Moreover, also in the fully connected case, $\gamma = 1/K$ (in case of FedAvg) and $K' = 1$ and $\pi = 0$ (for MM), leads to the same method, apart from involvement of the parameter server at FedAvg. Finally, if we limit $\sigma = 1$ for MM, we get the sMM algorithm.

Uncertain Convergence. To carry out a thorough analysis of convergence of NN training in a FL setting involves many variables which make giving meaningful guarantees extremely hard. A lot of effort [17, 20, 21, 26, 30, 31, 33, 35, 39] have been carried out in this direction, however, due to the complexity of the problem, all of them make certain restrictions. Even for the case of convex optimization, there are assumptions such that iid data distribution across the nodes or all the devices being active (the latter is equivalent to FedSGD). The latter assumption was made in [17, 31, 35], while authors of [26, 30, 33, 39] build on both. [20] and [21] provides convergence analysis for true FedAvg with non-iid data, but for the case of a strongly convex optimization objective, not applicable for the case of NNs.

Convex Case. Following the reasoning of [21], for convex optimization, the convergence rate of FedAvg is $\mathcal{O}(\frac{1}{n})$, where n is the total number of data points which contribute to the optimization. According to their analysis, to achieve an accuracy of ϵ, the number of training round to execute is

$$\frac{n}{E} = \mathcal{O}\left[\frac{1}{\epsilon}\left(\left(1 + \frac{1}{K'}\right)E\overline{||\nabla l||^2} + \frac{\sum_{k=1}^{K}\gamma^{k^2}\overline{Var(\nabla l^k)} + \Gamma + \overline{||\nabla l||^2}}{E}\right)\right] \quad (5)$$

where $E = \beta|\mathcal{B}|$ is the number of data points, from which the updates are computed between two communication rounds, γ^k is the probability of selecting the node k for the update round, $\Gamma = F^* - \sum_{k=1}^{K}p^k F^{k^*}$ quantifies "non-iid-ness" that is the difference of globally optimal loss and the local optima. $\overline{Var(\nabla l^k)}$ is a bound for variance, while $\overline{||\nabla l||^2}$ stands for squared norm of local stochastic gradients.

4.2   Privacy

An important challenge in FL, and distributed machine learning (ML) in general, is the question of privacy of potentially sensitive data. In our setup we see the following two major points for potential attacks:

Forward Inference. If we use the method of Algorithm 2, it is feasible to combinatorically infer classes of data held at the candidate nodes, granting an additional vulnerability to our method compared to FedAvg. However, applying a random noise over the indicator function, apart from some extreme situations, considerably decreases the vulnerability of the proposed approach.

Backward Inference. According to our best knowledge, as it is summarized in [14], attacks on privacy in distributed ML build on regular and frequent update messages following the same routes. Peer-to-peer nature of the training process adds an additional complexity to deal with, making it necessary to have access to all communication channels of the attacked node to achieve similar effectiveness of an attack. Gradient leakage attacks [37] can be executed having access to a single update vector, currently however they work only on single batch updates, or multiple batches with very few examples included. For both cases the tracker can be used to ensure legitimacy of routes, for example, to avoid building loops around the target or even to redraw the connection graph from time to time.

5   Conclusion

We presented our approach to alleviate the challenges imposed by the federated learning setup and, in general, distributed machine learning systems. The key idea of the proposed approach is that, instead of the transmission of model updates, the models themselves travel to the location of the data, evening the communication needs across the network.

With this (almost complete) decentralization of the learning process, the synchronization problems and straggler effect can be bypassed as well as communication burden at the parameter servers is not present anymore.

Our experiments have shown that similar performance can be achieved compared to the federated averaging baseline, however, with less communication and computational cost (at the price of using ensembles of small number of models at prediction phase). Based on the results of our experiments and the fact that the proposed approach is a generalization of the popular federated averaging approach, the presented work is worth further research.

References

1. Aji, A.F., Heafield, K.: Sparse communication for distributed gradient descent. arXiv preprint arXiv:1704.05021 (2017)
2. Alistarh, D., Grubic, D., Li, J., Tomioka, R., Vojnovic, M.: QSGD: communication-efficient sgd via gradient quantization and encoding. In: Advances in Neural Information Processing Systems, pp. 1709–1720 (2017)
3. Bellet, A., Guerraoui, R., Taziki, M., Tommasi, M.: Personalized and private peer-to-peer machine learning. arXiv preprint arXiv:1705.08435 (2017)
4. Boyd, S., Parikh, N., Chu, E., Peleato, B., Eckstein, J., et al.: Distributed optimization and statistical learning via the alternating direction method of multipliers. Found. Trends Mach. Learn. **3**(1), 1–122 (2011)
5. Chen, J., Pan, X., Monga, R., Bengio, S., Jozefowicz, R.: Revisiting distributed synchronous sgd. arXiv preprint arXiv:1604.00981 (2016)
6. Chen, Y., Sun, X., Jin, Y.: Communication-efficient federated deep learning with layerwise asynchronous model update and temporally weighted aggregation. IEEE Trans. Neural Networks Learn. Syst. (2019)

7. Dean, J., et al.: Large scale distributed deep networks. In: Advances in Neural Information Processing Systems, pp. 1223–1231 (2012)
8. Dimakis, A.G., Kar, S., Moura, J.M., Rabbat, M.G., Scaglione, A.: Gossip algorithms for distributed signal processing. Proc. IEEE **98**(11), 1847–1864 (2010)
9. Dryden, N., Moon, T., Jacobs, S.A., Van Essen, B.: Communication quantization for data-parallel training of deep neural networks. In: 2016 2nd Workshop on Machine Learning in HPC Environments (MLHPC), pp. 1–8. IEEE (2016)
10. Du, W., Zeng, X., Yan, M., Zhang, M.: Efficient federated learning via variational dropout (2018)
11. Hegedűs, I., Danner, G., Jelasity, M.: Gossip learning as a decentralized alternative to federated learning. In: Pereira, J., Ricci, L. (eds.) DAIS 2019. LNCS, vol. 11534, pp. 74–90. Springer, Cham (2019). https://doi.org/10.1007/978-3-030-22496-7_5
12. Hinton, G., Vinyals, O., Dean, J.: Distilling the knowledge in a neural network. arXiv preprint arXiv:1503.02531 (2015)
13. Hinton, G.E., Srivastava, N., Krizhevsky, A., Sutskever, I., Salakhutdinov, R.R.: Improving neural networks by preventing co-adaptation of feature detectors. arXiv preprint arXiv:1207.0580 (2012)
14. Hitaj, B., Ateniese, G., Perez-Cruz, F.: Deep models under the GAN: information leakage from collaborative deep learning. In: Proceedings of the 2017 ACM SIGSAC Conference on Computer and Communications Security, pp. 603–618 (2017)
15. Hsu, T.M.H., Qi, H., Brown, M.: Measuring the effects of non-identical data distribution for federated visual classification. arXiv preprint arXiv:1909.06335 (2019)
16. Jaggi, M., et al.: Communication-efficient distributed dual coordinate ascent. In: Advances in Neural Information Processing Systems, pp. 3068–3076 (2014)
17. Khaled, A., Mishchenko, K., Richtárik, P.: First analysis of local gd on heterogeneous data (2019)
18. Kingma, D.P., Salimans, T., Welling, M.: Variational dropout and the local reparameterization trick. In: Advances in Neural Information Processing Systems, pp. 2575–2583 (2015)
19. Konečnỳ, J., McMahan, H.B., Ramage, D., Richtárik, P.: Federated optimization: Distributed machine learning for on-device intelligence. arXiv preprint arXiv:1610.02527 (2016)
20. Li, T., Sahu, A.K., Zaheer, M., Sanjabi, M., Talwalkar, A., Smith, V.: Federated optimization in heterogeneous networks (2018)
21. Li, X., Huang, K., Yang, W., Wang, S., Zhang, Z.: On the convergence of fedavg on non-iid data. arXiv preprint arXiv:1907.02189 (2019)
22. Masters, D., Luschi, C.: Revisiting small batch training for deep neural networks. arXiv preprint arXiv:1804.07612 (2018)
23. McMahan, H.B., Moore, E., Ramage, D., Hampson, S., et al.: Communication-efficient learning of deep networks from decentralized data. arXiv preprint arXiv:1602.05629 (2016)
24. Nishio, T., Yonetani, R.: Client selection for federated learning with heterogeneous resources in mobile edge. CoRR abs/1804.08333 (2018)
25. Seide, F., Fu, H., Droppo, J., Li, G., Yu, D.: 1-bit stochastic gradient descent and application to data-parallel distributed training of speech DNNs. In: Interspeech (2014)
26. Stich, S.U.: Local SGD converges fast and communicates little (2018)
27. Strom, N.: Scalable distributed DNN training using commodity GPU cloud computing. In: Sixteenth Annual Conference of the International Speech Communication Association (2015)

28. Vanhaesebrouck, P., Bellet, A., Tommasi, M.: Decentralized collaborative learning of personalized models over networks (2017)
29. Wang, H., Kaplan, Z., Niu, D., Li, B.: Optimizing federated learning on non-iid data with reinforcement learning. In: IEEE INFOCOM 2020-IEEE Conference on Computer Communications, pp. 1698–1707. IEEE (2020)
30. Wang, J., Joshi, G.: Cooperative SGD: a unified framework for the design and analysis of communication-efficient SGD algorithms (2018)
31. Wang, S., et al.: Adaptive federated learning in resource constrained edge computing systems. In: IEEE INFOCOM 2018-IEEE Conference on Computer Communications (2018)
32. Wei, E., Ozdaglar, A.: On the o (1= k) convergence of asynchronous distributed alternating direction method of multipliers. In: 2013 IEEE Global Conference on Signal and Information Processing, pp. 551–554. IEEE (2013)
33. Woodworth, B., Wang, J., Smith, A., McMahan, B., Srebro, N.: Graph oracle models, lower bounds, and gaps for parallel stochastic optimization (2018)
34. Yang, H.H., Liu, Z., Quek, T.Q., Poor, H.V.: Scheduling policies for federated learning in wireless networks. IEEE Trans. Commun. (2019)
35. Yu, H., Yang, S., Zhu, S.: Parallel restarted SGD with faster convergence and less communication: demystifying why model averaging works for deep learning. In: Proceedings of the AAAI Conference on Artificial Intelligence, vol. 33, pp. 5693–5700 (2019)
36. Yurochkin, M., Agarwal, M., Ghosh, S., Greenewald, K., Hoang, T.N., Khazaeni, Y.: Bayesian nonparametric federated learning of neural networks. arXiv preprint arXiv:1905.12022 (2019)
37. Zhao, B., Mopuri, K.R., Bilen, H.: idlg: improved deep leakage from gradients. arXiv preprint arXiv:2001.02610 (2020)
38. Zhao, Y., Li, M., Lai, L., Suda, N., Civin, D., Chandra, V.: Federated learning with non-iid data. arXiv preprint arXiv:1806.00582 (2018)
39. Zhou, F., Cong, G.: On the convergence properties of a k-step averaging stochastic gradient descent algorithm for nonconvex optimization. In: Proceedings of the Twenty-Seventh International Joint Conference on Artificial Intelligence (2018)

Graph Embedding and Mining

Workshop on Graph Embedding and Mining (GEM 2021)

Graphs of various types, such as plain networks, linked/semantic web/RDF data, attributed, dynamic, interaction graphs, and possible combinations of these, are increasingly used as versatile and practical models for data encountered in today's data-intensive research and in industry. In contrast to data tables, graphs allow the capture of information about entities (using attributes or properties), as well as the relational structure between entities.

The ability to discover knowledge from and make predictions about such network data has gained in importance quickly. However, both the formalization of new problem types that match well with practical use cases and the algorithmic, statistical, and information theoretic aspects of such problems require further scientific inquiry. Graph representation learning – resulting in intermediate real-valued representations – enables learning and mining algorithms devised for non-relational data to be applied to graphs. With rapid advances in this area in particular, trustworthy AI on graphs requires particular attention.

The third Workshop on Graph Embedding and Mining (GEM 2021) was held in conjunction with the European Conference on Machine Learning and Principles and Practice of Knowledge Discovery in Databases (ECML PKDD 2021). The conference and all workshops were held virtually due to the ongoing COVID-19 pandemic. The aim of this workshop was to be a discussion forum for the most recent advances on the above topics. To this end, we called for both theoretical and practical contributions to stimulate interactions between participants. And we encouraged both long mature contributions and short, open for discussion, ideas.

The workshop featured

- a keynote given by Danai Koutra,
- a poster session,
- short pitch presentations of the contributed research, and
- a tutorial on the scikit-network, a python library for large graph analysis.

Our workshop was geared towards interactive scientific exchange in small groups. By devoting most of the time of the workshop to an interactive poster session format, we wanted to foster the interactions that are so direly needed in the current situation. To allow the participants an educated choice on which poster sessions to attend, short pitch presentations were given in advance (no questions to be asked!). Subsequently Danai Koutra gave a great keynote on "Representation Learning Beyond Homophily and Proximity". The workshop concluded with a best paper award that was given on the basis of highest reviewer scores.

Overall, we attracted 13 submissions, six of which were selected for these workshop proceedings. Additionally, four submissions were presented with a pitch and poster at the workshop which are not included in these proceedings. On average, 20 people attended the live meetings. The best paper award was given to Tobias Schumacher, Hinrikus Wolf, Martin Ritzert, Florian Lemmerich, Martin Grohe, and

Markus Strohmaier for their paper "The Effects of Randomness on the Stability of Node Embeddings".

The tutorial was split into two parts: (1) introducing the python library *scikit-network* and (2) applying it to the analysis of a subset of Wikipedia pages. We attracted around 15 attendees for the whole duration of the tutorial. The two sessions were balanced between theoretical foundations and basic data structures, and analysis of a real-world graph using the usual tools (pageranking, clustering of the nodes, etc.)

Organization

GEM 2021 Organizers

Bo Kang	Ghent University, Belgium
Christine Largeron	Université Jean Monnet, France
Jefrey Lijffijt	Ghent University, Belgium
Tiphaine Viard	Telecom Paris, France
Pascal Welke	University of Bonn, Germany

GEM 2021 Program Committee

Martin Atzmüller	Osnabrück University, Germany
Anes Bendimerad	LIRIS, France
Rémy Cazabet	Université Claude Bernard Lyon 1, France
Tijl De Bie	Ghent University, Belgium
Raphaël Fournier-S'niehotta	CNAM, France
Stephan Günnemann	Technical University of Munich, Germany
Mark Heimann	Lawrence Livermore National Laboratory, USA
Baptiste Jeudy	Laboratoire Hubert Curien, France
Di Jin	University of Michigan, USA
Márton Karsai	ENS de Lyon, France
Charlotte Laclau	Laboratoire Hubert Curien, France
Renaud Lambiotte	University of Oxford, UK
Sandra Mitrovic	IDSIA
Amedeo Napoli	LORIA Nancy, France
Marc Plantevit	LIRIS - Université Claude Bernard Lyon 1, France
Till Schulz	University of Bonn, Germany
Florian Yger	LAMSADE - Université Paris Dauphine, France

The Effects of Randomness
on the Stability of Node Embeddings

Tobias Schumacher[1] , Hinrikus Wolf[1(✉)] , Martin Ritzert[2] ,
Florian Lemmerich[3] , Martin Grohe[1] , and Markus Strohmaier[1,4,5]

[1] RWTH Aachen University, Aachen, Germany
{tobias.schumacher,markus.strohmaier}@cssh.rwth-aachen.de,
{hinrikus,grohe}@cs.rwth-aachen.de
[2] Aarhus University, Aarhus, Denmark
ritzert@cs.au.dk
[3] University of Passau, Passau, Germany
florian.lemmerich@uni-passau.de
[4] GESIS - Leibniz Institute for the Social Sciences, Cologne, Germany
[5] Complexity Science Hub Vienna, Vienna, Austria

Abstract. We systematically evaluate the (in-)stability of state-of-the-art node embedding algorithms due to randomness, i.e., the random variation of their outcomes given identical algorithms and networks. We apply five node embeddings algorithms—HOPE, LINE, node2vec, SDNE, and GraphSAGE—to assess their stability under randomness with respect to their performance in downstream tasks such as node classification and link prediction. We observe that while the classification of individual nodes can differ substantially, the overall accuracy is mostly unaffected by the geometric instabilities in the underlying embeddings. In link prediction, we also observe high stability in the overall accuracy and a higher stability in individual predictions than in node classification. While our work highlights that the overall performance of downstream tasks is largely unaffected by randomness in node embeddings, we also show that individual predictions might be dependent solely on randomness in the underlying embeddings. Our work is relevant for researchers and engineers interested in the effectiveness, reliability, and reproducibility of node embedding approaches.

Keywords: Node embedding · Graph embedding · Node classification · Link prediction · Reliability · Representation learning · Embedding stability

---

T. Schumacher, H. Wolf and M. Ritzert—Equal contribution.

This work is supported by the German research council (DFG) Research Training Group 2236 UnRAVeL, the Federal Ministry of Education and Research (BMBF), and the Ministry of Culture and Science of the German State of North Rhine-Westphalia (MKW). We thank Jan Bachmann, Max Klabunde, and Florian Frantzen for their help with implementing and running the experiments.

© Springer Nature Switzerland AG 2021
M. Kamp et al. (Eds.): ECML PKDD 2021 Workshops, CCIS 1524, pp. 197–215, 2021.
https://doi.org/10.1007/978-3-030-93736-2_16

1 Introduction

Many state-of-the-art node embedding algorithms make explicit use of random-ness in parameter initialization, edge sampling, or through stochastic optimiza-tion. Thus, the application of the same algorithm with identical parameters on the exact same graph data can lead to different embeddings.

Recent research [19] has provided an initial assessment of such instabilities, in particular with respect to the geometry of the embedding spaces. Yet, the impact of these instabilities on the outcomes of downstream tasks such as node classification and link prediction has not been systematically evaluated.

Research Objective. We investigate the effects of randomness on the stability of node embeddings. Towards this end, we specifically focus on assessing the *downstream stability* of node embeddings, i.e., the stability of outcomes from tasks such as node classification and link prediction.

Approach. We conduct experiments with five state-of-the-art embedding algo-rithms on empirical network datasets. For each embedding algorithm, we com-pute multiple node embeddings with the same parameters on the same networks but with different random seeds. Specifically, we apply HOPE [12], LINE [17], node2vec [4], SDNE [20], and GraphSAGE [6]. On the resulting embeddings, we then perform node classification and link prediction to quantify downstream stability with respect to these tasks. In that regard, we consider both stability in overall performance and stability of individual predictions.

Results and Implications. We find that despite substantial geometric insta-bilities, which have been reported in previous work [19] as well as our own prelim-inary experiments, the overall accuracy in node classification and link prediction is almost constant. This indicates a surprising stability in downstream tasks. At the same time, we show that the actual predicted classes of individual nodes can—and often do—differ between classifiers trained on embeddings based on different random seeds. For link prediction, we observe similar trends, although the stability of the single predictions is much higher than for node classification. This higher stability is however likely due to a higher overall accuracy in the considered scenarios for this task, which leaves less room for different misclassi-fications.

Overall, our work contributes towards a more fundamental understanding of the stability of node embeddings, and thereby opens up ways for more informed deployments and a better understanding of the effects of randomness on embedding-based predictions.

2 Related Work

Our paper extends a recent study by Wang et al. [19], who conducted their research independently and in parallel to ours. They provide an initial assessment of the issue of instability of node embeddings with an emphasis on geometric stability. Next to finding significant instabilities over most algorithms for both global and node-based stability, they perform a factor analysis to identify the main sources of those instabilities. The factor analysis suggests that the impact of dataset-dependent features such as size and density, as well as node properties such as closeness centrality, have higher impact than algorithmic parameters. They identified a correlation between embedding stability and node classification accuracy with SVMs. However, they did not investigate to which extent the accuracy of repeated downstream tasks varies and how far individual predictions differ in these downstream tasks.

In a set of preliminary experiments (cf. Appendix B), we confirmed the fundamental geometric instabilities which Wang et al. [19] have reported. However, we could not confirm the impact of network size and density as well as node centrality. Our paper complements their work by providing a thorough analysis on the impact of instability on downstream predictions, i.e., predictions occurring when combining embeddings with other machine learning algorithms.

Aside from the study by Wang et al. [19], there has not been any additional previous study on the stability of node embeddings. However, the issue of embedding instability has been thoroughly studied in the context word embeddings, which has also influenced our work.

The first work to point out instabilities in word embeddings has been conducted by Hellrich and Hahn [7]. They discovered that neighborhoods of words in the embedding space change significantly even under fixed corpora. These instabilities have been confirmed and further investigated by Antoniak and Minmo [1]. Both studies [1, 7] report significant instabilities of skip-gram-based word embedding methods with respect to local neighborhood similarities. To investigate which word properties influence stability, Wendlandt et al. [22] and Pierrejean and Tanguy [13] conducted regression-based factor analyses. They correlated the stability of a word embedding with semantic features such as a word's part of speech, as well as algorithmic parameters such as the dimensionality of the embedding space. Finally, Leszczynski et al. [10] specifically analyzed the relationship between geometric stability of word embeddings and the resulting instabilities in downstream tasks. They introduced an Eigenspace instability measure to quantify geometric instability, and proved that this measure theoretically determines the expected downstream disagreement on linear regression tasks. In our study, we directly measure the variance of downstream tasks with non-linear classifiers, such that this instability measure is not applicable.

3   Experimental Framework

Our main set of experiments quantifies the downstream stability of five state-of-the-art node embedding algorithms. We start with a short description of the algorithms and datasets and then describe the experiments. The code for our experiments is published on GitHub.[1]

---

[1] All code available on https://github.com/SGDE2020/embedding_stability.

We consider the following five node embedding algorithms as representatives of the spectrum of currently existing approaches. The spectral embedding algorithm *HOPE* [12] factorizes the Katz similarity matrix. *LINE* [17] embeds the local and global neighborhood structures separately and combines the resulting embeddings. *node2vec* [4] applies the word embedding algorithm *word2vec* on random walks generated from the network. *SDNE* [20] computes embeddings based on the encoder-decoder principle. The inductive node embedding algorithm *GraphSAGE* [6] applies a GNN to compute its embeddings.

We investigate the downstream stability of node embeddings on four graph datasets, which cover a broad spectrum of commonly used empirical graphs: the social graph BlogCatalog [23], the citation graph Cora [16], as well as the datasets Protein [15] and Wikipedia [11]. Statistics for each graph can be found in Table 1 in Appendix A.1.

Overview of Experiments. To analyze the impact of randomness in node embeddings on the outcomes of downstream predictions, we computed for each dataset 30 embeddings with each embedding algorithm, all with the embedding dimension of 128 and mostly standard parameters. We consider the two most common downstream tasks, node classification and link prediction. We evaluate two types of downstream stability, first the *stability of performance* and second the *stability of single predictions*. In stability of performance, we measure the variance of general performance scores such as micro-F1 of the classification on a holdout set. To quantify the stability of single predictions, we train (i) multiple classifiers on the same embedding and (ii) multiple classifiers on multiple embeddings of the same network produced by the same embedding algorithm. Differences in the classifications in (i) indicate the stability of the classification algorithm itself due to random elements in the classification algorithm, independent of the embedding. Such random elements naturally occur in most learning algorithms. Comparing outcomes of classifiers trained on different embeddings (ii) provides an indication of the combined stability of the embedding algorithm and the classifier. Thus, the difference between the outcomes of (i) and (ii) corresponds to the influence of the instability of the embeddings on the stability of the classification. To measure differences in the outcome of classifiers, we use general performance scores (such as micro-F1 of the classification on a holdout set) as well as the *stable core*, i.e., the ratio of nodes that are assigned to the same class in at least 90% of the classifier runs.

Since different machine learning algorithms have very different characteristics, we use multiple classifiers, namely AdaBoost, decision trees, random forests, and feedforward neural networks. For node classification, we performed a 10-fold cross-validation with 10 repetitions. For link prediction, we were able to generate a sufficient amount of training data and thus left out the cross-validation. More details on the parameterization of both the embedding and classification algorithms can be found in Appendix A.2.

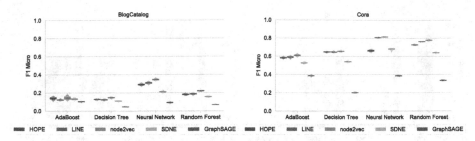

Fig. 1. Stability of classification performance of BlogCatalog and Cora.

4   Results

In a set of preliminary experiments (cf. Appendix B), we have found that all embedding algorithms except HOPE, which yields near-constant embeddings, display substantial geometric instabilities. These results are in line with results from Wang et al. [19]. In this section, we present the results from our experiments on the instability of downstream tasks as described in Sect. 3. We begin with the results for node classification and then continue with the stability of link prediction.

Node Classification. We first analyze the *stability of performance* in the node classification task. Due to limited space, we only present and discuss the results on BlogCatalog and Cora here. Results for the other datasets can be found in Appendix C. Figure 1 depicts the micro-F1 scores of the predictions. Each box in the figure aggregates the different micro-F1 scores of the repeated predictions on the 30 embedding per algorithm and dataset. We observe that the F1 scores of all classification tasks vary only marginally. Aside from stability, we observe a strong dependence of the micro-F1 scores on the classification algorithm, but not so much on the embedding algorithm, except for GraphSAGE which is always lower in performance.

Next, we investigate the stability of individual *node-wise predictions*. For that purpose, we determine the stable core of predictions over multiple classification runs, i.e., the ratio of nodes which are classified to have the same labels in 90% of all predictions. To distinguish between (i) instability originating from the classifiers and (ii) instability originating from the underlying embeddings, we compute the stable cores in two distinct settings. For (i), we train each classifier ten times on a fixed embedding and averaged the sizes of the resulting stable cores over five embeddings, for (ii), we trained each classifier once on all 30 embeddings. The results are shown in Fig. 2 where the stable cores from (i) are depicted in saturated colors and the stable cores from (ii) are shown in light colors. Compared to the stability of performance, the picture of stability in node-wise predictions is more mixed. Our first observation is that since the embeddings generated by HOPE are almost identical, also the stables cores

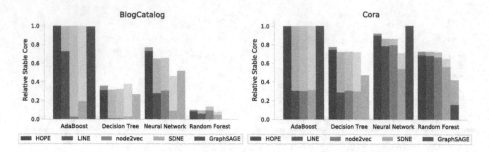

Fig. 2. Stability in node-wise predictions. We depict the mean stable core of predictions under varying embeddings in saturated colors, and the mean stable core of multiple predictions on fixed embeddings in lighter colors.

from (i) and (ii) are about the same size over all datasets and classifiers. For the remaining embedding algorithms, there is no clear trend recognizable. Next to the embedding algorithm, the choice of classifier seems to have a high impact on the stability of individual predictions. For AdaBoost, we observe almost stable predictions under fixed embeddings, while for varying embeddings, stability is highly dependent on the embedding algorithm. In contrast, the individual predictions of the decision trees are relatively unstable, in particular under varying embeddings. For random forests, we observe that the observed instabilities are mostly due to the classification algorithm itself. However, the degree of instability strongly varies over the datasets. Finally, for neural networks we observe that the degree of instability varies over both datasets and there is no clear trend on whether varying classifiers or varying embeddings have a stronger impact on the stability of individual predictions.

In general, we observe that both the chosen embedding algorithm as well as the selected classifier have a high influence on the stability of individual predictions. The impact of classifier and embedding algorithm varies over different datasets.

Link Prediction. For simplicity, we show the results of our link prediction experiments only on BlogCatalog, which are depicted in Fig. 3. The results on the other datasets can be found in Appendix C. The first observation is that in this binary task, the accuracies are naturally much higher than for multi-class and multi-label node classification. Further, we see low variances in those accuracies. In terms of the stability of individual link predictions, we also observe more stable individual predictions than in node classification. AdaBoost is almost perfect in repeating the task on the same input data, although the accuracy varies between 0.5 and 0.95 depending on the embedding. Decision trees also achieve highly reproducible predictions for varying embeddings, whereas for neural networks and in particular random forests there is a stronger dependence on the embeddings. For LINE and SDNE embeddings, we observe that most predictions stay the same independently of the underlying embedding, despite their geometric instability (see [19] and Appendix B). For node2vec and GraphSAGE,

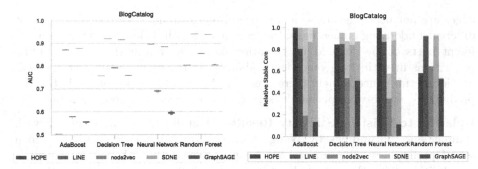

Fig. 3. Stability in link prediction on BlogCatalog. Left: Stability of accuracy. Right: Stability of individual predictions as difference between the mean stable core over all embeddings (saturated colors) and the mean stable core over repeatedly trained classifiers on fixed embeddings (lighter colors).

there is, however, a relatively high fluctuation in the predictions that results from instabilities in the embeddings. For HOPE, we confirm the stability of the predictions that is to be expected given its almost fully stable embeddings.

As expected, we see a strong dependence between performance and the individual predictions, i.e., higher performance in terms of accuracy corresponds to larger stable cores. Again, we observe a high impact of the embedding algorithm on the stability of individual predictions.

5   Discussion

Next, we discuss our experimental results, including potential explanations, relationship to prior research, implications, and limitations of our work.

Summary of Results. Our results show that the overall classification performance in both node embedding and link prediction is mostly unaffected by random variations in the embeddings, which were observed in preliminary experiments (cf. Appendix B) and previous work [19]. However, the actual predicted classes for single nodes vary depending on the embedding that the classifier was trained on, i.e., due to the randomness in the embeddings, different classifications are produced. To a lesser extent, this effect was also observed when analyzing the predictions of individual links.

Potential Explanations for Results. A potential explanation for the surprising stability in the overall classification performance is that classifiers seem to be able to extract and utilize local structural information from embeddings even if their global structure changes. This means that even in very different embeddings, the necessary information for a model that generalizes well is contained in each of those embeddings. Since the classifications of single nodes or edges are not nearly as stable as the overall performance, we conclude that for different underlying embeddings, the learning algorithm chooses to focus on different parts of the embedding. On other hand, the fluctuations in individual predictions fit with the geometric instabilities. Further, when the overall classification performance is high, there is not much room for variations in individual predictions, which we especially observed on the easier link prediction task.

Relation to Existing Stability Results. Overall, the results from our work complement the findings by Wang et al. [19] on embedding instability. In a set of preliminary experiments, we have confirmed the substantial geometric instabilities which they pointed out in their work. However, we did not observe a strong impact on downstream performance, which they have reported in a smaller experiment. Only when considering individual predictions, we observed substantial instabilities. For link prediction, we observed a relatively high downstream stability, again contrasting the results by Wang et al. [19].

Implications. In the authors' opinion, the outcomes of this paper have significant impact on the research of node embeddings. Since node embeddings vary just based on their internal random processes, great care must be taken in their evaluation and, if possible, experiments should be repeated several times in order to estimate and limit the influence of randomness and enable reproducibility of results. In settings in which unstable predictions are not problematic, for example for product recommendations, node embedding algorithms can safely be applied since the overall predictive performance is not influenced by the geometric stability. However, reproducibility of algorithms has emerged as a key factor for building trust in algorithmic decisions, which requires a high stability of predictions. This is especially important for high-stakes real-world decisions based on node embeddings. Practitioners should be aware that node embeddings add another level of uncertainty to individual (e.g., classification) decisions.

Limitations.   The stability of the investigated algorithms might be strongly influenced by their concrete implementations. In that regard, we picked reference implementations from the respective research papers or—if that was not possible—established code bases for the different algorithms. However, we cannot rule out that some (in-)stabilities we observed are a consequence of implementation details. Since the chosen implementations are widely used, our results are still highly relevant for researchers and practitioners. In our experiments, we did not aim for optimal performance, but for a comparable standard setting. Thus, we did not perform extensive hyperparameter optimization for each individual task, but relied on default parameters for each algorithm. We expect a slightly higher stability with optimized hyperparameters due to higher accuracies.

6   Conclusion

In this work, we analyzed the effects of instabilities in node embeddings on the predictions in downstream tasks. Despite substantial variations in the geometry of the embedding space, which have been pointed out in previous work [19] and confirmed in our own preliminary experiments (cf. Appendix B), we found that the overall performance in the downstream tasks node classification and link prediction only displays small deviations. However, we found considerable variations when looking the classifications of single nodes, and, to a smaller extent, in the prediction of single links.

In the future, we anticipate investigations of stability and robustness of node embedding algorithms towards an in-depth study of the effects of different embedding sizes and graph modifications such as deletions or additions of nodes or edges. Furthermore, we see an opportunity for developing measures that will allow to estimate the potential instability of an embedding without computing it multiple times.

A   Experimental Setup

A.1   Datasets

We provide some more details on the graphs datasets that were used in our experiments. Note that the Facebook dataset has only been used in our preliminary experiments on embedding geometry (cf. Appendix B), as it does not provide any node labels. Statistics for each graph can be found in Table 1.

- **BlogCatalog**: This graph models the relationships among the users of the BlogCatalog website. Each user is represented by a node and two nodes are connected if the respective users are friends. Each user additionally has one or more labels which correspond to the news category their blog belongs to.
- **Cora** [16]: In the well-known Cora citation network each scientific paper is represented by a node, and a directed edge indicates that the outgoing node cites the target node. Each paper is associated with a category that refers to its research topic.
- **Facebook** [14]: The Facebook government dataset models the social network structure of verified government sites on Facebook. Each site is represented by a node and nodes are connected by an edge if both sites like each other.
- **Protein** [15]: This biological network models protein interactions in human beings. Each node represents a protein and two nodes are connected if the corresponding proteins interact with each other. Additionally, each node is associated with one or more labels that represent biological states.
- **Wikipedia** [11]: This network represents the co-occurrence of words within a dump of Wikipedia articles. Each word corresponds to a node, and weighted edges represent the number of times two words occur in the same context. Additionally, each node has one or more labels that encode its part of speech.

We used the Cora dataset from the KONECT graph repository [8] and Blog-Catalog from the ASU Social computing repository [23]. The other empirical datasets were taken from the SNAP graph repository [9].

A.2   Implementations and Parameter Settings

To complement Sect. 3, in the following we give a more detailed overview on the chosen implementations and parameter settings of the node embedding algorithms, as well as the experimental setups of the downstream classification tasks that we used in our experiments.

Node Embedding Algorithms. For every algorithm from Sect. 3 we use the reference implementation except for HOPE, for which no reference implementation was published. Thus we resorted to the HOPE implementation from the GEM library [3]. We run the algorithms with default parameters from the given implementations whenever possible and compute embedding vectors of length $d = 128$. We adapted SDNE to use only a single intermediate layer and for larger graphs increased the weight on the reconstruction error and the regularization term, as otherwise SDNE maps all nodes onto the same vector.

Downstream Classification. For both node classification and link prediction, we use AdaBoost, decision trees, random forests, and feedforward neural networks as downstream classification algorithms. For all classifiers we used the standard methods with default parameters from scikit-learn (AdaBoost, decision tree, random forest) and TensorFlow (neural networks). In the case of neural networks, we use a network with a single hidden layer of width 100 with ReLu activation and an output layer with softmax or sigmoid activation depending on the classification type. Deeper and wider networks did not improve performance which is why we worked with this very simple architecture.

In node classification we predict either the class of a node, e.g., top-level research category in Cora, or a set of labels of a node, e.g., the news categories in BlogCatalog. In the latter case of multi label classification, we assume that we know the number l of labels and thus predict the l most probable labels. This approach leads to more stable predictions and is common in literature [18].

For the link prediction task, we considered subgraphs of each network where we removed 10% of the original edges at random while ensuring that the residual graph is still connected. For each reduced network, we computed 10 embeddings per algorithm. We then interpreted link prediction as a binary classification task

Table 1. *Statistics of empirical graph datasets.* We show number of nodes (—V—) and edges (—E—), density, and number of node labels. MC indicates multi class, ML multi label problems.

Data Set	$|V|$	$|E|$	Density	# Labels
BlogCatalog	10,312	333,983	0.00628	39 (ML)
Cora	23,166	91,500	0.00034	10 (MC)
Facebook	7,057	89,455	0.00359	-
Protein	3,890	76,584	0.01012	50 (ML)
Wikipedia	4,777	184,812	0.01620	40 (ML)

on the Hadamard product of two embedding vectors. The removed edges are then the positive examples for the link prediction, and we chose as many non-edges at random as negative examples for training the classifier.

For the stability of performance, we compute the variance of micro-F1 scores over one classifier computed on each of the 30 embeddings per graph and embedding algorithm in node classification, and each of the 10 embeddings per graph and embedding algorithm in link prediction. In both experiments, macro-F1 yields very similar results such that we only report micro-F1.

For the stability of single classifications, we have to separate inherent instability of the classifiers from the influence of different embeddings. We estimate the instability of a classifier by running it 10 times on a single embedding, averaged over 5 embeddings. The total variance in individual predictions is computed on the results of one classifier trained on each of the 30 embeddings using 75% of the nodes for training and 25% for evaluation.

B   Experiments on Geometric Stability

In this section, we present our preliminary experiments on the geometric stability of node embeddings. We first give a brief description of the measures for geometric stability, and then present the results.

B.1   Measures for Geometric Stability

To quantify geometric instability of node embeddings, we use two measures which have been introduced in related literature on word embeddings, namely *aligned cosine similarity* [5] and *k-NN Jaccard similarity* [1].

The aligned cosine similarity computes the node-wise cosine similarity between two embeddings after aligning the axes of the corresponding embedding spaces. To obtain the optimal alignment, we normalize all embedding vectors and solve the Procrustes problem: Given two embedding matrices $Z^{(1)}, Z^{(2)} \in \mathbb{R}^{N \times d}$, with N denoting the number of nodes in a given network, and d denoting the embedding dimension, we determine the transformation matrix $Q \in \mathbb{R}^{d \times d}$ by solving the minimization problem

$$Q := \underset{Q^T Q = I}{\operatorname{argmin}} \left\| Z^{(1)} Q - Z^{(2)} \right\|_F.$$

The k-NN Jaccard similarity measure compares the local neighborhoods of nodes between different embeddings. In both embedding spaces, we compute for a node u the k nearest neighbors with respect to cosine similarity. We then calculate the Jaccard similarity of the two nearest-neighbor sets of u.

Each of those two measures computes a score for a single node in two embeddings. In order to obtain a score for an embedding space to compare different algorithms, we average over all pairs of embeddings and all nodes.

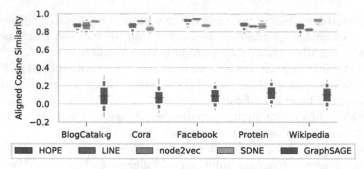

(a) Variability of aligned cosine similarity.

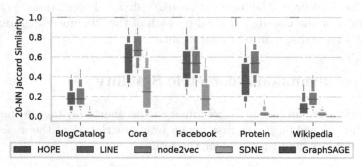

(b) Variability of k-NN Jaccard similarity.

Fig. 4. *Geometric stability.* Each letter-value plot shows the node-wise similarity values resulting from 30 runs per algorithm and graph. In (a) we use aligned cosine similarity, in (b) 20-NN Jaccard similarity.

B.2   Experimental Results

In our experiments on geometric stability, we used the same algorithmic parameter settings and datasets that have been introduced in Appendix A. Next to the overall stability of the embeddings, we also look into the *influence of node centrality*, and the *influence of network size and density* on the stability of node embeddings.

Geometric Stability. We start our analysis by computing 30 embeddings per dataset with every algorithm. We then compute node-wise stability measures averaged over all pairs of embeddings computed per graph and embedding algorithm. Figure 4 shows the distributions of (a) aligned cosine similarity and (b) k-NN Jaccard similarity over the nodes of each graph.

For the aligned cosine similarity, we observe that GraphSAGE achieves similarities that are generally only slightly above zero and sometimes even negative. Negative values correspond to angle differences of more than 90° between two embeddings of the same node. Thus, even after aligning axes, embedding vectors of the same node are mostly close to orthogonal to each other. In contrast,

HOPE yields near-constant embeddings (not shown) and shows hardly any instability. The algorithms SDNE, node2vec and LINE achieve aligned cosine similarities in the interval $(0.8, 0.9)$ with low variances. These values correspond to angles between $25°$ and $35°$ such that corresponding embedding vectors roughly point in the same direction after aligning the embedding spaces. Thus, the latter algorithms exhibit a moderate, but significant degree of instability in their embeddings.

Results for the k-NN Jaccard similarity, as shown in Fig. 4(b), generally confirm these findings. For HOPE, we observe perfectly matching neighborhoods, while for GraphSAGE the neighborhoods are completely disjoint. This matches our observations for aligned cosine similarity. For the other three algorithms, the resulting similarities seem to be highly dependent on the dataset, with quite large variances. Generally, node2vec appears most stable among these algorithms, though only by a slight margin over LINE. SDNE appears to the significantly less stable than node2vec and LINE with respect to Jaccard similarity, with similarity values close to zero on BlogCatalog, Protein and Wikipedia. This contrasts the results with respect to aligned cosine similarity, where SDNE appeared as stable as the other two algorithms.

Influence of Node Centrality. Now, we analyze whether nodes that are central in their graph have more stable embeddings. Closeness centrality has been identified to be one of the top influence factors for stability in the analysis of Wang et al. [19]. Also, from the definition of node2vec we expect this algorithm, among others, to produce more stable central node embeddings since central nodes occur more often in random walks. In Fig. 5, for the Cora and Facebook datasets we plot each node's closeness centrality against a moving average with window size 25 of their average node-wise (a) k-NN Jaccard similarity, and (b) k-NN angle divergence, aggregated over all 30 embeddings per network and algorithm. First of all, the (in)stability of the extreme cases HOPE and GraphSAGE appears invariant of the centrality of the node, both in (a) and (b). For SDNE, we observe that stability with respect to k-NN Jaccard similarity appears to increase with growing closeness centrality. This trend however is not visible when considering aligned cosine similarity. For LINE and node2vec, there is no simple trend visible with respect to any of the two measures, their similarity scores look rather arbitrary. Overall, we see that although closeness centrality is ranked high in the factor analysis of Wang et al. [19], there are no clear signs that more central nodes have more stable embeddings.

Influence of Graph Properties. To evaluate the impact of graph properties on the stability of the embeddings, we generated synthetic graphs with varying sizes and densities. More precisely, we utilized two network models, namely Barabasi-Albert networks [2] and Watts-Strogatz [21] networks. For each model, we generate two sets of networks, in which we either fixed the network's size at $n = 8000$ nodes and varied its density, or fixed the densities at $D = 0.01$ and varied their size. The results of this analysis can be found in Fig. 6, where we plot the average aligned cosine similarities over all nodes and embeddings per graph and algorithm against (a) graph size and (b) graph density. Figure 6(a) contains

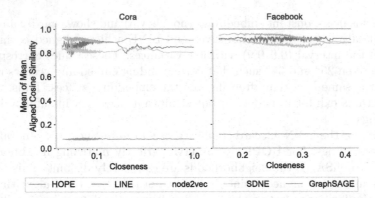

(a) Node-wise aligned cosine similarity against closeness centrality.

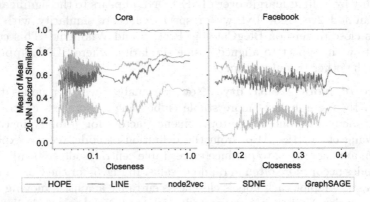

(b) Node-wise 20-NN Jaccard similarity against closeness centrality.

Fig. 5. *Influence of node centrality.* The moving averages of the node-wise (a) aligned cosine similarities and (b) 20-NN Jaccard similarities resulting from 30 embeddings per graph are plotted against each node's closeness centrality.

missing data points that result from terminating the embedding computation after a maximum of 72 h per embedding.

Considering the impact of network size, we see that for GraphSAGE, the already low stability rapidly drops with larger graph size on both synthetic models, whereas for HOPE, the near-perfect stability seems invariant of graph size. In between, LINE, SDNE and node2vec show similar stabilities like in our experiments on empirical graphs, however there is no consistent trend regarding the impact of network size on their stability. This finding contrasts results from Wang et al. [19], who stated that the stability of DeepWalk and node2vec primarily depends on the size of the input graphs.

For the dependence on network *density* plotted in Fig. 6(b), we see that the embedding stability of SDNE and node2vec seems to increase when graphs get more dense. HOPE is once again consistent in its high stability, whereas

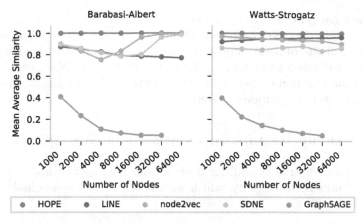

(a) Mean average aligned cosine similarity over varying sizes.

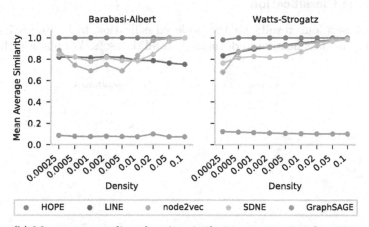

(b) Mean average aligned cosine similarity over varying densities.

Fig. 6. *Influence of graph properties.* In (a) synthetic graphs with varying size at fixed density 0.01 and in (b) synthetic graphs with varying density and 8000 nodes are used to measure the influence of those graph properties on stability. Each data point represents the average node-wise similarity over all nodes per graph and all 435 embedding pairs resulting from 30 runs of the corresponding algorithm.

GraphSAGE shows consistently low stability that is unaffected by network size. Finally, LINE does nor display any clear trend as it diverges between the two synthetic models.

Summary. Our results indicate clear differences in the geometric stability between the embedding algorithms, which is also in line with the results by Wang et al. [19]. HOPE consistently yields near-constant embeddings, whereas Graph-SAGE was shown to be very volatile. In between, the other algorithms (LINE, node2vec, and SDNE) exhibit a moderate, but significant degree of instability. When checking possible influence factors for stability, we found for none of them a strong and general trend. In particular, we observed that the influence of node centrality, graph size, and graph density have a rather small to negligible influence on the stability of node embeddings. This does not match the high ranking of the node and graph properties in the factor analysis by Wang et al. [19]. In contrast, stability is dominated by the choice of the embedding algorithm, which overshadows the aforementioned influences.

C   Additional Results on Downstream Stability

In the following we present additional plots from the experiments that we conducted on downstream stability, which we left out due to space limitations in the main part.

C.1   Node Classification

We first present our results on the node classification task. Figure 7 depicts the stability of classification performance on all datasets. We observe that over

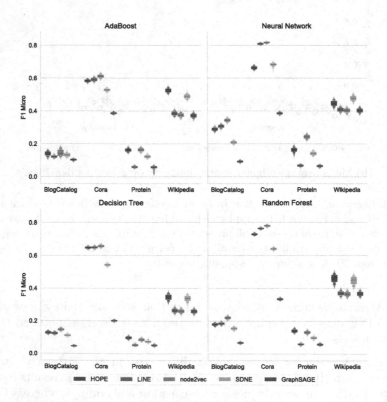

Fig. 7. *Stability of classification performance.* Stability of the micro-F1 score of the used classification methods is plotted against the used embedding algorithms. Each box corresponds to the prediction of 30 embeddings with 10 repetitions.

all algorithms and datasets, the resulting accuracies vary only marginally, and higher variances appear to depend on the datasets rather than embedding techniques (Fig. 8).

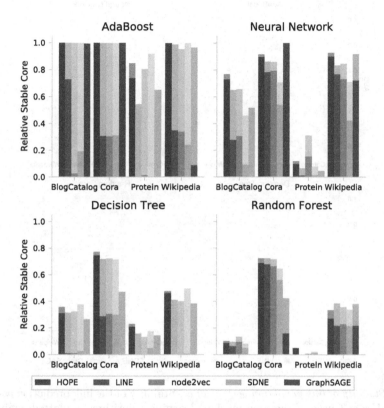

Fig. 8. *Stability in node-wise predictions.* This figure shows the stability of the classifiers as ratios of nodes which are always predicted to be in the same class. Saturated colors represent the mean stable core of all 30 embeddings and lighter colors the mean stable core of five randomly sampled embeddings with 10 repetitions each.

Our results regarding the *stability of single predictions* are shown in Fig. 7. The results on Wikipedia are mostly in line with the results that were obtained on BlogCatalog and Cora and discussed in the main part. For Protein, where we have already obtained the overall lowest accuracies in node classification, we observe an overall much lower stability in individual predictions compared to the other datasets.

C.2   Link Prediction

We close with the results regarding the stability of link prediction performance on all datasets, which are shown in Fig. 9. We observe that once again, the performance differences between different embeddings are negligible, except for neural networks on HOPE embeddings of the Protein network.

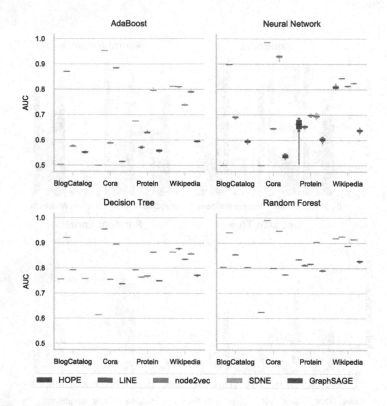

Fig. 9. *Stability of link prediction performance.* Stability of the link prediction accuracy in Area Under Curve of the used machine learning algorithms is plotted against the used embeddings algorithms. Each box corresponds to the prediction of 10 embeddings with 10 repetitions.

References

1. Antoniak, M., Mimno, D.: Evaluating the stability of embedding-based word similarities. Trans. Assoc. Comput. Linguist. **6**, 107–119 (2018)
2. Barabási, A.L., Albert, R.: Emergence of scaling in random networks. Science **286**(5439), 509–512 (1999)
3. Goyal, P., Ferrara, E.: GEM: a python package for graph embedding methods. J. Open Source Softw. **3**, 876 (2018)
4. Grover, A., Leskovec, J.: node2vec: scalable feature learning for networks. In: Proceedings of the 22nd ACM SIGKDD International Conference on Knowledge Discovery and Data Mining, pp. 855–864. ACM (2016)
5. Hamilton, W.L., Leskovec, J., Jurafsky, D.: Diachronic word embeddings reveal statistical laws of semantic change. In: Proceedings of the 54th Annual Meeting of the Association for Computational Linguistics (Volume 1: Long Papers), pp. 1489–1501 (2016)

6. Hamilton, W.L., Ying, Z., Leskovec, J.: Inductive representation learning on large graphs. In: Advances in Neural Information Processing Systems, pp. 1024–1034 (2017)
7. Hellrich, J., Hahn, U.: Bad company-neighborhoods in neural embedding spaces considered harmful. In: Proceedings of COLING 2016, the 26th International Conference on Computational Linguistics: Technical Papers, pp. 2785–2796 (2016)
8. Kunegis, J.: Konect: the koblenz network collection. In: Proceedings of the 22nd International Conference on World Wide Web, pp. 1343–1350 (2013)
9. Leskovec, J., Krevl, A.: SNAP Datasets: stanford large network dataset collection, June 2014. http://snap.stanford.edu/data
10. Leszczynski, M., May, A., Zhang, J., Wu, S., Aberger, C., Re, C.: Understanding the downstream instability of word embeddings. In: Proceedings of Machine Learning and Systems 2020, pp. 262–290 (2020)
11. Mahoney, M.: Large text compression benchmark (2011)
12. Ou, M., Cui, P., Pei, J., Zhang, Z., Zhu, W.: Asymmetric transitivity preserving graph embedding. In: Proceedings of the 22nd ACM SIGKDD International Conference on Knowledge Discovery and Data Mining, pp. 1105–1114. ACM (2016)
13. Pierrejean, B., Tanguy, L.: Predicting word embeddings variability. In: Proceedings of the Seventh Joint Conference on Lexical and Computational Semantics, pp. 154–159 (2018)
14. Rozemberczki, B., Davies, R., Sarkar, R., Sutton, C.: Gemsec: Graph embedding with self clustering. arXiv preprint arXiv:1802.03997 (2018)
15. Stark, C., Breitkreutz, B.J., Reguly, T., Boucher, L., Breitkreutz, A., Tyers, M.: Biogrid: a general repository for interaction datasets. Nucleic Acids Res. 34(suppl_1), D535–D539 (2006)
16. Šubelj, L., Bajec, M.: Model of complex networks based on citation dynamics. In: Proceedings of the 22nd International Conference on World Wide Web, pp. 527–530. ACM (2013)
17. Tang, J., Qu, M., Wang, M., Zhang, M., Yan, J., Mei, Q.: Line: large-scale information network embedding. In: Proceedings of the 24th International Conference on World Wide Web, pp. 1067–1077. International World Wide Web Conferences Steering Committee (2015)
18. Tang, L., Liu, H.: Relational learning via latent social dimensions. In: Proceedings of the 15th ACM SIGKDD International Conference on Knowledge Discovery and Data Mining, pp. 817–826. ACM (2009)
19. Wang, C., Rao, W., Guo, W., Wang, P., Liu, J., Guan, X.: Towards understanding the instability of network embedding. IEEE Trans. Knowl. Data Eng., 1 (2020)
20. Wang, D., Cui, P., Zhu, W.: Structural deep network embedding. In: Proceedings of the 22nd ACM SIGKDD International Conference on Knowledge Discovery and Data Mining, pp. 1225–1234. ACM (2016)
21. Watts, D.J., Strogatz, S.H.: Collective dynamics of 'small-world' networks. Nature 393(6684), 440 (1998)
22. Wendlandt, L., Kummerfeld, J.K., Mihalcea, R.: Factors influencing the surprising instability of word embeddings. In: Proceedings of the 2018 Conference of the North American Chapter of the Association for Computational Linguistics: Human Language Technologies, Volume 1 (Long Papers), pp. 2092–2102 (2018)
23. Zafarani, R., Liu, H.: Social computing data repository at ASU (2009). http://socialcomputing.asu.edu

Graph Homomorphism Features: Why Not Sample?

Paul Beaujean[✉], Florian Sikora, and Florian Yger

LAMSADE, CNRS, Université Paris-Dauphine, PSL Research University,
Paris, France
{Paul.Beaujean,Florian.Sikora,Florian.Yger}@lamsade.dauphine.fr

Abstract. Recent research in the domain of computed graph embeddings has shown that graph homomorphism numbers constitute expressive features that are well-suited for machine learning tasks such as graph classification. In this work-in-progress paper, we attempt to make this methodology scalable by obtaining additive approximations to graph homomorphism densities via a simple sampling algorithm. We show in experiments that these approximate homomorphism densities perform as well as homomorphism numbers on standard graph classification datasets. Moreover, we show that, unlike algorithms that compute homomorphism numbers, our sampling algorithm is highly scalable to larger graphs.

Keywords: Graph embedding · Graph homomorphism · Graph classification

1 Introduction

Graph embeddings are mappings from the set of all graphs to some well-behaved Euclidean space, which is the setting where most machine learning models operate. Approaches such as geometric deep learning [4] and more generally graph neural networks [17] attempt to learn these embeddings, but we focus instead on graph embeddings that are directly computed from the original graph data [10].

One particularly common approach in computed graph embeddings is the idea of representing graphs through statistics about particular substructures such as paths, trees, or walks. Graphlet kernels for example utilize the distribution of small induced subgraphs of size at most 5 to compare graphs [16].

In this work, we consider the notion of graph homomorphism, which can be seen as a relaxation of the notion of subgraph, that is easier to compute than induced subgraphs while being provably expressive. Statistics collected on these

Supported by Agence Nationale de la Recherche (ANR), projects STAP (ANR-17-CE23-0021) and ESIGMA (ANR-17-CE23-0010). F. Yger acknowledges the support of the ANR as part of the "Investissements d'avenir" program (ANR-19-P3IA-0001, PRAIRIE 3IA Institute).

© Springer Nature Switzerland AG 2021
M. Kamp et al. (Eds.): ECML PKDD 2021 Workshops, CCIS 1524, pp. 216–222, 2021.
https://doi.org/10.1007/978-3-030-93736-2_17

graph homomorphisms are then used as features to traditional machine learning models operating on Euclidean data. Novel ideas in computed embeddings have often led to advances in learned embeddings, see e.g. the Graph Substructure Network [3], and we are hopeful that graph homomorphisms may become a basis for new learned embeddings as well.

2   Graph Homomorphism Numbers

We will often consider a target simple undirected graph $G = (V, E)$ with n nodes and m edges from which we would like to obtain homomorphism information relative to some, usually smaller, pattern graph F with k nodes and l edges.

A graph morphism from F to G is a mapping from the node set $V(F)$ to $V(G)$. Graph homomorphisms are morphisms that preserve adjacency, i.e. $uv \in E(F)$ implies that $f(u)f(v) \in E(G)$, see e.g. Fig. 1. If a graph homomorphism is bijective then it also preserves non-adjacency and we call it a graph isomorphism. We write $G_1 \cong G_2$ when there exists an isomorphism between these two graphs.

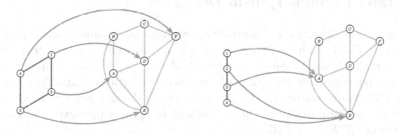

Fig. 1. Left: an homomorphism from a 4-cycle to a graph. Right: an homomorphism from a 4-path to a graph.

To see that a graph homomorphism from F to G is not the same notion as the fact that F is a subgraph of G notice that an homomorphism from a path graph on 4 nodes to G may represent a walk of size 4 in G and not only a path subgraph as can be seen in the right part of Fig. 1.

The number of graph homomorphisms from F to G is written $\hom(F, G)$ and ranges from 0 to n^k. It is convenient to combine multiple homomorphism numbers into a vector as follows: $\hom_{\mathcal{F}}(G) = (\hom(F, G))_{F \in \mathcal{F}}$. A celebrated result by Lovász [12] states the following:

Theorem 1. *Given two undirected simple graphs G_1 and G_2, both with n nodes. Denoting by $\mathcal{G}_n$ the set of all simple graphs with at most n nodes, we have:*

$$G_1 \cong G_2 \iff \hom_{\mathcal{G}_n}(G_1) = \hom_{\mathcal{G}_n}(G_2).$$

In particular this means that as long as two graphs have different homomorphism numbers for a single pattern F then they are not isomorphic to each other. Extensions of this result have been obtained by Lovász [13] and show that graphs that have close homomorphism vectors are similar in the sense that they have cuts of similar value.

NT & Maehara have proposed to use homomorphism numbers as features to machine learning models, hoping that the first few components of the homomorphism vectors would suffice in practice to classify real-life graphs [15]. Their approach was successful but remains limited in practice by the computational complexity of computing homomorphism numbers.

Indeed, the current best algorithm to compute $\hom(F, G)$ is an algorithm based on dynamic programming and tree decompositions given by Díaz et al. [8]. Its worst-case running time is $O(\text{poly}(k) \cdot n^{\text{tw}(F)+1})$ where $\text{tw}(F)$ is the treewidth of the pattern graph F. Hardness results in computational complexity show that there is no hope in finding a better algorithm, even if we settle for an approximate number that would be within $(1 \pm \varepsilon)$ of $\hom(F, G)$ [5,7].

3   Graph Homomorphism Densities

Homomorphism densities are normalized versions of homomorphism numbers. Formally, $t(F, G) = \hom(F, G)/n^k$, which means that densities live in the $[0, 1]$ interval. These quantities carry most of the properties of homomorphism numbers and constitute the basis of the theory of graph limits developed by Lovász [13]. More concretely, the homomorphism density from F to G can be interpreted as the probability that a random morphism preserve adjacency for every edge of $E(F)$ into $E(G)$.

This simple scaling allows us to use sampling to estimate $t(F, G)$ in the same manner as a coin of unknown bias, i.e. a Bernoulli distribution with unknown parameter p. Standard Chernoff bound arguments give conservative estimates to the number of morphisms that need to be drawn uniformly at random from $V(F) \to V(G)$ to obtain an additive approximation within $\pm\varepsilon$ of $t(F, G)$.

Algorithm 1 starts with drawing N morphisms uniformly at random, which is equivalent to drawing k nodes in G at random as they are the images of the k nodes of F and then checking for each edge $uv \in E(F)$ if their image is indeed in $E(G)$. The first set membership query that fails will short-circuit the condition as the morphism is not going to be an homomorphism. The total running time of Algorithm 1 is $O\left((k \log n + l) \cdot N\right)$ where $O(N \cdot k \log n)$ corresponds to sampling N morphisms and $O(N \cdot l)$ corresponds to testing whether each morphism is an homomorphism. This leads to a running time of $O\left((k \log n + l) \cdot \varepsilon^{-2} \log \delta^{-1}\right)$ to obtain a value that is guaranteed to be $t(F, G) \pm \varepsilon$ with probability $1 - \delta$.

Algorithm 1. sGHD: Sample Graph Homomorphism Density

Require: G an undirected graph on n nodes, F a pattern graph on k nodes and l edges, $\varepsilon > 0$ the requested additive precision, $1 - \delta \in (0, 1)$ the desired confidence.
Ensure: $\bar{t}$ such that $\mathbb{P}(|\bar{t} - t(F, G)| > \varepsilon) \leq \delta$
1: $N \leftarrow O(\varepsilon^{-2} \log \delta^{-1})$
2: **for** $i = 1$ to N **do**
3:    $\mathbf{f}^{(i)} \sim (\mathcal{U}([1 \mathrel{..} n]))_{[k]}$
4: **end for**
5: $\bar{t} \leftarrow \frac{1}{N} \sum_{i=1}^{N} \prod_{uv \in E(F)} \mathbb{1}\left[\mathbf{f}_u^{(i)} \mathbf{f}_v^{(i)} \in E(G)\right]$
6: **return** $\bar{t}$

The existence of this algorithm does not contradict the hardness results presented in the previous section. Indeed, if we assume that there is a single K_k in G then $t(K_k, G) = 1/n^k$. A fixed precision of ε would not be able to distinguish between $t(K_k, G) = 1/n^k$ and $t(K_k, G) = 0$. To put it simply, a fixed additive approximation would give accurate values for high $t(F, G)$ and inaccurate estimates for very low $t(F, G)$. However we will see in the next section that this theoretical behavior has a moderate impact on classification accuracy.

For more details on the scalability of Algorithm 1 and details on its implementation using adequate data structures, we invite the reader to refer to our workshop paper [1].

4   Graph Classification Using Homomorphism Features

We compare the performance of different homomorphism features on standard graph classification datasets from the TUDataset collection [14]. We note that these standard datasets mostly contain small graphs. The datasets containing the largest graphs still contain graphs with an average of a few hundred nodes. On the other hand, as Algorithm 1 has logarithmic complexity in the size of G, we would ideally use it on classification datasets containing very large graphs. However, to the best of our knowledge, there is no graph classification dataset which contains very large graphs. For example, the recently introduced Open Graph Database [11] offers a graph classification dataset of 3.8M graphs containing 14.5 nodes on average. Noting the scalability issues of exact homomorphism number algorithms like those used by NT & Maehara, we settle for well-studied datasets.

In Table 1 we collect test accuracy scores of a common cross-validation procedure [9]. SGHD models are vanilla logistic regressions trained on approximate homomorphism densities obtained by Algorithm 1 by choosing a set of pattern graphs $\mathcal{A}_{10}$ that corresponds to the first 10 connected simple graphs. $\overline{\text{SGHD}}$ models are trained on features computed from the same pattern graphs but to complement graphs instead.[1] To analyze the impact of topological information,

---

[1] Standard datasets contain relatively sparse graphs which means that the corresponding complement graphs are dense. Dense graphs have larger homomorphism densities which are easier to detect at fixed precision and more amenable for training machine learning models.

Table 1. Test accuracy scores. "-": experiment not provided in the corresponding study. *: details of experiment in Sect. 4

	MUTAG	NCI1	PROTEINS	DD	ENZYMES	REDDIT-B	COLLAB	IMDB-B
SGHD-$\mathcal{A}_{10}$ $\varepsilon = 0.1$	83.6 ± 8.7	62.6 ± 2.9	72.0 ± 4.1	**76.2 ± 3.2**	21.2 ± 4.1	73.6 ± 3.2	**68.1 ± 2.0**	63.3 ± 3.7
SGHD-$\mathcal{A}_{10}$ $\varepsilon = 0.01$	86.3 ± 7.9	62.7 ± 3.1	**72.3 ± 3.6**	76.1 ± 3.2	26.3 ± 4.5	73.8 ± 3.2	67.7 ± 2.1	68.7 ± 2.4
$\overline{\text{SGHD}}$-$\mathcal{A}_{10}$ $\varepsilon = 0.1$	83.8 ± 8.5	62.9 ± 2.9	72.1 ± 3.8	76.2 ± 3.4	23.3 ± 4.5	75.5 ± 2.7	67.3 ± 2.2	62.2 ± 3.5
GHC-$\mathcal{T}_{13}$ (NT&M.)	**88.2 ± 7.4**	**65.4 ± 2.5**	70.6 ± 4.7	75.3 ± 3.6	21.7 ± 3.6	**84.6 ± 2.3***	62.1 ± 1.9	**69.7 ± 4.4**
GHC-$\mathcal{C}_7$ (NT&M.)	-	-	-	76.1 ± 3.9	**29.5 ± 3.2**	-	-	-

we remove all weights and labels to obtain completely unlabeled undirected graphs. GHC models are SVM models trained on exact homomorphism numbers obtained by the algorithm of Díaz et al. [8]. GHC models also come with a variety of scaling algorithms in the data ingestion pipeline while SGHD do not. Because of scalability issues, GHC models are given homomorphism numbers from specific classes of graphs of low treewidth such as trees for the $\mathcal{T}_{13}$ model collecting homomorphism numbers from the first 13 trees, and cycles for $\mathcal{C}_7$. Note that Algorithm 1 is not bound by these limitations even if its complexity, like in the case of homomorphism numbers, favors small pattern graphs.

We notice that on most datasets, the performance of SGHD models is on par with exact homomorphism numbers. On some datasets there is a notable loss of accuracy depending on the precision requested e.g. in the case of IMDB-BINARY where a lower ε is required to reach higher test accuracy. In some other cases, SGHD models beat GHC ones such as COLLAB or achieve similar performance like DD. We focus our attention on the case of REDDIT-BINARY where there is a massive difference in test accuracy between SGHD and GHC. When looking closely at the supplementary material of the study by NT & Maehara [15], we notice that depending on the cross-validated scaler, test accuracy is 73.8% ± 2.8% with min/max or max/abs scaling while it's over 80% with other scaling techniques (standard, quantile, power). This shows that further preprocessing or models with higher capacity may extract more information than logistic regression or SVM models. We invite the reader to refer to our technical report for more details and comparisons to different machine learning models [2]. We conjecture that the cases where SGHD models beat GHC happen when approximate features operate as a regularizer which retaining most of the discriminating power of their exact counterparts. When the opposite happens, we suppose that information loss is responsible for most of the loss in accuracy.

5   Conclusion

We have proposed and implemented a simple randomized algorithm outputting an additive approximation of graph homomorphism densities. Our sampling algorithm is highly scalable and has a practically constant running time for a given fixed precision. This has to be contrasted with the theoretical and practical aspects of computing exact homomorphism numbers which rely on low treewidth pattern graphs and even then do not scale to large graphs.

We have shown in experiments on standard graph classification datasets that additively approximate homomorphism densities retain similar representational power compared to exact homomorphism numbers and sometimes to higher test accuracy even when they are used to train logistic regression models which are one of the simplest classifiers in the literature.

These preliminary results invite us to consider seeking larger datasets containing larger graphs to evaluate the performance of homomorphism densities in the setting that they were designed to excel. However, it remains to be seen how to train existing graph machine learning models on datasets of this scale. Another idea would be to consider learning the family of pattern graphs that best explains the information present in a given dataset.

Finally, we wish to compare the performance of models trained on homomorphism densities with alternative approaches such as the closely related graphlet kernels [6,16] or popular graph neural network architectures [17]. Moreover, we are interested in studying weighted and/or labeled variants of homomorphisms which would allow us to extract richer information from graph datasets.

References

1. Beaujean, P., Sikora, F., Yger, F.: Scaling up graph homomorphism features with efficient data structures. In: ICLR 2021 Workshop on Geometrical and Topological Representation Learning (2021). https://openreview.net/forum?id=EwT8NpZIth8
2. Beaujean, P., Sikora, F., Yger, F.: Scaling up graph homomorphism for classification via sampling. CoRR abs/2104.04040 (2021). https://arxiv.org/abs/2104.04040
3. Bouritsas, G., Frasca, F., Zafeiriou, S., Bronstein, M.M.: Improving graph neural network expressivity via subgraph isomorphism counting. CoRR abs/2006.09252 (2020). https://arxiv.org/abs/2006.09252
4. Bronstein, M.M., Bruna, J., LeCun, Y., Szlam, A., Vandergheynst, P.: Geometric deep learning: going beyond Euclidean data. IEEE Signal Process. Mag. **34**(4), 18–42 (2017)
5. Bulatov, A., Živný, S.: Approximate counting CSP seen from the other side. ACM Trans. Comput. Theory **12**, 1–19 (2020). https://doi.org/10.1145/3389390
6. Curticapean, R., Dell, H., Marx, D.: Homomorphisms are a good basis for counting small subgraphs. In: Proceedings of the 49th Annual ACM SIGACT Symposium on Theory of Computing, pp. 210–223 (2017)
7. Dalmau, V., Jonsson, P.: The complexity of counting homomorphisms seen from the other side. Theoret. Comput. Sci. **329**(1–3), 315–323 (2004)
8. Díaz, J., Serna, M., Thilikos, D.M.: Counting H-colorings of partial k-trees. Theor. Comput. Sci. **281**(1–2), 291–309 (2002)
9. Errica, F., Podda, M., Bacciu, D., Micheli, A.: A fair comparison of graph neural networks for graph classification. In: ICLR (2019)
10. Grohe, M.: word2vec, node2vec, graph2vec, x2vec: towards a theory of vector embeddings of structured data. In: PODS (2020)
11. Hu, W., Fey, M., Ren, H., Nakata, M., Dong, Y., Leskovec, J.: OGB-LSC: a large-scale challenge for machine learning on graphs. arXiv preprint arXiv:2103.09430 (2021)

12. Lovász, L.: Operations with structures. Acta Mathematica Academiae Scientiarum Hungarica (1967)
13. Lovász, L.: Large networks and graph limits. American Mathematical Soc. (2012)
14. Morris, C., Kriege, N.M., Bause, F., Kersting, K., Mutzel, P., Neumann, M.: TUDataset: A collection of benchmark datasets for learning with graphs. In: ICML Workshop on Graph Representation Learning and Beyond (GRL+) (2020). www.graphlearning.io
15. NT, H., Maehara, T.: Graph homomorphism convolution. In: ICML (2020)
16. Shervashidze, N., Vishwanathan, S., Petri, T., Mehlhorn, K., Borgwardt, K.: Efficient graphlet kernels for large graph comparison. In: AIStat (2009)
17. Wu, Z., Pan, S., Chen, F., Long, G., Zhang, C., Philip, S.Y.: A comprehensive survey on graph neural networks. IEEE Trans. Neural Networks Learn. Syst. (2020)

Neural Maximum Independent Set

Thomas Pontoizeau[✉], Florian Sikora, Florian Yger, and Tristan Cazenave

LAMSADE, CNRS, Université Paris-Dauphine, PSL Research University, Paris, France
{Thomas.Pontoizeau,Florian.Sikora,Florian.Yger,
Tristan.Cazenave}@lamsade.dauphine.fr

Abstract. The emergence of deep learning brought solutions to many difficult problems and has recently motivated new studies that try to solve hard combinatorial optimization problems with machine learning approaches. We propose a framework based on Expert Iteration, an imitation learning method that we apply to solve combinatorial optimization problems on graphs, in particular the Maximum Independent Set problem. Our method relies on training GNNs to recognize how to complete a solution, given a partial solution of the problem as an input. This paper emphasizes some interesting findings such as the introduction of learned nodes features helping the neural network to give relevant solutions. Moreover, we represent the space of good solutions and discuss the ability of GNN's to solve the problem on a graph without training on it.

1 Introduction

Given a graph, finding a set of pairwise non-adjacent nodes of maximum size is probably one of the most classical hard algorithmic graph problem [20,25] denoted MAXIMUM INDEPENDENT SET. Such a problem finds natural applications to non overlapping problems, including automatic label placement for instance [2].

MAXIMUM INDEPENDENT SET can also be seen as a 1-player game where the player chooses iteratively a new node to select in order to form the largest possible independent set. Considering such an optimization problem as a game opens the possibility of applying machine learning methods such as neural networks and deep reinforcement learning. Neural networks came back to the light when it made it possible to solve difficult and large scale classification tasks in computer vision. Nowadays, the development of graph convolution layers paves the way to the application of neural networks to graph problems. If the Operational Research community studied the resolution of such graph problems for 50 years, *learning* how to solve combinatorial optimization problems has been a recent interest in the machine learning community with the emergence of scalable methods like Graph Neural Networks (GNN) [13,15,28,36].

Deep Reinforcement Learning gave state of the art results on complex two-player complete information games such as Go, Chess and Shogi [30]. In those approaches, the underlying reinforcement learning paradigm is Expert Iteration [5]. It is a general

The authors acknowledge the support of the ANR as part of the "Investissements d'avenir" program (ANR-19-P3IA-0001, PRAIRIE 3IA Institute) and through the project DELCO (ANR-19-CE23-0016).

© Springer Nature Switzerland AG 2021
M. Kamp et al. (Eds.): ECML PKDD 2021 Workshops, CCIS 1524, pp. 223–237, 2021.
https://doi.org/10.1007/978-3-030-93736-2_18

algorithm that uses Monte Carlo Tree Search to leverage the learning of a policy and a value. Inspired by its successes on difficult games, we borrow similar ideas to apply it to combinatorial optimization problems on graphs. In this work, we focus on learning a policy for the MAXIMUM INDEPENDENT SET problem. To do so, we use Monte Carlo Search to generate diversified and well-explored states of this problem for each instance and use a Graph Neural Network to learn the resulting policy.

In this paper, we investigate several ingredients giving the ability to GNN to learn how to solve combinatorial optimization problems. Our main contributions are three-fold, first, we show that learning additional node features boosts the performance of GNNs. Secondly, we show how such a neural network can be used to generate a diversi-fied pool of good solutions, giving useful informations about the distribution of the best solutions and potentially indicating how hard it is to find the optimal solution. Finally, we show how to transfer knowledge between graphs, training a GNN on a small graph and applying it successfully on a larger graph. Although we primarly focus on MAX-IMUM INDEPENDENT SET problem, it is merely a case study and our framework is versatile enough to be adapted to other cases.

First, we introduce the relevant literature about solving combinatorial optimization problems with neural networks in Sect. 2. Then, in Sect. 3, we present the basic concepts and notations used throughout this paper. In Sect. 4, we provide the details about our method and how the results were obtained. The results of our numerical experiments are discussed in Sect. 5. Finally, we end the paper with concluding remarks and a roadmap for future works.

2 Related Work

Combinatorial optimization problems in graphs like MAXIMUM INDEPENDENT SET are widely studied in many fields and can be viewed from many different perspectives.

Our paper focuses on MAXIMUM INDEPENDENT SET which is a hard problem to solve. In terms of theoretical complexity, the problem is known to be NP-hard [20], *i.e.* it is not believed that there exists any polynomial algorithm for efficiently find-ing its optimal solution. Even from the approximation point of view, the problem is hard to approximate [25]. However it is possible to solve MAXIMUM INDEPENDENT SET by exact exponential approaches like divide-and-conquer methods [32] or branch-and-bounds methods [8,26]. Some efficient heuristics, although coming without any guarantee on the quality of the solution, have been devised [14,22].

Recently, several approaches tried to solve combinatorial optimization problems by using machine learning techniques [7,21,24]. It gained momentum with the emer-gence of Graph Neural Networks and their expressive power [13,36] and it motivated researchers to investigate how to apply GNN to graph combinatorial optimization prob-lems [15,18].

For example, Abe et al. applied Alpha Zero for several combinatorial optimiza-tion problems, including MAXIMUM INDEPENDENT SET [1]. Other works combined branch-and-bound methods with GNNs [16]. In [19], the authors introduced an unsu-pervised learning framework inspired by Erdős probabilistic method. Reinforcement learning has also been investigated [12]. In [21], the authors proposed a framework combining reinforcement learning and graph embeddings in order to incrementally construct the solution.

In our work, we build a framework that solves combinatorial optimization problem based on Expert Iteration [5], a promising reinforcement algorithm based on imitation learning. Our approach is innovative as it introduces learned features that help to improve the performance of the GNNs.

On the other hand, it has to be noted that GNNs are not always working perfectly. More precisely, it has been shown that stacking too many layers was detrimental as it lead to over-smoothing, *i.e.* the features of the nodes are propagating too far in the network and all information on each node vanishes [10]. In addition, when solving a combinatorial optimization problem in a supervised manner, some difficulties have been highlighted for generating unbiased representative labels for large scale instances [35]. In that way, we investigate transferring knowledge from small instances to larger instances and give encouraging results for future work.

3   Preliminaries

The Maximum Independent Set Combinatorial Optimization Problem. In the remainder of this paper, we denote by $G = (V, E)$ an undirected graph in which V (resp. E) is the set of nodes (resp. edges). The distance between two nodes is the number of edges in the shortest path connecting them.

Given a graph $G = (V, E)$, an independent set is a set $S \subset V$ of nodes such that $\forall i, j \in S, i \neq j \Rightarrow \{i, j\} \notin E$. The problem MAX INDEPENDENT SET consists in finding an independent set of maximum size.

Seeing Max Independent Set as a Game. It is possible to interpret MAX INDEPENDENT SET as a game in which the goal is to find a set of vertices of maximum cardinality by iteratively selecting a node under the constraints that the current set of selected nodes always forms an independent set. The game ends when there is no more node that can be selected and the score is the number of selected nodes.

In order to make the reading clear, we introduce some notations. All following definitions are always considering a given undirected simple (without any multiple edges nor any self-loops) graph $G = (V, E)$.

A *state* is a partial solution for MAX INDEPENDENT SET in G, *i.e.* a set of nodes $S \subset V$ that forms an independent set, not necessarily of maximal size. We call *initial state* the state in which no node has been selected, and *final state* a state of maximal size, *i.e.* a state in which no more nodes can be added to the solution without compromising the constraint of being an independent set. Sometimes, final states are naturally called *solutions*.

Given a state, a *legal move* is a node that does not belong to the state and can be added to it such that the whole set still forms an independent set.

Given a state, a *sequence* is a set of nodes not belonging to the state that can be added to it and form a final state. We call *labeled state* a couple {state, sequence}.

Given a state, a *policy* for the legal moves is a distribution of their probability to be selected.

Given a state and a method to obtain a policy for any state, a *rollout* is a sequence obtained by choosing nodes iteratively according to probability provided by the method.
Graph Neural Networks. A Graph Neural Network (GNN) is a neural network that contains graph convolutional layers that leverages the structure of a graph [36]. A graph convolutional layer takes the features of a node as an input and usually aggregates transformed features from the node and its neighbors into a set of output features. One of the major advantages of GNNs compared to other standard neural networks is that a GNN layer can be applied to any graph of any size.

In our neural networks, we used convolutional layers from *Graph Isomorphic Network* (GIN) [34] that have been shown to have good empirical performances. Note that we also tested Graph Attention Networks (GAT) [31] layers with a substantial drop in performance.
Expert Iteration. In imitation learning, an *expert* (usually represented by a tree search) creates a dataset of states with target moves drawn by him. Then, a *learner* (usually represented by a neural network) tries to imitate the behavior of the expert by predicting the policy among the moves. Expert Iteration [5] is a reinforcement learning algorithm based on imitation learning that repeats the process and improves the tree search at each iteration by using the learner policy to guide search during the dataset creation and the tree search, increasing the performance of the expert.

4   Method

4.1   Using Expert Iteration to Solve Max Independent Set

Labeling Phase. The first step of our algorithm consists in produces a set of labeled solutions that we attempt to predict in the learning phase.

First, the expert generates 500 random states that are not final with a uniform policy (*i.e.* probability distribution among the current legal moves). Then, for each randomly generated state, we designed the expert to make 50 random rollouts and keep the one with the best score as the sequence in the labeled state. Those settings allow the expert to provide a training set in short time and still give meaningful results for our observations, and turns out to be enough to reach the state-of-art in terms of quality of the solution for the studied datasets.

This naïve Monte Carlo search approach has the convenience to be very quick to execute and still to allow to highlight the quality improvement of the learning.

Learning Phase. The set of labeled states constitutes the training set of the GNN. Our learner is a graph neural network that we train on the training set, allowing to compute outputs for any state. Given a state, the learner tries to predict the nodes to select to complete the partial solution by mimicking the expert plays. Therefore, each input of the GNN represents a state of the game. To enrich the input, each node can be described by the following four possible groups of features:

- a $0-1$ feature *state* indicating if the node belongs to the current state or not.
- a $0-1$ feature *legal_moves* indicating if the node is a legal future move considering the current state.

- two random features *footprints* belonging to $[0, 1]$ that characterizes uniquely each node. This relates to the use of random features that have been proved to be helpful for GNN learning for solving other combinatorial problems like MINIMUM DOMINATING SET [29] by singularizing the nodes and breaks symmetries.
- two features *distances* indicating the number of nodes at distance 1 (resp. 2) from it.

Each group of feature is normalized according to the ℓ_2-norm.

In our architecture, we added the possibility to concatenate two additional features to the input for each nodes before the first layer that are learned during the training (*i.e.* those features are considered as weights of the neural network). We call them *node embeddings*. Thus, when we put all features for each node, each input tensor has a shape $(|V|, 8)$.

Our neural network has the following architecture. One block consists of one Conv1D layer (linear transformation that expands each node features separately), one GIN layer (containing ReLu layers and batchnorm layers), one dropout layer, and one Conv1D layer. The network consists of k blocks, adding the initial residual (copy of the initial input) to the result between each block, and finally apply a final sigmoid. In our experiments, GAT layers instead of GIN ones did not really give better performance and two blocks were usually enough. Adding the initial residual has been motivated in [11]. The target is a one-hot-vector of the nodes of the sequence given by the labeling phase. We then naturally evaluate the loss between the target and the output using the Binary Cross Entropy loss.

Note that at the beginning of each learning phase, we reset the parameters of the neural network. It makes sense since the expert plays are better each iteration. Once the GNN is trained, it can be used as a policy generator by applying a softmax function to the output for any given state, defined as: $softmax(z, t)_i := e^{\frac{z_i}{t}} / \sum_{k \in \{1...n\}} e^{\frac{z_k}{t}}$, with $z = (z_k)_{k \in \{1...n\}}$ the output of the model, t the temperature (which is a parameter). Notice that the greater t is, the closest to uniform policy the softmax function is. When t is close to 0, the softmax function tends to give the maximum value of the output.

Our approach consists in repeating the labeling phase and the learning phase, using the learner to perform rollouts instead of using a uniform policy for the labeling phase. Except for `dimacs-frb30-15-1`, our results were obtained within actually only one iteration (one labeling phase and one learning phase).

4.2 Evaluation of the Neural Network

In order to control how relevant is our neural network, we introduce two methods to evaluate it.

- **Argmax sequence method.** Starting with the initial state, compute the output of the neural network on the current state and choose the node with the highest activation among the legal moves. Iterate until the state is final. We call such a sequence the *argmax sequence*.
- **Stochastic exploration method.** To produce one solution, start with the initial state (*i.e.* the state in which no node has been chosen yet in the solution.), compute the output of the neural network on the current state and apply the softmax function with

a low temperature that we call the *exploration temperature* (in our experiments, we used a temperature of 0.106). Then, the next node is chosen according to the distribution of probabilities given by the softmax function among legal moves. Iterate until the state is final. Repeat the previous procedure k times to obtain a pool of k solutions as the result of the evaluation.

The latter gives richer information about the neural network performance, but is more than k times slower then the former. We sometimes use the first method as a simple and quick way to observe the quality of the neural network at each moment of the learning phase.

We tuned the hyper-parameters with Optuna [3]. The optimized algorithm to do so consisted in one labeling phase, one learning phase and one stochastic exploration for evaluation. The simplest objective function we could use is the score of the best found solution during the stochastic exploration, but we wanted to refine it by adding a notion of diversity in the pool of best solutions. In order to do so, we optimized the quality q of the set of the solution with the best found score, according to the following definition: $q = (b + 1) - 1/(\sqrt{|S|} \cdot trace(covariance(S^T)))$, with b the best score found during the stochastic exploration and S the set of found solutions with score b. In this way, the best found score is always highlighted by the objective function, but some bonus is given if the set of solutions with such score is large and diverse.

Our hyper-parameters suggested by Optuna comprise a very low exploration temperature (0.106), a dropout value of 0.2, and 145 epochs.

5   Experimental Results

In this section, we present interesting observations about learning how to solve MAX INDEPENDENT SET with GNNs. We highlight our experimental results among the following instances of graphs:

Instance	$|V|$		$|E|$	best known score
ba200_5	200	975	82	
er200_10	200	1957	41	
dimacs-frb30-15-1	450	17827	30	
bio-SC-LC	2004	20452	968	

er200_10 is a standard generated Erdős-Renyi graph that have edge probability $p = 0.10$ and ba200_5 is a Barabási-Albert generated graph in which 5 edges has been added per node [6].

dimacs-frb30-15-1[1] is a difficult instance from the Benchmarks with Hidden Optimum Solutions for Graph Problems [33] provided in [1] bio-SC-LC is a real-world instance from [27]. These four instances are a subset of instances used in [1].

---

[1] We kept the same name used in [1] but it is actually an instance of BHOSLIB [33].

5.1   Boosting the Learning Phase with Good Features

In this subsection we emphasize two major points. First, GNNs can be trained on a set of quickly labeled states and perform well to solve MAX INDEPENDENT SET. Second, adding embeddings to node features boosts the learning phase and helps the GNN to learn more efficiently.

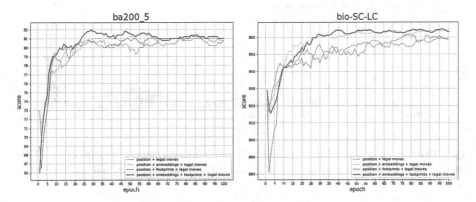

Fig. 1. The score of the argmax playout each epoch of the learning on each instance. Learning curves with embeddings are highlighted by a plain curve.

In Fig. 1, our GNN learns the label of a set of 500 labeled states given by the labeling phase for 100 epochs on each instance of our dataset (results for other instances can be found in appendix). At each epoch, we evaluate the GNN by calculating the score of the argmax sequence. We represent some curves with different ways to select footprints and embeddings among the features we introduced in last section. In order to make the graphic more readable, the curves have been smoothed by representing the average score of the 5 last scores for each epoch.

Since adding randomized features to the nodes has been observed quite efficient to singularize the nodes and to break symmetries [29], we included those new features as *footprints*, and also suggested to use the same idea allowing such footprint to be learned by the neural network (calling them *embeddings*). It appears clearly that all curves describing the performance of a network involving embeddings give better performance than ones without, and perform more efficiently than standard randomized features (see Fig. 1). On dimacs-frb30-15-1, our GNN showed more difficulty to learn but eventually succeeded on this difficult instance.

Those observations are based on a quick evaluation but can be made more precisely by using the second way to evaluate the GNN. In Fig. 2, we represented all the scores found in the exploring phase (in gray), the ones found by making playouts from the initial state with a GNN that trained on inputs with the state as a unique feature in light blue (resp. all features in dark blue). Resultas for other instances can be found in the appendix. The three sets contain 500 solutions each. Our observation show that the GNN does not succeed in beating the scores of the expert on dimacs-frb30-15-1.

We give some interpretation in next subsections based on the cluster of best found solutions. Overall, using all features clearly appears to be a good setting and outperform the expert plays and the stochastic exploration with no extra features.

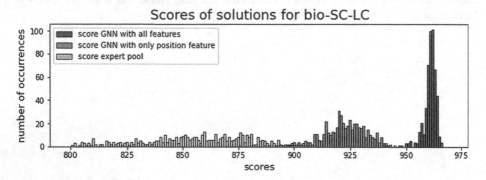

Fig. 2. Distribution of scores of 500 solutions from the training set and an exploration of 500 rollouts with the GNN for instance `bio-SC-LC`, first with only the state, and then with all the features. (Color figure online)

Once a neural network has learned on a small graph, it performs a little better on other bigger instances it never encountered, removing the footprint features and the embedding features during the learning phase since they had sense only in a particular instance. However, the quality of the learning is quite less effective than with all features. A discussion about this observation can be found in the appendix.

6 Conclusion

In this paper, we combined GNN with expert iterations to learn how to solve MAX INDEPENDENT SET problems and improved the overall network architecture with learned node features. We brought empirical evidences of the soundness of our approach and gave some qualitative results about the space of solutions explored by our approach.

On standard instances, GNNs seem to learn correctly how to solve MAX INDEPENDENT SET and are able to give good solutions with some easy generated training set. Our method seems to performs well on real-world instances but shows some difficulties on artificially created hard instances. Besides, we emphasized some interesting observations about the improvement made by adding embedding features to nodes. This could be investigated further in future work. In particular, we do not know yet what those embeddings are exactly learning, how meaningful they are, and how much they can be exported for other tasks.

Furthermore, our framework is general and could fit many problems such as MIN VERTEX COVER, MAX CLIQUE, or the weighted versions of them (by simply putting the weight instead of a $0 - 1$ activation in the state feature).

A   Results on other instances

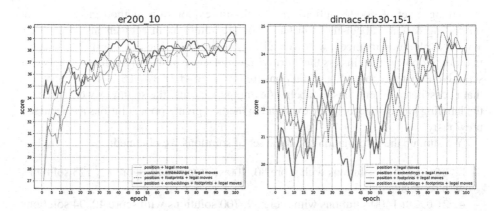

Fig. 3. The score of the argmax playout each epoch of the learning on each instance. Learning curves with embeddings are highlighted by a plain curve.

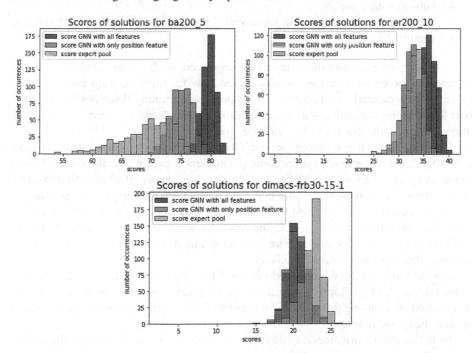

Fig. 4. Distribution of scores of 500 solutions from the training set and an exploration of 500 rollouts with the GNN for each instance, first with only the state, and then with all the features.

B   Stochastic exploration method allows to find various good solutions

By giving a set of good solutions, the stochastic exploration method gives us some interesting insights about the distribution of the solutions in a given instance. After several exploration methods, we saved all good solutions of the three best found scores in order to observe the clusters they form (*i.e.* in the sense of hamming distance between sets). Note that the pool of best solutions for dimacs-frb30-15-1 is the only one that necessitated 5 iterations of labeling/learning phase and the solutions were found in the labeling phase, the stochastic exploration method giving poor solutions as discussed previously.

We sum up the information about the set of solutions we obtained:

- ba200_5: 1400 solutions with score 80, 700 solutions with score 81, 123 solutions with score 82.
- er200_10: 1400 solutions with score 39, 700 solutions with score 40, 94 solutions with score 41.
- dimacs-frb30-15-1: 499 solutions with score 26, 77 solutions with score 27, 4 solutions with score 28.
- bio-SC-LC: 21 solutions with score 966, 5 solutions with score 967, 1 solution with score 968.

In order to observe how the solutions are organized, we first computed all hamming distances between each pair of solution to see how far they are from each other. In Fig. 5, we represented the distribution of all possible hamming distances for the three best found scores for each instance. For er200_10 and dimacs-frb30-15-1, we also represented the graphic in log scale for more readability.

For ba200_5, the solutions are well organized along a Gaussian curve. For er200_10, we observe almost the same phenomenon except that the best solutions (with score 41) are split into three clusters: one big cluster of close solutions, one sparse cluster of partially close solutions, and a small sparse cluster of solutions. For dimacs-frb30-15-1, all best found solutions look very sparse and have very few nodes in common. For bio-SC-LC, the solutions seem to be well organized and not so far from each other. Note that there is no hamming distance for the score 968 since we only obtained one solution for this score.

For each instance, we embedded our pool of solutions into a three dimensional space with a t-SNE [23] in Fig. 6, that allows to observe our previous remarks on the distribution of hamming distances. We can notice that the scatterplots are always nested around the best solutions.

Note that for the instances ba200_5 and er200_10, we had to sample the second and third best found solutions (in green and orange) in order to highlight the pool of best solutions (red).

With the instance dimacs-frb30-15-1, we observe that all best found solutions look very sparse in the sense that the solutions have very few nodes in common. This could explain why GNN cannot converge properly into a good solution. The latter

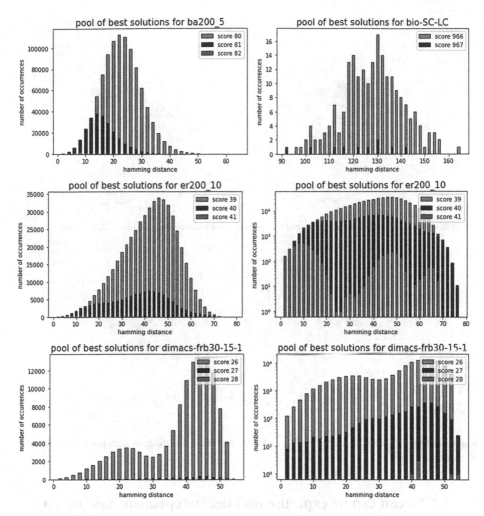

Fig. 5. The distribution of hamming distances between best found solutions for each instance of the dataset. Sometimes we also give the same graphic in log scale for more readability.

remark could explain why our neural network has so much difficulty to learn how to solve the problem on this instance and converge to good solutions.

Since enumerating maximal or maximum independent sets has also been studied in the literature [4,9,17], our method provides some interesting tools to obtain good various solutions with the help of a GNN.

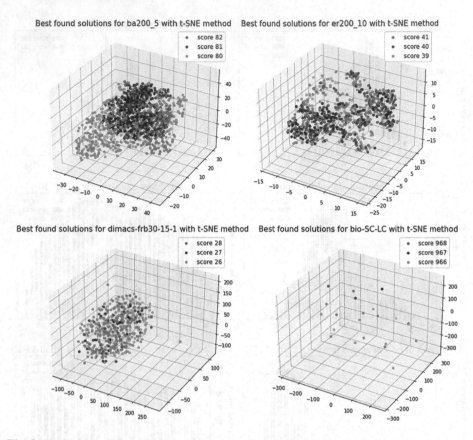

Fig. 6. The projection of the set of best found solutions for each instance with the t-SNE method. (Color figure online)

C   GNNs can export expertise on Max Independent Set from a small graph to a larger graph

The last observation we want to highlight is that once a neural network has learned on a small graph, it performs a little better on other bigger instances it never saw.

In order to do so, we had to remove the footprint features and the embedding features during the learning phase since they had sense only in a particular instance. Thus, the quality of the learning is quite less effective than with all features.

In our experiment, we compare training on ba200_5 and er200_10 and see how much the GNN performs on the other instances (including two other instances: ba100_5 and ba1000_5 constructed in the same way than ba200_5).

After a labeling phase, we make a learning phase and compute the score of the argmax sequence for each other instance using the trained model at each epoch. We then smoothed the curves by computing the average score of the last 5 scores at each epoch, and then represented the approximation ratio between the score of the argmax sequence and the best known result on the instance for each epoch. The resulted curves are represented in Fig. 7.

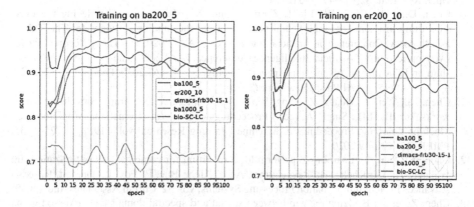

Fig. 7. Evolution of approximation ratio (compared to the best known score) on the argmax playout on other graph training on ba200_5 and er200_10.

When training on `ba200_5`, the quality of the argmax sequence improves quickly at the beginning and then stagnates. When training on `er200_10`, the qualitywork of the argmax sequence increases slower than with `ba200_5` but increases all along the learning. On the other hand, not surprisingly, note that the learning phase did not make any improvement on `dimacs-frb30-15-1`. Those observations indicate that our GNN was able to transfer some knowledge about MAXIMUM INDEPENDENT SET on brand new graphs, and is promising for future work.

References

1. Abe, K., Xu, Z., Sato, I., Sugiyama, M.: Solving NP-hard problems on graphs with extended AlphaGo zero (2020)
2. Agarwal, P.K., van Kreveld, M., Suri, S.: Label placement by maximum independent set in rectangles. Comput. Geom. **11**(3), 209–218 (1998)
3. Akiba, T., Sano, S., Yanase, T., Ohta, T., Koyama, M.: Optuna: a next-generation hyperparameter optimization framework. In: Proceedings of the 25th ACM SIGKDD International Conference on Knowledge Discovery & Data Mining, KDD 2019, pp. 2623–2631. Association for Computing Machinery, New York (2019)
4. Alexander, J., Mink, T.: A new method for enumerating independent sets of a fixed size in general graphs. J. Graph Theory **81**(1), 57–72 (2016)
5. Anthony, T., Tian, Z., Barber, D.: Thinking fast and slow with deep learning and tree search. In: Advances in Neural Information Processing Systems, pp. 5360–5370 (2017)
6. Barabási, A.L., Albert, R.: Emergence of Scaling in Random Networks. Science **286**(5439), 509–512 (1999)
7. Bengio, Y., Lodi, A., Prouvost, A.: Machine learning for combinatorial optimization: a methodological tour d'horizon. Eur. J. Oper. Res. **290**(2), 405–421 (2021)

8. Bourjolly, J.M., Laporte, G., Mercure, H.: A combinatorial column generation algorithm for the maximum stable set problem. Oper. Res. Lett. **20**(1), 21–29 (1997)
9. Byskov, J.M.: Enumerating maximal independent sets with applications to graph colouring. Oper. Res. Lett. **32**(6), 547–556 (2004)
10. Chen, D., Lin, Y., Li, W., Li, P., Zhou, J., Sun, X.: Measuring and relieving the over-smoothing problem for graph neural networks from the topological view. In: The Thirty-Fourth AAAI Conference on Artificial Intelligence, AAAI 2020, The Thirty-Second Innovative Applications of Artificial Intelligence Conference, IAAI 2020, The Tenth AAAI Symposium on Educational Advances in Artificial Intelligence, EAAI 2020, February 7–12, 2020, pp. 3438–3445. AAAI Press, New York (2020)
11. Chen, M., Wei, Z., Huang, Z., Ding, B., Li, Y.: Simple and deep graph convolutional networks. In: III, H.D., Singh, A. (eds.) Proceedings of the 37th International Conference on Machine Learning. Proceedings of Machine Learning Research, vol. 119, pp. 1725–1735. PMLR, 13–18 July 2020
12. Chen, X., Tian, Y.: Learning to perform local rewriting for combinatorial optimization. In: Wallach, H., Larochelle, H., Beygelzimer, A., d' Alché-Buc, F., Fox, E., Garnett, R. (eds.) Advances in Neural Information Processing Systems, vol. 32. Curran Associates, Inc. (2019)
13. Chen, Z., et al.: Bridging the gap between spatial and spectral domains: a survey on graph neural networks (2020)
14. Das, K.N., Chaudhuri, B.: Heuristics to find maximum independent set: An overview. In: Deep, K., Nagar, A., Pant, M., Bansal, J.C. (eds.) Proceedings of the International Conference on Soft Computing for Problem Solving (SocProS 2011) 20–22 December, 2011, pp. 881–892. Springer, India (2012)
15. Dwivedi, V.P., Joshi, C.K., Laurent, T., Bengio, Y., Bresson, X.: Benchmarking Graph Neural Networks. arXiv e-prints (2020)
16. Gasse, M., Chételat, D., Ferroni, N., Charlin, L., Lodi, A.: Exact combinatorial optimization with graph convolutional neural networks. In: Wallach, H.M., Larochelle, H., Beygelzimer, A., d'Alché-Buc, F., Fox, E.B., Garnett, R. (eds.) Advances in Neural Information Processing Systems 32: Annual Conference on Neural Information Processing Systems 2019, NeurIPS 2019, 8–14 December, 2019, Vancouver, BC, Canada, pp. 15554–15566 (2019)
17. Gurski, F., Rehs, C.: Counting and enumerating independent sets with applications to combinatorial optimization problems. Math. Methods Oper. Res. **91**(3), 439–463 (2020)
18. Hu, W., et al.: Open graph benchmark: Datasets for machine learning on graphs (2021)
19. Karalias, N., Loukas, A.: Erdos goes neural: an unsupervised learning framework for combinatorial optimization on graphs. In: Larochelle, H., Ranzato, M., Hadsell, R., Balcan, M.F., Lin, H. (eds.) Advances in Neural Information Processing Systems, vol. 33, pp. 6659–6672. Curran Associates, Inc. (2020)
20. Karp, R.M.: Reducibility among Combinatorial Problems, pp. 85–103. Springer, US (1972)
21. Khalil, E., Dai, H., Zhang, Y., Dilkina, B., Song, L.: Learning combinatorial optimization algorithms over graphs. In: Guyon, I., Luxburg, U.V., Bengio, S., Wallach, H., Fergus, R., Vishwanathan, S., Garnett, R. (eds.) Advances in Neural Information Processing Systems, vol. 30. Curran Associates, Inc. (2017)
22. Lamm, S., Sanders, P., Schulz, C., Strash, D., Werneck, R.F.: Finding near-optimal independent sets at scale. J. Heuristics **23**(4), 207–229 (2017)
23. Van der Maaten, L., Hinton, G.: Visualizing data using t-sne. J. Mach. Learn. Res. **9**(11) (2008)
24. Mazyavkina, N., Sviridov, S., Ivanov, S., Burnaev, E.: Reinforcement learning for combinatorial optimization: a survey (2020)
25. Piotr, B., Marek, K.: On some tighter inapproximability results. Technical report (1999)
26. Rossi, F., Smriglio, S.: A branch-and-cut algorithm for the maximum cardinality stable set problem. Oper. Res. Lett. **28**(2), 63–74 (2001)

27. Rossi, R.A., Ahmed, N.K.: The Network Data Repository with Interactive Graph Analytics and Visualization. In: Bonet, B., Koenig, S. (eds.) Proceedings of the Twenty-Ninth AAAI Conference on Artificial Intelligence, January 25–30, 2015, Austin, Texas, USA, pp. 4292–4293. AAAI Press (2015)

28. Sato, R., Yamada, M., Kashima, H.: Approximation ratios of graph neural networks for combinatorial problems. In: Wallach, H., Larochelle, H., Beygelzimer, A., d' Alché-Buc, F., Fox, E., Garnett, R. (eds.) Advances in Neural Information Processing Systems, vol. 32. Curran Associates, Inc. (2019)

29. Sato, R., Yamada, M., Kashima, H.: Random Features Strengthen Graph Neural Networks. CoRR abs/2002.03155 (2020). https://arxiv.org/abs/2002.03155

30. Silver, D., et al.: A general reinforcement learning algorithm that masters chess, shogi, and go through self-play. Science 362(6419), 1140–1144 (2018)

31. Velickovic, P., Cucurull, G., Casanova, A., Romero, A., Liò, P., Bengio, Y.: Graph attention networks. In: 6th International Conference on Learning Representations, ICLR 2018, Vancouver, BC, Canada, April 30–3 May 2018, Conference Track Proceedings (2018)

32. Xiao, M., Nagamochi, H.: Exact algorithms for maximum independent set. Inf. Comput. 255, 126–146 (2017)

33. Xu, K.: BHOSLIB: Benchmarks with Hidden Optimum Solutions for Graph Problems (Maximum Clique, Maximum Independent Set, Minimum Vertex Cover and Vertex Coloring). http://sites.nlsde.buaa.edu.cn/~kexu/benchmarks/graph-benchmarks.htm

34. Xu, K., Hu, W., Leskovec, J., Jegelka, S.: How Powerful are Graph Neural Networks? In: 7th International Conference on Learning Representations, ICLR 2019, New Orleans, LA, USA, 6–9 May 2019 (2019)

35. Yehuda, G., Gabel, M., Schuster, A.: It's not what machines can learn, it's what we cannot teach. In: III, H.D., Singh, A. (eds.) Proceedings of the 37th International Conference on Machine Learning. Proceedings of Machine Learning Research, vol. 119, pp. 10831–10841 (2020)

36. Zhou, J., et al.: Graph neural networks: a review of methods and applications (2019)

Fea2Fea: Exploring Structural Feature Correlations via Graph Neural Networks

Jiaqing Xie[1($\boxtimes$)] and Rex Ying[2]

[1] University of Edinburgh, Edinburgh, Scotland
s2001696@ed.ac.uk
[2] Stanford University, Stanford, USA
rexying@stanford.edu

Abstract. Structural features are important features in a geometrical graph. Although there are some correlation analysis of features based on covariance, there is no relevant research on structural feature correlation analysis with graph neural networks. In this paper, we introduce graph feature to feature (**Fea2Fea**) prediction pipelines in a low dimensional space to explore some preliminary results on structural feature correlation, which is based on graph neural network. The results show that there exists high correlation between some of the structural features. An irredundant feature combination with initial node features, which is filtered by graph neural network has improved its classification accuracy in some graph-based tasks. We compare differences between concatenation methods on connecting embeddings between features and show that *the simplest is the best*. We generalize on the synthetic geometric graphs and certify the results on prediction difficulty between structural features.

Keywords: Graph neural networks · Feature engineering

1 Introduction

Designing graph neural networks (GNN) with various message passing architectures has been a trend recently. Many powerful graph convolution methods such as GraphSAGE [8], GCN [12], GIN [33] and GAT [31] have been proposed, with a wide range of applications including social networks [23,32,40], molecules [10,11,14,34], natural language processing [22,24,29] and physics simulations [25,27]. Meanwhile, some issues about how to add extra node features reasonably are rising with the development of GNN, which are mainly discussed in this paper.

Some works emphasized the importance of adding structural feature information: pagerank [13,21], node degree [36], clustering coefficient [8] or adding different graph features [38], including one hot vector, constant scalar, clustering coefficient and pagerank to perform node or graph classifications. Shortest path length is also covered when considering the importance of distant nodes [7,35]. However, previous works ignored the importance of structural features'

© Springer Nature Switzerland AG 2021
M. Kamp et al. (Eds.): ECML PKDD 2021 Workshops, CCIS 1524, pp. 238–257, 2021.
https://doi.org/10.1007/978-3-030-93736-2_19

correlations of each node. In addition, they do not take the advantage of graph feature's correlation information to add features to the original node feature but arbitrarily add these node features, which might include repetitive or highly related information, leading to data redundancy.

Supervised tasks for selecting features via correlation based algorithm have been proposed [3,15]. One work shows whether there exists a strong correlation between node features and node labels [6] and another work uses node correlation information to perform convolution pooling on graph classification tasks [9]. However, most correlation metrics are not based on deep or graph neural network models but based on simple analytical solutions or analysis based on covariance matrices, which might ignore the role of graph neural network on generating correlation between features with enriched message passing information from neighbours. A graph neural network based model to achieve node structural feature correlations has been not investigated.

Present Work. In this paper, we propose a framework for processing graph *feature* to *feature* prediction (feature mainly refers to a graph's structural feature in this paper), called **Fea2Fea**, which includes two important components: single feature to single feature prediction (Fea2Fea-single) and multiple features to single feature prediction (Fea2Fea-multiple): 1) Fea2Fea-single returns the correlation matrix implemented by graph neural networks 2) Fea2Fea-multiple takes the use of the correlation matrix achieved in Fea2Fea-single to summarize some possible combinations of features with graph neural network based filter. Finally it is transferred to real world applications, such as combination of initial features with embedded filtered structural features to perform node or graph classification on some benchmark graph datasets, which is proved to be worked on some graph classification task based datasets such as PROTEINS and NCI1. Details of the framework and model architectures are described in part 3.

Based on the experiments implemented according to the framework, our works have reached several important **findings**: 1) Graph neural networks are superior to deep neural networks without graph embedding layers in mutual prediction of graph features 2) The expressive power of the graph neural network in achieving the correlation of graph features, such as the ease with which other structural features to predict *node degree* and general difficulty to predict *pagerank* 3) A feature combination of predicting another single feature via graph neural networks with different concatenation methods provides enriched embedding options but overall the simplest concatenation is the best 4) For some benchmark data sets, the initial node identity feature combined with additional structural embedded information leads to higher classification accuracy. The main **advantages** of our works are that they 1) illustrate feature correlations with low dimensionality, where the number of input dimensions is controlled to a maximum of 5 in Fea2Fea-multiple pipcline 2) filter additional redundant node features via graph neural network based models instead of covariance explained models 3) require a less rigorous constructed models to perform comparisons which means that we introduce the importance of adding non-correlated features other than defeating the state of the arts graph embedding methods.

2 Related Works

GNN Expressiveness. Some works that have emphasized the importance of adding structural features to input features are mentioned in the *Introduction* part, as well as the work [16] that uses graph structure combined with node features to perform predictions. There are also some works mentioning the importance of adding specific features to make graph neural network more expressive. GIN [33] has shown that identity features are not powerful as it does not pass WL tests possibly which is to check symmetric graphs in the datasets. A recent work called identity-awared GNN [37] emphasizes the self versus neighbour features, which proves that adding neighbourhood information as feature augmentation will distinguish symmetric graphs and pass WL tests more easily which is more expressive than GIN. Another work [26] also discusses the expressiveness of GNN.

Feature Correlation. Many useful analytical methods for measuring feature correlation have been proposed. Besides the covariance based methods that are mentioned in part 1, there are some other useful methods, such as a cross-entropy model with the conception of symmetrical uncertainty metrics [39] and computing a tree-decomposition based correlation graph to select unredundant features [20]. However, the lack of research on graph neural network based model on achieving feature correlation/selection is the potential motive of this paper. We'd like to present some feature correlation results with GNN based filters.

Other Graph Embedding Methods. Recent years, researches have been conducted on learning graph embeddings. Many useful graph convolution methods have been proposed besides GIN, GAT, GCN and GraphSAGE which are the four main graph embedding methods that are implemented in our paper. For example, the graph convolution with ARMA filters [2], graph neural network with attention [30], with gated recurrent units [18] and with chebyshev spectral graph convolutional operator [4].

3 Methods

3.1 Preliminary Graph Feature Extraction

Graph Annotation. Given a graph $\mathcal{G}(\mathcal{V}, \mathcal{E})$ where $\mathcal{V}$ is the vertex set of the graph and $\mathcal{E}$ is the edge set of the graph . We want to achieve a full-scale structural feature matrix $\mathbf{x}_\mathcal{G} \in \mathbb{R}^{|\mathcal{V}| \times \mathcal{D}}$ from $\mathcal{G}$ that requires an adjacency matrix $\mathcal{A}_g \in \mathbb{R}^{|V| * |V|}$ of $\mathcal{G}$, where V is the number of vertexes in the graph and $\mathcal{D}$ is the total input dimensions of structural features. In this paper we mainly choose five of all structural features to show exploratory results, which are *constant feature*(**Cons**), *node degree*(**Deg**), *clustering coefficient*(**Clu**), *average path length* (**AvgLen**) and *Pagerank*(**PR**). In this case $\mathcal{D}$ is equal to 5.

Feature Extraction. Constant feature of node u $\underset{u \in \mathcal{V}}{Cons}(u)$ is given by c, where $c \in \mathbb{R}^+$. Each node's constant feature is set to 1 in this paper for standardization. Degree of node u $\underset{u \in \mathcal{V}}{Deg}(u)$ is equal to the number of node u's neighbours.

Clustering coefficient of node u $\underset{u \in \mathcal{V}}{Clu(u)}$ is given by $\frac{2e_{jk}}{k_i*(k_i-1)}$, where $j, k \in \mathcal{V}$ and e_{jk} represents the total possible edges between node u's neighbours and k_i is the number of node u's neighbours. Pagerank of a node u $\underset{u \in \mathcal{V}}{PR(u)}$ is given by: $\frac{1-q}{|\mathcal{V}|} + q \times \sum_{v \in \mathcal{N}(u)} \frac{PR(v)}{\mathcal{L}(v)}$, where $\mathcal{N}(\cdot)$ represents the node u's neighbours and $\mathcal{L}(\cdot)$ is the number of outbound links from node u to its neighbours. In the traditional analysis of pagerank algorithm, the object graph is directed. However we can treat undirected graph as bidirectional graph, where outbound link is equal to the number of edges from node u to its neighbours. q is the residual probability and equal to 0.85 as default which can be regarded as a hyper-parameter. Average path length of a node $\underset{u \in \mathcal{V}}{Avglen(u)}$ is given by: $\frac{1}{\mathcal{V}'} \sum_{v \in \mathcal{V} \neq u} I(u,v) * d_{min}(u,v)$, where $I(u,v)$ indicates whether there's a path from node u to node v. If node v is reachable for node u, then $I(u,v)$ is equal to 1 otherwise it is 0. $\mathcal{V}'$ is equal to the total number of reachable nodes for node u, more specifically, $\mathcal{V}' = \sum_{v \in \mathcal{V} \neq u} I(u,v)$. $d_{min}(u,v)$ determines the shortest path length from node u to node v. Therefore, an entry $\mathbf{I}(v), v \in \mathcal{V}$ can be written as: $I(v) = Cons(v) \oplus Deg(v) \oplus Clu(v) \oplus PR(v) \oplus Avglen(v)$ where $\oplus$ is the direct concatenation (Fig. 1).

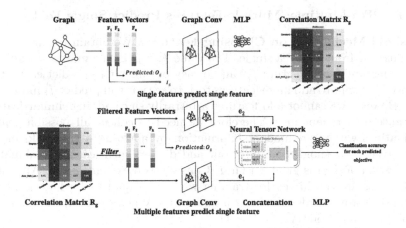

Fig. 1. Pipeline model for feature mutual prediction task, where the neural tensor network is originated in NLP domain [28] and sketch of the NTN block comes from *SimGNN* [1]. In the Fea2Fea-simple model, only single feature is regarded as the input while multiple features which are un-correlated are regarded as the input when it comes to Fea2Fea-multiple model, which is presented at the bottom part of the image. Neural Tensor Network, an example of feature concatenation methods is not necessary but just taken as an example to be familiar with our pipeline.

3.2 Fea2Fea-Single: Single Feature Predicts Single Feature

Obtain a Correlation Matrix by GNN. We have obtained a feature matrix $\mathbf{x}_\mathcal{G} \in \mathbb{R}^{|\mathcal{V}| \times 5}$. Specifically, graph features have been indexed to simplify analysis. The feature order is given by constant feature, degree, clustering coefficient, pagerank and average path length, which are indexed from 1 to 5 respectively. Suppose we have built a model $\mathcal{M}$ which is based on one of the four graph embedding methods together with multi-layer perceptrons. We extract two features each time, one for input $\mathbf{I}_\mathcal{G}$ and the other for output $\mathbf{O}_\mathcal{G}$. We set bins for the output since we focus more on classification task rather than regression task. After outputs are classified into their own bins, we use model $\mathcal{M}$ to predict $\mathbf{O}_\mathcal{G}$. The trained model will generate predictive output $\mathbf{O}_\mathcal{G}'$. We use mean classification *accuracy* on test datasets and *negative loss likelihood loss*(NLLloss) as two main evaluators in this paper. A higher accuracy indicates that it is easier to predict from input feature to output feature. We will implement a feature correlation matrix $\mathcal{R} \in \mathbb{R}^{5*5}$ since each feature can be taken both as input and output, except for the case of constant feature. We should note that constant feature can be only taken as input but not as output. It is the most basic feature for all graphs and will not lead to a classification problem. We can set the element in the matrix to infinity or let it symmetric to the diagonal under this circumstance. Algorithm 1 in appendix A shows the complete process of how to construct a correlation matrix for feature mutual prediction.

3.3   Fea2Fea-Multiple: Mutiple Features Predict Single Feature

Threshold Mechanism on Choosing Features. We obtain feature correlation matrix $\mathcal{R}$ from Fea2Fea-single. The value $\mathcal{R}(i, j)$ in the matrix $\mathcal{R}$ indicates whether feature f_i and feature f_j can be easy or difficult to predict each other. Moreover, we want to add more features to f_i to see if it will predict f_j more accurately. However we cannot add features arbitrarily since adding similar features may cause data redundancy. Therefore we need to collect all possible feature combinations without redundant information by applying a threshold mechanism. Given a threshold t, we filter out and discard such feature combination: (f_i, f_j) where $\mathcal{R}(i, j)$ is greater than t. Consider two extreme cases. All feature combinations are considered for training when t is equal to 1 while no feature combinations are considered when t is equal to 0. When $t \in (0,1)$, the number of possible concatenations $\mathcal{N} \in [1, 2^d - 1)$.

We initialize an array called *Comb* which includes all concatenation between features. The time complexity of this generation is $\Omega(2^d)$, where d is equal to the input feature dimension. It works efficiently when input feature dimension is small. A filter is applied after generating the array. For each f_i and f_j in each element of *Comb*, if both $\mathcal{R}(i, j)$ and $\mathcal{R}(j, i)$ are less than threshold t, the combination of f_i and f_j is valid, otherwise this combination is moved from *Comb*. Finally we achieve a filtered *Comb*.

Feature Concatenation. Given a valid combination f_1, $f_2,...,f_k$, where k $\in \{2,3,4\}$. We implement graph convolution layers for each feature to map them into graph embedding space: e_1, $e_2,...,e_k$. We provide three methods to concatenate those features after graph embeddings, which are simple concatenation,

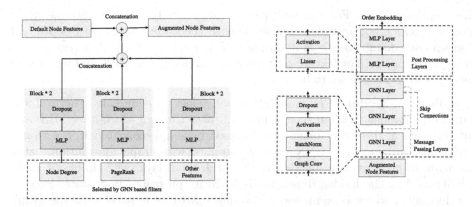

Fig. 2. Left: Features which are selected by GNN filter will first go through two MLP layers individually, then concatenate with other features. Right: irredundant features are concatenated with initial node features as augmented node features, which will go through three GNN layers with skip connection and a two-layer MLP.

bilinear concatenation and neural tensor network (NTN) [28] based concatenation. Assume that the graph embedding space of each feature is d. NTN concatenation contains both direct(simple) concatenation and bilinear concatenation, which is given by:

$$g_{(t)} = \mathbf{u}^T \mathbf{h}(\underbrace{g_{(t-1)}^T \mathbf{W}_{(t)} e_t}_{\text{bilinear concatenation}} + \underbrace{g_{(t-1)} \oplus e_t}_{\text{simple concatenation}} + \mathbf{b}), g_{(0)} = e_1 \in \mathbb{R}^d \quad (1)$$

where $g_{(t)} \in \mathbb{R}^{td}$ represents the temporal embedding after t-th concatenation, $\oplus$ represents the function that connects two features together with the summation of their dimensions, $g_{(0)}$ is initialized by e_1, the weight matrix $\mathbf{W}_{(t)} \in \mathbb{R}^{(t-1)d*td*d}$ with t$\geq$2, the obtained graph embedding $g_{(t)} \in \mathbb{R}^{td}$, weight vector $\mathbf{u}^T \in \mathbb{R}^{td}$ and $\mathbf{h}$ represents the activation function: *tanh* function as it is stated in original paper. However, we cannot conclude that NTN concatenation is superior to the other two concatenation methods according to Eq. (1). We should implement experiments with all three concatenation methods separately. Fea2Fea-Multiple will finally achieve the average accuracy of each combination, as well as the average accuracy of different numbers of input dimensions, which is the base of real world applications.

3.4 Model Architecture

Fea2Fea. For Fea2Fea-single pipeline, we implement two GNN layers followed by two multilayer-perceptron (MLP) layers. Hidden embedding size $h = 64$. The number of bins $\mathcal{B}$ is set to 6 as default. Model depth $\mathcal{D}$ is set to 2 as default. We add batch-norm layers when performing model depth hyperparameter tests. We add a two-layer MLP as our baseline model without graph convolution layers for comparison. For Fea2Fea-multiple pipeline, we observe the optimal graph convolution method from Fea2Fea-single with the same graph model depth and hidden embedding size. Default threshold $\mathcal{T}$ is set to 0.85, which is regarded as a hyperparameter. For feature concatenation, hidden dimension and number of neurons for NTN are set to 64.

Application Model. GNN based filters will select unredundant structural feature combinations. Each of the features will pass through a two-layer MLP and perform one of three concatenations, followed by simple concatenation with initial node features. Three GNN layers are applied with graph convolution, batch normalization, activation function(*relu*) and dropout layer with droupout probability of 0.6. The hidden dimension is 64 in graph embedding space. Preprocessing layer, which is composed of two-layer MLP is applied after the graph embedding, which is shown in Fig. 2.

Binning Methods. According to the distribution of structural features for *Planetoid* dataset, *degree* and *clustering coefficient* features will cause class inbalance (Appendix C). Therefore, we divide the unbalanced data into one bin, and the other data into remaining bins, especially more than half of the nodes' *clustering coefficient* are zero. *NCI1* dataset has a worse situation where most of the nodes in the graph have a zero *clustering coefficient*. One solution is to remove this feature from the feature set.

4   Experiments

For node datasets, we split the whole datasets into training, validation and test dataset with a fixed ratio in original setting [8]. For graph datasets, we split graph dataset into training and test datasets with a ratio of 8:1 and follow a traditional 10-fold cross validation in training splits [5].

Datasets. Six benchmark datasets are selected. For node datasets, we choose *Planetoid* [8], which includes Cora, CiteSeer and PubMed. For graph datasets, we choose *TUDataset* [19], which is a collection of benchmark datasets including data from small molecules, bioinformatics, social networks, computer vision and some synthetic datasets. In this paper, we choose two datasets Proteins and Enzymes from bioinformatics domain and Nci1 from molecule domain. After generating graph feature matrix for each dataset, we plot the distribution for each graph feature to search for imbalance (Appendix C).

4.1 Results

Table 1. Feature to feature prediction on cora and proteins datasets (bins = 6)

Task	CORA					PROTEINS				
	GCN	GIN	SAGE	GAT	MLP	GCN	GIN	SAGE	GAT	MLP
$Cons \rightarrow Deg$	0.509	**1.000**	0.213	0.202	0.206	0.560	**0.662**	0.469	0.469	0.469
$Deg \rightarrow Deg$	0.741	1.000	0.967	0.514	**1.000**	0.633	**0.662**	0.688	0.470	0.640
$Clu \rightarrow Deg$	0.423	**1.000**	0.504	0.474	0.285	0.537	**0.652**	0.482	0.469	0.467
$PR \rightarrow Deg$	0.308	**1.000**	0.311	0.223	0.197	0.430	**0.662**	0.446	0.469	0.469
$Avglen \rightarrow Deg$	0.409	**1.000**	0.499	0.228	0.274	0.522	**0.657**	0.482	0.469	0.469
$Cons \rightarrow Clu$	0.523	**0.533**	0.461	0.461	0.461	0.299	**0.436**	0.202	0.202	0.236
$Deg \rightarrow Clu$	**0.550**	0.542	0.548	0.506	0.531	0.435	0.387	**0.454**	0.312	0.346
$Clu \rightarrow Clu$	0.724	0.765	**0.968**	0.668	0.968	0.570	0.610	**0.707**	0.549	0.723
$PR \rightarrow Clu$	0.490	**0.538**	0.486	0.461	0.461	0.247	**0.330**	0.278	0.219	0.236
$Avglen \rightarrow Clu$	0.508	**0.538**	0.498	0.460	0.465	**0.276**	0.258	0.328	0.202	0.273
$Cons \rightarrow PR$	0.639	**0.756**	0.160	0.160	0.160	0.645	**0.648**	0.170	0.170	0.169
$Deg \rightarrow PR$	0.573	**0.792**	0.750	0.392	0.603	0.622	0.699	**0.722**	0.239	0.461
$Clu \rightarrow PR$	0.427	**0.695**	0.403	0.199	0.395	**0.575**	0.465	0.467	0.224	0.349
$PR \rightarrow PR$	0.345	**0.714**	0.323	0.185	0.160	0.170	**0.403**	0.251	0.170	0.176
$Avglen \rightarrow PR$	0.450	**0.741**	0.490	0.202	0.247	**0.579**	0.453	0.395	0.170	0.175
$Cons \rightarrow Avglen$	0.357	**0.384**	0.169	0.169	0.169	**0.182**	0.171	0.171	0.171	0.171
$Deg \rightarrow Avglen$	0.420	0.435	**0.440**	0.340	0.199	**0.212**	0.184	0.200	0.171	0.175
$Clu \rightarrow Avglen$	0.286	**0.310**	0.263	0.219	0.202	0.227	0.196	**0.254**	0.228	0.216
$PR \rightarrow Avglen$	0.215	**0.421**	0.266	0.185	0.169	0.171	**0.172**	0.171	0.171	0.171
$Avglen \rightarrow Avglen$	0.503	0.445	0.774	0.490	**0.958**	0.549	0.483	**0.612**	0.545	0.513

Model Performance. In terms of node datasets, *GIN* overall performs the best among all graph embedding method based models and baseline MLP model when performing Fea2Fea-single. Especially when predicting *degree*, *GIN* can reach 100% average test accuracy. *GraphSAGE* and *GCN* can predict features at an acceptable level but they are far from the performance of *GIN*. The reason might be that GIN is focused on structural features, which leads to higher accuracy in graph classification problems. *GAT* does not perform well in most of the tasks. *MLP* can achieve self-prediction, but it is poor at mutual prediction. On the view of graph datasets, *GIN* performs the best while performance of *GraphSAGE* and *GCN* is much better than the performance on node datasets. Results of *GAT* and *MLP* are close to the optimal value at some tasks but generally can not reach the performance provided by *GIN*. When it comes to the real world applications, improvements are made in graph datasets after we add irredundant features (Table 2), with an average improved accuracy of 4.8%. It does not improve a lot in node datasets. The reason might be attributed to the high dimensionality of CORA and CITESEER dataset.

Difficulty of Prediction. From Table 1 or Table 5, we figure out that *degree* is the easiest to predict. Predicting *clustering coefficient* and *average path length* is hard generally for both kinds of datasets. From the difficulty of prediction, we can also explain the property of each feature. For example, *degree* feature is a basic feature other than *constant feature* which is easier to predict from others. However, *clustering coefficient* is sparse while *average path length* and *pagerank* require the entire information from the whole graph which may bring redundant information. The choices of Neural Tensor Networks or just simple or bilinear concatenation are discussed in Appendix C. Overall it depends on the classification tasks.

Average Accuracy of Multi-to-One Pediction. We take the simple concatenation to illustrate an example. When predicting *clustering coefficient* from other features in Planetoid dataset, possible combination number of features is only equal to 2 or 3. For Citeseer dataset, a 3-set input is better than a 2-set input. A possible 3-set concatenation for Cora dataset is {*Cons, PR, Avglen*}, which is much better than a 1-set prediction for each of the feature in this 3-set. When predicting *Pagerank*, the prediction accuracy is not stable. A 3-set feature combination is worse than a 2-set feature combination for PUBMED and CITESEER and even worse than Fea2Fea-single. In *TUdataset*, predicting *pagerank* and *average path length* might exist the situation of 4-set feature set, but this 4-set feature set does not perform well and the average accuracy is descending. Figures are shown in appendix C (Fig. 9).

Table 2. Performance on node and graph classification problems (average test accuracy in %), where s, b, n are the abbreviations of "simple", "bilinear" and "NTN" methods respectively, followed by the number of irredundant features. A baseline Fea2Fea-null is based on GCN on node datasets and on GIN when it comes to graph datasets.

Model	CORA	CITESEER	PUBMED	ENZYMES	PROTEINS	NCI1
MLP	59.0 ± 1.0	59.6 ± 0.5	70.7 ± 2.2	32.7 ± 5.2	65.9 ± 3.6	58.0 ± 0.8
GraphSAGE	76.0 ± 4.4	66.8 ± 2.8	73.5 ± 1.5	37.8 ± 3.0	66.2 ± 2.5	64.7 ± 2.3
GCN	80.0 ± 0.9	68.1 ± 0.9	74.2 ± 1.1	36.0 ± 5.0	66.2 ± 0.8	61.3 ± 0.9
GAT	79.7 ± 1.2	69.2 ± 0.9	74.0 ± 1.4	31.0 ± 5.6	65.9 ± 2.4	60.9 ± 2.2
Fea2Fea-null	80.0 ± 0.9	68.1 ± 0.9	74.2 ± 1.1	47.2 ± 3.2	67.9 ± 1.2	71.8 ± 0.6
Fea2Fea-s2	79.7 ± 0.8	65.0 ± 2.4	77.4 ± 0.8	**48.5 ± 4.5**	**77.8 ± 0.9**	74.2 ± 0.8
Fea2Fea-b2	77.3 ± 3.1	64.7 ± 4.0	77.3 ± 1.1	45.8 ± 3.2	76.4 ± 1.2	70.8 ± 2.6
Fea2Fea-n2	77.0 ± 1.6	62.6 ± 4.0	75.7 ± 2.6	42.8 ± 3.4	74.9 ± 2.7	68.5 ± 0.7
Fea2Fea-s3	79.6 ± 0.9	66.2 ± 1.8	**78.5 ± 1.8**	48.0 ± 4.4	76.8 ± 1.4	**74.9 ± 0.9**

5   Conclusions and Future Works

In this paper, we introduce Fea2Fea-single to analyze correlations between structural features. We implement Fea2Fea-multiple to remove redundant features. We take the advantage of enriched graph embedding methods and feature concatenation to complete the experiments. Experiments show that adding structural features with GNN based pre-selection is necessary before training. In our future works, we plan to add more graph features and large-scale Open Graph Benchmark (OGB) datasets to enrich feature sets and to sum up a commonality on structural feature correlations. Graph embedding methods will also be enriched to make sure that *GIN* is the best graph embedding method in Fea2Fea-single. The conclusion of low self-prediction on pagerank is still under controversy which requires further explorations.

A   Algorithm Pseudocode

Algorithm 1: Get Feature Correlation Matrix

  Input: full-scale feature matrix $\mathbf{x}_\mathcal{G}$; model architecture $\mathcal{M}$; metrics $\mathcal{P}$
  Output: Feature correlation matrix $\mathcal{R}$
1 $\mathcal{R} \leftarrow 0, \mathcal{K} \leftarrow 5$
2 **for** $i \leftarrow 1$ *to* $\mathcal{K}$ **do**
3 $\quad$ $\mathbf{I}_\mathcal{G} \leftarrow \mathbf{x}_\mathcal{G}(:, i)$
4 $\quad$ **for** $j \leftarrow 2$ *to* $\mathcal{K}$ **do**
5 $\quad\quad$ $\mathbf{O}_\mathcal{G} \leftarrow \text{Binning}\,(\mathbf{x}_\mathcal{G}(:, j))$
6 $\quad\quad$ $\mathbf{O}_\mathcal{G}' \leftarrow \mathcal{M}\,(\mathbf{I}_\mathcal{G}, \mathbf{O}_\mathcal{G})$
7 $\quad\quad$ $\mathcal{R}(i, j) \leftarrow \mathcal{P}(\mathbf{O}_\mathcal{G}', \mathbf{O}_\mathcal{G})$
8 $\quad$ **end**
9 $\quad$ **if** $i == j == \mathcal{K}$ **then**
10 $\quad\quad$ **return** $\mathcal{R}$;
11 $\quad$ **end**
12 **end**

The case when j is equal to 1 is not shown in the description of pseudocode, but as we have mentioned in the main paper, we should not ignore the case of predicting constant feature. In real practice, the first column is symmetrical to the first row to simplify analysis. For example, predicting constant feature from node degree is equivalent to predicting node degree from constant feature.

Algorithm 2: Get $\mathcal{A}_\mathcal{G}$ which records multiple feature to single feature prediction results

---

Input: Feature Correlation Matrix $\mathbf{x}_\mathcal{G}$; model architecture $\mathcal{M}'$; metrics $\mathcal{P}$;
output feature index **idx**; threshold $\mathcal{T} = 0.85$

Output: Array $\mathcal{A}_\mathcal{G}$

1 Initialize $Comb \leftarrow \{\{0,1\}, ..., \{0,1,2,3,\mathcal{K}\}\}$
2 **for** $comb \in Comb$ **do**
3 | **for** $\forall\ \mathcal{F}_i, \mathcal{F}_j \in comb$ **do**
4 | | **if** $\mathcal{R}(F_i, F_j) \geq \mathcal{T}$ **or** $\mathcal{R}(F_j, F_i) \geq \mathcal{T}$ **or** **idx** in $comb$ **then**
5 | | | Remove $comb$ from $Comb$;
6 | | **end**
7 | **end**
8 **end**
9 **for** $comb \in Comb$ **do**
10 | $\mathbf{I}_\mathcal{G} \leftarrow \mathbf{x}_\mathcal{G}(:, comb)$
11 | $\mathbf{O}_\mathcal{G} \leftarrow \mathbf{x}_\mathcal{G}(:, \mathbf{idx})$
12 | $\mathbf{O}'_\mathcal{G} \leftarrow \mathcal{M}'(\mathbf{I}_\mathcal{G}, \mathbf{O}_\mathcal{G})$
13 | $\mathcal{A}_\mathcal{G}(comb|\mathbf{idx}) \leftarrow \mathcal{P}(\mathbf{O}'_\mathcal{G}, \mathbf{O}_\mathcal{G})$
14 **end**

Concatenation method is included in model $\mathcal{M}'$, which is not shown in the pseudocode but is instead shown in Fig. 2. We should also notice that the generation of total combination is not efficient if we consider adding large more structural features. One possible solution is to use reservoir sampling and do add-drop iterations until all pairs of feature correlation are under the threshold.

B   Hyper-parameter Tunning

Three hyperparameters are the number of bins, depth of graph convolution layers and threshold. We evaluated the influence of hyperparameters on *Citeseer* and *ENZYMES* datasets. Tasks are: from *pagrank* predict *average path length* and from *average path length* predict *clustering coefficient*. From the results of experiments, we find out that average accuracy is high when there're only two bins since the partition of classes is not strict. When the number of bins increases, the average prediction accuracy is descending. When model depth is shallow, the prediction effect is not good. Increasing the number of graph embedding layers with batch normalization will result in a better test generalization. We implement *SkipLayerGNN* [17] when layers are deep(≥ 3) (Table 3).

Table 3. Hyper-parameter tests on node and graph datasets

Tasks	Param order	CITESEER		ENZYMES	
		PR→AvgLen	AvgLen→Clu	PR→AvgLen	AvgLen→Clu
Bins	2	**0.774 ± 0.008**	**0.836 ± 0.003**	**0.856 ± 0.028**	**0.550 ± 0.001**
	3	0.727 ± 0.006	0.781 ± 0.003	0.751 ± 0.007	0.433 ± 0.058
	4	0.649 ± 0.010	0.754 ± 0.008	0.642 ± 0.016	0.327 ± 0.053
	5	0.616 ± 0.009	0.722 ± 0.004	0.551 ± 0.004	0.284 ± 0.023
	6	0.584 ± 0.006	0.719 ± 0.002	0.400 ± 0.157	0.294 ± 0.044
	7	0.558 ± 0.008	0.711 ± 0.003	0.390 ± 0.130	0.246 ± 0.002
	8	0.505 ± 0.006	0.703 ± 0.003	0.306 ± 0.158	0.251 ± 0.021
	9	0.491 ± 0.010	0.696 ± 0.002	0.293 ± 0.122	0.262 ± 0.026
	10	0.478 ± 0.008	0.694 ± 0.003	0.180 ± 0.119	0.249 ± 0.024
Depth	2	0.579 ± 0.005	0.720 ± 0.003	**0.505 ± 0.007**	0.281 ± 0.038
	4	0.387 ± 0.057	0.713 ± 0.004	0.237 ± 0.021	0.330 ± 0.050
	6	0.512 ± 0.055	0.707 ± 0.007	0.275 ± 0.032	0.330 ± 0.038
	8	0.638 ± 0.059	0.709 ± 0.006	0.348 ± 0.041	0.336 ± 0.030
	10	**0.651 ± 0.042**	**0.722 ± 0.007**	0.370 ± 0.059	**0.347 ± 0.013**

We set two thresholds: 0.6 and 0.8, which are based on Fea2Fea-multiple. Experiments are executed 10 times. We obtain average accuracy from the experiments. When threshold value is equal to 0.6, valid combined input feature sets are: (*Cons,AvgLen*). Mean accuracy is higher than the four combinations when threshold is equal to 0.8. For *ENZYMES* dataset, a higher threshold helps to obtain better results. Therefore, it infers that more feature combinations are not always the best which shows that we must try all combinations (Table 4).

Table 4. Threshold tests on node and graph datasets on predicting pagerank, the number in combination is in-ordered feature serial number as we mentioned before.

Tasks	Param order	CITESEER		ENZYMES	
		Valid combination	Accuracy	Valid combination	Accuracy
Threshold	0.6	(1,5)	**0.611 ± 0.039**	(1,3), (1,5), (3,5), (1,3,5)	0.405 ± 0.081
	0.8	(1,3), (1,5), (3,5), (1,3,5)	0.534 ± 0.067	(1,3), (1,5), (2,3), (3,5), (1,3,5))	**0.440 ± 0.079**

C   Visualization

We plot the distribution for each structural feature in each dataset. *Pagerank* and *Average Path Length* is normally distributed for most datasets. Degree and clustering coefficient have shown imbalance of data for most datasets. Therefore, binning method plays an important role in partitioning reasonable classes.

D   Supplementary Results on Feature Prediction

Concatenation Comparison. *NTN* method is worse than simple concatenation in the process of training the predicted *pagerank* of Cora dataset, and it generalizes worse on test datasets. The *bilinear* method faces the problem of over-fitting. In general, the *simple concatenation* has the best performance on test datasets. When predicting *average path length* on *Cora* dataset, *NTN* is far better than *bilinear* and *simple concatenation* but they show similar results on test datasets. More specifically, *Bilinear* and *NTN* have a serious fluctuation problem on accuracy, overall *NTN* is the optimal method according to Fea2Fea-multiple. These experiments indicate that complex concatenation method such as *NTN* does not always have the best performance. Sometimes the simplest way is the best way.

Table 5. Feature to feature prediction on planetoid datasets (bins = 6)

Aim	CITESEER					PUBMED				
	GCN	GIN	SAGE	GAT	MLP	GCN	GIN	SAGE	GAT	MLP
$Cons \rightarrow Deg$	0.587	**1.000**	0.379	0.379	0.379	0.678	**0.996**	0.478	0.478	0.478
$Deg \rightarrow Deg$	0.899	**1.000**	0.994	0.395	1.000	0.725	**1.000**	0.958	0.477	1.000
$Clu \rightarrow Deg$	0.628	**1.000**	0.631	0.492	0.670	0.643	**1.000**	0.603	0.471	0.574
$PR \rightarrow Deg$	0.466	**1.000**	0.409	0.376	0.379	0.539	**1.000**	0.600	0.478	0.478
$Avglen \rightarrow Deg$	0.536	**0.997**	0.686	0.476	0.473	0.696	**1.000**	0.785	0.478	0.524
$Cons \rightarrow Clu$	0.671	**0.693**	0.658	0.658	0.658	0.799	**0.805**	0.780	0.780	0.780
$Deg \rightarrow Clu$	0.688	0.714	**0.736**	0.658	0.726	0.794	**0.804**	**0.804**	0.780	0.805
$Clu \rightarrow Clu$	0.857	0.882	0.980	0.846	**0.992**	0.831	0.839	**0.939**	0.762	0.932
$PR \rightarrow Clu$	0.672	**0.687**	0.670	0.658	0.658	0.792	**0.805**	0.780	0.780	0.780
$Avglen \rightarrow Clu$	0.681	**0.696**	0.684	0.658	0.676	**0.794**	0.785	0.790	0.780	0.780
$Cons \rightarrow PR$	**0.702**	0.671	0.202	0.190	0.190	**0.669**	0.529	0.161	0.141	0.161
$Deg \rightarrow PR$	0.637	0.750	**0.752**	0.266	0.443	0.564	**0.629**	0.617	0.175	0.565
$Clu \rightarrow PR$	0.549	**0.575**	0.435	0.279	0.315	0.437	**0.559**	0.409	0.196	0.326
$PR \rightarrow PR$	0.192	**0.635**	0.415	0.263	0.190	0.478	**0.554**	0.336	0.161	0.161
$Avglen \rightarrow PR$	0.529	**0.691**	0.602	0.312	0.333	0.541	**0.591**	0.537	0.274	0.263
$Cons \rightarrow Avglen$	0.442	**0.503**	0.178	0.166	0.173	0.294	**0.394**	0.168	0.168	0.168
$Deg \rightarrow Avglen$	0.528	0.542	**0.553**	0.285	0.330	0.415	**0.443**	0.437	0.153	0.313
$Clu \rightarrow Avglen$	**0.420**	0.377	0.387	0.248	0.246	0.296	**0.330**	0.300	0.171	0.184
$PR \rightarrow Avglen$	0.268	**0.466**	0.207	0.310	0.173	0.316	**0.459**	0.198	0.197	0.168
$Avglen \rightarrow Avglen$	0.734	0.597	0.937	0.596	**0.979**	0.450	0.378	0.860	0.270	**0.984**

E   Graph Embedding Analysis

We extract graph embedding vectors and linear layer embedding vectors to perform tsne visualization and comparsion. When predicting *pagerank* from node *degree* on CITESEER dataset, the graph embedding is similar to the mlp embedding. On PROTEINS dataset, graph embedding cannot separate classes well compared with mlp embedding. When performing multiple features to single feature prediction, input embedding indicates that input features are mixed which cannot classify the class well but mlp embedding can separate the class well. It emphasizes the importance of a MLP after the GNN block (Figs. 3, 4, 5, 6, 7, 8, 10, 11, 12 and 13).

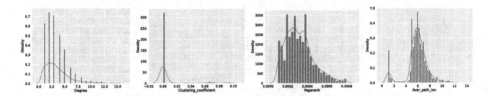

Fig. 3. Distribution of graph feature of Cora Dataset

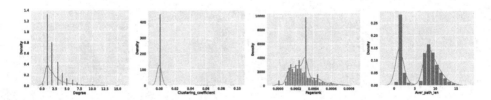

Fig. 4. Distribution of graph feature of Citeseer Dataset

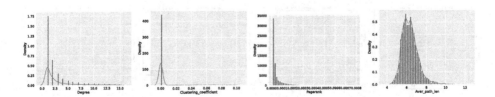

Fig. 5. Distribution of graph feature of PubMed Dataset

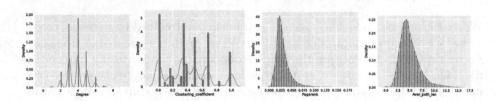

Fig. 6. Distribution of graph feature of ENZYMES Dataset

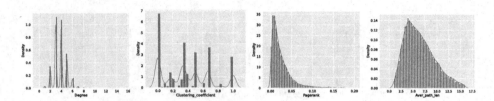

Fig. 7. Distribution of graph feature of PROTEINS Dataset

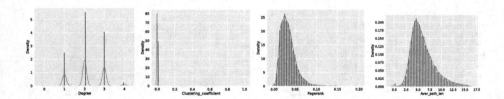

Fig. 8. Distribution of graph feature of NCI1 Dataset

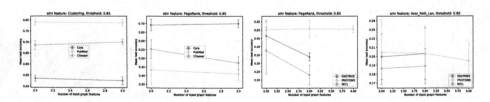

Fig. 9. Different numbers of features contribute to different prediction results for node and graph datasets. We perform 10 experiments and take the average accuracy with standard deviation shown in the figures.

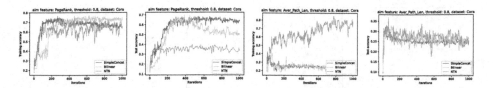

Fig. 10. Training and test accuracy versus epochs for different concatenation results when predicting clustering coefficient and pagerank in Fea2Fea-multiple

F   Generalization

We extend the results that we reach from benchmark datasets on synthetic dataset. We set up five new graph with number of nodes $n \in \{50, 200, 400, 800, 1000\}$. We use the same model with the same parameter settings in Fea2Fea-single. We found that the feature clustering coefficient is still the most difficult to predict and degree is the easiest to predict, which is the same as what we've previously found in Planetoid and TUdataset. As the number of nodes or features in the network increases, feature mutual prediction becomes more accurate (Table 6).

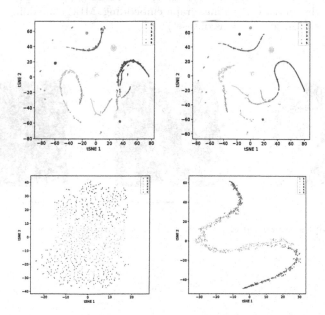

Fig. 11. tSNE on graph and MLP embeddings with Degree predicting PageRank, test on CiteSeer and Proteins datasets (from left to right: graph embedding, MLP embedding, from top to bottom: Citeseer dataset, Proteins dataset)

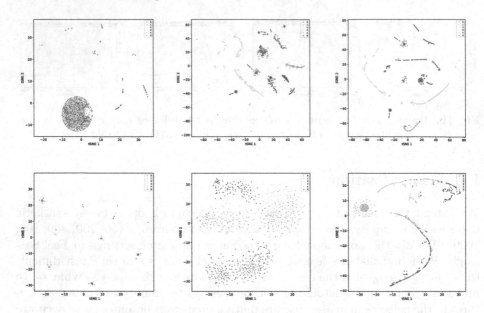

Fig. 12. tSNE on initial, graph and MLP embeddings with Constant feature together with Clustering Coefficient predicting Degree, test on CiteSeer and Proteins datasets (from left to right: initial embedding, graph embedding, MLP embedding, from top to bottom: Citeseer dataset, Proteins dataset)

Fig. 13. Examples of generated geometric dataset from networkx with nodes 200, 400 and 800

Table 6. Feature to feature prediction on planetoid datasets (bins = 6)

Aim	Number of nodes				
	50	200	400	800	1000
$Cons \rightarrow Deg$	0.800	0.950	**1.000**	0.975	**1.000**
$Deg \rightarrow Deg$	0.800	0.950	**1.000**	0.975	**1.000**
$Clu \rightarrow Deg$	**1.000**	**1.000**	**1.000**	0.975	**1.000**
$PR \rightarrow Deg$	0.800	0.950	**1.000**	**1.000**	0.990
$Avglen \rightarrow Deg$	**1.000**	**1.000**	**1.000**	**1.000**	**1.000**
$Cons \rightarrow Clu$	**0.800**	0.750	0.550	0.650	**0.800**
$Deg \rightarrow Clu$	0.800	**0.900**	0.525	0.775	0.780
$Clu \rightarrow Clu$	**1.000**	0.850	0.675	0.700	0.850
$PR \rightarrow Clu$	0.800	0.800	0.500	0.762	**0.810**
$Avglen \rightarrow Clu$	**0.800**	0.700	0.625	0.688	0.720
$Cons \rightarrow PR$	**1.000**	0.850	0.850	0.912	0.910
$Deg \rightarrow PR$	0.800	0.900	0.850	0.912	**0.950**
$Clu \rightarrow PR$	0.600	0.850	0.825	0.825	**0.930**
$PR \rightarrow PR$	**1.000**	0.850	0.875	0.850	0.930
$Avglen \rightarrow PR$	0.600	0.850	0.825	0.863	**0.890**

References

1. Bai, Y., Ding, H., Bian, S., Chen, T., Sun, Y., Wang, W.: Simgnn: a neural network approach to fast graph similarity computation (2020)
2. Bianchi, F.M., Grattarola, D., Livi, L., Alippi, C.: Graph neural networks with convolutional arma filters. IEEE Trans. Pattern Anal. Mach. Intell., 1 (2021)
3. Blessie, E.C., Karthikeyan, E.: Sigmis: a feature selection algorithm using correlation based method. J. Algorithms Comput. Technol. **6**(3), 385–394 (2012)
4. Defferrard, M., Bresson, X., Vandergheynst, P.: Convolutional neural networks on graphs with fast localized spectral filtering (2017)
5. Do, M.T., Park, N., Shin, K.: Two-stage training of graph neural networks for graph classification (2021)
6. Duong, C.T., Hoang, T.D., Dang, H.T.H., Nguyen, Q.V.H., Aberer, K.: On node features for graph neural networks (2019)
7. Dwivedi, V.P., Joshi, C.K., Laurent, T., Bengio, Y., Bresson, X.: Benchmarking graph neural networks (2020)
8. Hamilton, W., Ying, Z., Leskovec, J.: Inductive representation learning on large graphs. In: Advances in Neural Information Processing Systems, pp. 1024–1034 (2017)
9. Jiang, J., Lei, F., Dai, Q., Li, Z.: Graph pooling in graph neural networks with node feature correlation. In: Proceedings of the 3rd International Conference on Data Science and Information Technology, pp. 105–110. Association for Computing Machinery (2020)
10. Jin, W., Yang, K., Barzilay, R., Jaakkola, T.: Learning multimodal graph-to-graph translation for molecular optimization (2019)

11. Kearnes, S., McCloskey, K., Berndl, M., Pande, V., Riley, P.: Molecular graph convolutions: moving beyond fingerprints. J. Comput. Aided Mol. Des. **30**(8), 595–608 (2016)
12. Kipf, T.N., Welling, M.: Semi-supervised classification with graph convolutional networks. arXiv preprint arXiv:1609.02907 (2016)
13. Klicpera, J., Bojchevski, A., Günnemann, S.: Predict then propagate: Graph neural networks meet personalized pagerank (2019)
14. Knyazev, B., Lin, X., Amer, M.R., Taylor, G.W.: Spectral multigraph networks for discovering and fusing relationships in molecules (2018)
15. Kumar, G., Jain, G., Panday, M., Das, A.K., Goswami, S.: Graph-based supervised feature selection using correlation exponential. In: Mandal, J.K., Bhattacharya, D. (eds.) Emerging Technology in Modelling and Graphics. AISC, vol. 937, pp. 29–38. Springer, Singapore (2020). https://doi.org/10.1007/978-981-13-7403-6_4
16. Lerique, S., Abitbol, J.L., Karsai, M.: Joint embedding of structure and features via graph convolutional networks (2019)
17. Li, G., Müller, M., Thabet, A., Ghanem, B.: Deepgcns: Can GCNs go as deep as CNNs? (2019)
18. Li, Y., Tarlow, D., Brockschmidt, M., Zemel, R.: Gated graph sequence neural networks (2017)
19. Morris, C., Kriege, N.M., Bause, F., Kersting, K., Mutzel, P., Neumann, M.: Tudataset: a collection of benchmark datasets for learning with graphs (2020)
20. Ouali, A., Juniarta, N., Maigret, B., Napoli, A.: A feature selection method based on tree decomposition of correlation graph (2019)
21. Page, L., Brin, S., Motwani, R., Winograd, T.: The pagerank citation ranking: bringing order to the web. Technical Report 1999–66, Stanford InfoLab, November 1999
22. Palm, R.B., Paquet, U., Winther, O.: Recurrent relational networks (2018)
23. Qiu, J., Tang, J., Ma, H., Dong, Y., Wang, K., Tang, J.: Deepinf. In: Proceedings of the 24th ACM SIGKDD International Conference on Knowledge Discovery & Data Mining (2018)
24. Rahimi, A., Cohn, T., Baldwin, T.: Semi-supervised user geolocation via graph convolutional networks. In: Proceedings of the 56th Annual Meeting of the Association for Computational Linguistics (Volume 1: Long Papers), pp. 2009–2019. Association for Computational Linguistics (2018)
25. Sanchez-Gonzalez, A., Godwin, J., Pfaff, T., Ying, R., Leskovec, J., Battaglia, P.W.: Learning to simulate complex physics with graph networks (2020)
26. Sato, R.: A survey on the expressive power of graph neural networks (2020)
27. Seo*, S., Meng*, C., Liu, Y.: Physics-aware difference graph networks for sparsely-observed dynamics. In: International Conference on Learning Representations (2020)
28. Socher, R., Chen, D., Manning, C.D., Ng, A.: Reasoning with neural tensor networks for knowledge base completion. In: Advances in Neural Information Processing Systems, vol. 26. Curran Associates, Inc. (2013)
29. Sorokin, D., Gurevych, I.: Modeling semantics with gated graph neural networks for knowledge base question answering. In: Proceedings of the 27th International Conference on Computational Linguistics, pp. 3306–3317. Association for Computational Linguistics (2018)
30. Thekumparampil, K.K., Wang, C., Oh, S., Li, L.J.: Attention-based graph neural network for semi-supervised learning (2018)
31. Veličković, P., Cucurull, G., Casanova, A., Romero, A., Lio, P., Bengio, Y.: Graph attention networks. arXiv preprint arXiv:1710.10903 (2017)

32. Wang, H., et al.: Mcne. In: Proceedings of the 25th ACM SIGKDD International Conference on Knowledge Discovery & Data Mining (2019)
33. Xu, K., Hu, W., Leskovec, J., Jegelka, S.: How powerful are graph neural networks? arXiv preprint arXiv:1810.00826 (2018)
34. Xu, N., Wang, P., Chen, L., Tao, J., Zhao, J.: Mr-gnn: multi-resolution and dual graph neural network for predicting structured entity interactions. Proceedings of the Twenty-Eighth International Joint Conference on Artificial Intelligence (2019)
35. Yin, H., Wang, Y., Li, P.: Revisiting graph neural networks and distance encoding from a practical view (2020)
36. Ying, R., You, J., Morris, C., Ren, X., Hamilton, W.L., Leskovec, J.: Hierarchical graph representation learning with differentiable pooling (2019)
37. You, J., Gomes-Selman, J., Ying, R., Leskovec, J.: Identity-aware graph neural networks (2021)
38. You, J., Ying, R., Leskovec, J.: Design space for graph neural networks (2020)
39. Yu, L., Liu, H.: Feature selection for high-dimensional data: a fast correlation-based filter solution. In: ICML (2003)
40. Zhang, M., Chen, Y.: Link prediction based on graph neural networks (2018)

Web Image Context Extraction with Graph Neural Networks and Sentence Embeddings on the DOM Tree

Chen Dang[1], Hicham Randrianarivo[1(✉)], Raphaël Fournier-S'niehotta[2], and Nicolas Audebert[2]

[1] Qwant SAS, Paris, France
h.randrianarivo@qwant.com
[2] CEDRIC (EA4629), CNAM Paris, HESAM Université, Paris, France
{fournier,nicolas.audebert}@cnam.fr

Abstract. Web Image Context Extraction (WICE) consists in obtaining the textual information describing an image using the content of the surrounding webpage. A common preprocessing step before performing WICE is to render the content of the webpage. When done at a large scale (*e.g.*, for search engine indexation), it may become very computationally costly (up to several seconds per page). To avoid this cost, we introduce a novel WICE approach that combines Graph Neural Networks (GNNs) and Natural Language Processing models. Our method relies on a graph model containing both node types and text as features. The model is fed through several blocks of GNNs to extract the textual context. Since no labeled WICE dataset with ground truth exists, we train and evaluate the GNNs on a proxy task that consists in finding the semantically closest text to the image caption. We then interpret importance weights to find the most relevant text nodes and define them as the image context. Thanks to GNNs, our model is able to encode both structural and semantic information from the webpage. We show that our approach gives promising results to help address the large-scale WICE problem using only HTML data.

Keywords: Web image context extraction · Information retrieval · Graph neural network · Natural Language Processing

1 Introduction

Searching for images on the Web is essential for Internet users. Then, there is a need for efficient indexing methods to process large quantities of images. A useful step for the indexation in an image retrieval system consists in identifying the part of a webpage's text that best describes the image. The problem of extracting this *context* from the webpage is called Web Image Context Extraction (WICE) cf. Fig. 1. Visually rendering the webpage facilitates the extraction of an image's context, by loading and placing all structural elements of the page at the

© Springer Nature Switzerland AG 2021
M. Kamp et al. (Eds.): ECML PKDD 2021 Workshops, CCIS 1524, pp. 258–267, 2021.
https://doi.org/10.1007/978-3-030-93736-2_20

cost of evaluating several scripts. On a large scale, visual rendering and content extraction from a webpage is not tractable. We investigate how the HTML data structure may help in extracting images' contexts.

Fig. 1. An example of the WICE setting: an image, and its textual description (*context*) in the webpage.

Many approaches to WICE have been proposed. [17] use metadata related to the image as the textual context. Defining the context as the text in a "window surrounding the image" in the HTML is common, and some works try to find an optimal number of words to extract around the image [2,5]. [6] consider multiple sources of text, *e.g.*, title and meta information, as the context. These text-based methods often result in incomplete sentences, and do not provide accurate context when the context and image are not close in the HTML file.

Structure-based approaches focus more on the structure of the HTML document. An HTML document can be describe as a tree structure where each tag or text an object and the nested objects are "children" of the enclosing one wich is called the Document Object Model (DOM) [21]. Relying on the DOM tree, some works [7,8] measure similarities between the alternative text of the image and other texts, or develop precise webpage segmentation rules. [12] propose a *broadcast model* which combines the text blocks around images and information from other webpages linked to the webpage. [4] classify webpage's structures into three categories and handcraft rules to extract context. With a high focus on the page structure, DOM-tree based approaches often ignore or fail to fully use textual content, and recent evolutions of webpage programming have rendered many methods based on hard-coded rules no longer applicable.

Finally, some approaches focus on the webpage's visual layout. [1] use visual information to perform the segmentation using a set of predefined rules. [19] propose to extract all text which includes a caption and an alternative text at the same level as the image in the DOM tree. They also keep the texts around the image within a radius of 0.3 of the webpage rendering height. [18] compare visual and semantic clustering and find that visual based clustering performs much better in extracting information of web images

Besides, many graph-based information retrieval methods have been recently proposed. [14] use a graph-based framework to capture non-local and non-sequential context in sets of sentences. [3] introduce a Graph Neural Network model for multi-step reasoning. [13] study relation extraction for semi-structured

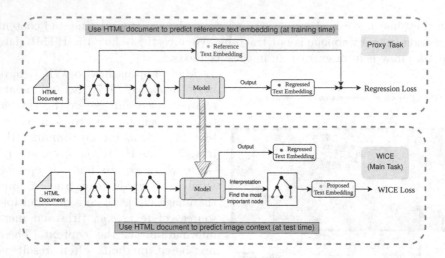

Fig. 2. Our pipeline. A graph neural network is first trained in the proxy task to predict the input HTML document's reference text (red dot), and then the importance score of the nodes (blue dots) in the prediction process is computed by interpreting the GCN model during the main task. The most predominant textual node (green dot) is then used as the context of the image. (Color figure online)

websites. [18] compares visual and semantic clustering and find that visual based clustering performs much better in extracting information of web images.

However, browser-based rendering is costly due to CSS and Javascript processing. It is nonetheless required by the state-of-the-art visual WICE approaches that rely on the full styled layout. We model webpages as graphs whose features are node types and text content. More specifically, our contributions to the problem of large-scale WICE are:

- Inject semantics into the DOM tree using state-of-the-art language models to generate sentence embeddings for each text node,
- Model webpages as graphs and use sentence embeddings as node features to train a graph neural network combining structural and semantic information,
- Use graphical models for large-scale processing of highly diverse news websites.

2   Method

Our goal is to identify nodes in the DOM tree that may contain part of an image's context. Since the DOM tree is a graph, we may use graph convolution networks (GCNs) [9] to do so. However, to bypass the lack of labeled datasets, we propose a *proxy task*, on unlabeled HTML documents to train our model. The results of this task may be interpreted to solve the WICE problem. Figure 2 illustrates the sequence of steps in our method.

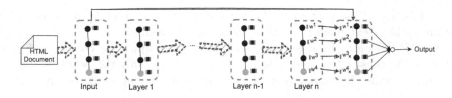

Fig. 3. An illustration of the wGCN (weight-GCN) architecture. The nodes in the last layer have a feature dimension of 1. We interpret it as the importance score of each node. The output is the weighted average of the text embeddings of all nodes.

For each HTML document that contains an image, we use the longest text between the alternative text[1] , the caption[2] , and the image title[3] as the *reference text*. We assume here that the reference text always describes the image. This assumption may potentially cause bias.

While crawling news websites, we empirically found that approximately 50% of them provide a caption or a reference text for their main illustrative images. This allows us to train on a relatively large corpus and also underlines the need for WICE in the wild: many websites do not provide clear textual contexts for their images.   Training models on a proxy task instead of directly predicting the context has many advantages. Firstly, we may rely on unlabeled datasets to train the model and solve the WICE problem, avoiding the need for annotations. It is easier and cheaper to crawl webpages with captions than manually annotate context sentences in thousands of articles. Moreover, at test time, our model will be used to infer the missing reference text: this means we can perform WICE even when the reference text is not present, which is where it is the most needed.

2.1   Framework

To extract the textual context of an image, we want to assign a weight w_i to all text nodes n_i from the DOM graph $\mathcal{G}$. Since we cannot learn w_i directly, we use a proxy task that consists of regressing a global embedding $\hat{\mathbf{z}}$ for the whole page. We use as a supervised target the sentence embedding $\mathbf{z}^*$ of the reference text (alt-text or caption) extracted for the image.

For most of our models, we will assume that it is possible to reconstruct the reference text's embedding using a linear combination of the other text embeddings, *i.e.*, we consider outputs in the form $\hat{\mathbf{z}} = \sum_i w_i f(\mathbf{z_i})$ where $\mathbf{z_i}$ is the embedding for the i-th node. Since the regressed text embedding is obtained by averaging weighted node embeddings, we assume that the largest contributor, *i.e.* the text node with maximum weight, is the most relevant for indexing. The image context can thus be obtained by taking the text node n_{target} that verifies $n_{\text{target}} = \arg\max_i w_i$.

---

[1] "alt" attribute of the `<img>` tag, which provides descriptive information of the image.

[2] text in `<figcaption>` that usually displays a short explanation besides the image.

[3] "title" attribute of the `<img>` tag.

In this work, the context is the most important text node, although this notion could be extended to all nodes over some threshold.

As a WICE metric for the extracted context, we compute the cosine similarity between the chosen text node's sentence embedding with the reference text's embedding, which is commonly used to measure document similarity in NLP.

2.2 Baselines

Inspired by the literature on WICE, we define various baselines for both the proxy and the WICE tasks. We define the `distance` baseline, which weights a text node by the inverse of its distance d_i to the image node in the graph:

$$\hat{\mathbf{z}} = \sum_i w_i \mathbf{z_i} = \sum_i \frac{1}{d_i} \cdot \mathbf{z_i} \tag{1}$$

Implicitly, this baseline also defines an equivalent WICE baseline, which we call `text after image`, where the textual context of the image is the closest text node in the graph. This is known as "window-based context extraction" in the WICE literature [2]. As a simpler rule-based baseline, we also consider the `title` WICE heuristic that uses the `title` attribute of the webpage as context.

We define a `blind` WICE baseline as selecting the node with the most similar text (in the embedding space) across all text nodes, compared to the embedding regressed by the network ($w_i = 1$ if $i = \arg\max_j(\cos(\hat{\mathbf{z}}, \mathbf{z_j}))$ and 0 otherwise). Intuitively, it can be interpreted as "looking for the missing caption". In this baseline, the WICE is "blind" since the reference text is completely unseen. This defines a lower-bound for our method: as a sanity check, the most important node we find should at least describe the image better than the predicted embedding.

Finally, we define an `oracle` baseline that uses the actual reference text to find the most similar node. The resulting cosine similarity gives us an upper bound of what is achievable using our proxy model. It may be used as an indicator representing the model's performance potential: the smaller the gap between a model's results and the `oracle`, the better its performance.

2.3 Models

Two base models are studied to perform the text regression: Graph Convolution Network (GCN) [9] and Graph Attention Network (GAT) [20]. GCNs are well-known for their promising performance on graph data. However, it is difficult to explain the prediction because the fusion of graph structure and feature dimension achieved by GCN is an explicitly irreversible process [22]. In comparison, GAT is a well-performing model, interpretable thanks to its attention mechanism which can be used as the weight of the text block. Its multi-headed mechanism can also be utilized to stabilize the performance of the model.

We study two different approaches to help interpret the GCNs and produce the node weights vector $\mathbf{w}$. First, we propose a GCN model that explicitly assigns weights to the nodes to facilitate the model's explanation, referred to as weight-GCN (wGCN). A traditional GCN would map the entire graph $\mathcal{G}$ to the target embedding: $\hat{\mathbf{z}} = \Psi_\theta(\mathcal{G})$ where Ψ is the GCN with parameters θ. However the information of which nodes contributed the most to the predictions is lost. Instead, we make the GCN produce one weight per node. The regression result is then the weighted average embedding of all nodes in the graph. Figure 3 illustrates the principle of the wGCN. Formally, let Ψ_θ be the GCN, $\mathbf{w} = \Psi_\theta(\mathcal{G})$ the output vector of node weights, and $\mathbf{z_i}$ be the text embedding of the node i of graph $\mathcal{G}$. Then the regressed text embedding $\hat{\mathbf{z}}$ can be denoted as:

$$\hat{\mathbf{z}} = \sum_i \Psi_\theta(\mathcal{G})[i] \cdot \mathbf{z_i} = \sum_i w_i \mathbf{z_i} \tag{2}$$

The wGCN is then trained using backpropagation and stochastic gradient descent: $\Psi_\theta^* = \arg\min_\theta \mathcal{L}_\Psi(\theta)$. $\mathcal{L}$ is the proxy task loss function. We minimize the negative cosine similarity between $\hat{\mathbf{z}}$ and the reference text $\mathbf{z}^*$ $i.e.$:

$$\mathcal{L}_\Psi(\hat{\mathbf{z}}, \mathbf{z}^*) = 1 - \cos(\hat{\mathbf{z}}, \mathbf{z}^*) = 1 - \cos\left(\sum_i w_i \mathbf{z_i}, \mathbf{z}^*\right) \tag{3}$$

Our second approach uses the GAT attention scores as the weights $\mathbf{w}$. The key difference between the GAT and wGCN is that wGCN learns the relationships between nodes and produces one weight per text node. The regression embedding is then an average of the embeddings weighted by the wGCN scores. In comparison, attention scores are only indirectly linked to the output embedding.

We also used the DeeperGCN (DGCN) architecture [10, 11] to create deeper GCN models. DGCN aims at solving common problems affecting deep GCNs, such as vanishing gradients, over-smoothing, and overfitting issues. DGCN also uses recent deep learning tricks such as residual learning or dilated aggregation.

In our study, deeper neural networks make more sense because more neighbors can be explored. Deeper networks lead to more nodes visited by the central node, and therefore more information collected. This way, the image node receive information from all the article's nodes, even for pages with complicated DOM structures; thus improving the representation capacities of the model.

3   Experiments and Results

The dataset for our study was constructed using webpages from the Qwant News search index[4]. It consists of 242 247 webpages from 1341 different websites, crawled mainly from French news websites. Some Italian, German, and Spanish

---

[4] https://www.qwant.com/?t=news.

Table 1. Proxy task regression performance for each model (average cosine similarity loss between the predicted embedding and the reference text embedding, lower is better).

Model	Split by webpages			Split by websites		
	Train	Valid.	Test	Train	Valid.	Test
distance	0.587	0.589	0.587	0.618	0.479	0.609
wGCN	**0.303**	**0.342**	**0.339**	**0.357**	**0.355**	**0.417**
GCN	0.365	0.407	0.406	0.426	0.482	0.569
GAT	0.367	0.441	0.439	0.423	0.563	0.623
DGCN	0.362	0.492	0.493	0.366	0.618	0.697

Table 2. WICE performance (average cosine similarity loss between the context node embedding and the reference text embedding, lower is better).

Model	Split by webpages			Split by websites		
	Train	Valid.	Test	Train	Valid.	Test
Oracle	0.293	0.297	0.293	0.334	0.264	0.259
Random	0.780	0.779	0.779	0.792	0.736	0.800
Title	0.834	0.835	0.833	0.834	0.861	0.814
Text after image	0.671	0.672	0.670	0.701	0.571	0.705
Blind WICE	0.654	0.658	0.653	0.715	0.508	0.680
WGCN	**0.381**	**0.386**	**0.381**	**0.415**	**0.404**	**0.441**

websites are also included. Both international and regional websites from France are included in order to maximize diversity.

We preprocess the HTML documents as follows: the content in `<main>` , `<body>` or `<article>` is first extracted from the webpage. Some tags pertaining to layout, such as `<style>` or `<button>` , are then removed to clean up the DOM tree from unnecessary nodes. Then, we extract the biggest image (in pixels) of a webpage with its reference text. Webpages without such an image are removed from the dataset. After preprocessing, the datasets contains 119 550 webpages in 805 websites. Texts are encoded using the multilingual sentence-BERT [15,16] that achieves state-of-the-art sentence embedding generation in several languages.

Node types are also considered useful and are one-hot encoded into 22 groups based on their HTML tag's semantics (lists, headers, paragraphs, etc.). There are two ways to split the dataset: splitting per document regardless of the original website and split by website using all the pages of one website. The second is more difficult because the data is not homogeneous: the test data may differ in structure and topic. In the first setting, the ratio of webpages in the training set, validation set and test set are 5:2:3, $i.e.$59 775 webpages in 723 websites for the training set, 23 910 webpages in 622 websites for the validation set and 35 865 webpages in 677 for the test set. In the second setup, the proportions of websites in the training set, validation set and test set are also 5:2:3, *i.e.*53 978 webpages from 402 websites in the training set, 28 872 webpages from 162 websites in the validation set and 36 700 webpages from 242 websites in the test set. The optimal cosine similarity regression losses for each model of the two settings are shown in Table 1. As can be seen, the explicit weight-GCN model performs better on both settings and generalizes significantly better on unknown websites. We use only the wGCN architecture for the WICE task.

The average cosine similarity loss between the proposed text, which is the text with the highest score, and the image's reference text is shown in Table 2. We see that naive WICE heuristics mostly fail on such a diverse dataset: `title` performs even worse than `random`, while `text after image` (which can be view as window-based WICE) rarely picks the best text node. In theory, older works managed WICE using more complex heuristics. However, defining a comprehensive ruleset does not scale up to more than 1000 websites and is unpractical in real applications. We do not compare with visual-based WICE, either, because the rendering step using requires at least 1 s per webpage[5], *i.e.*, more than three days for our whole dataset, not even including the segmentation algorithm. The preprocessing of our approach, *i.e.* generating text embeddings and graphs, takes ≈0.429 s per webpage, totaling ≈21 h.

For comparison, we add a random model that randomly choose a text node in the HTML page. Any model that has learnt anything should be better than this baseline.

The result shows that our model significantly outperforms WICE based on heuristics by at least 25%. The wGCN-extracted context is closer to the lower bound (`oracle`) than any other WICE approach we considered, thus validating our approach's relevance. In practice, we found that a cosine similarity ≥ 0.6 (*i.e.* a cosine loss ≤ 0.4) is enough, considering that two sentences have the same topic. Our wGCN does not always reach this threshold in average but is significantly closer than other approaches that average around 0.3 similarity.

We found a strong correlation between regression losses and WICE losses during the exploration of the results (Pearson correlation coefficient: 0.96), suggesting that better models for proxy tasks are better models for the main task. We also found that texts with lower WICE losses are often semantically very close to the images and the reference texts or their topics. This observation may be summarized as a correlation between lower regression loss and the relevance of the image for the extracted text (the original WICE problem's objective). This also shows that semantics may help in selecting nodes with similarly-named entities or dates, thus having a better chance to describe a given image.

We also observed that the model sometimes generalizes poorly to unknown websites, because of the heterogeneous webpages of each set.

However, this is less of a problem for closed set news crawler which parses a list of known websites regularly updated, with only the occasional introduction of a new domain.

---

[5] With *e.g.*, headless Chromium.

4   Conclusion

In this work, we address the WICE problem by modeling and learning webpages using language models and GNNs, making large-scale automatic WICE easier, and not bounded by hard-coded rules. We train a model on a large unlabeled news webpage corpus by learning to mimick the alt-text when it exists. Our weight-GCN model assigns a weight to each text node and we then extract the most important node and define it as the image context. This approach may blindly extract context sentences using semantic similarity between sentences and structural information learned from the DOM tree. By working directly with the HTML, we avoid the need for rendering the webpage into an image and cut the preprocessing time by a factor 3, making large-scale WICE more tractable.

References

1. Cai, D., Yu, S., Wen, J.R., Ma, W.Y.: VIPS: a vision-based page segmentation algorithm. Technical Report, MSR-TR-2003-79, Microsoft (2003). https://www.microsoft.com/en-us/research/publication/vips-a-vision-based-page-segmentation-algorithm/
2. Coelho, T.A.S., Calado, P.P., Souza, L.V., Ribeiro-Neto, B., Muntz, R.: Image retrieval using multiple evidence ranking. IEEE Trans. Knowl. Data Eng. **16**(4), 408–417 (2004). https://doi.org/10.1109/TKDE.2004.1269666
3. De Cao, N., Aziz, W., Titov, I.: Question answering by reasoning across documents with graph convolutional networks. In: Proceedings of the 2019 Conference of the North American Chapter of the Association for Computational Linguistics: Human Language Technologies. pp. 2306–2317. Association for Computational Linguistics, Minneapolis (2019). https://doi.org/10.18653/v1/N19-1240, https://www.aclweb.org/anthology/N19-1240
4. Fauzi, F., Hong, J.L., Belkhatir, M.: Webpage segmentation for extracting images and their surrounding contextual information. In: Proceedings of the 17th ACM International Conference on Multimedia, MM '09, pp. 649–652. Association for Computing Machinery, New York (2009). https://doi.org/10.1145/1631272.1631379
5. Feng, H., Shi, R., Chua, T.S.: A bootstrapping framework for annotating and retrieving WWW images. In: Proceedings of the 12th Annual ACM International Conference on Multimedia, Multimedia '04, pp. 960–967. Association for Computing Machinery, New York (2004). https://doi.org/10.1145/1027527.1027748
6. Gong, Z., Uu, R.L.H., Cheang, C.: Web image indexing by using associated texts. Knowl. Inf. Syst. **10**, 243–264 (2006). https://doi.org/10.1007/s10115-006-0011-0
7. Hattori, G., Hoashi, K., Matsumoto, K., Sugaya, F.: Robust web page segmentation for mobile terminal using content-distances and page layout information. In: Proceedings of the 16th International Conference on World Wide Web, WWW '07, pp. 361–370. Association for Computing Machinery, New York (2007). https://doi.org/10.1145/1242572.1242622
8. Joshi, P.M., Liu, S.: Web document text and images extraction using DOM analysis and natural language processing. In: Proceedings of the 9th ACM Symposium on Document Engineering, DocEng '09, pp. 218–221. Association for Computing Machinery, New York (2009). https://doi.org/10.1145/1600193.1600241

9. Kipf, T.N., Welling, M.: Semi-supervised classification with graph convolutional networks. In: 5th International Conference on Learning Representations, ICLR 2017, Toulon, France, 24–26 April 2017, Conference Track Proceedings. OpenReview.net, Toulon, France (2017). https://openreview.net/forum?id=SJU4ayYgl
10. Li, G., Muller, M., Thabet, A., Ghanem, B.: Deepgcns: can GCNs go as deep as CNNs? In: Proceedings of the IEEE/CVF International Conference on Computer Vision, pp. 9267–9276 (2019)
11. Li, G., Xiong, C., Thabet, A., Ghanem, B.: DeeperGCN: all you need to train deeper GCNs. Computing Research Repository arXiv:2006.07739 (2020)
12. Li, J., Liu, T., Wang, W., Gao, W.: A broadcast model for web image annotation. In: Zhuang, Y., Yang, S.-Q., Rui, Y., He, Q. (eds.) PCM 2006. LNCS, vol. 4261, pp. 245–251. Springer, Heidelberg (2006). https://doi.org/10.1007/11922162_29
13. Lockard, C., Shiralkar, P., Dong, X.L.: OpenCeres: when open information extraction meets the semi-structured web. In: Proceedings of the 2019 Conference of the North American Chapter of the Association for Computational Linguistics: Human Language Technologies, pp. 3047–3056. Association for Computational Linguistics, Minneapolis (2019). https://doi.org/10.18653/v1/N19-1309, https://www.aclweb.org/anthology/N19-1309
14. Qian, Y., Santus, E., Jin, Z., Guo, J., Barzilay, R.: GraphIE: a graph-based framework for information extraction. In: Proceedings of the 2019 Conference of the North American Chapter of the Association for Computational Linguistics: Human Language Technologies, pp. 751–761. Association for Computational Linguistics, Minneapolis (2019). https://doi.org/10.18653/v1/N19-1082, https://www.aclweb.org/anthology/N19-1082
15. Reimers, N., Gurevych, I.: Sentence-BERT: sentence embeddings using siamese BERT-networks. In: Proceedings of the 2019 Conference on Empirical Methods in Natural Language Processing. Association for Computational Linguistics, Asia World Expo, Hong Kong (2019). https://arxiv.org/abs/1908.10084
16. Reimers, N., Gurevych, I.: Making monolingual sentence embeddings multilingual using knowledge distillation. CoRR abs/2004.09813, https://arxiv.org/abs/2004.09813 (2020)
17. Shen, H.T., Ooi, B.C., Tan, K.L.: Giving meanings to WWW images. In: Proceedings of the Eighth ACM International Conference on Multimedia, MULTIMEDIA '00, pp. 39–47. Association for Computing Machinery, New York (2000). https://doi.org/10.1145/354384.376098
18. Tryfou, G., Tsapatsoulis, N.: Extraction of web image information: semantic or visual cues? In: Iliadis, L., Maglogiannis, I., Papadopoulos, H. (eds.) AIAI 2012. IAICT, vol. 381, pp. 368–373. Springer, Heidelberg (2012). https://doi.org/10.1007/978-3-642-33409-2_38
19. Tsapatsoulis, N.: Web image indexing using WICE and a learning-free language model. In: Iliadis, L., Maglogiannis, I. (eds.) AIAI 2016. IAICT, vol. 475, pp. 131–140. Springer, Cham (2016). https://doi.org/10.1007/978-3-319-44944-9_12
20. Veličković, P., Cucurull, G., Casanova, A., Romero, A., Liò, P., Bengio, Y.: Graph attention networks. In: 6th International Conference on Learning Representations (2017)
21. Wood, L., et al.: Document object model (dom) level 1 specification. W3C recommendation 1 (1998)
22. Xie, S., Lu, M.: Interpreting and understanding graph convolutional neural network using gradient-based attribution methods. Computing Research Repository arXiv:1903.03768 (2019). http://arxiv.org/abs/1903.03768

Towards Mining Generalized Patterns from RDF Data and a Domain Ontology

Tomas Martin[1(✉)], Victor Fuentes[1,2], Petko Valtchev[1],
Abdoulaye Baniré Diallo[1,2], René Lacroix[3], Maxime Leduc[2],
and Mounir Boukadoum[1]

[1] CRIA, Département d'informatique, UQÀM, Montréal, Canada
martin.tomas@cournes.uqam.ca
[2] LACIM, Département d'informatique, UQÀM, Montréal, Canada
[3] Lactanet, Sainte-Anne-de-Bellevue, Canada

Abstract. Nowadays, linked data (LD) are ubiquitous and mining them
for knowledge, e.g. frequent patterns, needs not be argued for.

A domain ontology (DO) on top of a LD dataset enables the discovery
of abstract patterns, a.k.a. *generalized*, capturing –rather than identical sub-structures–conceptual regularities in data. Yet with the resulting
ontologically-generalized graph patterns (OGP), a miner faces the combined challenges of graph topology and a label hierarchy, which amplifies well-known difficulties with graphs such as support counting or non
redundant pattern listing. As OGP mining is yet to be addressed in its
generality, we propose a formalization and study two workaround methods that avoid tackling it head-on, i.e. deal with each aspect separately.
Both perform pure graph mining with adapted label sets: *gSpan-OF*
merely strips labels of hierarchical structure while *Tax-ON* first mines
frequent graph topologies with only root classes as labels, then successively refines labels on each topology.

Keywords: Pattern mining · Ontologies · Graph data · Generalized
patterns

1 Introduction

Linked data (LD) are nowadays produced and published in ever increased numbers, hence mining them for knowledge about the underlying domain has been
recognized as an important research topic [19]. As a special case of such knowledge, structural patterns, either frequent [3] or rare [21], represent important
(a)typical trends and regularities that might, for instance, reflect previously
unknown phenomena or provide the explanation for observed behaviours.

The advantage of LD is they often come with, or might be fitted *a posteriori* to, a domain ontology (DO) [9] which is a prime source of descriptive domain
knowledge [15]. A DO, through its class hierarchy, makes possible the discovery
of more abstract patterns, a.k.a. *generalized patterns* (GP) [20] in the data mining

© Springer Nature Switzerland AG 2021
M. Kamp et al. (Eds.): ECML PKDD 2021 Workshops, CCIS 1524, pp. 268–278, 2021.
https://doi.org/10.1007/978-3-030-93736-2_21

(DM) field. GP refer to abstractions in places where data records refer to individual objects, or *items*. In this way, they go beyond the detection of identical sub-records (as in plain patterns) to capture the shared conceptual structure. GP are agnostic to data record topology, e.g. graph, sequence or flat set of items). We tackle pattern mining with a DO in a precision farming [6] context: DOs, typically designed in OWL, have gained significant popularity in life sciences [8], in particular, in agriculture. In a project revolving around dairy production optimization [11], we are looking after interpretable patterns [16] that might help compare and contrast populations of dairy cows and entire herds within the Canadian dairy livestock. To that end, we designed a DO [10] in OWL. In these settings, the resulting *ontologically-generalized graph patterns* (OGP), as we named them, are labelled multi-digraphs with vertex/edge labels being DO classes/properties.

An OGP miner faces significant challenges due to interplay between graph topology and hierarchy on labels, e.g. in typical pattern mining concerns like support counting and non redundant traversal of the pattern space. Since the OGP mining problem has yet to be addressed in its generality, we propose here a formalization thereof. Then, as a means to assess the need for a dedicated OGP miner, we study two workaround methods that avoid tackling the problem head-on. Instead, our methods deal with a single facet of the problem at a time. Both perform pure graph mining using *gSpan* [22], a reference graph miner, with adapted ontological label sets: *gSpan-OF* merely strips DO classes of their hierarchical structure while *Tax-ON* uses the most generic classes to discover frequent graph topologies then puts each through a sequence of label specializations.

When assessing these methods on a sample of our dairy control data, we observed that both incur high computational costs due to the combinatorial nature of the underlying pattern spaces. We see this as an argument in favour of a more subtle approach that mixes topology extensions with label refinements within a unique mining step.

The remainder of the paper is structured as follows: Sect. 2 summarizes related prior work. Next, Sect. 3 states the OGP mining problem and describes both workarounds. Then, Sect. 4 reports on their respective performances and observed limitations. Finally, Sect. 5 concludes the paper.

2   Related Work

Graphs are among the most difficult data structures to mine as basic operations involved are akin to the costly (sub-)graph isomorphism. *gSpan* [22] is arguably the reference method: To mine collections of undirected labelled graphs, it moves down a spanning tree of the pattern space each time extending a parent with a new edge. It exploits a canonical form, the depth-first-search (DFS) encoding, to prune redundant tree branches, which may require extensive graph comparisons. *gSpan* is the basis for a number of RDF graph miners. *Gaston* [18] is another popular graph pattern miner.

Taxonomies have been used as a source of domain knowledge in DM from its onset [4] thus leading to the *generalized patterns* (GP) where categories from a domain taxonomy replace some of the individual items, e.g. in sequential pattern mining [3]. Our own brand of patterns arise from labelled (multi-)graphs, i.e. named RDF graphs, as data records. Moreover, pattern nodes are labelled by OWL classes and edges by OWL properties at various abstraction levels (see definitions in Sect. 3).

The generalized graph pattern mining was introduced in [12] which proposes adapting *AGM* [13] to vertex/edge label taxonomies. While the intended pattern generalizing/specializing operator(s) is unclear, by paper's admission, vertex taxonomies alone make the task way more challenging and, without effective pruning strategies, the output grows prohibitively large. In [1,2], a framework for mining ontology-based patterns from click-streams is presented: Their *xPMiner* method outputs sequences of ontology classes linked by object properties. While patterns are basically graphs of classes/properties, they have handy sequence backbones simplifying both pattern space traversal and support computing w.r.t. to our unrestricted settings. In [5], a *Gaston*-based method is proposed for mining abstract graph patterns from RDF. They proceed as follows: In the pre-processing step, the RDF graphs are transformed by replacing individual resource nodes by one of its abstract types from the ontology. However, since no generalization step is included, the method is unable to discover patterns involving classes on different abstraction levels than those explicitly assigned as vertex labels.

Taxogram [7] is arguably the first method to mine frequent generalized graph patterns (over a mere vertex label taxonomy, though). At step one, it runs *gSpan* [22] to discover all pattern topologies using the most general concept label for every vertex in data graphs. Then, each of the resulting most general patterns is gradually specialized: The method goes down the taxonomy for every vertex up till reaching an infrequent specialization. Since no order is assumed on vertices, duplicate patterns might be generated, hence the need for (expensive) isomorphism checks.

GP-Close [14] mines GP from RDF datasets with a schema: It splits graphs into triples and mines those as mere transactions of triple-shaped items. Yet resulting GPs might not constitute connected graphs. Later on, [23] adapted *gSpan* to RDF with a DO, yet with no label refinement step. Recently, the extraction of ontology-based *path-shaped* frequent patterns, i.e. sequences, was studied in [17]. Their method focuses on scalability issues related to blending graph combinatorics and DO hierarchy traversals.

3   Ontology-Based Graph Pattern Mining

Below, we state the problem and present two simple mining methods. In that, we use our dairy cattle performance ontology (DCPO) [10]: An excerpt thereof is given in Fig. 1.

3.1   Problem Statement

A pattern mining task [3] is defined by two languages (data records and patterns) and a quality criterion. Let $\Omega = \langle O, C, R, \leq_C, \leq_R, \in_C, \rho \rangle$ be an ontology where O, C and R are its sets of objects, classes, and object properties, respectively. Both classes and properties are organized into hierarchies $H_C = \langle C, \leq_C \rangle$ and $H_R = \langle R, \leq_R \rangle$ with $\leq_C$ denoting the rdfs:subClassOf relation and $\leq_R$ the rdfs:subPropertyOf one. The instantiation relation $\in_C \subseteq O \times C$ is the translation of rdf:type. The incidence relation $\rho \subseteq C \times R \times C$ is made of triples $c_1 \times r \times c_2$ denoting a property r between classes c_1 (*domain*) and c_2 (*range*). Observe that, from RDF/OWL point of view, we admit only object properties: Data ones are assumed encoded, prior to the analysis, into a suitable class hierarchy where classes model value ranges. For instance, in Fig. 1, HerdLeaveReason and subclasses translate a data property whereby leaf classes model original values and the remainder expert-provided abstractions.

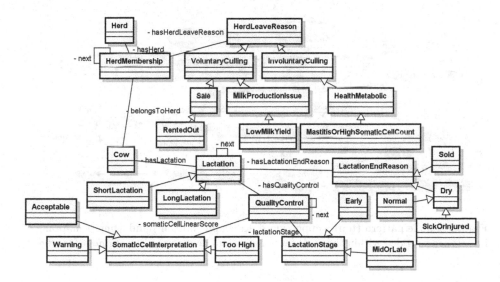

Fig. 1. Excerpt of our DCPO [10].

Our data language $\mathcal{L}_d$ is akin to RDF named graphs: A graph data record $g_d \in \mathcal{L}_d$ (see Fig. 2 on the bottom) represents a doubly-labelled directed multi-graph. Formally, it is a tuple $g_d = \langle V_d, E_d, \lambda_o, \lambda_r \rangle$ where V_d is a set of vertices, E_d is a bag of pairs of vertices and λ_o, λ_r are two labelling functions based on Ω. Moreover, $\lambda_o : V_d \to O$ maps each vertex to an object while $\lambda_r : E_d \to R$ maps an edge to a property. Intuitively, a pair of adjacent vertices in g_d exists iff the corresponding RDF triple exists in the triple store. Next, a pattern $g_p \in \mathcal{L}_p$ is also a doubly-labelled directed multi-graph $g_p = \langle V_p, E_p, \lambda_c, \lambda_r \rangle$. $\lambda_c : V_p \to C$ sends vertices into ontology classes while λ_r works the same way as above. We call such patterns *ontologically-generalized graphs patterns* (OGP), yet the term will only be used whenever ambiguity can arise.

Figure 2 shows an OGP and a matching data graph (labels from Fig. 1). Albeit of similar composition, $\mathcal{L}_p$ is not easily mapped to RDFS. It is rather akin to an RDF format where resources are *exemplars* of the corresponding DO entities.

Next, a relation $\dashv_\Omega \subseteq \mathcal{L}_d \times \mathcal{L}_p$ reflects the fact that a data graph $g_d = \langle V_d, E_d, \lambda_o, \lambda_r \rangle$ matches a pattern $g_p = \langle V_p, E_p, \lambda_c, \lambda_r \rangle$. Formally speaking, it is akin to a sub-graph isomorphism extended by *is-a* links. Thus, we note $g_d \dashv_\Omega g_p$ whenever an injective graph morphism $\mu : g_p \to g_d$ exists s.t. $\forall v_p \in V_p$, $\lambda_o(\mu(v_p)) \in_C \lambda_c(v_p)$ and $\forall e_p \in E_p$, $\lambda_r(\mu(e_p)) \leq_R \lambda_r(e_p)$. As an example, in Fig. 2, consider the μ mapping (dashed line) of the edge labelled *&hasLactation*[1] (top) to `hasFirstLactation` (bottom). In DCPO, there is a `rdfs:subPropertyOf` link between both. Similarly, *&InvoluntaryCulling* refers to a super-class of `MastitisOrHighSCC`, the most specific type of `Mastitis_1`.

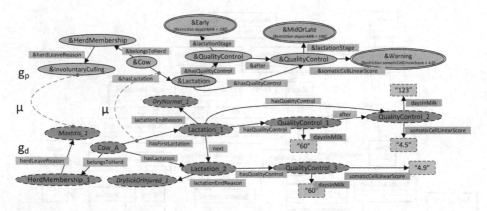

Fig. 2. Sample pattern (top), supporting data graph (bottom) and some μ-mappings. Labels are provided within the ovals for vertices and next to the line segment for edges.

In a similar way, we define the generality between two patterns, $\sqsubseteq_\Omega \subseteq \mathcal{L}_p \times \mathcal{L}_p$, i.e. via a subgraph isomorphism extended by *subclass* links from the ontology (the major difference w.r.t. $\dashv_\Omega$ is `rdf:type` is replaced by `rdfs:subClassOf*`). Thus, we note $g_p \sqsubseteq_\Omega g'_p$ whenever an injective graph morphism $\eta : g'_p \to g_p$ exists s.t. $\forall v'_p \in V'_p$, $\lambda_c(\eta(v_p)) \leq_C \lambda_c(v'_p)$ and $\forall e'_p \in E'_p$, $\lambda_r(\eta(e_p)) \leq_R \lambda_r(e'_p)$. A proper illustration of η is not in our figures: The pattern in Fig. 2 trivially generalizes the one in Fig. 3 as a subgraph thereof. For a non trivial example one could replace *&InvoluntaryCulling* in Fig. 3 by *&HealthMetabolic*. $\sqsubseteq_\Omega$, induces a hierarchy on $\mathcal{L}_p$, $\langle \mathcal{L}_p, \sqsubseteq_\Omega \rangle$. OGP miners have to traverse it while using an interestingness criterion, e.g. *support*.

---

[1] Pattern labels in $\mathcal{L}_p$ will be prefixed by & to differentiate them from ontology entities.

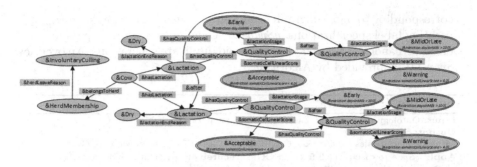

Fig. 3. An interesting pattern too far down the pattern space ($\sigma = 138$, $|D| = 7425$).

3.2   An OGP from the Dairy Dataset

The pattern in Fig. 3 –found via a SPARQL query and deemed useful by our experts– reflects the fact that 138 cows culled for reasons beyond farmer's control (`Involuntary` class) had prior health issues. Thus, for at least two lactations, they recorded a period of acceptable somatic cell levels followed by another one with more worrisome values. Such scenarios are plausible as higher scores are major signals for *mastitis* (udder inflammation). Consequently, recurrent health issues late in the lactation period could very well be the trigger for the involuntary culling and hence deserve closer monitoring.

Noteworthily, OGPs capture shared structure in data plain patterns would miss as the matching elements in data graphs might differ. For instance, the cows whose graphs match our OGP might have been culled for a variety of reasons, e.g. udder breakdown. Yet by referring to the common `Involuntary` class, the OGP helps factor out higher-order commonalities only available through the DO. This effect of label hierarchy, known from GP, spreads over edge labels also in our DO-based settings. Computationally, though, the OGP proved hardly reachable since located 32-deep within $\mathcal{L}_p$.

3.3   Two Workaround Approaches

As a first approach, we designed two workaround solutions that avoid mixing concept hierarchy- and graph topology-related aspects during $\langle \mathcal{L}_p, \sqsubseteq_\Omega \rangle$ traversal. Here, both methods produce $\mathcal{F}^\Omega$, the set of all frequent OGPs. The goal was to assess the alternatives to a fully-blown OGP miner going down $\langle \mathcal{L}_p, \sqsubseteq_\Omega \rangle$ while using a yet-to-define canonical form and a dedicated refinement operator (e.g. as in [2]).

First, *gSpan-OF* (for *gSpan* with Ontology Flattened), runs *gSpan* with a flat set of ontological labels, i.e. ignoring the relations $\leq_C$ and $\leq_R$ at pattern generation time. Only during support check for a pattern ($\dashv_\Omega$), do $\leq_C$, $\leq_R$, and $\in_C$ interfere in mining. Precisely speaking, the traversal follows a subset of

$\sqsubseteq_\Omega$ corresponding to pure subgraph isomorphism, thus ignoring links in $\langle \mathcal{L}_p, \sqsubseteq_\Omega \rangle$ representing label specializations. *gSpan-OF* benefits from *gSpan*'s parsimonious traversal: Its canonical form checks help avoid duplicate generations. An overview of *gSpan-OF* is provided by Algorithm 1.

Algorithm 1: *gSpan-OF*

Input: Ontology Ω, graph database D, minimal support threshold ς
Output: Set of frequent OGPs $\mathcal{F}^\Omega$
1 **Mine** frequent triples $t \in C \times \mathcal{R} \times C$ into $\mathcal{F}_1^\Omega$, the set of frequent *single-edge* OGPs
2 **Apply** *gSpan* to **discover** all frequent OGPs, **recursively** extending OGPs in $\mathcal{F}_1^\Omega$
3   **Extend** a frequent OGPs g_p using a frequent $t \in \mathcal{F}_1^\Omega$ to produce $g_{p'}$
4   For each $g_{p'}$, **test** its DFS encoding for canonicity
5   **Compute** $\sigma(g_{p'})$ (find data graphs $g_d \dashv_\Omega g_{p'}$)
6   If $\sigma(g_{p'}) > \varsigma$, $g_{p'}$ is a frequent OGP, then **recursively** extend $g_{p'}$
7 **end**

Second, *Tax-ON* (for _Tax_ogram with _O_ntology and _N_on-redundant output) untangles topology extensions and label specialization differently (Algorithm 2). Loosely following [7], it tackles them separately, in subsequent steps: (1) graph mining on suitably re-labelled data graphs and (2) label refinements on each of the patterns found at step (1). Thus, *Tax-ON* first mines $\mathcal{F}^{_G}$, the most generic frequent graph patterns, using plain *gSpan*. To that end, data graphs are cloned and then vertex/edge labels replaced by the most generic super-entities in H_C and H_R, respectively. In step two, an extension of *Taxogram* refines vertex labels. Thus, each OGP undergoes a sequence of individual label specializations, i.e. substitutions of a class by a direct subclass, to go down H_C.

To provide additional context, the pattern in Fig. 3 is a 21-pattern within the *gSpan-OF*'s pattern space (same as *gSpan*), while it can be obtained by refining a more realistic 10-pattern belonging to $\mathcal{F}^{_G}$ with *Tax-ON* (striping out all white vertices).

Algorithm 2: *Tax-ON*

Input: Ontology Ω, graph database D, minimal support threshold ς
Output: Set of frequent OGPs $\mathcal{F}^\Omega$
1 **Mine** $\mathcal{F}^{_G}$ the set of frequent most-generic OGPs
2   **Build** D' a copy of D, where vertex labels are replaced by their most generic class in $C_0 = roots(H_C)$
3   **Mine** frequent triples $t \in C_0 \times \mathcal{R} \times C_0$ into $\mathcal{F}_1^G$, the single-edge subset of $\mathcal{F}^G$ over D'
4   **Apply** *gSpan* to **discover** all OGPs in $\mathcal{F}^{_G}$, from all OGPs in $\mathcal{F}_1^G$
5 **end**
6 **Mine** all other frequent OGPs, from $\mathcal{F}^{_G}$ over D
7   **Apply** *Taxogram* to **recursively** label-refine OGPs in $\mathcal{F}^{_G}$
8   **Replace** one vertex label by one of its direct descendants in H_C to generate $g_{p'}$
9   **Compute** $\sigma(g_{p'})$ (same as in *gSpan-OF*), if $g_{p'}$ is a frequent OGP, then **recursively** label-refine $g_{p'}$
10   **end**
11   **Eliminate** duplicate OGPs using graph isomorphism checks
12 **end**

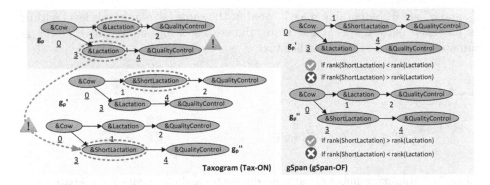

Fig. 4. Hierarchy-centered exploration vs Vertex-centered exploration.

4 Evaluation of the DO-Powered Pattern Miners

We put both methods through some preliminary experiments. In practice, the main limiting factor, shared by both, is the number of automorphisms in data graphs and patterns. In *gSpan*, this slows down candidate generation, canonical check and support-computing steps. Also, it creates a prohibitive number of embeddings of a candidate pattern to the data graphs (up to 10^6 for one data graph). This prevents, in turn, the efficient pruning during candidate generation based on explicit embedding management. That same issue prevents *Taxogram* from building a vertical database, which forces expensive support counting over the graph database (via $\dashv_\Omega$).

Conceptually, the key point is that both methods forgo part of the available structure: *gSpan-OF* ignores the hierarchies in the DO, while *Tax-ON* strips a pattern from its graph information bringing it down to vertex sequence. Simply put, neither flattening the conceptual hierarchy nor ignoring the graph topology is efficient enough in practice to allow for an in-depth exploration of $\langle \mathcal{L}_p, \sqsubseteq_\Omega \rangle$. On one hand, *gSpan*'s non-redundant exploration is burdened by a large number of candidates induced by flat label sets since. On the other, *Taxogram*'s loose refinement approach outputs lots of duplicate patterns.

Figure 4 compares *gSpan*'s vertex-centered exploration to *Taxogram*'s hierarchy-centered one on their way to deal with an OGP g_p and its two (iso-morphic) specializations g'_p and g''_p. On the left, *Taxogram* produces both g'_p and g''_p from g_p by refining *&Lactation* into *&ShortLactation* on positions 1 and 3, respectively. This exemplifies a key shortage: By specializing all positions unrestrictedly, it allows for duplicates to arise. In contrast, *gSpan* –through its canonical form-driven exploration– avoids either g'_p or g''_p since exactly one of them will comply to that form constraints. This capacity is independent from the order on labels that underlies the canonical form, as shown on the right of the figure (for additional details, see Fig. 6 in Appendix).

Conversely, by traversing every subspace of $\langle \mathcal{L}_p, \sqsubseteq_\Omega \rangle$ induced by a OGP topology while following $\sqsubseteq_\Omega$, *Taxogram* benefits from support anti-monotony. Thus, g'_p and g''_p will be tested only if g_p proves frequent. For *gSpan*, though, g_p,

g'_p, and g''_p are incomparable since located at the same (size-induced) level within its flattened pattern space. Therefore, the status of any of the three OGPs is immaterial while testing the other two.

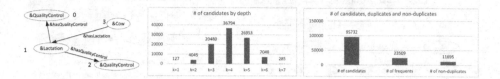

Fig. 5. *Taxogram*: number of patterns, duplicates and candidates for a 3-pattern.

Figure 5 clarifies the number of candidates *Taxogram* examines while testing all possible specializations of a specific 3-pattern (left). Here, up to seven specializations are required to reach a most specific pattern while the peak number of candidates is generated at depths four and five below. The worrying aspect is among the ca. 100k specializations tested, some 50% were duplicates (ratio increases with the pattern size).

A suitable canonical form is essential for the efficient exploration of $\langle \mathcal{L}_p, \sqsubseteq_\Omega \rangle$. Yet current forms such as *gSpan*'s are not designed to work with label specializations. A critical property to guarantee for a refinement operator is the anti-monotonicity of the canonical form: A canonical OGP code must only have canonical ancestors. The highly irregular structure of $\langle \mathcal{L}_p, \sqsubseteq_\Omega \rangle$ is a major challenge faced by such operators.

5   Conclusion

We presented here a first attempt at efficiently mining frequent generalized patterns from an RDF dataset while using a DO as a generalization source. As the corresponding DM task is beyond the reach of state-of-the-art methods, we designed two workaround solutions: (1) pure graph pattern mining with a flattened set of labels and (2) graph mining with only root class/property labels followed by recursive pattern specialization.

Having put them through experimental evaluation, we observed that both approaches suffer on high computational overhead, most likely due to the highly combinatorial nature of the resulting pattern spaces. This clearly warrants a more subtle blend of topology extension and label refinement in a uniform descend in $\langle \mathcal{L}_p, \sqsubseteq_\Omega \rangle$. Therefore, we are currently investigating the design of a dedicated OGP miner exploiting a tailor-made canonical form and support computing mechanisms.

Appendix

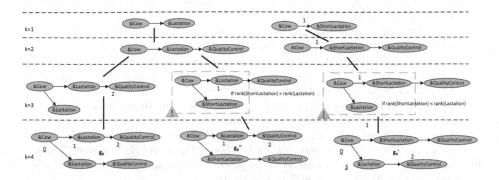

Fig. 6. *gSpan*'s flattened exploration of $\mathcal{L}_p$

References

1. Adda, M., et al.: On the discovery of semantically enhanced sequential patterns. In: 4th ICMLA, p. 8. IEEE (2005)
2. Adda, M., et al.: A framework for mining meaningful usage patterns within a semantically enhanced web portal. In: 3rd C* Conference on Computer Science and Software Engineering, pp. 138–147 (2010)
3. Aggarwal, C.C., Han, J. (eds.): Frequent Pattern Mining. Springer, Cham (2014). https://doi.org/10.1007/978-3-319-07821-2
4. Anand, S., et al.: The role of domain knowledge in data mining. In: CIKM, pp. 37–43 (1995)
5. Berendt, B.: Using and learning semantics in frequent subgraph mining. In: Nasraoui, O., Zaïane, O., Spiliopoulou, M., Mobasher, B., Masand, B., Yu, P.S. (eds.) WebKDD 2005. LNCS (LNAI), vol. 4198, pp. 18–38. Springer, Heidelberg (2006). https://doi.org/10.1007/11891321_2
6. Brett, D., et al.: A survey of semantic web technology for agriculture. Inf. Process. Agric. **6**, 487–501 (2019)
7. Cakmak, A., Ozsoyoglu, G.: Taxonomy-superimposed graph mining. In: 11th EDBT, pp. 217–228. ACM (2008)
8. Cannataro, M., Santos, R.D., et al.: Biomedical and bioinformatics challenges to computer science. Procedia Comput. Sci. **1**(1), 931–933 (2010)
9. Dou, D., et al.: Semantic data mining: a survey of ontology-based approaches. In: IEEE ICSC, pp. 244–251 (2015)
10. Fuentes, V., et al.: Dairy ontology to support precision farming. In: 12th ICBO (2021)
11. Gonçalves Frasco, C., et al.: Towards an effective decision-making system based on cow profitability using deep learning. In: 12th ICAART, pp. 949–958 (2020)
12. Inokuchi, A.: Mining generalized substructures from a set of labeled graphs. In: Fourth IEEE International Conference on Data Mining (ICDM 2004), pp. 415–418. IEEE (2004)

13. Inokuchi, A., Washio, T., Motoda, H.: An apriori-based algorithm for mining frequent substructures from graph data. In: Zighed, D.A., Komorowski, J., Żytkow, J. (eds.) PKDD 2000. LNCS (LNAI), vol. 1910, pp. 13–23. Springer, Heidelberg (2000). https://doi.org/10.1007/3-540-45372-5_2

14. Jiang, T., et al.: Mining generalized associations of semantic relations from textual web content. IEEE Trans. Knowl. Data Eng. **19**(2), 164–179 (2007)

15. Kramer, F., Beißbarth, T.: Working with ontologies. In: Keith, J.M. (ed.) Bioinformatics. MMB, vol. 1525, pp. 123–135. Springer, New York (2017). https://doi.org/10.1007/978-1-4939-6622-6_6

16. Martin, T., et al.: Leveraging a domain ontology in (neural) learning from heterogeneous data. In: CIKM (Workshops) (2020)

17. Monnin, P.: Matching and mining in knowledge graphs of the web of data-applications in pharmacogenomics. Ph.D. thesis, Université de Lorraine (2020)

18. Nijssen, S., Kok, J.: A quickstart in frequent structure mining can make a difference. In: 10th ACM KDD, pp. 647–652 (2004)

19. Rettinger, A., et al.: Mining the semantic web. DMKD **24**(3), 613–662 (2012)

20. Srikant, R., Agrawal, R.: Mining generalized association rules. Futur. Gener. Comput. Syst. **13**(2–3), 161–180 (1997)

21. Szathmary, L., et al.: Towards rare itemset mining. In: 19th IEEE ICTAI, vol. 1, pp. 305–312, October 2007

22. Yan, X., Han, J.: gSpan: graph-based substructure pattern mining. In: IEEE ICDM, pp. 721–724 (2002)

23. Zhang, X., et al.: Mining link patterns in linked data. In: 13th WAIM, pp. 83–94 (2012)

Machine Learning for Irregular Time Series

Workshop on Machine Learning for Irregular Time Series (ML4ITS 2021)

Time series data are ubiquitous. The broad diffusion and adoption of the Internet of Things (IoT), as well as major advances in sensor technology are examples of why such data have become pervasive. These technologies have applications in several domains, such as healthcare, finance, meteorology, and transportation, where, for instance, the following tasks have high importance: prediction of the health status of patients, stock market analysis, prediction of the weather, and logistics. Deep Neural Networks (DNNs) have recently been used to create models that improve on the state of the art for some of these tasks. In time series classification and forecasting, Deep Learning (DL) has been beneficial for avoiding heavy data pre-processing and feature engineering. Time series data influences both political and industrial decisions every day, yet there is, surprisingly, very limited research in Machine Learning (ML) for time series - especially in situations where data is scarce or of low-quality.

In many real-world applications, we have the following two scenarios: 1) the amount of available training data is limited, or 2) there is a huge amount of available data which is scarcely labeled or unlabeled due to high costs of data collection and annotation. As a result, the future of Artificial Intelligence (AI) will be about "doing more with less". There is a need for focusing on modern AI techniques that can extract value from such challenging datasets. These considerations can also contribute to the increasing need to address sustainability and privacy aspects of ML and AI. Furthermore, there is a need to overcome the issue of limited availability of data and scarcity of labeled data for (multivariate) time series modeling. In this context, heterogeneity of the data (e.g. non-stationarity, multi-resolution, irregular sampling), as well as noise, pose further challenges.

The main scope of this workshop is to advance the state of the art in time series analysis for "irregular" time series. We define time series to be "irregular" if they fall under one or several of the following categories: a) Short: univariate and multivariate time series with a limited amount of data and history, b) Multiresolution: multivariate time series where each signal has a different granularity or resolution in terms of sampling frequency, c) Noisy: univariate/multivariate time series with some additional perturbation appearing in different forms (including time series with missing data), d) Heterogeneous: multivariate time series, usually collected by many physical systems, that exhibit different types of embedded, statistical patterns and behaviours, e) Scarcely labeled and unlabeled: univariate/multivariate time series where only a small part of the data is labeled or the data is completely unlabeled.

The first international workshop on Machine Learning for Irregular Time Series (ML4ITS 2021) intended to offer the ideal context for dissemination and cross-pollination of novel ideas in designing machine learning models suitable for dealing with irregular time-series

We want to thank all the authors who submitted their papers to ML4ITS 2021, all presenters and participants to the workshop, and the invited speakers Claudio Alippi and Boris Oreshkin for their inspiring talks. We are also grateful to the Program

Committee for their exceptional work in reviewing the submitted papers and the ECML Workshop Chairs and ECML PKDD organizers who made this event possible.

Full information about the workshop can be found at the workshop's website at https://ml4its.github.io/ml4its2021/.

October 2021

Massimiliano Ruocco
Erlend Aune
Claudio Gallicchio

Organization

Workshop Co-chairs

Massimiliano Ruocco — Sintef/Norwegian University of Science and Technology, Norway

Erlend Aune — BI/Norwegian University of Science and Technology, Norway

Claudio Gallicchio — University of Pisa, Italy

Program Committee

Sara Malacarne — Telenor Research, Norway

Pierluigi Salvo Rossi — Norwegian University of Science and Technology, Norway

Bjorn Magnus Mathisen — Sintef, Norway

Per Gunnar Auran — Sintef, Norway

Jo Eidsvik — Norwegian University of Science and Technology, Norway

Leif Anders Thorsrud — BI, Norway

Gard Spreeman — Telenor Research, Norway

Pablo Ortiz — Telenor Research, Norway

Vegard Larsen — BI/Norges Bank, Norway

Stefano Nichele — Oslomet/Simula, Norway

Filippo Maria Bianchi — UiT the Arctic University of Norway, Norway

Juan-Pablo Ortega — St. Gallen University, Switzerland

Azarakhsh Jalalvand — Ghent University, Belgium, and Princeton University, USA

Benjamin Paaßen — Humboldt University of Berlin, Germany

Petia Koprinkova-Hristova — Institute of Information and Communication Technologies, Bulgarian Academy of Sciences, Bulgaria

Homological Time Series Analysis
of Sensor Signals from Power Plants

Luciano Melodia$^{(\boxtimes)}$ and Richard Lenz

Professorship for Evolutionary Data Management, Friedrich-Alexander University
Erlangen-Nürnberg, 91058 Erlangen, Germany
{luciano.melodia,richard.lenz}@fau.de

Abstract. In this paper, we use topological data analysis techniques
to construct a suitable neural network classifier for the task of learning
sensor signals of entire power plants according to their reference designation system. We use representations of persistence diagrams to derive
necessary preprocessing steps and visualize the large amounts of data.
We derive deep architectures with one-dimensional convolutional layers
combined with stacked long short-term memories as residual networks
suitable for processing the persistence features. We combine three separate sub-networks, obtaining as input the time series itself and a representation of the persistent homology for the zeroth and first dimension. We
give a mathematical derivation for most of the used hyper-parameters.
For validation, numerical experiments were performed with sensor data
from four power plants of the same construction type.

Keywords: Power plants · Time series · Signal processing · Geometric
embedding · Persistent homology · Topological data analysis

1 Introduction

Power plants, regardless of their construction, must be intensively maintained
and monitored to ensure constantly efficient power generation and to minimize
the risk of damage. Sensors measure pressure, temperature, enthalpy, electrical
resistance, etc., and their readings are recorded for this reason in order to monitor
them. Since power plant operators operate at an international level, there is
a need to evaluate information from power plants whose measured values are
organized and stored in accordance with the norms that are standard for the
specific country. Both the identifiers and the storage structure rarely resemble
each other, so signals from power plant sensors must be manually assigned to
the appropriate identifiers. These identifiers are called the *power plant reference
designation system*, which is defined as an international standard [13].

The code can be found at: https://github.com/karhunenloeve/TwirlFlake.
This project has been partially supported by Siemens Energy AG.
The authors thank Leonie Rumi, Noah Becker, Philipp Gäbelein and the anonymous
reviewers for useful suggestions for improvement and proofreading.

© Springer Nature Switzerland AG 2021
M. Kamp et al. (Eds.): ECML PKDD 2021 Workshops, CCIS 1524, pp. 283–299, 2021.
https://doi.org/10.1007/978-3-030-93736-2_22

Unfortunately, this standard is not supported by all countries, so seamless mapping is not possible for the time being. In addition, there are problems such as the choice of acronyms for the identifiers, the language and the lack of uniqueness, so that the engineer often performs the mapping manually based on the measured values and previously calculated statistics, since (s)he cannot rely on the predefined identifiers. Classifiers that assign the measured values to the appropriate identifier based on some of their features are suitable for this purpose. Of particular importance are the periodicities or quasi-periodicities occurring within the sensor's signal, which encode certain recurring events within the power plant. We use persistent homology on an embedding of these signals that lies on or dense within an N-dimensional torus to encode (quasi-)periodicities. We train neural networks with the raw signal, the zero-dimensional and the one-dimensional homology groups of a filtered toroidal embedding of the signal.

Our work is structured as follows:

Section 2 We introduce the theory of persistent homology on triangulable topological spaces. Specifically, we introduce simplicial complexes, filtrations and the associated persistence module, and define some representations of persistent Betti numbers – and hence the persistence diagram.

Section 3 We derive the assumption to describe a time series as a smooth manifold.

Section 4 We discuss Takens' embedding and the topological and geometrical properties of the sliding window point cloud, which encodes (quasi-) periodicities that can be detected in its persistence diagrams.

Section 5 We detail the heuristics used to determine an ideal embedding dimension and a time delay, and compute these quantities for our data.

Section 6 We present the results of examining the data using Betti curves and persistence silhouettes. We evaluate our proposed architecture based on accuracy, F_1-score, precision and recall.

Section 7 Finally, we discuss the results and summarize our experiments. We state two issues that arose from our work, particularly with respect to the applicability to other power plants of the same construction type.

2   Primer: Persistent Homology

Homology groups are Abelian groups attached to a topological space, counting, in an intuitive sense, the holes of the very same object in a particular dimension. We assume for the persistent homology theory the triangulability of the underlying topological space. This is not essential for homology theories in general, but it is extremely useful for the persistent one. We have no prior knowledge of the underlying space and consequently we must approximate it by some construction on the given points.

Let said topological space contain all points in the data set. We first consider points in general position $\{v_0, \cdots, v_k\} \subset \mathbb{R}^n$, such that the vectors $\{v_1 - v_0, \cdots, v_k - v_0\}$ are linearly independent with $k < n$. Thus, the points do not lie on a hyperplane of dimension less than k.

Their convex hull is the simplex

$$[v_0, \cdots, v_k] := \left\{ \sum_{i=1}^{k} \lambda_i(v_i - v_0) \;\middle|\; \sum_{i=1}^{k} \lambda_i = 1, \; \lambda_i \geq 0 \right\}, \tag{1}$$

with dimension k. The ith face of a simplex can be written as

$$d_j[v_0, \cdots, v_k] = [v_0, \cdots, \hat{v}_i, \cdots, v_k], \tag{2}$$

where $\hat{v}_i$ denotes the removal of the element v_i from the simplex. Note, that the ith face is a $(k-1)$-simplex. We call a finite union of such simplices in Euclidean space a simplicial complex $\mathfrak{K}$, if every face of a simplex in $\mathfrak{K}$ is also in $\mathfrak{K}$, and every intersection of two simplices from $\mathfrak{K}$ is either empty or a common face of both. A filtration is a nested sequence

$$\mathbb{K}: \quad \emptyset = \mathfrak{K}_0 \subseteq \mathfrak{K}_1 \subseteq \cdots \subseteq \mathfrak{K}_q = \mathfrak{K}, \tag{3}$$

where we assume without restriction of generality that $\mathfrak{K}_i = \mathfrak{K}_q$ for $i \geq q$. The expression $\mathfrak{K}_i$ denotes a one-parameter family of simplicial complexes. The parameter can be interpreted as a particular time point on the filtration and determines how the realization of the simplicial complex looks on the set of points. There are several ways to construct simplicial complexes. For example, the Vietoris-Rips complex $\mathfrak{R}_i$ with $i \in \mathbb{R}$, is defined for a set of points as:

$$\mathfrak{R}_i(X) := \left\{ U \subseteq X \;\middle|\; ||x - y|| \leq i \text{ for all } x, y \in U \right\}. \tag{4}$$

We are looking for an ideal set of points that contains all the points in our data set. For this, we assume that we can find a suitable set on which we define a topology. This is precisely the assumption of an underlying topological space, which was made at the beginning of the section. If we assume more structure for the topological space we are looking for, such as smooth coordinate maps and a Riemannian metric, we can assume a Riemannian manifold underlying the data.

A Vietoris-Rips complex over a Riemannian manifold (see Sect. 3.2) is homotopy equivalent to the manifold itself for sufficiently small i [10, §3.5]. Under mild conditions, its homotopy groups have an isomorphism into its respective homology groups by the Hurewicz homomorphism [8,16, p. 390,§3]. Thus, its homology theory provides an isomorphism to the cohomology theory of a smooth manifold, a suitable description due to the quality of data [17, §6].

2.1 Homology Groups

The kth chain group C_k on $\mathfrak{K}$ is the free Abelian group on the set of k-simplices. An element $c \in C_k$ is called k-chain and can be written as $c = \sum_{i=1}^{k} \lambda_i \sigma_i$ with $\sigma := [v_0, \cdots, v_k] \in \mathfrak{K}$ with coefficients λ_i in any ring. The group becomes a vector space if the coefficients are chosen to be within some field $\mathbb{F}$. The ring of integers $\mathbb{Z}$ modulo a maximal prime ideal – $\mathbb{Z}/(p\mathbb{Z})$ – gives us such a field.[1]

We can study the chain groups on the filtration considering a chain complex, a pair $(C_\star, \partial)$, where $C_\star = \bigoplus_{k\in\mathbb{Z}} C_k$ and $\partial = \bigoplus_{k\in\mathbb{Z}} \partial_k$, with C_k as $\mathbb{F}$-vector space and $\partial_{k+1} : C_{k+1} \to C_k$ as $\mathbb{F}$-linear maps with $\partial_k \circ \partial_{k+1} = 0, [v_0, \cdots, v_k] \mapsto \sum_{i=0}^k (-1)^i [v_0, \cdots, \hat{v}_i, \cdots, v_k]$. Elements from $\ker \partial_k$ will be called k-cycles and to elements from $\operatorname{im} \partial_{k+1}$ we will refer to as k-boundaries. Thus, each boundary is a cycle. Similarly, cohomology can be defined on simplicial complexes, which is exploited in the implementation of the algorithm for the computation of persistent homology [7, §4.1]. Having the chain complex $(C_\star, \partial)$, the kth homology group of chain complexes is defined as a quotient over $\mathbb{F}$ – or just over some ring – such that $H_k(C_\star; \mathbb{F}) := \ker \partial_k (C_\star) / \operatorname{im} \partial_{k+1} (C_\star)$. The module is defined as a family of $\mathbb{F}$-vector spaces V_i for a real number i together with $\mathbb{F}$-linear maps $f_{ij} : V_i \to V_j$, for $i \leq j$, which for a $k \leq i \leq j$ satisfy the equation $f_{kj} = f_{ij} \circ f_{ki}$. We can define persistent homology by considering a family $\{\mathfrak{K}_i\}_{i\in\mathbb{R}}$ of simplicial complexes, with simplicial maps $f_{ij} : \mathfrak{K}_i \to \mathfrak{K}_j$.

The *persistence module* are the k-dimensional homology groups $H_\star(\mathfrak{K}_i; \mathbb{F})$ together with the maps $H_\star(f_{ij}) : H_\star(\mathfrak{K}_i; \mathbb{F}) \to H_\star(\mathfrak{K}_j; \mathbb{F})$ induced by f_{ij}.

Multi-sets. *Persistence diagrams* encode the ranks of a persistence module and are *multi-sets*, a pair consisting of a set and a function (Z, φ), where $\varphi : Z \to \mathbb{R} \cup \{+\infty\}$. For $z \in Z$, $\varphi(z)$ denotes a multiplicity. The union of two multi-sets (Z, φ) and (Z', φ') is the multi-set $(Z \cup Z', \varphi \cup \varphi')$, with

$$(\varphi \cup \varphi')(z) = \begin{cases} \varphi(z), & \text{if } z \in Z, z \notin Z', \\ \varphi'(z), & \text{if } z \notin Z, z \in Z', \\ \varphi(z) + \varphi'(z), & \text{if } z \in Z, z \in Z'. \end{cases} \tag{5}$$

2.2 Persistent Homology

The persistence module is defined as the collection of all $\mathbb{F}$-vector spaces V_i for a real number i together with the $\mathbb{F}$-linear maps $f_{ij} : V_i \to V_j$ such that for each pair i, j it holds that $i \leq j$. Moreover, for a $k \leq i \leq j$, it holds that $f_{kj} = f_{ij} \circ f_{ki}$, under the condition that all but finitely many such maps are isomorphisms. Let us choose a filtration of the Vietoris-Rips complex $\{\mathfrak{R}_i(X)\}_{i\in\mathbb{R}}$ together with its simplicial maps $f_{ij} : \mathfrak{R}_i(X) \to \mathfrak{R}_j(X)$ for each pair $i \leq j$, so that the above mentioned conditions hold. We write the *persistent (simplicial) homology groups* with $\mathbb{F}$-coefficients as persistence module

$$H_\star(\mathfrak{R}_i(X); \mathbb{F}), \text{ with } H_\star(f_{ij}) : H_\star(\mathfrak{R}_i(X); \mathbb{F}) \to H_\star(\mathfrak{R}_j(X); \mathbb{F}). \tag{6}$$

---

[1] For our work, we use the Mersenne prime $p = 6972593$, because of the efficiency of memory allocation, since it fits into an integer data type and does not cause overflow. One could also use $p = 2$, but since about 3% of the data yields homological coefficients other than the one in $\mathbb{Z}_2$, we use the largest possible coefficients field.

The persistence diagram is a multi-set of points in $\mathbb{R} \times (\mathbb{R} \cup \{+\infty\})$. We consider the case of finite persistence modules, which are represented by persistence diagrams. Such diagrams are realized for a finite and discrete set $I \subset \mathbb{R}$ as real open intervals $\{(b_i, d_j)\}_{i,j \in I}$. The *birth points* b_i and *death points* d_j satisfy for a persistence module $b_i \leq i \leq j < d_j$. The multiplicity of points (b_i, d_j) in the multi-set is equal to $\mathrm{rank}(f_{ij})$. We represent a persistence diagram as a finite multi-set $\mathfrak{P} := \{(\frac{b_i + d_j}{2}, \frac{d_j - b_i}{2})\}_{i,j \in I}$ and $i \leq j$ [3, §2]. We use *persistence landscapes*, a functional representation of *persistent points* $(\frac{b_i + d_j}{2}, \frac{d_j - b_i}{2})$, which are elements of the *persistence diagram* $\mathfrak{P}$, given by the following function [2, §2.2]:

$$
\Lambda_i(t) = \begin{cases} t - b_i, & \text{if } t \in [b_i, \frac{b_i + d_j}{2}], \\ d_j - t, & \text{if } t \in (\frac{b_i + d_j}{2}, d_j], \\ 0, & \text{otherwise.} \end{cases} \tag{7}
$$

Silhouettes and Betti Curves. For points in the multi-set $\mathfrak{P}$ weights $\{w_i = |d_j - b_i|^p \mid 0 < p \leq \infty\}_{i,j \in I}$ with $i \leq j$ exist, such that the *w-weighted silhouette* of $\mathfrak{P}$ is a representation of a persistence diagram within the vector space of real-valued functions [3, §2.3]:

$$
\xi : \mathbb{R} \to \mathbb{R}, \quad t \mapsto \frac{\sum_{i \in I} w_i \Lambda_i(t)}{\sum_{i \in I} w_i}. \tag{8}
$$

Let the multi-set $\mathfrak{P}$ be a persistence diagram, then we define the *Betti curve* as a function $\beta_{\mathfrak{P}} : \mathbb{R} \to \mathbb{N}$, whose values on $s \in \mathbb{R}$ are given by the number of points $(b_i, d_j) \in \mathfrak{P}$ – counted with multiplicity – such that we satisfy $b_i \leq s < d_j$.

We have now introduced persistent homology and the persistence module, and defined persistence diagrams and their representations, which we intend to use for our classifiers. Persistent homology is able to encode (quasi-)periodicities of signals. However, this requires an embedding that encodes the signals as a curve on a compact geometric object, a torus. We discuss this embedding next.

3 Time Series Embedding

For the embedding of the time series, which is itself a subset of some topological space T, we call a mapping $f : T \to R$ an embedding in a topological space R if f is a homeomorphism from T to the subspace $f(T)$ of its image. Thus f is said to be continuous and injective such that every open set $O \subseteq T$ is open again as an image $f(O) \subseteq f(T)$. Two reasons lead us to the embedding of a time series:

1. The stability of the persistence diagrams we want to compute is guaranteed to the best of our knowledge for tame functions [6, §2]. The function f is tame if it is continuous, all sub-level-sets have homology groups of finite rank, and there are finitely many critical values at which these ranks change.
2. We want to relate homology groups with periodicity. This is done naturally by embedding on a compact geometric object, the N-torus.

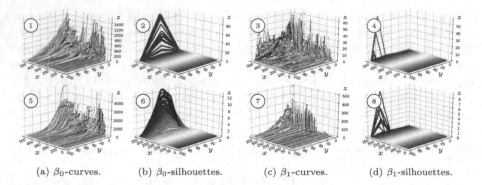

(a) β_0-curves.   (b) β_0-silhouettes.   (c) β_1-curves.   (d) β_1-silhouettes.

Fig. 1. Betti curves and silhouettes of power plant signals. The signals are ordered by decreasing persistence entropy. The x-axis displays the resulting order. The y-axis indicates the parameterization of the simplicial complex normalized to $[0, 100]$ – for visualization purposes. The z-axis indicates the number of representatives of the zeroth/first homology group. (1–4) are zeroth/first Betti curves and persistence silhouettes from the heating medium system of a combined cycle gas turbine power plant with one steam and two gas turbines. (5–8) show the main condensate pumping plant. The sample size per signal varies between 10^3 and up to 10^5.

A time series is a, possibly finite, strictly totally ordered sequence $(t_i) := \{t_i\}_{i=0}^n$ of real numbers and corresponds to the measured quantity by a sensor.[2] It can be constructed choosing a point t_0 and a time step size s, such that $t_i = t_0 + s \cdot i$. Taking the sequence $(t_i) \subset T \subset \mathbb{R}$ and a function $f : T \to \mathbb{R}$ we get another time series $f \circ (t_i) = (f(t_i))$. The total order for a single sequence gives a one-dimensional manifold with smooth structure. We motivate this choice rigorously.

3.1 Polynomial Approximation

Let $\mathcal{T} := \{T_j\}_{j=0}^m$ be a finite family of sets and $T_j := \{(t_i) \mid t_i \in [a, b], i \in \{1, \ldots, n\}\}$ be a finite strictly totally ordered set of points. Our domain is a compact Hausdorff space since every point in $[a, b]$ can be obtained by the intersection of its closed neighborhoods. We can even obtain a k-times differentiable function $f : M \to \mathbb{R} \in C^k(\mathbb{R})$ for a compact Hausdorff space M and for any $k \in \mathbb{Z}$, which in turn can be approximated arbitrarily exactly by a polynomial function $p(t)$.

Thus, we can choose polynomial functions as coordinate functions to describe our time series as a smooth manifold, without restriction of generality.

Theorem 1. *For a strictly totally ordered sequence $T_j := \{(t_i) \mid t_i \in [a, b], i \in \{1, \ldots, n\}\}$ there exists a well defined polynomial function $p : M \to \mathbb{R}$ on a closed interval $M \subset \mathbb{R}$ – approximating T_j arbitrarily well.*

---

[2] Usually, arbitrary n-tuples are written as $(t_i)_{i=0}^n$, but we denote explicitly strictly totally ordered ones. We introduce here non-standard notation to ensure readability.

Proof. To see this, we use Heine-Borel's theorem, which says that any closed interval of the real line is compact [27, §9.1.24]. We choose $M = [a, b]$. Let $C \subset [a, b]$ be compact, then $f(C)$ is compact in $\mathbb{R}$. Now, we take a sequence (t_i) in the range set $f(C)$. Let (h_i) be a sequence in the domain of f. Then, let for $(h_i) \subseteq f(C)$ and every $i \in \mathbb{N}$ be at least one $t_i \in C$ with $f(t_i) = h_i$. Thus, $(t_i) \subseteq C$ is a sequence. Since C is compact, there exists a convergent sub-sequence (t_{i_j}) whose limit for $j \to \infty$ is also contained in C. Suppose that f is continuous in $[a, b]$, hence, continuous in $t_i \in [a, b]$. Given the fact that $(t_{i_j}) \to t_i$, we conclude that $(h_{i_j}) \to f(t_i)$. Due to the fact that t_i is in C, we have that $f(t_i) \in f(C)$, and thus $f(C)$ is compact by continuity. By [25, §21], there exists a polynomial function $p(t_i)$ to approximate $f(t_i)$ with arbitrary error $|f(t_i) - p(t_i)| < \epsilon$ on compact Hausdorff spaces.   □

3.2 Smooth Manifold Construction

As $p(t)$ is smooth, we can use an argument for smooth functions to yield a smooth manifold and its atlas. Roughly speaking a manifold of dimension n is a topological space locally homeomorphic to $\mathbb{R}^n$. A requirement that provides a manifold would be that $\mathbb{M} \subset \mathbb{R}^n$ and for each $t_i \in \mathbb{M}$ there exists an open ball $B_\epsilon(t_i) = \{m \in \mathbb{M} \mid d(t_i, m) < \epsilon\}$ given the Euclidean metric d and a smooth function, with a smooth diffeomorphism $\phi_{t_i} : B_\epsilon(t_i) \to \{z \in \mathbb{R}^n \mid ||z|| < 1\}$ [18, §1]. For any smooth function $f : \mathbb{R} \to \mathbb{R}$ its graph $\mathcal{G}f := \{(t_i, f(t_i)) \mid t_i \in \mathbb{R}\}$ is a smooth manifold diffeomorphic to $\mathbb{R}$ with an embedding into $\mathbb{R}^2$, given by inclusion. This also holds for functions, that are smooth on a compact domain. In this example, the polynomial is the special case of such a smooth function. An atlas is given by the one map $\varphi : \mathcal{G}f \to \mathbb{R}, \varphi(t_i, f(t_i)) \mapsto t_i$. Thus, f is a transition map from $\mathbb{R}$ into $\mathcal{G}f$ with $t_i \mapsto (t_i, f(t_i))$. This means that we can assume without restrictions that the manifold underlying our data is smooth.

The graph of the polynomial is a connected topological space, thus the manifold is also connected. In this construction also the higher homology groups are trivial, as $\mathcal{G}f \cong \mathbb{R}$ are isomorphic as manifolds.

But we want to make the homology groups utilizable to detect patterns in the form of (quasi-)periodicities in sequences. Vice versa, their analysis can provide information about outliers. We discuss the intended embedding, which establishes a connection between periodic signals and non-trivial higher homology groups by embedding the underlying 1-manifold in an N-torus.

4 Sliding-Window Embedding

For the analysis of time series the expected behavior of the frequencies and amplitudes are of importance, thus, the behavior of a physical entity measured by sensors over some time. Subsets of the time series can be (quasi-)periodically recurring or single events. We take advantage of this fact and use Takens' embedding to make this behavior topologically and geometrically explicit.

4.1   Takens' Embedding

We need a representation of the signal in a space that makes (quasi-)periodicities and outliers visible. Mathematically, we obtain an embedding into a manifold M using a map $\varphi : \mathbb{M} \times I \to \mathbb{M}$, with $\mathbb{M} \subseteq \mathbb{R}^d$. We write the embedded sequence for a function $f : \mathbb{M} \to \mathbb{R}$ as $(f(\varphi(t_i, l)))$ for each $t_i \in \mathbb{M}, l \in I$. We embed the time series into a compact manifold whose homology groups as well as its dimension can be inferred.

The dimension is a useful invariant that can be used to determine a variety of machine learning algorithm parameters, such as the maximum number of neurons required in a neural network layer.

We thus interpret the time series as a smooth dynamical system, a pair $(\mathbb{M}, \Psi)$ where $\mathbb{M}$ denotes a smooth manifold and $\Psi : \mathbb{M} \times \mathbb{R} \to \mathbb{M}$ is a flow such that $\Psi(t_i, 0) = t_i$ and $\Psi(\Psi(t_i, r), k) = \Psi(t_i, r + k)$ for all $t_i \in \mathbb{M}$ and $r, k \in \mathbb{R}$.

Takens' embedding theorem makes this clear if we choose $\mathbb{M}$ as a compact Riemannian manifold:

Let $\tau \in \mathbb{R}_+ \setminus \{0\}$ and $M \geq 2 \dim \mathbb{M}$ be an integer. Furthermore, if $\Psi \in \mathcal{C}^\infty(\mathbb{M} \times \mathbb{R}, \mathbb{M})$ is a function and $F \in \mathcal{C}^\infty(\mathbb{M}, \mathbb{M})$ is generic, then $\psi_{t_i} : \mathbb{M} \to \mathbb{R}, r \mapsto F \circ \Psi(t_i, r)$ is called delay map. Thus, $\psi : \mathbb{M} \to \mathbb{R}^{M+1}, t_i \mapsto (\psi_{t_i}(0), \psi_{t_i}(\tau), \psi_{t_i}(2\tau), \cdots, \psi_{t_i}(M\tau))$ is the desired smooth embedding [19,26, §2]. Let $f : [a, b] \to \mathbb{R}$ be a function.

Further, let M be the *embedding dimension* and τ be called *time delay*. The *sliding-window embedding* of $f : \mathbb{M} \cong \mathbb{R} \to \mathbb{R}$ into $\mathbb{R}^{M+1}$ can then be written as [22, §2]:

$$\mathrm{SW}_{M,\tau} f(t_i) = [f(t_i), f(t_i + \tau), \cdots, f(t_i + M\tau)]^\top . \tag{9}$$

The product $M\tau$ is called the *window size* of the sliding-window embedding. For different values of $t_i \in T := [a, b] \subset \mathbb{R}$ we get a set of points that we call *sliding-window point cloud associated with* T [21, §3.2]:

$$\mathbb{SW}_{M,\tau} f := \{\mathrm{SW}_{M,\tau} f(t_i) \mid t_i \in T\} . \tag{10}$$

We have anticipated an embedding which is visibly smooth, but whose geometry and topology we do not know yet. To make the geometry and topology of the associated sliding window point cloud directly exploitable, we treat the periodic and quasi-periodic cases next.

4.2   Periodic Signals

A 2π-periodic signal can be expressed as a sum of cosine or sine functions together with its Fourier coefficients. A function $f : [t_i, t_i + 2\pi] \to \mathbb{R}$ is 2π-periodic if $f(t_i + 2\pi) = f(t_i)$. We pick up the result that the chosen embedding for L-periodic functions is dense within a torus. A function f is L-periodic on $[0, 2\pi]$, iff $f(t_i + \frac{2\pi}{L}) = f(t_i)$ for all $t_i \in \mathbb{R}$.[3] Under our main assumption, point-

---

[3] Recall, if our signals were L-periodic for some natural number L, we could, by definition, treat them as functions on some torus $\mathbb{T} \cong \mathbb{R}/L\mathbb{Z}$.

wise convergence is also guaranteed since the approximation is allowed to be smooth.

We use a theorem proven by Perea to understand the topological and geometric structure of the *sliding window point cloud* [22, §5.6].

First, we define a centering map $Z : \mathbb{R}^{M+1} \to \mathbb{R}^{M+1}$,

$$Z(\mathbf{x}) = \mathbf{x} - \frac{\langle \mathbf{x}, \mathbf{1} \rangle}{||\mathbf{1}||^2}, \quad \text{where} \quad \mathbf{1} = [1, \dots, 1]^\top \in \mathbb{R}^{M+1}, \tag{11}$$

and we use the sliding window embedding on the truncated Fourier transform,

$$\mathrm{SW}_{M,\tau} S_N f(t_i) = \sum_{n=0}^{N} \cos(nt_i)(a_n \mathbf{u}_n + b_n \mathbf{v}_n) + \sin(nt_i)(b_n \mathbf{u}_n - a_n \mathbf{v}_n), \tag{12}$$

where $S_N f(t_i)$ is the *N-truncated Fourier series expansion* of f with some remainder, such that $f(t_i) = S_N f(t_i) + R_N f(t_i)$, $j^2 = -1$ and

$$S_N f(t_i) = \sum_{n=0}^{N} a_n \cos(nt_i) + b_n \sin(nt_i) = \sum_{n=-N}^{N} \hat{f}(n) e^{jnt_i},$$

$$\hat{f}(n) = \begin{cases} \frac{1}{2} a_n - \frac{j}{2} b_n, & \text{if } n > 0, \\ \frac{1}{2} a_{-n} - \frac{j}{2} b_{-n}, & \text{if } n < 0, \\ a_0, & \text{if } n = 0. \end{cases}$$

If we take f to be L-periodic, with $L(M+1)\tau - 2\pi$, then we yield $\mathrm{SW}_{M,\tau} S_N f(t_i)$ $= \hat{f}(0) \cdot \mathbf{1} + Z(\mathrm{SW}_{M,\tau} S_N f(t_i))$. For its norm we obtain $||Z(\mathrm{SW}_{M,\tau} S_N f(t_i))||$ $= \sqrt{M+1}(||S_N f||_2^2 - \hat{f}(0)^2)^{1/2}$. Constructing the orthonormal set $\{\tilde{\mathbf{x}}_n, \tilde{\mathbf{y}}_n \in \mathbb{R}^{M+1} \mid 1 \le n \le N, n \equiv 0 \mod L\}$ gives, as proven in [22, §5.6]:

$$\varphi_\tau(t_i) := \frac{\sqrt{M+1} \left(||S_N f||_2^2 - \hat{f}(0)^2 \right)^{\frac{1}{2}}}{|| \sqrt{M+1} \left(||S_N f||_2^2 - \hat{f}(0)^2 \right)^{\frac{1}{2}} ||} \tag{13}$$

$$= \sum_{\substack{n=1 \\ n \equiv 0 \mod L}}^{N} \frac{2|\hat{f}(n)|}{\sqrt{||S_N f||_2^2 - \hat{f}(0)^2}} (\cos(nt_i)\tilde{\mathbf{x}}_n + \sin(nt_i)\tilde{\mathbf{y}}_n) \tag{14}$$

$$= \sum_{\substack{n=1 \\ n \equiv 0 \mod L}}^{N} \tilde{r}_n (\cos(nt_i)\tilde{\mathbf{x}}_n + \sin(nt_i)\tilde{\mathbf{y}}_n), \quad \text{with} \quad \sum_{n=1}^{N} \tilde{r}_n^2 = 1. \tag{15}$$

As studied in [22, §5. 6], a clear geometric picture of the centered and normalized sliding window point cloud arises for $S_N f$, since if we consider $S^1(r) \subset \mathbb{C}$ as a circle centered around zero with radius r in the complex plane, then $t_i \to \varphi_\tau(t_i)$ is a curve on an N-torus $\mathbb{T} = S^1(\tilde{r}_1) \times \cdots \times S^1(\tilde{r}_N)$, as shown in Eq. 15.

Thus, we can use the zeroth persistent homology group - as a Euclidean metric dependent quantity between the connected components of Takens' embedding of the signal samples - and the first persistent homology group as features for our classifier to describe the torus on which the data lie. The dependence on the Euclidean metric is due to the choice of the embedding space of the simplicial complex. Another Riemannian manifold would also be conceivable, but we use the ordinary Riemannian metric of $\mathbb{R}^n$.

For the N-torus the Betti numbers are computed as $\beta_k = \binom{N}{k}$ for the kth persistent homology group. It follows that rank $H_1(\varphi_\tau; \mathbb{F})$ corresponds to the number of 1-spheres forming the product space $\mathbb{T}$ [17]. Thus, the underlying manifold can be determined.

4.3   Quasi-periodic Signals

For the second case, we are interested in functions $f : [a, b] \to \mathbb{C}$, which are quasi-periodic signals of the form $f(t_i) = \sum_{l=0}^{L} \lambda_l e^{j\omega_l t_i}$, where $n \in \mathbb{N}$, λ_l are non-negative complex numbers, and ω_l are incommensurate non-negative reals. As such, they are functions which also have incommensurate periods. Such signals naturally occur in the context of power plants.

Perea proved [20, §2], that for points

$$p_f(t_i) = (\lambda_0 e^{j\omega_0 t_i}, \lambda_1 e^{j\omega_1 t_i}, \cdots, \lambda_L e^{j\omega_L t_i}) \tag{16}$$

and for $\lambda_l \in \mathbb{C}$, the generated set $T_f = \{p_f(t_i) \mid t_i \in \mathbb{Z}\}$ is dense within an $(N+1)$-torus $\mathbb{T}^{N+1} = S_{\lambda_0}^1 \times S_{\lambda_1}^1 \times \cdots \times S_{\lambda_N}^1$, by Kronecker's Theorem [14]. More succinct: if $0 < \tau < 2\pi/\max(\omega_l)$, then Ω_f has full rank. Moreover, if $M \geq N$, then the sliding window point cloud is given by

$$\mathrm{SW}_{M,\tau}f := \{\mathrm{SW}_{M,\tau}f(t_i) \mid t_i \in \mathbb{Z}\}, \tag{17}$$

and is dense within a space homeomorphic to $\mathbb{T}^{N+1}$ [20, §2.1].

We address a rather trivial observation that makes all reasoning amenable to computations in the real number field: The complex plane has an isomorphism of sets given by the bijection $\wp : \mathbb{R} \times \mathbb{R} \to \mathbb{C}, (t_i, s) \to t_i + js$, with $j^2 = -1$, where $\times$ denotes the direct product of sets. The vector space structure can be defined on both, $\mathbb{R} \times \mathbb{R}$ as well as on $\mathbb{C}$, with basis $B_{\mathbb{R} \times \mathbb{R}} = \{(1,0),(0,1)\}$ and $B_{\mathbb{C}} = \{(1,j)\}$. In the category of rings and fields the two objects differ, but for our purposes we use the bijection $\wp$ as an isometry between $\mathbb{C}$ and $\mathbb{R}^2$ – allowing us to apply the above arguments to real vector spaces. Since these $\mathbb{R}$-vector spaces are both of dimension two, they are isomorphic as $\mathbb{R}$-modules.

5   Heuristic Choice of Parameters

Some of the hyper-parameters have to be estimated in advance by heuristic approaches. We describe them next and determine the ideal embedding dimension and time-delay of our particular data set.

5.1 Time Delay τ

We determine an optimal value for the time delay creating a partition for each time series T_j. For this approach, we write the interval $[t_1^{\min}, t_n^{\max}] \subset T_j$ as the interval between the smallest and largest value for $t_i \in T_j$ of a discrete multi-set of time series values, i.e., $i \in \mathbb{Z}$. Accordingly, $t_1^{\min}$ is a global minimum and $t_n^{\max}$ is a global maximum. The interval is split into a partition of s-size:

$$\mathcal{P}_{T_j} := \left\{ \underbrace{[t_1^{\min}, t_s^{\min}]}_{:=[1]}, \underbrace{[t_{s+1}, t_{2s+1}]}_{:=[2]}, \underbrace{[t_{2s+2}, t_{3s+2}]}_{:=[3]}, \cdots, \underbrace{[t_{n-s}^{\max}, t_n^{\max}]}_{:=[n/s]} \right\}. \tag{18}$$

We choose s as the smallest divisor of n for the time steps whose particular partitioning contains the fewest turning points. We justify this choice by arguing that a large number of elements are defined in $\mathcal{P}_{T_j}$, but each interval most likely contains a single period. Hereafter we will refer to all intervals with $k \in \{1, 2, \cdots, n/s\}$ as *bins*.

Let $\mathsf{P}([k]) := \mathsf{P}(t_i \in [k])$ denote the probability that t_i is contained in the kth bin. Let $\mathsf{P}([k], [l]) := \mathsf{P}(t_i \in [k], t_{i+\tau} \in [l])$ denote the probability that t_i is contained in the kth bin, while $t_{i+\tau}$ is contained in the lth bin, with $k \neq l$. The mutual information $\mathsf{I}([k], [l]|\tau)$ for these two probability distributions is then computable as follows:

$$\mathsf{I}([k], [l]|\tau) = -\sum_{k=1}^{n/s} \sum_{l=1}^{n/s} \mathsf{P}([k], [l]|\tau) \, \mathrm{ld} \, \frac{\mathsf{P}([k], [l]|\tau)}{\mathsf{P}([k]|\tau)\mathsf{P}([l]|\tau)}. \tag{19}$$

To gain intuition, recall the entropy for a finite alphabet $T_j = \{t_1, t_2, \ldots, t_n\}$ as $\mathsf{H}(T_j) = -\sum_{t_i \in T_j} \mathsf{P}(t_i) \, \mathrm{ld} \, \mathsf{P}(t_i)$, which gives the mean value of bits for each element to encode it uniquely. Hence, the mutual information indicates how many bits are needed to generate a second random variable, given one at hand, with both functions running over the alphabet.

The optimal period τ is obtained for $\min_\tau \mathsf{I}([k], [l]|\tau)$, since we get the most information when we add another value $t_{i+\tau}$ to this particular subset. We minimize $\mathsf{I}([k], [l]|\tau)$. We get as a result $\tau = 1$ for over 98% of the data. The size of one bin ranges from 463 to 624. Accordingly, we choose a sample size of 500.

5.2 Embedding Dimension M

We examined about $18 \cdot 10^3$ different signals coming from four different combined cycle gas turbine power plants with a total of two gas turbines, two boilers for steam generation and one steam turbine of the same construction type.

A constant embedding dimension must be specified for our model, which isn't optimal for the individual signal. However, since the toroidal embedding is also to be performed for the preprocessing of a new signal, a constant embedding dimension is a prerequisite. To determine a suitable M, we use the *false nearest neighbor* algorithm. We apply the assumption from Sect. 3 that the embedding of a deterministic system in higher dimensions is smooth, relying on the generality of Takens' embedding. Thus, points that are close as measured values on the graph of their function will be close in their embedding with respect to the induced Euclidean norm in data space.

For two points $t_i, t_k \in T_j$ there is an $\epsilon^{(ik)}$ such that:

$$\epsilon^{(ik)} = \frac{|t_{i+M\tau} - t_{k+M\tau}|}{\left(\frac{1}{n}\sum_{l=1}^{n}(t_l - \frac{1}{n}\sum_{l=1}^{n}t_l)^2\right)^{\frac{1}{2}}} > \epsilon. \tag{20}$$

The L^1-norm is chosen for the numerator because it is more robust to outliers in the sense that the values are not higher powers of the absolute value. Whenever the inequality in Eq. 20 is satisfied, t_i and t_k are called *false nearest neighbors*. As suggested by the authors of the original algorithm [12], the difference $|t_{i+M\tau} - t_{k+M\tau}|$ is very likely to be as large as $\sqrt{2}\sigma_{T_j}$, if the data were white noise of standard deviation σ. We set $\epsilon = \sqrt{2}\sigma_{T_j}$, for some time series $T_j \in \mathcal{T}$. The optimal embedding dimension is the optimization problem that minimizes the number of such false neighbors.

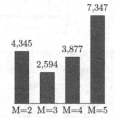

Fig. 2. Cardinality of the set of time series with their optimal embedding dimension M.

We choose $M = 5$, since this is the ideal embedding for the majority of the time series according to Fig. 2, as we can always immerse all other signals with lower optimal embedding dimension.

6   Results

To compute simplicial complexes, filtrations and persistence representations, we used GUDHI v.3.4.1 [15] and giotto-tda v.0.4.0 [28]. Neural networks were written in Keras v.2.4.0 [5] and trained using Tensorflow v.2.4.0 [1] as backend on NVIDIA Quadro RTX 4000 graphics cards. cuDNN v.8.4.0 [4] is used to enable graphics acceleration for LSTMs.

In a first step, we removed all time series from the data that do not have persistent features of the zeroth and first homology group in their persistence diagrams. Thus, we removed all time series with persistence entropy

$$H\left(\mathfrak{P}_{\mathrm{SW}_{M,\tau}f}\right) = -\sum_{(d_j,b_i)\in\mathfrak{P}_{\mathrm{SW}_{M,\tau}f}} \frac{d_j - b_i}{\kappa} \, \mathrm{ld}\left(\frac{d_j - b_i}{\kappa}\right) \geq 0.98,$$

$$\text{where} \quad \kappa = \sum_{(d_j,b_i)\in\mathfrak{P}_{\mathrm{SW}_{M,\tau}f}} d_j - b_i.$$

The goal is to get rid of erroneous measurements, as well as (nearly) uniformly distributed signals or constants. We then replaced all NaN values in the time series and Betti curves with their corresponding median, respectively.

6.1   Persistence Representations

Each individual colored curve in Fig. 1(a), (c) represents a recorded signal of a power plant within the respective component over one year.

Figure 1 (b), (d) show the corresponding persistence silhouettes.

β_0-curves. The β_0-curves in Fig. 1(a) count the number of representatives of the zeroth homology group at each parameter y of the filtration. Plateaus indicate persistent properties, i.e., connectivity is preserved for some time. The curve for β_0 is monotonically decreasing. Initially, each measurement counts itself as a connected component – as it is an embedded point, so the amount of points available is trivially the largest value of each β_0-curve. Thus, the plot of Betti curves gives information about the number of points per signal. It visualizes the class balance of the data set. We can measure this balance considering the persistence entropy of all β_0-curves from Fig. 1(a), restricted to $y = 0$, of about 67%. For $y \in \{1, \ldots, 20\}$, i.e., the first twenty recorded steps of the filtration, the entropy of the β_0-curves in Fig. 1(a) – of all signals – varies on average between about 67% and 45%, but decreases as y increases. This means that there is a large difference in the number of connected components in the first 20% of the computed filtration. Entropy gives a measure of the distribution of connected components among the signals at some specific parameter y of the filtration.

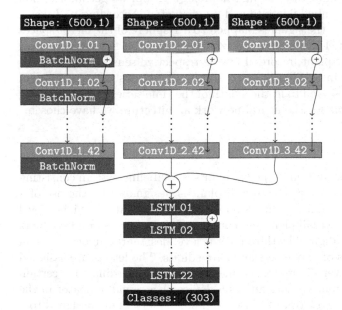

ConV1D	Parameters
Activation	$\frac{t}{1+e^{-\beta t}}$ [23]
#-Layer	42
Filters	64
Kernel-size	3
Padding	Causal
Kernel init.	Glorot normal
Bias init.	Zeros
Residual	C^1 [9]
L^1-regularization	0.001
L^2-regularization	0.01

LSTM	Parameters
Activations	See [11]
#-Layer	22
Units	32
Kernel init.	Glorot normal
Bias init.	Zeros
Residual	C^1 [9]
L^1-regularization	0.001
L^2-regularization	0.01

Fig. 3. An illustration of the derived architecture. We trained on sigmoid focal cross-entropy loss [24], with a batch size of 128, a learning rate of 10^{-4} with stopping patience for 5 epochs, and a minimum learning rate of 10^{-6} for about $6 \cdot 10^3$ epochs. The task was to classify $303 \cdot 10^3$ samples into 303 classes, with 10^3 samples per class. We used 10% of the data for validation and did no further tuning of hyper-parameters.

Thus, we can assume that most of the data differs with regard to its density distribution in Euclidean space. Further, Fig. 1(a) shows β_0-curves, restricted to $x = 500$, which are almost diagonal. This means that connected components are present in large number throughout the filtration for this particular signal.

β_1-curves. Similarly, the persistence entropy of the β_1-curves in Fig. 1(c), restricted to $y = 50$, is about 47%. Therefore, the signals differ by the number of cyclic elements. This can also be easily seen by looking at the scale of the z-axes in Fig. 1(a).

Recall that the signals are toroidally embedded. If no connected components are merged, it means that along the filtration we can expect an S^1-factor of the torus. In other words, the representatives of the first homology group, which we can see in the diagrams Fig. 1(c), (d), count the periodic elements within the signal. The toroidal embedding allows us to interpret that certain modes of the original signal lie on different tori. This claim is directly observable in Fig. 1(c) (3); the diagram shows numerous counts of holes within the embedding per signal, even for higher filtration parameters.

Since the dimension N of the torus corresponds to the rank of its first homology group, we can also read the dimension of the toroidal embedding of a signal from the persistence diagrams and thus from its β_1-curve in Fig. 1(c) (d). As β_0-curves contain the information about connected components and the dimension of the torus is determined by the first homology group, we get information about the object we are embedding into and its geometry. According to Takens' embedding, each period in the signal is mapped to a 1-sphere of the factor spaces of the N-torus, see Sect. 4.2 and Sect. 4.3, which is why the β_1-curves 'encode events' within the power plant, recorded by the respective sensor.

The Betti curves lie in a Hilbert space, allowing statistics to be computed, and can themselves be resolved to fit the length of the time series sample, making them particularly suitable for the neural network architecture we have chosen.

6.2   Classification

The reference designation system for power plants is a labeling system consisting of four levels of structure. First, the overall plant is designated by the use of a letter (L) or a digit (D). The second level of detail designates a higher-level functional system in the overall plant and consists of three letters and two digits with an optional leading digit. The third outline level designates an aggregate in the sub-plant. It consists of two letters and three digits. The letters are assigned to the aggregates in power plants (e.g. a measurement) according to a certain key. The fourth subdivision level designates a device or a signal indicator in the aggregate. It consists of two letters and two digits. The letters are assigned to a piece of equipment (e.g. a drive) according to a given key:

Overall system (OS)	Function (F)	Aggregate (A)	Operating resources (OR)
L or D	(D)LLLDD	LLDDD(L)	LLDD

Table 1. Classification results. The used, corresponding parts of the reference designation system are marked as such (✓/✗). We tested with 10-fold cross-validation and report the mean values of the measurements with associated standard deviation. The experiments with complete/incomplete identifiers are colored (•/•), respectively.

OS	F	A	OR	Accuracy	F1	Precision	Recall
\multicolumn{8}{}{$\mathcal{C}^0$-CONVNET WITHOUT TOPOLOGICAL FEATURES SIMILAR TO FIG. 3:}
✓	✓	✓	✓	0.4821 ±0.0031	0.5677 ±0.0033	0.6912 ±0.0029	0.4816 ±0.0037
✓	✗	✗	✗	0.7129 ±0.0102	0.7904 ±0.0092	0.9010 ±0.0097	0.7041 ±0.0088
✓	✓	✗	✗	0.5691 ±0.0037	0.6830 ±0.0058	0.8699 ±0.0065	0.5622 ±0.0052
✓	✓	✓	✗	0.5426 ±0.0055	0.6681 ±0.0036	0.8682 ±0.0048	0.5429 ±0.0029
\multicolumn{8}{}{$\mathcal{C}^0$-CONVNET AS IN FIG. 3:}
✓	✓	✓	✓	0.6142 ±0.0047	0.6212 ±0.0077	0.7681 ±0.0082	0.5216 ±0.0073
✓	✗	✗	✗	0.8316 ±0.0121	0.8511 ±0.0063	0.9327 ±0.0163	0.7827 ±0.0039
✓	✓	✗	✗	0.7024 ±0.0091	0.7567 ±0.0101	0.8756 ±0.0109	0.6663 ±0.0094
✓	✓	✓	✗	0.6291 ±0.0078	0.7376 ±0.0065	0.8726 ±0.0056	0.6389 ±0.0077
\multicolumn{8}{}{$\mathcal{C}^1$-CONVNET AS IN FIG. 3:}
✓	✓	✓	✓	**0.6383** ±0.0085	**0.6566** ±0.0055	**0.7849** ±0.0074	**0.5597** ±0.0076
✓	✗	✗	✗	0.8221 ±0.0028	0.8497 ±0.0023	0.9267 ±0.0033	0.7846 ±0.0018
✓	✓	✗	✗	**0.7284** ±0.0019	**0.7670** ±0.0027	**0.8826** ±0.0017	**0.6782** ±0.0066
✓	✓	✓	✗	**0.6524** ±0.0009	**0.7276** ±0.0028	**0.8821** ±0.0032	**0.6192** ±0.0025

In each case, we classify all training samples, each drawn without replacement from one sensor signal recorded over one year. After transforming the sample with the parameters determined from Sect. 5.1 and Sect. 5.2 using Takens' embedding, we compute the Betti curves for β_0 and β_1 of its persistence diagram. Afterwards, we use them as input to the sub-nets without batch normalization as shown in Fig. 3. Thus, we enrich the neural network input with information about connectedness and the underlying N-dimensional torus of the embedded signal.

We summarize the results from Table 1 as follows:

1. The best classification results for the power plant reference designation system have been measured with an accuracy of approximately 64% (OS F A OR), about 83% for the *overall system* (OS), 73% for the *functional level* (OS F), and 65% for the *aggregate* (OS F A) on the validation data.
2. We observe for all experiments that precision is higher than recall. We interpret the high precision as a solid exactness of our classifiers. Recall, on the other hand, is significantly lower, as an indicator of the completeness of the classifier. The higher the recall value, the less accurate our classifiers are at assigning signals to the corresponding labels. This can already be anticipated

from the Betti curves shown in Fig. 1, which reveal similar features for the zeroth/first persistent homology group for some of the signals, symptomatic for their poor distinctiveness.

3. We have shown, that residual networks improve the classification results for all labels except for the assignment to the *overall system* (OS). Further investigation is needed to find an explanation for this behavior.

4. Moreover, using β_0 and β_1-curves we could improve the expected value of the classification results for all label variants studied, see Table 1.

7   Summary

In this work, we used Betti curves of the zeroth and first persistent homology group as feature vectors for neural networks. Using recent research, we have shown how these two persistent homology groups naturally '*encode events*' in (quasi-)periodic time series and justified their suitability mathematically. We used a variant of Perea's framework for the analysis of (quasi-)periodicity, and applied it as a feature generator for machine learning on irregular time series of measurements taken from power plants. The topological features obtained through Takens' delay embedding and persistent homology are used to enrich the input of a deep neural network. We have shown that these extra features significantly improve the network's performance in a classification task.

Moreover, we have shown that the experiments also reveal an improvement in classification through β_0 and β_1-curves for all hierarchy levels of the power plant reference designation system. We designed the architecture using residual connections and were able to confirm their usefulness compared to the same architecture without such connections.

Our future work will address the following open questions:

- How well does our model perform when trained on a larger and more complete data set with signals from multiple entire power plants, but validated with sensor data from an independent new power plant?
- Can we improve the classifier by first training it at the coarsest hierarchical level of the power plant reference designation system (OS) and then using the resulting weights to initialize training for more detailed levels of labels (such as F,A) down to the operating resource (OR) in a recursive manner?

References

1. Abadi, M., et al.: TensorFlow: Large-scale machine learning on heterogeneous systems (2015). http://tensorflow.org/, software available from tensorflow.org
2. Bubenik, P.: Statistical topological data analysis using persistence landscapes. J. Mach. Learn. Res. **16**(1), 77–102 (2015)
3. Chazal, F., et al.: Stochastic convergence of persistence landscapes and silhouettes. In: Annual Symposium on Computational Geometry, pp. 474–483 (2014)
4. Chetlur, S., et al.: cuDNN: efficient primitives for deep learning. arXiv:1410.0759 (2014). https://developer.nvidia.com/cudnn
5. Chollet, F., et al.: Keras (2015). https://github.com/fchollet/keras
6. Cohen-Steiner, D., Edelsbrunner, H., Harer, J.: Stability of persistence diagrams. Discrete Comput. Geom. **37**(1), 103–120 (2007)

7. De Silva, V., Morozov, D., Vejdemo-Johansson, M.: Dualities in persistent (co) homology. Inverse Probl. **27**(12), 124003 (2011)
8. Hatcher, A.: Algebraic Topology. Cambridge University Press, Cambridge (2005)
9. Hauser, M., Gunn, S., et al.: State-space representations of deep neural networks. Neural Comput. **31**(3), 538–554 (2019)
10. Hausmann, J.C.: On the vietoris-rips complexes and a cohomology theory for metric spaces. Ann. Math. Stud. **138**, 175–188 (1995)
11. Hochreiter, S., Schmidhuber, J.: Long short-term memory. Neural Comput. **9**(8), 1735–1780 (1997)
12. Kennel, M., Abarbanel, H.: False neighbors and false strands: a reliable minimum embedding dimension algorithm. Phys. Rev. E **66**, 059903 (2002)
13. Königstein, H., Müller, H., Kaiser, J.: Das RDS-PP: Übergang vom KKS zu einer internationalen Norm, vol. 8. VGB Powertech (2007)
14. Kronecker, L.: Näherungsweise ganzzahlige Auflösung linearer Gleichungen. Monatsberichte der Königlich Preussischen Akademie der Wissenschaft (1884)
15. Maria, C., Boissonnat, J.-D., Glisse, M., Yvinec, M.: The Gudhi library: simplicial complexes and persistent homology. In: Hong, H., Yap, C. (eds.) ICMS 2014. LNCS, vol. 8592, pp. 167–174. Springer, Heidelberg (2014). https://doi.org/10.1007/978-3-662-44199-2_28
16. Melodia, L., Lenz, R.: Persistent homology as stopping-criterion for voronoi interpolation. In: Lukić, T., Barneva, R.P., Brimkov, V.E., Čomić, L., Sladoje, N. (eds.) IWCIA 2020. LNCS, vol. 12148, pp. 29–44. Springer, Cham (2020). https://doi.org/10.1007/978-3-030-51002-2_3
17. Melodia, L., Lenz, R.: Estimate of the neural network dimension using algebraic topology and lie theory. In: Image Mining. Theory and Applications VII (2021). https://doi.org/10.1007/978-3-030-68821-9_2
18. Milnor, J., Weaver, D.: Topology from the Differentiable Viewpoint. Princeton University Press, Princeton (1997)
19. Packard, N., Crutchfield, J., Farmer, D., Shaw, R.: Geometry from a time series. Phys. Rev. Lett. **45**(9), 712 (1980)
20. Perea, J.: Persistent homology of toroidal sliding window embeddings. In: IEEE International Conference on Acoustics, Speech and Signal Processing, pp. 6435–6439 (2016)
21. Perea, J.: Topological time series analysis. Not. Am. Math. Soc. **66**, 686–694 (2019)
22. Perea, J., Harer, J.: Sliding windows and persistence: an application of topological methods to signal analysis. Found. Comput. Math. **15**, 799–838 (2015)
23. Ramachandran, P., Zoph, B., Le, Q.V.: Swish: a self-gated activation function. arXiv:1710.05941 (2017)
24. Ross, T.Y., Dollár, G.: Focal loss for dense object detection. In: IEEE Conference on Computer Vision and Pattern Recognition, pp. 2980–2988 (2017)
25. Stone, M.H.: The generalized Weierstrass approximation theorem. Math. Mag. **21**, 237–254 (1948)
26. Takens, F.: Detecting strange attractors in turbulence. In: Rand, D., Young, L.-S. (eds.) Dynamical Systems and Turbulence, Warwick 1980. LNM, vol. 898, pp. 366–381. Springer, Heidelberg (1981). https://doi.org/10.1007/BFb0091924
27. Tao, T.: Analysis I. SLB. Springer, Heidelberg (2006). https://doi.org/10.1007/3-540-33278-2
28. Tauzin, G., et al.: giotto-tda: : A topological data analysis toolkit for machine learning and data exploration. J. Mach. Learn. Res. **22**, 39:1–39:6 (2021). https://github.com/giotto-ai/giotto-tda

Continuous-Discrete Recurrent Kalman Networks for Irregular Time Series

Mona Schirmer[1,2]($\boxtimes$), Mazin Eltayeb[2], and Maja Rudolph[3]

[1] ENSAE Paris, Palaiseau Cedex, France
mona.schirmer@ensae.fr
[2] Bosch Center for Artificial Intelligence, Renningen, Germany
mazin.eltayeb@de.bosch.com
[3] Bosch Center for Artificial Intelligence, Pittsburg, USA
maja.rudolph@us.bosch.com

Abstract. In this paper we address the problem of modelling time series with irregular intervals by incorporating a continuous-time version of the Kalman filter into a neural network architecture. Building on the idea of Recurrent Kalman Networks (RKNs) we use an encoder-decoder structure to learn a latent observation space and latent state space in which the dynamics of the data can be approximated linearly. Here, a recurrent Kalman component alternates between continuous latent state propagation and Bayesian updates from incoming observations. This allows us to model and react instantaneously to observations as they come at arbitrary time steps while ensuring sufficient expressive power to model nonlinear dynamics. Experiments on synthetic data show that the model is indeed able to capture continuous, nonlinear dynamics.

Keywords: Irregular time series · Kalman filter · Neural networks

1 Introduction

Time series are omnipresent in various fields such as healthcare, climate forecast and finance. In many application areas, time series naturally come in an irregular fashion. In healthcare for example, lab measurements may be taken at irregular time intervals. This results in sparse observations - a characteristic that challenges well established models for time series such as Recurrent Neural Networks (RNNs). A common heuristic to apply RNNs for this type of data is to impose discrete dynamics by splitting the timeline into equally sized bins.

Recently, steps have been taken towards a more natural modelling of irregular time series. Notably, the work on Neural Ordinary Differential Equations (Chen et al. 2018) introduced continuous formulations of neural networks allowing to propagate hidden states in continuous time. However, this approach can not react to new observations as their state trajectory is fixed once the parameters are learned. Rubanova et al. (2019) address this issue by adding RNN updates at each observation time, Kidger et al. (2020) by proposing Neural Controlled

© Springer Nature Switzerland AG 2021
M. Kamp et al. (Eds.): ECML PKDD 2021 Workshops, CCIS 1524, pp. 300–305, 2021.
https://doi.org/10.1007/978-3-030-93736-2_23

Differential Equations and De Brouwer et al. (2019) by combining a continuous formulation of GRUs with a Bayesian update network.

Updating a trajectory based on incoming information while considering noise and prior knowledge is the essence of Kalman filters (Kalman 1960). In his celebrated paper, Kalman showed that the optimal solution for the linear filtering problem has closed form updates. The updates balance prior belief based on previous observations with inherently noisy new observations in an optimal manner.

We propose to embed a continuous-discrete version of the Kalman filter into a neural network architecture thereby combining continuous states with the flexibility of neural networks. As the underlying state dynamics of the continuous-discrete Kalman filter are governed by a Stochastic Differential Equation (SDE), our model can deal with irregular time intervals by simply integrating the continuous state dynamics up to any point in time. Since the Kalman filter can be seen as a Bayesian update model, our network is responsive to the data as it comes in. Our contributions are as follows:

- We present the Continuous-Discrete Recurrent Kalman Network (CDRKN), a time series model that combines the power of neural networks for feature extraction with the advantages of a probabilistic state-space model, specifically the continuous-discrete Kalman filter (Sect. 2.1).
- The resulting neural architecture (Sect. 2.2) can process time series data like a recurrent neural network by sequentially processing observations and internally updating its latent state. Due to the properties of the continuous-discrete Kalman filter, the model deals with irregular time intervals between observations in a principled manner.
- In Sect. 3, we study the CDRKN on a spiral and pendulum data set and find that it is able to capture the nonlinear dynamics of the data, even as the data is observed irregularly.

2   Method

The CDRKN addresses the challenge of modeling a time series $\mathbf{x}_{\mathcal{T}} = [\mathbf{x}_t | t \in \mathcal{T} = \{t_0, t_1, \cdots t_N\}]$ whose observation times $\mathcal{T} = \{t_0, t_1, \cdots t_N\}$ can occur at irregular intervals. It assumes a continuous latent state $\mathbf{z} \in \mathbb{R}^M$ whose dynamics are governed by a linear SDE. Figure 1 illustrates the network architecture: An encoder and decoder relate observation space with a latent state space. A continuous-discrete Kalman filter alternates between observation updates and continuous state propagation. We first describe the continuous discrete Kalman filter for the latent state (Sect. 2.1) and then develop the CDRKN in Sect. 2.2.

2.1   Continuous-Discrete Kalman Filter

The continuous-discrete Kalman filter (see (Jazwinski 1970)) assumes a continuous latent state $\mathbf{z} \in \mathbb{R}^M$ that evolves according to the SDE

$$d\mathbf{z} = \mathbf{A}\mathbf{z}dt + \mathbf{G}d\beta \tag{1}$$

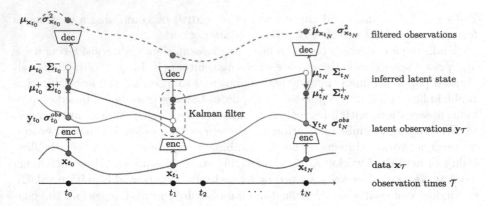

Fig. 1. CDRKN: An encoder network maps the observation $\mathbf{x}_t$ to a latent observation $\mathbf{y}_t$ and returns an elementwise uncertainty estimate $\boldsymbol{\sigma}_t^{obs}$. Both are used to update the latent state prior $\boldsymbol{\mu}_t^-, \boldsymbol{\Sigma}_t^-$ to the latent state posterior $\boldsymbol{\mu}_t^+, \boldsymbol{\Sigma}_t^+$ (red arrows), which the decoder maps back to the observation space. The latent state posterior is propagated to the next observation time (red line). (Color figure online)

where $\mathbf{A} \in \mathbb{R}^{M \times M}$ is a time-invariant transition matrix, $\boldsymbol{\beta} \in \mathbb{R}^B$ a Brownian motion process with diffusion matrix $\mathbf{Q} \in \mathbb{R}^{B \times B}$ and diffusion coefficient $\mathbf{G} \in \mathbb{R}^{M \times B}$. At times $t \in \mathcal{T}$ an observation $\mathbf{y}_t \in \mathbb{R}^D$ is sampled from the latent state

$$\mathbf{y}_t \sim \mathcal{N}(\mathbf{H}\mathbf{z}_t, \boldsymbol{\Sigma}_t^{obs}) \tag{2}$$

where $\mathbf{H} \in \mathbb{R}^{D \times M}$ denotes the observation model. The noise components of $\mathbf{y}_t$, $\mathbf{z}_t$ and the initial state $\mathbf{z}_{t_0}$ are assumed to be independent. This version of the Kalman filter allows modelling discrete observations at potentially arbitrary time-intervals as samples from a continuous trajectory of $\mathbf{z}$. Posterior estimates of the latent state can be obtained by Bayes' theorem in a recurrent two-step procedure.

Prediction Step. Between observations the prior density describes the evolution of the latent state. It is governed by the SDE in (1). For linear, time-invariant systems - as it is the case here - the SDE has an analytical solution. After an observation at time t the solution of the SDE for $t' > t$ is given by

$$\mathbf{z}_{t'} = \exp(\mathbf{A}(t' - t))\mathbf{z}_t + \int_t^{t'} \exp(\mathbf{A}(t' - \tau))\mathbf{G}d\boldsymbol{\beta}_\tau \tag{3}$$

From here, the prior density at time t' is $p(\mathbf{z}_{t'}|\mathbf{y}_t) = \mathcal{N}(\boldsymbol{\mu}_{t'}^-, \boldsymbol{\Sigma}_{t'}^-)$ with

$$\boldsymbol{\mu}_{t'}^- = \exp(\mathbf{A}(t' - t))\boldsymbol{\mu}_t^+ \tag{4}$$

$$\boldsymbol{\Sigma}_{t'}^- = \exp(\mathbf{A}(t' - t))\boldsymbol{\Sigma}_t^+ \exp(\mathbf{A}(t' - t))^T$$
$$+ \int_t^{t'} \exp(\mathbf{A}(t' - \tau))\mathbf{G}\mathbf{Q}\mathbf{G}^T \exp(\mathbf{A}(t' - \tau))^T d\tau \tag{5}$$

where $\exp(\cdot)$ denotes the matrix exponential and $(\boldsymbol{\mu}_t^+, \boldsymbol{\Sigma}_t^+)$ the posterior mean and variance of the latent state at the last observation time t. Note that the integral in (5) can be computed analytically using matrix fraction decomposition (Axelsson and Gustafsson 2014). We summarize the prediction step for the parameters of the prior $p(\mathbf{z}_{t'}|\mathbf{y}_t) = \mathcal{N}(\boldsymbol{\mu}_{t'}^-, \boldsymbol{\Sigma}_{t'}^-)$ as

$$[\boldsymbol{\mu}_{t'}^-, \boldsymbol{\Sigma}_{t'}^-] = \text{predict}_\mathbf{A}(\boldsymbol{\mu}_t^+, \boldsymbol{\Sigma}_t^+ | \mathbf{G}, \mathbf{Q}) \tag{6}$$

where the transition matrix $\mathbf{A}$ is learned and $\mathbf{G}$ and $\mathbf{Q}$ are given. The initial conditions are hyperparameters and set to $\boldsymbol{\mu}_{t_0}^- = \mathbf{0}$ and $\boldsymbol{\Sigma}_{t_0}^- = 10^4 \cdot \mathbf{I}$

Update Step. At the time of a new observation t' the prior is updated by Bayes' theorem.

$$p(\mathbf{z}_{t'}|\mathbf{y}_{t'}) = \frac{p(\mathbf{y}_{t'}|\mathbf{z}_{t'})p(\mathbf{z}_{t'}|\mathbf{y}_t)}{p(\mathbf{y}_{t'}|\mathbf{y}_t)} \tag{7}$$

Thanks to the Gaussian assumption all quantities are known and the posterior density at time $t \in \mathcal{T}$ is given by $p(\mathbf{z}_t|\mathbf{y}_t) = \mathcal{N}(\boldsymbol{\mu}_t^+, \boldsymbol{\Sigma}_t^+)$ with

$$\boldsymbol{\mu}_t^+ = \boldsymbol{\mu}_t^- + \mathbf{K}_t(\mathbf{y}_t - \mathbf{H}\boldsymbol{\mu}_t^-) \tag{8}$$

$$\boldsymbol{\Sigma}_t^+ = (\mathbf{I} - \mathbf{K}_t\mathbf{H})\boldsymbol{\Sigma}_t^- \tag{9}$$

$$\mathbf{K}_t = \boldsymbol{\Sigma}_t^-\mathbf{H}^T(\mathbf{H}\boldsymbol{\Sigma}_t^-\mathbf{H}^T + \boldsymbol{\Sigma}_t^{obs})^{-1} \tag{10}$$

We summarize these updates as

$$[\boldsymbol{\mu}_t^+, \boldsymbol{\Sigma}_t^+] = \text{update}(\boldsymbol{\mu}_t^-, \boldsymbol{\Sigma}_t^-, \mathbf{y}_t, \boldsymbol{\Sigma}_t^{obs} | \mathbf{H}) \tag{11}$$

2.2 Continuous-Discrete Recurrent Kalman Networks

Recent work has been done on integrating the Kalman filter in neural networks to expand its expressiveness (Fraccaro et al. 2017; Becker et al. 2019). In this paper, we built on the concept of (discrete) RKNs (Becker et al. 2019).

CDRKNs consist of three parts: an encoder f_θ, a recurrent continuous-discrete Kalman filter and a decoder g_ϕ. At each time point $t \in \mathcal{T}$, an observation is propagated through the three components: A nonlinear function f_θ parametrized by a neural network maps the observation $\mathbf{x}_t$ to a latent observation space and outputs a transformed observation $\mathbf{y}_t$ along with an elementwise uncertainty estimate $\boldsymbol{\sigma}_t^{obs}$.

$$\text{encoder:} \qquad [\mathbf{y}_t, \boldsymbol{\sigma}_t^{obs}] = f_\theta(\mathbf{x}_t) \tag{12}$$

The Kalman filter updates the latent state prior using the latent observation and its uncertainty estimate (update step (8)–(10)). It propagates the resulting posterior up to the next observation time t' (prediction step (4) and (5)).

$$[\boldsymbol{\mu}_t^+, \boldsymbol{\Sigma}_t^+] = \text{update}(\boldsymbol{\mu}_t^-, \boldsymbol{\Sigma}_t^-, \mathbf{y}_t, \boldsymbol{\sigma}_t^{obs}\mathbf{I}|\mathbf{H}) \qquad [\boldsymbol{\mu}_{t'}^-, \boldsymbol{\Sigma}_{t'}^-] = \text{predict}_\mathbf{A}(\boldsymbol{\mu}_t^+, \boldsymbol{\Sigma}_t^+ | \mathbf{G}, \mathbf{Q}) \tag{13}$$

The decoder maps the posterior estimate back to the observation space and returns either a filtered observation or a prediction of future time steps along with an elementwise uncertainty estimate.

$$\text{decoder:} \qquad [\boldsymbol{\mu}_{\mathbf{x}_t}, \sigma^2_{\mathbf{x}_t}] = g_\phi(\boldsymbol{\mu}_t^+, \boldsymbol{\Sigma}_t^+) \qquad (14)$$

Our objective function is the Gaussian negative log-likelihood of observations $\mathbf{x}_{\mathcal{T}}$

$$\mathcal{L}_{\theta,\phi,\mathbf{A}}(\mathbf{x}_{\mathcal{T}}) = -\frac{1}{N} \sum_{t \in \mathcal{T}} \log \mathcal{N}(\mathbf{x}_t | \boldsymbol{\mu}_{\mathbf{x}_t}, \sigma^2_{\mathbf{x}_t} \mathbf{I}) \qquad (15)$$

The model parameters $(\theta, \phi, \mathbf{A})$ can be trained with backpropagation in an end-to-end manner. In the current version of the model, the Brownian motion β, the latent state $\mathbf{z}$ and the latent observation $\mathbf{y}$ operate in the same space and $\mathbf{H} = \mathbf{G} = \mathbf{I}$. For now, $\mathbf{Q}$ is also kept fix at $0.1 \cdot \mathbf{I}$.

3   Empirical Study

We studied the ability of the CDRKN to model nonlinear, irregular time series on a prediction and a state estimation task. To assess the benefits of continuous state transitions, we compared it to the discrete RKN. In the first experiment, we generated 1000 2-dimensional spiral trajectories with different starting points, irregular time steps and additive Gaussian noise. We keep the overall length of the spiral (i.e. its maximum angle) fixed and experiment with different granularity for $N = \{50, 100, 200\}$. The model was tasked to predict another N random observations on each spiral trajectory.

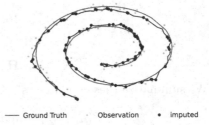

— Ground Truth   Observation   • imputed

Fig. 2. Test sample of the imputation experiment for $N = 100$. Blue lines connect imputed with filtered observations. Our method is able to model the dynamics of the spiral despite time periods with few or no observations. (Color figure online)

In the second experiment, we created highly noise corrupted image sequences of a pendulum and trained the model to estimate the pendulum angle $\theta_t = (sin(\theta_t), cos(\theta_t))$ as in Becker et al. (2019). To create irregular time intervals we subsample $1/3$ of the images.

Table 1. Average Gaussian negative log-likelihood on test set: Despite the strict linearity assumption in the transition model of CDRKN, our method outperforms RKN in granular prediction scenarios.

Model	Spiral			Pendulum
	$N = 200$	$N = 100$	$N = 50$	$N = 25$
RKN (Becker et al. 2019)	3.5016	3.5565	**3.0710**	−4.2533
CDRKN (ours)	**0.8288**	**3.3090**	3.4096	**−4.7423**

We use the same encoder/decoder architecture for RKN and CDRKN. Note that the RKN uses more parameters in its transition model allowing for locally linear dynamics. It also has twice the latent state dimensionality of our model. In the pendulum experiment, we also allow locally linear transitions for the CDRKN and impose the same approximations on the state covariance as in Becker et al. (2019). Table 1 shows CDRKN better accounts for granular time steps in the prediction task despite the more complex parameterization of RKN. In the pendulum experiment, when both models employ locally linear transitions, CDRKN outperforms RKN.

4   Conclusion

We proposed CDRKNs, a network class that processes data as samples of continuous dynamics providing a principled framework to model irregular time series. Embedding a continuous version of the Kalman filter into a neural network allows modelling nonlinear systems, adjusting to new observations and providing uncertainty estimates. In future work, we like to further investigate locally linear transitions of the latent state, experiment with partially observable time series and explore potential computational savings.

References

Chen, T.Q., Rubanova, Y., Bettencourt, J., Duvenaud, D.: Neural ordinary differential equations. In: Advances in Neural Information Processing Systems 31, pp. 6571–6583. CurranAssociates, Inc. (2018)

Rubanova, Y., Chen, T.Q., Duvenaud, D.: Latent ordinary differential equations for irregularly-sampled time series. In: Advances in Neural Information Processing Systems 32, pp. 5320–5330. CurranAssociates, Inc. (2019)

Kidger, P., Morrill, J., Foster, J, Lyons, T.: Neural controlled differential equations for irregular time series. arXiv preprint arXiv:2005.08926 (2020)

De Brouwer, E., Simm, J., Arany, A., Moreau, Y.: GRU-ODE-Bayes: continuous modeling of sporadically-observed time series. In: Advances in Neural Information Processing Systems 32, pp. 7379–7390. Curran Associates, Inc. (2019)

Kalman, R.E.: A new approach to linear filtering and prediction problems. Trans. ASME J. Basic Eng. **82**, 35–45 (1960)

Jazwinski, A.H.: Stochastic Processes and Filtering Theory. Academic Press, New York (1970)

Axelsson, P., Gustafsson, F.: Discrete-time solutions to the continuous-time differential Lyapunov equation with applications to Kalman filtering. IEEE Trans. Autom. Control **60**(3), 632–643 (2014)

Fraccaro, M., Kamronn, S., Paquet, U., Winther, O.: A disentangled recognition and nonlinear dynamics model for unsupervised learning. In: Advances in Neural Information Processing Systems, pp. 3601–3610. NeurIPS (2017)

Becker, P., Pandya, H., Gebhardt, G., Zhao, C., Taylor, C. J., Neumann, G.: Recurrent Kalman networks: factorized inference in high-dimensional deep feature spaces. In: International Conference on Machine Learning, pp. 544–552. PMLR (2019)

Adversarial Generation of Temporal Data: A Critique on Fidelity of Synthetic Data

Ankur Debnath[✉], Nitish Gupta, Govind Waghmare, Hardik Wadhwa, Siddhartha Asthana, and Ankur Arora

Mastercard AI Garage, Gurugram, India
{ankur.debnath,nitish.gupta,govind.waghmare,hardik.wadhwa,
siddhartha.asthana,ankur.arora}@mastercard.com

Abstract. Generative modelling for temporal data has seen a paradigm shift from autoregressive to adversarial models. Adversarial generation algorithms have proven to be more efficient in capturing the complex temporal correlations that the simplistic autoregressive model could not. Albeit, high-fidelity remains a concern even for adversarial modelling. The generation of high-fidelity data requires the model to have three strengths: capture feature correlations, model long-term dependencies, and scalability in dimensions. This paper analyzes these strengths on the existing methods of adversarial temporal generation regarding the fidelity of synthetic data. Towards this, we evaluate different algorithms for adversarial temporal generation on five different datasets of varying dynamics (long-term vs. short-term dependency) and dimensionality. We conclude by discussing gaps in the literature and future directions for high fidelity temporal data generation through adversarial methods.

Keywords: Adversarial temporal generation · Fidelity · Time series

1 Introduction

Generative modelling of time series is an active area of research. In particular, generating high-quality synthetic time series data is a challenging problem. The recent success of Generative Adversarial Networks (GAN) in other modalities like text and images has led researchers to explore generative methods for synthetic temporal data. GAN-based approaches have attempted to model time-series distributions and have shown a better efficacy. We discuss the ability of GANs to generate realistic sequences and present a detailed analysis of the most relevant GANs existing in the literature. We evaluate the fidelity of the synthetic data based on three characteristics: *Modelling long-term patterns*, *capturing feature correlations*, and *scalability in dimensions*. Capturing long-term patterns is crucial in data that have periodicity over an extended time frame. To model the joint distribution of the features, learning the complex correlations between them is essential even with high dimensionality. Thus, we perform experiments on five publicly available multivariate time-series datasets and examine the performance

© Springer Nature Switzerland AG 2021
M. Kamp et al. (Eds.): ECML PKDD 2021 Workshops, CCIS 1524, pp. 306–321, 2021.
https://doi.org/10.1007/978-3-030-93736-2_24

of these algorithms. Further, we explain the different evaluation metrics for temporal data generation algorithms. Various metrics evaluate different aspects of the data fidelity, leading to a comprehensive assessment. The structure of this paper is as follows: an introduction of the temporal data generation and GANs (Sect. 2). Then, we present a thorough explanation and analysis of the different variants of GANs. After that, we discuss the evaluation metrics and experimental results. Finally, we conclude this work by identifying relevant literature gaps concerning the three critical factors, i.e., modelling long-term patterns, capturing feature correlations, and scalability in dimensions.

In literature, many works concentrate on GANs focused on images, and to some extent, on audio and text [10,11,14]. To the best of our knowledge, this is the first attempt to extensively analyze adversarial time series generation. This work aims to evaluate different algorithms for adversarial temporal generation and highlight their advantages and limitations.

2   Background and Notations

GANs [22] fall in the broad category of deep generative models that take input samples and create a model that produces new samples from the same distribution as that of the original data. Consider samples $S_1, S_2, ...S_n$ where $S_i \in \mathbb{R}^d$, and each sample drawn i.i.d from some distribution $S_i \sim P_S$. A GAN uses these samples to create a model that can sample from the distribution P_S. A typical GAN comprises of a Generator G and a Discriminator D, which are trained in an adversarial manner. Both G and D are deep neural networks. In the most basic form, a GAN maps a noise vector $z \in \mathbb{R}^p$ to a sample $S \in \mathbb{R}^d$ where $d \gg p$. z is usually drawn from a predefined Gaussian or uniform distribution P_z. Simultaneously, We train the discriminator to map from latent space to $[0, 1]$, which takes both real samples and samples generated by the generator and classifies them as real or fake. Both G and D use this classification error to train. Analogous to game theory, both players are trying to beat each other in a min-max game. This game theory approach ensures adversarial learning across the networks and models the data distribution effectively. For generating sequences, each sample S^i is a sequence $X_{1:T}$, where T is the sequence length and $X \in \mathbb{R}^d$. The goal of the GAN is to learn both the conditional density $p(X_t|X_{1:t-1})$ and P_S. In this section, we discuss generative methods for temporal data.

2.1   Sequence Generation with RNNs

The initial works on sequence generation focused on density modelling using recurrent neural networks [25,26]. The objective was to learn the conditional densities of the sequences, where the subsequent sample is generated conditionally using the past samples. This approach was extended to seq2seq tasks and has been successful in many applications such as machine translation, handwriting generation, etc. However, such sampling-based methods cannot effectively generate high fidelity and diverse samples to be utilised by the downstream tasks. In particular, finding a good predictive distribution (usually referred to as density modelling) for high-dimensional real-valued data can be challenging. In general, RNNs and their variants are inefficient in modelling long-term temporal dependencies and complex feature correlations [13]. Therefore, supervised methods of generating sequences are insufficient to generate high-fidelity synthetic data. A sequence $x = (x_1, x_2, ..x_T)$ is passed through a stacked layer of RNNs, and generate an output sequence $y = (y_1, y_2, ..y_T)$. The goal is to learn $Pr(x_{t+1}|y_t)$ using the previous time output y_t. Sampling from this conditional distribution can help generate sequences with moderate fidelity [25]. However, this sampling is usually non-trivial and often requires beam-search to generate synthetic sequences [26].

2.2   Adversarial Generation

Apart from density estimation methods of generating synthetic sequences using RNNs, there has been a growing interest in utilizing GANs for synthetic temporal data generation. The success of GANs in other domains such as CV [10], NLP [14] has motivated researchers to formulate sequence generation in an adversarial manner. The works in this area have focussed majorly on two aspects: architectural design and training strategies. The most relevant works in this space are C-RNN-GAN [7], RCGAN [1], TimeGAN [9] and DoppelGANger (DG) [5]. The three key aspects in generating synthetic multivariate time series sequences that we observe throughout this literature are: capturing long-term temporal dependencies, feature correlations, and dimensional scalability. In the following section, we explain each algorithm on the basis of these three aspects.

3   Adversarial Generation Algorithms: Variants

In this section, we describe the works in adversarial generation for temporal data. In the next section, we present a taxonomy for the GAN based architectures and analyse them.

3.1   C-RNN-GAN

C-RNN-GAN architecture (Fig. 1a) uses recurrent neural networks as its backbone for modelling sequential data in an adversarial fashion. The generator (G) and discriminator (D) are deep recurrent networks. The RNNs used in these networks are LSTMs. Moreover, the discriminator consists of deep bi-directional LSTMs, whereas the generator has unidirectional LSTMs. The purpose of choosing such a configuration is to ensure that the discriminator is trained to classify real and fake data better. In such a way, the training becomes a zero-sum game for which the Nash equilibrium is a state of the discriminator where it can no longer distinguish between real and fake data. In C-RNN-GAN, one measurement of a sequence is generated simultaneously, similar to the non-adversarial counterparts (RNN based sequence generation). Also, the generator output from

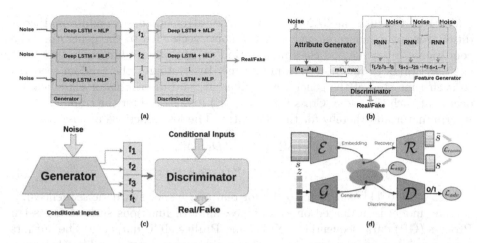

Fig. 1. Architectures of GAN variants a) C-RNN-GAN b) DoppelGANger c) RCGAN d) TimeGAN (zoom-in for visibility)

the previous time step is fed to the next RNN cell of the generator concatenated with noise. The loss functions of the model are:

$$\mathcal{L}_G = \frac{1}{n} \sum_{i=1}^{n} \log(1 - D(G(\boldsymbol{z}^i))) \tag{1}$$

$$\mathcal{L}_D = \frac{1}{n} \sum_{i=1}^{n} [-\log(D(S^i)) - \log(1 - D(G(z^i)))] \tag{2}$$

(where S^i is the ith sequence, z_i is the random vector).

The model is designed for a music generation task. The training data is publicly available classical music data. The features of the data are four real-valued scalars at every data point: *tone length, frequency, intensity,* and *time spent* since the previous tone. The model output is evaluated on *Polyphony, Scale consistency, Repetitions* and *tone span.* The model can generate realistic examples of music but nowhere close to the actual training data. Overall, the paper claims that adversarial learning leads to higher fidelity data generation than density estimation techniques.

3.2   RCGAN

This architecture is proposed to generate multivariate real-valued sequences using adversarial training. Moreover, they take motivation from Conditional GANs, where the generation of samples is conditioned on additional inputs. The original work proposes LSTMs as generator and discriminator. Unlike C-RNN-GAN, the previous time-step output of the generator is not fed to the next time step. Also, both the networks are unidirectional LSTMs. The paper establishes two versions of their architecture, namely RGAN (Recurrent GAN) and RCGAN

(Recurrent Conditional GAN) (Fig. 1c). The latter one has an additional, conditional input c_i, concatenated with the noise vector fed to the generator. The conditional input may be class information or any other additional input that the user wants to retain over training time. Interestingly, the random seed fed to each time step of the generator is different, and discriminator classification occurs at each time step. Cross-Entropy (CE) loss is used for the training of the discriminator and thereby for the generator. The loss functions are as follows:

$$\mathcal{L}_D(S^i) = -CE(D(S^i), y_i) \tag{3}$$

$$\mathcal{L}_G(z^i) = -CE(D(G(z^i)), \mathbf{1}) \tag{4}$$

(where $y_i \in \{0, 1\}^T$ is a vector of 1's for real and 0's for synthetic sequences).

The model is evaluated on Sine waves, smooth functions such as Gaussian Process (GP) data, sequential MNIST, and Philips eICU database. The authors suggest using specialized metrics curated for sequence generation like Maximum Mean Discrepancy (MMD) and Train on Synthetic Test on Real (TSTR) instead of general GAN evaluation metrics used in other domains. The paper also highlights the privacy aspects and proposes a differentially private version of the GANs, which is out of scope for this paper. The paper, however, claims to generate reasonably realistic labelled sequences and proves the utility of such architecture in the medical domain. Additionally, the paper discusses the evaluation and motivates more accurate measures to evaluate sequential GANs.

3.3   TimeGAN

TimeGAN is the first paper that primarily targets temporal modelling in sequential GANs. Also, it is the first work to combine representation learning and unsupervised GAN into a single framework. TimeGAN consists of four networks (Fig. 1d): embedding function ($\mathcal{E}$), recovery function ($\mathcal{R}$), generator ($\mathcal{G}$) and discriminator ($\mathcal{D}$). The first two are autoencoding networks, which are trained jointly alongside the adversarial components. This architecture reduces the high dimensionality of the adversarial learning space as fewer and lower-dimensional factors of variation often drive the temporal dynamics of complex systems. The paper introduces a stepwise-supervised loss to encourage the model to learn the stepwise conditional distribution. In this model, all four networks are recurrent models (LSTM and GRU). The key contribution of this paper is the supervised loss that is used to aid the generator. This ensures that the generated samples contain the same temporal information as that of the original data. Unlike other GANs discussed in this paper, the adversarial learning space is not the original data space but a lower-dimensional space learned by the autoencoder. The different loss functions used by this paper are as follows: The first two components are trained on:

$$\min(\lambda \mathcal{L}_S + \mathcal{L}_R) \tag{5}$$

and the generator and discriminator are trained adversarially as:

$$\min(\eta \mathcal{L}_S + \max \mathcal{L}_U) \tag{6}$$

Where, $\mathcal{L}_R$, $\mathcal{L}_S$ and $\mathcal{L}_U$ are defined as follows:

$$\mathcal{L}_R = \mathbb{E}_{S \sim P_S}[\|S - \tilde{S}\|_2] \tag{7}$$

$$\mathcal{L}_U = \mathbb{E}_{S \sim P_S}[\log(\mathcal{D}(\mathcal{E}(S))) + (1 - \log(\mathcal{D}(\mathcal{G}(z))))] \tag{8}$$

$$\mathcal{L}_R = \mathbb{E}_{S \sim P_S} \sum_{t=1}^{T}[\|(\mathcal{E}(S_t) - \mathcal{G}(\mathcal{E}(S_{t-1}), z_t)\|_2] \tag{9}$$

The paper evaluates the proposed model on real-valued multivariate time series data such as autoregressive multivariate Gaussian data, Sines, Stocks, Energy, and Events datasets. It benchmarks across different sequential generative algorithms using *Discriminative* and *Predictive* scores. These scores are trainable metrics, post-hoc RNN classifiers, and prediction networks to assess the quality of the synthetic data. The paper also visually analyze through PCA and t-SNE plots to validate its performance further. The authors claim that TimeGAN, by virtue of its *supervised* loss, generates high-quality synthetic sequences for datasets with high temporal correlations and low periodicity. However, it performs decently for datasets with high periodicity as well. It is important to note that although the algorithm supports metadata (static feature) generation, the authors do not show results on such datasets. In original work, the sequence lengths of the synthetic samples are around 30.

3.4 DoppelGANger

This paper aims to create high fidelity, easily generalizable synthetic data for multivariate measurement data for networking applications. To this end, the paper targets two critical aspects of the problem: fidelity and privacy. Fidelity refers to (a) capturing complex correlations between measurements and their associated metadata and (b) long-term correlations within time series. Additionally, they address the mode-collapse of GANs in generating synthetic data with a high dynamic range. For privacy, they highlight that existing privacy-preserving training techniques may sacrifice the utility of the data. They explain the privacy-quality trade-off of the proposed algorithm.

Architecturally, DG (Fig. 1b) consists of two sets of generators and discriminators for measurements (time-varying) and metadata (static) features. To model the correlations between metadata and the time-varying measurements, metadata is fed to the time series generator at each time step. This is possible due to the decoupled generation of metadata from the time-varying measurements. The time series generator is a recurrent network, while the metadata generator and both the discriminators are MLP networks. Unlike its counterparts, it does not rely on joint modelling both temporal and static features. DG also addresses mode collapse by generating normalized time series and rescale to the original range using generated max and min limits. The authors also introduce the concept of *Batch Generation*, in which RNNs output batched samples rather than singletons. To better model the feature correlations, DG makes use of conditional generation of temporal features from metadata. The architecture uses Wasserstein loss functions and combines the loss as follows:

$$\min_{\mathcal{G}} \max_{\mathcal{D}_1, \mathcal{D}_2} \mathcal{L}_1(\mathcal{G}, \mathcal{D}_1) + \alpha \mathcal{L}_2(\mathcal{G}, \mathcal{D}_2) \tag{10}$$

where $\mathcal{L}_i$, $i \in \{1, 2\}$ is the Wasserstein loss of the original and second discriminator, respectively and :

$$\mathcal{L}_i = \mathbb{E}_{S \sim P_S}[T_i(\mathcal{D}_i(S)) - T_i(\mathcal{D}_i(\mathcal{G}_i(z)))] - \mathbb{E}_{\hat{S} \sim P_{\hat{S}}}[(\|\nabla_{\hat{S}} \mathcal{D}_i(T_i(\hat{S}))\|_2 - 1)^2] \tag{11}$$

where, $T_1(S) = S$, $T_2(S) = $ metadata part of S. $\hat{S} := tS + (1 - t)\mathcal{G}(z)$ where $t \sim Unif[0, 1]$

The model is evaluated on three network datasets: Wikipedia Web Traffic (WWT), Measuring Broadband America (MBA), and Google Cluster Usage Traces (GCUT). The paper uses auto-correlation plots to compare the quality of the generated data and evaluate the model's performance on capturing temporal correlations. Pearson correlations measure the performance in regards to capturing feature correlations. Further, distributions of temporal and metadata features are evaluated to compare the performance against the baselines on fidelity of generated data. The datasets used in the experiments are diverse in terms of sequence lengths and dimensionality. The results appear to be promising across the metrics. Additionally, the paper discusses membership inference and differential privacy and elaborates on the quality-privacy trade-off in such GANs.

4   Analysis of GAN Variants

One of the first attempts in generating sequences with GANs started with C-RNN-GAN. The primary objective of this algorithm was to model sequences with LSTMs. However, the LSTM framework was not able to generate realistic sequences and failed to model temporal correlations. Following that, RCGAN was proposed, which combined the concept of conditional GANs and sequence generation. Instead of feeding the previous time step generator out as input to the next time step, the model generated each time step with a different random seed. The conditional version of this architecture could generate class-labelled synthetic data and helped model the feature correlations better than an unconditional generation. However, both C-RNN-GAN and RCGAN could not model long sequences and high dimensionality due to the RNNs. To this end, TimeGAN was proposed to solve the dimensionality problem by introducing supervised and representation learning into the GAN setup. TimeGAN's architecture allowed it to model the temporal correlations and variable interactions in a much better way. However, all of these algorithms were proposed to generate sequences on the scale of a few tens of samples. DG highlighted that these algorithms are unable to model long annual or diurnal patterns in the data as it requires the learning of large sequences. It also addressed the issue of the low receptive field

Fig. 2. A Taxonomy of architectures: Green region shows long-term modelling which is exhibited by RCGAN, TimeGAN, and C-RNN-GAN. DoppelGANger captures both feature correlations and long-term dependencies while none of them is addressing all the three properties (long-term modelling, feature correlations, and scalability) (Color figure online)

of such RNNs by incorporating the concept of *Batch Generation*. However, none of these algorithms showed results on extreme cases of temporal modelling and dimensionality. We study the limitations of these algorithms in terms of sequence lengths, dimensionality, and feature correlations. We perform extensive experiments on RCGAN, TimeGAN, and DG and discuss them in the following sections. Figure 2 shows the taxonomy of the different architectures proposed and placed according to their respective research proposal. The placement of each algorithm is further established through the set of experiments performed.

5   Evaluation Metrics

The qualitative and quantitative assessment of synthetic data generated by the GANs needs to be performed extensively. Several metrics have been proposed for temporal GANs. Interestingly, not all metrics from GAN literature are suitable to measure the fidelity of synthetic sequences. Hence, metrics that can assess the synthetic data based on capturing temporal dynamics, modes captured, the performance on tail-heavy distributions, and diversity are required in order to examine such GANs thoroughly. In this work, we restrict our experimentation to the following metrics:

- **Discriminative Score**: For a quantitative measure of similarity, the original and synthetic sequences are classified by a post-hoc RNN classifier (2-layer LSTM). Classification accuracies are reported.
- **Predictive Score**: A post-hoc sequence-prediction model (2-layer LSTM) is trained on synthetic to predict next-step temporal vectors over each input sequence and tested on original data. Performance is measured in terms of the mean absolute error (MAE).
- **Kullback-Leibler (KL) divergence**: It is a measure of distance between two probability density functions in terms of entropy. This measures the similarity between two distributions.
- **Pairwise Correlation Difference (PCD)**: It is a metric that captures the difference in variable correlations in two data matrices.
- **Training Time**: Training time of the algorithm (in hours).

6   Datasets

In this paper, we perform experiments on five public benchmark datasets[1], described as follows:

Stocks[2]: It contains daily historical Google stocks data from 2004 to 2019. **Solar-Energy** (see footnote 1): The solar power production records in the year of 2006 in Alabama State. **Electricity** (see footnote 1): The electricity consumption in kWh recorded every 15 min from 2012 to 2014, for N = 321 clients. **Exchange-Rate** (see footnote 1): The collection of the daily exchange rates of eight countries ranging from 1990 to 2016. **Traffic** (see footnote 1): This data is a col-

Table 1. Dataset Statistics, where N is the number of variables, L is the length of the time series

Datasets	N	L
Electricity	321	26,304
Exchange Rate	8	7,588
Stocks	6	3,685
Solar-Energy	137	52,560
Traffic	500	14,035

lection of 48 months (2015–2016) hourly data from the California Department of Transportation. Financial datasets such as Exchange Rate and Stocks are inherently aperiodic and have high temporal correlations across dimensions, as shown in Fig. 3. On the other hand, Electricity, Solar, and Traffic are more periodic and contain seasonalities. The different statistics associated with each dataset are mentioned in Table 1.

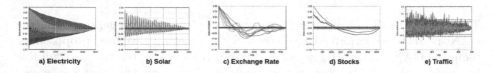

a) Electricity   b) Solar   c) Exchange Rate   d) Stocks   e) Traffic

Fig. 3. Auto-correlation graphs of sampled variables (zoom-in for visibility)

7   Experiments and Results

In this section, we consider the three most recent GAN algorithms for sequence generation and perform experiments on five public benchmark datasets (see footnote 1). We do not perform experiments on C-RNN-GAN as it known to be inferior to all the other three algorithms. The supplementary code for the base experiments of RCGAN [1][3], TimeGAN [9][4] and DG [5][5] provided by the authors are used in this work. Sequence lengths, dimensionality, and periodicity in the datasets ensure that these algorithms are tested extensively for long-term dependencies, feature correlations, and scalability.

---

[1] https://github.com/laiguokun/multivariate-time-series-data.
[2] https://finance.yahoo.com/quote/GOOG/history?p=GOOG.
[3] https://github.com/ratschlab/RGAN.
[4] https://github.com/jsyoon0823/TimeGAN.
[5] https://github.com/fjxmlzn/DoppelGANger.

7.1 Performance Evaluation

Data Distribution. One of the key evaluation criteria of GANs is through comparison of data distribution of original data and synthetic data. For this, we have used plots like PCA (Fig. 4), t-SNE (Fig. 5), density plots (Fig. 6), along with KL Divergence metric to examine the performance of each model. We also show data distribution plots to visualize the effectiveness of these models to capture different modes in the data.

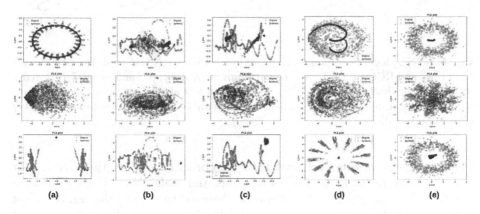

Fig. 4. PCA plots of real and synthetic data for TimeGAN (1st row), DoppelGANger (2nd row) and RCGAN (3rd row). Each column provides the visualization for Electricity (a), Stocks (b), Exchange Rate (c), Solar (d) and Traffic (e) datasets. Synthetic data is represented using blue colour and Real data using red colour (zoom-in for visibility) (Color figure online)

The results obtained for TimeGAN, RCGAN, and DG are shown in Table 2, 3, and 4 respectively. We observe that RCGAN has the worst performance in terms of modelling data distribution as compared to TimeGAN and DG. The performance of all these models tend to deteriorate as the complexity and scale of the datasets are increased. For smaller datasets such as Stocks and Exchange Rate, the synthetic data distribution is captured well by TimeGAN and DG. For the other three datasets, due to larger dimensionality, the performance is decreased significantly. This can be explained by the dynamic range of these RNNs [18,23]. The dynamic range is greatly impacted by the normalization strategies used and can lead to mode collapse if not performed correctly [12]. RCGAN and TimeGAN do not focus on the normalization and hence suffer from mode collapse when dynamic ranges in the datasets increase. Whereas, special attention is given to normalization and mode-collapse by DG and hence able to manage the dynamic ranges in the data and reduce mode-collapse. As we increase sequence length, the GANs become more susceptible to the shortcoming of RNNs as it needs to model longer sequences.

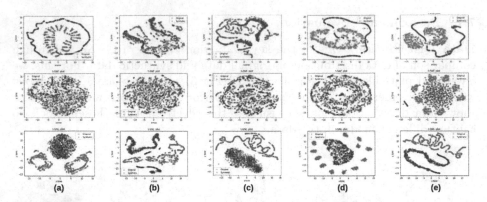

Fig. 5. t-SNE plots of real and synthetic data for TimeGAN (1st row), DoppelGANger (2nd row) and RCGAN (3rd row). Each column provides the visualization for Electricity (a), Stocks (b), Exchange Rate (c), Solar (d) and Traffic (e) datasets. Synthetic data is represented using blue colour and Real data using red colour (zoom-in for visibility) (Color figure online)

Feature Correlations and Scalability. In order to evaluate these models for their capacity to model feature correlations and scalability, we use PCD as a metric. From Table 2, 3, and 4, it can be observed that when the cardinality of the dataset is low, such as in Stocks and Exchange Rate, the performance of TimeGAN and DG is better. For each algorithm, as dimensionality is increased by changing datasets, PCD increases, implying that it becomes more difficult for the GANs to capture variable interactions. Even though DG seems to perform decently with increased dimensions, but the time complexity of the model increases with increasing dimensionality. The worst performance is observed in the Traffic dataset having 500 dimensions and hence empirically indicating the limitations of such models in terms of scalability and capturing variable interactions. Interestingly, even though TimeGAN incorporates dimensionality reduction through its autoencoder, it still suffers in high dimensions and indicates a need for better representations in lower-dimensional space. Results show that the overall complexity of the models in terms of training time seems to be significantly impacted with higher sequence lengths and dimensionality[6]. Figure 9 shows the training times of each algorithm on the different datasets (in increasing order of dimensions). We infer that the complexity of the dataset concerning temporal correlations and dimensionality increases, training time simultaneously rises for each algorithm.

Long Term Dependencies. The reported lengths of sequences by RCGAN and TimeGAN are in the order of tens. Therefore, we used high sequence lengths

---

[6] The original Traffic dataset has 862 dimensions. None of the experiments could be completed on that due to memory overflow. Hence, all experiments on traffic were done on 500 dimensions only.

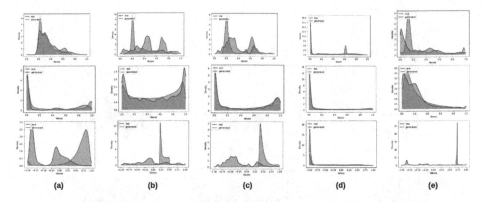

Fig. 6. Probability density functions of synthetic and real data for TimeGAN (1st row), DoppelGANger (2nd row) and RCGAN (3rd row). Each column provides the visualization for Electricity (a), Stocks (b), Exchange Rate (c), Solar (d) and Traffic (e) datasets. Synthetic data is represented using orange colour and Real data using blue colour (zoom-in for best view) (Color figure online)

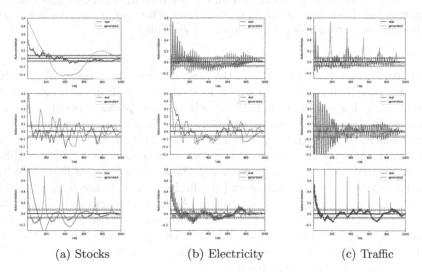

    (a) Stocks   (b) Electricity   (c) Traffic

Fig. 7. Auto-correlation graphs of real and synthetic data for TimeGAN (1st row), DoppelGANger (2nd row) and RCGAN (3rd row). These are plotted for lag = 1000, which show short term periodicity. (zoom-in for visibility)

for each dataset as a benchmark. Note that DG claims to model longer sequence lengths (in order of 100 s) for network datasets but seems to not scale in terms of other parameters simultaneously. Here, we increase sequence length for testing long-term modelling along with high-dimensional datasets (with and without periodic patterns). We use Discriminative Score, Predictive Score, and Autocorrelation plots to assess the quality of synthetic data.

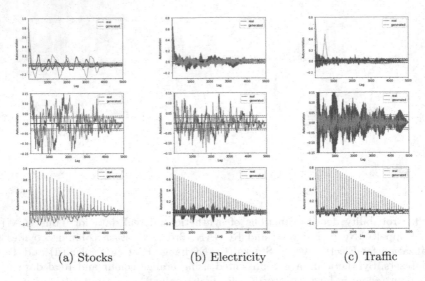

(a) Stocks   (b) Electricity   (c) Traffic

Fig. 8. Auto-correlation graphs of real and synthetic data for TimeGAN (1st row), DoppelGANger (2nd row) and RCGAN (3rd row). These are plotted for lag = 5000, which show long term periodicity. (zoom-in for visibility)

Figure 7 and 8 show that DG is the best performing in terms of modelling short and long-term patterns. However, its performance drops as dimensionality is increased. This indicates that all models tend to perform poorly on longer sequence lengths of high dimensional data. RNNs used in all of these models suffer from memory bottleneck and are not scalable in high dimensions [13]. Although techniques like *Batch Generation* and representation learning have been used, they are not entirely resolving these issues. Interestingly, Tables 2, 3, and 4 show that discriminative and predictive scores do not follow the trend as other metrics, which is counter-intuitive. As these metrics are trainable, they have a fixed network structure (constant parameters). Therefore, they cannot be generalised enough to distinguish or assess the quality of synthetic data from all the models. Hence, it is more intuitive to use generalisable metrics like KLD and PCD which focus on data fidelity.

8 Discussion

In this section, we discuss on results and limitations of the generative models. This paper examines the most recent works in the field of generative modelling for temporal data. We also discuss motivation and approaches taken by these algorithms (refer Sect. 3) and analyze them critically. We perform an evaluation of these algorithms on five different datasets to highlight the gaps in the literature. We examine the fidelity of the synthetic data on three principal attributes: long-term modelling, feature correlations, and scalability.

Table 2. Results for TimeGAN which include Discriminative Score (Disc. score), Predictive Score (Pred. score), KL Divergence (KLD), Pairwise correlation difference (PCD). (↑) means higher, the better and Training time. (↓) means lower, the better.

Details	Electricity	Stocks	Exchange Rate	Solar	Traffic
Size	26,304	3,685	7,588	52,560	14035
Cardinality	321	6	8	137	500
Disc. score (↓)	0.4999	0.1834	0.2377	0.4626	0.5
Pred. score (↓)	0.0347	0.0406	0.0981	0.0412	0.108
KLD (↓)	64992.33	6847.39	8256.62	32319.53	499035.91
PCD (↓)	48.44	1.073	1.397	10.587	160.433
Training Time (↓)	18.25 h	6.2 h	8.8 h	13.5 h	22.5 h

Table 3. Results for RCGAN which include Discriminative Score (Disc. score), Predictive Score (Pred. score), KL Divergence (KLD), Pairwise correlation differnce (PCD) and Training time. (↑) means higher, the better. (↓) means lower, the better.

Details	Electricity	Stocks	Exchange Rate	Solar	Traffic
Size	26,304	3,685	7,588	52,560	14035
Cardinality	321	6	8	137	500
Disc. score (↓)	0.5	0.4971	0.5	0.5	0.5
Pred. score (↓)	0.5447	0.4259	0.6442	0.5528	0.8722
KLD (↓)	874814.27	28807.85	44684.76	288776.67	1924776.35
PCD (↓)	95.268	3.0349	3.402	91.36	198.7723
Training Time (↓)	15.25 h	4.7 h	6.5 h	8.3 h	20 h

Table 4. Results for DG which include Discriminative Score (Disc. score), Predictive Score (Pred. score), KL Divergence (KLD), Pairwise correlation differnce (PCD) and Training time. (↑) means higher, the better. (↓) means lower, the better.

Details	Electricity	Stocks	Exchange Rate	Solar	Traffic
Size	26,304	3,685	7,588	52,560	14035
Cardinality	321	6	8	137	500
Disc. score (↓)	0.4927	0.2788	0.4289	0.4737	0.4998
Pred. score (↓)	0.0692	0.1015	0.1959	0.0445	0.0938
KLD (↓)	41879.94	358.86	878.89	4338.29	191267.65
PCD (↓)	41.27	0.50	0.654	10.22	189.77
Training Time (↓)	8.5 h	0.4 h	0.9 h	4.7 h	22.6 h

From results obtained in Sect. 7, we observe that modelling sequences become challenging with increasing length. Since these algorithms are based on RNNs, this limits the models' capacity to remember long sequences. Additionally, in terms of feature correlations and scalability, we observe that if both sequence lengths and dimensionality are increased, the performance of these models decreases, along with increase in training time. Apart from

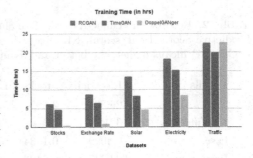

Fig. 9. Comparison of training time (in hrs). A detailed explanation is given in Sect. 7.1

these architectural gaps, there is a need for generalised metrics that can thoroughly evaluate synthetic data quality. Further, the problem of mode-collapse and the high dynamic range of datasets need robust and stable GAN frameworks.

This paper aims to highlight the research gaps and design considerations of such generative models and motivate further research. It points out that even though these models are good enough for shorter sequences and moderate dimensionality, the research gap on modelling long sequences in high dimensions remains open and can be explored in future research.

References

1. Esteban, C., Hyland, S.L., Rätsch, G.: Real-valued (medical) time series generation with recurrent conditional GANs. arXiv preprint arXiv:1706.02633 (2017)
2. Fawaz, H.I., et al.: Data augmentation using synthetic data for time series classification with deep residual networks. arXiv preprint arXiv:1808.02455 (2018)
3. Goodfellow, I., Bengio, Y., Courville, A.: Deep Learning. MIT Press, Cambridge (2016)
4. Iwana, B.K., Uchida, S.: An empirical survey of data augmentation for time series classification with neural networks. arXiv preprint arXiv:2007.15951 (2020)
5. Lin, Z., et al.: Using GANs for sharing networked time series data: challenges, initial promise, and open questions. In: Proceedings of the ACM Internet Measurement Conference (2020)
6. Lyu, X., et al.: Improving clinical predictions through unsupervised time series representation learning. arXiv preprint arXiv:1812.00490 (2018)
7. Mogren, O.: C-RNN-GAN: continuous recurrent neural networks with adversarial training. arXiv preprint arXiv:1611.09904 (2016)
8. Wen, Q., et al.: Time series data augmentation for deep learning: a survey. arXiv preprint arXiv:2002.12478 (2020)
9. Yoon, J., Jarrett, D., van der Schaar, M.: Time-series generative adversarial networks. In: Advances in Neural Information Processing Systems, pp. 5508–5518 (2019)
10. Wang, Z., She, Q., Ward, T.E.: Generative adversarial networks in computer vision: a survey and taxonomy. ACM Comput. Surv. (CSUR) **54**(2), 1–38 (2021)
11. Jabbar, A., Li, X., Omar, B.: A survey on generative adversarial networks: variants, applications, and training. arXiv preprint arXiv:2006.05132 (2020)

12. Wiatrak, M., Albrecht, S.V., Nystrom, A.: Stabilizing generative adversarial networks: a survey. arXiv preprint arXiv:1910.00927 (2019)
13. Collins, J., Sohl-Dickstein, J., Sussillo, D.: Capacity and trainability in recurrent neural networks. arXiv preprint arXiv:1611.09913 (2016)
14. Yu, L., et al.: SeqGAN: sequence generative adversarial nets with policy gradient. In: Proceedings of the AAAI Conference on Artificial Intelligence, vol. 31, no. 1 (2017)
15. Donahue, C., McAuley, J., Puckette, M.: Adversarial audio synthesis. arXiv preprint arXiv:1802.04208 (2018)
16. Arora, S., Zhang, Y.: Do GANs actually learn the distribution? An empirical study. arXiv preprint arXiv:1706.08224 (2017)
17. Abadi, M., et al.: Deep learning with differential privacy. In: Proceedings of the 2016 ACM SIGSAC Conference on Computer and Communications Security, pp. 308–318. ACM (2016)
18. Arjovsky, M., Chintala, S., Bottou, L.: Wasserstein GAN. arXiv preprint arXiv:1701.07875 (2017)
19. Choi, E., Biswal, S., Malin, B., Duke, J., Stewart, W.F., Sun, J.: Generating multi-label discrete patient records using generative adversarial networks. arXiv preprint arXiv:1703.06490 (2017)
20. Mirza, M., Osindero, S.: Conditional generative adversarial nets. arXiv preprint arXiv:1411.1784 (2014)
21. Goodfellow, I.: NIPS 2016 tutorial: generative adversarial networks. arXiv preprint arXiv:1701.00160 (2016)
22. Goodfellow, I., et al.: Generative adversarial nets. In: Advances in Neural Information Processing Systems, pp. 2672–2680 (2014)
23. Gulrajani, I., Ahmed, F., Arjovsky, M., Dumoulin, V., Courville, A.C.: Improved training of Wasserstein GANs. In: Advances in Neural Information Processing Systems, pp. 5767–5777 (2017)
24. Cai, T.T., Ren, Z., Zhou, H.H., et al.: Estimating structured high-dimensional covariance and precision matrices: optimal rates and adaptive estimation. Electron. J. Stat. **10**(1), 1–59 (2016)
25. Graves, A.: Generating sequences with recurrent neural networks. arXiv preprint arXiv:1308.0850 (2013)
26. Sutskever, I., Vinyals, O., Le, Q.V.: Sequence to sequence learning with neural networks. In: Advances in Neural Information Processing Systems (2014)

IoT, Edge, and Mobile for Embedded Machine Learning

Workshop on IoT, Edge, and Mobile
for Embedded Machine Learning (ITEM 2021)

The second Workshop on IoT, Edge, and Mobile for Embedded Machine Learning (ITEM 2021) was held in conjunction with the European Conference on Machine Learning and Principles and Practice of Knowledge Discovery in Databases (ECML PKDD 2021) on September 17, 2021. Due to the COVID-19 pandemic, the conference and its workshops were held fully virtually. The previous edition of the workshop was also held jointly with ECML PKDD in 2020.

Local and embedded machine learning (ML) is a key component for real-time data analytics in upcoming computing environments like the Internet of Things (IoT), edge computing, and mobile ubiquitous systems. The goal of the ITEM workshop is to bring together experts, researchers, and practitioners from all relevant communities, including ML, hardware design and embedded systems, IoT, edge, and ubiquitous/mobile computing.

There is an increasing need for real-time intelligent data analytics, driven by a world of big data, and society's need for pervasive intelligent devices, such as wearables for health and recreational purposes, smart city infrastructure, e-commerce, Industry 4.0, and autonomous robots. Most applications share facts like large data volumes, real-time requirements, limited resources including processor, memory, network, and, possibly, battery life. Data might be large but possibly incomplete and uncertain. Notably, often powerful cloud services can be unavailable, or not an option due to latency or privacy constraints. For these tasks, local and embedded ML is among the most promising approaches to address learning and reasoning under uncertainty. Examples include image and speech processing, such as image recognition, segmentation, object localization, multi-channel speech enhancement, speech recognition, and signal processing such as radar signal denoising, with applications as broad as robotics, medicine, autonomous navigation, recommender systems, etc.

The call for papers included the following non-exclusive list of topics:

- Compression of neural networks for inference deployment
- Learning on edge devices, including federated and continuous learning
- Trading among prediction quality, efficiency of representation, and computational efficiency
- Automatic code generation from high-level descriptions, including linear algebra and stencil codes, targeting existing and future instruction set extensions
- Tool-driven optimizations from ML model level down to instruction level, automatically adapted to the current hardware requirements
- Understanding the difficulties and opportunities using common ML frameworks with marginally supported devices
- Exploring new ML models designed for use on designated device hardware
- Future emerging processors and technologies for use in resource-constrained environments, e.g. RISC V, embedded FPGAs, or analogue technologies
- Applications and experiences from deployed use cases using embedded ML

- Energy efficiency of ML models created with distinct optimization techniques
- Security/privacy of embedded ML
- New benchmarks suited to edge and embedded devices

The workshop papers were selected through a peer-reviewed process in which each submitted paper was assigned to several members of the Program Committee. The main selection criteria were the novelty and technical quality of the approach. Our Program Committee chair, Günther Schindler from Heidelberg University, was responsible for managing the review and selection process. Eight papers were accepted for presentation. The program included two keynotes:

- "Machine Learning + Logic Synthesis For Fun and Profit" by Sat Chatterjee from Google
- "Composable domain specific compiler abstractions for ML and beyond" by Jacques Pienaar, also from Google

More information about the workshop can be found on the workshop website: https://www.item-workshop.org.

Many people contributed to making this workshop a successful event. We would like to thank our keynote speakers for their inspiring talks in the early hours of their day, our Program Committee members and additional reviewers for their detailed and constructive reviews, our Program Committee chair for his tireless work to get the best possible program selected, our authors for their well-prepared presentations, all workshop attendees for their engagement and participation, and of course the team of ECML PKDD for accepting us at the conference.

November 2021

Gregor Schiele
Franz Pernkopf
Michaela Blott
Holger Fröning

Organization

ITEM 2021 Chairs

Gregor Schiele   University of Duisburg-Essen, Germany
Franz Pernkopf   Graz University of Technology, Austria
Michaela Blott   Xilinx Research, Ireland
Holger Fröning   Heidelberg University, Germany

Program Committee

Günther Schindler   Heidelberg University, Germany
  (Program Co-chair)
Costas Bekas   Citadel Securites, Switzerland
Herman Engelbrecht   Stellenbosch University
Giulio Gambardella   Xilinx, Ireland
Tobias Golling   Geneva University, Switzerland
Domenik Helms   OFFIS e.V. - Institut für Informatik, Germany
David King   Air Force Institute of Technology, USA
Benjamin Klenk   NVIDIA Research, USA
Manfred Mücke   Materials Center Leoben Forschung GmbH,
                Germany
Marco Platzner   Paderborn University, Germany
Thomas B. Preußer   ETH Zurich, Switzerland
Wei Shao   Royal Melbourne Institute of Technology,
                Australia
Yannik Stradmann   Heidelberg University, Germany
Jürgen Teich   Friedrich-Alexander-Universität
                Erlangen-Nürnberg, Germany
Nicolas Weber   NEC Labs Europe, Germany

Towards Precomputed 1D-Convolutional Layers for Embedded FPGAs

Lukas Einhaus[✉], Chao Qian, Christopher Ringhofer, and Gregor Schiele

University Duisburg-Essen, Duisburg, Germany
{lukas.einhaus,chao.qian,christopher.ringhofer,gregor.schiele}@uni-due.de

Abstract. We present a new type of 1D-convolutional block allowing us to precompute large parts of a 1D-CNN. The block combines quantization with depthwise-separable convolutions to reduce the overhead for precomputation, making the approach feasible. We present two proof of concept architectures and evaluate them on a Xilinx Spartan-7 S15 low power embedded FPGA. This way we are able to detect atrial fibrillation from 42 s ECG samples. Classifying a sample takes us 0.052 ms, while consuming 0.004 mJ of energy. The networks achieve a classification accuracy of 82.37% and 94.22% respectively. The implementations do not use any block RAM or DSP slices.

Keywords: Embedded machine learning · FPGA · Quantized neural networks

1 Introduction

In this paper we present initial results for a new optimization approach for 1D-convolutional neural networks (1D-CNNs) on embedded Field Programmable Gate Arrays (FPGAs). 1D-CNNs are often used to process time series signals e.g. to detect damage to building structures, analyze medical data or monitor complex machines [7]. Offloading them to the Cloud is often not an option in these scenarios, e.g. due to communication latency, privacy concerns, or unreliable network connections. Therefore, they must be executed in-situ, i.e. on the embedded device, meeting strict requirements regarding execution speed and energy consumption.

We use an embedded low-power FPGA as our embedded device [10]. FPGAs are widely used to accelerate deep learning and AI applications, due to their ability to instantiate highly efficient and tailor-made circuit designs. It allows to take all the knowledge about the network architecture or even a specific model into account when applying optimizations. Because of their reconfigurability, developers can update models to adapt to changing requirements or to incorporate new techniques and optimizations after system deployment.

To realize local 1D-CNNs on FPGAs, many modern approaches focus on the implementation of optimized multiply-accumulate operations to reduce the cost of computations [3,12,14].

© Springer Nature Switzerland AG 2021
M. Kamp et al. (Eds.): ECML PKDD 2021 Workshops, CCIS 1524, pp. 327–338, 2021.
https://doi.org/10.1007/978-3-030-93736-2_25

However, this still leads to high costs due to memory accesses since weight matrices have to be read from block RAM or external memory [12]. In contrast, the core of our approach is to avoid costly computations by performing most of the necessary calculations before deployment and storing the results in truth tables. To avoid memory accesses, including block RAM, we convert these truth tables directly to look-up tables (LUTs). These LUTs are one of the essential core building blocks on the FPGA. In addition to potentially extremely short inference times and low energy cost per inference, this approach also enables us to work with full resolution weights during training and inference even for networks where activations are quantized aggressively.

But this comes at the cost of an increased resource utilization. Without further optimizations the size of the needed lookup tables – and therefore the number of used LUTs – grows very quickly. For our approach to be useful for relevant problems, it is essential to reduce the table growth and keep the table sizes small enough to be deployable on small low-power FPGAs.

We will present a method to keep these sizes low enough to make the method feasible and present a proof of concept architecture to back our result. For our proof or concept, we use a 1D-CNN for atrial fibrillation detection and deploy it on a Xilinx Spartan-7 S15 FPGA.

The remainder of this paper is structured as follows. First, we discuss related work in Sect. 2. Then, we present our current approach in Sect. 3. We evaluate it in Sect. 4 before we conclude the paper in Sect. 5.

2   Related Work

Before we present our own approach for optimizing neural networks on FPGAs in more detail, we first discuss closely related optimization approaches.

A 1D-convolution basically moves a filter over an input vector with a specific stride, i.e. step size, until the input vector is exhausted. A dot product is calculated for each step. With multiple input channels and filters, each of these steps can be considered a matrix multiplication. On an FPGA these matrix multiplications are typically implemented as a set of multiply-accumulate operations, in short MAC. One well known approach is to use binary neural networks (BNN) and to implement the MAC operations on the model level with XNOR-Popcount operations in hardware [2,3,9]. Although this leads to faster calculations and less memory usage, weights have to be binarized, contrary to our approach using full resolution weights.

In [14] the authors present their LUTNet approach to compensate that. They replace the weighted sum of an input vector by a sum over arbitrary boolean functions each consuming a subset of the input vector's coefficients. Instead of calculating the dot product by using XNOR followed by a Popcount operation

$$\mathbf{w} \cdot \mathbf{x} = \sum_{n=1}^{N} w_n x_n \tag{1}$$

for a weight vector $\mathbf{w}$ and an input vector $\mathbf{x}$, they find a set of functions $(g_n)_{\{1,\ldots,\tilde{N}\}}$ that behaves like the dot product in the sense that

$$\mathbf{w} \cdot \mathbf{x} = \sum_{n=1}^{\tilde{N}} g_n \left((x_{n_1}, \ldots, x_{n_K}) \right) \tag{2}$$

The approach develops its real strength when combined with network pruning. The final goal is to obtain $\tilde{N} \ll N$ with only small accuracy loss. The authors present a three-stage network construction process of training a conventional BNN, pruning that network and finally replacing the dot products by the functions g_n. Each of the two latter steps is followed by a retraining phase. The idea is that by using higher arity g_n the network regains most of the accuracy that was lost during pruning. This is possible due to the increased network capacity when using g_n instead of XNOR-Popcount on (w_n, x_n). To find the g_n the authors solve Eq. 2 for each g_n, wherein g_n has been expanded to a Lagrange interpolating polynomial. The resulting expressions for g_n are then used during the last retraining phase. This approach still uses online calculations and block RAM. We precompute operations and do not use block RAM.

Most closely related to our approach, in [13] the authors present an approach called LogicNets. They also precompute operations but focus on dense networks instead of 1D-CNNs. Each neuron is represented by one LUT with six inputs and one output (6:1 LUT). Thus, each neuron is precomputed individually and the mapping to LUTs is straightforward. This however imposes a restriction to the network topology since a neuron can never have more than six input bits. To enforce this, the authors apply random sparsity and use binary activations.

In contrast, our approach precomputes a whole layer and maps it to several LUTs. Thus, estimating the resulting resource consumption (i.e. the number of LUTs) is more difficult in our case. On the plus side, we are more flexible with regards to network topology and quantization, since we do not need to map to LUTs directly in a one-to-one fashion.

3   Our Approach: Precomputed 1D-CNNs

After discussing related approaches, we now present our proposed approach in more detail. First, we give a brief overview and describe the optimization goal. Then, we describe our two optimization steps, namely (i) quantizing activations and (ii) splitting convolutions.

Our approach is based on constructing a specialized neural network model for a specific application use case. This tailored – and pretrained – model is then deployed on the FPGA. The model is structured in multiple consecutive layers, e.g. 1D-convolutions and fully connected layers, that realize a problem-specific 1D-CNN. In this paper we focus mainly on 1D-convolutions but we believe that our techniques can be applied to other layer types as well. Note, that we do not train our neural networks on the FPGA but restrict ourselves to inference.

We realize each 1D-convolutional layer as a distinct hardware block that can be wired to preceding and following layers. We consider convolutions as arbitrary logical functions that we represent as precomputed truth tables that hold all input-output pairs necessary to fully define the corresponding function. As such, each 1D-convolution is represented by its own unique truth table that is then implemented in the LUTs of the FPGA.

Following this idea however poses a major challenge: the size of the truth tables depends directly on the codomain of the convolution. We assume a codomain consisting of vectors that are comprised of coefficients with a bit-width of b. Each of these vectors is built from coefficients. We obtain a table with $M = 2^{b \cdot n}$ entries where n denotes the kernel size of the corresponding convolution. Taking into account that a 1D-convolution typically has to process several input channels c the truth table will have

$$M = f \cdot 2^{b \cdot n \cdot c} \tag{3}$$

entries, where f denotes the number of convolutional filters. Note that for an output bit-width of b_{out} we need b_{out} such truth tables. For demonstration purposes we will continue to refer to an example 1D-convolutional block. This block consists of the sequence (Conv1D(3, 5, 4)-Batchnorm1D-Activation), where the corresponding convolution has a kernel size of $n = 3$, $c = 5$ input channels and $f = 4$ filters. For 16bit fixed-point decimal numbers this convolutional block will already result in $M = 2^{16 \cdot 5 \cdot 3} \approx 7.07 \cdot 10^{72}$ table entries. It is obvious that this number needs to be drastically reduced to make the approach feasible.

One way to do so would be to compress the truth tables directly with a compression algorithm. That however induces the overhead of decompressing at runtime and in case of a lossy compression could introduce errors into the network that have not been taken into account during training.

Therefore, we decided to design the network topology in such a way that it already reduces the value of M. This has the advantage that errors introduced by the optimizing techniques can already be taken into account during training. Additionally, the network designer knows the final accuracy that the system will yield on hardware before deploying it. From Eq. 3 we can see that we have different ways to achieve smaller table sizes. We can

1. reduce one or multiple of the involved variables, e.g. bit-width, number of input channels, kernel size, or number of filters.
2. try to remove as many of them as possible from the exponent in the formula by changing the internal structure of our 1D-convolutional block.

We use a combination of both of these possibilities as described in the following.

3.1 Quantizing Activations

Our first optimization focuses on reducing variables. Kernel size, number of filters and input channels typically are quite essential to a network's architecture and depend on the size and number of features that the layer is supposed to detect as well as the spatial or temporal resolution. Instead, we focus on reducing the bit-width.

This leads to a quantization scheme that should quantize the activations, i.e. the input values to our convolutions aggressively. Note that we do not need to quantize weights since they are not stored during deployment due to our precomputations.

We use a binarization as shown in [2] and two-level residual binarizations [3]. Using just one bit we can reduce the truth table size of our example convolution with $n = 3$, $c = 5$ and $f = 4$ compared to full resolution activations approx. by the factor $\approx 0.33 \times 10^5$ to 131072. To realize the training with quantized activations, we replace the derivative of the sign function used for the (residual) binarization with $\mathbf{1}_{\leq 1}$ (sometimes also denoted rect_1) [2]. This prevents the gradient propagated backwards to be cancelled.

However, the techniques in the following sections will allow our use case scenario to be implemented with a ternarization to achieve higher accuracy.

$$\pi \colon \mathbb{R} \longrightarrow \{-1, 0, 1\}, \quad x \longmapsto \begin{cases} 0 & \text{if } |x| \leq 0.5 \\ 1 & \text{else} \end{cases} \tag{4}$$

Analogue to the binarization we replace the gradient during the backward pass with $\mathbf{1}_{\leq 1}$. Using a batch normalization layer with a trainable scaling factor [6] in front of every ternarization, we can learn effectively during training which intervals should be mapped to zero, therefore we can use a fixed threshold.

3.2 Splitting Convolutions

In addition to quantizing activations, we also want to remove as many of the involved variables from the exponent in Eq. 3. To do so, we can change the internal structure of the 1D-convolution block, e.g. by splitting the convolution into several consecutive ones. This is the basic idea of depthwise separable convolutions [5,11].

Mathematically speaking, depthwise separable convolutions are convolutions that are representable by two consecutive matrix multiplications. This leads to applying two consecutive MAC operations, which can be represented by an expression of the following form:

$$\sum_{i=1}^{f} \mathbf{e}_i \sum_{j=1}^{c} w'_{ij} \sum_{k=1}^{n} x_{jk} w_{jk} \tag{5}$$

For each channel j the sum over the n inputs x_{jk} weighted with w_{jk} is calculated. Each of the resulting numbers is then weighted with an additional set of weights w'_{ij} per filter i. Finally this is multiplied with the unit vector $\mathbf{e}_i$ for each filter to form an output vector with one coefficient per filter. That means that this convolution can also be subdivided into two steps: (i) depthwise and

(ii) pointwise convolution. The depthwise convolution is applied to each channel separately, in our case using one filter per channel. Then the pointwise convolution is applied to the result of (i) using a kernel size of one with an arbitrary number of filters. This technique has been shown to produce good results with respect to model accuracy and resource consumption [1,5]. It is used especially in Edge and embedded contexts to reduce the number of weights and computations of a model. Depthwise separable convolutions, however, also allow us to decouple the different input channels from each other and calculate a separate truth table for each of them. This way we can consider each channel separately for our truth tables instead of having to combine the input across all channels in one huge truth table. Figure 1 visualizes that principle. The dotted and solid lines represent a different channel respectively. For the *Convolution* block on the left side (a), we need to consider all inputs across all channels for the single big truth table. For the *Split Convolution* on the right side (b), we can instead obtain an intermediate result for each of the input channels separately using one truth table per channel. The two intermediate results are then used to find the final result using another truth table. Thus combining several smaller truth tables.

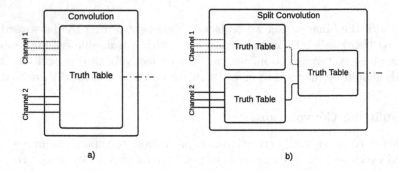

Fig. 1. Realizing the convolution via truth tables.

Generalizing that principle, we can describe the new consecutively applied MAC operations with the help of the composition of $c + 1$ arbitrary functions

$$\sum_{i=1}^{f} e_i t'_{ic}(t_{n1}(x_{11}, \ldots, x_{1n}), \ldots, t_{nc}(x_{c1}, \ldots, x_{cn})) \tag{6}$$

where each of the t_{nj} takes n parameters and each of the t'_{ic} takes c parameters. Therefore we can realize the resulting block with a total of

$$f \cdot 2^{b \cdot c} + c \cdot 2^{b \cdot n} \tag{7}$$

table entries. This corresponds to a reduction in size by the factor

$$2^{b \cdot c(1-n)} + \frac{c}{f} \cdot 2^{b \cdot n(1-c)}. \tag{8}$$

We combine it with quantizing activations by adding a quantization after each individual convolution. For our example with $n = 3$, $c = 5$, $f = 4$ the new resulting set of tables only make up 0.12% of the original table's size for the binary case.

Note that we can choose to implement arbitrary functions as long as they are representable by the form given in Eq. 6. This especially applies to the batch normalization that we use before every quantization. This is possible since the batch normalization operation does not recombine any of the inputs, i.e. it operates element wise.

4 Evaluation

In this section we present an end-to-end implementation of a 1D-CNN as a proof of concept. All convolutions in the network except the first are precomputed and the network detects atrial fibrillation from ECG signals with an accuracy of 94%. First, we introduce the problem and data set. Then we compare two versions of our network, both using quantized activations and convolution splitting: one with binary activations, another one with ternary activations.

Our evaluation task is the detection of atrial fibrillation on ECG sample data. We derive our dataset from the MIT-BIH arrhythmia database that is publicly available [4,8]. The ECG samples contain two channels that have been recorded 250 Hz. Additionally, each sample is provided with metadata describing which sections are showing atrial fibrillation or sinus rhythm respectively. To obtain our training, validation and test data sets, we divide 16 records of 10h into 42 s examples, that are labeled as either AF (atrial fibrillation) or N (normal sinus rhythm). The resulting set was split into 9278 training samples and 2285 samples for validation and testing respectively. Additionally we down-sampled the data 125 Hz by simply picking only every second sample point, since we experienced that this is enough to already yield good results. Figure 2 shows three example windows of ECG signals for each of these classes.

In our evaluation we compare two different example architectures, both using our optimization techniques. The only difference between them is that the first one (split_bin) uses binary activations. The second one (split_ter) uses ternary activations. Both use convolution splitting and precomputation with full resolution weights. The base architecture has been hand-crafted for the application problem, independently of precomputation. The structure of both architectures is presented in Table 1.

Note that every convolution is followed implicitly by a quantizing activation function. The first convolution binConv processes the raw 16bit fixed point numbers that the ECG signal consists of. Consequently this layer is not subject to precomputation, but will instead remain to be dynamically calculated during runtime using binary weights. By using a stride of 3, we reduce the length of the output this layer produces by a factor of 3. That way, less data needs to be shifted through the remainder of the network and the dense layer in the end will be correspondingly smaller. After that, we perform a maxPool operation with a stride

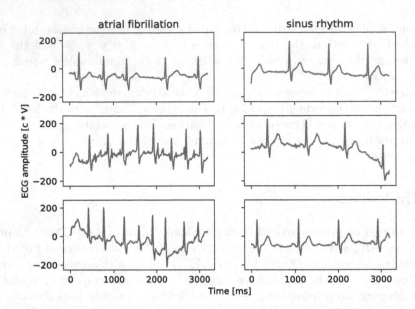

Fig. 2. The six plots show six 3.2 s excerpts of channel I of an ECG signal. The left side shows atrial fibrillation while the right side shows a normal sinus rhythm.

smaller than the pooling window size, i.e. the pooling regions overlap. That way, less temporal information is lost compared to disjoint pooling regions, because it will produce more variation in the output compared to using a stride equal to the pooling window size of 3. It is followed by two of our precomputed *splitConv* 1D-convolutional layers. Then, we do another *maxPool*, another *splitConv* and two consecutive *maxPool*. Finally, we flatten the output and apply a fully

Table 1. The 1D-CNN architecture used for atrial fibrillation detection with precomputed convolutions

Layer	Stride	Kernel size k	Channels c (in)	Filters f (out)
binConv	3	10	1	6
maxpool	6	8	6	6
splitConv	1	6	6	5
splitConv	1	6	5	5
maxpool	2	3	5	5
splitConv	1	6	5	6
maxpool	4	4	6	6
maxpool	3	3	6	6
flatten	–	–	6	1
binDense	–	–	–	–

Table 2. Detailed view of the first *splitConv* block.

Layer	Stride	Kernel size k	Channels c (in)	Filters f (out)
quantize	–	–	6	6
batchNorm	–	–	6	6
depthwiseConv	1	6	6	6
batchNorm	–	–	6	6
quantize	–	–	6	6
pointwiseConv	1	1	6	5
batchNorm	–	–	5	5

binarized dense layer *binDense* to it. The dense layer is calculated dynamically using XNOR-Popcount. For the activation preceding the dense layer we chose a two-bit residual binarization [3]. The method finds a binary encoded approximation of the input by recursively applying the sign functions to its residuals. This procedure effectively doubles the number of output channels since every incoming integer is encoded as a two-dimensional bit-vector. Table 2 shows a detailed view of the first *splitConv* block. In this block as well as for the rest of the architecture, each quantizing function is preceded by a batch normalization layer. While enabling the trainable scaling factor for the batch normalization layers [6] does not have any impact when using binary activations [2], it is essential for our ternarizations with fixed thresholds.

We implemented both versions of our architecture and deployed them on a Xilinx Spartan-7 S15 FPGA. We operated the FPGA with 10 MHz (although our model could operate with up to 200 MHz) to reduce the power consumption. We trained with the Adamax optimizer using initial learning rates between 0.0043 and 0.0045, using a batch size of 512. Furthermore we chose binary cross-entropy with logits for our loss function. We trained each architecture 24 times over 400 epochs.

Table 3 shows the resulting test accuracy, the total number of table entries, the total LUT utilization on the FPGA, as well as the energy consumption and latency per sample. Note that no block RAM or DSP slices are used.

Our binary network *split_bin* achieves an accuracy of approx. 82% with a latency of 0.052 ms and a corresponding energy consumption of 0.004 mJ per sample. The ternary version *split_ter* improves the accuracy noticeably to 94%.

Table 3. Evaluation results on a Xilinx Spartan-7 S15 FPGA. The table shows the mean and standard deviation of the accuracy across all experiment runs.

Version	Accuracy	Table entries	LUT Utiliz.	Latency	Energy Consum.
split_bin	$82.37 \pm 3.19\%$	1600	12.33%	0.052 ms	0.004 mJ
split_ter	$94.22 \pm 1.59\%$	92160	29.39%	0.052 ms	0.004 mJ

Table 4. Results for architectures without convolution splitting. The table shows the mean and standard deviation of the accuracy across all experiment runs.

Version	Accuracy	Table entries
non_split_bin	$96.72 \pm 0.74\%$	3.52×10^{11}
non_split_ter	$97.94 \pm 0.67\%$	2.36×10^{22}

Interestingly, it retains the same latency. This is because both use mainly precomputed components that can achieve up to 200 MHz, way more than the 10 MHz dictated by the speed of the flash chip the data was read from during the experiments. To understand better the achieved accuracy, we conducted additional experiments with binary and ternary activations and standard 1D-convolutions instead of our *splitConv* layers. Weights still used full resolution. The results of these experiments are shown in Table 4. We kept the hyperparameters from the previous experiments and again performed 24 training runs. Comparing both tables we observe an interesting difference with respect to the quantization scheme. Using standard 1D-convolutions with binary activations results in an accuracy of approx. 97%. For ternary activations the accuracy increases only very slightly to approx. 98%. Apparently, for binary activations our convolution splitting impacts the accuracy a lot. For ternary activations this effect – while still present – is much smaller. We believe that this is because the ternary case has a neutral element with respect to addition.

Both of our precomputed networks fit comfortably on our FPGA, which features a total of 8000 LUTs. The sum of all table entries that are needed for *split_bin* is just 1600, leading to it using only 12% of the available LUTs. For *split_ter* we get a total number of table entries of 92160, 57.6 times more. However, the LUT utilization increases only to approx. 29%, a factor of around 2.4. This is likely due to a number of different reasons. First, not all LUTs are used for representing the truth tables. In fact, for *split_ter* only about 8% of them are. The rest is used for connecting the layers as well as for the first and last layers, which together make up approx. 80% of the consumed LUTs.

Since for our ternary network we do not use all numbers that are representable with 2 bit, only a small margin of table entries needs to be explicitly precomputed and defined. In fact building input vectors with only three possible coefficients $(-1, 0, 1)$ will yield a total of only 16767 table entries for the whole network. Therefore 81.8% of all table entries can be arbitrarily chosen, allowing the synthesizing tool (in our case Vivado) to apply a large set of optimization strategies.

Comparing the number of table entries for *non_split_bin* and *split_bin* as well as for *non_split_ter* and *split_ter*, we can conclude that convolution splitting reduces the size of the precalculated truth tables by a factor of 0.87×10^7 in the binary case (*split_bin*) and by a factor of 0.26×10^{18} for the ternary case (*split_ter*).

5   Conclusion and Future Work

Precomputing operations in neural networks and mapping them directly to LUTs on FPGAs is an exciting new approach to construct fast and energy efficient deep learning models. In this paper we showed our current approach to use this idea for 1D-CNNs. We use a combination of quantization for binary or ternary activations with splitting convolutions based on depthwise separable convolutions. We applied this to a 1D-CNN for detecting atrial fibrillation in ECG signals and were able to implement a model that achieved a competitive accuracy of 94.22% while using only 29.39% of the available LUTs on a small Xilinx Spartan-7 S15. In future work we plan to apply our approach to other use cases, e.g. for audio processing, and to other layer types, e.g. 2D-convolutions. We also want to examine further techniques to reduce the truth table sizes, e.g. by pooling over input channels. Additionally we think that the accuracy of the networks can be improved by further exploiting the fact that our approach allows for the implementation of arbitrary boolean functions instead of being restricted to convolutions. Finally, we plan to use the remaining LUTs on the FPGA to instantiate parallel convolutions, speeding up inference even more.

Acknowledgements. The authors acknowledge the financial support by the Federal Ministry of Education and Research of Germany in the KI-Sprung LUTNet project (project number 16ES1125).

References

1. Chollet, F.: Xception: deep learning with depthwise separable convolutions (2017)
2. Courbariaux, M., Hubara, I., Soudry, D., El-Yaniv, R., Bengio, Y.: Binarized neural networks: training deep neural networks with weights and activations constrained to +1 or −1 (2016)
3. Ghasemzadeh, M., Samragh, M., Koushanfar, F.: ReBNet: residual binarized neural network. arXiv:1711.01243 (2018)
4. Goldberger, A.L., et al.: PhysioBank, PhysioToolkit, and PhysioNet: components of a new research resource for complex physiologic signals. Circulation **101**(23), e215–e220 (2000)
5. Howard, A.G., et al.: MobileNets: efficient convolutional neural networks for mobile vision applications. arXiv:1704.04861 (2017)
6. Ioffe, S., Szegedy, C.: Batch normalization: accelerating deep network training by reducing internal covariate shift. CoRR, abs/1502.03167 (2015). http://arxiv.org/abs/1502.03167
7. Kiranyaz, S., Avci, O., Abdeljaber, O., Ince, T., Gabbouj, M., Inman, D.J.: 1D convolutional neural networks and applications: a survey. arXiv:1905.03554 (2019)
8. Moody, G.B., Mark, R.G.: The impact of the MIT-BIH arrhythmia database. IEEE Eng. Med. Biol. Mag. **20**(3), 45–50 (2001). https://doi.org/10.1109/51.932724
9. Rastegari, M., Ordonez, V., Redmon, J., Farhadi, A.: XNOR-Net: ImageNet classification using binary convolutional neural networks. arXiv:1603.05279 (2016)
10. Schiele, G., Burger, A., Cichiwskyj, C.: The elastic node: an experimentation platform for hardware accelerator research in the Internet of Things. In: 2019 IEEE International Conference on Autonomic Computing (ICAC), pp. 84–94 (2019). https://doi.org/10.1109/ICAC.2019.00020

11. Sifre, L.: Rigid-motion scattering for image classification. Ph.D. thesis, Ecole Polytechnique, CMAP, October 2014
12. Umuroglu, Y., et al.: FINN. In: Proceedings of the 2017 ACM/SIGDA International Symposium on Field-Programmable Gate Arrays, February 2017. https://doi.org/10.1145/3020078.3021744. http://dx.doi.org/10.1145/3020078.3021744
13. Umuroglu, Y., Akhauri, Y., Fraser, N.J., Blott, M.: LogicNets: co-designed neural networks and circuits for extreme-throughput applications. In: 2020 30th International Conference on Field-Programmable Logic and Applications (FPL), pp. 291–297. IEEE (2020)
14. Wang, E., Davis, J.J., Cheung, P.Y.K., Constantinides, G.A.: LUTNet: learning FPGA configurations for highly efficient neural network inference. arXiv: 1910.12625 (2020)

Embedded Face Recognition for Personalized Services in the Assistive Robotics

Iris Walter[1]([ID]), Jonas Ney[2], Tim Hotfilter[1], Vladimir Rybalkin[2], Julian Hoefer[1], Norbert Wehn[2], and Jürgen Becker[1]

[1] Karlsruhe Institute of Technology, Karlsruhe, Germany
{iris.walter,hotfilter,julian.hoefer,becker}@kit.edu
[2] Technical University of Kaiserslautern, Kaiserslautern, Germany
{ney,rybalkin,wehn}@eit.uni-kl.de

Abstract. Recently, the field of assistive robotics has drawn much attention in the health care sector. In combination with modern machine learning-supported person recognition systems, they can deliver highly personalized services. However, common algorithms for person recognition such as convolutional neural networks (CNNs) consume high amounts of power and show low energy efficiency when executed on general-purpose computing platforms.

In this paper, we present our hardware architecture and field programmable gate array (FPGA) accelerator to enable on-device person recognition in the context of assistive robotics. Therefore, we optimize a neural network based on the SqueezeNet topology and implement it on an FPGA for a high degree of flexibility and reconfigurability. By pruning redundant filters and quantization of weights and activations, we are able to find a well-fitting neural network that achieves a high identification accuracy of 84%. On a *Xilinx Zynq Ultra96v2*, we achieve a power consumption of 4.8 W, a latency of 31 ms and an efficiency of 6.738 FPS/W. Our results outperform the latency by 1.6x compared to recent person recognition systems in assistive robots and energy efficiency by 1.7x for embedded face recognition, respectively.

Keywords: Ambient assisted living · Assistive robotics · Convolutional neural networks · Face recognition · Field programmable gate array · Quantization

1 Introduction

Assistive robotics promise major enhancements in the health and elderly care sector. In an ambient assisted living (AAL) where several people live at the same place, assistive services include, for example, reminder and help with medication, support in rehabilitation, calling for help in emergency cases or needs-based help with daily living. For this purpose, assistive robots must be able to understand

© Springer Nature Switzerland AG 2021
M. Kamp et al. (Eds.): ECML PKDD 2021 Workshops, CCIS 1524, pp. 339–350, 2021.
https://doi.org/10.1007/978-3-030-93736-2_26

their environment, assess interaction possibilities with persons in their vicinity and adapt to their personal needs. While movements and environment perception are still big challenges to overcome, the tremendous improvements of deep neural networks (DNNs) in recent years might help to address them. The emergence of convolutional neural networks (CNNs), which take spatial information into account and hence show very good performance in image processing, vision and perception, confirms this trend further. Their performance and prediction results even surpass human vision as well as traditional computer vision algorithms, for instance in cancer detection [3]. Although Machine Learning (ML) algorithms are considered enablers for those challenging use cases, they come with a very high computational complexity, typically resulting in high power consumption when implemented on general purpose processors. In addition, those throughput oriented architectures fail to meet the low latency requirements. Thus, many currently available off the shelf solutions such as central processing units (CPUs) or graphics processing units (GPUs) cannot satisfy the requirements for embedded applications.

For this reason, a shift towards dedicated DNN hardware accelerators can be observed. With dedicated accelerators, co-design in accordance with the algorithm is conceivable [9]. For instance, optimization strategies like quantization [11] or pruning [8] can improve neural network inference performance 100-fold and energy efficiency 1000-fold compared to a CPU baseline implementation. Quantization aims at a reduction of the bit precision of data and weights, thus it reduces the computational complexity. Pruned neural network topologies can reduce the total amount of operations through smart bypass techniques of superfluous operations during execution or by removing redundant connections in the network beforehand, respectively.

While hardware accelerators for DNNs can achieve vast improvements in the inference, dedicated hardware structures usually lack flexibility. The trade-off between accuracy and energy efficiency cannot be fully exploited when the number format and bitwidth are fixed. Accelerators on Field programmable gate arrays (FPGAs) [2] on the other hand, are highly customizable and offer great potential for optimization to balance numerical precision and desired accuracy for a specific use case.

In this paper, we present an FPGA-based accelerator for face recognition focusing on privacy and low power consumption to enable personalized services in assistive robotics. In summary, our contributions are threefold:

- For the application in the field of personalized assistive robotics, we choose and derive a hardware-aware CNN topology. This model emphasizes low power for mobile deployment and low latency for high user acceptance.
- We implement the network with an HLS-based tool-flow on a *Xilinx* FPGA and compare different configurations in terms of accuracy, resource usage and energy efficiency.
- We select and evaluate the optimal configuration regarding the trade-off between resource utilization and accuracy. Based on this model, we demonstrate low power person recognition with a latency of 31 ms and an energy effi-

ciency of 6.7 FPS/W, outperforming recent work by 1.6x regarding latency [5] and 1.7x regarding energy efficiency [32].

2   Related Work

2.1   Personalization in Assistive Robotics

AAL methods, targeting elderly and other people who are dependent on care, support and help people in daily life. Exemplary, they can improve the safety of a person or assist rehabilitation. Personalization of the provided service as well as continuous adaption to the constantly changing needs of the patient are identified as key factors for user acceptance and satisfaction with an assistive robot [20].

Ghiţă et al. [7] present a people detection and recognition of a *Pepper* robot to show reminders to people in an AAL. Locally captured images of the environment of the robot are transmitted to a remote machine where people are recognized using FaceNet [22]. Their person recognition takes 300 ms.

Duque et al. [5] implemented a gaze control for a robot head to improve interaction with persons. To determine the person to look at, they combine face recognition, pose recognition and speech detection. For face recognition, they use a ResNet with 29 convolutional layers inferencing in less than 0.05 s on a *GPU GeForce RTX 2080 Ti*.

2.2   Face Recognition

Face recognition tasks have been widely studied for several decades. In general, the common algorithms for face recognition can be divided into three different categories: holistic methods, local methods and deep learning (DL) methods. Most holistic methods are based on Eigenfaces, i.e., a set of eigenvectors representing images in a lower dimensional space, and were first proposed by Sirovich and Kirby [23]. Local or feature-based methods for face recognition aim to find similar local features from the face or from special regions of the face. As a result, they are more robust to variations like pose, viewpoint and expression, compared to holistic methods [31]. DL methods are based on DNNs achieving state-of-the-art results in various image processing tasks, and have also drawn attention in the field of face recognition in recent years. By making use of a large set of training samples, DNNs are able to learn how and which features to extract from an input image to increase recognition accuracy significantly compared to holistic and local methods. The first breakthroughs of DL-based algorithms for face recognition were achieved by DeepFace [25] and Deep ID [24] in 2014. Since then, the state-of-the-art performance has been improved drastically, mainly driven by the advancements made in DL, for instance, Google's FaceNet accomplished an accuracy of over 99% on the labeled faces in the wild (LFW) benchmark. Overall, the accuracy has been increased from around 90% for the classical methods, up to around 99.8% [18,21] in just three years [27].

2.3   Low-Complexity Neural Networks

In recent years, one main focus of DNN research was increasing accuracy without considering the size and complexity of the network. This becomes a problem, especially in embedded devices where both latency and power requirements need to be fulfilled. Thus, there is an increasing demand for neural network architectures that are suitable for low power embedded devices.

One of the first presented DNNs optimized for the embedded domain is SqueezeNet, introduced in 2016 [13]. It achieves an accuracy on the same level as AlexNet [15] for the ImageNet [4] dataset with 50 times less trainable parameters. SqueezeNet consists of 18 convolutional layers with 1.25 M parameters. The main building block of the network is the fire module that consists of squeeze layers with 1×1 convolutional kernels and expand layers with a mix of 1×1 and 3×3 kernels.

In 2017, Howard et al. presented MobileNets which are CNNs for mobile vision application, especially focusing on low latency and low power applications [10]. The architecture is based on depthwise separable convolutions to reduce computational complexity while increasing the model depth. MobileNets accomplish a top-1 accuracy of 70.6% for ImageNet with only 4.2 M trainable parameters.

Shortly later, in 2018, ShuffleNet was proposed [30], which is an extremely efficient neural network designed for mobile devices. It makes use of pointwise group convolution and channel shuffle operations to reduce computation cost while maintaining accuracy. A ShuffleNet unit is used to enable information flow across feature channels by shuffling input data of different groups. As a result, ShuffleNet obtains a three percentage points higher top-1 accuracy than MobileNet on ImageNet while, at the same time, achieving a 13x speedup compared to AlexNet. However, ShuffleNet is composed of 50 layers with 5.4 M parameters and therefore requiring 4.3x more memory space than SqueezeNet.

2.4   FPGA Implementations of Face Recognition

Various FPGA implementations of DNN-based face recognition systems have been proposed recently.

Zhunge et al. implemented a DNN for face recognition based on FaceNet using FaceNet using high-level synthesis (HLS) on a *Xilinx Virtex Ultrascale+* device [32]. They explored multiple fast convolution algorithms to find an optimal strategy to apply them to the different types of convolutions of the model. Their FPGA accelerator accomplishes a single face recognition in 23.7 ms with a power consumption of 10.6 W. As a result, they outperform the *Nvidia GTX 1080 GPU* with respect to latency by a factor of 3.75 and achieve an energy efficiency of 3.981 FPS/W.

In 2021, Liu et al. proposed a heterogeneous computing system with a hardware accelerator based on MobileNet that is used as face tracking platform [17]. The FPGA accelerator is integrated into a CPU and GPU based system that is aided by a delay-aware, energy-efficient scheduling algorithm. Their design, implemented on the *Intel Stratix 10*, achieves a latency of only 1.3 ms per image. They claim to provide a low power solution, although power consumption never is reported but only the power efficiency of their system. Derived from this value and the complexity of MobileNet, the power consumption can be estimated to be around 26 W.

In summary, state-of-the-art face recognition implementations based on DNNs achieve latencies that are sufficient for real-time applications. However, none of the works provide solutions that are suitable for battery-powered embedded devices with respect to power consumption, while still fulfilling real-time constraints. They have either very high power consumption [17] or provide only moderate latency [32]. Thus, we focus on a DNN-based face recognition implementation that is real-time suitable as well as consuming low power for the CASIA-Webface dataset [28].

2.5 FINN

To reduce development time and to guarantee compatibility to other designs, the hardware implementation presented in this work is based on the open-source hardware library [16] which is part of Xilinx FINN framework [26]. FINN is a tool to explore the design space of DNN inference accelerators on FPGAs. The DNN layers of the FINN hardware library are designed as dataflow architecture, resulting in high-throughput and low latency implementations. Furthermore, the framework provides an adjustable degree of parallelization for each layer, which can be used to find an optimal trade-off between resource usage and latency. Additionally, FINN uses fixed-point format that allows for a higher level of flexibility and enables customizable quantization.

3   Context and Methodology

In a typical AAL, several people live at the same place, hence an assistive robot needs a real-time scene perception to locate and interact with persons in a room. To perform personalized assistive services, the robot must first and foremost identify the people in its workspace. Recent work already proposed face recognition techniques for assistive robots, but lack either privacy due to remote computation [7] or adaption to low power requirements of embedded systems [5].

We use the assistive robot *ARMAR-6* [1] which is equipped with one depth camera and two stereo cameras for peripheral and foveal vision, respectively. The camera images are forwarded to a face detection unit, e.g. realized with an MTCNN [29] that is a commonly used CNN to detect bounding boxes of objects or faces. A typical face detection with an MTCNN on a GPU takes 122 ms [6]. Afterwards the detected faces are forwarded to a face recognition unit.

In this paper, we focus on the real-time face recognition and its efficient implementation on the robot with low power consumption to address limited power supply in the embedded, battery powered application. We implement the accelerator entirely on an FPGA board to enable on-device execution of face recognition and thereby enhance privacy. This platform furthermore facilitates updates and allows reconfiguration during runtime to supply other algorithms that provide the assistive services from the FPGA.

3.1 Model Topology

As DL methods achieve very good accuracy results on the task of face recognition, our approach is based on a DNN. Selecting the DNN topology is a crucial design decision for embedded face recognition, as it determines the initial architecture and performance optimization potential of the later hardware implementation. SqueezeNet [13], MobileNet [10] and ShuffleNet [30] are good candidates for embedded face recognition, since they provide a relatively small computational complexity as well as sufficient accuracy on the ImageNet dataset. As SqueezeNet has the smallest memory footprint with 1.25 M parameters and has no residual paths or skip connections that increase implementation complexity, this topology suites excellent for the hardware implementation of face recognition in assistive robotics. In addition, it has a high reduction potential which was shown in the original paper [13].

Starting from the original SqueezeNet v1.1 topology [12], we apply different optimizations: First, we reduce the number of filters of the last convolution to 256 instead of 1,000. This decreases the output dimensionality and still enables face recognition in environments with a relatively small group of persons for instance in the aforementioned application. For regularization and to avoid overfitting, we insert a dropout layer before the last convolution. Additionally, we replace the final softmax layer with a convolutional layer with 128 filters and train with triplet loss [22] to learn a multi-dimensional embedding for faces. We implement this variant of SqueezeNet in PyTorch. To reduce the model size and computational complexity, we apply an unconditional filter pruning by factor two that is proved to be as effective as structured filter pruning [19], i.e., half the number of filters compared to the original SqueezeNet v1.1 is applied per convolutional layer. This pruning quarters the model size with a negligible accuracy degrada-

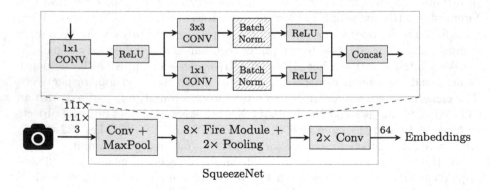

Fig. 1. Overview of our SqueezeNet, changes are highlighted by a dashed pattern.

tion of less than one percentage point for face recognition on the CASIA-Webface dataset [28].

Moreover, we insert batch normalization layers after each expand convolutional layer of the fire module to stabilize training, see Fig. 1. During inference, we fold the batch normalization parameters into the previous convolutional layer as proposed in [14] to reduce computational complexity further.

3.2 Hardware Architecture

We use the hardware library of FINN [16] to implement an HLS model of SqueezeNet. For synthesis and deployment of the HLS model on the FPGA board, we use *Xilinx Vivado (HLS)*. To avoid offloading intermediate feature maps between the layers, we implement the entire GNN as dataflow architecture on the chip. A dataflow architecture can be considered as task-level pipelining, i.e., the results of one layer are forwarded to the next before the computation of the entire layer has finished. Therefore, the subsequent layer starts processing as soon as the data for the first calculation step is available and the need for buffering intermediate results between the layers decreases.

The total inference latency of the CNN is determined by the first convolutional layer which we observed being the bottleneck layer and thus has to be maximally unrolled. We unroll the other layers by the minimum factor that does not increase this latency. Additionally, we reduce computational complexity, hence the resource utilization, by quantization of the weights and activations of the SqueezeNet with fixed point datatypes instead of floating point. To determine an appropriated quantization scheme which covers the whole range of values, we profile the weight and activation floating point values of our trained model and choose minimal integer bit widths for each layer. As the fraction bit width influences accuracy, it is part of the design space exploration (DSE). We further add a user-level driver for communication between the hardware and software on the board.

4   Evaluation and Discussion

4.1   Experimental Setup and Training

Initially, we train our CNN with 32-bit floating point on the CASIA-Webface dataset [28]. $2,115$ of $10,575$ identities are assigned to the test dataset and the images of the remaining $8,460$ identities are split by 80% into training data and by 20% into validation data. The learning rate is set to 0.05, dropout to 0.5, momentum to 0.9, batch size to 400 and our pruned SqueezeNet is trained for 500 epochs. To evaluate the identification accuracy in the presented use case, we create a small test dataset from the unseen test data including ten identities with ten images each. We calculate a reference embedding for each identity corresponding to the mean of the embedding of five different images. For identification of a person, an image passes the SqueezeNet and the euclidean

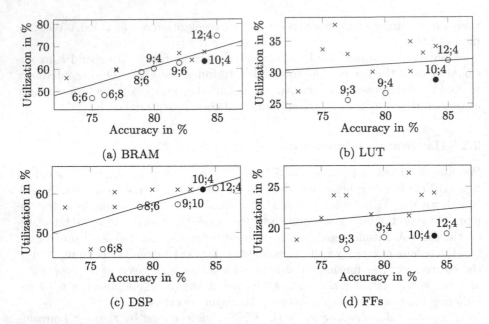

Fig. 2. Utilization of BRAMs, LUTs, DSPs and Flip-Flops of the *ZCU102* in relation to the accuracy. Pareto optimal configurations are shown as circles and are labeled with the quantization step: The first and latter number represents the fraction bit width of weights and activations, respectively. A trend line indicates the correlation between primitive utilization and accuracy.

distance to the reference embeddings of each identity is determined. The identity related to the closest reference embedding is recognized by the network. On our test dataset, the software model achieves 89.6% accuracy with full precision.

After batch normalization folding, our CNN has 222, 940 parameters and 17.47 M MAC operations are executed to process one image. Compared to the original SqueezeNet v1.1, the computational complexity of our variant decreases by factor 20 and the model size by factor 5.

The *Xilinx ZCU102* with large FPGA resources allows for a wide range of variation in the optimization parameters of the accelerator. This is why we choose the *ZCU102* as target platform for the DSE. For later deployment of a low power accelerator for face recognition, we choose the well-fitting *Xilinx Ultra96v2* which provides four times less resources and lower static power consumption.

4.2   Design Space Exploration

For embedded face recognition, we analyze the influence of different quantization schemes on the trade-off between accuracy and resource utilization which directly affects energy efficiency. Therefore, we apply individual quantization schemes for each layer, i.e., we use minimal integer bit widths as described in Subsect. 3.2 and apply different fixed fraction bit widths for weights and activations. For

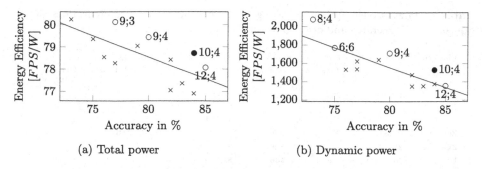

(a) Total power   (b) Dynamic power

Fig. 3. Total and dynamic energy efficiency on the *ZCU102* in relation to the accuracy. Pareto optimal configurations are shown as circles and are labeled with the quantization level. A trend line indicates the correlation between the efficiency and the accuracy.

example, we denote 10;4 as the quantization scheme with ten fraction bits for all weights and four fraction bits for all activation values.

The design is unrolled to the degree which accomplishes a minimal latency of 27, 225 *cycles*. For this configuration, we report the accuracy, resource utilization and energy efficiency for different quantization schemes on the *ZCU102*, see Fig. 2 and Fig. 3. For less precision in the number representation of weights and activations, less resources are required, energy efficiency increases and the identification accuracy decreases. As it is shown in Fig. 2a, the BRAM utilization dominates the total resource utilization and is effected most by the varying quantization scheme. Hence, BRAMs are the limiting resource which determine feasibility of the accelerator on a certain FPGA.

The configuration 10;4 with ten weight fraction bits and four activation fraction bits is a Pareto-optimal design with respect to all hardware costs, which achieves a moderate accuracy drop of five percentage points compared to the software model with full precision and without batch normalization folding. This model requires 63.4% of the BRAM resources of the *ZCU102* and is selected for further considerations.

For power evaluation, we first measure static power of the FPGA board with a digital multimeter and then the total power consumed to process face recognition. Dynamic power consumption is reported as the difference between these measurements. At a clock frequency of 300 MHz, the accelerator executes a single face recognition in 0.481 ms and draws 26.423 W in total. As a result, the accelerator achieves a total energy efficiency of 78.713 FPS/W, see Table 1. Compared to [17], our unrolled accelerator on a *ZCU102* is 2.7 times faster and more energy efficient, respectively. If the HLS model is implemented as a full sequential architecture, the model requires 19.1% of the BRAM resources of a *ZCU102* and realizes a single face recognition in 25.338 ms at the same clock frequency. Therefore, the total energy efficiency of the rolled model is 1.667 FPS/W on a *ZCU102*.

For very high throughput, the unrolled implementation might be preferred. However, on the *ZCU102*, we could show that the real-time constraints of a robot

Table 1. Power consumption and energy efficiency of different configurations of our SqueezeNet accelerator and other embedded face recognition.

Platform	Runtime [ms]	Power consumption		Throughput [FPS]	Energy efficiency	
		Total [W]	Dynamic [W]		Total [FPS/W]	Dynamic [FPS/W]
Stratix 10 [17]	1.310	26	–	763.5	29.586	–
VU9P [32]	23.7	10.6	–	42.194	3.981	–
ZCU102[a]	**0.481**	26.423	1.360	**2,079.824**	**78.713**	**1,529.282**
ZCU102[b]	25.338	23.679	0.279	39.467	1.667	141.457
Ultra96v2[c]	30.712	**4.832**	**0.236**	32.560	6.738	137.967

[a]unrolled at 300 MHz, [b]rolled at 300 MHz, [c]rolled at 250 MHz

scene perception can also be satisfied with the rolled model. As this model has a smaller resource footprint, we intend the implementation on a smaller FPGA to reduce power and energy consumption further. An *Ultra96v2* has four times less BRAMs than a *ZCU102* which suites well for the implementation of the rolled model. We synthesize the rolled model for the *Ultra96v2* at a maximum clock frequency of 250 MHz resulting in a BRAM resource utilization of 80.8%. On the *Ultra96v2*, our implementation inferences in 30.71 ms which is 1.6 times faster than the face recognition of [5]. Moreover, we measure a dynamic power consumption of 0.236 W and a total power consumption of 4.832 W which is five times lower than on the *ZCU102* and less than half the power consumption of [32]. The total energy efficiency on the *Ultra96v2* is four times better than on the *ZCU102* and 1.7 times better than in [32], respectively, see Table 1.

Considering 122 ms of processing time for the face detection unit [6], the entire person identification is realized within 152 ms using the *Ultra96v2* for low power face recognition. This is in the range of human reaction time and thus natural interaction with the *ARMAR-6* is enabled.

5   Conclusion and Future Work

In this paper, we present an embedded low power face recognition, which enables real-time person recognition for assistive robotics. Due to on-device execution, personalized services can be provided at the edge and privacy is assured. Therefore, we optimize a CNN topology for the design of an FPGA accelerator and analyze the performance-accuracy trade-off on a *Xilinx* FPGA depending on the quantization level. Our implementation on an *Ultra96v2* accomplishes a single face recognition in 31 ms consuming only 4.8 W. We achieve 6.738 FPS/W energy efficiency which outperforms the state-of-the-art 1.7-fold. In the future, we plan to implement the whole processing chain of scene perception, including face detection and the execution of personalized assistive services.

References

1. Asfour, T., et al.: ARMAR-6: a high-performance humanoid for human-robot collaboration in real-world scenarios. IEEE Robot. Autom. Mag. **26**, 108–121 (2019)
2. Baehr, S., et al.: Low latency neural networks using heterogenous resources on FPGA for the Belle II trigger. CoRR abs/1910.13679 (2019)
3. Brinker, T.J., et al.: Deep learning outperformed 136 of 157 dermatologists in a head-to-head dermoscopic melanoma image classification task. Eur. J. Cancer **113**, 47–54 (2019)
4. Deng, J., Dong, W., Socher, R., Li, L.J., Li, K., Fei-Fei, L.: ImageNet: a large-scale hierarchical image database. In: IEEE Conference on Computer Vision and Pattern Recognition (2009)
5. Duque-Domingo, J., Gómez-García-Bermejo, J., Zalama, E.: Gaze control of a robotic head for realistic interaction with humans. Front. Neurorobot. **14**, 34 (2020)
6. Esler, T.: Face recognition using PyTorch. https://github.com/timesler/facenet-pytorch. Accessed 06 July 2021
7. Ghiţă, ŞA., Barbu, M.Ş, Gavril, A., Trăscău, M., Sorici, A., Florea, A.M.: User detection, tracking and recognition in robot assistive care scenarios. In: Giuliani, M., Assaf, T., Giannaccini, M.E. (eds.) TAROS 2018. LNCS (LNAI), vol. 10965, pp. 271–283. Springer, Cham (2018). https://doi.org/10.1007/978-3-319-96728-8_23
8. Han, S., Pool, J., Tran, J., Dally, W.J.: Learning both weights and connections for efficient neural networks. CoRR abs/1506.02626 (2015)
9. Hotfilter, T., Kempf, F., Becker, J., Reinhardt, D., Baili, I.: Embedded image processing the European way: a new platform for the future automotive market. In: World Forum on Internet of Things, pp. 1–6 (2020)
10. Howard, A.G., et al.: MobileNets: efficient convolutional neural networks for mobile vision applications. CoRR abs/1704.04861 (2017)
11. Hubara, I., Courbariaux, M., Soudry, D., El-Yaniv, R., Bengio, Y.: Quantized neural networks: training neural networks with low precision weights and activations. CoRR abs/1609.07061 (2016)
12. Iandola, F.N.: Github forresti/squeezenet/squeezenet_v1.1/ (2016). https://github.com/forresti/SqueezeNet/tree/master/SqueezeNet_v1.1. Accessed 01 Feb 2021
13. Iandola, F.N., Moskewicz, M.W., Ashraf, K., Han, S., Dally, W.J., Keutzer, K.: SqueezeNet: AlexNet-level accuracy with 50x fewer parameters and < 1MB model size. CoRR abs/1602.07360 (2016)
14. Jacob, B., et al.: Quantization and training of neural networks for efficient integer-arithmetic-only inference. In: IEEE Conference on Computer Vision and Pattern Recognition (2018)
15. Krizhevsky, A., Sutskever, I., Hinton, G.E.: ImageNet classification with deep convolutional neural networks. In: Pereira, F., Burges, C.J.C., Bottou, L., Weinberger, K.Q. (eds.) Advances in Neural Information Processing Systems, vol. 25. Curran Associates, Inc. (2012)
16. Labs, X.R.: FINN HLS library. https://github.com/Xilinx/finn-hlslib. Accessed 22 Feb 2021
17. Liu, X., et al.: Collaborative edge computing with FPGA-based CNN accelerators for energy-efficient and time-aware face tracking system. IEEE Trans. Comput. Soc. Syst., 1–15 (2021). https://ieeexplore.ieee.org/document/9363321
18. Liu, Y., Li, H., Wang, X.: Rethinking feature discrimination and polymerization for large-scale recognition. CoRR abs/1710.00870 (2017)

19. Liu, Z., Sun, M., Zhou, T., Huang, G., Darrell, T.: Rethinking the value of network pruning. In: International Conference on Learning Representations (2018)
20. Matarić, M.J., Scassellati, B.: Socially assistive robotics. In: Siciliano, B., Khatib, O. (eds.) Springer Handbook of Robotics, pp. 1973–1994. Springer, Cham (2016). https://doi.org/10.1007/978-3-319-32552-1_73
21. Ranjan, R., Castillo, C.D., Chellappa, R.: L2-constrained softmax loss for discriminative face verification. CoRR abs/1703.09507 (2017)
22. Schroff, F., Kalenichenko, D., Philbin, J.: FaceNet: a unified embedding for face recognition and clustering. In: IEEE Conference on Computer Vision and Pattern Recognition (2015)
23. Sirovich, L., Kirby, M.: Low-dimensional procedure for the characterization of human faces. J. Opt. Soc. Am. A **4**(3), 519–524 (1987)
24. Sun, Y., Wang, X., Tang, X.: Deep learning face representation from predicting 10,000 classes. In: IEEE Conference on Computer Vision and Pattern Recognition (2014)
25. Taigman, Y., Yang, M., Ranzato, M., Wolf, L.: DeepFace: closing the gap to human-level performance in face verification. In: IEEE Conference on Computer Vision and Pattern Recognition, pp. 1701–1708 (2014)
26. Umuroglu, Y., et al.: FINN: a framework for fast, scalable binarized neural network inference. In: ACM/SIGDA International Symposium on Field-Programmable Gate Arrays, pp. 65–74. ACM (2017)
27. Wang, M., Deng, W.: Deep face recognition: a survey. CoRR abs/1804.06655 (2018)
28. Yi, D., Lei, Z., Liao, S., Li, S.Z.: Learning face representation from scratch. CoRR abs/1411.7923 (2014)
29. Zhang, K., Zhang, Z., Li, Z., Qiao, Y.: Joint face detection and alignment using multitask cascaded convolutional networks. IEEE Signal Process. Lett. **23**(10), 1499–1503 (2016)
30. Zhang, X., Zhou, X., Lin, M., Sun, J.: ShuffleNet: an extremely efficient convolutional neural network for mobile devices. CoRR abs/1707.01083 (2017)
31. Zhou, H., Mian, A., Wei, L., Creighton, D., Hossny, M., Nahavandi, S.: Recent advances on singlemodal and multimodal face recognition: a survey. IEEE Trans. Hum.-Mach. Syst. **44**(6), 701–716 (2014)
32. Zhuge, C., Liu, X., Zhang, X., Gummadi, S., Xiong, J., Chen, D.: Face recognition with hybrid efficient convolution algorithms on FPGAs. CoRR abs/1803.09004 (2018)

FLight: FPGA Acceleration
of Lightweight DNN Model Inference
in Industrial Analytics

Hassan Ghasemzadeh Mohammadi[1(✉)], Felix Paul Jentzsch[1],
Maurice Kuschel[1], Rahil Arshad[1], Sneha Rautmare[1], Suraj Manjunatha[1],
Marco Platzner[1], Alexander Boschmann[2], and Dirk Schollbach[3]

[1] Computer Science Department, Paderborn University, Paderborn, Germany
hassan.ghasemzadeh.mohammadi@uni-paderborn.de
[2] Weidmüller Interface GmbH & Co. KG, Detmold, Germany
Alexander.Boschmann@weidmueller.com
[3] Weidmüller Monitoring Systems GmbH, Dresden, Germany
Dirk.Schollbach@weidmueller.com

Abstract. Deep Neural Networks (DNNs) have been recently in the
focus of interest owing to their high-quality results in various application
domains of big data analytics. The lack of end-to-end toolchains
that can automatically optimize and translate a DNN algorithm, usually
implemented in high-level languages, on FPGA-based platform hinders
software developers to exploit such platforms. A few available toolchains,
namely Xilinx DPU, have been optimized for computational expensive
applications, e.g., real-time image and video processing, and therefore
they introduce large performance overhead on lightweight applications,
e.g., sensor-driven decision analytics. In this paper, we introduce the
FLight framework to fully automatize the acceleration of lightweight
DNN algorithms on FPGA-based embedded platforms. The framework
takes a trained model in TensorFlow/Keras as input and generates an
optimized synthesizable C++ version and maps it on the target platform
using HLS tools. FLight is an easy-to-use framework and does not require
deep knowledge of embedded system design since it automatically performs all the required mapping steps. FLight supports the acceleration
of various deep learning algorithms, e.g., Feed-Forward Neural networks
(FFNNs), Convolutional Neural Networks (CNNs), and sequential models like Recurrent Neural Networks (RNNs). We evaluated the applicability of FLight with various DNN models on both academic and industrial datasets. As a case study, we exploited the FLight framework for
wind turbine condition monitoring using an industrial dataset provided
by *Weidmüller Monitoring Systems GmbH*. The experimental results
revealed the 40 times speedup compared to the Xilinx DPU framework.

Keywords: Deep learning · Accelerator · Embedded systems ·
FPGA · Industrial analytics · Condition monitoring · Wind turbine ·
Approximate computing

© Springer Nature Switzerland AG 2021
M. Kamp et al. (Eds.): ECML PKDD 2021 Workshops, CCIS 1524, pp. 351–362, 2021.
https://doi.org/10.1007/978-3-030-93736-2_27

1   Introduction

In recent years, Deep Neural Networks (DNNs) have been exploited as a versatile and potent tool in various application domains, such as human behavior analytics, computer vision, condition monitoring, smart control systems, and system security, owing to their proven capability as universal approximators of functions that originate from unknown non-linear systems. With the growing popularity of edge computing and availability of large amounts of sensor data, predictive DNN models are also employed in *industrial analytics* for tasks such as performance monitoring, anomaly detection, and predictive maintenance. Deploying DNN models in industrial analytics settings is challenging since the embedded compute platforms used exhibit rather constrained memory and computing resources. The training of DNNs is usually performed offline on high-performance GPU-based computing platforms, but the inference needs to run on the embedded compute platform. While capturing of sensor data, signal conditioning, pre-preprocessing, and feature extraction can often be done on embedded CPUs, the DNN-based prediction is computationally much heavier and can benefit from hardware acceleration.

Mapping DNN models efficiently to FPGAs poses a problem to industrial analytics experts. There exists a lack of comprehensive and automated frameworks to accelerate various DNN models with limited or no knowledge of embedded system and FPGA design. DNN algorithms are typically realized in high-level languages, e.g., Python, with the help of libraries such as TensorFlow, Keras, Pytorch, or Caffe. Although High-Level Synthesis (HLS) tools intend to mitigate the hardware design challenge by synthesizing C/C++ based designs into Hardware Description Language (HDL) designs and, ultimately, to FPGA configuration bitstreams, they are still far from being satisfying solutions due to the huge number of tuning knobs and compiler directives that create significant obstacles for the automated synthesis and optimization of such accelerators.

Most of the available frameworks for accelerating DNN inference on FPGAs focus on computationally expensive models and algorithms mainly used for real-time computer vision and video analytics. These frameworks support many features suited to the required huge bandwidth of image/video data processing and provide the possibility of large convolution operations, batch normalization, and large, fully-connected layer networks. For instance, the Xilinx Deep Learning Processor Unit (DPU) employs a configurable IP core and parallelizes the given input DNN according to the available resources and required computations. DPU is a programmable computation engine controlled by an embedded ARM processor.

In this paper, we introduce the FLight framework to efficiently accelerate *lightweight* DNN model inference on platform FPGAs. We focus on industrial analytics applications where input data comes from a set of sensors that provide typically 1-dimensional data at a sampling rate of a few kHz at most. While the resulting DNN models still can be computationally demanding and ask for hardware acceleration over an embedded CPU, the implementations are less throughput and memory bandwidth oriented than, for example, in real-time computer vision and video analytics.

FLight introduces a TensorFlow/Keras-to-FPGA (TFPGA) engine that analyzes and fine-tunes the given input DNN and generates a synthesizable C++ implementation, that is then fed into Xilinx SDSoC to create the bitstream needed for the FPGA and the associated software components. FLight can map a broad range of DNN models, i.e., Feed-Forward Neural Networks (FFNN), Convolutional Neural Networks (CNNs), and sequential DNNs such as Recurrent Neural Networks (RNNs), and is easily expandable due to its modular structure. FLight can employ quantization of weights and activations to reduce hardware requirements, but requires no special training or re-training of the given DNN. As a result, the given network models can be accelerated easily based on their original specification.

The remainder of this paper is organized as follows: Sect. 2 briefly reviews the most related works. Section 3 details the FLight framework and the TFPGA engine. In Sect. 4, we present condition monitoring for wind turbines as an industrial case study. Section 5 discusses experimental results and, finally, Sect. 6 concludes the paper.

2   Related Work

In recent years, there has been a significant effort by the research community and industry to implement the inference phase of DNNs on FPGA-based embedded platforms. Since FPGAs offer a flexible design space via reconfigurable hardware resources, and high bandwidth on-chip memory, they are preferable to be used as the hardware platform of the deep learning algorithms. In the following, we briefly review some recently proposed techniques to implement DNNs on FPGAs.

One category of research starts from a Register Transfer Level (RTL) description of the given DNN and directly synthesizes it into an FPGA [9,13]. Despite their satisfactory quality and performance, they require deep knowledge of hardware design and embedded platforms. Moreover, converting the given trained model, usually developed with high-level languages, i.e., python, into an equivalent RTL description is still cumbersome and very time-consuming even for experienced designers. Last but not least, analyzing the target model to optimize the model parameters for hardware implementation, i.e., quantization of data elements and operations, is not straightforward.

Another category consists of HW/SW decomposition of the target DNN, and benefits a hardware accelerator/or architecture controlled by software running on the embedded CPU [7,12,14–17]. This category benefits offloading the details of hardware design overheads, i.e., accelerator architecture, communication among hardware components, and development of hardware drives. Here, the main benefit is the reduction off-chip memory accesses. However, the model optimization is somehow limited and the designer cannot perform a fine-grain optimization on different parts of the DNN model. Possible optimizations are usually restricted to a few supported parameter optimizations such as homogeneous bit-width scaling, i.e., 8-bit and/or 16-bit. The other drawback is that most of the available platforms in this category have been optimized for computationally expensive tasks, i.e., real-time image and video processing, and they are not appropriate for accelerating lightweight DNN models owing to their considerable performance overhead.

3   FLight Framework

This section presents FLight by first giving an overview over the framework and then elaborating on its two core components, the repository of customizable HLS functions and the TensorFlow/Keras-to-FPGA (TFPA) engine. Finally, the architecture of the created hardware accelerators is sketched.

3.1   Overview

Figure 1 illustrates the FLight framework and its associated workflow. It starts with DNN model training in TensorFlow and Keras [2], which are among the most popular Machine Learning (ML) APIs and enable easy development and prototyping in Python. The target DNN model resulting from training is fed into TFPGA that, relying on a repository of customizable HLS functions for performance tuning on the FPGA, analyses the DNN model and outputs a corresponding target model in synthesizable C++ as well as the set of optimized weights and biases. Target model and optimized parameters are split in this phase since they are used independently to assemble the hardware accelerator (cmp. Subsect. 3.4). Both the model and the optimized parameters are exploited in Xilinx SDSoC to create configuration data for a Xilinx platform FPGA comprising the bitstream for reconfigurable logic and binaries for the embedded ARM CPU cores.

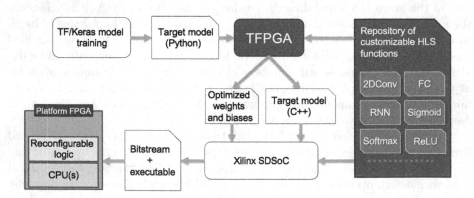

Fig. 1. Overview of the FLight framework

The structure of FLight facilitates the decoupling of the development of DNN models with the standard TensorFlow/Keras training process in Python and the mapping of the resulting models to a platform FPGA for inference. It was a design goal that the conversion of the model to a form that supports hardware acceleration should add no extra effort to the original model development, for example, for re-training. In particular, FLight is agnostic to any optimizations and techniques used during training and takes as inputs only the resulting network model and associated parameters.

3.2 Repository of Customizable HLS Functions

To ensure the decoupling of model training and accelerator synthesis, we have developed a repository of functions, components, and DNN layers in synthesizable C++ that have the same behavior as their TensorFlow counterparts. This engineering step required a detail analysis of the TensorFlow source code, in particular the behavior of the different DNN layers. FLight currently includes customizable HLS templates for implementing TensorFlow-compatible fully connected layers, two-dimensional convolution layers, and recurrent neural network layers. These three layers are utilized in the per-layer-accelerator architecture (cmp. Sect. 3.4) and allow us to build a huge variety of DNNs depending on the given input model. The HLS templates cover functionality for DNN inference, features related to model training such a back-propagation are not required.

Fully Connected Layer. The dense layer is the most basic and simple layer in DNN models. The TensorFlow implementation is also quite similar to the definitions in the literature. Flight follows the Tensorflow way of organizing the weights of the layers and storing them to get correct results after loading. The array has the shape $[NumInputs, NumNeurons]$ and we store it by iterating over the inputs and saving their weights to the neurons.

2D Convolution Layer. The two-dimensional convolutional layer is more complex, but still similar in structure to the dense layer. The most notable difference lies in the input buffering requirement. Because the convolution operates on a window that is sliding along the input image, it is not sufficient to read input pixels only once and in order. Instead, we use a sliding window buffer that stores the last $2 \times image\_width + kernel\_width$ pixels. This allows us to continue the practice of sequential input data access, which is very convenient for our data motion technique while minimizing the buffering requirements. The buffer is implemented as a LUT-based shift register and requires no costly BRAM cells. This approach very well-fitted with small dimension models. For larger models, BRAM cells are utilized for buffering purposes.

Recurrent Layer. The RNN layer is implemented in the same way as proposed in [3]. Instead of concatenating the recurrent result with the new input, both are handled independently. Therefore, the RNN layer consists of two independent dense layers whose results are summed up and then are added to the bias. Consequentially, here we have two kernels to care about, the kernel for the input with the shape $[NumInputs, NumStates]$ and the kernel for the recurrent state with the shape [*NumStates*, *NumStates*]. An RNN-specific layer argument, that can also be found in the TF implementation of the RNN layer, is the "return sequences" flag. That flag decides if the cell outputs values just after a complete sample or after every time step to produce a new sequence. In the Flight framework, we implemented the standard with just one output per sample, not every time step, to able to feed the output directly into the dense layers.

3.3 TFPGA Engine

Figure 2 shows the internal structure of the TFPGA engine that transforms a TensorFlow/Keras model to synthesizable C++ code. The process starts with a parsing step, in which the input model is split into a list of layers. TFPA then parses each layer and creates a layer dictionary that stores information such as a layer identifier, the type of the layer, the layer's parameters including number of neurons, dimensions of input and output, a string with the trained weights/biases, and the type of activation function. In the following model analysis step, we identify corresponding HLS functions and templates for the model layers and set the data types for the inputs, weights/biases, and outputs of each layer. We perform conversion from the Python floating-point data type to a fixed-point data format and subsequent quantization. Generally, the corresponding loss of information may result in reduced accuracy at the benefit of a faster and more resource-efficient hardware implementation. TFPGA uses `AP_fixed` as fixed-point data type with the total bit width as a hyper parameter, defaulting to 16 bits.

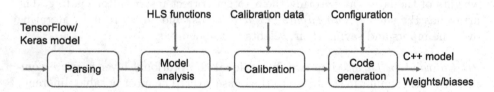

Fig. 2. Overview of the main steps in TFPGA engine.

In the calibration step, TPFGA defines the position of the binary points of the different fixed-point variables, i.e., determines the integer and fractional parts. Since the trained weights/biases are fixed, we can easily analyze their value range and determine corresponding integer bit widths. Typically, the data types for weights, biases, and layer inputs/outputs will differ. Finally, the coder generation step creates the DNN model in synthesizable C++ as well as the weights/biases.

3.4   Architecture of Hardware Accelerator

For our framework, we exploit heterogeneous streaming architecture. This brings us the benefits of mapping each neural network layer to a dedicated and customized accelerator. Using this architecture, we can easily generate I/O and memory interfaces for various components and create required control logic via SDSoC. This prevents the burden of understanding hardware details for mapping DNN layers on FPGA and enables us to automate the mapping process via scripting. Buffers among various layers are utilized to adjust the different latencies of the layers in a DNN model. Figure 3 for a high-level representation of our architecture.

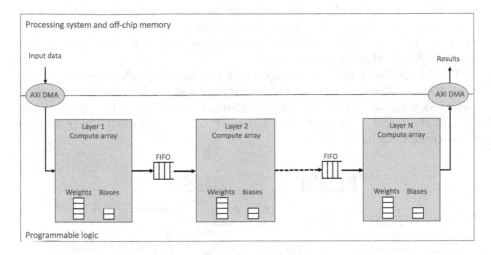

Fig. 3. Conceptual structure of a per-layer-accelerator architecture.

4   Industrial Analytics Using FLight: A Case Study

To demonstrate the applicability of the FLight framework, we have used it to build an end-to-end condition monitoring system for Wind Turbines in collaboration with *Weidmüller Monitoring Systems GmbH*[1] [1]. The system monitors the outcome of accelerometer sensors of a wind turbine and performs condition monitoring to identify faults on rotor blades. Fault detection using machine learning has received considerable interest for various wind turbine components including rotor blades, gearbox, shaft, generator, and the hydraulic system, e.g., [4]. While the majority of research in condition monitoring and fault detection for wind turbines focuses on gearbox and bearing subsystems [5,8,10], our focus is on automated fault detection for sensors of rotor blades [6].

In each blade, a pair of accelerometer sensors with the sampling frequency in a range of several kHz monitor the vibration and movements of the blades in two orthogonal directions, namely edge-wise, and flap-wise. The former captures blade vibrations in the direction of rotation and the latter scans oscillations in the reverse direction of airflow. The captured data is then processed and given to a machine learning model for fault detection.

---

[1] https://www.weidmueller.com/int/company/markets_industries/wind/index.jsp.

Figure 4 illustrates the main steps of the fault detection system for wind turbines as implemented by the FLight framework. The captured raw data from the sensors is processed into a cleanly formatted dataset that can be used later for fault detection purposes. The first preprocessing step downsamples the time signal via the Mode-Median-Bucket algorithm [11] to reduce the complexity of the sensor data. Then, the resulting signal is divided into several segments, so-called windows, in such a way that each window contains the same number of samples. Finally, the frequency spectrum for each window is calculated by a Fast Fourier Transform (FFT). Every window simply adopts the label of its data frame as *faulty* or *non-faulty*. Finally, the obtained spectra are reshaped and fed into an CNN to identify fault occurrences.

FLight has been utilized to implement the inference phase of the above-mentioned condition monitoring system. Preprocessing tasks, e.g., downsampling and windowing, as well as frequency spectrum computation are assigned to the CPU of the target platform FPGA and the CNN computations are performed in programmable hardware.

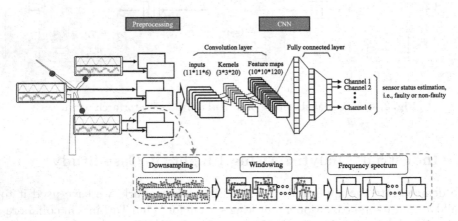

Fig. 4. Wind turbine sensor fault detection based on a CNN model. The input of each channel is reshaped into a 2D array of 11×11. The Convolution layer includes 20 kernels with the size of 3×3 and leads to a set of feature maps with the size of $10 \times 10 \times 120$.

5 Experimental Results

We chose the Xilinx ZCU102 development board as the hardware platform for our experiments. All experiment runs were automatically generated by FLight and tested against the included software model. The accelerators in this baseline implementation were not specially optimized by hand, nor do they support multithreading. However, we did use the highest level of compiler optimization available, which improved their performance significantly. We evaluated the applicability of the FLight framework with various DNN models on both academic and industrial datasets. For the former one, we utilized Fashion-MNIST provided by Zalando Research for the classification of grayscale images of clothing articles coming from 10 classes. For the latter one, we used a dataset from *Weidmüller Monitoring Systems GmbH* for fault detection in WTs.

We evaluated the applicability of the FLight for condition monitoring as explained in Sect. 4. The training and testing sets contain 4000 and 1025 samples provided by *Weidmüller Monitoring Systems GmbH*. Each sample represent a piece of time-series data with the length of 1 s labeled as *faulty* or *non-faulty* respectively. In the training set, we applied bootstrapping with replacement to be sure that the number of samples from both faulty and non-faulty classes is comparable. This is a very vital step later to train a high-quality classifier that provides high accuracy and recall on the testing set.

Figure 5 represents the accuracy for various quantization settings and for the reference software implementation which utilizes double-precision floating-point. The baseline technique, provided by *Weidmüller Monitoring Systems GmbH*, uses a Support Vector Machine (SVM), which results in an accuracy of 0.64 on average for all the input channels. For all of the six channels, the proposed CNN provides better classification results in comparison with SVM, making our model a successful technique to capture the most discriminative features for the sensor blade fault detection problem. The results show an accuracy penalty of 6% for a 16-bit quantization, which is acceptable without model retraining. When only the weights are quantized further to 8-bit, we even observe a slight increase in accuracy to 87%. We attribute this to the inherent regularization characteristic of the quantization. We also evaluated the FPGA resource usage for different quantization settings.

Figure 6 depicts the results generated with a parallelization factor of 32. The LUT savings of 12%, when going from 16- to 8-bit, are not very significant. This is expected though since most internal computations still operate on higher bit width registers. Using 8-bit weight quantization, we achieved a 34% decrease in BRAM usage, while a further quantization of activations does not affect, as our architecture stores activations only in small buffer registers and not in BRAM. These results show that weight quantization is a successful measure to reduce the memory footprint of a FLight-based accelerator. We evaluated the speed of the condition monitoring system for different degrees of parallelization of the CNN model. The parallelization factor here refers to the loop unrolling rate enabled by pragmas in the HLS tool. We reach a speedup of 90x, which equals about 52 μs per inference in absolute time, for the maximum parallelization factor of 100 (see Fig. 7).

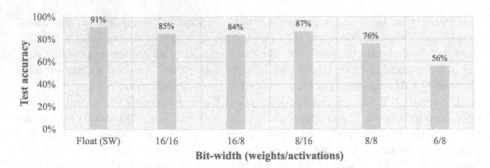

Fig. 5. Accuracy of condition monitoring system with various quantization in FLight framework. The numbers in X/Y format denote the bitwidth of weights (X) and activations (Y) respectively.

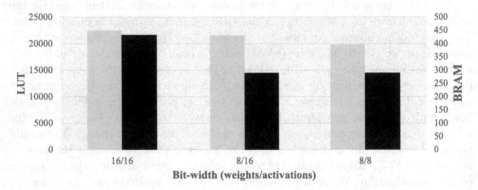

Fig. 6. Impact of quantization on the FPGA resources. The numbers in X/Y format denote the bitwidth of weights (X) and activations (Y) respectively.

We also compared our FLight-based implementation of the condition monitoring model with an existing off-the-shelf solution for inference acceleration. We selected the Xilinx Deep Learning Processor Unit (DPU) that offers an HW/SW architectural accelerator that is different from our approach. The accuracy of the DPU model, which is compressed to an 8-bit integer representation, experienced a drop similar to our experiments. 5% were lost, resulting in a 91% accuracy on hardware. Using 2.1 μs per inference, it is about 40x slower than the best FLight implementation.

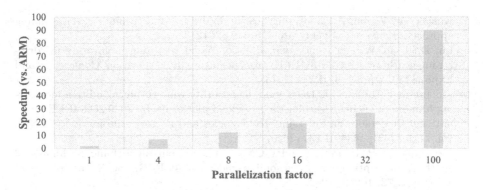

Fig. 7. Speedup via parallelization factor in comparison with SW implementation on ARM

6 Conclusion and Future Work

In this paper, we presented the FLight framework as a fully automated toolflow that accelerates the inference phase of lightweight DNN algorithms on FPGA-based embedded platforms. The main motivation for such a framework was the lack of end-to-end toolflow that can automatically optimize and translate a DNN algorithm, usually implemented in high-level languages, e.g., Python, on an FPGA-based platform. The framework takes a trained model in TensorFlow/Keras as input and generates an optimized synthesizable C++ version. It analyzes the parameters of the algorithms, i.e., weights and biases, for further compression, i.e., possible quantization, and maps it on the target platform using HLS tools. Without any requirement on hardware design knowledge, FLight supports the acceleration of various deep learning algorithms, e.g., Feed-Forward Neural networks (FFNNs), Convolutional Neural Networks (CNNs), and sequential models like Recurrent Neural Networks (RNNs). We evaluated the applicability of FLight with various DNN models on both academic and industrial datasets.

As a case study, we exploited the FLight framework for wind turbine condition monitoring using an industrial dataset provided by *Weidmüller Monitoring Systems GmbH*. The condition monitoring system identifies the fault occurrences on the rotor blades using a CNN-based model. The experimental results revealed the 40 times speedup compared to the Xilinx DPU framework.

References

1. BLADEcontrol condition monitoring system. Weidmüller Monitoring Systems GmbH. https://mdcop.weidmueller.com/mediadelivery/asset/900_87890
2. Keras: The Python Deep Learning API. https://keras.io
3. Goodfellow, I., Bengio, Y., Courville, A., Bengio, Y.: Deep Learning, vol. 1. MIT Press, Cambridge (2016)

4. Kusiak, A., Verma, A.: Analyzing bearing faults in wind turbines: a data-mining approach. Renewable Energy **48**, 110–116 (2012)
5. Lu, D., Qiao, W.: Frequency demodulation-aided condition monitoring for drivetrain gearboxes. In: 2013 IEEE Transportation Electrification Conference and Expo (ITEC), pp. 1–6. IEEE (2013)
6. Mohammadi, H.G., et al.: DeepWind: an accurate wind turbine condition monitoring framework via deep learning on embedded platforms. In: 2020 25th IEEE International Conference on Emerging Technologies and Factory Automation (ETFA), vol. 1, pp. 1431–1434. IEEE (2020)
7. Mousouliotis, P.G., Petrou, L.P.: CNN-grinder: from algorithmic to high-level synthesis descriptions of CNNs for low-end-low-cost FPGA SoCs. Microprocess. Microsyst. **73**, 102990 (2020)
8. Oyague, F., Butterfield, C., Sheng, S.: NREL gearbox reliability collaborative analysis round robin. Technical report, National Renewable Energy Lab. (NREL), Golden, CO, United States (2009)
9. Qiu, J., et al.: Going deeper with embedded FPGA platform for convolutional neural network. In: Proceedings of the 2016 ACM/SIGDA International Symposium on Field-Programmable Gate Arrays, pp. 26–35 (2016)
10. Schlechtingen, M., Santos, I.F., Achiche, S.: Wind turbine condition monitoring based on SCADA data using normal behavior models. Part 1: system description. Appl. Soft Comput. **13**(1), 259–270 (2013)
11. Steinarsson, S.: Downsampling time series for visual representation. Ph.D. thesis (2013)
12. Stone, J.E., Gohara, D., Shi, G.: OpenCL: a parallel programming standard for heterogeneous computing systems. Comput. Sci. Eng. **12**(3), 66 (2010)
13. Venieris, S.I., Bouganis, C.S.: fpgaConvNet: mapping regular and irregular convolutional neural networks on FPGAs. IEEE Trans. Neural Netw. Learn. Syst. **30**(2), 326–342 (2018)
14. Wirbel, L.: Xilinx SDAccel: a unified development environment for tomorrow's data center. The Linley Group Inc, p. 24 (2014)
15. Xing, Y., et al.: DNNVM: end-to-end compiler leveraging heterogeneous optimizations on FPGA-based CNN accelerators. IEEE Trans. Comput. Aided Des. Integr. Circuits Syst. **39**(10), 2668–2681 (2019)
16. Zhang, C., Sun, G., Fang, Z., Zhou, P., Pan, P., Cong, J.: Caffeine: toward uniformed representation and acceleration for deep convolutional neural networks. IEEE Trans. Comput. Aided Des. Integr. Circuits Syst. **38**(11), 2072–2085 (2018)
17. Zhu, J., Wang, L., Liu, H., Tian, S., Deng, Q., Li, J.: An efficient task assignment framework to accelerate DPU-based convolutional neural network inference on FPGAs. IEEE Access **8**, 83224–83237 (2020)

Exploring Cell-Based Neural Architectures for Embedded Systems

Ilja van Ipenburg, Dolly Sapra$^{(\boxtimes)}$, and Andy D. Pimentel

University of Amsterdam, Amsterdam, Netherlands
i.vanipenburg@student.uva.nl, {d.sapra,a.d.pimentel}@uva.nl

Abstract. *Neural Architectures Search* (NAS) methodologies, which automatically discover state-of-the-art neural networks, have seen a growing interest in recent years. One particular group of NAS methodologies searches for small sub-networks called cells, which are then linearly connected to form the complete neural network. The composition of the final neural network, established through the width of the cells and the depth of the connections, is manually designed while being influenced by the available GPU memory. Typically, the hardware architectures targeted in NAS research are powerful, high-end GPUs. Hence, the attention is on creation of a large neural network that will still fit in the GPU, in turn leading to a very high accuracy for the given task. In direct contrast, we exploit the inherent flexibility of cells to create smaller neural networks, with the intention to study their behaviour on resource-constrained embedded systems. We use the cells discovered from *Stochastic Neural Architecture Search* (SNAS), to explore the effect that the composition of the cell has on various metrics, namely, the number of parameters, accuracy, latency and power usage. The last two metrics are measured on NVIDIA Jetson Nano, an embedded AI computing platform with a small GPU with mere 4GB on-chip memory. When comparing results of our exploration to the original SNAS architecture's with 97.02% accuracy for the CIFAR-10 dataset, one particular architecture, with only a tenth of original parameters, achieved an accuracy of 96.14%, notably with 15% lower power consumption and ≈3x faster inference time. Furthermore, this model outperforms other architectures, which are designed for edge devices, specifically to reduce the model size. Thus demonstrating that cell-based architectures, with adequate composition, provide efficient models to be deployed on resource-constrained edge devices.

1 Introduction

Recently, there has been a growing interest in *Neural Architecture Search* (NAS), the automation of architecture engineering for efficient neural networks. The automatically discovered neural architectures have routinely outperformed the hand-designed ones in a variety of domains such as language processing and image classification tasks [1]. There is an important and increasingly popular

© Springer Nature Switzerland AG 2021
M. Kamp et al. (Eds.): ECML PKDD 2021 Workshops, CCIS 1524, pp. 363–374, 2021.
https://doi.org/10.1007/978-3-030-93736-2_28

subgroup of NAS methodologies, consisting of algorithms focused on cell-based neural architectures. As discussed in [2], cell-based NAS has several advantages over other methodologies. Firstly, the search space of the NAS algorithm is reduced as the algorithm only searches for a small sub-network called *cell*, which is a small part of a complete neural architecture. Secondly, these cells can be transferred and re-used in different datasets and domains. Thirdly and most importantly in this work, the architectures created by repeating building blocks are a useful design principle in general.

The cell-based NAS typically discovers two types of cells, namely, a normal cell and a reduction cell [3]. The normal cell is designed to maintain the feature map size of the input, whereas the reduction cell reduces the feature map size. The complete neural architecture is generated by forming a linear connection of the normal cells, interrupted by a few reduction cells at regular intervals. The neural architectures created by repeating the same cell possess an inherent flexibility to be able to form neural networks of different sizes. Individual cells can be wide or narrow (depending on the number of filters it has), and the variable frequency of cell repetition in the neural network further adds to their flexible nature.

The GPUs targeted in NAS research typically are state-of-the-art hardware, usually resulting in the creation of a large neural network, containing a stack of many wide cells. However, most NAS methodologies do not consider hardware limitations during engineering. The neural networks designed for a high-end GPU may not be a viable option for an edge device. Oftentimes, the edge devices are cheap, able to fit in small spaces, and run on an internal battery. Consequentially, they have constraints on memory, processing power, speed, and energy. Figure 1 illustrate the enormous difference between the performance of neural networks on a high-end GPU (NVIDIA Tesla T4) and a resource-constrained GPU (NVIDIA Jetson Nano). The graphs are drawn for latency vs parameters for various cell-based architectures. In a glance they do look similar, however the scale of the latency for these two devices are in stark contrast. An architecture designed for a high-end GPU, providing inference in 0.1 s can take more than 2 s on an embedded device, driving the point that the best model for one device can not always be the optimal choice for another device. This also demonstrates the importance of designing efficient neural network architectures for embedded systems.

Taking the flexibility of the cell-based neural architectures into consideration, the question arises if it is possible to alter and organize the cells, such that it is feasible to be deployed on resource-constrained edge devices. This question motivates the current work, where we first analyze the performance of the original architecture created by a popular NAS cell on our target hardware. Subsequently, this information is utilized to design a grid search, which evaluates various architectures created from the chosen cell. The objective of this search is to discover good architecture(s), composed from the same cell, which have better performance in terms of power and latency on an embedded system while retaining an acceptable accuracy. In other words, the aim is to investigate the

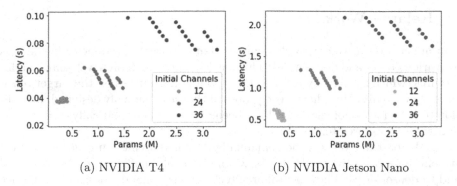

(a) NVIDIA T4   (b) NVIDIA Jetson Nano

Fig. 1. Latency of various architectures on the NVIDIA T4 and the NVIDIA Jetson Nano, illustrating the latency difference between embedded systems and high-end GPUs.

cell-based architecture's suitability to an embedded system. As far as we know, this is the first experimental and exploratory study that systematically analyses cell-based architectures for deployment on embedded devices.

Specifically, in this work, the cells chosen were those discovered from a NAS methodology called *Stochastic Neural Architecture Search* (SNAS). We studied the effects of the composition of the cells with a variable number of parameters, on its accuracy along with latency, and power usage on NVIDIA Jetson Nano, which is an embedded AI computing platform with a small GPU with mere 4 GB on-chip memory. The results of this exploration illustrate that the architectures, composed of the same cells that were originally designed for high efficiency and a large memory footprint, are able to achieve competitive performance on another hardware with resource limitations.

Since the search is based on multiple objectives, selection of the best candidates is concluded through *Pareto optimization*, where any objective cannot be improved without worsening some of the other objectives. The set of candidates selected in such a fashion are collectively called as a *Pareto Front*. The Pareto Front obtained upon convergence, presents the various possible architectures that can be deployed on the edge device. It allows the designer to be aware of the trade-offs that exist between different evaluation parameters. For instance, a highly efficient model generally has many convolutional kernels, thereby having a large memory footprint and a longer processing time. On the other hand, a smaller model with fewer parameters is highly likely to have less prediction accuracy, but might be the best option for a resource-constrained device.

The remainder of this paper is structured as follows. Firstly, in Sect. 2, related works in the domain of neural architectures for embedded systems is discussed. Secondly, in Sect. 3, we discuss the tradition composition of cell based architecture and its behaviour on resource-constrained Jetson-Nano. Subsequently, Sect. 4 presents our search methodology and the evaluation objectives. Next, the results from experiments are described in Sect. 5. Lastly, Sect. 6 concludes the paper.

2   Related Work

Neural networks are increasingly being used in resource-constrained edge devices for various tasks and domains [4]. This has led to research on novel neural architectures, which specifically cater to the resource limitations on the target hardware. New architectures have been proposed that are manually designed, in addition to NAS methodologies, that have been defined to automatically discover the hardware aware architectures.

In the past few years, various manually-designed architectures for embedded systems have been proposed, such as, MobileNets [5], ShuffleNet [6], DenseNet [7] and CodenseNet [8]. These neural architectures are generally designed to reduce the resource usage, however, they do not evaluate the performance on a specific target hardware. They require significant design time, in addition to the human expertise. Moreover, they are not always optimal for a new application or hardware, and may require further manual effort to fine-tune.

Early NAS methodologies [2] were focused only on improving the accuracy and took many days to converge [9,10]. With the introduction of faster differentiable search algorithms, which only takes a few hours to converge on a cell-based search space, the cell-based NAS methodologies have become mainstream [11–13]. Consequently, there are many efficient known cells, however, once a cell is discovered, all architectures are constructed in a similar manner (as first proposed in [9]).

Specific hardware aware NAS methodologies, such as MnasNet [14], PPP-Net [15], are efficient in searching for neural networks that achieve high accuracy, low computation cost, and low latency on a specific device. However, most of them do not generate flexible architectures (or cells) that can be adapted after the NAS algorithm has finished execution. For example, PPP-Net and MnasNet, both generate architectures with pre-defined number of blocks.

The inherent flexibility of a cell however, allows it to be used in various configurations to construct the whole architecture. Motivated by this thought, the current work is an exploratory study that investigates the architectures constructed from one such cell (SNAS [13]). To the best of our knowledge, no other study has been proposed to systematically analyse the effects of cell-based architecture composition for embedded systems.

3   Cell-Based Architecture Analysis

In this section, we first explain the baseline neural architecture, originally constructed by the cells discovered during the SNAS (Stochastic Neural Architecture Search) [13] work. Next, we analyse the baseline model for the CIFAR-10 dataset, on NVIDIA Jetson Nano and further utilize the details from this analysis to search for architectures that are suitable to resource-constrained devices.

The SNAS methodology discovers two types of cells: a *Normal Cell*, which preserves the feature map size of the input, and a *Reduction Cell*, which reduces the feature map size by half. For the CIFAR-10 dataset, the original baseline

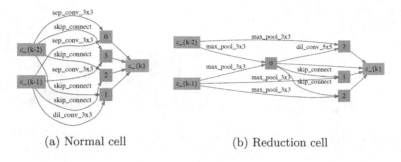

(a) Normal cell   (b) Reduction cell

Fig. 2. Cells found by SNAS (mild constraint) [13].

model has linearly connected 18 normal cells, partitioned into three blocks by
two additional reduction cells at 1/3 and 2/3 of the total depth of the network.

The cell itself is designed to take the output of the two previous cells, and as
depicted in Fig. 2, consists of an acyclic graph of various nodes. In the figure, blue
nodes are input/output nodes of the cell and orange nodes are the intermediate
nodes. Each edge between an input node and an intermediate node is either a
convolutional operation or a skip connection. The output of all the intermediate
nodes is then concatenated to produce the final cell output.

In the SNAS work of [13], resource constraints were weakly considered during
the search, by placing limits on the size of the cell. Three levels of resource
constraints were used: mild, moderate, and aggressive. For each of these levels,
a normal cell and a reduction cell were discovered. The mild constraint allows
for relatively large cells to be discovered, whereas, the aggressive mode leads to
discovery of cells with fewer parameters. However, the number of parameters in
the aggressive-cell were still too high, the final model has ≈80% of the size of
the model constructed from the mild-cell. This still may not be considered to be
of suitable size to be deployed on many edge devices with memory limitations.

Furthermore, most NAS approaches construct the neural network in the same
standard manner. The classic architecture was first introduced by [9] and has
been deployed by many other cell-based NAS works [11–13]. These works spend
considerable time and effort to discover the cells and subsequently construct
the architecture as originally suggested. In the classic architecture, blocks of
the normal cells are partitioned by the reduction cells at regular intervals. The
standard composition for the eventual neural network is a balanced architecture,
with an equal number of cells in each block. Figure 5a illustrates the baseline
architecture composition commonly used for the CIFAR-10 dataset. In the figure,
FxN to the right of each block represents the number of channels per cell in the
block × number of cells per block. In the classic architectures, the number of
cells per block is the same for all blocks, however, the number of channels of the
cell gets doubled in subsequent blocks. The number of channels of a cell refers
to the number of kernels in every convolutional operator edge in the cell.

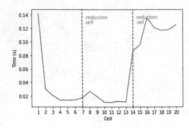
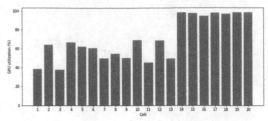

(a) Average latency per cell per epoch   (b) Average GPU utilization per cell per epoch

Fig. 3. Analysis of cell behaviour on Jetson-Nano. The last block of the architecture is compute intensive, relying heavily on GPU based computation and is taking longer time to complete all its operations.

We analysed the behaviour of the baseline SNAS architecture on Jetson-Nano and one of the chief observations was, that the last block consumes most computational resources. It is unsurprising though, considering the fact that the last block has very wide cells, with 4x as many channels as compared to the cells in the first block. Figure 3 shows the behaviour of individual cells during an inference cycle. Cells 1 to 20 are consecutive cells in the order of execution during the inference. Figure 3a shows execution time of each cell and Fig. 3b shows GPU utilization during execution of each cell. It is evident from these graphs that the cells in the last block are utilizing the GPU to the maximum, as a direct consequence of the large number of computational operations required. Additionally, it takes between 5x–10x longer for each cell (in the last block) to finish execution as compared to cells in first and second block.

From this analysis, it is obvious that to achieve a suitable neural architecture composition for an embedded device, utilization of narrow cells should be ensured. Certain design principles from the classic architecture may still be retained, such as, the pattern of doubling the number of channels after every reduction cell can be maintained, by reducing the number of initial channels in the first block. Another alternate approach may be to reduce the number of cells only in the last block. The total number of cells can also be retained by varying the numbers of cells in different blocks and thereby creating an unbalanced architecture. We eventually employ these strategies in a grid search methodology to explore the architecture composition for embedded systems.

4   Architecture Search

In this section, we explain the methodical neural architecture search approach using the SNAS cells and their evaluation process on the target hardware. We expect a cell from any other NAS methodology will also provide a similar exploration result on an embedded device, since all existing NAS algorithms favor a similar cell with wide and shallow structure [3].

4.1 Search Space

To indicate different architectures, we use the following notation: $(C@K-L-M)$, where C is the number of channels in the first block and K,L,M are the number of cells in the first, second and third block respectively. Thus, the original SNAS architecture in Fig. 5a is denoted as $(36@6-6-6)$.

The neural architectures generated by our search algorithm maintain a constant depth of 20 cells, with reduction cells placed at $K+1$ and $K+L+2$ depth of the network, for which $K \geq 2, L \geq 2, M \geq 2$. The amount of initial channels is sampled at an interval of 12, with a maximum of 36 channels.

4.2 Grid Search

To evaluate various neural architecture compositions, a list of all possible K-L-M meta-architectures is generated and sorted using radix sort, which prioritizes values in the order K, L, M. The distance measure between architectures is defined as a three-dimensional Manhattan distance. For meta-architectures NN_1 and NN_2:

$$d = |K_1 - K_2| + |L_1 - L_2| + |M_1 - M_2| \tag{1}$$

In order to prevent architectures that are too unbalanced, the first stipulation on architectures is that the distance between baseline $(6-6-6)$ architecture and K, L, M can not differ from each other by more than a balance factor b. The value $b = 6$ was chosen for this research. Architectures in the list that do not adhere to this rule were removed.

Next, architectures that are similar to each other were removed after it was observed that architectures that were close to each other had comparable evaluated metrics, including accuracy. Moving through the sorted list, architectures with a distance $d < 4$, between itself and the last valid architecture were removed. This step was performed to reduce the number of architectures that would need to undergo a resource expensive training process.

For the final step, architectures with a trainable parameter size of 3.3 million parameters or more are removed, as to be able to fit the networks on a single GPU and reduce training times. The final sample consists of 18 unique $K-L-M$ iterations, which were each trained with 12, 24, and 36 initial channels, resulting in a total of 54 architectures that were trained and evaluated.

4.3 Architecture Evaluation

For evaluation, all architectures are trained following the evaluation settings of SNAS; all networks are trained from scratch for 600 epochs with batch size 96 on the CIFAR-10 data set for image classification. CIFAR-10 consists of $60,000$ labeled images of dimensions $32 \times 32 \times 3$, comprising of $50,000$ training and $10,000$ testing images. The images are divided into 10 classes. Standard data augmentation techniques [16] with small translations, cropping, rotations and horizontal flips along with cutout [17] were utilized during the training.

Latency is measured on the NVIDIA Jetson Nano developer kit for embedded applications. This hardware, as previously discussed, is extremely constrained in resources, when compared to high-end GPUs. This has such a significant impact, that the power measurements must be done separately from the latency measurements, as they slow down the network by at least 33%. Both latency and power usage are measured and averaged over 50 individual runs of 50 batches and a batch size of 64 images.

5   Exploration Results

In this section, we present the evaluations and results of the cell-based architecture exploration. All the generated architectures are first trained and then the performance is measured on the target hardware specific metrics.

All training settings were the same as followed in [13], where every neural network was trained using stochastic gradient descent with initial learning rate 0.1, weight decay 3×10^{-5} and batch size 128. The learning rate was decayed by a factor of 0.97 after each epoch and auxiliary towers with weight 0.4 were used as additional enhancements. All neural networks were trained on NVIDIA Tesla T4 and on an average took 1.5 days to train completely.

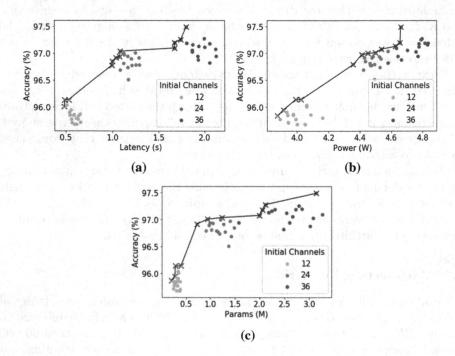

Fig. 4. Pareto fronts for accuracy of different architectures w.r.t. (a) latency, (b) power, and (c) parameter size.

Table 1. The multi-dimensional Pareto front for cell-based architectures on CIFAR-10 with accuracy, number of parameters, inference time, and power usage on Jetson-Nano as evaluation objectives. All architectures use the SNAS mild constraint cells with cutout. SNAS^{orig} refers to the original baseline architecture. SNAS^c refers to other architectures composed using the SNAS-cell.

Architecture	Init-Channels	Accuracy (%)	Params (M)	Latency (s)	Power-usage (W)
SNAS^{orig} (6 - 6 - 6)	36	97.02	2.9	1.87	4.70
SNAS^c (6 - 4 - 8)	36	**97.49**	3.14	1.80	4.66
SNAS^c (7 - 7 - 4)	36	97.28	2.12	2.00	4.80
SNAS^c (3 - 9 - 6)	36	97.20	2.85	1.68	4.66
SNAS^c (4 - 10 - 4)	36	97.19	2.31	1.81	4.82
SNAS^c (5 - 7 - 6)	24	97.04	1.24	1.08	4.58
SNAS^c (4 - 10 - 4)	24	96.94	1.05	1.08	4.52
SNAS^c (8 - 8 - 2)	24	96.92	0.73	1.29	4.43
SNAS^c (4 - 10 - 4)	12	96.14	0.28	**0.53**	4.00
SNAS^c (9 - 3 - 6)	12	95.84	0.30	0.64	**3.88**
SNAS^c (9 - 5 - 4)	12	95.72	**0.25**	0.66	3.95

Once all the neural networks were trained, they were evaluated on four metrics, namely, accuracy, number of parameters, latency and power usage. The last two metrics were measured on the NVIDIA Jetson Nano developer board. Next, the Pareto Front was selected based on all evaluated metrics, which is presented in Table 1. All the models in the Pareto set are considered to be equally adequate to be marked as a good model. This set is a handy tool for system designers, as it provides the quantitative trade-offs between different objectives.

Considering it is not easy to draw and understand four-dimensional plots, Fig. 4 shows the two-dimensional Pareto fronts of different sets of parameters, comparing latency, power, and number of parameters to the accuracy of an architecture. Beside the visualization, these graphs also provide insight into the impact of initial channels on different evaluation metrics. In each graph, three clusters are clearly visible, which indicate the three different settings for the number of initial channels of the architectures.

Looking at the whole pareto front (Table 1), surprisingly, the $(36@6 - 4 - 8)$ architecture performs the best in terms of accuracy, achieving 97.49% compared to 97.02% achieved by the original balanced SNAS architecture with a comparable parameter size. This difference is more than what can be attributed to statistical difference. This architecture strongly suggests that the composition of the cells in a neural network plays an important role in the eventual performance of the model. Beyond the search for a cell, there is little manual effort required in training of some of the architecture compositions, and this can be a recommended segment of the underlying NAS methodology.

For the resource-constrained devices, sacrificing some precision for efficiency is often acceptable, since not all systems require perfect accuracy. From the exploration results, it is evident that the cell based architectures are still very

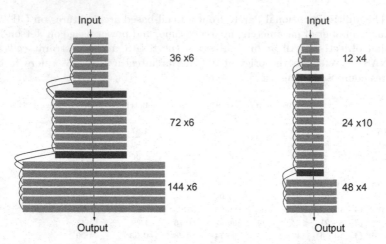

(a) SNAS original architecture (36@6-6-6) (b)   SNAS^c Architecture   (12@4-10-4)
(2.9M parameters, 97.02% accuracy)   (0.3M parameters, 96.14% accuracy)

Fig. 5. Cell-based architectures for CIFAR-10 with three blocks of normal cells partitioned by two reduction cells (in red). FxN to the right of each block represents the number of channels per cell in the block × number of cells per block. (Color figure online)

efficient with fewer parameters. For example, the architecture $(24@5 - 7 - 6)$ with less than half the parameters of the baseline model, has a similar accuracy.

As expected, the models with the fewest parameters, fastest inference time and lowest power usage were all the architectures with 12 initial channels. Most research only takes the number of parameters into account when optimizing for an edge device. However, latency and power do not always have a linear relationship with the model size. One notable architecture, $(12@4 - 10 - 4)$, as shown in Fig. 5b, is up to three times faster and consumes about 15% less power, while still achieving an accuracy of 96.14%. This architecture has narrow cells, as well as fewer cells in the last block.

We compared the models on pareto front that had less than 1M parameters, to other state of the art neural architectures that were designed specifically for embedded systems. We use the number of parameters for the comparison, since latency and power usage cannot be fairly compared as either the target hardware is different or not included in the search. The results are presented in Table 2.

The neural networks explored in this work outperform all other small architectures when it comes to prediction accuracy. This result strongly indicates the advantages of constructing neural architectures for an edge device by utilizing cells that were discovered for high-end GPUs. Not only are they flexible and highly efficient, but can also expedite the design process of an application for a specific target hardware, specially when fast turn-around time is desired.

Table 2. Accuracy of various architectures for CIFAR-10 with <1M parameters.

Architecture	Type	Accuracy (%)	Params (M)
DenseNet-BC (k = 12) [7]	Non-cell	95.49	0.8
LEMONADE Cell 9 [18]	Cell	95.43	0.5
CondenseNet-86 [8]	Non-cell	95.0	0.52
CondenseNet^{light}-94 [8]	Non-cell	95.0	0.33
PPP-Net-A [15]	Non-cell	94.72	0.45
PPP-Net-B [15]	Non-cell	95.42	0.52
One-Shot Top (F = 16) [19]	Cell	94.6	0.7
One-Shot Small (F = 16) [19]	Cell	94.6	0.4
SNAS^c (24@8-8-2)	Cell	**96.92**	0.73
SNAS^c (12@4-10-4)	Cell	96.14	0.28
SNAS^c (12@9-3-6)	Cell	95.84	0.3
SNAS^c (12@9-5-4)	Cell	95.72	0.25

6   Conclusion

The aim of this research was to explore cell-based architectures and their composition strategies for an embedded system. In order to do this, a grid search was defined and a carefully truncated list of architectures was trained and evaluated. A Pareto Front was presented, describing the trade-offs between different evaluation metrics, namely, accuracy, number of parameters, along with the latency and power usage on the NVIDIA Jetson Nano developer board. Among these architectures several noteworthy architectures were found, highlighting the importance of the composition of cells, geared towards different situations.

As an extension to this work, we aim to explore more connection patterns to create the architectures from a cell. In the current work, there were many restrictions put on the search space, such as, the connection pattern of doubling the number of channels after every reduction cell was maintained. In the future, the aim is to create a search space where highly unbalanced architectures as well as different connection patterns can be explored through a search algorithm. We further aim to utilise a population based meta-heuristic algorithm [20] to cover this large search space to achieve a better pareto front.

Acknowledgements. This project has received funding from the EU Horizon 2020 Research and Innovation programme under grant agreement No. 780788.

References

1. He, X., Zhao, K., Chu, X.: AutoML: a survey of the state-of-the-art. Knowl.-Based Syst. **212**, 106622 (2021)
2. Elsken, T., Metzen, J.H., Hutter, F.: Neural architecture search: a survey. arXiv preprint arXiv:1808.05377 (2018)

3. Shu, Y., Wang, W., Cai, S.: Understanding architectures learnt by cell-based neural architecture search. In: International Conference on Learning Representations (2020)
4. Wang, X., Han, Y., Leung, V.C., Niyato, D., Yan, X., Chen, X.: Convergence of edge computing and deep learning: a comprehensive survey. IEEE Commun. Surv. Tutor. **22**, 869–904 (2020)
5. Howard, A.G., et al.: MobileNets: efficient convolutional neural networks for mobile vision applications. arXiv preprint arXiv:1704.04861 (2017)
6. Zhang, X., Zhou, X., Lin, M., Sun, J.: ShuffleNet: an extremely efficient convolutional neural network for mobile devices. In: Proceedings of the IEEE Conference on Computer Vision and Pattern Recognition (2018)
7. Huang, G., Liu, Z., Van Der Maaten, L., Weinberger, K.Q.: Densely connected convolutional networks. In: Proceedings of the IEEE Conference on Computer Vision and Pattern Recognition (2017)
8. Huang, G., Liu, S., Van der Maaten, L., Weinberger, K.Q.: CondenseNet: an efficient DenseNet using learned group convolutions. In: Proceedings of the IEEE Conference on Computer Vision and Pattern Recognition (2018)
9. Zoph, B., Vasudevan, V., Shlens, J., Le, Q.V.: Learning transferable architectures for scalable image recognition. In: Proceedings of the IEEE Conference on Computer Vision and Pattern Recognition (2018)
10. Real, E., Aggarwal, A., Huang, Y., Le, Q.V.: Regularized evolution for image classifier architecture search. In: Proceedings of the AAAI Conference on Artificial Intelligence (2019)
11. Pham, H., Guan, M., Zoph, B., Le, Q., Dean, J.: Efficient neural architecture search via parameters sharing. In: International Conference on Machine Learning (2018)
12. Liu, H., Simonyan, K., Yang, Y.: DARTS: differentiable architecture search. In: International Conference on Learning Representations (2019)
13. Xie, S., Zheng, H., Liu, C., Lin, L.: SNAS: stochastic neural architecture search. In: International Conference on Learning Representations (2019)
14. Tan, M., et al.: MnasNet: platform-aware neural architecture search for mobile. In: Proceedings of the IEEE/CVF Conference on Computer Vision and Pattern Recognition (2019)
15. Dong, J.-D., Cheng, A.-C., Juan, D.-C., Wei, W., Sun, M.: PPP-Net: platform-aware progressive search for pareto-optimal neural architectures (2018)
16. Springenberg, J.T., Dosovitskiy, A., Brox, T., Riedmiller, M.: Striving for simplicity: the all convolutional net. arXiv preprint arXiv:1412.6806 (2014)
17. DeVries, T., Taylor, G.W.: Improved regularization of convolutional neural networks with cutout. arXiv preprint arXiv:1708.04552 (2017)
18. Elsken, T., Metzen, J.H., Hutter, F.: Efficient multi-objective neural architecture search via Lamarckian evolution. In: International Conference on Learning Representations (2019)
19. Bender, G., Kindermans, P.-J., Zoph, B., Vasudevan, V., Le, Q.: Understanding and simplifying one-shot architecture search. In: International Conference on Machine Learning. PMLR (2018)
20. Beheshti, Z., Shamsuddin, S.M.H.: A review of population-based meta-heuristic algorithms. Int. J. Adv. Soft Comput. Appl. **5**(1) (2013)

Design Space Exploration of Time, Energy, and Error Rate Trade-offs for CNNs Using Accuracy-Programmable Instruction Set Processors

Armin Schuster[✉], Christian Heidorn, Marcel Brand, Oliver Keszocze, and Jürgen Teich

Friedrich-Alexander-Universität Erlangen-Nürnberg (FAU), 91058 Erlangen, Germany
armin.schuster@fau.de

Abstract. We proclaim the use of application-specific instruction set processors with programmable accuracy called Anytime Instruction Processors (AIPs) for Convolutional Neural Network (CNN) inference. For a floating-point operation, the number of correctly computed mantissa result bits can be freely adjusted, allowing for a fine-grained trade-off analysis between accuracy, execution time and energy. We propose a Design Space Exploration (DSE) technique in which the accuracy of CNN computations is determined layer-by-layer. As one result, we show that reductions of up to 62% in energy consumption are achievable for a representative ResNet-18 benchmark in comparison to a solution where each layer is computed at full accuracy according to the IEEE 754 single precision floating-point format.

1 Introduction

Deep learning practices are present throughout many application fields for cyber-physical systems, ranging from robotics and autonomous driving to smart cameras. In the context of embedded systems, fulfilling non-functional requirements is just as essential as fulfilling functional requirements. Therefore, this work focuses on exploring trade-offs between the objectives of achieving a low error rate R, a low execution time T as well as a low energy consumption E for inference for a class of neural networks called Convolutional Neural Networks (CNNs). For these networks, a range of optimization strategies has previously been proposed. These techniques can be divided into the categories parameter pruning, quantization, low-rank factorization, transferred/compact convolutional filters and knowledge distillation. Much research work has been performed in precision tuning under detection rate constraints, e.g., in proposing hardware designs using unconventional floating-point formats. For a survey of these methods, see, e.g., [5]. This paper focuses exclusively on the exploitation of quantization for a novel application-specific instruction processor (ASIP) called Anytime Instruction Processor (AIP) [2,3]. For a given precision, e.g., half, single or double

© Springer Nature Switzerland AG 2021
M. Kamp et al. (Eds.): ECML PKDD 2021 Workshops, CCIS 1524, pp. 375–389, 2021.
https://doi.org/10.1007/978-3-030-93736-2_29

precision, an AIP supports the execution of floating-point instructions with a user-programmable number a of mantissa bits to be correctly computed. This allows to provide execution time, energy and accuracy trade-offs in a very flexible way. Whereas neural network inference using a variable bitwidth per layer is a well-explored field, this paper a) proclaims the use of AIPs for realizing a layer-programmable evaluation accuracy rather than proposing layer-specific hardware optimizations, and b) proposes a multi-objective evolutionary algorithm (MOEA)-based Design Space Exploration (DSE) to c) systematically explore trade-offs, particularly the opportunity for execution time and energy reductions. For the ResNet-18, we show that energy reductions of up to 62% are achievable without any increase of error rates or execution time compared to an implementation using the standard IEEE 754 single precision floating-point format.

2 Related Work

Particularly in the context of CNNs, much research work has been performed in precision tuning under detection rate constraints e.g., in proposing hardware designs using unconventional floating-point format, or even layer-parallel implementations on massively parallel processor arrays, e.g., [9,11]. For CNNs with more than one bit of precision, Wang et al. showed that computing each CNN layer at a different fixed-point precision may reduce energy consumption while maintaining the classification accuracy compared to a homogeneous fixed 8-bit quantization for all layers [22]. Their approach optimizes hardware implementations using a Bit-serial Matrix Multiplication Overlay (BISMO) that was previously proposed in [21]. To determine the optimal configuration, reinforcement learning is used. De la Parra et al. have investigated hardware designs of CNNs using approximate arithmetic units including multipliers and adders [17]. They investigated how the error rate of the resulting CNNs is affected when utilizing approximate multipliers and adders for CNN computations.

In our proposed approach, DSE is performed using AIP processors by optimizing the number of mantissa bits a in each layer by exploiting their support of floating-point instructions at programmable accuracy. Also different to [22], our approach uses the NSGA-II evolutionary algorithm [6] for DSE instead of reinforcement learning. Moreover, our approach is a multi-objective one: We provide a DSE of the three objectives error rate R, energy E and execution time T, by individually selecting a certain accuracy a for each layer. With respect to the aspect of DSE, ALWANN [16] is the approach closest to the methodology presented in this paper. ALWANN also uses an evolutionary algorithm, but explores the selection of pre-synthesized approximate multipliers from a library. Our work is different in two aspects: First, we do not consider hardware designs, but rather explore the programmable accuracy of floating-point units of Anytime Instruction Processors [2,3] for each layer. Second, ALWANN only explores a two-dimensional objective space including energy and error rate but not execution time. As will be shown, AIPs provide the flexibility to go from a highest accuracy, i.e., single precision accuracy to just one single mantissa bit ($a = 1$) being computed, thereby allowing for execution time and energy reductions as will be shown.

Finally, looking at other processor-based approaches, i.e., not targeting or requiring the design of specific hardware, enables the highest degree of flexibility for the inference of quantized neural networks (QNN) at the edge. The PULP-NN project [8] proposes DSPs with ISA extensions for the RISC-V processor to enable the efficient inference of Quantized Neural Networks for integer arithmetic with $8, 4, 2$ and one bit(s). But contrary to this approach, AIPs support a) floating-point computations and b) arbitrary accuracy $a \in \{1, ..., 23\}$. Finally, the PULP-NN project does not provide an exploratory approach to the optimization of quantization in each individual layer. Many further approaches exploring quantization such as $[1, 4, 7]$ exist, however, none of these is investigating floating-point formats.

In summary, contrasting to all of the aforementioned works, we exploit floating-point processor targets called AIPs [2,3] with variable mantissa accuracy, as they allow for a wider dynamic range than integer arithmetic and, at the same time, not requiring any application-specific hardware design. Moreover, we explore a three-dimensional design space spanned by error rate, energy as well as execution time.

3   Fundamentals

3.1   Anytime Instruction Processors

The concept of Anytime Instructions [2,3] is based on the idea of processing basic arithmetic operations such as additions, subtractions, multiplications, and divisions at a programmable accuracy. By encoding the number of mantissa bits a of a floating-point operation that shall be computed within the instruction word itself, instructions can be executed more efficiently regarding energy than at full accuracy (i.e., $a = 23$ for single, resp. $a = 10$ for half precision), but with possible errors in the less significant mantissa bits. Implementations of such operations have been proposed in [3] for floating-point division which is classically computed Most Significant Bit (MSB) first. Recently, Brand et al. [2] suggested also implementations for floating-point additions and multiplications. The problem of conventional Least Significant Bit (LSB) first arithmetic is solved by proposing an implementation using *online arithmetic* and one alternative solution called *bitmasking*. In online arithmetic, MSB first computations are enabled using a *redundant signed digit radix-2* number format. In case of bitmasking, the computation of the least significant part is just disabled by masking corresponding input and intermediate result bits. For more details, we refer to [2,3]. The latency formulæ for these instructions in dependence on a will be introduced in Sect. 4.2 as well as formulæ for estimating the average energy (see Sect. 4.3) of such instructions.

3.2   Deep Learning and CNNs

Modern CNNs can consist of more than 1,000 layers of different types [10,20]. In this work, the considered CNNs consist of convolutional layers, fully-connected layers, pooling layers, and activation layers. The most computationally expensive layers are the convolutional layers, such that the overall energy demands may be reduced by computing these layers at diminished accuracy.

A convolutional layer i transforms a so-called input feature map, a 3-dimensional tensor of shape $R_i \times C_i \times N_i$, into a 3-dimensional tensor of shape $R_{i+1} \times C_{i+1} \times M_i$, called output feature maps. Hereby, R_i and C_i denote the number of rows and columns of one input feature map. N_i represents the number of input feature maps, whereas R_{i+1}, C_{i+1} are the respective rows and columns of one output feature map and M_i represents the number of output feature maps. The transformation is done by applying M_i 3-dimensional convolutional filters of shape $K_i \times K_i \times N_i$ on the input feature maps.

In this context, K_i denotes the kernel size of the ith filter. One output element of a convolutional layer is computed as

$$O[m][r][c] = \sum_{n=0}^{N-1} \sum_{k_1=0}^{K-1} \sum_{k_2=0}^{K-1} I[n][r+k_1][c+k_2] \times W[m][n][k_1][k_2]. \qquad (1)$$

In Eq. (1), O, I, and W denote the output, input and weight tensors, respectively. The number of Multiply-Accumulate (MAC) operations $MACS_i$ of convolutional layer i is given by

$$MACS_i = R_{i+1} \cdot C_{i+1} \cdot M_i \cdot N_i \cdot K_i^2. \qquad (2)$$

For fully-connected layers, the parameters K_i, C_{i+1}, R_{i+1} are one ($K_i = C_{i+1} = R_{i+1} = 1$), thus included in Eq. (2).

4   Design Space Exploration

The objective of our following DSE is to minimize the error rate R, the execution time T as well as the energy consumption E of the inference computations for a given CNN when individually choosing the number $a_i \in \{1, \ldots, M\}$ of accurately computed mantissa result bits of floating-point operations in each layer i, $0 \leq i < |V|$, with $|V|$ denoting the number of the convolutional and fully-connected layers and $M = 23$ for single precision floating-point numbers. The resulting 3-objective optimization problem, introducing a three-objective cost function f to be minimized and making use of a variable x_i per layer (i.e., $0 \leq i < |V|$), is given by:

$$\min \quad f : \mathbb{N}^{|V|} \to \mathbb{R}^3$$
$$\mathbf{x} \mapsto (R(\mathbf{x}), T(\mathbf{x}), E(\mathbf{x})) \qquad (3)$$
$$s.t. \quad 1 \leq x_i \leq M \quad \text{for all } 0 \leq i < |V|.$$

The entries x_i of the vector $\mathbf{x}$ encode the number of mantissa bits a used for all floating-point operations in layer i of the CNN. The functions R, T and E evaluate the error rate, the execution time and the energy, respectively. We call a vector $\mathbf{x}$ also a *configuration* or *design point*. In the following, the formulæ for evaluating R, T, and E are derived and explained.

4.1 Error Rate R

The error rate R of a CNN inference is defined as

$$R = \frac{\#\text{incorrectly classified validation images}}{\#\text{ all validation images}} \cdot 100\%. \qquad (4)$$

In our work, R is evaluated using Eq. (4) when passing a validation set of 10,000 images through one network configuration. The fraction of incorrectly classified images thus determines the error rate. The images for the validation set are taken from the same data set (CIFAR-10 [13]) on which the CNN (ResNet-18 [10]) has been trained on.

4.2 Execution Time T

Online Arithmetic. The latency of each anytime instruction depends on the number of accurately computed mantissa bits a_i for each layer i, $0 \leq i < |V|$. With bpc_{op} being the number of bits computed per cycle and ΔFP_{op} representing the latency of the additional steps of a floating-point computation, the latency $L_{\mathrm{op}}(a_i)$ for one instruction is obtained according to [2] as

$$L_{\mathrm{op}}(a_i) = \lceil \frac{a_i + \delta_{\mathrm{op}}}{bpc_{\mathrm{op}}} \rceil + \Delta FP_{\mathrm{op}} \quad (5) \quad \text{with} \quad bpc_{\mathrm{op}} = \lceil \frac{\delta_{\mathrm{op}} + m + 1}{p_{\mathrm{op}}} \rceil. \quad (6)$$

The latency $L_{\mathrm{op}}(a_i)$ must be computed for the two types of operations used in the convolutional and fully-connected layer, i.e., op $\in \{\mathrm{mul}, \mathrm{add}\}$, with $\Delta FP_{\mathrm{mul}} = 1$ and $\Delta FP_{\mathrm{add}} = 2$. bpc_{op}, depending on the word length wl, the so-called *online delay* δ_{op}, and the number of pipeline stages p_{op}, is determined during synthesis time and assumed to be constant. For the DSE, we first choose the online arithmetic AIP implementation proposed in [2], and select $\delta_{\mathrm{mul}} = 3$, $\delta_{\mathrm{add}} = 0$, $p_{mul} = 7$ and $p_{add} = 6$, thus, with $m = 23$ (single precision), $bpc_{\mathrm{mul}} = \lceil \frac{3+m+1}{7} \rceil = 4$ and $bpc_{\mathrm{add}} = \lceil \frac{m+1}{6} \rceil = 4$. The reason for this choice will get clear later when explaining Fig. 3 (see Sect. 5.4). As a result, we obtain

$$L_{\mathrm{add}}(a_i) = \lceil \frac{a_i}{4} \rceil + 2 \quad (7) \quad \text{and} \quad L_{\mathrm{mul}}(a_i) = \lceil \frac{a_i + 3}{4} \rceil + 1. \quad (8)$$

The latency of a complete layer i can now be calculated as the product of the latency required for each instruction $L_{op}(a_i)$ and the number of MACs $MACS_i$ in Eq. (2). As the multiplications and additions are pipelined in each CNN layer, the latency of one layer computes as

$$L_i(a_i) = \max(L_{\mathrm{mul}}(a_i), L_{\mathrm{add}}(a_i)) \cdot MACS_i. \qquad (9)$$

Under the assumption of non-overlapping layer processing, the overall latency for the $|V|$-layer CNN then computes to L (Eq. (10)). Finally, the overall execution time T can be computed for a given clock frequency f (Eq. (11)).

$$L = \sum_{i=0}^{|V|-1} L_i(a_i) \qquad (10) \qquad\qquad T = \frac{L}{f} \qquad (11)$$

Bitmasking Approach. We decided to evaluate a second variant of AIP called bitmasking according to [3]. Here, the least significant operand bits of an addition are simply set to zero. For the multiplication, this approach is applied on the partial products, which are then summed up. Contrary to the online arithmetic approach, the latency of each operation is given according to Eq. (12) and thus not varying with a, as the operations can be performed by applying simple AND-operations on the input operands.

$$L_{\mathrm{op}}^{\mathrm{bm}} = 1 + \Delta F P_{\mathrm{op}}. \qquad (12)$$

With $\Delta F P_{op}$ being identical as for the online arithmetic approach (see Sect. 4.2), the latency for an addition, respectively multiplication is given as $L_{\mathrm{add}}^{\mathrm{bm}} = 3$, $L_{\mathrm{mul}}^{\mathrm{bm}} = 2$, respectively. Analogously to the online arithmetic approach, the latency for a complete layer evaluation utilizing the bitmasking implementation is obtained using Eq. (9). Similar, the latency computation L of the whole CNN according to Eq. (10) and execution time T according to Eq. (11) apply.

4.3   Energy E

For the evaluation of the energy consumption of each design point $(a_0, \cdots, a_{|V-1|})$, anytime functional units for multiplications and additions in single precision called `Anytime32`, have been individually synthesized. Subsequently, the average power of each operation op $\in$ {mul, add} has been characterized based on 10^6 random input combinations for each value of a_i from which an average energy consumption $E_{\mathrm{op}}(a_i)$ is obtained. The energy for each layer i is then estimated as

$$E_i(a_i) = (E_{\mathrm{mul}}(a_i) + E_{\mathrm{add}}(a_i)) \cdot MACS_i. \qquad (13)$$

The mentioned power values have been obtained by conducting a synthesis using Cadence Genus v2019.1 and targeting the Synopsys 28 nm Educational Library [19]. Evaluation of the generated design has been performed in Modelsim v10.6c. As the power estimate directly depends on the calculations conducted in the system, the switching activity has been recorded for 10^6 random calculations. For place&route, Cadence Innovus v2019.1 has been used. This process consists of creating an initial floorplan, loading the switching activity from the simulations, elaborating the place&route steps, and reporting back the estimated dynamic power dissipation. It should be noted that, independent of the choice of the number of mantissa bits a, full 32-bit words are fetched from main memory. This means that the average energy consumption related to just memory accesses is a constant offset to the values shown in Fig. 9 and therefore discarded in the following considerations. Thus, the overall energy E of a design point including all convolutional and fully-connected layers can be calculated as

$$E = \sum_{i=0}^{|V|-1} E_i(a_i). \tag{14}$$

5   Experiments and Results

5.1   Comparison of Anytime Instructions Against IEEE 754 Single Precision and Variable Precision Formats

Table 1. Floating-point number formats as analyzed and compared in this paper.

Design	s	e	m	wl
IEEE 754 single precision	1	8	23	32
Anytime32	1	8	a	32
float<8, a>	1	8	a	$1+8+a$

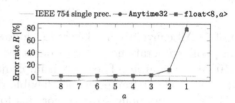

Fig. 1. Error rate R of LeNet5-base for varying numbers a of mantissa bits for the three floating-point formats in Table 1.

Where as the rounding mode in IEEE 754 single precision is "round-to-nearest, ties-to-even", Anytime Instruction Processors (AIPs) use "round-to-zero", as only the a most significant mantissa bits are computed, while the remaining ones simply set to zero. In the following first experiment, we investigate the influence of a as well as the different rounding modes on the error rate of the inference process of a simple CNN named LeNet5-base that has been trained in IEEE 754 single precision format. For comparison, three different floating-point formats according to Table 1 are introduced. The first format is the standard IEEE 754 single precision floating-point format with $s = 1$ sign bit, $e = 8$ exponent bits, and $m = 23$ mantissa bits. This format is used as the reference. The second format is the proposed Anytime32 format, is the AIP 32 bit format with $s = 1$ sign bit, $e = 8$ exponent bits, and also $m = 23$ mantissa bits but from which only $1 \leq a \leq m$ bits are correctly computed. The third format, denoted as float<8, a>, corresponds to a variable precision IEEE floating-point number presentation in which the word length wl is given as $wl = 1 + 8 + a$, where numbers are rounded using the "round to nearest, ties to even" mode. The arithmetic operations are performed as per the IEEE 754 standard, i.e., no approximations are performed. As CNN, LeNet5-base has been trained on the MNIST-benchmark [14] for hand-written digit recognition. The CNN consists of two convolutional layers, having 6 and 16 filters, respectively, with a filter kernel

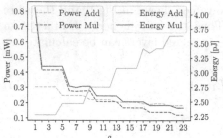 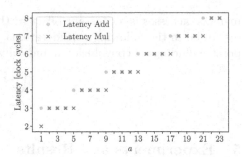

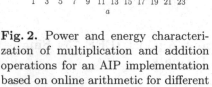

Fig. 2. Power and energy characterization of multiplication and addition operations for an AIP implementation based on online arithmetic for different values of a.

Fig. 3. Latency of AIP addition (blue) and multiplication (red) operations according to Eq. (7) and Eq. (8) for an AIP implementation based on online arithmetic. (Color figure online)

size of 5. All convolutional layers are followed by ReLU activations and maximum pooling operations. A fully-connected layer at the end gives the result for the 10 output neurons. The CNN is implemented and trained in TypeCNN [18], a C++ development framework for flexible datatypes, using the single precision IEEE 754 floating-point format. For this format, an error rate of 2.06% on the 10,000 test images provided by the MNIST-benchmark is achieved, see Fig. 1. For the evaluation of the other two formats in Table 1, no retraining or fine-tuning has been performed. For each, Fig. 1 also plots the achieved error rate for each value a of mantissa bits computed. In case of the lower bitwidth IEEE format float<8, a>, all a mantissa bits are computed correctly as the mantissa is stored in a bits only. In case of the anytime instructions, only the a most-significant bits of the mantissa of size $m = 23$ are computed, while fixing the remaining mantissa bits to zero. We use the arbitrary precision number library *Aarith* [12] for the simulation of the non-standard floating-point formats as well as the Anytime32 format. As can be seen, the error rate of the CNN stays constantly small for values of $a \geq 3$. Only starting from an accuracy of $a \leq 2$, the error rate rises significantly when compared to the IEEE 754 single precision floating-point format. Comparing anytime instructions with the variable precision float<8, a>, we can only observe deviations in error rates for $a \leq 2$. The deviations amount to less than 1% due to the use of different rounding modes.

5.2 DSE Setup

The DSE problem introduced formally in Sect. 4 is solved using the NSGA-II [6] multi-objective evolutionary algorithm (MOEA), embedded in the Opt4J [15] framework. The MOEA was configured with a population size of 100, 25 parents per generation, an offspring of 25 per generation and a crossover rate of 0.95, and the DSE performed over 1000 generations. For the exploration, a larger CNN benchmark called ResNet-18 has been trained on the CIFAR-10 dataset [13]. After training, the CNN using IEEE 754 single precision, an error rate R of 7.99% has been obtained.

5.3 Search Space Analysis

Similar to the LeNet5-base benchmark as shown in Fig. 1, we found that no more than $a_i = 8$ correct mantissa bits are required in each layer to obtain the above desirable error rate of 7.99% as achievable using IEEE 754 single precision computations. We therefore decided to set the maximum value of a_i to 9 for every layer i. With ResNet-18 consisting of $|V| = 21$ layers, the resulting search space therefore consists of 9^{21} design points to be explored[1].

5.4 Experimental Results

In the following, we present the exploration results for the ResNet-18 benchmark only due to space limitations. For each individual explored during the DSE, the execution time T and the energy consumption E are evaluated analytically according to Eqs. (7–11, 13–14), respectively. The only time-consuming part of the DSE is the evaluation of the error rate R of each explored solution. Here, 10,000 validation images need to processed to calculate R, resulting in overall exploration times of about 1 day per exploration of 1000 generations. We provide exploration results for the presented online arithmetic as well as the bitmasking approach and finally compare the explored Pareto-fronts with the design point of an IEEE 754 single precision reference implementation.

Online Arithmetic Approach. The results on the dynamic power characterization per operation according to Sect. 4.3 are shown in Fig. 2. As can be seen, the average power per operation $P_{op}(a_i)$ is highest for $a_i = 1$ for both multiplication and addition operations. The reason for this can be explained using Fig. 3 which shows the latencies $L_{op}(a_i)$ for each value of a_i according to Eq. (7) and Eq. (8). The values for p_{op} in Eq. (6), and hence the value for bpc_{op} in Eq. (5), have been chosen in such a way that the

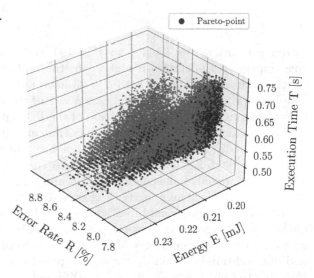

Fig. 4. 3D objective space including error rate R, energy consumption E, and execution time T with Pareto-optimal CNN configurations shown in blue. (Color figure online)

---

[1] Note that $|V| = 21$ for ResNet-18 when also counting three 1×1 convolutional layers.

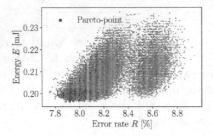
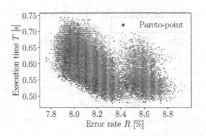

Fig. 5. 2D projection of the 3D objective space visualizing error rate R against energy consumption E for an AIP implementation based on online arithmetic with Pareto-optimal CNN configurations shown in blue. (Color figure online)

Fig. 6. 2D projection of the 3D objective space visualizing error rate R against execution time T for an AIP implementation based on online arithmetic with Pareto-optimal CNN configurations shown in blue. (Color figure online)

latency of additions and multiplications within the MAC-operations is equalized. More power is consumed if new instructions are introduced more frequently which is the case for the values of smaller a_i. From the power characterization, the values $E_{op}(a_i)$ can then be computed as

$$E_{op}(a_i) = \frac{P_{op}(a_i) \cdot L_{op}(a_i)}{f}. \tag{15}$$

In our evaluations, we fixed f to $f = 400$ MHz. As can be seen, the energy consumption per operation is highest for the multiplication for $a = 1$, whereas for the addition, operations with $a = 23$ have the highest energy consumption. The high switching power for the case $a = 1$ of the online multiplication is due to two reasons: First, the computation of the first bit is only possible after computing a minimal number of three δ-bits that are later discarded from the final result. Secondly, due to the higher rate of computed operations (the computation of one bit takes fewer cycles than the computation of two or more bits) and because we only issue one instruction in the functional unit's pipeline in our evaluations, the switching activity is considerably higher. By adding the estimated energy consumption of an addition and a multiplication instruction, we obtain the average energy consumption of a MAC operation. Interestingly, an average energy minimum of a MAC operation consisting of one multiplication and one addition is obtained at $a = 7$. Figure 4 depicts the explored 3D Pareto front for the objectives R, E and T. Different two-dimensional projections of this point set are also given in Fig. 5, Fig. 6, and Fig. 7. In the three-dimensional plot, two clusters of solutions can be distinguished, separated by different error rates, which is further illustrated in Fig. 5. The left cluster of solutions is characterized by design points with the majority of layers quantized with $a_i = 7$.

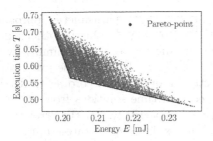

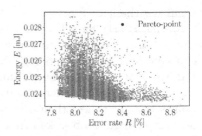

Fig. 7. 2D projection visualizing execution time T against energy consumption E for an AIP implementation based on online arithmetic with Pareto-optimal CNN configurations shown in blue. (Color figure online)

Fig. 8. 2D objective space for an AIP variant using bitmasking according to [3] visualizing error rate R against energy E with Pareto-optimal CNN configurations shown in blue. (Color figure online)

The Pareto-points (blue) inside this cluster are characterized by as well low error rates R as low energy E. In contrast, Pareto-points in the right cluster with higher error rates are points with lowest execution time T. Interestingly, these design points quantize several layers at much lower value of a_i, i.e., many low precision quantizations use $a_i < 5$. Overall, design points containing many layers with $a_i = 5$ or $a_i = 6$ are rare. Thus, the diversity of accuracy values between those two point clouds is sparse. The total spread of error rate among the evaluated points is also rather small and values from 7.8% up to 8.8% are observed. Another 2D projection of the 3D objective space is shown in Fig. 6, which plots the execution time T over the error rate R. Regarding the execution time, there is a clear trend that solutions with a small execution time T have an increased error rate R. Such solutions contain layers with a_i close to one. Another aspect worth mentioning is that minimum error rate R solutions show an execution time T in the range between $0.60\,\text{s} < T < 0.62\,\text{s}$, where obviously allocating more bits for a_i does not lead to a lower error rate R. Finally, in Fig. 7, a 2D projection of the 3D objective space along the error rate is shown. It can be clearly observed that Pareto-optimal points are located along two lines. On the right line, the Pareto-points are solutions tending to lower values of a_i with increasing energy E. Solutions on the left line correspond to higher values of a_i with execution time T increasing accordingly. The triangular shape can be explained using Fig. 3. Obviously, the execution time of a layer increases step-wise only with increasing accuracy values a_i. The change of slope is given exactly at the point where a decrease in energy E requires an increase of a_i to values higher than 5 that, in turn, leads to a higher execution time T. Table 2 reports the concrete values for some of these points. Apart from a Pareto-point with lowest error rate R that is even lower than the IEEE 754 reference point, a solution with lowest energy consumption E (almost all layers processed at $a_i = 7$ or $a_i = 8$) and one with lowest execution time T. For the latter, all layers are processed at an accuracy of $a_i \leq 4$ except for one layer. The configuration requiring the least amount of energy E using the online arithmetic approach is choosing $a_i = 7$ for all layers i of the CNN (see Fig. 2). On the other side of the spectrum, configurations resulting in the shortest execution time T would choose $a_i = 1$ for all layers i of the CNN (see Fig. 3). Even if these two extreme Pareto-points have not been found directly by our DSE according to Table 2, the deviation of best energy and best execution time solutions from these two points amount to less than 0.19% and 1.9%, respectively. For the error rate R, no best configuration is known a priori. Our explored error rate-minimal solutions, however, provide interesting insights into the importance of the individual layers of the CNN: For the best error rate solution in Table 2, obviously the first layers (close to the inputs) and the last layers (close to the outputs) are more important to the overall correctness of classification than the middle layers. This can be observed also when inspecting other Pareto-optimal points close to this point in the objective space.

Bitmasking Approach. In analogy to Fig. 2 for the online arithmetic approach, a characterization of average energy of additions and multiplications using the bitmasking approach is visualized in Fig. 9. The bitmasking addition requires less energy than the referenced IEEE 754 single precision operation for $1 \leq a \leq 4$. It uses less energy for the multiplication for $1 \leq a \leq 23$, i.e. in all possible configurations. The implementation of the bitmasking multiplication masks the partial products. It requires in average considerably less energy than an IEEE 754 single precision operation,

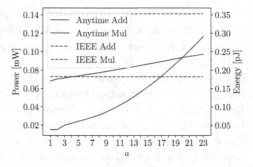

Fig. 9. Power and energy characterization of multiplication and addition operations for an AIP implementation based on bitmasking according to [3] for different values of a.

because for $a = 23$, half of the sum of partial products is not computed. Finally, based on the power models, we also explored the Pareto front for the bitmasking approach as depicted in Fig. 8. When comparing the online arithmetic approach with the bitmasking approach, Table 2 reveals a huge disparity in energy consumption between the two approaches. The best energy solution of the online arithmetic approach requires a factor of 7.8× more energy than using the bitmasking approach. Finally, we compare our solutions to a reference implementation of the IEEE 754 single precision standard. Interestingly, the best energy solution of the bitmasking approach requires just 62% of the IEEE 754 reference (see Table 2). Finally, the minimal error rate is obtained when using the online arithmetic approach with a reduction of approximately 7% compared to the IEEE 754 single precision reference point. We conclude that the use of AIPs enables to save a considerable amount of energy for CNN inference using the bitmasking approach whereas the implementation based on online arithmetic

Table 2. Comparison of different ResNet-18 design points

Design point	x	R [%]	E [mJ]	T [s]
IEEE 754 single prec.	$(23, 23, \quad\quad \ldots \quad\quad , 23, 23)$	**8.57**	**0.060**	0.469
— Online Arithmetic —				
Reference point	$(23, 23, \ldots, 23, 23)$	8.57	0.226	1.500
Best error rate	$(6, 8, 6, 2, 3, 3, 3, 5, 2, 2, 2, 4, 2, 2, 2, 7, 4, 6, 4, 3, 8)$	7.76	0.201	0.608
Best energy	$(7, 7, 8, 7, 7, 7, 7, 4, 7, 7, 2, 7, 2, 7, 7, 7, 8, 8, 7, 8)$	7.95	0.196	0.743
Best execution time	$(1, 1, 1, 1, 1, 4, 1, 8, 1, 1, 1, 1, 4, 1, 1, 4, 1, 2, 1, 1, 4)$	8.33	0.237	0.478
— Bitmasking —				
Reference point	$(23, 23, \ldots, 23, 23)$	8.57	0.061	0.469
Best error rate	$(3, 7, 7, 2, 5, 2, 2, 2, 2, 2, 2, 1, 3, 7, 1, 1, 1, 1, 1, 1, 8)$	**7.80**	0.025	0.469
Best energy	$(1, 1, 1, 1, 1, 2, 1, 3, 1, 1, 1, 1, 2, 1, 1, 1, 1, 1, 1, 1, 1)$	8.65	**0.023**	0.469

provide reductions in error rate when not computing all layers at an accuracy of $a = 23$ such as the IEEE 754 single precision reference design.

6   Conclusion

In contrast to approaches proposing specialized hardware solutions for CNNs, this paper proclaims the use of a class of accuracy-configurable processors called Anytime Instruction Processors (AIPs). Here, for a given precision, e.g., half, single or double precision floating-point format, the number of most significant mantissa bits to be accurately determined is encoded in the instruction word itself and can thus be adjusted from instruction to instruction in order to realize energy and execution time savings. For AIPs, we considered the DSE problem of energy, execution time and error rate optimization for CNNs in which the number a of mantissa bits to be computed for all neurons of each layer is explored using an evolutionary algorithm. It was shown that configurations can be found that lead to as much as 62% in energy reduction for a representative ResNet-18 benchmark using an AIP in comparison to an IEEE 754 single precision processor reference implementation. For a given CNN, the exploration needs to be carried out only once. Finally, exploration times of less than a day may still be reduced by performing the time-consuming evaluation of the error-rate per solution in parallel on a server of GPUs or by running the evaluations on an AIP FPGA implementation.

Acknowledgments. This work was partially funded by the Deutsche Forschungsgemeinschaft (DFG, German Research Foundation)—Project Number 146371743—TRR 89 Invasive Computing and the German Federal Ministry for Education and Research (BMBF) within project KISS (01IS19070B).

References

1. Alizadeh, M., Behboodi, A., van Baalen, M., Louizos, C., Blankevoort, T., Welling, M.: Gradient ℓ_1 regularization for quantization robustness. In: 8th International Conference on Learning Representations, ICLR (2020)

2. Brand, M., Witterauf, M., Bosio, A., Teich, J.: Anytime floating-point addition and multiplication-concepts and implementations. In: 31st IEEE International Conference on Application-Specific Systems, Architectures and Processors, ASAP, pp. 157–164. IEEE (2020)
3. Brand, M., Witterauf, M., Hannig, F., Teich, J.: Anytime instructions for programmable accuracy floating-point arithmetic. In: Proceedings of the 16th ACM International Conference on Computing Frontiers, CF, pp. 215–219. ACM (2019)
4. Cai, Y., Yao, Z., Dong, Z., Gholami, A., Mahoney, M.W., Keutzer, K.: ZeroQ: a novel zero shot quantization framework. In: 2020 IEEE/CVF Conference on Computer Vision and Pattern Recognition, CVPR 2020, pp. 13166–13175 (2020)
5. Cheng, Y., Wang, D., Zhou, P., Zhang, T.: Model compression and acceleration for deep neural networks: the principles, progress, and challenges. IEEE Sig. Process. Mag. **35**(1), 126–136 (2018)
6. Deb, K., Agrawal, S., Pratap, A., Meyarivan, T.: A fast and elitist multiobjective genetic algorithm: NSGA-II. IEEE Trans. Evol. Comput. **6**(2), 182–197 (2002)
7. Fu, Y., et al.: FracTrain: fractionally squeezing bit savings both temporally and spatially for efficient DNN training. In: Advances in Neural Information Processing Systems 33: Annual Conference on Neural Information Processing Systems 2020, NeurIPS, pp. 12127–12139 (2020)
8. Garófalo, A., Rusci, M., Conti, F., Rossi, D., Benini, L.: PULP-NN: accelerating quantized neural networks on parallel ultra-low-power RISC-V processors. Phil. Trans. R. Soc. A **378**, 20190155 (2019)
9. Genc, H., et al.: Gemmini: an agile systolic array generator enabling systematic evaluations of deep-learning architectures. CoRR abs/1911.09925 (2019)
10. He, K., Zhang, X., Ren, S., Sun, J.: Deep residual learning for image recognition. In: 2016 IEEE Conference on Computer Vision and Pattern Recognition, CVPR, pp. 770–778. IEEE Computer Society (2016)
11. Heidorn, C., Witterauf, M., Hannig, F., Teich, J.: Efficient mapping of CNNs onto tightly coupled processor arrays. J. Comput. **14**(8), 541–556 (2019)
12. Keszocze, O., Brand, M., Witterauf, M., Heidorn, C., Teich, J.: Aarith: an arbitrary precision number library. In: ACM/SIGAPP Symposium On Applied Computing (2021). Aarith is publicly available at https://github.com/keszocze/aarith
13. Krizhevsky, A., Nair, V., Hinton, G.: The CIFAR-10 dataset (2010). https://www.cs.toronto.edu/~kriz/cifar.html. Accessed 20 Feb 2020
14. LeCun, Y., Cortes, C., Burges, C.J.: MNIST handwritten digit database (1998). http://yann.lecun.com/exdb/mnist/. Accessed 18 Nov 2020
15. Lukasiewycz, M., Glaß, M., Reimann, F., Teich, J.: Opt4j: a modular framework for meta-heuristic optimization. In: Proceedings of the 13th Annual Conference on Genetic and Evolutionary Computation, pp. 1723–1730 (2011)
16. Mrazek, V., Vasícek, Z., Sekanina, L., Hanif, M.A., Shafique, M.: ALWANN: automatic layer-wise approximation of deep neural network accelerators without retraining. In: Proceedings of the International Conference on Computer-Aided Design, ICCAD 2019, pp. 1–8. ACM (2019)
17. la Parra, C.D., Guntoro, A., Kumar, A.: ProxSim: GPU-based simulation framework for cross-layer approximate DNN optimization. In: 2020 Design, Automation & Test in Europe Conference & Exhibition, DATE 2020, pp. 1193–1198. IEEE (2020)
18. Rek, P., Sekanina, L.: TypeCNN: CNN development framework with flexible data types. In: Design, Automation & Test in Europe Conference & Exhibition, DATE 2019, pp. 292–295 (2019)

19. Synopsys: Synopsys 32/28 nm and 90 nm generic libraries (2020). https://www.synopsys.com/community/university-program/teaching-resources.html. Accessed 06 Aug 2020
20. Sze, V., Chen, Y.H., Yang, T.J., Emer, J.S.: Efficient processing of deep neural networks: a tutorial and survey. Proc. IEEE **105**(12), 2295–2329 (2017)
21. Umuroglu, Y., Rasnayake, L., Själander, M.: BISMO: a scalable bit-serial matrix multiplication overlay for reconfigurable computing. In: 28th International Conference on Field Programmable Logic and Applications, FPL 2018, pp. 307–314. IEEE Computer Society (2018)
22. Wang, K., Liu, Z., Lin, Y., Lin, J., Han, S.: HAQ: hardware-aware automated quantization with mixed precision. In: IEEE Conference on Computer Vision and Pattern Recognition, CVPR 2019, pp. 8612–8620 (2019)

Ultra-low Power Machinery Fault Detection Using Deep Neural Networks

Sven Nitzsche[1(✉)], Moritz Neher[1], Stefan von Dosky[2], and Jürgen Becker[3]

[1] FZI Research Center for Information Technology, Karlsruhe, Germany
{nitzsche,neher}@fzi.de
[2] Siemens AG, Digital Industries, Process Automation, DI PA TI OTP,
Munich, Germany
[3] Karlsruhe Institute of Technology, Karlsruhe, Germany

Abstract. This work in progress investigates artificial and spiking neural networks for machinery fault detection on extremely resource-limited devices. We evaluate various architectures on classifying vibration data of centrifugal pumps and on two public bearing fault datasets. The target system is a sensor with an integrated ARM Cortex M4F microcontroller with 80 kB of main memory powered only by standard AA batteries. Compared to state of the art approaches the developed neural networks are optimized to run with extremely limited resources (i.e. a model size of less than 15 kbytes and a power consumption of few hundred milliwatts). We target an overall run time of 10 years without changing batteries. Current results show a run time of roughly 7.5 years.

Keywords: Machinery fault detection · Bearing fault detection · Artificial neural network · Spiking neural network

1 Introduction

Machinery faults are a major failure cause in process industry. A catastrophic failure of an electrical machine (e.g. due to faulty pumps or bearings) may trigger subsequent errors in the production line and poses a high economic risk [10]. Predictive maintenance based on machine learning can be used to identify such faults early and take appropriate countermeasures [9,12]. Typically, such machine learning approaches require substantial computational power to process sensor data in real time. Consequently, they run on powerful workstations that gather sensor data from the observed machines via network. Depending on the application, transmitting this data requires a lot of energy on the sensor side, which in turn makes wired connections or a sensor with connection to the power grid necessary. This complicates the placement of sensors and makes predictive maintenance retrofits costly and inflexible.

In this work, we investigate various machine learning approaches for machinery fault detection to run directly in a battery-powered sensor, where only prediction results are transmitted via network so that appropriate action can be

© Springer Nature Switzerland AG 2021
M. Kamp et al. (Eds.): ECML PKDD 2021 Workshops, CCIS 1524, pp. 390–396, 2021.
https://doi.org/10.1007/978-3-030-93736-2_30

taken. The system should be able to run for multiple years without battery replacement, making a retrofit simple and flexible. As use case, a low cost multi-sensor from Siemens is used to detect anomalies in centrifugal pumps based on vibrations. The resulting neural network architectures are also trained on public datasets to investigate their suitability for bearing fault detection. In a first step, mainly various artificial neural networks were used. Spiking neural networks are considered as well, however since this is still work in progress, only preliminary simulation results are available.

2   Approach and Limitations

The main focus of this work is to find suitable neural networks architectures to run in a very resource-limited environment. We considered various types of Artificial Neural Networks (ANNs), specifically Feed Forward Neural Networks (FFNNs) as well as Convolutional Neural Networks (CNNs). Details on the exact architecture and training methods can be found in Sect. 4. Furthermore, we investigate Spiking Neural Networks (SNNs) by converting the best performing ANNs to rate-coded SNNs using NengoDL [8]. Preliminary results based on simulations are given in Sect. 4.2.

In a first step, recorded vibration data from a KSB Etabloc centrifugal pump in four different operating states is used to test various architectures for the neural networks and find suitable candidates. The four operating states are a nominal state, weak and strong cavitation and hydrolock. The data was collected in intervals of 30 s over the course of approximately one hour per operating state. Each sample contains 512 data points at a sample rate of 6.66 kHz, resulting in a duration of 76.8 ms per measurement. More details on the recorded data can be found on GitHub[1]. The best networks are then fine-tuned and retrained for two common bearing fault datasets. The first one is the CaseWestern Reserve University (CWRU) Dataset [5], which uses artificially damaged bearings that exhibit three different fault patterns and three fault sizes. Together with data from healthy bearings a classification dataset with 10 classes is formed similar to related work presented in Sect. 3 [6,13]. As a second dataset, the Paderborn University Dataset [11] is used. It provides data obtained from bearings with real damages caused by accelerated life tests. A dataset with three classes based on the fault location is used comparable to the approach by Pandhare et al. [7]. However, in our case all three operating states are combined to form a more diverse dataset. The dataset has a sample rate of 64 kHz but is subsampled by a factor of four to be similar in sample rate to the two other datasets. Results for all datasets are discussed in Sect. 5.

During development of the models, various limitations imposed by the target platform had to be taken into account. The sensor system is based on an ARM Cortex M4F microcontroller with operating system and Bluetooth stack, leaving 30 kB of main memory for the neural network and its runtime library. For ANNs, we chose TensorFlow Lite Micro as runtime library, since it offers a seamless

---

[1] https://github.com/nitzsche-fzi/machinery-fault-detection.

workflow from training to deployment and the possibility to use ARM's CMSIS-NN library for optimized kernel implementations. The runtime library requires about 15 kB. Consequently, all network models were limited to a size of 15 kB. The whole system is powered by two standard AA batteries and consumes 20 μW during sleep. It wakes up from sleep mode every 15 s for measuring and inference. The target run time with this energy budget is ten years, including energy for the transmission of results.

3   State of the Art

Since the generated data is one-dimensional and originates from a centrifugal pump, it can be referred to as vibration analysis for machinery fault detection. Combinations of both high and low frequencies present in this kind of data pose a major challenge for conventional CNNs. To solve this issue, Zhang et al. [13] propose Wide First-layer Kernels Deep CNNs (WDCNNs). These networks have a convolutional layer with very large kernels as their first layer, which is followed by several layers with conventional small kernels. This way the receptive field of the network gets increased while maintaining good resolution.

A specialized subcategory of machinery fault detection is the so called bearing fault detection, where several attempts have been made to effectively detect and classify common fault types of bearings using deep learning techniques. In their work, Xiaojie et al. [3] use a hierarchical approach with a two-stage classification framework on the CWRU dataset. In the first stage, the fault pattern is determined by means of a CNN. Next, a separate convolutional network is trained for each fault pattern to evaluate the fault size. Lu et al. [6] compare CNNs to conventional bearing fault detection methods like Stacked Denoising Autoencoders (SAEs), support vector machines and Softmax Regression Classifiers (SRs) and show their superior accuracy under different noise levels. They also use the CWRU dataset, but in contrast to the work mentioned above, they define ten distinct classes as combinations of fault pattern and size. Pandhare et al. [7] use multiple input data formats in their work on the Paderborn dataset in combination with a multilayer CNN. A comprehensive review of recent works using various public datasets is given by Zhang et al. [12]. They show different deep learning methods and their application to bearing fault detection. So far, related work focused on achieving high accuracy instead of small model size and energy efficiency.

With respect to Spiking Neural Network, there has been little related work to date. Zuo et al. [14] simulate a shallow SNN with no hidden layer that is trained to on the CWRU dataset, reaching 98.95% accuracy with data preprocessing. Dennler et al. [2] present an SNN to detect vibration anomalies using preprocessing based on a cochlea model. Besides simulation in Brian2 they also run their model on analog neuromorphic hardware and test it using IBF and R2F datasets. Again, both works focus on accuracy and do neither report model size nor required energy.

4   Neural Network Models

The deep neural networks presented in this work have been trained using the TensorFlow framework by Google. All networks have a batch normalization layer before each dense or convolutional layer. Additionally a max pooling layer is present after each convolutional layer. To prevent overfitting, a dropout of 0.1 is used. The training and test data is oversampled to increase the total number of samples. For training, the Adam stochastic optimization algorithm is applied. Hyperparameter scan is done using the Python package Talos [1]. In case of the public bearing fault datasets a ten fold cross validation is performed. The source code for training and evaluation and a detailed overview of the hyperparameters can be found on GitHub[2].

4.1   Artificial Neural Networks

In this work, three different variants of small sized CNNs are proposed. The first variant only has one convolutional layer. The second one is an adapted version of the previously discussed WDCNNs with three convolutional layers. In the third variant with two convolutional layers called Depth-wise Separable Convolutional Neural Network (DS-CNN) the second layer is replaced by a separable convolution, which enables more space and time efficient inference [4]. We use a wide kernel for the first layer of all variants, while following layers have conventional small kernels. To provide a benchmark and show the suitability of CNNs for the task of machinery and bearing fault detection, a FFNN with two layers of varying size is included as well.

4.2   Spiking Neural Networks

Spiking Neural Networks potentially provide a drastically increased energy efficiency compared to ANNs. Therefore, we investigate the feasibility of using SNNs in the given use case by converting a two-layer CNN model via NengoDL using rate-coding and leaky integrate-and-fire neurons. Preliminary results are presented in this section, however the SNN implementation is still work in progress. With a spike rate scaling factor of at least 20, we achieved a sufficient accuracy in simulation. Figure 1 shows the effect of spike rates on accuracy for different values of synaptic smoothing s. Synaptic smoothing is a low pass filtering of generated spikes to prevent fast, erratic switching of the network outputs. Besides spike rates, the accuracy mainly depends on the time a specific input is presented to the network. We achieved good results with a presentation time of at least 50 ms. However, as Fig. 2 shows, the required energy per inference scales linearly with presentation time, hence it should be as short as possible.

---

[2] https://github.com/fzi-neuromorphic/machinery-fault-detection.

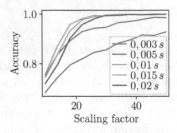

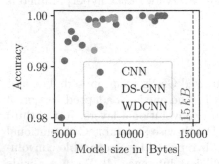

Fig. 1. Accuracy depending on spike rate scaling and synaptic smoothing

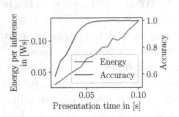

Fig. 2. Influence of presentation time on energy per inference and accuracy

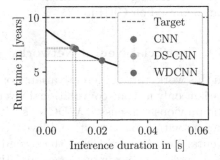

Fig. 3. Accuracy of different network architectures compared to model size

Fig. 4. Run time over inference duration for measuring period $\tau = 15\,s$

5   Evaluation

For a first selection of models we evaluated multiple candidates for each ANN architecture with respect to accuracy in relation to the model size. Accuracy and size refer to the converted and 8-bit quantized networks in ".tflite" binary format. This first step was done using our self-recorded centrifugal pump dataset, with results shown in Fig. 3. Only models smaller than 15 kB were considered. Normal CNNs perform best in this regard, reaching an accuracy of up to 100% with a model size of about 11 kB. DS-CNNs also perform well. They reach an accuracy of 99.9% with just 8.5 kB. WDCNNs result in comparably large model sizes. A similar picture emerges for the other two datasets, detailed data can be found on GitHub.

However, this statement is put into perspective when model complexity in form of FLOPs is taken into account. Table 1 shows size, number of FLOPs and accuracy for the best candidates for each architecture using our pump dataset (Table 1a) and for equal sized models using the CWRU dataset (Table 1b). Results for the Paderborn dataset are similar and can be found on GitHub. Note that no results for SNNs are given in these tables since their development is still in a very early stage and so far they have only been implemented in a simulation framework. Hence, the numbers would not be comparable with the microcontroller implementations of the ANNs.

Normal CNNs require heavy computation with over ten times as many FLOPs as for DS-CNNs for the smallest possible models (Table 1a). With roughly equal size, CNNs still require almost two times as many FLOPs (Table 1b). The large amount of FLOPs leads to an increased inference time and consequently higher energy requirements. This, in turn, drastically decreases battery life of the sensor as can be seen in Fig. 4. It visualizes sensor run time in years depending on the inference time. Inference for CNNs takes roughly twice as long as for other architectures, decreasing run time by about 1.5 years.

Table 1. Comparison of ANN architectures

(a) Pump dataset				(b) CWRU dataset			
Model	Size (B)	FLOPs	Acc. (%)	Model	Size (B)	FLOPs	Acc. (%)
FFNN	12 320	17 716	83.59	FFNN	14 128	19 314	84.11
CNN	11 056	269 348	100	CNN	14 256	180 290	99.72
DS-CNN	8 400	21 076	99.93	DS-CNN	14 320	101 690	99.54
WDCNN	12 256	73 908	100	WDCNN	14 144	116 946	99.68

6   Conclusion and Future Work

Based on two public and one self-recorded dataset, we investigated various neural network architectures regarding size and energy requirements when run on an embedded microcontroller. We achieved accuracies of over 99% in all scenarios with models smaller than 15 kB. Especially standard CNNs performed good in terms of accuracy and model size. However, since they require a high number of calculations, the overall system run time was comparably short. As an alternative, CNNs with depth-wise separable convolutions can be used to achieve similar accuracy and a long run time of about 7.5 years using two standard AA batteries. Compared to the state of the art, we managed to keep a high accuracy but significantly reduced model size and energy requirements.

With artificial neural networks we were not able to achieve the target run time of 10 years. However, preliminary results using more efficient spiking neural networks show their suitability for the given tasks. For future work, we therefore plan to optimize our SNN implementation and port it to the target system by extending it with suitable hardware-based acceleration. Furthermore, a native implementation instead of the converted SNN may help to further reduce energy requirements. For ANNs, a more robust domain adaptability has to be achieved for the solution to be viable in an industrial context. More diverse datasets and additional methods can be explored to reach that goal.

References

1. Autonomio: Talos [computer software] (2019). http://github.com/autonomio/talos
2. Dennler, N., Haessig, G., Cartiglia, M., Indiveri, G.: Online detection of vibration anomalies using balanced spiking neural networks. CoRR abs/2106.00687 (2021)
3. Guo, X., Chen, L., Shen, C.: Hierarchical adaptive deep convolution neural network and its application to bearing fault diagnosis. Measurement **93**, 490–502 (2016)
4. Howard, A.G., et al.: MobileNets: efficient convolutional neural networks for mobile vision applications. arXiv abs/1704.04861 (2017)
5. Loparo, K.: Case western reserve university bearing data center. Case Western Reserve University, Bearing Data Center, Cleveland, OH, USA, Technical report (2012)
6. Lu, C., Wang, Z., Zhou, B.: Intelligent fault diagnosis of rolling bearing using hierarchical convolutional network based health state classification. Adv. Eng. Inform. **32**, 139–151 (2017)
7. Pandhare, V., Singh, J., Lee, J.: Convolutional neural network based rolling-element bearing fault diagnosis for naturally occurring and progressing defects using time-frequency domain features. In: 2019 Prognostics and System Health Management Conference (PHM-Paris), pp. 320–326. IEEE (2019)
8. Rasmussen, D.: NengoDL: combining deep learning and neuromorphic modelling methods. Neuroinformatics **17**, 611–628 (2019)
9. Saufi, S.R., Ahmad, Z., Leong, M., Lim, M.H.: Challenges and opportunities of deep learning models for machinery fault detection and diagnosis: a review. IEEE Access **7**, 122644–122662 (2019)
10. SKF Group: Bearing damage and failure analysis. Technical report, SKF Group, June 2017
11. University of Paderborn: Bearing datacenter (2020). https://mb.uni-paderborn.de/kat/forschung/datacenter/bearing-datacenter
12. Zhang, S., Zhang, S., Wang, B., Habetler, T.G.: Deep learning algorithms for bearing fault diagnostics - a comprehensive review. IEEE Access **8**, 29857–29881 (2020)
13. Zhang, W., Peng, G., Li, C., Chen, Y., Zhang, Z.: A new deep learning model for fault diagnosis with good anti-noise and domain adaptation ability on raw vibration signals. Sensors **17**(2), 425 (2017)
14. Zuo, L., Zhang, L., Zhang, Z., Luo, X., Liu, Y.: A spiking neural network-based approach to bearing fault diagnosis. J. Manuf. Syst. **61**, 714–724 (2020)

SPNC: Fast Sum-Product Network Inference

Lukas Sommer[1]([✉]) [iD], Cristian Axenie[2], and Andreas Koch[1] [iD]

[1] Embedded Systems and Applications Group Technical University Darmstadt,
Darmstadt, Germany
{sommer,koch}@esa.tu-darmstadt.de
[2] Intelligent Cloud Technologies Laboratory Huawei Munich Research Center,
Munich, Germany
cristian.axenie@huawei.com

Abstract. Sum-Product Networks have received increasing attention from academia and industry alike, but the software ecosystem is comparably sparse. In this work, we enhance the ecosystem with an open-source, domain-specific compiler that allows to easily and efficiently target CPUs and GPUs for Sum-Product Network inference. The implementation of the compiler is based on the open-source MLIR framework. Using a real-world application of Sum-Product Networks, a robust speaker identification model, we showcase the performance improvements our compiler can achieve for SPN inference on CPUs and GPUs.

Keywords: Sum-Product Networks · Compiler · MLIR · LLVM

1 Introduction

Probabilistic models are receiving increasing attention from both academia and industry, being a complementary alternative to more widespread machine learning approaches such as (deep) neural networks (NN). Probabilistic models can handle the *uncertainty* found in real-world scenarios better, and are also, in contrast to NNs, able to *express uncertainty* over their output.

However, in contrast to neural networks, for which a rich ecosystem with a variety of frameworks, libraries and compilers, such as Tensorflow's XLA, or Facebook's Glow, is available, the ecosystem for probabilistic models such as Sum-Product Networks (SPN) is comparatively sparse, due to them being a relatively young class of models.

One of the most popular libraries for research with Sum-Product Networks is SPFlow by Molina et al. [3], which provides a programmatic representation of Sum-Product Network models and allows to learn their structure and parameters

Calculations for this research were conducted on the Lichtenberg high performance computer of TU Darmstadt. This research was funded by the German Federal Ministry for Education and Research (BMBF) with the funding ID ZN 01|S17050.

© Springer Nature Switzerland AG 2021
M. Kamp et al. (Eds.): ECML PKDD 2021 Workshops, CCIS 1524, pp. 397–408, 2021.
https://doi.org/10.1007/978-3-030-93736-2_31

from data. SPFlow also allows to perform inference on the models obtained through learning, but is implemented in pure Python and can therefore not leverage the full feature set of CPUs or GPUs. However, exploiting all available hardware features is crucial for the deployment of SPN models on embedded-grade devices and for efficient inference in real-world applications, e.g., to fulfill real-time requirements.

Therefore, in this work, we enhance the SPN ecosystem by developing *SPNC*, an *open-source* domain-specific, multi-platform compiler for performing fast Sum-Product Network inference on CPUs and GPUs, and present the following contributions:

- Based on the MLIR framework [2], we develop custom high-level intermediate representations capturing the semantics of Sum-Product Networks in the compiler (Sect. 3.1).
- We define efficient strategies to map Sum-Product Network inference to CPUs with vector extensions and CUDA GPUs. The mapping strategies make use of the underlying hardware's specific features for efficient inference (Sect. 3.2 and Sect. 3.3).
- Using a real-world application of Sum-Product Networks, we evaluate our approach in detail, and compare it to the currently available framework (Sect. 5).
- We develop a Python interface to our compiler, which seamlessly integrates with SPFlow [3] and allows to target CPUs and GPUs with ease (Sect. 4).

Furthermore, we provide necessary background information on Sum-Product Networks and the open-source MLIR framework in the next section, and discuss related works in Sect. 6.

2   Background

2.1   Sum-Product Networks

SPNs [7] are a relatively young class of probabilistic graphical models (PGM). In principle, such class of models can be considered a unified approach combining Bayesian network representation formalism and Markov random field computation. Such a computational configuration enables SPNs to efficiently reason under incompleteness and uncertainty, which is a challenging task in many real-world scenarios [11]. Unfortunately, inference requires intense computation that introduces a long delay. This is a core motivation of our work.

Additionally, in contrast to most neural network architectures, SPNs are also able to quantify uncertainty over the output. An example for this property can be found in [6], where SPNs, when confronted with out-of-domain images, indicate this through a low likelihood for the output class, in contrast to the multi-layer perceptron undergoing the same test. An overview of other practical usage examples of SPNs can be found in the survey by Paris et al. [5].

Sum-Product Networks capture the joint probability of a set of variables (i.e., features) in the form of a directed acyclic graph (DAG). Regardless of the

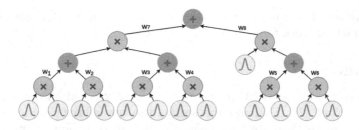

Fig. 1. Example of a sum-product network graph.

application and the underlying data, the DAG is always composed from three different types of nodes. At the bottom of the DAG, so-called *leaf nodes* capture the univariate probability distribution of a single variable/feature. Depending on the type of data (e.g., continuous vs. discrete), and underlying distribution of the data, different probability distributions can be used, e.g., Gaussian distribution for continuous variables, or a categorical distribution for discrete values. The so-called *scope* of a leaf node is the single variable associated with it. Further up in the graph, a combination of product nodes and weighted sum nodes is used to capture the joint probability distribution. Product nodes represent factorizations of independent variables. For the SPN to be valid, the scopes of the different child nodes of a product node must be *disjoint*. Weighted sum nodes, on the other hand, represent a mixture of distributions and the scopes of all child nodes must be *identical* in a valid SPN. The structure of the SPN depends on the distribution of the underlying data, and can either be learned from data, or be hand-crafted, just followed by parameter learning. An overview of SPN learning algorithms can be found in [5]. A small example of an SPN graph is shown in Fig. 1. As a core functional principle, the SPN decomposes complex multivariate "global" functions by exploiting the way in which the global function factors into a product of simpler "local" functions of a subset of the variables [1]. SPN can be used to solve machine learning tasks, such as classification, by performing inference on the underlying DAG. In general, SPNs support multiple different types of inference. In this study, we are focusing on two of these types, namely joint probability inference and marginal inference. Joint probability inference is used to obtain the joint probability given *full* evidence (i.e., a value for each variable). To this end, the evaluation of the SPN DAG starts by evaluating the distribution of the leaf nodes, given the value of the variable associated with each of them. After that, the values are propagated upwards through the DAG, performing multiplication or weighted addition at the product and sum nodes, until a final probability value is obtained at the root node of the SPN. Marginal inference, on the other hand, is used when only *partial* evidence is available. Leaf nodes for which no evidence is available are set to 1, the remaining ones are evaluated just as in joint inference, and the propagation of values through the SPN is performed analogously to the description above.

The compiler developed in this study aims to accelerate the *inference* in Sum-Product Networks by efficiently mapping them to different hardware targets.

Learning of the SPN is assumed to have taken place beforehand, using a standard Sum-Product Network framework such as SPFlow [3].

2.2 MLIR

The implementation of SPNC in this work is heavily based on the open-source MLIR framework [2][1]. Therefore this section presents a brief overview of MLIR.

MLIR aims to facilitate the implementation of compilers by providing an *extensible* framework for the implementation of multi-level IRs. The main reason for adding multiple levels of abstractions into compilers is that an early lowering to a *low-level* intermediate representation such as LLVM IR loses too much of the high-level structure of the program, which later on must be reconstructed using often fragile approaches based on the low-level IR in order to perform transformations, e.g., on loops. Capturing additional information and potentially domain-specific semantics in one or multiple high-level IRs enables the compiler to perform more powerful program transformations.

Because MLIR provides common components for the implementation of IRs, such as pass managers and common transformations, users can focus on the design of the IR itself.

In order to not impose too many constraints on the semantics of different IRs, MLIR defines a minimal set of *generic* abstractions that must be used by all IRs. Similar to most modern compilers, MLIR uses the static single assigment (SSA) form, with *operations* (short: Ops) consuming and producing *values*. All values are typed, with the type system being extensible, while also defining a number of common types. So-called *attributes*, which are also typed, can additionally be used to attach compile-time information to operations.

Operations, types, and attributes are organized in so-called *dialects*, which do not add any semantics, but are a mere logical unit for the organization of the IR. Dialects can be mixed in the same logical unit, and so-called *lowerings* translate *between* different dialects, with intra-dialect transformations also being available. Typically, a progressive, step-wise lowering from a high-level dialect to lower-level dialects is used to compile for a specific target, e.g., a CPU. To enable the implementation of common transformations, MLIR uses the notion of *traits* and *interfaces* that can be attached to operations, and provides generic interfaces for transformations such as constant folding.

MLIR's extensible nature allows us to design *custom* high-level IRs. We use these to represent Sum-Product Networks in the compiler developed in this work, while we can rely on the *common* infrastructure and dialects provided with MLIR to efficiently target different hardware platforms.

3 Approach

The aim of SPNC is to automatically compile Sum-Product Networks and probabilistic queries operating on them to executable kernels. Compiling individual

---

[1] https://mlir.llvm.org.

SPNs allows to employ *all* hardware features available on the target platform for fast inference, for example vector extensions present on most modern CPUs. Currently, SPNC supports two main targets:

- **CPUs:** Being based on MLIR and LLVM, SPNC can target any CPU for which a backend is present in LLVM. Vector extensions are currently supported on x86 (AVX, AVX2, AVX-512) and Arm (Neon Advanced SIMD) CPUs.
- **GPUs:** The flow currently supports Nvidia CUDA GPUs, but the generic GPU abstractions of MLIR would allow to target other GPUs with comparably few changes.

The compilation flow for both targets is based on MLIR. To this end, two SPN-specific MLIR dialects have been designed and implemented, which will be described in Sect. 3.1. Starting from these dialects, the target-specific lowerings will create an executable, using the flows described in Sect. 3.2 and Sect. 3.3. The user interface and some implementation details are described in Sect. 4.

3.1 MLIR Dialect Design

The first of the two SPN-specific dialects that SPNC employs during compilation, called **HiSPN**, captures the DAG structure of a Sum-Product Network and the information about the query to perform on a high level of abstraction. It was designed to closely match the representation used internally by the SPFlow framework [3], similar to how an abstract syntax tree captures a general-purpose programming language on a high level of abstraction.

In the HiSPN dialect, an abstract probability type is used for values inside the SPN DAG, allowing SPNC to delay the decision on the actual data type used for computation and take graph characteristics into account.

The second SPN-specific dialect, called **LoSPN**, represents the actual computation that needs to be performed to process the requested query on the SPN. The top-level unit in this dialect is a *Kernel*, comprising one or multiple *Tasks*. A Task does not only represent (parts of) the SPN DAG structure, including weighted sum, product, and leaf nodes, but also contains information about which inputs values will need to be accessed and which outputs will be produced as intermediate or final result. In contrast to HiSPN, LoSPN uses a concrete type for the values. To represent the computation in log-space, which commonly used to avoid arithmetic underflow in Sum-Product Network inference, a SPN-specific data type was added to the LoSPN dialect.

The lowering from HiSPN to LoSPN is currently identical for both flows, targeting CPU and GPU. In this step, the necessary computation for the query is derived from the SPN DAG structure and query information captured by HiSPN and is the lowered into the operations of the LoSPN dialect. After lowering, the LoSPN representation undergoes a number of transformations, including steps such as common subexpression elimination (CSE), followed by the target-specific lowerings to dialects provided by the MLIR framework described in the next two sections.

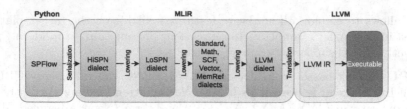

Fig. 2. CPU Compilation Flow.

3.2   CPU Compilation Flow

The compilation flow for the CPU starts with the serialized SPN model (cf. Sect. 4), an overview is shown in Fig. 2. After deserialization to the HiSPN dialect, lowering to LoSPN, and the transformations on the LoSPN dialect have been performed, the IR is again lowered to dialects provided as part of the MLIR framework. The Kernel and the Tasks in the LoSPN dialect are lowered to functions, with the Kernel function calling the functions for the individual Tasks and the Task functions iterating multiple inputs for batch processing.

The operations contained inside each Task are lowered to a combination of different dialects (Note that MLIR allows to mix operations from different dialects in the same function/module):

- **Standard dialect:** Contains operations such as simple addition or multiplication on arbitrary data-types, including vectors.
- **Math dialect:** Elementary math functions, such as the `exp` and `log` function, are represented by operations from this dialect.
- **SCF dialect:** Operations from this dialect represent structured control flow, e.g. for-loops.
- **MemRef dialect:** Contains facilities to handle memory, e.g., allocation or store/load operations.
- **Vector dialect:** Vector specific operations, e.g., vector lane shuffling.

The combination of the Vector dialect and the Standard operation's ability to handle vector data-types lets the compiler exploit the CPU's SIMD extensions, if present, for maximum efficiency. In contrast to a generic loop vectorization, a domain-specific compiler such as SPNC can, thanks to the MLIR framework, leverage high-level information to generate more efficient code, e.g., by employing a combination of simple vector loads and shuffles instead of expensive gather loads.

After some transformation passes provided by the MLIR framework, all dialects are lowered to the LLVM dialect and then translated to LLVM IR, so LLVM can produce the final executable. As part of this process, the executable is also linked with vector libraries, providing optimized implementations of elementary math functions (e.g., `exp`) for vector code. The currently supported vector libraries are Intel SVML and Libmvec for x86 CPUs, and ARM Optimized Routines for ARM Neon.

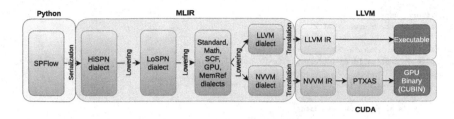

Fig. 3. GPU Compilation Flow.

Although it is technically possible to perform within the MLIR framework, we have decided to implement *multi-threading* in the runtime-component (cf. Sect. 4) rather than directly in the generated code. This allows to adopt the threading behavior dynamically, e.g., when executing multiple compiled kernels concurrently.

3.3   GPU Compilation Flow

Similar to the CPU compilation flow, the GPU compilation flow also starts from the serialized SPN model and performs the same steps up to the lowering of LoSPN to dialects from the MLIR framework. Here, for the GPU, the Kernel is lowered into a function, which will remain on the host CPU and will be responsible for GPU/CPU data transfers and the invocation of the Tasks, which, in contrast to the CPU flow, are lowered into *device* functions executing on the GPU.

For the operations inside the Tasks, a similar combination of MLIR-provided dialects is used, with one notable difference: Instead of the Vector dialect, the **GPU** dialect is used to represent the SIMT execution model, with operations for access to block and thread identifiers and for representation of GPU device functions and runtime functions for memory & execution management.

After that step, the GPU- and host portion of the IR are separated into two compilation units. While the flow for the host portion via LLVM is very similar to the CPU flow, eventually resulting in an executable, the GPU portion of the code is translated in multiple steps to NVVM IR, PTX assembly and a GPU binary (CUBIN format). This GPU binary is then loaded at runtime by the host function to execute inference on the GPU.

An overview of the overall compilation flow for the GPU is shown in Fig. 3.

4   Python Interface and Implementation

In order to make the compiler and the execution of the compiled binaries via the runtime component accessible to machine learning experts working with the SPFlow library, SPNC offers a Python-based interface to the compiler and runtime. In this manner, machine learning experts can create the SPNs using their familiar tools from the SPFlow library and feed their results to the compiler.

```
1   import numpy as np
2   from spn.structure... import ...
3
4   # Create an example SPN
5   p0 = Product(children=[Categorical(p=[0.3, 0.7], scope=1), Categorical(p=[0.4, 0.6], scope=2)])
6   ...
7   spn = Sum(weights=[0.4, 0.6], children=[p2, p4])
8
9   # Create some random test data
10  ...
11  test_data = np.c_[a, b, c].astype("float32")
12
13  # Perform inference using SPFlow
14  from spn.algorithms.Inference import log_likelihood
15  spflow_results = log_likelihood(spn, test_data)              # Location (1)
16
17  # Compile for CPU and perform inference
18  from spnc.cpu import CPUCompiler
19  cpu_results = CPUCompiler().log_likelihood(spn, test_data)   # Location (2)
20
21  # Compile for CUDA GPU and perform inference
22  from spnc.gpu import CUDACompiler
23  gpu_results = CUDACompiler().log_likelihood(spn, test_data)  # Location (3)
```

Fig. 4. Python interface usage example.

Figure 4 shows an usage example of the Python interface, the example SPN is taken from SPFlow's documentation. Location (1) in the code shows how inference is usually performed in SPFlow, by invoking log_likelihood.

The other two locations show the invocation of SPNC for compilation and execution on the CPU (2) or CUDA GPU (3). In both cases, the invocation of log_likelihood on the compiler will first compile the SPN using the respective flow described in Sect. 3.2 and Sect. 3.3 and then execute inference using the compiled kernel. A small runtime component part of SPNC is responsible for loading the compiled kernel and executing inference. In case of CPU execution, the runtime is also responsible for multi-threaded execution using OpenMP. The Python interface also supports separate compilation and execution, so an SPN only needs to be compiled once to repeatedly perform inference.

Similar to SPFlow, the compiled kernels also support marginalized inference by passing NaN as input value for marginalized variables.

The Python interface is implemented using Pybind11[2]. As Pybind11 has full support for numpy arrays, input data for execution can simply be provided as numpy arrays, and the result data will likewise be returned as a numpy array. For efficient exchange of SPN models between the Python interface and the compiler, implemented in C++, a binary serialization based on the open-source Cap'n Proto[3] library was implemented.

---

[2] https://github.com/pybind/pybind11.
[3] https://capnproto.org/.

5   Evaluation

To demonstrate SPNC's ability to target different heterogeneous systems, we are evaluating it on two different systems: As an example of an embedded-grade device, a Nvidia Jetson AGX Xavier device with 6-core ARM v8 CPU and Volta GPU will be used. As a non-embedded device, a machine with an AMD Ryzen 9 3900XT CPU equipped with 32 GB RAM and an Nvidia RTX 2070 Super GPU with 8 GB RAM will be used. As the Ryzen processor does not support AVX-512, experiments for AVX-512 will be performed on a dual-socket system with two Intel Xeon Platinum 9242 CPUs and 384 GB RAM.

As a real-world application of SPNs, an SPN-based automatic speaker identification from [4] is used as example application. Based on the open-source release by Nicolson et al.[4], we evaluate two different scenarios, namely the clean speech samples (245567 samples) and noisy speech samples with marginalization (1227835 samples). A sample comprises 26 features, each encoded as single-precision floating point value. We use computation in log-space to avoid deviation from the original result, using single-precision floats as the underlying data type. The implementation by Nicolson et al. contains an SPN *per speaker*, so a set of 628 different SPNs is used for evaluation.

In all experiments using our compiler, we measure the execution time from Python, i.e., the execution time always also includes the invocation overhead of the Python interface in addition to the actual execution time. We track compilation time and execution time separately (also for Tensorflow). The average compilation time across all platforms for CPU is 7 s (max. 33 s) and for GPU 2 s (max. 5 s). The translation of the SPFlow graph to a Tensorflow graph, which is provided by the SPFlow framework, takes 18 s on average (max. 61s).

5.1   Non-embedded Systems

Figure 5 shows the performance comparison for the non-embedded systems, the numbers are given as speedup over the inference execution with SPFlow.

The speedup achieved by translating the SPFlow graph to a Tensorflow graph is relatively low on both CPU (geo.-mean 1.5x) and GPU (1.38x), as the graph is still broken down into individual operations that are launched through the Tensorflow runtime. Marginalization is currently not supported by the Tensorflow translation in SPFlow, therefore no bars are shown for Tensorflow in Fig. 5b.

SPNC on the other hand achieves speedups of 564x and 482x by compiling for the CPU and multithreaded execution, without employing vector extensions. If the vector extensions and vector libraries for elementary functions (Libmvec for AVX-2 and Intel SVML for AVX-512) are used additionally, the speedup increases to 801/814x and 976/935x, respectively. The compilation for the GPU also achieves a significant speedup of 352x and 524x, but data movements between host and device in both cases make up for more than 60% of the execution time, so even though the execution on the GPU itself is very fast, the data movement overhead, which is not present when compiling for CPU, limits the speedup.

---

[4] https://github.com/anicolson/SPN-ASI.

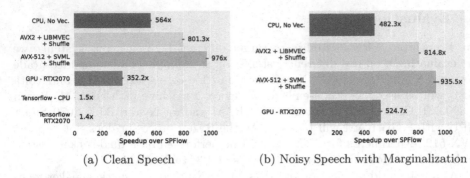

(a) Clean Speech   (b) Noisy Speech with Marginalization

Fig. 5. Performance comparison on non-embedded systems, given as speedup over execution in SPFlow.

5.2 Embedded System

Figure 6 shows the same comparison for the embedded-grade Jetson Xavier platform. As there are fewer CPU cores available than on the Ryzen/Xeon CPU, the speedup achieved by compilation for CPU is smaller compared to Fig. 5, but still reaches 124x (clean) and 58x (noisy) compared to SPFlow. When using the Neon Advanced SIMD extensions, the speedup increases by 2.9x/2.3x to 369x and 133x. In contrast to Fig. 5, the GPU compilation on the Xavier platform provides better performance than the CPU compilation. This is due to the fact that GPU and CPU *physically* share the same memory and no memory transfers between host CPU and GPU are necessary. With the memory transfers eliminated, our GPU compilation achieves speedups of 1004x and 784x.

Another important aspect on embedded systems is memory usage: For the noisy speech samples, it is not possible to process all samples in one batch with SPFlow, as the SPFlow inference runs out of memory (16 GB) and the input has to be processed in multiple blocks sequentially. The compiled kernels are much more memory-efficient and allow to process all samples in a single invocation.

For the Tensorflow comparison on this platform, the Tensorflow package officially provided by Nvidia for Jetson platforms is used. Similar to Fig. 5, the translation provides a speedup over SPFlow (2.36x), but is still significantly slower than the compiled executables.

6 Related Work

To the best of our knowledge, the compiler presented in this work is the *first* compiler for Sum-Product Networks, enabling efficient inference on multiple hardware platforms.

For creation, training, inference, and experimentation with Sum-Product Networks, a number of libraries have been proposed over the years. The two most popular ones, according to the survey conducted by Paris et al. [5], are SPFlow [3] and libspn [8].

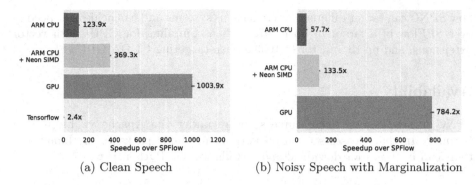

(a) Clean Speech   (b) Noisy Speech with Marginalization

Fig. 6. Performance comparison on embedded systems, given as speedup over execution in SPFlow.

SPFlow allows users to either programmatically create an SPN or learn it, including its structure, from data. It also supports inference on the obtained SPN, either in pure Python, or, for a limited number of cases, through a translation to a Tensorflow graph and execution of that graph. As our evaluation has shown, our compiler significantly outperforms both variants.

Libspn also allows to perform parameter learning and inference for SPNs, again through translation to a Tensorflow graph, which has yielded suboptimal performance in our evaluation in Sect. 5.

Another interesting approach to efficient training and inference for SPNs is through tensorization of the SPN graph, as shown in [6] or [12]. However, these implementations are limited to weight learning, with the structure of the SPNs being subject to additional constraints, whereas our compiler can process SPNs with *arbitrary* DAG structure.

In previous work [9,10], we have developed a custom, FPGA-based inference accelerator for Sum-Product Networks. However, as the automatically generated accelerator uses a fully spatial hardware layout, the maximum size of SPNs that can be mapped to the FPGA is limited by the available hardware resources to sizes significantly smaller than the SPNs evaluated in this work, and the flow currently does not support Gaussian distributions.

7   Conclusion

In this work, we have presented SPNC, a domain-specific compiler for fast inference in Sum-Product Networks. The implementation of SPNC is based on the open-source MLIR framework, which facilitates the implementation of domain-specific compilers.

SPNC was designed to seamlessly integrate with SPFlow, a popular open-source library for SPN construction, learning, and representation, through its Python interface.

In our evaluation, using an SPN-based robust automatic speaker identification as a real-world example of Sum-Product Networks, we have demonstrated how SPNC can target different heterogeneous systems and can achieve a speedup over SPFlow of a factor of up to 978x when compiling for CPUs with vector extensions, and up to a factor of 1003x when targeting CUDA GPUs.

Availability

SPNC is available as open-source software under the Apache v2 License on Github[5]. In the releases section on Github, pre-built packages for Linux systems can be found for download and installation via Python `pip`.

References

1. Kschischang, F.R., Frey, B.J., Loeliger, H.A.: Factor graphs and the sum-product algorithm. IEEE Trans. Inf. Theory **47**(2), 498–519 (2001)
2. Lattner, C., et al.: Mlir: scaling compiler infrastructure for domain specific computation. In: CGO 2021 (2021)
3. Molina, A., et al.: Spflow: an easy and extensible library for deep probabilistic learning using sum-product networks (2019)
4. Nicolson, A., Paliwal, K.K.: Sum-product networks for robust automatic speaker identification (2020)
5. Paris, I., Sanchez-Cauce, R., Diez, F.J.: Sum-product networks: A survey (2020)
6. Peharz, R., et al.: Random sum-product networks: a simple but effective approach to probabilistic deep learning. In: Proceedings of UAI (2019)
7. Poon, H., Domingos, P.: Sum-product networks: a new deep architecture. In: 2011 IEEE International Conference on Computer Vision Workshops (ICCV Workshops) (2011)
8. Pronobis, A., Ranganath, A., Rao, R.P.: Libspn: a library for learning and inference with sum-product networks and tensorflow. In: Principled Approaches to Deep Learning Workshop (2017)
9. Sommer, L., Oppermann, J., Molina, A., Binnig, C., Kersting, K., Koch, A.: Automatic mapping of the sum-product network inference problem to fpga-based accelerators. In: IEEE International Conference on Computer Design (ICCD), IEEE (2018)
10. Sommer, L., Weber, L., Kumm, M., Koch, A.: Comparison of arithmetic number formats for inference in sum-product networks on fpgas. In: 2020 IEEE 28th Annual International Symposium on Field-Programmable Custom Computing Machines (FCCM) (2020)
11. Sugiarto, I., Axenie, C., Conradt, J.: Fpga-based hardware accelerator for an embedded factor graph with configurable optimization. J. Circuits Syst. Comput. **28**(02), 1950031 (2019)
12. van de Wolfshaar, J., Pronobis, A.: Deep Generalized Convolutional Sum-Product Networks for Probabilistic Image Representations. arXiv:1902.06155 (September 2019)

---

[5] https://github.com/esa-tu-darmstadt/spn-compiler.

Towards Addressing Noise and Static Variations of Analog Computations Using Efficient Retraining

Bernhard Klein[1]([✉])[iD], Lisa Kuhn[1][iD], Johannes Weis[2][iD], Arne Emmel[2][iD], Yannik Stradmann[2][iD], Johannes Schemmel[2][iD], and Holger Fröning[1][iD]

[1] Institute of Computer Engineering, Heidelberg University, Heidelberg, Germany
{bernhard.klein,holger.froening}@ziti.uni-heidelberg.de,
lisa.kuhn@stud.uni-heidelberg.de
[2] Kirchhoff-Institute for Physics, Heidelberg University, Heidelberg, Germany
{johannes.weis,arne.emmel,yannik.stradmann,
schemmel}@kip.uni-heidelberg.de

Abstract. One of the most promising technologies to solve the energy efficiency problem for artificial neural networks on embedded systems is analog computing, which, however, is fraught with noise due to summations of unwanted or disturbing energy, and static variations related to manufacturing. While these inaccuracies can have a negative effect on the accuracy, in particular for naively deployed networks, the robustness of the networks can be significantly enhanced by a retraining procedure that considers the particular hardware instance. However, this hardware-in-the-loop retraining is very slow and thus often the bottleneck hindering the development of larger networks. Furthermore, it is hardware-instance-specific and requires access to the instance in question.

Therefore, we propose a representation of a hardware instance in software, based on simple, parallelization-friendly software structures, which could replace the hardware for the major fraction of retraining. The representation is based on lookup tables, splines as interpolated functions and additive Gaussian noise to cover static variations together with electrical noise of the multiplier array and column-wise integrators. The combined approach using the proposed representation together with some final epochs of hardware-in-the-loop retraining reduces the overall training time from over 10 h to less than 2 h compared to a full hardware-in-the-loop retraining, while notably increasing accuracy. This work highlights that including device-specific static variations and noise in the training process is essential for a time-efficient hardware-aware network training for analog computations, and that major parts can be extracted from the hardware instance and represented with simple and efficient software structures. This work is the first step towards hardware-specific but hardware-inaccessible training, addressing speed and accuracy.

Keywords: Analog hardware representation · Hardware-aware training · Static variations · Electrical noise · Analog computations

© Springer Nature Switzerland AG 2021
M. Kamp et al. (Eds.): ECML PKDD 2021 Workshops, CCIS 1524, pp. 409–420, 2021.
https://doi.org/10.1007/978-3-030-93736-2_32

1 Introduction

Machine learning (ML) has quickly established as key component for various applications, ranging from image processing over robotics and natural language processing to signal processing, and is pervasively deployed on devices including mobile, embedded and edge devices. However, there is a substantial gap in between ML's compute and memory requirement and hardware capability, amplified by a staggering Moore's law and the end of Dennard scaling.

As in the Post-Dennard performance scaling era performance in operations per second is best defined as the product of the power budget in Watts and the energy efficiency in operations per Joule, energy efficiency of arithmetic operations and their associated data movements are becoming first-class citizens, and limited resources for mobile, embedded and edge devices due to physical footprint and battery life are amplifying the problem. While digital CMOS processing is still the most common option for ML and various other computational tasks, the interest in alternatives is increasing. There is a plethora of options in the context of *Beyond Moore*, such as quantum computing, neuromorphic computing, and several more. Among those, possibly analog CMOS computing is least disruptive to the existing compute stack. While analog computations can substantially improve the energy efficiency of computations, it comes at the cost of uncertainty in computations, both with regard to *static variations* such as non-linearities as well as *dynamic noise*. While digital computing relies on binarization to increase the resilience towards such imperfections, analog computing is much more sensitive. Still, there are tradeoffs, for instance by choosing a suitable bit width one can avoid a dominating influence of otherwise prohibitive thermal noise in analog computing [19].

There exists a notable set of related work on analog computing, which usually is due to its uncertainty rather application-specific. In the context of machine learning, the work by Murmann describes a mixed-signal processor architecture for artificial neural networks (ANN) [9], which proposes a processing array that allows for much denser processing elements than its digital counterpart, and leverages this in combination with a reduced power budget per element for an increase in parallelism. Another analog computing platform for machine learning applications is BrainScaleS-2 (BSS-2), which utilizes analog operations for the computation of spiking networks as well as ANNs [15]. While this is another example for mixed-signal computation, the implementation details differ substantially.

Furthermore, ANN architectures are apparently quite tolerant against uncertainties in the computation, as there exists a rich body of previous work on quantization and pruning [7,14]. Essentially, both techniques are unsafe optimizations [22] and introduce noise in the computation, however, previous work has shown that retraining is an effective measure to counter a potential loss in prediction quality [1]. Examples for work that enhances the robustness of ANNs for analog computing are based on knowledge distillation [23] and noise injecting training [12], and demonstrate the potential robustness of ANNs.

The main idea of this work stems from the observation that neural networks can be tolerant against inaccurate computations, but this inaccuracy has to be represented in the training process. While including the uncertain hardware in the forward path of the training process can be an option, so-called *hardware-in-the-loop training*, it is often slowing down this process substantially as such analog hardware is not optimized for peak throughput but rather peak energy efficiency. Based on this observation, the present work is concerned with finding a suitable *representation* of the hardware including non-idealities found in analog computing, such that the required hardware involvement can be reduced to a minimum, thereby saving training time. Although being precise enough in representing the hardware peculiarities, this representation has to be compact and efficient to compute, as otherwise training time penalties can occur. Last, as every analog computing instance differs, it has to be trainable in an automated manner to a particular hardware instance.

We therefore propose a representation based on lookup tables and splines as interpolated functions in combination with a configurable amount of additive noise, to represent static variations as well as dynamic noise. The lookup tables represent non-linearities, the splines describe saturation effects, while the additive noise mimicks the various noise sources found in analog computations. We show that this representation can be derived from an automated characterization of a particular hardware instance. We base our characterization on BSS-2 as a real-world prototype of analog computing, fabricated in a generic 65nm CMOS technology and available in sufficient quantities. We demonstrate the effectiveness of our abstract yet precise representation by comparing the accuracy of a keyword spotting task to the accuracy of an ANN trained directly on hardware as well as full-precision standard computations. In detail, this work makes the following contributions:

- Proposing a sufficiently precise and compute efficient representation of hardware non-idealities, suitable to be included in the training framework.
- Designing a training methodology to compensate all major hardware imperfections, reducing time-costly hardware-in-the-loop training.
- Comparing the performance of various representations in terms of accuracy and training time, including various baselines such as full-precision accuracy.

As a result, the training of neural networks on inaccurate hardware is substantially simplified and accelerated, furthermore allowing researchers to reason in an abstract way about the implications of uncertain computations on ML tasks. We see the resulting methodology as key for further research on robust network architectures and explorations in the context of HW/SW-Codesign with uncertain hardware.

2   Background and Related Work

While in principle analog computing exists also based on optical [6], photonic [16] or phase-change-memory [11] technology, most commonly found and focus of the present work is electronic CMOS technology [9]. In such, computations such as multiplication and addition can be easily represented using the physical laws of network analysis. The key operation of most ANNs is the dot product. As one option among various alternatives, the input operands of a multiply-accumulate (MAC) operation can be represented as current pulses, with a pulse's length Δt_i (time) being the input activation and its amplitude the synaptic weight $I_{i,j}$. A single multiplication is then the time integral over this pulse, therefore charge, whilst the result of the MAC over all inputs equals $Q_j = \sum_i I_{i,j} \cdot \Delta t_i$. In particular, BSS-2 operates in exactly this scheme.

While in theory analog computations following Kirchhoff's laws require no energy, in practice charging and de-charging capacitors results in inefficiencies, the required translation from digital signaling to analog quantities requires conversion, and leakage currents and conductances are not ideal, which all contribute to a notable amount of energy. Still, as long as thermal noise is not the dominant non-ideality, analog electrical computing is considered orders of magnitude more energy-efficient than its digital counterpart [9].

2.1   BrainScaleS-2

The presented framework is tested using BrainScaleS-2—a mixed-signal neuromorphic platform based on a custom ASIC manufactured in 65nm CMOS technology [15]. It supports the accelerated emulation of spiking neural networks (SNN) as well as the processing of MAC operations within its analog core [21]. The ASIC consists of 512 neuron circuits, each of which is connected to a column of 256 synapses as well as a dedicated ADC channel for activation readout. The synapses are of 6bit precision. As its sign can be selected row-wise, to achieve signed weights we combine two hardware synapses to a single virtual synapse (Fig. 1). When computing a matrix-vector multiplication,

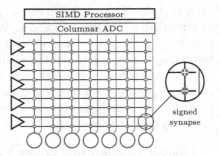

Fig. 1. Block diagram of the BSS-2 analog core, showing synapse drivers (triangles), neurons (large circles), and synapses (small circles in matrix) [2].

the input is encoded as a vector of 5bit unsigned integers. Its entries are multiplied inside the synapse array and accumulated on a capacitor representing the neuron's activation, which is digitized using columnar ADCs with 8bit precision. Each vector may be sent multiple times N_{sends} with an adjustable pause between individual vector entries T_{pause} to optimize signal amplitudes for the circuits' dynamic range.

The process of analog computation is affected by multiple sources of uncertainties: As the analog matrix-vector multiplication is embedded in a full digital setting, multiple *digital-analog and analog-digital conversions* are required. The involved circuitry does not maintain perfect linearity throughout its full dynamic range; especially apparent through offsets, saturation and range-specific amplification effects. Most notably, low vector inputs are exaggerated within the respective digital-to-analog conversion. Each computational element involves multiple *amplifiers*, all of which are subject to different degrees of saturation. Most prominently, saturation effects are observable for the circuits driving synaptic signals to the neuronal accumulators, being strongest for negative summands if the accumulated activation from preceding vector entries is already low. Moreover, the signals emitted by the *synapses* weaken if multiple high-valued summands of the same sign arrive within a short time interval. Last, a driver's signal deteriorates with physical distance, most notably for small values. In addition, all modules of an ASIC are subject to device mismatch due to imprecisions within the manufacturing process. For BrainScaleS-2, effects of these *static variations* primarily emerge for the analog circuits and result in a random distribution of the computational characteristics of all neurons and synapses. The hardware therefore provides calibration mechanisms that can be used to reduce the deviations between its computational elements, even though static variations cannot be eliminated completely in practice. Also, the result from any analog computation is superimposed by *electrical noise* of various sources, including coupled noise from neighboring circuitry and thermal noise. Integrating this noise results in deviations of the output activations that can be observed after digitization, ultimately reducing the obtainable resolution. Models for analog hardware therefore need to be robust against non-linearities, post-calibration mismatch and electrical noise.

2.2 Noisy Computations in the Context of Machine Learning

Most often, the dynamic noise of analog computations is modeled as additive zero-mean Gaussian noise, while the variance differs among the related works. For instance, in Rekhi et al. [13] the variance is a function of the number of bits of the output, in Joshi et al. [5] it depends on range of values a non-volatile memory device can store, while in Zhou et al. [23] the amount of additive noise is a configurable parameter to explore the robustness of network architectures against such. Static variations have also been considered before, but rather with regard to interactions among multiple devices as found in arrays of analog processing elements [3,4].

An early work on robust neural networks reports the benefits of noise injection during training on the example of MLPs [10]. Noisy machines [23] focuses on the additive Gaussian noise and neglects static variations to propose a methodology based on knowledge distillation to highlight and combat the reduced learning capacity of ANNs when executed on noisy hardware. A more general treatment of fault-tolerance of ANNs can be found in [18].

While we are also concerned about finding more robust ANNs, our work distinguishes from related work by the fact that we are not only concerned with *dynamic noise*, but also *static variations* and the inevitable amount of variance found across multiple hardware instances. While hardware-in-the-loop training is suitable to address the noise and variances, it comes with a substantial slowdown.

3 Hardware Representation

The provided method is designed to represent chip specific peculiarities of inaccurate matrix multiply operations in the training process and is applicable for uncertain accelerators which are based on a multiplier array and accumulate with column-wise integrators. Although the proposed hardware representation requires that the columns are largely independent of each other and that there are no major time dependencies besides electrical noise, it is a very general concept that can be applied to a wide variety of accelerators. The peculiarities of the accelerator are distilled from measurements and mapped in a hardware representation, which is suitable to train machine learning models for this specific hardware as quickly as possible. Figure 2 illustrates the three distinct parts of the

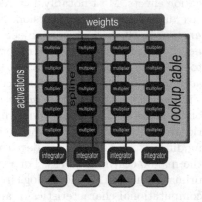

Fig. 2. Schematic overview of the representation: *lookup table* for the static variations of the multiplier array, *splines* for the column and integrator specific variations, and additive *Gaussian noise* that models the electrical noise, among others.

model which simulate the hardware and capture multiplier-array- and integrator-specific static variations together with electrical noise, which are a *lookup table* to represent the multiplier non-linearities, *splines* as piecewise polynomial functions to represent integrator imperfections, and *Gaussian additive noise* to represent the electrical noise observed during analog computations. In the following, we describe these components in more detail.

Lookup Table. All multiplier effects are covered by storing the output for all possible inputs for each multiplier in a lookup table. On condition that the effects are reproducible and do not change over time, the lookup table enables to model complex behavior of the multipliers with highest possible generality. For the BSS-2 hardware such a column-row-activation-weight-based lookup table can be created by performing *row-wise measurements* to capture the row and column specific details with minimal interference. Since only one row is active at the same time the electrical signal strength is low, therefore saturation effects are negligible. The signal is amplified with $N_{sends} = 20$, $T_{pause} = 8$ to a medium high level to be measured with good quality. All other measurements and hardware-in-the-loop training uses $N_{sends} = 1$, $T_{pause} = 8$ to minimize saturation effects.

Splines. While the lookup table can model multiplier specific variations, it is not capable to model effects which are based on interactions between them or belong to the *integrator*. For BSS-2, saturation effects dominate these *column-wise variations*. On condition that only multipliers connected to the same integrator can

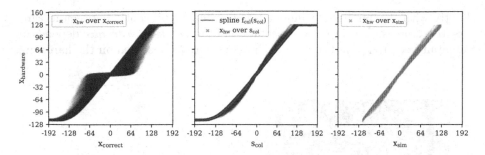

Fig. 3. Mismatch between values expected from a correct matrix multiply $x_{correct}$ to measured results $x_{hardware}$ for a random column (left). A relevant part of this errors can be corrected by using the sums of the lookup table s_{col} (middle), together with a interpolated spline to represent column specific saturation effects. On the right: comparison of representation x_{sim} and measurement $x_{hardware}$ results. An optimal representation would lead to a straight line broadened by electrical noise.

interact and interaction between integrators is negligible, properly designed functions can model this inter-multiplier and integrator specific behavior. The sum of values s_{col} in a column c extracted from the lookup table e_{lookup},

$$s_{col} = \sum_{r}^{I} e_{lookup}[r, c, a_r, w_r] \tag{1}$$

is mapped with column-specific function f_{col},

$$x_{sim} = f_{col}(s_{col}) \tag{2}$$

to the predicted hardware values x_{sim}, with I as input vector size, activation a_r and weight w_r of row r. Splines—piecewise polynomial functions—can be interpolated to act as this functions. A *full range measurement* is performed for all possible activations, weights and input vector sizes to include the limits of the value space and thus extract the saturation effects of the accelerator. All measured values $x_{hardware}$ whose summed lookup table values s_{col} are in the same interval $\Delta s_i = [i\Delta s, (i+1)\Delta s]$ are averaged to get a stable mapping from Δs_i to $\overline{x_{hardware}}$. Figure 3 shows exemplary the operation of lookup table and spline.

Electrical Noise. With all *static variations* modeled by the *lookup table* and *splines* there are still sources of *electrical noise* which can not be covered with a deterministic model. However, due to their very statistical properties, they can be represented by adding noise sampled from a zero-mean *Gaussian distribution*. The variance is determined by a measurement where activation and weight matrices are chosen randomly and the same random input matrices are used to computed the outputs many times to measure the standard deviation of exactly the same operation. As Fig. 4a illustrates, the *electrical noise* increases slightly with the number of summands and differs significantly between integrators, therefore the hardware representation of the *Gaussian variance* depends on the input vector size and column in the same way as observed on the hardware.

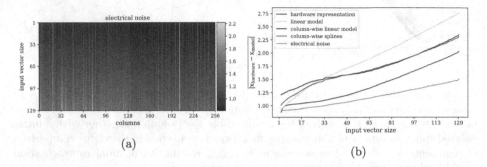

(a)   (b)

Fig. 4. Left: Electrical noise measured as standard deviation of repeated calculations with identical inputs. With a larger input vector the average noise increases (vertical gradient), since more noisy values are accumulated, but also the variances between columns are significant (vertical lines). Right: Model performance as difference to hardware measurements for random inputs. A perfect model would be align with the electrical noise (gray) of the hardware. The proposed hardware representation outperforms other linear-regression and splines based models. Noise and model imperfections increase with input vector size due to more involved components and saturation effects.

Performance of Various Models to Represent BSS-2. Figure 4b compares linear models together with a column-wise spline-based model to the proposed hardware representation. The better a model is, the closer it comes to the baseline given by electrical noise. Due to more involved inaccurate and noisy components, a larger input vector size increases the error of all models. With increasing input vector size also the error of column-specific variations becomes more dominant since more values are amplified by them. Therefore for larger input vectors the column-wise models perform better than their non-specific counterparts. Although larger input vectors lead to more hidden saturation effects and thus increase the mismatch between representation and hardware, including multiplier-specific variations by its lookup table enables the proposed hardware representation to outperform all other models.

While the proposed general hardware representation models saturation effects with columnar splines, the BSS-2 hardware accumulates positive and negative summands separately. This leads to *hidden saturation effects* which occur in the sign intrinsic accumulation process and due to positive and negative cancellation are not visible in the final sum. However, the representation has to be compact and of high throughput, thus it cannot model all hardware imperfections in detail, particularly since accurate hardware emulations mostly are notably slower than execution on the hardware itself, which is in direct conflict to the design goals of a representation used to speed up the training process.

While PyTorch was used to be interchangeable with the BSS-2 interface *hxtorch* [17], the concept is universal and applicable with all major ML frameworks.

4   Training Methods and Results

Keyword Spotting Task. We consider the keyword spotting task of the Speech Commands Dataset V1 [20] as being both representative and sufficiently complex to unveil differences in between proposed hardware representation and hardware itself. While there are different options to model such time-series problems, including recurrent neural networks, LSTMs and transformers, this work focuses on a multi-layer perceptron with log-mel filter [8]—illustrated in Fig. 5. Other architecture options are considered to be beyond the scope of the present work.

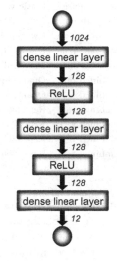

Fig. 5. Keyword spotting model

Training Methods Based on the Hardware Representation. For compression of ML models, retraining with the compressed model has proven to be very successful to compensate accuracy loss induced by the less accurate representations [1]. Here we follow a similar approach to compensate for hardware imperfections. In the retraining step, the already trained full-precision model is trained again but this time with the hardware or their representation used in the forward path. The intuition is, that then the backpropagation will automatically address the hardware imperfections by minimizing the errors injected in the forward path. Notably, complete hardware-in-the-loop retraining is the most precise way to integrate the hardware but also by far the slowest option.

Table 1 compares different training strategies reporting accuracy and retraining time, notably a *plain* training based on 300 epochs (which is also used as initialization for all other methods), a *quantized* retraining to achieve the required quantization (uint5 activations and int6 weights) of BSS-2, a *noise-only* retraining based on the plain method that adds injected zero-mean additive Gaussian noise, a *representation (rep.) no noise* retraining based on this work's lookup table and splines, a *representation (rep.) with noise* retraining based on this work's lookup table, splines and additive noise, a *representation (rep.) increasing (inc.) noise* retraining based on this work's lookup table, splines and a noise increasing over the epochs from 0 % for first epoch to 100 % of the original noise level on the last epoch, a *hardware-in-the-loop* retraining that uses a BSS-2 instance in the forward path, and a *combined* retraining that first relies on either the quantized model or on our representation with increasing noise, before finally retraining for a couple of epochs using hardware-in-the-loop.

Table 1. Comparison of various retraining methods

Method		Retraining		Accuracy	
Name	Description	Epochs	Time in minutes	SW	BSS-2
Plain	Original full precision model	0	0.0	80.8 %	12.3 %
Quantized	Uint5 activations, int6 weights	300	13.1	79.6 %	25.8 %
Noise only	Plain model & noise	300	10.4	80.5 %	18.4 %
Rep. no noise	Static variances without noise	300	83.4	76.5 %	26.4 %
Rep. with noise	Hw. rep. with noise	300	83.4	73.5 %	35.1 %
Rep. inc. noise	Hw. rep. increasing noise	300	83.6	76.0 %	41.0 %
Hw-in-the-loop	Full hardware-in-the-loop	300	652.3		66.8 %
Hw-in-the-loop	Full hardware-in-the-loop	350	769.5		66.7 %
Combined qt.	Quant. (300 ep.) & hw. (5 ep.)	305	13.1 + 11.5		62.1 %
Combined qt.	Quant. (300 ep.) & hw. (50 ep.)	350	13.1 + 117.5		67.3 %
Combined rep.	Rep. (300 ep.) & hw. (1 ep.)	301	83.6 + 2.2		64.9 %
Combined rep.	Rep. (300 ep.) & hw. (5 ep.)	305	83.6 + 11.5		67.4 %
Combined rep.	Rep. (300 ep.) & hw. (10 ep.)	310	83.6 + 23.5		69.7 %
Combined rep.	Rep. (300 ep.) & hw. (50 ep.)	350	83.6 + 117.5		70.1 %

Comparing Various Representations. While the plain model performs very poorly on BSS-2, including more hardware characteristics with quantization, additive noise and hardware representation leads to an increased accuracy of the retrained model. Although most related work focuses on counter measures against noise [5,12,23], the experiments show that *static variations* are equally important and have to be considered. Moreover we found that increasing the noise level during the retraining leads to better results. This suggests that increasing the difficulty, in our case hardware imperfections, slowly during retraining makes it easier for the model to adapt. This might also be the reason for the astonishing fact that combining representation retraining with only a few epochs hardware-in-the-loop retraining leads to higher test accuracies than full hardware-in-the-loop retraining.

These first experiments clarify that although the representation is not yet exact enough to replace hardware-in-the-loop completely, it covers enough hardware particularities during the training to be much more accurate than a naive deployment. Harnessing the best of both worlds with a combined approach outperforms hardware-in-the-loop retraining in terms of accuracy and training time.

Training Time. While the training with hardware in the loop is bound by BSS-2 throughput, the representation training is only limited by the available compute resources. For the experiments a single *NVIDIA Titan Xp* was used, however, the throughput scales with the resources and by using more or more powerful GPUs the required training time can be reduced easily. Although with this lightweight training resource, in contrast to typical high-end, multi-GPU training systems, a speedup of 7.8 for representation training and 6.9 for combined training (5 hw. epochs) over hardware-in-the-loop retraining can be reported.

5 Summary and Outlook

A representation is proposed to minimize the involvement of analog hardware during retraining, based on distilling all major hardware characteristics from measurements, and represented during training with a lookup table, splines and additive Gaussian noise, modeling particularities of the multiplier array, saturation effects of the integration, analog and digital conversion processes and inevitable electrical noise. Although the representation cannot replace the costly hardware-in-the-loop retraining step completely, it replaces most of it with much faster representation training, such that the amount of remaining epochs with the hardware is reasonable small compared to the rest and the overall training time is reduced from more than 10 h to less than 2 h, notably increasing accuracy. Overall, the proposed method demonstrates that including device-specific static variations and noise in the training process is essential to train hardware-aware robust neural networks for analog computations, and that major parts can be extracted from the hardware and represented with simple and parallelization-friendly software structures. This is the first step towards hardware-specific but hardware-inaccessible training, addressing speed and accuracy.

Acknowledgements. The development of BrainScaleS-2 has received funding from the German Federal Ministry of Education and Research under grant number 16ES1127, the EU (H2020/2014-2020: 720270, 785907, 945539 (HBP)) and the Lautenschläger-Forschungspreis 2018 for Karlheinz Meier. We also acknowledge the financial support from the COMET program within the K2 Center "Integrated Computational Material, Process and Product Engineering (IC-MPPE)" (Project No. 859480). This program is supported by the Austrian Federal Ministries for Transport, Innovation and Technology (BMVIT) and for Digital and Economic Affairs (BMDW), represented by the Austrian research funding association (FFG), and the federal states of Styria, Upper Austria and Tyrol.

References

1. Courbariaux, M., Bengio, Y., David, J.P.: BinaryConnect: training deep neural networks with binary weights during propagations. In: Advances in Neural Information Processing Systems, vol. 28. Curran Associates, Inc. (2015). https://dl.acm.org/doi/10.5555/2969442.2969588
2. Cramer, B., et al.: Training spiking multi-layer networks with surrogate gradients on an analog neuromorphic substrate (2020). https://arxiv.org/abs/2006.07239
3. Feinberg, B., Wang, S., Ipek, E.: Making memristive neural network accelerators reliable. In: 2018 IEEE International Symposium on High Performance Computer Architecture (HPCA), pp. 52–65 (2018). https://doi.org/10.1109/HPCA.2018.00015
4. Jain, S., Sengupta, A., Roy, K., Raghunathan, A.: Rx-caffe: Framework for evaluating and training deep neural networks on resistive crossbars (2018). http://arxiv.org/abs/1809.00072
5. Joshi, V., et al.: Accurate deep neural network inference using computational phase-change memory. Nature Commun. **11**(1), 2473 (2020). https://doi.org/10.1038/s41467-020-16108-9

6. Lin, X., et al.: All-optical machine learning using diffractive deep neural networks. Science **361**(6406), 1004–1008 (2018). https://doi.org/10.1126/science.aat8084
7. Liu, Z., et al.: Rethinking the value of network pruning. In: International Conference on Learning Representations (2019). https://arxiv.org/abs/1810.05270
8. Mermelstein, P.: Distance measures for speech recognition, psychological and instrumental. Pattern Recognit. Artif. Intell. **116**, 374–388 (1976)
9. Murmann, B.: Mixed-signal computing for deep neural network inference. IEEE Trans. Very Large Scale Integr. (VLSI) Syst. **29**(1), 3–13 (2021). https://doi.org/10.1109/TVLSI.2020.3020286
10. Murray, A., Edwards, P.: Enhanced MLP performance and fault tolerance resulting from synaptic weight noise during training. IEEE Trans. Neural Netw. **5**(5), 792–802 (1994). https://doi.org/10.1109/72.317730
11. Nandakumar, S.R., Le Gallo, M., Boybat, I., Rajendran, B., Sebastian, A., Eleftheriou, E.: A phase-change memory model for neuromorphic computing. J. Appl. Phys. **124**(15), 152135 (2018). https://doi.org/10.1063/1.5042408
12. Qin, M., Vucinic, D.: Noisy computations during inference: Harmful or helpful? CoRR abs/1811.10649 (2018). http://arxiv.org/abs/1811.10649
13. Rekhi, A.S., et al.: Analog/mixed-signal hardware error modeling for deep learning inference. In: 56th Annual Design Automation Conference. DAC, Association for Computing Machinery (2019). https://doi.org/10.1145/3316781.3317770
14. Roth, W., et al.: Resource-efficient neural networks for embedded systems. CoRR abs/2001.03048 (2020). http://arxiv.org/abs/2001.03048
15. Schemmel, J., Billaudelle, S., Dauer, P., Weis, J.: Accelerated analog neuromorphic computing. CoRR abs/2003.11996 (2020). https://arxiv.org/abs/2003.11996
16. Shen, Y., et al.: Deep learning with coherent nanophotonic circuits. Nat. Photonics **11**(7), 441–446 (2017). https://doi.org/10.1038/nphoton.2017.93
17. Spilger, P., et al.: `hxtorch`: PyTorch for BrainScaleS-2. In: Gama, J., et al. (eds.) ITEM/IoT Streams -2020. CCIS, vol. 1325, pp. 189–200. Springer, Cham (2020). https://doi.org/10.1007/978-3-030-66770-2_14
18. Torres-Huitzil, C., Girau, B.: Fault and error tolerance in neural networks: a review. IEEE Access **5**, 17322–17341 (2017). https://doi.org/10.1109/ACCESS.2017.2742698
19. Vittoz, E.: Future of analog in the VLSI environment. In: IEEE International Symposium on Circuits and Systems, vol. 2, pp. 1372–1375 (1990). https://doi.org/10.1109/ISCAS.1990.112386
20. Warden, P.: Speech commands: A dataset for limited-vocabulary speech recognition. CoRR abs/1804.03209 (2018). http://arxiv.org/abs/1804.03209
21. Weis, J., et al.: Inference with artificial neural networks on analog neuromorphic hardware. In: Gama, J., et al. (eds.) ITEM/IoT Streams -2020. CCIS, vol. 1325, pp. 201–212. Springer, Cham (2020). https://doi.org/10.1007/978-3-030-66770-2_15
22. Whatmough, P., Wei, G.Y., Brooks, D.: Deep Learning for Computer Architects. Morgan & Claypool Publishers, San Rafael (2017)
23. Zhou, C., et al.: Noisy machines: Understanding noisy neural networks and enhancing robustness to analog hardware errors using distillation. CoRR (2020). https://arxiv.org/abs/2001.04974

eXplainable Knowledge Discovery in Data Mining

Workshop and Tutorial on eXplainable Knowledge Discovery in Data Mining (XKDD 2021)

The 3rd International Workshop and Tutorial on eXplainable Knowledge Discovery in Data Mining (XKDD 2021) was held in conjunction with the European Conference on Machine Learning and Principles and Practice of Knowledge Discovery in Databases (ECML PKDD 2021) on Monday, September 13, 2021. Previous editions of the workshop were also held jointly with ECML PKDD 2019 and 2020. In line with the organization of ECML PKDD 2021, the XKDD 2021 workshop was a fully virtual event with live presentations which were also pre-recorded and are freely available on the workshop website[1].

In the past decade, machine learning-based decision systems have been widely used in a wide range of application domains such as, for example, credit scores, insurance risk, and health monitoring, in which accuracy is of the utmost importance. Although the support of these systems has a big potential to improve decisions in different fields, their use may present ethical and legal risks, such as codifying biases, jeopardizing transparency and privacy, and reducing accountability. Unfortunately, these risks arise in different applications and they are made even more serious and subtle by the opacity of recent decision support systems, which often are complex with internal logic that is usually inaccessible to humans.

Nowadays most of the Artificial Intelligence (AI) systems are based on machine learning algorithms. The relevance and need of ethics in AI is supported and highlighted by various initiatives arising from the research to provide recommendations and guidelines in the direction of making AI-based decision systems explainable and compliant with legal and ethical issues. These include the EU's General Data Protection Regulation (GDPR) which introduces, to some extent, a right for all individuals to obtain "meaningful explanations of the logic involved" when automated decision making takes place, the "ACM Statement on Algorithmic Transparency and Accountability", Informatics Europe's "European Recommendations on Machine-Learned Automated Decision Making", and "The ethics guidelines for trustworthy AI" provided by the EU High-Level Expert Group on AI.

The challenge to design and develop trustworthy AI-based decision systems is still open and requires a joint effort across technical, legal, sociological, and ethical domains.

The purpose of XKDD, eXaplaining Knowledge Discovery in Data Mining, is to encourage principled research that will lead to the advancement of explainable, transparent, ethical, and fair data mining and machine learning. XKDD is an event organized in two parts: a tutorial to introduce audience to the topic and a workshop to discuss recent advances in the research field. The tutorial will provides a broad overview of the state of the art on the major applications for explainable, and transparent

---

[1] https://kdd.isti.cnr.it/xkdd2021/.

approaches and their relationship with fairness and privacy. Moreover, it presents Python/R libraries that practically show how explainability and fairness tasks can be addressed. The workshop seeks top-quality submissions addressing uncovered important issues related to ethical, fair, explainable, and transparent data mining and machine learning. Papers should present research results in any of the topics of interest for the workshop as well as application experiences, tools, and promising preliminary ideas. XKDD asks for contributions from researchers, academics and practitioners from industry, working on topics addressing these challenges, primarily from a technical point of view but also from a legal, ethical, or sociological perspective.

Topics of interest include, but are not limited to, the following:

- Explainable Artificial Intelligence
- Interpretable Machine Learning
- Transparent Data Mining
- Explainability in Clustering Analysis
- Technical Aspects of Algorithms for Explanation
- Explaining Black Box Decision Systems
- Adversarial Attack-based Models
- Counterfactual and Prototype-based Explanations
- Causal Discovery for Machine Learning Explanation
- Fairness Checking
- Fair Machine Learning
- Explanation for Privacy Risk
- Ethics Discovery for Explainable AI
- Privacy-Preserving Explanations
- Transparent Classification Approaches
- Anonymity and Information Hiding Problems in Comprehensible Models
- Case Study Analysis
- Experiments on Simulated and Real Decision Systems
- Monitoring and Understanding System Behavior
- Privacy Risk Assessment
- Privacy by Design Approaches for Human Data
- Statistical Aspects, Bias Detection, and Causal Inference
- Explanation, Accountability, and Liability from an Ethical and Legal Perspective
- Benchmarking and Measuring Explanation
- Visualization-based Explanations
- Iterative Dialogue Explanations
- Explanatory Model Analysis
- Human-Model Interfaces
- Human-Centered Artificial Intelligence
- Human-in-the-Loop Interactions

The XKDD 2021 workshop was preceded by a tutorial which consisted of four parts:

- Motivations for Explainability and Links to Other Ethical Aspects by Anna Monreale,

- Explaining and Checking Fairness for Predictive Models by Przemyslaw Biecek,
- Useful Explanations and How to Find Them by Riccardo Guidotti, and
- Explaining with LIME and LORE Wrapped in the X-Lib Library by Salvatore Rinzivillo.

The XKDD 2021 workshop papers were selected through a single-blind peer-review process in which each submitted paper was assigned to at least three members of the Program Committee. The main selection criteria were the novelty of the proposal and its impact in explanation/privacy/fairness processes. XKDD 2021 received a total of 22 submissions. Six papers were accepted for presentation (an acceptance rate of 27%). The workshop was organized as a Zoom webinar with an accompanying slack channel, and we registered an audience of about 104 different users across the whole workshop.

The workshop started with an excellent keynote on Ingredients of future medical AI: explainability and robustness by Andreas Holzinger from the Human-Centered AI Lab at the Medical University of Graz, Austria.

More information about the workshop, including the videos of the talks, can be found on the workshop website: https://kdd.isti.cnr.it/xkdd2021/. We would like to thank all the participants for joining XKDD 2021, the authors for their interesting works and presentations, and all workshop attendees for their engagement and the questions. A special thank goes to the wonderful Program Committee for their effective and timing reviews.

The organization of XKDD 2021 was supported the European Community H2020 programme under the following funding schemes: INFRAIA-1-2014-2015 Res. Infr. G. A. 871042 *SoBigData++* (sobigdata), G.A. 952026 *HumanE AI Net* (humane-ai), G.A. 825619 *AI4EU* (ai4eu), G.A. 834756 *XAI,* (xai), and NCN Sonata Bis-9 grant 2019/34/E/ST6/00052 (HOMER).

October 2021

Riccardo Guidotti
Anna Monreale
Salvatore Rinzivillo
Przemysław Biecek

Organization

XKDD 2021 Program Chairs

Riccardo Guidotti   University of Pisa, Italy
Anna Monreale   University of Pisa, Italy
Salvatore Rinzivillo   ISTI-CNR, Italy
Przemysław Biecek   Warsaw University of Technology, Poland

XKDD 2021 Program Committee

Avishek Anand   Leibniz University, Germany
Umang Bhatt   University of Cambridge, UK
Francesco Bodria   Scuola Normale Superiore, Italy
Giuseppe Casalicchio   Ludwig Maximilian University of Munich, Germany
Chaofan Chen   University of Maine, USA
Miguel Couceiro   Inria, France
Josep Domingo-Ferrer   Universitat Rovira i Virgili, Spain
Thibault Laugel   AXA, France
Paulo Lisboa   Liverpool John Moores University, UK
Marcin Luckner   Warsaw University of Technology, Poland
Stan Matwin   Dalhousie University, Canada
Ramaravind Kommiya Mothilal   Everwell Health Solutions, India
Francesca Naretto   Scuola Normale Superiore, Italy
Roberto Prevete   University of Napoli, Italy
Antonio Rago   Imperial College London, UK
Jan Ramon   Inria, France
Xavier Renard   AXA, France
Mahtab Sarvmaili   Dalhousie University, Canada
Dominik Slezak   University of Warsaw, Poland
Myra Spiliopoulou   University Magdeburg, Germany
Vicenc Torra   Umea University, Sweden
Grigorios Tsoumakas   Aristotle University of Thessaloniki, Greece
Marco Virgolin   Chalmers University of Technology, Sweden
Wendy Hui Wang   Stevens Institute, USA
Guangyi Zhang   KTH Royal Institute of Technology, Sweden

The Next Frontier: AI We Can Really Trust

Andreas Holzinger[1,2]([✉]) [iD]

[1] Human-Centered AI Lab, Medical University Graz, Graz, Austria
`andreas.holzinger@human-centered.ai`
[2] xAI Lab, Alberta Machine Intelligence Institute, Edmonton, Canada

Abstract. Enormous advances in the domain of statistical machine learning, the availability of large amounts of training data, and increasing computing power have made Artificial Intelligence (AI) very successful. For certain tasks, algorithms can even achieve performance beyond the human level. Unfortunately, the most powerful methods suffer from the fact that it is difficult to explain why a certain result was achieved on the one hand, and that they lack robustness on the other. Our most powerful machine learning models are very sensitive to even small changes. Perturbations in the input data can have a dramatic impact on the output and lead to entirely different results. This is of great importance in virtually all critical domains where we suffer from low data quality, i.e. we do not have the expected i.i.d. data. Therefore, the use of AI in domains that impact human life (agriculture, climate, health, ...) has led to an increased demand for trustworthy AI. Explainability is now even mandatory due to regulatory requirements in sensitive domains such as medicine, which requires traceability, transparency and interpretability capabilities. One possible step to make AI more robust is to combine statistical learning with knowledge representations. For certain tasks, it can be advantageous to use a human in the loop. A human expert can - sometimes, of course not always - bring experience, domain knowledge and conceptual understanding to the AI pipeline. Such approaches are not only a solution from a legal point of view, but in many application areas the "why" is often more important than a pure classification result. Consequently, both explainability and robustness can promote reliability and trust and ensure that humans remain in control, thus complementing human intelligence with artificial intelligence.

Keywords: Artificial intelligence · Trust · Explainable AI · Robustness · Human-in-the-loop

1 Success in Machine Learning Enabled a New AI Spring

At the interface between politics, industry and consumers, artificial intelligence is experiencing unprecedented popularity. Politicians around the world are declaring AI a strategic goal, industry sees it as a huge growth engine, and many

ⓒ Springer Nature Switzerland AG 2021
M. Kamp et al. (Eds.): ECML PKDD 2021 Workshops, CCIS 1524, pp. 427–440, 2021.
https://doi.org/10.1007/978-3-030-93736-2_33

application areas that impact human life (e.g. agriculture, climate, health, ...) see it as a great opportunity for multiple improvements in predictive modelling [4], diagnostics [39] and therapy [9].

For example, in July 2017, China's State Council published the country's artificial intelligence (AI) development strategy, titled "New Generation Artificial Intelligence Development Plan", with the goal of becoming a world leader in AI by 2030 [41]. Many other countries followed, e.g. in 2018 the German Federal Government with "AI made in Germany 2030"[1].

This recent AI spring was triggered by three main drivers: 1) the worldwide trend towards digitization, and thereby the 2) availability of big data, and above all 3) the remarkable advances in statistical machine learning and computational power.

The spread of AI solutions is accelerated by current events, a very recent example is the health domain: The potential for medical AI-based systems in the near future has increased enormously after the sad 200 million COVID-19 cases and 4 million deaths worldwide (as of 6 August 2021)[2]. To give another very recent example: A graph-based machine learning method enables the identification of bioactive anti-COVID-19 molecules in foods based on their ability to target the SARS-CoV-2 host gene (protein-protein) interactome. Based on this work, a "food map" was created that estimates the theoretical anti-COVID-19 potential of each ingredient based on the diversity and relative content of antivirally active candidates. Such approaches will play an important role in future clinical trials of precise nutritional interventions against COVID-19 and other viral diseases [32].

Indeed, the increased availability of data has reignited interest in AI algorithms for the medical domain, especially convolutional neural networks in image analysis and specifically in radiology and pathology. However, in order to use AI to solve problems in medicine and life sciences beyond the laboratory and routine, there is an urgent need to go beyond simple benchmarking and improve the performance of methods that only work with independent and identically distributed (i.i.d.) data. Independently and identically distributed random variables have the same distribution and do not affect each other - however, this is rarely the case with real data. Machine learning is learning from observed data by constructing stochastic models that can be used to make predictions and decisions. It sounds simple, but when do we have i.i.d. in reality, real-world data is highly non-linear, non-stationary and high-dimensional and often noisy. Data quality is therefore a basic requirement for the correct functioning of our data-driven algorithms. This often requires great efforts of data pre-processing, data cleansing, because malfunctions due to "dirty data" can have dramatic effects. However, this data cleaning can also have a negative impact on data quality, especially if not done carefully [48].

---

[1] https://www.ki-strategie-deutschland.de.
[2] https://www.worldometers.info/coronavirus/, accessed 6 August 2021.

Unfortunately, even the best current machine learning models do not generalize well, have difficulty with *small* training datasets ("little data") and are sensitive to even small perturbations as we will see later on.

Above all, the most successful approaches are so complex, so nonlinear, and so high-dimensional that they are difficult or impossible for human experts to interpret and, above all, can no longer derive causal relationships. Robustness and explainability have therefore been declared by the European Union to be definitely the most important properties for future trustworthy AI [17].

In this paper, we first define the terms trust and trustworthy AI, then explain what explainability and causability are and why this is important, briefly look at robustness, and conclude with how a human-in-the-loop can contribute to robustness and explainability.

2   Trust and Trustworthy AI

Before we dive into our topic, we need some definitions so that we can develop a common understanding of the terms used. Incidentally, this is also a good example of the *quality of explanations* and the question of when an explanation is good, which we will discuss in more detail later under the terminus Causability. As Jean Piaget (1896–1980) advocated in his tradition of human-centered and trans-disciplinary science, we first need a *common framework to enable mutual understanding* (see e.g. [37,38]).

2.1   What Is Trust?

Trust is a multidisciplinary concept that is very difficult to define [8], similarly as human intelligence [5], and this is why AI is also very difficult to define. The most common definition of intelligence is given by cognitive science as mental capability, and includes, among others, the ability to think abstract, to reason, and to solve problems from the real world. A hot topic in current AI/machine learning research is to find out whether and to what extent algorithms are able to learn such abstract thinking and reasoning similarly as humans can do - or whether the learning outcome remains on purely statistical correlation [22].

The concept of trust is also linked to several disciplines and influenced by many diverse factors. It can be understood better when we consider that trust has evolved genealogically from some fundamental features of human social life with the aim that we can *rely on other people to act cooperatively* [47].

As a social psychological construct, trust is a belief or an assessment, i.e. it is always *subjective* and dependent on personal attitudes and expectations. Nevertheless, trust has some consistent characteristics: subjectivity, dynamism, context awareness (risk situations, perceived domain importance, e.g. fields or situations that impact human life), incomplete transitivity, time decay, asymmetry and measurability. The basic factors include security, dependability, integrity, predictability, and reliability [53].

Trust evaluation is the process of quantifying trust with attributes that influence individual trust. A number of machine learning methods have even been used for trust assessment [51].

It is obvious that trust is very important for Human-AI interaction because trust is an attributional process and perceived trust is an important aspect of developing and maintaining interpersonal relationships. Successful cooperation between human communicators occurs when ambiguity and uncertainty in social perceptions are reduced through the development of trust. In particular, it is important to emphasise that the upcoming human-AI interaction represents a new paradigm in which human communication is augmented or even generated by an intelligent system. Here, trust develops differently than in classical early human-computer interaction or in interactions between humans [18].

2.2   What Is Trustworthy AI?

Despite all the successes and the recurring euphoria about AI, recent work shows that AI can unintentionally harm humans and that it is precisely the large-scale and wide introduction of AI technologies that holds enormous and unimagined potential for new types of unforeseen threats [26].

For example, AI can make unreliable decisions in safety-critical scenarios (e.g. in the medical domain) or undermine fairness by inadvertently discriminating against a group [16].

For this reason, the international research community has recently paid much attention to so-called *trustworthy AI*. Dimensions of trustworthy AI include: security, safety, privacy, non-discrimination, fairness, accountability (re-traceability, replicability), auditability and environmental Well-being, and most of all *robustness and explainability* [34]. These dimensions have also been included into the European Commissions ethics guidelines for trustworthy AI[3] [13,17]. For all these reasons, Trustworthy AI is a strongly emerging field in the international research community [7].

3   Explainability and Causability

3.1   What Is Explainable AI?

Although explainable AI (xAI) only emerged through the DARPA initiative [15] it is in principle not a new field. The problem of explainability is at least as old as AI itself, in fact it is the result of AI itself. DARPA's Explainable Artificial Intelligence (XAI) program aimed to develop AI systems whose models and decisions can be understood and trusted by end users. To this end, a large number of diverse methods have been developed by the growing xAI community. While also interpretable methods (aka ante-hoc methods) have been used, which are also known as glass-box models (such as decision trees, or graph-based methods),

---

[3] https://ec.europa.eu/commission/presscorner/detail/en/IP_19_1893, last accessed on August, 31, 2021.

the emphasis has been put by the community on the interpretation of so-called black-box methods (see below). The holistic integrative approach of the DARPA program is worth noting, as the realization of xAI has included not only methods for learning more explanatory models, but also the design of effective explanatory interfaces, as well as the psychological requirements for understanding effective explanations. Meanwhile xAI is an established vibrating field [1,44].

3.2   What Is Explainability?

Explainability, in the sense of the machine learning community focuses mostly on post-hoc methods, i.e. to make so-called "black-box" models explainable by a human [31]. Such methods highlight technically decision-relevant parts of machine representations and machine models, i.e., parts that contributed to model accuracy during training or to a particular prediction. A typical example of a method that does this very well is Layer Wise Relevance Propagation (LRP) [33]. With this method, heat maps can be used to visualize the parts that contributed to the given explanation. Graph Neural Networks (GNNs) are an increasingly popular approach for predicting graph-structured data, however, the input graphs are tightly entangled with the neural network structure, making traditional xAI approaches inapplicable on such graphs. Therefore, "GNN-LRP" [45] has already been further developed to provide explainability with graphs, or trees, as well. These methods can be very helpful in the biology, medicine and the life sciences, e.g. [24,25]. However, this "explainability" is a technical approach and does *not* relate to a human model. However, in certain domains, especially in the medical field, there is a need for causability, introduced by Holzinger et al. (2019) [23].

3.3   What Is Causability?

Causability is the measurable extent to which an explanation - resulting from an explainable AI method to a human expert achieves a specified level of causal understanding. Causability refers to a human model and can be measured with the System Causability Scale [21]. Causability is not a synonym for causality, instead the term causa-bil-ity was introduced in reference to usa-bil-ity. Whilst explainability (represented by the field of xAI) is about the technical implementation of transparency and traceability in AI approaches, causability is about measuring and ensuring the quality of explanations.

So let's briefly summarize: Explainability technically highlights decision relevant parts of machine representations and machine models, i.e. parts that have contributed to model accuracy in training or to a specific prediction for a given observation. This is already an important step and this is where the xAI community has already developed a variety of successful methods. However, explainability does not relate to a human model. Causability is the measurable extent to which an explanation (resulting from explainability) achieves a certain level of causal understanding for a human expert (or layperson, of course). Causal in the sense of Judea Pearl as relationship between cause and effect [35].

Why is this important? Because human understanding, especially checking whether and to what extent something has been understood, can only be guaranteed if we can map Explainability with Causability. Consequently, successful mapping between Explainability and Causability requires new human-AI interfaces that allow domain experts to interactively ask questions and counterfactuals to gain insight into the underlying explanatory factors of an outcome [20].

In an ideal world, statements originating from both "human intelligence" and "artificial intelligence" would be identical and congruent with the "ground truth," which is that it must be defined equally for humans and AI. That this is not easy and often does not work becomes quickly clear in the complex domain of medicine: medicine is a good exemplar of real-world challenges: (i) ground truth cannot always be precisely defined, especially for medical diagnoses; and (ii) human (scientific) models are often based on causality in the sense of Judea Pearl as the ultimate goal for understanding the underlying explanatory mechanisms.

While correlation is accepted as the basis for decisions in medical AI for a long time [42], it can only be considered as an intermediate step for causal considerations, which is relevant due to the importance of validity and necessary to build human trust [6].

Currently, there is much debate in the xAI community about avoiding bias and how to ensure fairness in AI decisions [30]. Bias is a core issue in causality, and causability is one possible measure of it. Validation of causal effects under particular causal structures is especially necessary when such effects are estimated in limited arrays. Randomized controlled trials are a good example. Such studies allow causal hypotheses to be tested because randomization by design is guaranteed, even with limited knowledge about the domain. A particular generalizability problem has been described by (Bareinboim & Pearl, 2013) [2], referred to as transportability, which can be viewed as a "data fusion framework" for external validation of intervention models and counterfactual queries. Transportability allows causal effects learned in experimental studies to be transferred to a new setup in which only observational studies can be conducted. Transportable models can be integrated into clinical guidelines to augment subject matter experts with "actionable" predictions to achieve better precision medicine [36].

The domain of artificial intelligence has tremendous potential to contribute to a better understanding of disease, which can lead to more accurate diagnoses, more rational disease prevention strategies, better treatment selection, and the development of new therapies. In addition, a better understanding of disease can contribute to the long-term goal of personalized precision medicine, which seeks to redefine the understanding of disease development and progression, treatment response, and health outcomes by measuring as precisely as possible the molecular, genetic, environmental, and behavioral individual factors that contribute to health and disease. Here, it is imperative that AI decisions be fully traceable across all modalities involved so that the medical professional has the ability to i) understand, ii) confirm, or iii) reject them. Whatever future human-AI interfaces look like, they must enable a domain expert to understand causal pathways in order to compute meaningful counterfactuals [28]. This is where the use of graphs and learning graph representations can be beneficial [24].

4   Robustness

4.1   Robustness in General

Many machine learning models achieve amazing performance on standard i.i.d. data. However when working with real data (e.g. from the medical domain) they fail miserably, being perturbed very easily. Robustness is generally defined as the property of a model to produce unperturbed results even if the input data is perturbed [50].

For example, it has been observed that commonly occurring image corruptions, such as random noise, contrast change, and blurring, can lead to significant performance degradation. Consequently, improving distributional robustness is an important step towards safely deploying models in complex, real-world settings [54]. Robustness is a ubiquitously observed property of biological systems and is considered a fundamental feature of complex evolvable systems. It is achieved by several underlying principles that apply to biological organisms as well as to sophisticated technical systems [29].

4.2   Robustness in Interventional Studies

Biomedical observational studies are affected by confounding and selection bias, which makes causal inference to be unfeasible if robust assumptions are not made. These require a priori domain knowledge, as data-driven prediction models may be used for drawing causal effects, but neither their parameters nor their predictions necessarily have a causal interpretation. The healthcare informatics communities are recommended to employ causal approaches and learn causal structures by using the *linchpins* to develop and test intervention models [40]: 1) target trials, 2) transportability, and 3) prediction invariance.

Target trials refer to algorithmic emulation of randomized studies. Transportability is a *license* to "transfer causal effects learned in experimental studies to a new population, in which only observational studies can be conducted." Akin to transportability is prediction invariance, where a "true causal model is contained in all prediction models whose accuracy does not vary across different settings".

When a causal structure is available or a target trial design can be devised, the evaluation of model transportability for a given set of action queries (e.g., treatment options or risk modifiers) is recommended; while for exploratory analyses where causal structures are to be discovered, prediction invariance could be used. In this way, as advocated by [40] transportability and prediction invariance could become guideline core tools and part of reporting protocols for intervention models, for a better alignment with the standards for prognostic and diagnostic models of medicine and biomedical practice today.

4.3   Robustness to Adversarial Attacks

Technically, we are talking about performance on unseen examples from the underlying distribution, and the goal is to train models so that the expected loss reaches a minimum:

$$\mathop{\mathbb{E}}_{(x,y)\sim\mathcal{D}} = [\mathcal{L}(x,y;\theta)] \tag{1}$$

Szegedy et al. (2013) [49] made a fascinating discovery: several non-linear machine learning models, including state-of-the-art neural networks, misclassify well-known examples if they have been disturbed even slightly, e.g. with almost invisible salt and pepper noise. For example, a pig's face is suddenly classified as an airliner, a panda becomes a gibbon, or a benign melanoma becomes malignant (see picture) with extremely high confidence. Such total misclassifications can have dramatic effects in many application areas and do not contribute at all to trust building.

Several non-linear machine learning models, including state-of-the-art neural networks, also misclassify well-known examples if they have been disturbed even slightly, e.g. with almost invisible salt and pepper noise. The vehement however was that the output is completely misclassified, for example a pig face is suddenly classified as an airliner (see picture). This may be funny, however it can have dramatic effects in many application areas. For example, in medicine, such misclassifications can lead to serious consequences.

However, many non-linear ML models, particularly deep learning ones, falsely classify the so-called adversarial examples, i.e., inputs formed from small perturbations (e.g., salt-and-pepper noise) applied to training samples, which are usually not even visible for a time-limited human [10]. This results into dramatic effects, and completely wrong outputs with high confidence. A typical example is shown in Fig. 1.

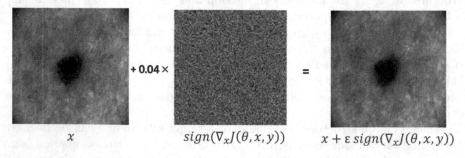

$$x \qquad\qquad sign(\nabla_x J(\theta,x,y)) \qquad x + \varepsilon\, sign(\nabla_x J(\theta,x,y))$$

Fig. 1. Example of the susceptibility of our currently best performing deep learning models: One of the maybe most prominent adversarial example, cf. with [12]. Perturbations of just a few pixels ("salt-and-pepper noise") can dramatically change the classification and turn a malign melanoma into benign and vice versa.

Determining the appropriate Δ to use is a domain specific question, and therefore a human-in-the-loop [19] can be of help because humans, even if they also make mistakes, can be considered a robust proxy in decision making (see next chapter).

In Fig. 1 the θ are the feature parameters, x is the the input to the model, y is the targeted output function associated with x, and $J(\theta, x, y)$ is the cost function (loss function) used to train the neural network. The cost function around the current value of θ can be linearized, obtaining an optimal max-norm constrained pertubation of

$$\epsilon = sign(\nabla_x J(\theta, x, y)) \tag{2}$$

This is called "fast gradient sign method" of generating adversarial examples, the required gradient can be computed using backpropagation. The $sign(x)$ as real sign function maps from $\mathbb{R}$ to a Bit and adds a sign to it. The Nabla Operator denotes the vector representation of the differential operators (gradient, divergence, rotation). The property to resist such disturbances is called robustness, and achieving it can mean training models with *low expected adversarial loss*:

$$\mathop{\mathbb{E}}_{(x,y)\sim\mathcal{D}} = [\max_{\delta\in\Delta} \mathcal{L}(x + \delta, y; \theta)]. \tag{3}$$

5 How Can a Human-in-the-Loop Help?

When we compare human learning and problem solving with the capabilities of our most advanced learning algorithms, some serious differences immediately stand out: Supervised learning requires a lot of labelled data while model-free reinforcement learning requires far too many trials. Humans, on the other hand, are able to generalise quickly and surprisingly well even in complex situations with little prior experience. Humans can generalise in a way that is different and more powerful than ordinary i.i.d. generalisation, namely these can correctly interpret novel combinations of existing concepts even when these combinations are extremely unlikely under training distribution, at least as long as they take into account higher-level syntactic and semantic patterns that have already been learned [46]. Humans are often very robust to change and can adapt quickly to change even with little training. Current Deep Learning is most successful in perceptual tasks and, more generally, in the previously mentioned System 1 tasks. The use of Deep Learning for System 2 tasks that require a deliberate sequence of steps is still in its infancy [3]. Humans are also very adept at inferring new causal relationships from even a few observations. Prior knowledge about the probability of occurrence of causal relationships of different kinds and the nature of the mechanisms linking causes and effects plays a crucial role in these inferences [14].

Let's stay again with an example from network medicine. Interactions of human experts on graphs clearly need to be based on the low-dimensional input features to efficiently discover, reject, or confirm causal links between biomedical modalities. Once these connections are computed and the structure of the input graphs is updated accordingly, methods for explainable GNN can be applied [52,55]. Human domain knowledge in the loop thereby enhances the model building process. Furthermore, human interactions can be realized here by "what-if" queries (counterfactuals) to the system, leading to a graph of counterfactuals in which features are defined as nodes and edges refer to combinations of features. Such a counterfactual graph can be generated in a purely data-driven manner: Given a test set that includes a sufficient number of samples, an algorithm traverses the feature space and exchanges feature values between nearest neighbors of a different class of results until the class of the instance itself changes. The nearest neighbor-based sampling leads to counterexamples of realistic patient profiles and is thus based on plausible counterfactuals. Of course, providing such counterfactuals, roughly based on the internals of a model, is not yet sufficient for explainability. The plausibility of the counterfactual change is therefore a must, i.e., the "counterfactual path" leading to the label change should have a real chance of occurring in practice for the counterfactual to be realistic. In this regard, recent attempts to find plausible counterfactuals for image classification should be extended to models for graph data. The sampled feature path leading to the class change is stored and forms a *contrafactual decision path*. Repeating this procedure results in a graph consisting of *multiple decision paths* that can be used as a communication channel back to the black box model. Recent work has shown how a DF can be efficiently reduced to a single decision tree [11,43], from which counterfactuals can be easily observed by the leaf nodes, so that it could be used as a model for *global* explanations. In such an approach, the *human-in-the-loop* will be able to study this consensus decision tree and thus adopt the changes to the counterfactual graph. Studying the impact of modifications to the counterfactual graph on the decision trees can facilitate the definition of symbolic rules to revise the internal structure of the input graph. Possible modifications include adding or deleting semantic links between modalities, however also adjusting their edge weights (reference) [24,36].

To implement such robust, explainable and thus trustworthy AI applications, we need an iterative, agile and human-centred AI design process. These processes have long been known in traditional software engineering as agile user-centred design methods [27] and now need to be taken to the next level of future AI engineers.

6   Conclusion

Explainability offers the great opportunity not only to meet the legal requirement for transparency and traceability of "black boxes", but also to promote trust in AI and, above all, to foster a deeper understanding of previously unknown connections - in other words, to contribute to the discovery of knowledge. Just think of the enormous support that doctors can draw from the combination of human intelligence and AI (e.g. in diagnosis): Humans show very good intuition in low-dimensional problems, can generalise amazingly well from a small amount of data, and recognise connections thanks to their everyday intelligence. For example, doctors can apply AI to "interesting" data and interrogate it interactively. Conversely, machine-generated results can be reconstructed from high-dimensional data spaces that no human could ever have found and checked for plausibility. The most important contribution of explainability is to clarify what is cause and what is effect in order to avoid falsely including artefacts and surrogates. This is desirable in many application domains and even mandatory in safety-critical domains.

The major challenges of our field for the future is to merge the two AI approaches, e.g. logic-based ontologies with probabilistic machine learning with one (or more or even many) humans in the loop into a hybrid multi-agent interaction model, used as a kind of "power steering for the brain". This would not only mean an extension (augmentation) of human intelligence by machine intelligence, but also, conversely, an extension of artificial intelligence by human intuition and thus an important contribution to making algorithms more robust.

Acknowledgements. The author declares that there are no conflicts of interests. This work does not raise any ethical issues. Parts of this work have been funded by the Austrian Science Fund (FWF), Project: P-32554, explainable AI.

References

1. Arrieta, A.B., et al.: Explainable artificial intelligence (XAI): Concepts, taxonomies, opportunities and challenges toward responsible AI. Inf. Fusion **58**, 82–115 (2020). https://doi.org/10.1016/j.inffus.2019.12.012
2. Bareinboim, E., Pearl, J.: A general algorithm for deciding transportability of experimental results. arXiv:1312.7485 (2013)
3. Bengio, Y., Lecun, Y., Hinton, G.: Deep learning for AI. Commun. ACM **64**(7), 58–65 (2021). https://doi.org/10.1145/3448250
4. Biecek, P.: Dalex: explainers for complex predictive models in r. J. Mach. Learn. Res. **19**(1), 3245–3249 (2018)
5. Binet, A.: L'étude expérimentale de l'intelligence. Schleicher frères and cie, Paris (1903)
6. Cabitza, F., Campagner, A., Balsano, C.: Bridging the "last mile" gap between AI implementation and operation: "data awareness" that matters. Ann. Transl. Med. **8**(7), 501 (2020). https://doi.org/10.21037/atm.2020.03.63
7. Chatila, R., et al.: Trustworthy AI. In: Braunschweig, B., Ghallab, M. (eds.) Reflections on Artificial Intelligence for Humanity. LNCS (LNAI), vol. 12600, pp. 13–39. Springer, Cham (2021). https://doi.org/10.1007/978-3-030-69128-8_2
8. Corazzini, J.G.: Trust as a complex multi-dimensional construct. Psychol. Rep. **40**(1), 75–80 (1977). https://doi.org/10.2466/pr0.1977.40.1.75
9. Donsa, K., Spat, S., Beck, P., Pieber, T.R., Holzinger, A.: Towards personalization of diabetes therapy using computerized decision support and machine learning: some open problems and challenges. In: Holzinger, A., Röcker, C., Ziefle, M. (eds.) Smart Health. LNCS, vol. 8700, pp. 237–260. Springer, Cham (2015). https://doi.org/10.1007/978-3-319-16226-3_10

10. Elsayed, G.F., et al.: Adversarial examples that fool both human and computer vision. In: Bengio, S., Wallach, H., Larochelle, H., Grauman, K., Cesa-Bianchi, N., Garnett, R. (eds.) Neural Information Processing Systems (NIPS 2018), pp. 1–11. NIPS Foundation (2018)

11. Fernández, R.R., De Diego, I.M., Aceña, V., Fernández-Isabel, A., Moguerza, J.M.: Random forest explainability using counterfactual sets. Inf. Fusion 63(11), 196–207 (2020). https://doi.org/10.1016/j.inffus.2020.07.001

12. Finlayson, S.G., Bowers, J.D., Ito, J., Zittrain, J.L., Beam, A.L., Kohane, I.S.: Adversarial attacks on medical machine learning. Science 363(6433), 1287–1289 (2019). https://doi.org/10.1126/science.aaw4399

13. Floridi, L.: Establishing the rules for building trustworthy AI. Nat. Mach. Intell. 1(6), 261–262 (2019). https://doi.org/10.1038/s42256-019-0055-y

14. Griffiths, T.L., Sobel, D.M., Tenenbaum, J.B., Gopnik, A.: Bayes and blickets: effects of knowledge on causal induction in children and adults. Cogn. Sci. 35(8), 1407–1455 (2011). https://doi.org/10.1111/j.1551-6709.2011.01203.x

15. Gunning, D., Aha, D.W.: Darpa's explainable artificial intelligence program. AI Mag. 40(2), 44–58 (2019). https://doi.org/10.1609/aimag.v40i2.2850

16. Hajian, S., Bonchi, F., Castillo, C.: Algorithmic bias: from discrimination discovery to fairness-aware data mining. In: Proceedings of the 22nd ACM SIGKDD International Conference on Knowledge Discovery and Data Mining. ACM (2016). https://doi.org/10.1145/2939672.2945386

17. Hamon, R., Junklewitz, H., Sanche, I.: Robustness and Explainability of Artificial Intelligence - From technical to policy solutions. Publications Office of the European Union, Luxembourg (2020). https://doi.org/10.2760/57493

18. Hohenstein, J., Jung, M.: Ai as a moral crumple zone: the effects of AI-mediated communication on attribution and trust. Comput. Hum. Behav. 106(2020). https://doi.org/10.1016/j.chb.2019.106190

19. Holzinger, A.: Interactive machine learning for health informatics: when do we need the human-in-the-loop? Brain Inf. 3(2), 119–131 (2016). https://doi.org/10.1007/s40708-016-0042-6

20. Holzinger, A.: Explainable ai and multi-modal causability in medicine. Wiley i-com J. Interact. Media 19(3), 171–179 (2020). https://doi.org/10.1515/icom-2020-0024

21. Holzinger, A., Carrington, A., Müller, H.: Measuring the quality of explanations: the system causability scale (SCS). KI - Künstliche Intelligenz 34(2), 193–198 (2020). https://doi.org/10.1007/s13218-020-00636-z

22. Holzinger, A., Kickmeier-Rust, M., Müller, H.: KANDINSKY patterns as IQ-test for machine learning. In: Holzinger, A., Kieseberg, P., Tjoa, A.M., Weippl, E. (eds.) CD-MAKE 2019. LNCS, vol. 11713, pp. 1–14. Springer, Cham (2019). https://doi.org/10.1007/978-3-030-29726-8_1

23. Holzinger, A., Langs, G., Denk, H., Zatloukal, K., Mueller, H.: Causability and explainability of artificial intelligence in medicine. Wiley Interdisc. Rev. Data Min. Knowl. Discov. 9(4), 1–13 (2019). https://doi.org/10.1002/widm.1312

24. Holzinger, A., Malle, B., Saranti, A., Pfeifer, B.: Towards multi-modal causability with graph neural networks enabling information fusion for explainable AI. Inf. Fusion 71(7), 28–37 (2021). https://doi.org/10.1016/j.inffus.2021.01.008

25. Holzinger, A., Mueller, H.: Toward human-AI interfaces to support explainability and causability in medical AI. IEEE Comput. 54(10) (2021). https://doi.org/10.1109/MC.2021.3092610

26. Holzinger, A., Weippl, E., Tjoa, A.M., Kieseberg, P.: Digital transformation for sustainable development goals (SDGs) - a security, safety and privacy perspective on AI. In: Holzinger, A., Kieseberg, P., Tjoa, A.M., Weippl, E. (eds.) CD-MAKE 2021. LNCS, vol. 12844, pp. 1–20. Springer, Cham (2021). https://doi.org/10.1007/978-3-030-84060-0_1

27. Hussain, Z., Slany, W., Holzinger, A.: Investigating agile user-centered design in practice: a grounded theory perspective. In: Holzinger, A., Miesenberger, K. (eds.) USAB 2009. LNCS, vol. 5889, pp. 279–289. Springer, Heidelberg (2009). https://doi.org/10.1007/978-3-642-10308-7_19

28. Kahneman, D.: Varieties of counterfactual thinking. In: Roese, N.J., Olson, J.M. (eds.) What might have been: The social psychology of counterfactual thinking. Taylor and Francis, New York (1995)

29. Kitano, H.: Biological robustness. Nat. Rev. Genet. 5(11), 826–837 (2004). https://doi.org/10.1038/nrg1471

30. Kusner, M.J., Loftus, J.R.: The long road to fairer algorithms. Nature 578, 34–36 (2020). https://doi.org/10.1038/d41586-020-00274-3

31. Lakkaraju, H., Arsov, N., Bastani, O.: Robust and stable black box explanations. In: Daumé, H., Singh, A. (eds.) International Conference on Machine Learning (ICML 2020), pp. 5628–5638. PMLR (2020)

32. Laponogov, I., et al.: Network machine learning maps phytochemically rich "hyperfoods" to fight covid-19. Human genomics 15(1), 1–11 (2021). https://doi.org/10.1186/s40246-020-00297-x

33. Lapuschkin, S., Binder, A., Montavon, G., Mueller, K.R., Samek, W.: The LRP toolbox for artificial neural networks. J. Mach. Learn. Res. (JMLR) 17(1), 3938–3942 (2016)

34. Liu, H., et al.: Trustworthy ai: A computational perspective. arXiv:2107.06641 (2021)

35. Pearl, J.: Causality: Models, Reasoning, and Inference, 2nd edn. Cambridge University Press, Cambridge (2009)

36. Pfeifer, B., Saranti, A., Holzinger, A.: Network module detection from multimodal node features with a greedy decision forest for actionable explainable AI. arXiv:2108.11674 (2021)

37. Piaget, J.: On the Development of Memory and Identity. Clark University Press, Worchester (1961)

38. Piaget, J., Inhelder, B.: Memory and Intelligence. Routledge, London (1973)

39. Ploug, T., Holm, S.: The four dimensions of contestable AI diagnostics-a patient-centric approach to explainable AI. Artif. Intell. Med. 107(2020). https://doi.org/10.1016/j.artmed.2020.101901

40. Prosperi, M., et al.: Causal inference and counterfactual prediction in machine learning for actionable healthcare. Nat. Mach. Intell. 2(7), 369–375 (2020). https://doi.org/10.1038/s42256-020-0197-y

41. Roberts, H., Cowls, J., Morley, J., Taddeo, M., Wang, V., Floridi, L.: The Chinese approach to artificial intelligence: an analysis of policy, ethics, and regulation. AI Soc. 36(1), 59–77 (2020). https://doi.org/10.1007/s00146-020-00992-2

42. Roque, F.S., et al.: Using electronic patient records to discover disease correlations and stratify patient cohorts. PLoS Comput. Biol. 7(8) (2011). https://doi.org/10.1371/journal.pcbi.1002141

43. Sagi, O., Rokach, L.: Explainable decision forest: transforming a decision forest into an interpretable tree. Inf. Fusion 61, 124–138 (2020). https://doi.org/10.1016/j.inffus.2020.03.013

44. Samek, W., Montavon, G., Vedaldi, A., Hansen, L.K., Müller, K.-R. (eds.): Explainable AI: Interpreting, Explaining and Visualizing Deep Learning. LNCS (LNAI), vol. 11700. Springer, Cham (2019). https://doi.org/10.1007/978-3-030-28954-6

45. Schnake, T., et al.: Xai for graphs: Explaining graph neural network predictions by identifying relevant walks. arXiv:2006.03589 (2020)

46. Shepard, R.N.: Toward a universal law of generalization for psychological science. Science **237**(4820), 1317–1323 (1987). https://doi.org/10.1126/science.3629243

47. Simpson, J.A.: Psychological foundations of trust. Curr. Dir. Psychol. Sci. **16**(5), 264–268 (2007). https://doi.org/10.1111/j.1467-8721.2007.00517.x

48. Stoeger, K., Schneeberger, D., Kieseberg, P., Holzinger, A.: Legal aspects of data cleansing in medical AI. Comput. Law Secur. Rev. **42**(2021). https://doi.org/10.1016/j.clsr.2021.105587

49. Szegedy, C., Zaremba, W., Sutskever, I., Bruna, J., Erhan, D., Goodfellow, I., Fergus, R.: Intriguing properties of neural networks. arXiv:1312.6199 (2013)

50. Tsipras, D., Santurkar, S., Engstrom, L., Turner, A., Madry, A.: Robustness may be at odds with accuracy. In: 7th International Conference on Learning Representations, ICLR 2019 (2019)

51. Wang, J., Jing, X., Yan, Z., Fu, Y., Pedrycz, W., Yang, L.T.: A survey on trust evaluation based on machine learning. ACM Comput. Surv. (CSUR) **53**(5), 1–36 (2020). https://doi.org/10.1145/3408292

52. Xu, K., Hu, W., Leskovec, J., Jegelka, S.: How powerful are graph neural networks? arXiv:1810.00826 (2018)

53. Yan, Z., Holtmanns, S.: Trust modeling and management: from social trust to digital trust. In: Subramanian, R. (ed.) Computer Security, Privacy and Politics: Current Issues, Challenges and Solutions, pp. 290–323. IGI Global (2008)

54. Yin, D., Lopes, R.G., Shlens, J., Cubuk, E.D., Gilmer, J.: A fourier perspective on model robustness in computer vision. arXiv:1906.08988 (2019)

55. Ying, Z., Bourgeois, D., You, J., Zitnik, M., Leskovec, J.: Gnnexplainer: generating explanations for graph neural networks. In: Wallach, H., Larochelle, H., Beygelzimer, A., d'Alche Buc, F., Fox, E., Garnett, R. (eds.) Advances in Neural Information Processing Systems, pp. 9244–9255 (2019)

This Looks Like That, Because ... Explaining Prototypes for Interpretable Image Recognition

Meike Nauta[1,2]($\boxtimes$), Annemarie Jutte[1], Jesper Provoost[1], and Christin Seifert[1,2]

[1] University of Twente, Enschede, The Netherlands
m.nauta@utwente.nl, {a.m.p.jutte,j.c.provoost}@student.utwente.nl
[2] University of Duisburg-Essen, Essen, Germany
christin.seifert@uni-due.de

Abstract. Image recognition with prototypes is considered an interpretable alternative for black box deep learning models. Classification depends on the extent to which a test image "looks like" a prototype. However, perceptual similarity for humans can be different from the similarity learned by the classification model. Hence, only visualising prototypes can be insufficient for a user to understand what a prototype exactly represents, and why the model considers a prototype and an image to be similar. We address this ambiguity and argue that prototypes should be explained. We improve interpretability by automatically enhancing visual prototypes with quantitative information about visual characteristics deemed important by the classification model. Specifically, our method clarifies the meaning of a prototype by quantifying the influence of colour hue, shape, texture, contrast and saturation and can generate both global and local explanations. Because of the generality of our approach, it can improve the interpretability of any similarity-based method for prototypical image recognition. In our experiments, we apply our method to the existing Prototypical Part Network (ProtoPNet). Our analysis confirms that the global explanations are generalisable, and often correspond to the visually perceptible properties of a prototype. Our explanations are especially relevant for prototypes which might have been interpreted incorrectly otherwise. By explaining such 'misleading' prototypes, we improve the interpretability and simulatability of a prototype-based classification model. We also use our method to check whether visually similar prototypes have similar explanations, and are able to discover redundancy. Code is available at https://github.com/M-Nauta/Explaining_Prototypes.

A. Jutte and J. Provoost—Both authors contributed equally to this work.

© Springer Nature Switzerland AG 2021
M. Kamp et al. (Eds.): ECML PKDD 2021 Workshops, CCIS 1524, pp. 441–456, 2021.
https://doi.org/10.1007/978-3-030-93736-2_34

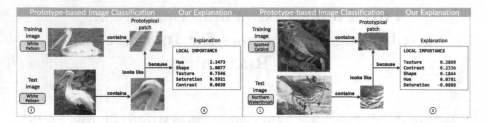

Fig. 1. (1) Prototype-based image classification (e.g. ProtoPNet [7]). (2) Our contribution: Quantifying the importance of visual characteristics to explain why the classification model deemed an image patch and prototype similar. Left: Logical explanation for the clear similarity between the patches. Right: a 'misleading' prototype: humans might expect these patches to be dissimilar, but our method explains that the classification model considers these patches alike because of similar texture.

1   Introduction

Convolutional Neural Networks (CNNs) are the de-facto standard for object detection due to their impressive performance in numerous automated image classification tasks. However, the black box nature of neural networks prevents a human to assess the model's decision making process, which is especially problematic in domains with high stakes decisions [21]. Following this demand on understanding automated decision making, explainable Artificial Intelligence (XAI) has been actively researched [10]. *Post-hoc* explanation methods learn a second, transparent model to approximate the first black box model [10], but these reverse-engineering approaches are not guaranteed to show the *actual* reasoning of the black box model [21]. *Intrinsically* interpretable models on the other hand, are faithful by design and allow *simulatability*: a user should be able to reproduce the model's decision making process based on the input data together with the explanations of the interpretable model and come to the same prediction [15]. One type of such models is prototypical learning, which has a transparent, built-in case-based decision making process. We focus on the problem of supervised image recognition where a machine learning model should label an image. Prototypes in this context are usually 'nearest neighbours', i.e., images from the training set that look similar to the image being classified [1,4]. The similarity between a prototype and an image is often measured in latent space, learned by the neural network, where images from the same class are close and dissimilar images are far apart with respect to a certain distance or similarity metric. Recently, the Prototypical Part Network (ProtoPNet) [7] and ProtoTree [17] were introduced which use prototypical *parts* and identify similar patches in an image. The classification depends on the extent to which *this* part of the image "looks like" *that* prototypical part, measured by a similarity score. An example of this reasoning is shown in Fig. 1.

Prototype Ambiguity. In this paper, we address the ambiguity that prototypes can have and present a method to *explain prototypes*. Consider the left part in Fig. 1, showing a prototypical patch ('prototype') of a white pelican. Although the similarity between this prototype and the patch in the test image is not surprising, it is unclear what this prototype exactly represents. Is the prototype looking for a white neck, an orange-coloured beak or is the shape of the beak specifically important? The similarity score between the two patches assigned by the model depends on its classification strategy, and hence its learned latent space. Explanations are especially needed when similarity is not so obvious. When seeing the two patches in the right part of Fig. 1, a human might argue that these patches are dissimilar because of the colour differences. The classification model however assigns these patches a high similarity score, and thus considers them alike, even though the test image is from a different class than the prototype. This shows that a human and the CNN might have different reasoning processes, despite using the same prototypes. The classification strategy of a neural network, dependent on the learned latent space, determines the reason for considering two patches as being similar or different. It has been shown that CNNs trained on ImageNet are strongly biased towards recognizing texture [8], although other work shows that CNNs can be biased towards shape [18] or colour [11]. Perceptual similarity for humans however is biased towards shape [3], but also based on e.g. colour, size, semantic similarity and complexity [13, 20]. It is also questionable whether humans and CNNs will ever follow the exact same similarity reasoning, since Rosenfeld et al. found that neural networks fall short on predicting human similarity perception [19]. Since a user is not aware of the underlying classification strategy of the trained CNN and might also be unaware of personal biases, only visualising prototypes is insufficient for understanding what a prototype exactly represents, and why a prototype and an image are considered similar. This issue may also arise with other explainability methods that show or highlight image parts, such as attention mechanisms [6], components [22] and other part-based models e.g. [26, 27]. Including our explanations can help users to increase the simulatability [15] and general understanding of the model.

Contribution. We improve the interpretability of a prototype-based CNN by automatically enhancing prototypes with extra quantitative information about visual characteristics used by the model. Specifically, we present a methodology to quantify the influence of colour hue, saturation, shape, texture, and contrast in a prototype. This clarifies what the model pays attention to and why a model considers two images to be similar. Hence, we disentangle localisation and explanation. Our method can extend any prototype-based model for image recognition, such as ProtoPNet [7] and ProtoTree [17]. In this paper, we show its applicability for the prototypical parts of ProtoPNet. For example, again considering the left part of Fig. 1, our explanation shows that ProtoPNet considers the prototype and patch from the test image to be similar because of the similar colour hue and shape of the beak in the test image. Our method is especially useful when similarity is not so obvious. It can explain potentially misleading prototypes such as the right prototype in Fig. 1. Whereas a human might look for something green, our explanation reveals that ProtoPNet considers these two patches similar because of texture, contrast and shape. The similarity is thus because of the dotted pattern and colour hue was not important. This explanation seems reasonable given that the prototype is from the class "Spotted Catbird".

Our method automatically modifies images to change their hue, shape, texture, contrast or saturation. We forward both the original image and the modified image through the prototype-based network and analyse the resulting similarity scores. Specifically, the similarity score between the prototype and the original image is compared with the similarity score of the prototype and a modified image. The intuition is that a visual characteristic is considered *unimportant* by the classification model when the difference between these two similarity scores is small (and will therefore get a low importance score), and is deemed *important* when the similarity scores differ sufficiently. For example, a blue bird is changed to a purple bird by changing the hue of the image. If hue would have been important for the specific prototype, it would be expected that the model assigns a low similarity between the prototype and the purple bird, whereas the similarity with the blue bird was high. As shown in Fig. 1, the prototypes can subsequently be explained by quantifying the importance of visual characteristics.

2 Prototypical Part Network

We apply the methodology presented in this paper to ProtoPNet, the Prototypical Part Network from Chen et al. [7] that follows the "*this* looks like *that*" reasoning. Prototypical parts learned by ProtoPNet are subsequently explained by our method. Key for presenting our explanation methodology is having a global understanding of the workings of ProtoPNet.

The ProtoPNet architecture consists of a standard CNN (e.g. DenseNet), followed by a prototype layer and a fully-connected layer. The prototype layer consists of a pre-determined number of class-specific prototypes, usually 10 prototypes per class [7]. The fully-connected layer learns a weight for each prototype. During training, prototypes are vectors in latent space that should learn discriminative, prototypical parts of a class. An input image is forwarded through the CNN, after which the prototype layer compares the resulting latent embedding with the prototype. A kernel slides over the latent image and at each location, the distance between the latent prototype and a patch in the latent image is calculated. This creates an activation map, containing the distance to the prototype at each location in the latent image. To ensure that the prototype can be visualised, the training procedure of ProtoPNet requires that each prototype is *identical* to some latent training patch such that it can be upsampled to the size of the original image and visualised as an image patch (Fig. 2).

After training, ProtoPNet classifies a test image k by calculating the similarity between a prototype and image k. The distance $d_{j,k}$ between the nearest patch in latent image k to the j-th prototype is converted to a similarity score:

$$g_{j,k} = \log\left(\frac{d_{j,k} + 1}{d_{j,k} + \epsilon}\right), \tag{1}$$

where ϵ is an arbitrarily small positive quantity to prevent zero division. To classify this image, the similarity scores of the image and each prototype are weighted by the fully-connected layer and summed per class, resulting in a final score for an image belonging to each class. The left part of Fig. 4 illustrates this reasoning process.

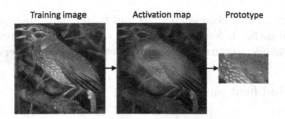

Fig. 2. A prototype from ProtoPNet is the nearest patch of a training image.

3 Methodology

In order to obtain importance scores for visual characteristics of prototypes, images are automatically modified. The characteristics in focus are contrast, colour hue and saturation, shape, and texture (cf. Sect. 3.1). Our approach for automatically modifying these characteristics is described in Sect. 3.2. Section 3.3 presents a methodology to explain prototypes by quantifying the importance of a visual characteristic.

3.1 Important Visual Characteristics

The perceptual and cognitive processing in the human visual system is influenced by various features. To determine which image modifications we need to effectively explain prototypes, we discuss important visual characteristics for the human perceptual system.

The data visualisation domain has a ranking of channels to control the appearance of so-called *marks* [16]. A 'mark' is a basic graphical element in an image, such as a black triangle or moving red dot. Important visual channels for marks are position, size, angle, spatial region, colour hue, colour luminance, colour saturation, curvature, motion and shape [16]. For static 2-dimensional natural images, motion is not applicable and we consider curvature related to shape. Furthermore, it is not necessary to modify the size, position, angle or spatial region of objects in images, since CNNs with pooling, possibly combined with suitable data augmentation, are invariant to these characteristics [9,23]. Moreover, research in neuroscience shows that the human eye can recognise objects independent of ambient light level during the day [24], whereas contrast (spatial variation in luminance) is needed for edge detection and delineation of objects [16]. The human visual system is thus more sensitive to contrast than absolute luminance [24]. We therefore will not modify the absolute luminance, but the contrast in an image. Thus, the visual characteristics from the data visualisation domain that we deem important for explaining a prototype are **hue**, **contrast**, **saturation** and **shape**.

The channels for marks mentioned in the previous paragraph are however too simplistic, because they do not include the texture or material of an object. Research in neuroscience also emphasises the importance of texture for classifying objects in the natural world [5,14]. Related to this, Bau et al. [2] disentangled visual representations by layers in a CNN and found that self-supervised models, especially in the earlier layers of the network, learn many texture detectors. We therefore also include **texture** as an important visual characteristic.

3.2   Image Modifications

Original   Contrast   Saturation   Hue   Shape   Texture

Fig. 3. Image modifications for corresponding visual characteristics.

For each of the visual characteristics, an image set is created. Each of these sets contains modified images, which are designed to be harder to classify based on the respective characteristic. For example, we generate a set of low-contrast images, such that contrast information can not (or hardly) be used by the model. Using these modified images, the importance of a characteristic for a specific prototype can be determined by comparing the differences between the prototype-image similarity of the original and modified image.

To create the modified images, we apply image transformations to reduce the intensity of each characteristic, i.e., we create images with reduced contrast, saturation, hue, shape and texture. Figure 3 shows an example image and its modified versions. We opt for automated image modifications instead of manual modifications used for experiments in psychology research (e.g. [18]), to be able to create a large number of modified images efficiently.

To create low **contrast** images, the original image is blended with the mean of its grayscale version. More concretely, we first create a grayscale version of the image and calculate its mean value. We then generate the modified image by pixel-wise averaging each channel (RGB) of the original image with the mean grayscale value. Similarly, the low-**saturation** image is created by averaging the original image with its grayscale counterpart. To generate an image with different **colour hues**, the RGB image is converted to the HSV colour space after which the H-dimension is modified for each pixel. In order to modify **shapes** in an image, we add a linear displacement by warping the image. Specifically, we shift pixels according to a sine wave in both the horizontal and vertical direction. To modify **texture**, we apply a non-local means denoising technique which removes small quantities of noise, and can therefore be used to blur the sophisticated texture of a bird while preserving its overall shape. Implementation details are presented in Sect. 4.1.

3.3   Importance Scores for Image Characteristics

We evaluate the importance of visual characteristics by calculating local and global importance scores. The **local** score measures the importance of the visual characteristics for a single image, and is therefore applied on previously unseen images, i.e., any image in the test dataset S_{test}. The **global** score measures the importance of visual characteristics for one prototype in general, and is independent of a specific input image. The global score is obtained from the training dataset S_{train}.

Let $i \in \{\text{contrast}, \text{saturation}, \text{hue}, \text{shape}, \text{texture}\}$ denote the type of modification, $j \in \{1, 2, ..., n\}$ the prototype index and k the image. Furthermore, as introduced in Sect. 2, let the similarity of the original image and the prototype be denoted as g, and the similarity of the modified image and the prototype be denoted as $\hat{g}$. Then the local importance score $\phi_{\text{local}}^{i,j,k}$ of characteristic i for test image $k \in S_{\text{test}}$ on the j-th prototype is the difference in similarity scores:

$$\phi_{\text{local}}^{i,j,k} = g_{j,k} - \hat{g}_{i,j,k}. \tag{2}$$

For the calculation, we fix the patch location such that the part of image k compared with prototype j is the same for both the original and modified image.

Whereas the local importance score indicates to what extent a visual characteristic influences the similarity score given by the prototype-based model *for a single image*, the *global* importance score gives a general impression of the importance of visual characteristics in a prototype. The global score of characteristic i on the jth prototype can be calculated by taking all training images into account. A naive approach would be to average over the local scores of all training images. However, prototype j might not be present in all images and modifying those images will therefore not (or barely) influence the resulting similarity score. For example, if prototype j represents a specific beak which is not present in original image k, ProtoPNet will give a low similarity score $g_{j,k}$. Since this prototype will also be absent in the modified image, the difference between the similarity scores, Eq. 2, is near zero. This result could indicate that a certain characteristic is not important, although the result is actually indicating that the prototype was simply not present. Therefore, we create a more informative global importance score by calculating a weighted arithmetic mean by weighing the local scores of all images in S_{train} by their similarity score with the prototype:

$$\phi_{\text{global}}^{i,j} = \frac{\sum_{k=1}^{|S_{\text{train}}|} \phi_{\text{local}}^{i,j,k} \cdot g_{j,k}}{\sum_{k=1}^{|S_{\text{train}}|} g_{j,k}}. \tag{3}$$

Hence, if unmodified image k gets a low similarity score with prototype j, it will get a low weight for the global importance calculation. In contrast, if prototype j is clearly present in image k, ProtoPNet will assign a high similarity score and hence k gets a high weight.

These importance scores can be used to create *global* explanations that explain a prototype, and *local* explanations that explain the similarity score between a given image and a prototype. The global explanation for the j-th prototype lists for each visual characteristic i its importance by showing the importance scores $\phi_{\mathrm{global}}^{i,j}$. This explanation is thus input independent and can be created before applying the prototype model to unseen images. The local explanation is of use during testing and explains a single prediction.

4   Experimental Setup

For our experiments, we use the Caltech-UCSD Birds dataset [25], a dataset for bird species identification also used by Chen et al. [7] for training their ProtoP-Net. It contains 200 different classes with approximately 60 images per class. The dataset provides a train-test split, leading to S_{train} with 5994 images and S_{test} with 5794 images. To evaluate our method for explaining prototypes, we first train a ProtoPNet [7] that results in an interpretable predictive model with prototypical parts for fine-grained image recognition. We apply the same data processing techniques as the original work [7]. We cropped the images according to the bounding boxes provided with the dataset and apply data augmentation on S_{train} as described in ProtoPNet's supplementary material [7]. All images are resized to 224×224. We opted for DenseNet-121 [12] as backbone of ProtoPNet, as this was reported to be the best-performing network on the Caltech-UCSD dataset [7]. The DenseNet-121 network has been pre-trained on ImageNet.[1] When forwarding the resized images through DenseNet, the input image dimensions, $H_{\mathrm{in}} = W_{\mathrm{in}} = 224$ and $D_{\mathrm{in}} = 3$, are transformed to the output dimensions $H = 7$, $W = 7$ and $D = 128$. Depth D is a hyperparameter in ProtoPNet determining the number of channels for the network output and the prototypes, and is set to 128 as in ProtoPNet [7]. As in the original paper [7] we use 10 prototypes per class, leading to 2000 prototypes in total. All other training parameters are also replicated from the implementation by Chen et al. [7].

We apply our method to the resulting prototypes for generating global and local explanations. Section 4.1 presents the implementation details for our image modifications. The design of our experiments to evaluate our explanations is presented in Sect. 4.2.

4.1   Modification Implementation

When implementing the image modifications as described in Sect. 3.2, we aim for a similar modification 'strength' for all characteristics in order to compare importance scores. Furthermore, the image modifications should be modest, since too extreme modifications can lead to out-of-distribution images that result in erratic behaviour of the underlying neural network of ProtoPNet. A similar modification degree depends on how ProtoPNet perceives the images. Therefore, we find a suitable modification degree by forwarding both the unmodified image and

---

[1] We use the same methodology as ProtoPNet [7] in order to reproduce results, although it is known that there is some overlap between Caltech-UCSD and ImageNet.

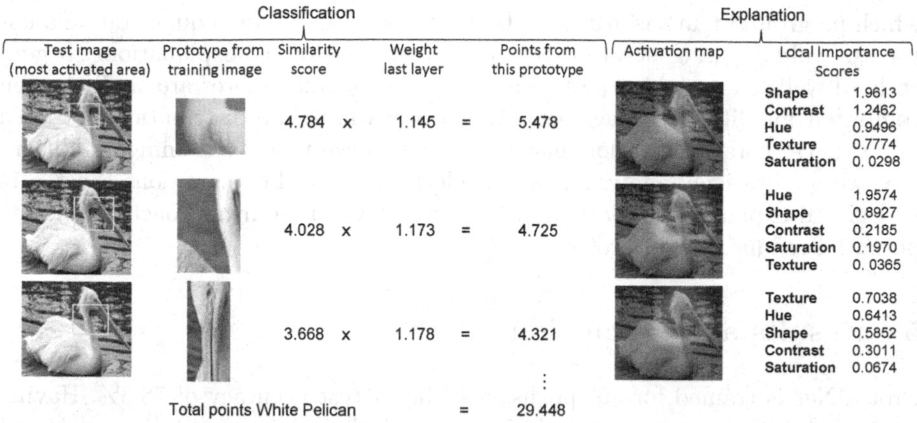

	Classification					Explanation	

Test image (most activated area)	Prototype from training image	Similarity score		Weight last layer		Points from this prototype		Activation map	Local Importance Scores	
		4.784	x	1.145	=	5.478			Shape	1.9613
									Contrast	1.2462
									Hue	0.9496
									Texture	0.7774
									Saturation	0. 0298
		4.028	x	1.173	=	4.725			Hue	1.9574
									Shape	0.8927
									Contrast	0.2185
									Saturation	0.1970
									Texture	0. 0365
		3.668	x	1.178	=	4.321			Texture	0.7038
									Hue	0.6413
									Shape	0.5852
									Contrast	0.3011
									Saturation	0.0674

$$\vdots$$

Total points White Pelican $\quad\quad\quad = \quad$ 29.448

Fig. 4. Left: ProtoPNet's reasoning with a subset of all prototypes of the White Pelican class. To classify a test image, ProtoPNet compares the class-specific prototypes of each class with the test image to calculate the total number of points for this class. An image is classified as the class with the most points. Right: The activation maps produced by ProtoPNet and our corresponding local explanations that explain which characteristics were important for a similarity score.

the modified image through the underlying CNN and compare their L1-norm distance in latent space. We tune the modifications parameters such that the mean distance between the unmodified latent training images and the latent modified images is exactly 0.0002 for all characteristics. This value is experimentally chosen such that it results in modifications that are clearly distinguishable for the human eye, while still being perceived by ProtoPNet as being close to the original images. For the colour modifications (contrast, saturation and hue), we use PyTorch's image transformations[2]. More specifically, we use the `ColorJitter` function where we set the contrast value to 0.45, saturation to 0.7 and hue to 0.1 for the respective modifications. The shape modification is manually implemented in Python. The texture modification is implemented with the Non-local Means Denoising algorithm for coloured images in OpenCV[3]. The filter strength of the denoising algorithm is set to 4 to get the correct mean latent distance.

4.2 Considerations for Evaluation

We would like to emphasise that we do not want to explain human perception, but the perception of the prototype-based model. Hence, we cannot ask users what they deem important, since we aim to explain the model's reasoning. Also, we cannot construct a ground-truth since we are opening up a "black-box" for

---

[2] https://pytorch.org/docs/stable/torchvision/, accessed June 2020.

[3] https://docs.opencv.org/3.4/d1/d79/group__photo__denoise.html# ga03aa4189fc3e31dafd638d90de335617, accessed June 2020.

which no ground-truth is available. However, we can still do a quantitative analysis to evaluate the generalizability and robustness of the explanations. If generalised well, one would expect that global importance scores are similar when computed for different image sets. We also evaluate the distribution of global importance scores to get more insight in the general model reasoning. Additionally, we qualitatively analyse a varied selection of local explanations (Sect. 5.1) and global explanations (Sect. 5.2). In Sect. 5.3 we use our approach to analyse potential redundancy of prototypes.

5   Results and Discussion

ProtoPNet is trained for 30 epochs, reaching a test accuracy of 78.3%. Having applied the same data and training process as the original work [7], we do not know why our accuracy is lower than the accuracy reported in the original work (80.2%). However, the aim of this paper is not to train the best ProtoPNet, but to find a reasonable well-performing model in order to explain its prototypes.

Figure 4 (left) shows a selection of prototypical patches ('prototypes') learned by ProtoPNet. ProtoPNet measures the similarity between a prototype and patches in a given test image. The resulting similarity scores are multiplied with learned weights resulting in a final score per class. An image is classified as the class with the highest score (i.e. most points).

5.1   Analysing Our Local Explanations

Figure 4 (right) shows how our local explanations complement the prototypical reasoning by explaining which visual characteristics were important for ProtoPNet's similarity score between a prototype and a specific image. We show the activation map as implemented by Chen et al. [7] and list the local importances for each visual characteristic. The importances identified by our local explanations for the test image shown in Fig. 4 seems reasonable given the typical white colour of the pelican and its long neck. Furthermore, our explanations enable a user to understand why ProtoPNet gave a high similarity score to a prototype and a patch in the test image. Whereas the prototypes of the white pelican look similar to the human eye, our local importance scores can differentiate between prototypes and estimate the prototype's purpose. The topmost prototype in Fig. 4 mostly focuses on shape and contrast, the second prototype deems hue important and the third prototype focuses on texture.

Our local explanations can also be useful to confirm the user's expectations, such as the importance of the yellow colour for the prototype in Fig. 5a. The explanations are however especially insightful when the given similarity score is in contrast with human perceptual similarity and an explanation is needed. Figure 5b explains why different test images received a high similarity score by ProtoPNet. Our local importance scores therefore serves as an extension to a trained prototype-based model, to explain a single prediction.

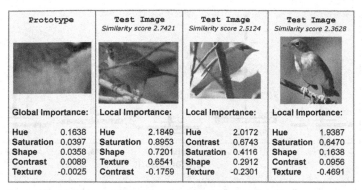

Prototype	Test Image *Similarity score 2.7421*	Test Image *Similarity score 2.5124*	Test Image *Similarity score 2.3628*
Global Importance:	**Local Importance:**	**Local Importance:**	**Local Importance:**
Hue 0.1638	Hue 2.1849	Hue 2.0172	Hue 1.9387
Saturation 0.0397	Saturation 0.8953	Contrast 0.6743	Saturation 0.6470
Shape 0.0358	Shape 0.7201	Saturation 0.4116	Shape 0.1638
Contrast 0.0089	Texture 0.6541	Shape 0.2912	Contrast 0.0956
Texture -0.0025	Contrast -0.1759	Texture -0.2301	Texture -0.4691

(a) As expected, the yellow hue is dominant, and the local explanations correspond with the global importance.

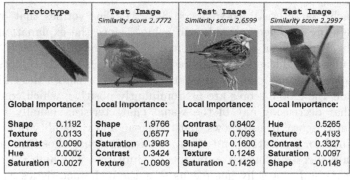

Prototype	Test Image *Similarity score 2.7772*	Test Image *Similarity score 2.6599*	Test Image *Similarity score 2.2997*
Global Importance:	**Local Importance:**	**Local Importance:**	**Local Importance:**
Shape 0.1192	Shape 1.9766	Contrast 0.8402	Hue 0.5265
Texture 0.0133	Hue 0.6577	Hue 0.7093	Texture 0.4193
Contrast 0.0090	Saturation 0.3983	Shape 0.1600	Contrast 0.3327
Hue 0.0002	Contrast 0.3424	Texture 0.1248	Saturation -0.0097
Saturation -0.0027	Texture -0.0909	Saturation -0.1429	Shape -0.0148

(b) Test images can get high similarity scores for different reasons.

Fig. 5. Similarity between a class-specific prototype and test images from different classes explained by our local importance scores.

5.2   Analysing Our Global Explanations

Local explanations are useful to explain an unexpected result, but do not give a coherent, overall explanation of the prototype-based model. Our methodology therefore also produces *global* explanations that give an average view regarding the importance scores for each prototype. Specifically, the global importance scores are computed for each prototype by taking the weighted mean of all training images, as introduced in Sect. 3.3. Hence, these explanations are independent of test input.

Quantitative Evaluation. To quantitatively evaluate our global explanations, we compute global importance scores not only for S_{train}, but for evaluation purposes also for S_{test}. We confirmed with a Shapiro-Wilk test that the importance scores for each characteristic are normally distributed, such that we could apply the Welch t-test to confirm that there is no significant difference between the global importance scores of all prototypes calculated from S_{train} and from

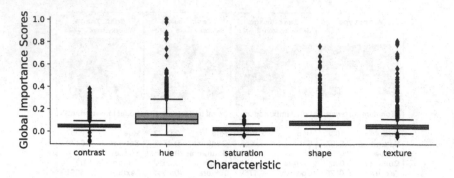

Fig. 6. Box plot of global importance scores across the training set.

S_{test} for each characteristic (p-values $< 3.5^{-11}$). This verifies that our global importance scores are generalisable and robust, since the explanations do not significantly change when computed from a different image set.

To get more insight as to how characteristics are used, Fig. 6 plots the distributions of global importance scores of all prototypes across the training set. It shows that the global scores are predominantly positive, which confirms our intuition that decreasing any of the visual characteristics usually leads to a lower similarity score. It also shows that the mean and variability of importance scores is small for saturation and contrast, meaning that those characteristics only have a moderate influence on prototype similarity. The high variability for shape, texture and especially hue means that these characteristics can be substantially important for some prototypes. This corresponds with the fact that hue and shape are considered more important and effective for humans in the data visualisation domain than saturation or contrast [16].

Qualitative Evaluation. Figure 7 shows a varied selection of prototypes with their global explanation. For the upper two rows, the importance of characteristics corresponds to the visually identifiable properties of the prototypes and hence, the explanations seem reasonable. However, the explanations in the bottom row might come as a surprise. A human might think that shape is important for the bottom left prototype and that the prototype resembles fin-footed birds. Our global explanation indicates that shape is of little importance and that colour hue is the dominant characteristic. Since a ground-truth is not available, we verify the correctness of the global explanation by analysing test images that had a high similarity with the prototype. Although a prototype is trained to be class-specific, Fig. 8a shows that images from a different class can still get assigned a high similarity score. These images confirm that the prototype deems hue important and therefore resembles *red* feet, instead of webbed feet. The reverse is true for the bottom right prototype of Fig. 7. Humans could think that the prototype resembles a red eye, and would be surprised by a high similarity score with a black-eyed bird, and hence might lower their trust in the model. Our global importance scores indicate that the importance for hue is rather low,

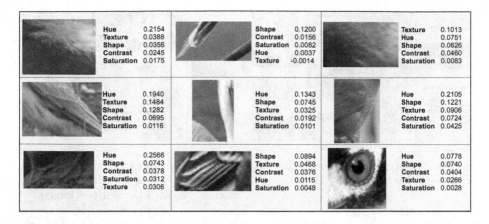

Fig. 7. Selection of prototypes explained with their global importance scores. Top row: Prototypes with predominantly a single important characteristic. Center row: Prototypes with intuitive explanations. Bottom row: ambiguous and potentially misleading prototypes.

and hue and shape are of similar importance. When analysing birds that get assigned a high similarity with this prototype as shown in Fig. 8b, it is easily verified that the prototype does indeed not represent a red eye. These examples show that our global explanations can clarify visual prototypes. Without our explanations, a user would not be aware of the meaning of a given prototype and correct *simulatability* [15] would not be guaranteed.

5.3 Redundant Prototypes

An interesting question is whether prototypes that are slightly different, deem the same visual characteristics important. Prototypes with different global importance scores complement each other, whereas similar explanations indicate prototype redundancy. We consider prototypes to be visually similar when they are close to each other in the latent space learned by ProtoPNet. We measure the Euclidean distance between the latent representation of two prototypes of the same class (a 'pair'). This gives $\binom{10}{2} = 45$ unique pairs per class, and $45 \cdot 200 = 9000$ pairs in total. Let P be the set of unique pairs of two prototypes from the same class, such that $|P| = 9000$, and $V \subset P$ the set of pairs with two visually similar prototypes. We consider a pair of two prototypes i and j identical when the Euclidean distance in latent space $d_{i,j} = 0$ and *visually similar but not identical* when $d_{i,j} < \tau$ and $d_{i,j} > 0$, where $\tau = 0.15$ is found to be a suitable threshold for perceptual similarity. This gives 63 pairs of identical prototypes and $|V| = 93$ unique pairs of 164 visually similar prototypes. To evaluate whether these visually similar prototypes also have similar global explanations, we consider the global importance scores of a prototype as a vector of length 5 and calculate the Euclidean distance between the global explanations of two prototypes. The orange plot in Fig. 9 shows that most pairs with visually similar

(a) The prototype indeed deems hue more important than shape.

(b) The global score explains that the red hue from the eye is not that important, which is validated by near test images.

Fig. 8. Test images from a different class than the prototype-class which have the highest similarity scores with the prototype.

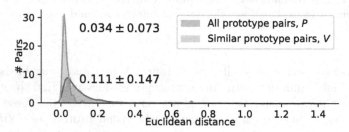

Fig. 9. Histogram with the distribution of Euclidean distances between the global explanations of two prototypes of the same class (a 'pair').

prototypes have a small distance between their global importance scores, which is not the case in general (blue). Therefore, these prototypes might be redundant and unnecessarily increase explanation size. Additionally, a few pairs in the orange plot have dissimilar explanations (distance of roughly 0.7) and therefore complement each other.

6   Conclusion and Future Work

A prototype-based image recognition model learns visual prototypes and localises a patch in a test image that looks alike a prototype to assign it a similarity score. We argue that these prototypes should be explained with respect to the model's reasoning and extend localisation with explanation. We presented an automated approach to explain visual prototypes learned by any prototypical image recognition model. Our method automatically modifies the hue, texture, shape, contrast or saturation of an image, to identify which visual characteristics of a prototype the model deems important. We applied our method to the prototypes learned by ProtoPNet [7]. The importance of visual characteristics identified by our explanations often corresponded to the visually perceptible properties of the prototypes, showing that our explanations are reasonable. We also showed that perceptual similarity for humans can be different from the similarity learned by the model, indicating the need for explaining the model's reasoning. Such 'misleading' prototypes will hinder correct simulatability and only visualising prototypes can be insufficient for understanding why the model considered a prototype and an image highly similar. To the best of our knowledge, we are the first to address such ambiguity of visual prototypes and the elegant simplicity of our approach makes it a suitable stand-alone solution. We think the extra computational complexity required is justifiable given the extra insights our method provides. Furthermore, because of the stand-alone nature of our method, it can be applied to any prototypical image recognition method, including ProtoPNet [7] and ProtoTree [17]. Our approach can also easily be extended with more visual characteristics or other image modifications a user is interested in.

Future work concerns the potential interactions between characteristics. Our importance scores assume that characteristics from image modifications are mutually exclusive. However, denoising the image to lower its texture could also slightly influence shape. We implemented the image modifications in such a way to limit interactions between characteristics as much as possible, but future analysis could determine to what extent visual characteristics are correlated.

References

1. Arik, S.Ö., Pfister, T.: Attention-based prototypical learning towards interpretable, confident and robust deep neural networks. CoRR abs/1902.06292 (2019)
2. Bau, D., Zhou, B., Khosla, A., Oliva, A., Torralba, A.: Network dissection: quantifying interpretability of deep visual representations. In: CVPR (2017)
3. Op de Beeck, H.P., Torfs, K., Wagemans, J.: Perceived shape similarity among unfamiliar objects and the organization of the human object vision pathway. J. Neurosci. **28**(40), 10111–10123 (2008)
4. Biehl, M., Hammer, B., Villmann, T.: Prototype-based models in machine learning. Wiley Interdisc. Rev. Cognitive Sci. **7**(2), 92–111 (2016)
5. Cavina-Pratesi, C., Kentridge, R., Heywood, C., Milner, A.: Separate channels for processing form, texture, and color: evidence from fMRI adaptation and visual object agnosia. Cereb. Cortex **20**(10), 2319–2332 (2010)
6. Chaudhari, S., Polatkan, G., Ramanath, R., Mithal, V.: An attentive survey of attention models. CoRR abs/1904.02874 (2019)
7. Chen, C., Li, O., Tao, D., Barnett, A., Rudin, C., Su, J.K.: This looks like that: deep learning for interpretable image recognition. In: NeurIPS, pp. 8928–8939 (2019)
8. Geirhos, R., Rubisch, P., Michaelis, C., Bethge, M., Wichmann, F.A., Brendel, W.: Imagenet-trained CNNs are biased towards texture; increasing shape bias improves accuracy and robustness. In: ICLR (2019)
9. Goodfellow, I., Bengio, Y., Courville, A.: Deep Learning. MIT Press, Cambridge (2016)
10. Guidotti, R., Monreale, A., Ruggieri, S., Turini, F., Giannotti, F., Pedreschi, D.: A survey of methods for explaining black box models. ACM Comput. Surv. (CSUR) **51**(5), 1–42 (2018)

11. Hosseini, H., Xiao, B., Jaiswal, M., Poovendran, R.: Assessing shape bias property of convolutional neural networks. In: Proceedings of the IEEE Conference on Computer Vision and Pattern Recognition (CVPR) Workshops (June 2018)
12. Huang, G., Liu, Z., Van Der Maaten, L., Weinberger, K.Q.: Densely connected convolutional networks. In: CVPR, pp. 4700–4708 (2017)
13. King, M.L., Groen, I.I., Steel, A., Kravitz, D.J., Baker, C.I.: Similarity judgments and cortical visual responses reflect different properties of object and scene categories in naturalistic images. NeuroImage **197**, 368–382 (2019)
14. Kourtzi, Z., Kanwisher, N.: Cortical regions involved in perceiving object shape. J. Neurosci. **20**(9), 3310–3318 (2000)
15. Lipton, Z.C.: The mythos of model interpretability. Queue **16**(3), 31–57 (2018)
16. Munzner, T., Maguire, E.: Visualization Analysis & Design. CRC Press, Boca Raton (2015)
17. Nauta, M., van Bree, R., Seifert, C.: Neural prototype trees for interpretable fine-grained image recognition. In: Proceedings of the IEEE/CVF Conference on Computer Vision and Pattern Recognition (CVPR), pp. 14933–14943 (June 2021)
18. Ritter, S., Barrett, D.G.T., Santoro, A., Botvinick, M.M.: Cognitive psychology for deep neural networks: a shape bias case study. In: ICML Proceedings of Machine Learning Research, vol. 70, pp. 2940–2949. PMLR (2017)
19. Rosenfeld, A., Solbach, M.D., Tsotsos, J.K.: Totally looks like - how humans compare, compared to machines. In: CVPR Workshops (2018)
20. Rossion, B., Pourtois, G.: Revisiting snodgrass and vanderwart's object pictorial set: The role of surface detail in basic-level object recognition. Perception **33**(2), 217–236 (2004). pMID: 15109163
21. Rudin, C.: Stop explaining black box machine learning models for high stakes decisions and use interpretable models instead. Nat. Mach. Intell. **1**(5), 206–215 (2019)
22. Saralajew, S., Holdijk, L., Rees, M., Asan, E., Villmann, T.: Classification-by-components: probabilistic modeling of reasoning over a set of components. In: NeurIPS, pp. 2792–2803 (2019)
23. Simonyan, K., Zisserman, A.: Very deep convolutional networks for large-scale image recognition. In: Bengio, Y., LeCun, Y. (eds.) ICLR (2015)
24. Striedter, G.F.: Neurobiology: a Functional Approach. Oxford University Press, Oxford (2016)
25. Welinder, P., et al.: Caltech-UCSD Birds 200. Technical Report. CNS-TR-2010-001 (2010)
26. Zhang, N., Donahue, J., Girshick, R., Darrell, T.: Part-based R-CNNs for fine-grained category detection. In: Fleet, D., Pajdla, T., Schiele, B., Tuytelaars, T. (eds.) ECCV 2014. LNCS, vol. 8689, pp. 834–849. Springer, Cham (2014). https://doi.org/10.1007/978-3-319-10590-1_54
27. Zheng, H., Fu, J., Zha, Z., Luo, J., Mei, T.: Learning rich part hierarchies with progressive attention networks for fine-grained image recognition. IEEE Trans. Image Process. **29**, 476–488 (2020)

Prototypical Convolutional Neural Network for a Phrase-Based Explanation of Sentiment Classification

Kamil Pluciński, Mateusz Lango$^{(\boxtimes)}$ (ID), and Jerzy Stefanowski (ID)

Faculty of Computing and Telecommunication, Institute of Computer Science,
Poznan University of Technology, ul. Piotrowo 2, 61-138 Poznan, Poland
kamil.plucinski97@gmail.com, {mlango,jstefanowski}@cs.put.edu.pl

Abstract. The attention mechanisms are often used to support an interpretation of neural network based classification of texts by highlighting words to which the network paid attention while making a prediction. Following recent studies, the attention technique does not always provide a faithful explanation of the model. Thus, in this paper we study another idea of prototype-based neural networks. Although for texts they obtain promising results, they may provide explanations in the form of comparisons of whole (potentially long) documents or also run into problems with providing reliable explanations. To overcome it, in this work a new prototype-based convolutional neural architecture for text classification is introduced, which provides predictions' explanations in the form of similarities to phrases from the training set. The experimental evaluation demonstrates that the proposed network obtains similar classification performance to the black-box convolutional networks while providing faithful explanations. Moreover, it is shown that a new method for dynamic tuning of the number of prototypes introduced in this paper offers performance gains against static tuning.

Keywords: Prototypes · Explainable neural networks · Text classification · Phrase-based explanation

1 Introduction

Intensive research in machine learning in the last decades has led to the development of many methods and their applications in many fields. The improvements in predictive capabilities, particularly while using deep neural networks, opened new challenges as well as encouraged practitioners to use them practically in business, industry and many other real life domains. Nevertheless, these applications exposed the importance of some aspects of machine learning models that the AI community has not previously paid much attention.

This is especially true for explainable AI, as the lack of it may slow down the prevalence of the new intelligent systems in some areas of application [16]. Recall that current approaches demonstrating the best predictive performance

© Springer Nature Switzerland AG 2021
M. Kamp et al. (Eds.): ECML PKDD 2021 Workshops, CCIS 1524, pp. 457–472, 2021.
https://doi.org/10.1007/978-3-030-93736-2_35

are usually based on complex, black-box models such as deep neural networks or gradient boosting trees. They do not provide direct information on the system's internal logic or how the prediction is produced. The ability to provide explanations in ways that are understandable to humans would support verification of even the most accurate prediction systems, help in detecting potential biases in data or algorithms, enable the assessment of causes of incorrect decisions, and the possibility of correcting them. Finally, it will improve human trust in AI systems and ensure their acceptance [16].

The need to develop explainable artificial intelligence solutions is particularly well seen in the modern natural language processing solutions that often use pre-trained word embeddings as the text representation. Such neural representations are trained on large text corpora to model the words' distributional semantics [9,19]. Even though using such representations improve performance on various NLP tasks, word embeddings also model undesirable human biases present in the training corpora such as gender, racial bias or social bias around disability [7].

The methods for explaining classifier prediction are usually divided into approaches finding post-hoc explanations for black-box models (agnostic ones) and approaches redesigning machine learning methods into more transparent ones [4]. For text data, one common way of shedding light on the inner workings of a neural network is using architectures with the attention mechanism [1]. Neural networks with attention implement decisions while looking at the particular fragments of the input sequence, which is selected by the mechanism itself. Therefore, the network's decision can be partially explained to the user by indicating phrases in the text that are taken into account while making predictions. Such a way of providing explanations for NLP models gained some popularity [1,5,10,20], among other things, due to the easiness of interpretation while demonstrating the actual operation of the model.

Unfortunately, the recent papers [8,21] demonstrated some severe issues with this approach. It was found that, for some tasks, it is possible to find different attention weights, i.e. pointing towards entirely different elements of the input, while returning the same decision. Moreover, the attention weights are often poorly correlated with feature importance measures, indicating problems with the faithfulness of the provided explanation.

In this work, we propose a new neural network architecture for text classification, which is not based on the attention mechanism but provides a faithful explanation in a similar form. More concretely, the proposed neural network builds upon *prototype-based architectures* that were initially proposed for image classification [3,11] and recently were also considered by a few researchers to text data [6,12]. This approach is based on memorizing the most representative examples (prototypes) during the training, and making the final prediction based on the similarity of them to the input example. Nevertheless, these prototype-based neural networks for text offer prototypes in the form of either whole sentences or full documents, whereas our approach will offer prototypes in the form of *phrases*, allowing explanations in a similar form to these being the result of the attention mechanism. Furthermore, the prototypes in the form of phrases better generalize to a higher number of examples and provide shorter, simpler and more intuitive explanations.

Moreover, our proposal is based on convolutional neural networks, whereas other prototypical architectures for texts use recurrent networks, with the representation build by a complex transformer architecture. We claim that it should make our approach easier to parallelize and enables the lower computation time.

Our approach also has fewer hyperparameters, and most importantly, the critical number of prototypes in the model is dynamically selected by the online training procedure. The performed experimental evaluation demonstrates that the proposed approach, whereas being simpler and easier to parametrize, provides competitive results in terms of classification accuracy and offers explanations in a more succinct form.

2   Related Works

Machine learning is applied in many tasks involving textual data in the area of natural language processing. Text classification, in particular in the context of sentiment analysis, document categorization, or topic labeling, is identified as one of the main problems where XAI methods are used [2]. XAI methods often provide decision explanation by highlighting which words or sentences are the most relevant for particular predictions.

Although the categorization of XAI approaches now covers many different methods [4], the main categories specialized for text data include various types of model distillation [15], identification of feature's importance in the text [1,2] and visualization [20] as well as intrinsic approaches that offer explainable modifications of models. The last group of methods, which is at the focus of the current paper, is primarily based on *attention* [1]. This mechanism, used among others in recurrent neural networks, makes predictions basing on the dynamically assigned weights to all generated hidden states. As successive states are generated based on successive words, by analyzing the weights assigned by the attention mechanism, one can obtain the influence of the word on the final prediction. This approach is straightforward to understand for users, as the words on which "the network is looking at" can be highlighted. However, the recent research suggests that these methods suffer from low fidelity, showing that a completely different distribution of attention weights leads to the same results, questioning the plausibility of this explanation [8,21].

More recently, other methods consisting of neural architectures based on prototypes were introduced. These approaches, firstly proposed for image classification [3], perform decisions basing on the similarity between the input and the selected instances from the datasets, called *prototypes*. Typically, the image is processed by various convolutional layers, then the similarity between the input and prototypes in the latent space is calculated. The final decision of the model is made basing on these similarities, e.g. by computing their weighted sum. The network can provide an explanation for the decision by stating that "this looks like that" [3], i.e. the decision was taken because the input is similar to a training example of which we know that belongs to a given class.

In some earlier works, the prototypes were simply data instances that were highly representational of many other examples [13] and selected automatically from the training set during model training. However, in a neural network, a prototype does not necessarily have to be associated with a whole training example but can represent only some part of it. For instance, considering the classification of bird species' images, a bounding box marking a characteristic part of the bird's body is an even more intuitive and human-like explanation than showing a whole similar image. Note that even if the latent representation of the example is constructed in a nontransparent manner, the decision of the network performed by finding similarities with items that are of a certain class is rather easy to explain. By simply showing the prototypes aligned with the input, one can present a faithful and plausible explanation of the real inner working of the system to the user.

The idea of neural networks based on prototypes was adapted for sequential text classification by the authors of ProSeNet [12]. In their work, the hidden representations of whole documents from the training set are used as prototypes. The model consists of LSTM layers generating the latent representation for each sample and a prototype layer that calculates the L2 distance between prototypes and the input. The prediction is made by a fully connected layer that relies only on similarities to prototypes.

This approach was further explored by [6], whose authors noticed that explaining predictions basing on the similarity between whole documents is not efficient when the input sequence is long. The proposed approach, ProtoryNet, uses prototypes in the form of the sentences, not whole documents. The network process the input sentence by sentence, generating its embeddings with the pre-trained DistilBERT model [17]. Then, each sentence is represented as a vector of similarities to the existing prototypes. This representation is additionally normalized and made sparse by applying softmax transform with a proper value of a temperature parameter. Such constructed representation of input sentences is fed into the LSTM layer, followed by a fully connected layer that performs the final decision.

3   Prototype-Based Convolutional Neural Network for Text

The previously proposed prototype-based neural networks for text classification suffer from some disadvantages. Firstly note that ProtoryNet, contrary to the earlier works in image classification, obtains the final prediction in a rather complex and nontransparent way. One of the main advantages of using prototype networks was that even if the latent representation is constructed by a blackbox model, the decision itself is performed by a relatively simple calculation involving the similarity with the prototypes. This feature made the prototype-based neural model so appealing, since the decision was performed by a white-box model working on the similarities that can be intuitively understood even by an inexperienced user. Nevertheless, if the similarities become an input to the next black-box model – an LSTM model – the possibility of a simple explanation becomes more difficult.

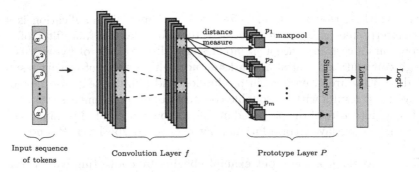

Fig. 1. Architecture of the prototypical convolutional neural network.

ProSeNet does not have this issue and classifies instances with a single linear layer but, on the other hand, provides a document level explanation. Unfortunately, comparing whole, especially long, documents can be hard to analyze and interpret. This is especially the case in sentiment classification, where even an overall positive review can mention some negative aspects of a service or a product. Therefore, it may be unclear for the user which fragments of the text really influenced the calculation of similarity and further the classifier's decision. We hypothesize that it could be overcome by indicating the most influential words in the text similarly to earlier solutions in the attention-based explanations.

In this work, we propose a new prototypical neural network that operates on the prototypes in the form of *phrases* that are automatically selected and extracted from the training set. Consequently, our model provides an explanation basing on similarity of phrases. Moreover, the proposed model calculates prediction with a white-box model, i.e. single linear layer, providing intelligible and faithful explanations.

In this section, we describe details of the proposed model. In Subsect. 3.1, we explain how the convolutional network was adapted to text data. In the following Subsect. 3.2, the loss function used to optimize the model parameters is explained. Further details on the model's training are provided in the two final subsections, where we show how network weights are converted into representations of phrases and propose a method to automatically select the number of prototypes that is an important parameter for prototype-based networks.

3.1 Architecture of the Proposed Solution

We have chosen a convolutional neural network as a basis of the proposed neural model that constructs prototypes in the form of phrases. This seems to be a natural choice for two reasons. First, convolutional networks were already successfully adopted to prototypical-based architectures in image classification [3], and our proposal will build upon these works. Moreover, kernels in a convolutional network have an intuitive interpretation in text classification. A kernel performs the convolution of several consecutive words, where their number depends directly

on the width of the kernel. Therefore, a kernel in text classification is sometimes interpreted as a soft n-gram detector. For example, one filter of length
5 represents a certain concept in latent space that consists of five words. Even
though convolutional neural networks do not achieve state-of-the-art results for
text data, they are often used in production and popular text libraries as they
achieve good results with reasonable computational requirements [18].

The architecture of the proposed model is shown in Fig. 1. The model consists
of an embedding layer, convolutional layer f, prototype layer P, and a linear
layer.

The processing of the input example begins by converting each token into
its dense representation in the embedding layer. Then, the input sequence is
applied into the convolutional block that processes it with a convolution layer
that contains d kernels of length k. A zero-padding of size $pad = floor(\frac{k}{2})$ has
been applied to the input so that the resulting length of the hidden representation
is equal to the length of the original sequence. The result of the convolution layer
at position i is computed on the word representations at positions $[i-pad; i+pad]$
which makes it straightforward to align representations with the words in the
input. Furthermore, batch normalization and ReLu activation are applied after
the convolutional layer. The result of this layer is later denoted as $f(x)$.

The prototype layer P contains m prototypes, each of them having dimensionality $[1, d]$, so it can be aligned to each element of the representation $f(x)$.
The number of prototypes m is one of the model's most important parameter;
if not specified otherwise, the default value is $m = 16$ and the layer is initialized
with values from the $\mathcal{U}(0, 1)$ distribution. Recall that each latent representation
$f(x)^{(i)}$ was created by the convolution of k tokens embeddings, so it is aligned to
a single phrase, thus the prototypes are also selected as *phrase representations
from the training set*.

While processing the latent representation, the squared L_2 distance between
the prototype and each element of $f(x)$ is calculated, obtaining a distance map
of all phrases to the prototype. Following the approach of [3], the calculated
distances are transformed into similarity values using the similarity transforma-
tion defined as $sim(x) = log(\frac{x+1}{x+\epsilon})$, where ϵ is a small number added for stable
computations. Then, the most similar phrase from the input is selected by a
max-pooling layer. This process is repeated for each prototype, resulting in a
vector containing information about the presence of a prototype in the input
text, expressed by the prototype similarity to the most similar input phrase.

A fully connected layer without bias subsequently processes the resulting vector of similarities, returning the logit used to calculate standard cross-entropy
loss. Weights in the hidden layer indicate the significance of the prototype occurrence for the prediction.

3.2   Loss Function

The loss function used to train the proposed model should both construct an
accurate classifier and enable the construction of suitable prototypes. A high-
quality set of prototypes should cover all the examples in the training set, since

each instance should be well represented by a close prototype. From the user's perspective, the set of prototypes should also be diversified and concise to avoid long explanations containing very similar phrases. The loss function used to train our model combines standard cross-entropy with several additional components that address these assumptions.

The first additional component of the loss function is responsible for constructing prototypes that are well aligned to the training set examples. More concretely, it ensures that the hidden representation of each example has a close representative in the set of prototypes in the sense of Euclidean distance. Therefore, for each example in the batch, the minimum distance between the prototypes and any input phrase is computed and minimized. Since this means that in a sense, the function is mapping a prototype to a cluster of matching examples, this component is denoted as *clustering* and defined as:

$$L_{clst} = \frac{1}{|D|} \sum_{x \in D} \min_{j,k} d(p_j, f(x)^{(k)})^2 \tag{1}$$

where $d(p_j, f(x)^{(k)})$ is the L_2 distance between the prototype and the hidden representation of the input example at k position and D is the set of examples.

The next component of the loss is responsible for creating a diverse representation. It maximizes the mutual distance between prototypes until they are different of at least a predefined margin. The component is expressed by:

$$L_{sep} = \frac{1}{m} \sum_{p_i \subset P} \min_{j,i \neq j} \max\{0, margin - d(p_j, p_i)\} \tag{2}$$

where m is the number of prototypes and $d(p_j, p_i)$ is the L_2 distance between prototypes. The margin in our experiments was set to 1.

Finally, to encourage the construction of sparse weight vectors, the L_1 regularization term is added for the fully connected layer weights. It further improves the simplicity of the final white-box model.

The complete form of the proposed loss function is defined as:

$$L = \alpha CrossEntropy + \beta L_{clst} + \theta L_{sep} + \gamma L_1 \tag{3}$$

where $\alpha, \beta, \theta, \gamma$ are parameters that can be tuned. Following our experiments, the best results are obtained for $\alpha = 0.99, \beta = 0.005, \theta = 0.005, \gamma = 0.01$.

3.3 Prototype Projection

During the model training with backpropagation, the weights of the prototype layer are updated according to the gradient and constitute an abstract representation located somewhere in the latent space. However, they do not necessarily point to a hidden representation of any particular training example.

Since we want to point the prototypes to the real phrases in the training set, we should require that the representation of the prototype is exactly equal to the latent representation of some real example. Therefore, between training and validation phases at each epoch, a special projection phase is performed. In detail, the abstract representation of each prototype is replaced by the closest latent representation of the phrase existing in the training dataset as it refers to a more human readable form. This distance is measured as L_2 norm. Aside from the weight replacement, the words that comprised this representation are saved for future explanations.

3.4 Dynamic Number of Prototypes

The number of prototypes is a key parameter of the model. Its too small values usually led to underfitting, whereas a large number of prototypes will result in many prototypes with minimal impact on the prediction and some prototypes semantically very similar to each other. For the sake of simplicity and clarity, providing the user with information about the occurrence of one important prototype is more desirable than presenting several very similar prototypes.

To overcome this problem, we propose an approach to select the number of prototypes during model training dynamically. The proposed method addresses the aforementioned issues in the following way: too similar prototypes are merged into one with the accumulated weight, while the prototypes with marginal impact on the prediction are removed.

More precisely, dynamic prototype tuning begins after completing the prototype projection phase. Prototypes whose weights in the fully connected layer are smaller than $\epsilon = 2 \cdot 10^{-3}$ are removed from the network, then pairs of prototypes whose L_2 distance is smaller than $s = 0.3$ are merged. Since the merging operation of a pair of very similar prototypes should not change the overall network behavior, the merging is done by removing one of the prototypes from the pair and adding its weight vector to the weight vector of the remaining prototype. Note that if identical prototypes were merged with this procedure, the network would return precisely the same results as before this operation.

In order to prevent underfitting, the model is allowed to improve its prediction by using an additional amount of new prototypes. After prototype trimming, a selected number $k = 2$ of new prototypes is added with random initialization. If increasing the number of prototypes does not improve classification performance, they will be automatically removed by the next trimming phase. This process is repeated until the result on the validation set is not improved during several consecutive epochs (early-stopping method). The pseudo-code of model training along with projection of prototypes and dynamic tuning of its number can be seen in Algorithm 1.

4   Experiments

This section describes an experimental evaluation of the proposed framework. In Sect. 4.2 the performance of the proposed model to related works has been compared. The impact of the number of prototypes on the prediction quality is

---

Algorithm 1: Training of the proposed model with automatic selection of the number of prototypes

---

Input: dataset D, neural network including convolutional layer f, prototype layer P and linear layer g; number of warm-up epochs - $warm\_up$, minimum distance between prototypes - s, minimum prototype weight - ϵ, number of epochs - $epochs$

for $i \leftarrow 1, 2, ..., epochs$ **do**

    if $i > warm\_up$ **then** add k new prototypes to P

    Training phase(D,f,P,g)

    Prototype projection:

        foreach $p_k \in P$ **do**

            $P_k \leftarrow \arg\min_{f(x)^{(j)} \,:\, x \in D} ||p_k - f(x)^{(j)}||_2$

    if $i \geq warm\_up$ **then**

        Remove prototypes from P associated with weights $\leq \epsilon$ in g

        Merge prototype pairs in P whose mutual L_2 distance is $\leq s$

    Validation phase()

---

studied in Sect. 4.3. A visualization example of the set of prototypes together with use-case of the model is presented in Sects. 4.4 and 4.5, respectively. Finally, the explanation quality has been examined in Sect. 4.6.

4.1   Experimental Setup

The models were optimized using AdamW algorithm with a weight decay of 0.1 and a learning rate of 10^{-3} which was divided by 10 after every 3 epochs without improvement in cross entropy loss for CNN, and every 10 epochs for our approach. The training was performed for 30 epochs with batches of size 32. In the embedding layer GloVe-42B [14] pre-trained word embeddings were used. The number of convolutional kernels was experimentally selected to 64 (due to best results of the predictive accuracy), while the length of the prototypes was set to 5.

Experiments were conducted using 5-fold cross-validation on five datasets for sentiment classification, which were considered in the related works. The datasets were preprocessed in the same manner[1] as in [6]. Typical training time per fold performed on Tesla K80 fluctuated between 2 min for the Hotel dataset and 12 min for IMDB dataset.

4.2   Comparison of the Proposed Model with Reference Methods

Table 1 reports the accuracy results of the proposed Prototype Convolutional Network, other reference prototype-based neural models from the literature, as

---

[1]   The purpose of the preprocessing was 1) to binarize the datasets where sentiment was expressed on a scale of 1–5, 2) to balance the size of the datasets, and 3) to balance the number of examples from the positive and negative classes through under-sampling. For more details, please refer to [6].

Table 1. Prediction quality (accuracy) of the prototype convolutional network, together with the results of the approaches presented in related works. The mean with standard deviation calculated on the 5-fold cross-validation is reported. The results of ProSeNet and ProtoryNet as reported in [6].

Dataset	ProSeNet	ProtoryNet	Ours	CNN (blackbox)
IMDB	0.835 (±0.008)	0.849 (±0.002)	0.853 (±0.003)	0.893 (±0.002)
Amazon Reviews	0.840 (±0.011)	0.882 (±0.004)	0.880 (±0.004)	0.911 (±0.003)
Yelp Reviews	0.868 (±0.008)	0.872 (±0.002)	0.834 (±0.004)	0.867 (±0.003)
Rotten Tomatoes	0.748 (±0.006)	0.762 (±0.007)	0.735 (±0.003)	0.776 (±0.004)
Hotel Reviews	0.909 (±0.007)	0.949 (±0.004)	0.904 (±0.006)	0.929 (±0.008)

well as a black-box CNN baseline. One can observe that addition of an explainable prototypical layer to the black-box CNN model, caused a relatively small loss in accuracy of about 3% points.

Comparing our results with ProSeNet, the proposed method achieved comparable classification performance with two better, two worse, and one tie result. However, our approach provides explanations in a shorter form, uses faster convolutional layers rather than recursive ones, and requires, as we will see, a smaller number of prototypes.

Although the results of our model do not outperform the ProtoryNet architecture, note that ProtoryNet uses a black-box recursive model to perform the final prediction, which, in our opinion, reduces trust in the provided explanations. The user knows that the decision was performed based on analyzing similarities of every input sentence to some reference ones, but he/she cannot fully understand how they were combined to obtain the final prediction. In particular, due to the problem of learning long dependencies, some of these similarities can be even discarded by ProtoryNet while making the prediction. Note also that our approach, by using the max-pooling layer, effectively performs the decision basing on only one selected occurrence of each prototypical phrase, rather than using a prototype for each sentence like ProtoryNet. Moreover, ProtoryNet uses a much more complex transformer network to construct input representations.

4.3 Analyzing the Impact of the Number of Prototypes

The experiment was planned to measure the influence of the number of prototypes on the predictive accuracy. The results are presented in Fig. 2. One can notice that the static approach with a fixed number of prototypes has a large variance of the accuracy results for a smaller number of prototypes. As the number of prototypes increases, the results become more stable and closer to the baseline i.e. black-box CNN. However, these results also indicate that when the number of prototypes increases, the diversity of prototypes decreases at the same time. Moreover, the prototypes with little influence on the prediction appear.

The dynamic tuning of the number of prototypes was proposed to solve these problems. Looking at the prediction results of our approach with the dynamic

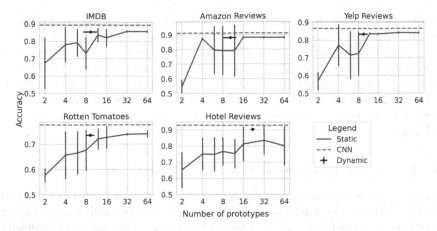

Fig. 2. The impact of the number of prototypes on the predictive accuracy. Each result is averaged over 5-folds and the bars indicate the standard deviation. CNN (green dashed line) was included for reference as a black-box baseline. Dynamic prototype selection (orange dot) was initialized with 16 prototypes, bars in both the X and Y axes present the standard deviation. (Color figure online)

tuning, we can say that they are no worse than the results of the static approach, even with the highest number of prototypes. Moreover, for the Hotel dataset, the dynamic tuning offers significantly better results in terms of the accuracy. At the same time, the deviation of the results on individual folds is small compared to the static approach. Using the proposed dynamic tuning results in constructing fewer and more relevant prototypes, additionally eliminating the need for tuning the number of prototypes.

4.4 Visualization of Learned Prototypes

Recall that during prototype projection, the prototype is mapped to the latent representation of the most similar phrase in the training set. The tokens of this phrase are remembered, which allows later visualization of phrases that

Table 2. Visualization of phrases that form latent representations of prototypes of a model learned on the IMDB dataset.

Weight	Phrase
0.974	The best filmgoing experiences i
0.784	Is a wonderful film full
0.445	Lots of great comedy from
0.379	Crawford is excellent as the
0.373	I really enjoyed this debut
−0.560	Was the worst film i
−0.861	A waste of film
−0.962	Unbelievable rambling nonsense that should

compose the prototype representation. An example list of the prototypes in human-readable form is given in Table 2 for one of the datasets. The weight of the prototype comes from the fully connected layer. Note that the higher the weight, the more positive the phrase is. The prototypes with weights close to zero become increasingly neutral, where prototypes with the lowest weights represent the most negative phrases. Each prototype consists of 5 tokens, but the special tokens representing zero-padding and beginning or end of the document are hidden.

4.5 Explaining Text Classification with Prototypes

During the prediction phase, the interpretation of the model is provided according to prototypes. The model's prediction is performed by finding the most similar phrase from the input and then taking the linear combination of similarities and prototype's weights. The result of such a linear combination for the purpose of explanation is called an *evidence score*. When such evidence is greater than zero, it indicates the prediction to a positive class, while negative score results in negative class assignment.

An example of explaining the correct prediction is presented in Table 3. The prototypes were associated with the corresponding classes based on their weight in the fully connected layer. Positively-weighted prototypes represent phrases associated with positive sentiment, and negative-weighted ones represent negatively charged examples. For each class, three crucial phrases impacting the most the prediction result were selected and showed to the user. In the presented example, the user receives information why the input example has been classified as negative. It resulted that according to the neural network, the phrase *"was mediocre overall and considering"* is similar to *"unbelievable rambling nonsense that should"*, which is a phrase strongly representative of the negative class. Furthermore, the user is informed that this prototype added 2.08 to the negative class evidence, while all prototypes gave the evidence for negative class of 5.49. The user may notice that the key phrases associated to the evidence of positive class were not similar to the prototypes. The similarity of the most similar phrases to the positive prototypes was 0.6 when for the negative classes it was over 2. For example, the most similar phrase to *"is a wonderful movie full"*, was *"the first ninety minutes so"* with a similarity of only 0.61, which should be interpreted that the network did not find evidence for this prototypical phrase occurrence in the text. Finally, by the large difference of evidences between the classes (5.49 versus 2.01) the user can deduce that the network was confident in its decision.

Let us consider another case, where it is necessary to explain to the user the behavior of the model when it made a wrong prediction. An example of such a situation is presented in Table 4. The user can observe that the sums of the evidence for each of the two classes are quite close to each other – 4.56 compared to 4.40. The decision margin is relatively small, so the model was not confident in making this decision. When analyzing key phrases, the user can notice that the model focused too much attention on the phrase *'is silly it was*

Table 3. Explanation for a correct prediction.

Input example: about twenty minutes into this movie i was already bored quite simply these characters were fairly dull occasionally something enjoyable would happen but then things would slow down again fortunately my patience was eventually rewarded and the ending to this movie was n't bad at all however it was by no means good enough to justify sitting through the first ninety minutes so i would say that the movie was mediocre overall and considering all of the talent in the cast i 'd call this a disappointment.

Prediction: **Negative**, Gold standard: **Negative**

Evidence for negative sentiment:

Prototype	Most similar phrase in input text	Similarity * Weight
unbelievable rambling nonsense that should	was mediocre overall and considering	2.17 * 0.96 = **2.08**
a waste of film	characters were fairly dull occasionally	2.08 * 0.86 = **1.79**
was the worst film i	the movie was mediocre overall	2.89 * 0.56 = **1.62**

Sum of evidence: **5.49**

Evidence for positive sentiment:

Prototype	Most similar phrase in input text	Similarity * Weight
the best filmgoing experiences i	occasionally something enjoyable would happen	0.60 * 0.97 = **0.58**
is a wonderful film full	the first ninety minutes so	0.61 * 0.78 = **0.48**
lots of great comedy from	my patience was eventually rewarded	0.82 * 0.45 = **0.36**

Sum of evidence: **2.01**

Table 4. Explanation for an incorrect prediction.

Input example: this is a pleasant film even if the premise is silly it was sort of a guilty pleasure to watch meg ryan seems to be able to pull off roles in this kind of film another example is joe vs. the volcano that 's what makes her a star in part walter matthau of course had that ability too and he really puts himself into the role making an amusing good hearted einstein i suppose you could say they 're both good at portraying loveable characters though loveable in different ways loveable young women vs. loveable curmudgeon.

Prediction: **Negative**, Gold standard: **Positive**

Evidence for negative sentiment:

Prototype	Most similar phrase in input text	Similarity * Weight
a waste of film	is silly it was sort	2.87 * 0.86 = 2.47
was the worst film i	is silly it was sort	1.89 * 0.56 = 1.06
unbelievable rambling nonsense that should	premise is silly it was	1.07 * 0.96 = 1.03

Sum of evidence: **4.56**

Evidence for positive sentiment:

Prototype	Most similar phrase in input text	Similarity * Weight
the best filmgoing experiences i	is a pleasant film even	1.80 * 0.97 = 1.75
is a wonderful film full	good at portraying loveable characters	1.47 * 0.78 = 1.15
i really enjoyed this debut	this is a pleasant film	1.72 * 0.37 = 0.64

Sum of evidence: **4.40**

sort', whose similarity was high to the two prototypes. However, looking at the input example, the user observes that this fragment was taken out of context in a positive sentence. On the other hand, the user, noticing that the network operates on short phrases, may find such a mistake justified, or conclude that in such a situation changing this model to another one operating on longer phrases can be beneficial. Other examples of model's explanations are available in the repository with source code[2].

4.6   Examining the Fidelity of the Model

Finally, we conducted tests to measure the fidelity of provided explanations. High-fidelity characterizes an explanation that indicates the crucial information for the model while making a prediction. On the other hand, an explanation with low fidelity can point to random words in the example. To verify how the explanation mimics the model's behavior, we investigated the impact of removing phrases indicated as crucial on the classifier's prediction.

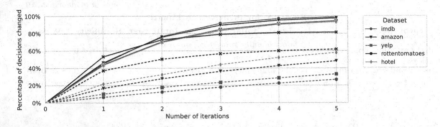

Fig. 3. Percentage of examples whose prediction was changed by masking the phrases indicated as crucial by the explanation of the decision (solid lines). The dashed lines demonstrate the result for a random baseline, i.e. an approach removing from the example random phrases at each iteration.

More precisely, for each input example, we performed a model prediction followed by constructing an explanation of that prediction. Then, we selected the most similar phrase to each prototype, indicating the predicted class. All tokens of selected phrases were converted to unknown tokens, and the prediction of the example was repeated. If the model's decision changed, the test was considered a success, otherwise, the explanation construction and selection of key phrases were repeated. The process was continued up to 5 iterations or until all tokens were converted to unknown.

Figure 3 shows the percent of successful tests in each iteration. One can observe that masking only the key phrases indicated by the explanation resulted in changing the decision in over 40% of the examples. In the third test iteration, the decision has changed in at least 80% of the examples for all datasets. On Rotten Tomatoes dataset, further iterations do not result in additional flipped decision because the test examples are short texts that after several iterations has almost all words hidden (they are classified to the "default" class).

---

[2] https://github.com/plutasnyy/ProtoCNN.

5   Final Remarks

In this paper, we have introduced a novel approach to explain sentiment predictions of texts. It is based on a new proposed prototypical convolutional neural network that uses prototypes in the form of automatically selected phrases from the training set, performs prediction by calculating similarities in the learned latent representation, and is trained with specially designed loss functions. Such phrase prototypes should provide shorter, simpler, more readable explanations and better generalize to the larger datasets than earlier considered solutions with document-level or sentence-level prototypes. Furthermore, the other changes in the output layer of this network also support obtaining more faithful explanations than ones offered by the two earlier, most related recurrent network approaches. We have also proposed the dynamic method for tuning the number of prototypes.

The performed experiments demonstrated that our proposed network obtains predictive accuracy similar to other reference networks while providing more faithful explanations. Moreover, the other experiments have clearly shown that a new method for dynamic tuning of the number of prototypes introduced in this paper offers performance gains against static tuning. Furthermore, we have demonstrated a user scenario with an explanation provided by the network and have shown that the provided explanations are characterized by high fidelity.

Nevertheless, in future research we are going to explore additional approaches to improve the structure, understandability and legibility of phrases extracted from texts, in particular for obtaining more meaningful prototypes.

Acknowledgments. The authors are grateful to the Poznan Supercomputing and Networking Center for computational resources. The research by Kamil Pluciński and Jerzy Stefanowski was supported by TAILOR, a project funded by EU Horizon 2020 research and innovation programme under GA no. 952215. Mateusz Lango was supported by the Polish National Science Centre grant no. 2016/22/E/ST6/00299.

References

1. Bahdanau, D., Cho, K., Bengio, Y.: Neural machine translation by jointly learning to align and translate. In: Bengio, Y., LeCun, Y. (eds.) 3rd International Conference on Learning Representations ICLR (2015)
2. Bodria, F., Giannotti, F., Guidotti, R., Naretto, F., Pedreschi, D., Rinzivillo, S.: Benchmarking and Survey of Explanation Methods for Black Box Models. arXiv e-prints arXiv:2102.13076 (February 2021)
3. Chen, C., Li, O., Tao, D., Barnett, A., Rudin, C., Su, J.K.: This looks like that: Deep learning for interpretable image recognition. In: Wallach, H., Larochelle, H., Beygelzimer, A., d'Alché-Buc, F., Fox, E., Garnett, R. (eds.) Advances in Neural Information Processing Systems, vol. 32, pp. 8928–8939 (2019)
4. Guidotti, R., Monreale, A., Ruggieri, S., Turini, F., Giannotti, F., Pedreschi, D.: A survey of methods for explaining black box models. ACM Comput. Surv. (CSUR) **51**(5), 1–42 (2018)

5. He, R., Lee, W.S., Ng, H.T., Dahlmeier, D.: Effective attention modeling for aspect-level sentiment classification. In: Proceedings of the 27th International Conference on Computational Linguistics, pp. 1121–1131 (2018)
6. Hong, D., Baek, S., Wang, T.: Interpretable sequence classification via prototype trajectory (July 2020). https://arxiv.org/abs/2007.01777
7. Hutchinson, B., Prabhakaran, V., Denton, E., Webster, K., Zhong, Y., Denuyl, S.: Social biases in NLP models as barriers for persons with disabilities. In: Proceedings of the 58th ACL, pp. 5491–5501 (2020)
8. Jain, S., Wallace, B.C.: Attention is not Explanation. In: Proceedings of the NAACL, pp. 3543–3556 (2019)
9. Lampridis, O., Guidotti, R., Ruggieri, S.: Explaining sentiment classification with synthetic exemplars and counter-exemplars. In: Appice, A., Tsoumakas, G., Manolopoulos, Y., Matwin, S. (eds.) DS 2020. LNCS (LNAI), vol. 12323, pp. 357–373. Springer, Cham (2020). https://doi.org/10.1007/978-3-030-61527-7_24
10. Letarte, G., Paradis, F., Giguère, P., Laviolette, F.: Importance of self-attention for sentiment analysis. In: Proceedings of the 2018 EMNLP Workshop BlackboxNLP: Analyzing and Interpreting Neural Networks for NLP, pp. 267–275 (2018)
11. Li, O., Liu, H., Chen, C., Rudin, C.: Deep learning for case-based reasoning through prototypes: a neural network that explains its predictions. In: AAAI (2018)
12. Ming, Y., Xu, P., Qu, H., Ren, L.: Interpretable and steerable sequence learning via prototypes. In: Proceedings of the 25th ACM SIGKDD International Conference on Knowledge Discovery and Data Mining (July 2019)
13. Molnar, C.: Interpretable Machine Learning (2019). https://christophm.github.io/interpretable-ml-book/
14. Pennington, J., Socher, R., Manning, C.: GloVe: global vectors for word representation. In: Proceedings of the EMNLP, pp. 1532–1543 (2014)
15. Ribeiro, M.T., Singh, S., Guestrin, C.: Model-agnostic interpretability of machine learning. In: Workshop on Human Interpretability in Machine Learning at International Conference on Machine Learning (2016)
16. Samek, W., Müller, K.-R.: Towards explainable artificial intelligence. In: Samek, W., Montavon, G., Vedaldi, A., Hansen, L.K., Müller, K.-R. (eds.) Explainable AI: Interpreting, Explaining and Visualizing Deep Learning. LNCS (LNAI), vol. 11700, pp. 5–22. Springer, Cham (2019). https://doi.org/10.1007/978-3-030-28954-6_1
17. Sanh, V., Debut, L., Chaumond, J., Wolf, T.: DistilBERT, a distilled version of BERT: smaller, faster, cheaper and lighter. In: 5th Workshop on Energy Efficient Machine Learning and Cognitive Computing @ NeurIPS 2019 (2019)
18. Strubell, E., Verga, P., Belanger, D., McCallum, A.: Fast and accurate entity recognition with iterated dilated convolutions. In: Proceedings of EMNLP, pp. 2670–2680 (2017)
19. Sundararajan, M., Taly, A., Yan, Q.: Axiomatic attribution for deep networks. In: International Conference on Machine Learning, pp. 3319–3328. PMLR (2017)
20. Wang, Y., Huang, M., Zhu, X., Zhao, L.: Attention-based LSTM for aspect-level sentiment classification. In: Proceedings of the EMNLP, pp. 606–615 (2016)
21. Wiegreffe, S., Pinter, Y.: Attention is not explanation. In: Proceedings of the EMNLP-IJCNLP, pp. 11–20 (2019)

Explanations for Network Embedding-Based Link Predictions

Bo Kang[✉], Jefrey Lijffijt, and Tijl De Bie

IDLab, Department of Electronics and Information Systems, Ghent University,
Technologiepark-Zwijnaarde 122, 9052 Ghent, Belgium
{bo.kang,jefrey.lijffijt,tijl.bie}@ugent.be

Abstract. Graphs (also called *networks*) are powerful data abstractions, but they are challenging to work with, as many machine learning methods may not be applied to them directly. Network Embedding (NE) methods resolve this by learning vector representations for the nodes, for subsequent use in downstream machine-learning tasks. Link Prediction is one such important downstream task, used for example in recommender systems. NE methods perform exceedingly well in accuracy for Link Prediction, but predictions following from the embeddings, whose dimensions have no intrinsic meaning, are not straightforward to understand. Explaining why predictions are made can increase trustworthiness, help understand the underlying models and give insight into what features of the network are important in light of the predictions, and answer posed regulatory requirements on the ability to explain machine-learning-based decisions. We study the problem of providing explanations for NE-based link predictions and introduce ExplaiNE, an approach to derive counterfactual explanations by identifying links in the network that explain link predictions. We show how ExplaiNE can be used generically on NE-based methods and consider ExplaiNE in more detail for Conditional Network Embedding, a particularly suitable state-of-art NE method. Extensive experiments demonstrate ExplaiNE's accuracy and scalability.

Keywords: Explainability · Network embedding · Link prediction · XAI

1 Introduction

Network embeddings (NEs) have exploded in popularity in both the machine learning and data mining communities. By mapping a network's nodes into a vector space, NEs enable the application of a variety of machine learning methods on networks for important tasks such as link prediction and classification. Link prediction (LP) is the task to predict whether nodes are likely to be or become connected in partially observed or evolving networks. LP has wide-ranging applications, for friendship recommendations, recommender systems, knowledge graph completion, etc. While there are numerous conventional LP methods that predict links based on heuristic statistics computed over networks (e.g. based on the number of common neighbors) (see, e.g. [17]), recently proposed NE-based methods typically outperform those heuristic approaches (e.g. [9, 11]).

A major disadvantage of NE-based LP methods is that they do not immediately provide intelligible explanations of the predicted links. The ability to understand link

© Springer Nature Switzerland AG 2021
M. Kamp et al. (Eds.): ECML PKDD 2021 Workshops, CCIS 1524, pp. 473–488, 2021.
https://doi.org/10.1007/978-3-030-93736-2_36

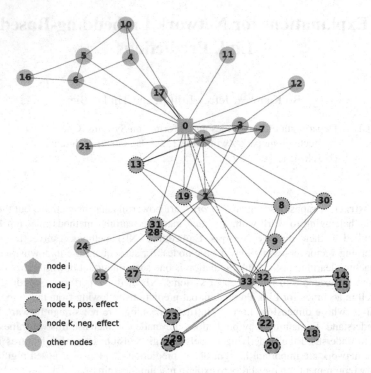

Fig. 1. Visualization of Zachary's karate club.

predictions is important and useful for several reasons: (a) recommender systems that provide explanations are more easily trusted and more effective (b) it allows data analysts to have a better understanding of the network characteristics such as node features and network dynamics, (c) transparency of automated processing systems is required in a growing number of regulations, and explanations can increase transparency.

To address these needs, we present ExplaiNE, a principled counterfactual reasoning approach for explaining NE-based LPs. In its simplest form, ExplaiNE quantifies how the probability of a predicted link is affected by considering the non-existence (weakening) of an existing link. Links that after weakening most strongly reduce the probability of the predicted link then serve as counterfactual explanations.

Example. We illustrate the idea behind ExplaiNE with an example. Zachary's karate club network [29] is a network of 34 karate club members connected through 78 friendship links. We used a recent probabilistic NE-based LP method [11], embedded the data in two-dimensional Euclidean space (so that we can also visualize it; see Fig. 1) and asked the NE-based LP method for a recommendation of a link for node $i = 33$.

The LP method suggests the most probable link for 33 is node $j = 0$. Although node 0 is not the closest unlinked node to 33, it has high degree, making a link likely under the model used by this embedding method. Figure 1 visualizes the embedding and also shows which existing links incident to i ExplaiNE details as positive (orange circle with dotted edge) and negative (blue circle with dashed edge) contributions, towards the prediction of link $\{0, 33\}$. It concludes this because weakening links to the orange nodes would reduce the link probability $\{i, j\}$, whereas weakening links to the blue nodes would increase it. These explanations visually appear to make sense: the orange nodes 'pull' node 33 closer to 0, while the blue nodes pull node 33 away from 0.

The paper is organized as follows. In Sect. 2, we first derive ExplaiNE generically, allowing for explanations not only in terms of links incident to the predicted link, but also in terms of other links as well as non-links. We then reduce its scope to explanations in terms of only incident links (as in the example above), and make an approximation (which we justify empirically) to obtain a still generic but highly scalable approach. Next we apply ExplaiNE to Conditional Network Embedding (CNE) [11], a recent state-of-the-art NE method. The application of ExplaiNE to CNE is particularly transparent, thanks to its straightforward use in LP, requiring no training once the embedding is found. We also outline how ExplaiNE can be applied to NE methods based on skip gram with negative sampling-based such as LINE [25], DeepWalk [19], PTE [24], and node2vec [9], with further details provided in the supplementary material. Experiments covering quantitative and qualitative analysis of the performance and scalability are described in Sect. 3. Related work is discussed in Sect. 4 and the conclusions in Sect. 5.

The **main contributions** of this paper are:

- We introduce ExplaiNE, a generic counterfactual reasoning approach for explaining LPs based on NEs (Sect. 2.3).
- We provide a scalable tight approximation of ExplaiNE (Sect. 2.4).
- We present a detailed application of ExplaiNE to CNE (Sect. 2.5)
- An outline of applying ExplaiNE to skip gram with negative sampling based NE methods (Sect. 2.6).
- We discuss quantitative and run-time analyses, finding that the approximation is stable and scalable (Sect. 3).
- We study qualitative and quantitative empirical results from realistic case studies, which indicate ExplaiNE achieves its goal (Sects. 3.2, 3.3).

2 Methods

2.1 Notation

We first introduce basic notation, similar to, e.g. [4] and [8]. An *undirected network* is denoted $\mathcal{G} = (V, E)$ where V is a set of $n = |V|$ *nodes* and $E \subseteq \binom{V}{2}$ is the *set of links* (or *edges*). A *link* is denoted by an unordered node pair $\{i, j\} \in E$. We use A to denote an *adjacency matrix*, with element $a_{ij} = 1$ for $\{i, j\} \in E$ and $a_{ij} = 0$ otherwise. The symbol $\hat{A}$ will be used to denote the adjacency matrix of a particular observed network. NE methods find a mapping $f : V \to \mathbb{R}^d$ from nodes to d-dimensional real vectors. An *embedding* is denoted as $X = (x_1, x_2, \ldots, x_n)' \in \mathbb{R}^{n \times d}$, where $x_i \triangleq f(i)$ for $i \in V$ is the embedding of each node. Finally, we write X^* for a locally optimal embedding of an adjacency matrix A, and likewise write $\hat{X}^*$ for $\hat{A}$.

2.2   NE-Based Link Predictions

All well-known NE methods aim to find an embedding X^* for given graph $\mathcal{G}$ (with adjacency matrix A) that maximizes a continuously differentiable[1] objective function $\mathcal{O}(A, X)$ for the given adjacency matrix A. Thus X^* must satisfy the following necessary condition of local optimality:

$$\nabla_X \mathcal{O}(A, X^*) = 0. \tag{1}$$

Defining $F(A, X) \triangleq \nabla_X \mathcal{O}(A, X)$, a locally optimal embedding X^* is thus a solution to $F(A, X^*) = 0$.

Based on an embedding X, it is common to predict the existence of a link between any pair of nodes i and j by computing a link probability (or other score) $g_{ij}(X)$, using a differentiable function $g_{ij} : \mathbb{R}^{nd} \to \mathbb{R}$. In practice, g_{ij} often only depends on the embeddings x_i and x_j of i and j, and often it can be written as $g_{ij}(X) = g(x_i, x_j)$ for some function $g : \mathbb{R}^d \times \mathbb{R}^d \to \mathbb{R}$. It is often found by training a classifier (e.g. logistic regression) on a set of known linked and unlinked node pairs (see Sect. 2.6), but sometimes it follows directly from the NE model (e.g. for CNE). We also introduce the function $g_{ij}^* : \mathbb{R}^{n \times n} \to \mathbb{R}$ defined as $g_{ij}^*(A) \triangleq g_{ij}(X^*)$ where X^* is locally optimal w.r.t. A. I.e., g_{ij}^* directly computes the link probability w.r.t. the locally optimal embedding (assumed here as given) for a specified adjacency matrix.

2.3   ExplaiNE as a Generic Approach

ExplaiNE uses a counterfactual reasoning approach to explain link predictions based on a NE. Namely, it quantifies the change of the link probability (or other score) of a node pair $\{i, j\}$ if the presence of a link between a given pair of nodes $\{k, l\}$ were to be altered.

Consider first the situation where $\{k, l\} \in E$: If removing the link $\{k, l\}$ strongly decreases the probability of a link between i and j, the link $\{k, l\}$ is a good counterfactual explanation of this predicted link. Conversely, consider the situation where $\{k, l\} \notin E$: If adding a link $\{k, l\}$ strongly decreases the probability of a link between i and j, it is the absence of a link between k and l that is a good counterfactual explanation of the predicted link $\{i, j\}$.

Intuitively, adding or removing an existing link $\{k, l\}$ will alter the probability of a link between i and j because it will alter the optimal embedding, which in turn will change the link probability of the target pair $\{i, j\}$. For the ExplaiNE strategy to be effective, we must be able to compute and combine these two effects in an efficient manner.

A naive approach would be to recompute the embedding with a link added or removed, and to quantify how much this changes the probability of a link between i and j. However, recomputing the embedding is computationally demanding, and is

---

[1] Note that, although NE methods are often described for unweighted networks (i.e., a binary adjacency matrix), the objective $\mathcal{O}(A, X)$ is often continuously differentiable also w.r.t. the adjacency matrix A. This is required for ExplaiNE to be applicable, but as we will see this requirement is often satisfied.

practically impossible to do even for a moderate number of pairs $\{k, l\}$. Moreover, even adding or removing a single link can dramatically change the optimization landscape, and as there are potentially many local optima, we can end up with a very different embedding—even if initialized with the same original embedding—, making a change in link probability erratic and hard to interpret.

Instead, ExplaiNE investigates the effect of an *infinitesimal* change to a_{kl} around its observed value $\hat{a}_{kl}$, on the link probability as computed by g_{ij}^*. Specifically, ExplaiNE seeks as explanations node-pairs $\{k, l\}$ ($k \neq l$ and $\{k, l\} \neq \{i, j\}$) for which $\frac{\partial g_{ij}^*}{\partial a_{kl}}(\hat{A})$ is large in absolute value, and positive if $\hat{a}_{kl} = 1$ (as then decreasing a_{kl} down from $\hat{a}_{kl} = 1$ by a small amount would maximally decrease g_{ij}^*), and negative if $\hat{a}_{kl} = 0$ (as then increasing a_{kl} up from $\hat{a}_{kl} = 0$ by a small amount would maximally decrease g_{ij}^*). This can be done analytically. Indeed, applying the chain rule, we find:

$$\frac{\partial g_{ij}^*}{\partial a_{kl}}(\hat{A}) = \nabla_X g_{ij} \left(\hat{X}^*\right)^T \cdot \frac{\partial X^*}{\partial a_{kl}}(\hat{A}). \tag{2}$$

For many NE methods the first factor can be computed analytically from the expression for g_{ij}, as we will see in the next subsections. The second factor can be computed using the *implicit function theorem* (see, e.g. [3]). Rephrased for our specific setting and overloading the symbol X^* here to also signify a function, this theorem states:

Theorem 1 (Implicit function theorem). *Let $F : \mathbb{R}^{n \times n} \times \mathbb{R}^{n \times d} \to \mathbb{R}^{n \times d}$ be a continuously differentiable function with arguments denoted $A \in \mathbb{R}^{n \times n}$ and $X \in \mathbb{R}^{n \times d}$. Moreover, let $\hat{A}$ and $\hat{X}^*$ be such that $F(\hat{A}, \hat{X}^*) = 0$. If the Jacobian matrix $\nabla_X F(\hat{A}, \hat{X}^*)$ is invertible, then there exists an open set $S \subset \mathbb{R}^{n \times n}$ with $\hat{A} \in S$ such that there exists a continuously differentiable function $X^* : S \to \mathbb{R}^{n \times d}$ with:*

$$X^*(\hat{A}) = \hat{X}^*, \text{ and}$$
$$F(A, X^*(A)) = 0 \text{ for all } A \in S, \text{ and}$$
$$\frac{\partial X^*}{\partial a_{kl}}(A) = -(\nabla_X F(A, X^*(A)))^{-1} \frac{\partial F}{\partial a_{kl}}(A, X^*(A)).$$

It is the latter expression, evaluated at $\hat{A}$, that we need in in order to evaluate Eq. (2). Note that the Jacobian $\nabla_X F$ is in fact the Hessian of $\mathcal{O}$ with respect to X. This means that $\nabla_X F(\hat{A}, \hat{X}^*)$ is negative definite (as $\hat{X}^*$ is optimal for $\hat{A}$). While for some NE-methods it may not be *strictly* negative definite and thus not invertible as required by the theorem (because, e.g. any translation of $\hat{X}^*$ may be equally optimal according to $\mathcal{O}$), this situation can be avoided by adding a regularizer to $\mathcal{O}$ on, e.g. the Frobenius norm of $\hat{X}^*$ with very small weight. Without going into detail, we note that as this regularization constant approaches zero, this becomes equivalent with using the pseudo-inverse of the Hessian, instead of its inverse. This is the approach we have taken whenever this situation arose. Denoting this Hessian evaluated at $\hat{A}$ and $\hat{X}^*$ as H, can thus write:

$$\frac{\partial X^*}{\partial a_{kl}}(\hat{A}) = -H^{-1} \cdot \frac{\partial F}{\partial a_{kl}}(\hat{A}, \hat{X}^*). \tag{3}$$

Putting Eqs. (2) and (3) together, we now can compute the derivative of g_{ij}^* with respect to a_{kl}:

$$\frac{\partial g_{ij}^*}{\partial a_{kl}}(\hat{A}) = -\nabla_X g_{ij}\left(\hat{X}^*\right)^T \cdot H^{-1} \cdot \frac{\partial F}{\partial a_{kl}}(\hat{A}, \hat{X}^*). \tag{4}$$

For efficiency, one can compute the partial derivatives for a given predicted link $\{i, j\}$ and for all pairs $\{k, l\}$ by pre-computing the vector $\nabla_X g_{ij}\left(\hat{X}^*\right)^T \cdot H^{-1}$: solve a linear system with nd variables and equations, and right multiplying it with the vectors $\frac{\partial F}{\partial a_{kl}}(\hat{A}, \hat{X}^*)$ that depend on k and l. Unfortunately, the computational cost of solving this linear system is $O((nd)^3)$, limiting scalability both in network size and dimensionality. Thus, while this is a clear improvement over the naive approach, it is not sufficient for realistic network sizes. The next subsection describes how to make ExplaiNE tractable also for large networks and dimensionalities.

2.4   Making ExplaiNE Scalable

First, we choose to focus on explanations in terms of linked pairs $\{k, l\}$, rather than in terms of unlinked pairs. Such positive explanations are arguably more insightful than negative ones, and especially in sparse networks.

Second, experiments show that the best explanation for a predicted link $\{i, j\}$ for a node i, tends to be a link $\{k, l\}$ that is incident to node i, i.e., for which $l = i$. This is arguably because links adjacent to node i affect the link probability $g_{ij}^*(\hat{A})$ by directly affecting the embedding x_i^*, whereas links not incident to i are likely to have a secondary effect only. Besides this, we also believe that nodes incident to i are likely to be more meaningful from node i's perspective than other links, in practical applications. Thus, we can restrict ourselves to seeking an explanation for a predicted link from node i to node j in terms of an existing link $\{i, k\}$ for which $\frac{\partial g_{ij}^*}{\partial \hat{a}_{ik}}(\hat{A})$ is large and positive.

Third, we consider only NE methods where $g_{ij}(X^*)$ only depends on x_i^* and x_j^* (true for all NE methods we are aware of). Thus, Eq. (2) can be written as:

$$\frac{\partial g_{ij}^*}{\partial a_{ik}}(\hat{A}) = \nabla_{x_i} g_{ij}\left(\hat{X}^*\right)^T \cdot \frac{\partial x_i^*}{\partial a_{ik}}(\hat{A})$$
$$+ \nabla_{x_j} g_{ij}\left(\hat{X}^*\right)^T \cdot \frac{\partial x_j^*}{\partial a_{ik}}(\hat{A}). \tag{5}$$

Finally, we make an approximation inspired by the fact that changing a_{ik} will have a direct effect on the optimal embeddings x_i^* and x_k^*, but only indirectly (and thus typically less so) on the embedding of the other nodes—including on x_j^*. Hence, we ignore the second term in Eq. (5).

What remains to be computed is thus $\frac{\partial x_i^*}{\partial a_{ik}}(\hat{A})$. To do so, we consider the optimality condition of the embedding w.r.t. x_i^* alone, considering all other node embeddings fixed to their optimum in $\hat{X}^*$ for the observed $\hat{A}$. Letting $\hat{X}_{(i)}^*$ denote the set of $\hat{x}_l$ with $l \neq i$, this optimality condition is:

$$\nabla_{x_i} \mathcal{O}(\hat{A}, x_i, \hat{X}_{(i)}^*) = 0 \tag{6}$$

Writing $\hat{F}_i(A, x_i) \triangleq \nabla_{x_i} \mathcal{O}(A, x_i, \hat{X}^*_{(i)})$ for conciseness, optimality of $\hat{x}^*_i$ given the observed network $\hat{A}$ then requires that $\hat{F}_i(\hat{A}, \hat{x}^*_i) = 0$. We can use the implicit function theorem on this optimality condition to approximate $\frac{\partial x^*_i}{\partial a_{ik}}$ as:

$$\frac{\partial x^*_i}{\partial a_{ik}}(\hat{a}_{ik}) = -H_i^{-1} \cdot \frac{\partial \hat{F}_i}{\partial a_{ik}}. \tag{7}$$

Here, $H_i = \nabla_{x_i} \hat{F}_i(\hat{A}, \hat{x}^*_i)$ is the Jacobian of $\hat{F}_i$ or equivalently the Hessian of $\mathcal{O}$ w.r.t. x_i, evaluated at $(\hat{A}, \hat{X}^*)$. Putting Eqs. (7) and (5) (ignoring the second term as discussed) together, this yields:

$$\frac{\partial g^*_{ij}}{\partial a_{ik}}(\hat{A}) = -\nabla_{x_i} g_{ij} \left(\hat{X}^*\right)^T \cdot H_i^{-1} \cdot \frac{\partial \hat{F}_i}{\partial a_{ik}}(\hat{A}, \hat{x}^*_i). \tag{8}$$

Comparing Eq. (4) with Eq. (8) reveals the dramatic complexity reduction achieved: Inverting $H_i \in \mathbb{R}^{d \times d}$ has a practical complexity of only $O(d^3)$, which is entirely feasible given common dimensionalities used in the literature (often 128).

2.5 ExplaiNE for CNE

We now apply the generic ExplaiNE approach to the Conditional Network Embedding method (CNE) [11]. Detailed derivations are deferred to the supplementary Section.1. CNE proposes a probability distribution for the network conditional on the embedding, and finds the optimal embedding by maximum likelihood estimation. Specifically, the objective function $\mathcal{O}$ in CNE is the log-probability of the network conditioned on the embedding:

$$\mathcal{O}(\hat{A}, X) = \log(P(\hat{A}|X))$$
$$= \sum_{\{i,j\}:\hat{a}_{ij}=1} \log P_{ij}(a_{ij} = 1|X)$$
$$+ \sum_{\{i,j\}:\hat{a}_{ij}=0} \log P_{ij}(a_{ij} = 0|X).$$

Here, the link probabilities P_{ij} conditioned on the embedding are defined as follows:

$$P_{ij}(a_{ij} = 1|X) = 1 - P_{ij}(a_{ij} = 0|X)$$
$$= \frac{P_{\hat{A},ij}\mathcal{N}_{+,\sigma_1}(\|x_i - x_j\|)}{P_{\hat{A},ij}\mathcal{N}_{+,\sigma_1}(\|x_i - x_j\|) + (1 - P_{\hat{A},ij})\mathcal{N}_{+,\sigma_2}(\|x_i - x_j\|)}, \tag{9}$$

where $\mathcal{N}_{+,\sigma}$ denotes a half-Normal distribution [14] with spread parameter σ, $\sigma_2 > \sigma_1 = 1$, and where $P_{\hat{A},ij}$ is a prior probability for a link to exist between nodes i and j as inferred from the degrees of the nodes (or based on other information about the structure of the network [1,13]). CNE, being based on a probabilistic model for the graph conditioned on the embedding, naturally allows for LP using the probabilities $P_{ij}(\hat{a}_{ij} = 1|X)$. In other words, $g_{ij}(X) = P_{ij}(a_{ij} = 1|X)$ as shown in Eq. (9). Note that it depends on x_i and x_j alone, as required for the approximate version of ExplaiNE to be applicable (third assumption).

Next we show how to apply approximated ExplaiNE to CNE (in the remainder of the text, we drop the modifier 'approximated'; exact ExplainNE applied to CNE is derived in supplementary Section.2. First, we derive the optimality condition:

$$\hat{F}_i(\hat{A}, \hat{x}_i^*) = \nabla_{x_i^*} \log(P(\hat{A}|\hat{X}^*))$$
$$= \gamma \sum_{j \neq i} (\hat{x}_i^* - \hat{x}_j^*) \left(P\left(a_{ij} = 1|\hat{X}^* \right) - \hat{a}_{ij} \right)$$
$$= \mathbf{0}.$$

Denoting $\gamma = \frac{1}{\sigma_1^2} - \frac{1}{\sigma_2^2}$, and $\hat{P}_{ij}^* \triangleq g_{ij}^*(\hat{A}) = P_{ij}(a_{ij} = 1|\hat{X}^*)$ (the probability of a link between i and j given the optimal embedding $\hat{X}^*$ for $\hat{A}$), we can now derive the three factors in Eq. (8) (see Supplementary Section 1 for detailed derivations):

$$\nabla_{x_i} g_{ij}\left(\hat{X}^*\right) = -\gamma(x_i^* - x_j^*)\hat{P}_{ij}^*(1 - \hat{P}_{ij}^*),$$
$$H_i = \nabla_{x_i} \hat{F}_i(\hat{A}, \hat{x}_i^*)$$
$$= \gamma I \sum_{l \neq i} (P_{il}^* - \hat{a}_{il})$$
$$- \gamma^2 \sum_{l \neq i} (x_i^* - x_l^*)(x_i^* - x_l^*)' \hat{P}_{il}^*(1 - \hat{P}_{il}^*),$$
$$\frac{\partial \hat{F}_i}{\partial a_{ik}}(\hat{A}, \hat{x}_i^*) = \gamma(x_k^* - x_i^*),$$

This means:

$$\frac{\partial g_{ij}^*}{\partial a_{ik}}(\hat{A}) = (x_i^* - x_j^*)^T \left(\frac{-H_i}{\gamma^2 \hat{P}_{ij}^*(1 - \hat{P}_{ij}^*)} \right)^{-1} (x_i^* - x_k^*).$$

Note that the Hessian should be invertible and negative definite, if $\hat{x}_i^*$ is indeed a local maximum. Interestingly, this expression has an intuitive interpretation: without the inverted Hessian, it would be an inner product between the distance of x_i^* to the embeddings of both nodes x_j^* and x_k^*, indicating that the best explanation is as far as possible in the direction of x_j^* as seen from x_i^*. Yet, the Hessian modulates the metric and reduces the explanatory power in directions with lots of embedded nodes l for which $\hat{P}_{il}^*(1 - \hat{P}_{il}^*)$ is large, i.e. for which the model is undecided whether there should be a link.

2.6 ExplaiNE for Other NE Methods

Here we illustrate the generic applicability of ExplaiNE by outlining the steps of applying it to NE methods based on skip gram with negative sampling (SGNS) (e.g. LINE, PTE, DeepWalk, node2vec). In supplementary Section 3, we derive a concrete example for LINE [25].

In those methods, $g_{i,j}(\boldsymbol{X}) = g(\boldsymbol{x}_i, \boldsymbol{x}_j)$, where $g \triangleq \sigma \circ h$ with $\sigma : \mathbb{R}^d \to \mathbb{R}$ a linear classifier (often logistic regression) applied to edge embeddings, whereby the embedding $h(\boldsymbol{x}_i, \boldsymbol{x}_j)$ of an edge $\{i,j\}$ is computed by applying an edge embedding operator $h : \mathbb{R}^d \times \mathbb{R}^d \to \mathbb{R}^d$ (e.g. element-wise product) to the embeddings of the nodes at its end-points.

[15] and [21] found that SGNS-based NE methods all share the same objective:

$$\mathcal{L} = \sum_{i=1}^{|V|} \sum_{j=1}^{|V|} \log \sigma(\boldsymbol{x}_i \cdot \boldsymbol{y}_j) + b \sum_{i=1}^{|V|} \mathbb{E}_{j' \sim P_N} \left[\log \sigma(-\boldsymbol{x}_i \cdot \boldsymbol{y}_{j'}) \right],$$

where $\boldsymbol{x}_i$ is the target embedding of node i, $\boldsymbol{y}_i$ is the embedding of node j as context (usually discarded, node2vec does not differentiate target and context), $\sigma(\cdot)$ is a sigmoid function, P_N is known as the noise that generates negative samples, and b is the number of negative samples. Moreover, [21] showed that $\mathcal{L}$ often has a closed form representation (or converges to one in probability). This makes it possible to obtain an analytical expression of the NE optimality condition, and thus of the function $F(\boldsymbol{A}, \boldsymbol{X})$. Given this, both exact and approximated ExplaiNE can be derived.

3   Experiments

We investigated the following questions: **Q1** How does the approximation compare to the exact version (Sect. 3.1)? **Q2** Does ExplaiNE give sensible explanations (Sect. 3.2–Sect. 3.3)? **Q3** Does the proposed method scale (Sect. 3.4)?. All experiments are based on CNE with parameters $\sigma_1 = 1$, σ_2 is tuned on a 90%–10% train-validation split with values from $\{2, 8, 16, 32, 64\}$. Any weights associated to links are ignored. Due to the lack of space, we summarize the experimental results in this section, and discuss the results more extensively in supplementary material. Code to reproduce all the experiments as well as supplementary material are available at: https://github.com/aida-ugent/ExplaiNE. We used the following networks.

The **Game of Thrones' (GoT) network**[2] consisting of 796 characters (nodes) and 2823 links between characters that are mentioned within 15 words of one another in books 1–5. We used a 2-dimensional embedding of this network to assess the quality of the approximated ExplaiNE approach.

The **DBLP co-authorship network** [26] (DBLP dataset V10[3]) with papers published up to year 2017, from which we selected all papers published at ICML, NeurIPS, ICLR, JMLR, MLJ, KDD, ECML-PKDD, and DMKD. This results in 23,359 authors (nodes) and 20,545 papers, converted into 66,597 links between authors who co-authored at least one paper. We conducted both qualitative and quantitative evaluations on a 8-dimensional embedding of this network.

The **MovieLens dataset**[4] [10] with 100,000 ratings by 943 users on 1,682 movies. The network is thus bipartite and consists of 943+1,682 nodes and 100,000 edges. The

---

[2] https://github.com/mathbeveridge/asoiaf.
[3] https://aminer.org/citation.
[4] https://grouplens.org/datasets/movielens/100k/.

dataset also contains metadata such as title and genre, which we have used as external validation sources. We conducted qualitative and quantitative experiments on a 8-dimensional embedding of this network.

3.1 Quality of the ExplaiNE Approximation

Before applying approximated ExplaiNE to the real-world datasets, we first evaluate the quality of the approximation (**Q1**). We will assess the extent to which the top K explanations for a predicted link $\{i, j\}$ incident to a given node i, as given by approximated ExplaiNE, overlap with the top-K explanations given by exact ExplaiNE. Relevant parameters here are (1) the value of K and (2) the number of neighbors. As we consider only links to neighbors as candidate explanations, K must be smaller than the number of neighbors of i. Moreover, if the number of neighbors is not much larger than K, a substantial overlap in the top-K explanations of the exact and approximate method is not surprising. Indeed, if i has m neighbors, two random subsets of K neighbors would share l elements with probability $\binom{K}{l}\binom{m-K}{K-l}/\binom{m}{K}$, which is large for large l if m is not much larger than K.

Thus, we performed a stratified analysis, computing the size of the overlap of the top-K explanations, aggregated in a histogram over nodes with a specific degree. We did this on the GoT dataset for K from 1 to 5. This experiment revealed that the top-1 is always identical between the approximated and exact versions, while the elements further in the ranked list very rarely swapped positions (2 to 3 differences out of 796 on ranks 2–4, and 7 differences out of 796 for rank 5).

We also compared the complete ranking of the neighbors between the approximated and exact ExplaiNE versions, and this simply for the most probable link for every node (which empirically meant seeking explanations for links that are also present in the network). We computed the normalized Kendall tau distance between the ranked explanations given by approximated and exact ExplaiNE. The average normalized Kendall tau distance is 0.008 ± 0.03. For comparison, the average Kendall tau distance between a random ranking and exact ExplaiNE is 0.47 ± 0.28, so the observed rank distance between the exact and approximate ranking is indeed very small. Having established confidence in the accuracy of the approximation, we can now evaluate the behavior of approximated ExplaiNE on two larger networks.

3.2 Qualitative Evaluation

As a qualitative evaluation, we applied ExplaiNE to two real world networks to assess whether ExplaiNE gives sensible explanations for predicted links (**Q2**).

DBLP Network. In this co-authorship network, a predicted link between authors i and j suggests two non-linked authors would network-wise form a sensible collaboration.

While ExplaiNE uses no external information to provide its explanations for such suggested collaborations, our experiments indicate that such explanations tend to be existing collaborators working on a topic on which the suggested collaborator is active as well. As an example, we predict links for Eric P. Xing (node i), and compute the

Table 1. Predicted/recommended collaborations for Eric P. Xing.

Rank	Recommendations	Explain: 'Zoubin Ghahramani'
1	Zoubin Ghahramani	Yee Whye Teh
2	John D. Lafferty	Mário A. T. Figueiredo
3	Tong Zhang	Willie Neiswanger
4	Jeff G. Schneider	Andrew Gordon Wilson
5	Dale Schuurmans	Ruslan Salakhutdinov

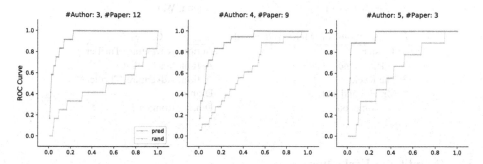

Fig. 2. ROC curves of co-author predictions for i = 'Eric P. Xing', with author-list lengths 3, 4, and 5 (orange = rand., blue = ExplaiNE). (Color figure online)

explanations for his top recommendation[5]: Zoubin Ghahramani (node j). We find that the existing co-authors of Eric P. Xing identified by ExplaiNE as top-5 explanations for this recommendation are either colleagues or co-authors of Zoubin Ghahramani (see Table 1), with a shared interest in graphical models and Bayesian machine learning.

MovieLens Network. In this network, a predicted link between a user i and movie j amounts to recommending movie j to user i. To do this, CNE is not given access to meta-data of the users or movies, and neither does ExplaiNE to identify explanations. Yet, we can make use of this meta-data to qualitatively assess whether the explanations make sense: We computed the recommendation for the first user (uid = 0) in the user list. The top recommended movie is 'Mission: Impossible' with genre tags 'Action, Adventure, Mystery'. The genres of the top explanations given by ExplainNE arguably have strongly overlapping genre tags (Table 2).

More results from these two data sets are given in the supplementary material. Based on these and other qualitative experiments (not included), our general impression is that ExplaiNE gives sensible explanations. The next subsection aims to quantify these findings.

---

[5] The recommended authors are not co-authors of the querying author according to the constructed co-author network.

Table 2. Recommended movie to user uid = 0. The top movie recommended by CNE (Mission: Impossible) is explained through movies already seen by user uid = 0. The top-ranked explanations have genres that overlap with the recommended movie.

j	Recommendations	Genres
1	Mission: Impossible	Action, Adventure, Mystery
2	E.T. the Extra-Terrestrial	Children's, Drama, Fantasy, Sci-Fi
3	Scream	Horror, Thriller
4	Liar Liar	Comedy
5	Schindler's List	Drama, War

k	Explanations for 'Mission: Impossible'	Genres
1	Twister	Action, Adventure, Thriller
2	Independence Day	Action, Sci-Fi, War
3	The Rock	Action, Adventure, Thriller
4	Mr. Holland's Opus	Drama
5	Phenomenon	Drama, Romance

3.3   Quantitative Evaluation

Objectively evaluating the quality of an explanation is conceptually non-trivial, due to a lack of datasets with ground-truth explanations for LP. Yet, as we show here, it is possible to use metadata to derive reasonable ground-truth proxy explanations, and compare with those.

DBLP Network. For this data, we can construct ground truth explanations for *existing* links (as opposed to *predicted* ones). While this is not the intended use case of ExplaiNE, it is a possible use case and justified here given our intention to objectively validate the quality of the explanations. Our approach is based on the intuition that a one-time co-author j of a given author i could have been introduced to that author i by another co-author k on the same paper, thus explaining the link $\{i, j\}$. While this is not always true, we postulate it is sufficiently common.

Given an author i and a one-time co-author j of i, we used ExplaiNE to rank the other co-authors of i, from more to less explanatory (according to Eq. 8). We then took the top-r of this ranked list as predicted co-authors on the paper i co-authored with j, and created a confusion matrix. Clearly, the hardness of this prediction task is different for papers with different numbers of authors. Thus, in order to get a more aggregate assessment, we summed the top-r confusion matrices for all one-time co-authors of node i on papers with a given number of co-authors L, and this for different L between 3 and 5. For a given author-list length, the confusion matrices with different r were then used to create precision-recall curves or ROC curves. Figure 2 shows the ROC curves for Eric P. Xing as node i for each author-list length. For comparison, also ROC curves computed based on randomly ranked lists are shown (as the size of the data is

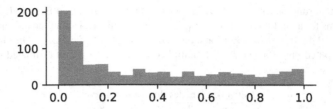

Fig. 3. Histogram of p-values that indicates the significance of the correlation between the genre recommended and the genres in the explanation. Each p-value is computed against 50 random explanations. Those explanations are drawn from user's watched movies. The empirical distribution has Kolmogorov-Smirnov test statistic 0.26 and a p-value against uniform distribution that is smaller than the numpy floating point machine accuracy 5.5e–58. This shows the significance of positive correlation between the recommended movies and the explanations made by ExplaiNE.

rather small, these are not always close to the diagonal). All results indicate that the explanations are remarkably effective, indicating that ExplaiNE performs well. Results for other nodes are similar.

MovieLens Network. The idea behind our second quantitative evaluation is that a good explanation k of a predicted link between a movie-user pair $\{i, j\}$ should often have a similar list of genres as j. To test this, we computed the top-5 explanations for user i and her top recommended movie j. Then we averaged the Jaccard similarity between the set of genres for movie j and the set of genres of each of the 5 explanations. To assess the significance of this average, we computed an empirical p-value for it by randomly sampling 50 sets of 5 'explanations' drawn from the watched movies of i, resulting in 50 random average Jaccard similarities to compare with the one obtained by ExplaiNE. Thus we obtained an empirical p-value for each user i, indicating the significance of the overlap between the set of genres of the recommended movie j and the top-5 explanations.

A histogram of these p-values is shown in Fig. 3. While p-values are uniformly distributed under the null hypothesis—that the explanations have genres unrelated to those of j—, here this is not the case, which indicates the null hypothesis is false. The Kolmogorov-Smirnoff test indeed rejects the null (with test statistic 0.26 and p-value is smaller than the numpy floating point machine accuracy 5.5e–58).

3.4 Scalability and Runtime

To address **Q3**, we measured the runtime of exact and approximated ExplaiNE when computing $\frac{\partial g_{ij}^*}{\partial a_{ik}}(\hat{A})$ for all $k \notin \{i, j\}$, as per Eqs. (4) and (8), on average over random pairs of nodes $\{i, j\}$. The runtime was measured on a PC with Intel quad Core i5 2.7 GHz and 16 GB RAM. Results shows that approximated ExplaiNE is efficient and applicable to large networks with higher dimensionality (e.g. on average 0.02 s on DBLP network with dim $= 8$), while exact ExplaiNE is not (exact method ran out of memory) (Table 3).

Table 3. Average runtime (in sec., 10 trials) of exact and approximated ExplaiNE in computing the explanations for a random pair of nodes $\{i, j\}$. Note that the exact method also has substantial memory cost: 10.4 Gb for MovieLens and on DBLP it went out of memory. On MovieLens, the time was computed only for one k, and multiplied by $n - 2$ to get an estimated total time for all k.

Network	#nodes	dim	exact	approx.
Karate	34	2	0.03	1.8e−4
GoT	796	2	64.1	4.1e−4
GoT	796	8	1490	9.8e−4
MovieLens	2625	8	3.4e5	6.3e−4
DBLP	23359	8	—	0.02

4   Related Work

There are a few existing works that aim to explain link predictions (see, e.g. [2,7,27]). More specifically, they explicitly encode specific edge types (e.g. topological roles, whether it is social or topic link) in the model, thus the explanations are limited to the edge types that the model induced.

In parallel, the importance of accountability of AI has sparked growing research interest in interpretable ML. Interpretable ML research can be categorized into model-based and post-hoc approaches [6, 18]. The first category focuses on incorporating interpretability while constructing the ML model. ExplaiNE belongs to the second category of interpretable ML methods: it is a post-hoc method that focuses on interpreting the local structure of ML models (here, NE models). Most directly related (although not for LP) are [22] and [16], who provide a model-agnostic explanation via local approximation of the model. Closely related, [23] and [12] compute the gradient of the loss of a (black-box) model w.r.t. the input to gauge the relevance of the input features. The former computes the gradient using back-propagation, while the second approximates the gradient using a Taylor series expansion. In a more recent work [28], the explanation of a prediction made by a graph neural network is represented by the subgraph and sub-features that largely affect the prediction.

A related research line concerns adversarial attacks on graphs, as such attacks tend to search for graph modification with a large impact on a specific task. [5] and [30], for example, consider attacks that target node classifications based on graph convolutional neural networks (instead of LPs based on NE, such as ExplaiNE). [20] investigate the robustness of LP in knowledge graphs via adversarial modifications, for the specific case of knowledge graph embeddings using multiplicative models. The authors show how their approach also enables interpretations of the knowledge graph in terms of length-2 Horn rules (e.g. isMarriedTo$(a, b) \wedge$ hasChild$(b, c) \Rightarrow$ hasChild(a, c)). ExplaiNE on the other hand is derived as a generic approach, with explaining the presence of individual links as a primary focus.

ExplaiNE is the first generic approach (and, as far as we know, the first approach at all) for explaining link predictions based on a NE. Unlike the previous attempts of explaining link prediction results, ExplaiNE is agnostic to the edge types. Moreover, to the best of our knowledge, ExplaiNE is the first method that uses the implicit function theorem for explainability. This proved to be a crucial element for computing the gradient of the link probability w.r.t. the network structure, as it allowed us to rigorously track the optimal embedding given an infinitesimal change in the input network. We believe this theorem can prove valuable also for other tasks, particularly those where an intermediate representation is obtained by optimizing an unsupervised objective function (e.g. an autoencoder), to be fed into a subsequent model that is trained in a supervised manner.

5   Conclusions

Link prediction (LP) is an important task, with numerous applications. State-of-the-art approaches are based on first embedding the network in a vector space, followed by a LP step. Unfortunately, these approaches offer no insight in their predictions. To remedy this, we introduced ExplaiNE, a generic approach to explain Network Embedding (NE)-based LPs based in terms of existing links in the network that contribute most to a link being predicted (or not predicted), and we presented several means to make it scalable. We applied ExplaiNE to CNE, a state-of-the-art NE method. Through extensive qualitative and quantitative evaluations, we evaluated the usefulness of ExplaiNE, and its ability to scale to large networks, indicating an affirmative answer to both. As further work, it may be considered how to exploit the principles behind ExplaiNE for detecting and mitigating adversarial modifications, and studying robustness, similar to [5,20,30].

Acknowledgements. The research leading to these results has received funding from the European Research Council under the European Union's Seventh Framework Programme (FP7/2007–2013) (ERC Grant Agreement no. 615517), and under the European Union's Horizon 2020 research and innovation programme (ERC Grant Agreement no. 963924), from the Flemish Government under the "Onderzoeksprogramma Artificiële Intelligentie (AI) Vlaanderen" programme, and from the FWO (project no. G091017N, G0F9816N, 3G042220).

References

1. Adriaens, F., Lijffijt, J., De Bie, T.: Subjectively interesting connecting trees. In: Ceci, M., Hollmén, J., Todorovski, L., Vens, C., Džeroski, S. (eds.) ECML PKDD 2017. LNCS (LNAI), vol. 10535, pp. 53–69. Springer, Cham (2017). https://doi.org/10.1007/978-3-319-71246-8_4
2. Barbieri, N., Bonchi, F., Manco, G.: Who to follow and why: link prediction with explanations. In: KDD, pp. 1266–1275. ACM (2014)
3. Chiang, A.C.: Fundamental Methods of Mathematical Economics. McGraw-Hill, Aukland (1984)
4. Cui, P., Wang, X., Pei, J., Zhu, W.: A survey on network embedding. TKDE **31**, 833–852 (2018)
5. Dai, H., et al.: Adversarial attack on graph structured data. In: ICML, pp. 1115–1124 (2018)
6. Du, M., Liu, N., Hu, X.: Techniques for interpretable machine learning. arXiv preprint arXiv:1808.00033 (2018)

7. van Engelen, J.E., Boekhout, H.D., Takes, F.W.: Explainable and efficient link prediction in real-world network data. In: Boström, H., Knobbe, A., Soares, C., Papapetrou, P. (eds.) IDA 2016. LNCS, vol. 9897, pp. 295–307. Springer, Cham (2016). https://doi.org/10.1007/978-3-319-46349-0_26

8. Goyal, P., Ferrara, E.: Graph embedding techniques, applications, and performance: a survey. Knowl.-Based Syst. **151**, 78–94 (2018)

9. Grover, A., Leskovec, J.: node2vec: Scalable feature learning for networks. In: KDD. pp. 855–864. ACM (2016)

10. Harper, F.M., Konstan, J.A.: The movielens datasets: history and context. TIIS **5**(4), 19 (2016)

11. Kang, B., Lijffijt, J., De Bie, T.: Conditional network embeddings. In: ICLR (2019)

12. Koh, P.W., Liang, P.: Understanding black-box predictions via influence functions. In: ICML, pp. 1885–1894 (2017)

13. van Leeuwen, M., De Bie, T., Spyropoulou, E., Mesnage, C.: Subjective interestingness of subgraph patterns. Mach. Learn. **105**(1), 41–75 (2016). https://doi.org/10.1007/s10994-015-5539-3

14. Leone, F., Nelson, L., Nottingham, R.: The folded normal distribution. Technometrics **3**(4), 543–550 (1961)

15. Levy, O., Goldberg, Y.: Neural word embedding as implicit matrix factorization. In: NeurIPS, pp. 2177–2185 (2014)

16. Lundberg, S.M., Lee, S.I.: A unified approach to interpreting model predictions. In: NeurIPS, pp. 4765–4774 (2017)

17. Martínez, V., Berzal, F., Cubero, J.C.: A survey of link prediction in complex networks. CSUR **49**(4), 69 (2017)

18. Murdoch, W.J., Singh, C., Kumbier, K., Abbasi-Asl, R., Yu, B.: Interpretable machine learning: definitions, methods, and applications. arXiv preprint arXiv:1901.04592 (2019)

19. Perozzi, B., Al-Rfou, R., Skiena, S.: Deepwalk: online learning of social representations. In: KDD, pp. 701–710. ACM (2014)

20. Pezeshkpour, P., Tian, Y., Singh, S.: Investigating robustness and interpretability of link prediction via adversarial modifications. arXiv pre-print arXiv:1905.00563 (2019)

21. Qiu, J., Dong, Y., Ma, H., Li, J., Wang, K., Tang, J.: Network embedding as matrix factorization: unifying deepwalk, line, pte, and node2vec. In: WSDM, pp. 459–467. ACM (2018)

22. Ribeiro, M.T., Singh, S., Guestrin, C.: Why should i trust you?: explaining the predictions of any classifier. In: KDD, pp. 1135–1144. ACM (2016)

23. Simonyan, K., Vedaldi, A., Zisserman, A.: Deep inside convolutional networks: visualising image classification models and saliency maps. arXiv preprint arXiv:1312.6034 (2013)

24. Tang, J., Qu, M., Mei, Q.: PTE: predictive text embedding through large-scale heterogeneous text networks. In: KDD, pp. 1165–1174. ACM (2015)

25. Tang, J., Qu, M., Wang, M., Zhang, M., Yan, J., Mei, Q.: Line: large-scale information network embedding. In: WWW, pp. 1067–1077 (2015)

26. Tang, J., Zhang, J., Yao, L., Li, J., Zhang, L., Su, Z.: Arnetminer: extraction and mining of academic social networks. In: KDD, pp. 990–998. ACM (2008)

27. Xu, L., Wei, X., Cao, J., Yu, P.S.: On exploring semantic meanings of links for embedding social networks. In: WWW, pp. 479–488 (2018)

28. Ying, Z., Bourgeois, D., You, J., Zitnik, M., Leskovec, J.: Gnnexplainer: generating explanations for graph neural networks. In: NeurIPS, pp. 9240–9251 (2019)

29. Zachary, W.W.: An information flow model for conflict and fission in small groups. J. Anthropol. Res **33**(4), 452–473 (1977)

30. Zuegner, D., Akbarnejad, A., Guennemann, S.: Adversarial attacks on neural networks for graph data. In: IJCAI, pp. 6246–6250 (2019)

Exploring Counterfactual Explanations for Classification and Regression Trees

Suryabhan Singh Hada$^{(\boxtimes)}$ ⓘ and Miguel Á. Carreira-Perpiñán ⓘ

Department of Computer Science and Engineering,
University of California, Merced, CA 95343, USA
{shada,mcarreira-perpinan}@ucmerced.edu

Abstract. The problem of counterfactual explanations is that of minimally adjusting attributes in a source input instance so that it is classified as a target class under a given classifier. They answer practical questions of the type "what should my annual income be for my loan to be approved?", for example. We focus on classification and regression trees, both axis-aligned and oblique (having hyperplane splits), and formulate the counterfactual explanation as an optimization problem. Although this problem is nonconvex and nondifferentiable, an exact solution can be computed very efficiently, even with high-dimensional feature vectors and with both continuous and categorical features. We also show how the counterfactual explanation formulation can answer a range of important practical questions, providing a way to query a trained tree and suggest possible actions to overturn its decision, and demonstrate it in several case studies. The results are particularly relevant for finance, medicine or legal applications, where interpretability and counterfactual explanations are particularly important.

Keywords: Interpretability · Counterfactual explanations · Decision trees

1 Introduction

In the last decade, deep learning and machine learning models have become widespread in many practical applications. This has been beneficial as these models provide intelligent, automated processing of tasks that up to now were hard for machines. However, at the same time, concerns related to the ethics and safety of these models have arisen as well. One is the problem of interpretability, i.e., explaining how these models function. This is an old problem, which has been studied (possibly under different names, such as explainable AI) since decades ago in statistics and machine learning (e.g. [2,11,14,21]). A second problem is explaining why the model made a decision or how to change it [28]. This problem is more recent and has become more pressing due to concerns over the opaqueness of current AI systems. That said, related problems have been studied in data mining or knowledge discovery from databases [1,9,23,30], in particular in applications such as customer relationship management (CRM).

© Springer Nature Switzerland AG 2021
M. Kamp et al. (Eds.): ECML PKDD 2021 Workshops, CCIS 1524, pp. 489–504, 2021.
https://doi.org/10.1007/978-3-030-93736-2_37

Much of the recent works focus on a specific version of the second problem that focuses on changing a classifier's decision in a prescribed way [28]. This problem is called *counterfactual explanation* [28]. Formally, a counterfactual explanation seeks a minimal change to a given input's features that will change the classifier's prediction in the desired way. These explanations are very helpful in understanding the behavior of a model for a given instance, as they can answer questions like what input feature the model focuses on to make a certain prediction. Mathematically, the problem of counterfactual explanations can be formulated as an optimization problem. The objective is to minimize the change in input features subject to classified as user-defined class. This similar formulation has been used in deep nets for adversarial examples and model inversion [10,13,16,17,22,25,26,31]. There are also formulations that are specific for linear models [6,24,27]. Most algorithms to solve the optimization assume differentiability of the classifier with respect to its input instance so that gradient-based optimization can be applied. However, none of these algorithms apply to decision trees, which define nondifferentiable classifiers.

Decision trees have long been widely used in applications and are regularly highly ranked in user surveys such as KDnuggets.com or data mining and machine learning reviews such as [29]. This is particularly owing to their ease of interpretability compared to more accurate classifiers such as neural nets or decision forests. However, the prediction accuracy of decision trees learned using the recently introduced *Tree Alternating Optimization (TAO)* algorithm [3,7,38] is much higher than using traditional algorithms such as CART or C5.0. This applies both to the traditional, axis-aligned trees, but also to the far more accurate oblique trees (having hyperplane splits) and sparse oblique trees (having hyperplane splits with few nonzero weights), as well as to trees of more complex forms [34,37], and to forests using bagging, boosting or other ensembling mechanisms, but where each tree is trained with TAO [8,12,33,35,36]. Finally, the stronger predictive power of sparse oblique decision trees together with their interpretability and fast inference makes them useful for other uses, such as in understanding deep neural networks [18,19] or compressing deep neural networks [20].

For these reasons, solving counterfactual explanations for decision trees is important in practice. Our recent work from [4,5] shows that the counterfactual problem can be solved exactly and efficiently in a decision tree. We showed that the counterfactual problem in decision trees is equivalent to solving the counterfactual problem in each leaf region and picking the best among them (see Sect. 2). This exact formulation for solving the counterfactual problem in decision trees allows us to extend the problem, which can answer some very interesting questions. For instance, what is the closest class to change from the original class, what is the closest boundary if changing the only feature, which feature has the lowest cost to change the class to a target class, and more. We describe these extended problems in detail in Sect. 3. These problems are very difficult and probably very expensive to answer in other models like a neural network. However, we show that in decision trees using the [4,5] framework, we can solve these extended problems exactly and efficiently.

Next, we first briefly describe our formulation of the counterfactual problem and how to solve it exactly (Sect. 2); this follows closely the main results of [4,5]. Then we describe each extended problem in detail (Sect. 3). Finally, in Sect. 4 we discuss these extended problems along with real-life use cases.

2   Solving Counterfactual Exactly in Decision Trees

2.1   Leaf Region: Definition

Assume we are given a classification tree that can map an input instance $\mathbf{x} \in \mathbb{R}^D$, with D real features (attributes), to a class in $\{1, \ldots, K\}$. Assume the tree is rooted, directed and binary (where each decision node has two children) with decision nodes and leaves indexed in the sets $\mathcal{D}$ and $\mathcal{L}$, respectively, and $\mathcal{N} = \mathcal{D} \cup \mathcal{L}$. We index the root as $1 \in \mathcal{D}$. For example, in Fig. 1 we have $\mathcal{N} = \{1, \ldots, 15\}$, $\mathcal{L} = \{5, 8, 9, 11, \ldots, 15\}$ and $\mathcal{D} = \mathcal{N} \setminus \mathcal{L}$. In oblique decision trees each decision node $i \in \mathcal{D}$ has a real-valued decision function $f_i(\mathbf{x})$ defined by a hyperplane (linear combination of all the features) $f_i(\mathbf{x}) = \mathbf{w}_i^T \mathbf{x} + b_i$, with fixed weight vector $\mathbf{w}_i \in \mathbb{R}^D$ and bias $b_i \in \mathbb{R}$. For axis-aligned trees, $\mathbf{w}_i$ is an indicator vector (having one element equal to 1 and the rest equal to 0). The decision function $f_i(\mathbf{x})$ send down an input instance $\mathbf{x} \in \mathbb{R}^D$ to i's right child if $f_i(\mathbf{x}) \geq 0$ and down i's left child otherwise. Each leaf $i \in \mathcal{L}$ is labeled with one class label $y_i \in \{1, \ldots, K\}$. For an input instance $\mathbf{x}$ the tree predicts its label by sending down $\mathbf{x}$ via the decision nodes, to exactly one leaf and outputting its label. The parameters $\{\mathbf{w}_i, b_i\}_{i \in \mathcal{D}}$ and $\{y_i\}_{i \in \mathcal{L}}$ are estimated by TAO [3,7] (or another algorithm) when learning the tree from a labeled training set.

The tree partitions the input space into $|\mathcal{L}|$ regions, one per leaf, as shown in Fig. 1 (bottom panel). Each region is an axis-aligned box in case of axis-aligned trees and a polytope for oblique trees. This region is defined by the intersection of the hyperplanes found in the path from the root to the leaf. Specifically, define a linear constraint $z_i(\mathbf{w}_i^T \mathbf{x} + b_i) \geq 0$ for decision node i where $z_i = +1$ if going down its right child and $z_i = -1$ if going down its left child. Then we define the constraint vector for leaf $i \in \mathcal{L}$ as $\mathbf{h}_i(\mathbf{x}) = (z_j(\mathbf{w}_j^T \mathbf{x} + b_j))_{j \in \mathcal{P}_i \setminus \{i\}}$, where $\mathcal{P}_i = \{1, \ldots, i\}$ is the path of nodes from the root (node 1) to leaf i. We call $\mathcal{F}_i = \{\mathbf{x} \in \mathbb{R}^D : \mathbf{h}_i(\mathbf{x}) \geq 0\}$ the corresponding feasible set, i.e., the region in input space of leaf i. For example, in Fig. 1 (top) the path from the root to leaf 14 is $\mathcal{P}_{14} = \{1, 3, 6, 10, 14\}$ and its region is given by:

$$\mathbf{h}_{14}(\mathbf{x}) = \begin{pmatrix} f_1(\mathbf{x}) \\ -f_3(\mathbf{x}) \\ -f_6(\mathbf{x}) \\ -f_{10}(\mathbf{x}) \end{pmatrix} = \begin{pmatrix} \mathbf{w}_1^T \mathbf{x} + b_1 \\ -\mathbf{w}_3^T \mathbf{x} - b_3 \\ -\mathbf{w}_6^T \mathbf{x} - b_6 \\ -\mathbf{w}_{10}^T \mathbf{x} - b_{10} \end{pmatrix} \geq \mathbf{0}.$$

2.2   Counterfactual Problem in Decision Trees

In this section we briefly describe the counterfactual problem in a decision tree. Assume we are given a *source input instance* $\overline{\mathbf{x}} \in \mathbb{R}^D$ which is classified by the tree as class $\overline{y}$, i.e., $T(\overline{\mathbf{x}}) = \overline{y}$, and we want to find the closest instance $\mathbf{x}^*$ that would be classified as another class $y \neq \overline{y}$ (the *target class*). We define the *counterfactual explanation* for $\overline{\mathbf{x}}$ as the (or a) minimizer $\mathbf{x}^*$ of the following problem:

$$\min_{\mathbf{x} \in \mathbb{R}^D} E(\mathbf{x}; \overline{\mathbf{x}}) \quad \text{s.t.} \quad T(\mathbf{x}) = y, \ \mathbf{c}(\mathbf{x}) = \mathbf{0}, \ \mathbf{d}(\mathbf{x}) \geq \mathbf{0} \tag{1}$$

where $E(\mathbf{x}; \overline{\mathbf{x}})$ is a cost of changing attributes of $\overline{\mathbf{x}}$, and $\mathbf{c}(\mathbf{x})$ and $\mathbf{d}(\mathbf{x})$ are equality and inequality constraints (in vector form). The fundamental idea is that problem (1) seeks an instance $\mathbf{x}$ that is as close as possible to $\overline{\mathbf{x}}$ while being classified as class y by the tree and satisfying the constraints $\mathbf{c}(\mathbf{x})$ and $\mathbf{d}(\mathbf{x})$.

The constraint $T(\mathbf{x}) = y$ makes the problem severely nonconvex, nonlinear and nondifferentiable because of the tree function $T(\mathbf{x})$. However as described in [4,5] this problem can be solved exactly and efficiently. In [4] we show that problem (1) is equivalent to:

$$\min_{i \in \mathcal{L}} \min_{\mathbf{x} \in \mathbb{R}^D} E(\mathbf{x}; \overline{\mathbf{x}}) \quad \text{s.t.} \quad y_i = y, \ \mathbf{h}_i(\mathbf{x}) \geq \mathbf{0}, \ \mathbf{c}(\mathbf{x}) = \mathbf{0}, \ \mathbf{d}(\mathbf{x}) \geq \mathbf{0}. \tag{2}$$

In English, what this means is that solving problem (1) *over the entire space can be done by solving it within each leaf's region (defined by* $\mathbf{h}_i$, *Sect. 2.1) and then picking the leaf with the best solution.* This is shown in Fig. 1 (bottom panel). That is, the problem has the form of a mixed-integer optimization where the integer part is done by enumeration (over the leaves ($\mathcal{L}$)) and the continuous part (within each leaf) by other means to be described later.

Hence, the problem we still need to solve is the problem over a single leaf $i \in \mathcal{L}$ (having the desired label $y_i = y$), and henceforth we focus on this. We write it as:

$$\min_{\mathbf{x} \in \mathbb{R}^D} E(\mathbf{x}; \overline{\mathbf{x}}) \quad \text{s.t.} \quad \mathbf{h}_i(\mathbf{x}) \geq \mathbf{0}, \ \mathbf{c}(\mathbf{x}) = \mathbf{0}, \ \mathbf{d}(\mathbf{x}) \geq \mathbf{0}. \tag{3}$$

here, $\mathbf{h}_i(\mathbf{x})$ is the set of hyperplanes that represents decision rule of the nodes in the path from root to leaf i (Sect. 2.1). If the function $E(\mathbf{x}; \cdot)$ is convex over $\mathbf{x}$ and the constraints $\mathbf{c}(\mathbf{x})$ and $\mathbf{d}(\mathbf{x})$ are linear, then this problem is convex (since for oblique trees $\mathbf{h}_i(\mathbf{x})$ is linear). In particular, if E is quadratic then the problem is convex quadratic program (QP), which can be solved very efficiently with existing solvers. See [4,5] for more details.

2.3   Separable Problems: A Special Case for Axis-Aligned Trees

Our earlier work [4,5] provides following result, which vastly simplifies the problem for axis-aligned trees.

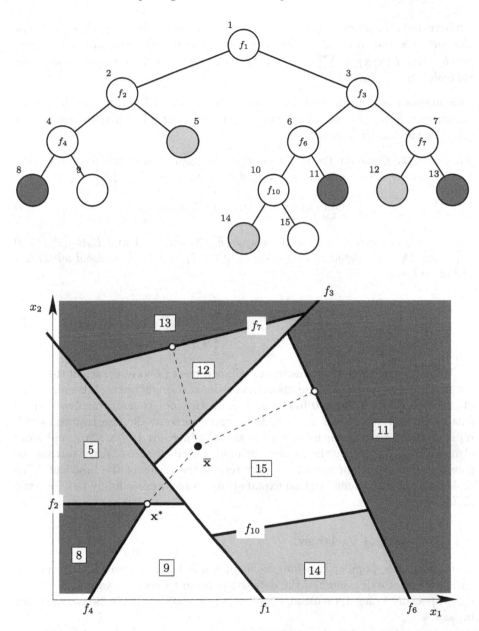

Fig. 1. *Top*: an oblique classification tree with $K = 3$ classes (colored white, light gray and gray) from [4]. A decision node i sends an input instance $\mathbf{x}$ to its right child if $f_i(\mathbf{x}) \geq 0$ and to its left child otherwise. The decision function $f_i(\mathbf{x}) = \mathbf{w}_i^T \mathbf{x} + b_i$, where $\mathbf{w}$ and b are weights and the bias respectively. *Bottom* the space of the input instances $\mathbf{x} \in \mathbb{R}^2$, assumed two-dimensional, partitioned according to each leaf's region in polytopes (the region boundaries are labeled with the corresponding decision node function). The source instance $\overline{\mathbf{x}}$ is in the white class and the counterfactual one (using the ℓ_2 distance) subject to being in the gray class is $\mathbf{x}^*$.

Theorem 1. *In problem* (1)*, assume that each constraint depends on a single element of* $\mathbf{x}$ *(not necessarily the same) and that the objective function is separable, i.e.,* $E(\mathbf{x}; \overline{\mathbf{x}}) = \sum_{d=1}^{D} E_d(x_d; \overline{x}_d)$. *Then the problem separates over the variables* $x_1, \ldots, x_D$.

This means that, within each leaf, we can solve for each x_d independently, by minimizing $E_d(x_d; \overline{x}_d)$ subject to the constraints on x_d. Further, the solution is given by the following result.

Theorem 2. *Consider the scalar constrained optimization problem, where the bounds can take the values* $l_d = -\infty$ *and* $u_d = \infty$:

$$\min_{x_d \in \mathbb{R}} E_d(x_d; \overline{x}_d) \quad s.t. \quad l_d \leq x_d \leq u_d. \tag{4}$$

Assume E_d *is convex on* x_d *and satisfies* $E_d(\overline{x}_d; \overline{x}_d) = 0$ *and* $E_d(x_d; \overline{x}_d) \geq 0$ $\forall x_d \in \mathbb{R}$. *Then* x_d^*, *defined as the median of* $\overline{x}_d$, l_d *and* u_d, *is a global minimizer of the problem:*

$$x_d^* = median(\overline{x}_d, l_d, u_d) = \begin{cases} l_d, & \overline{x}_d < l_d \\ u_d, & \overline{x}_d > u_d \\ \overline{x}_d, & otherwise \end{cases} . \tag{5}$$

We can apply these theorems to axis-aligned trees (assuming each of the extra constraints $\mathbf{c}(\mathbf{x})$ and $\mathbf{d}(\mathbf{x})$ depends individually on a single feature), because each of the constraints $\mathbf{h}_i(\mathbf{x}) \geq \mathbf{0}$ in the path from the root to leaf i involve a single feature of $\mathbf{x}$. Within each leaf $i \in \mathcal{L}$ we can represent the constraints (which represents the path from the root to i) as bounding box $\mathbf{l}_i \leq \mathbf{x} \leq \mathbf{u}_i$, and solve elementwise by applying the median formula described above. After solving the counterfactual problem in each leaf, we return the result of the best leaf. This makes solving the counterfactual explanation problem exceedingly fast for axis-aligned trees.

2.4   Categorical Variables

Although many popular benchmarks in machine learning use only continuous variables, in practice, most of the datasets contain categorical variables. This is true especially in legal, financial, or medical applications, for instance, use cases in Sect. 4.

In this work we handle categorical variables as described in [4,5]. That is we encode the categorical variables as one-hot. This means, if an original categorical variable can take C different categories, we encode it using C dummy binary variables jointly constrained so that exactly one of them is 1 (for the corresponding category): $x_1, \ldots, x_C \in \{0, 1\}$ s.t. $\mathbf{1}^T \mathbf{x} = 1$.

Since we only need to read the values of dummy variables during training, we treat them as if they were continuous and without the above constraints. However, when solving the counterfactual problem, we modify those variables, so we need to respect the above constraints. This makes the problem a mixed-integer optimization, where some variables are continuous and others binary (the dummy variables). While these problems are NP-hard in general, in many practical cases, we can expect to solve them exactly and quickly using modern mixed-integer optimization solvers, such as CPLEX or Gurobi [15].

3 Exploring Different Types of Counterfactual Explanation Questions

The counterfactual problem (2) accommodates a variety of useful, practical questions about the source instance ($\overline{\mathbf{x}}$). We list each below and explain how that can be solved exactly.

1. *Finding the closest boundary. The minimum-distance change to $\overline{\mathbf{x}}$ that changes its original class k.* Solve the problem (3) in every leaf except the ones with label k, and pick the solution with the lowest cost.
2. *Critical attribute for change to the target class y. Which attribute has the lowest cost to change the class of $\overline{\mathbf{x}}$ to a target class y, if changing only one attribute?* For given a attribute d, we add all other attributes to the equality constraint ($\mathbf{c}(\mathbf{x}) = \mathbf{0}$) and solve the counterfactual problem (2). We repeat this process for each attribute in $\overline{\mathbf{x}}$ and pick the attribute for which the counterfactual ($\mathbf{x}^*$) has the lowest cost.
3. *Critical attribute for changing the class. Which attribute has the lowest cost to change the class of $\overline{\mathbf{x}}$ to any other class if changing only one attribute?* For each attribute, we solve the *finding the closest boundary* problem, where other attributes are in the equality constraint; and pick the attribute for which the counterfactual ($\mathbf{x}^*$) have the lowest cost.
4. *Robust counterfactuals. Here, we want to find the counterfactuals that are well inside a leaf region rather than on the boundary, so they are more robust to flipping their class due to small changes.* This problem can easily be solved by shrinking the leaf region size in problem 3. That is in problem 3, the constraint "$\mathbf{h}_i(\mathbf{x}) \geq \mathbf{0}$" becomes "$\mathbf{h}_i(\mathbf{x}) \geq \epsilon$", where $\epsilon > 0$.

We can also use the counterfactual problem (2) to explore more practical problems that are related to the regression trees.Consider a regression tree T, where $T(\overline{\mathbf{x}})$ and $T(\mathbf{x}^*)$ represent the predicted value of the source instance ($\overline{\mathbf{x}}$) and the counterfactual ($\mathbf{x}^*$). Similar, to above we list each problem below and explain how that problem can be solved exactly.

1. *$T(\mathbf{x}^*) > T(\overline{\mathbf{x}})$: find the minimum change in $\overline{\mathbf{x}}$ that increase its predicted value.* For this we only consider the leaves whose label is larger than the $T(\overline{\mathbf{x}})$, and solve problem (3) in each of them and pick the $\mathbf{x}^*$ with the lowest cost.
2. *$T(\mathbf{x}^*) \geq T(\overline{\mathbf{x}}) + \beta$ find the minimum change in $\overline{\mathbf{x}}$ that increase its predicted value atleast by β.* Same as above, the only difference is the leaves we consider have the label greater than $T(\overline{\mathbf{x}}) + \beta$.

3. $\alpha \geq T(\mathbf{x}^*) \geq \beta$ *find the minimum change in* $\overline{\mathbf{x}}$ *that change its predicted value between* α *and* β. It is again same as the previous two, but this time we only consider the leaves with label between α and β.

All these extended problems can be applied to any type of decision tree with hard thresholds, but here we only focus on oblique trees.

4   Experiments

Since we show all extension problems (Sect. 3) can be solved exactly by converting them into the problem (2), we do not need to assess the optimality and validity of the generated counterfactual examples experimentally. Instead, we apply each problem to a real-life dataset [32] and explain their usage with three use case studies.

- Our first use case deals with the students' grades in secondary education of two Portuguese schools. For each student, we have social, demographic, and school-related attributes and their final grades. In this study, we focus on the students who have failing grades, and we suggest the changes that can improve students' grades by applying our extension problems.
- Next, we focus on the loan applicants as credit risk from the German credit dataset. We focus on the applicants that are a bad credit risk. Our goal here is not only to generate counterfactuals that can classify these candidates as good credit risk but also to suggest more changes to have a stronger credit profile. We do this by generating counterfactuals that are more robust to small changes.
- Our last use case study focuses on the median value of owner-occupied homes in multiple suburbs of Boston. We treat this as a regression problem whose goal is to predict the value of the home. Then we apply our formulations to explore what factors affect the value of a home at different price points.

In all the case studies, our model is an oblique decision tree that we train using a recently proposed algorithm called tree alternating optimization (TAO) [3,7]. To generate counterfactuals we use ℓ_2 distance as our cost function (E). Also, these datasets contain categorical variables, so to generate counterfactuals, we use Gurobi [15] to solve the mixed-integer optimization problem.

Next, we describe each dataset in brief and then discuss each use case study in detail.

4.1   Dataset Information

All our datasets are from UCI [32], and they are described in the same order as the user case study. Since none of the datasets contains a separate test set, we randomly divide the instances into training (80%) and test (20%) sets.

Student Portuguese Grades. Each instance have three grades, but we use only the final grades, and remove the other two from the attributes. The final grades are in the range of 1 to 20. So, we divide these grades into five categories. These categories are as follows:
- If the grades are greater or equal to 16, then *excellent*.
- If grades are either 15 or 14, then *good*.
- If grades are either 14 or 13, then *satisfactory*.
- If grades are in range between 12 and 10, then *sufficient*.
- If grades are 9 or lower, then *fail*.

The prediction task is to determine the final grades of a student. There are 649 instances, and each has 30 attributes (after removing the two grades). Out of 30 attributes, 11 are integer, and the rest are categorical. We convert each categorical attribute into a one-hot encoding attribute. Thus each instance of the dataset has 64 attributes.

German Credit. The prediction task is to determine whether an applicant is considered a *good* or a *bad* credit risk for 1000 loan applicants. There are 20 attributes, out of which 7 are integer, and the rest are categorical. Similarly to the above dataset, we convert each categorical attribute to a one-hot encoding attribute. Thus each instance of the dataset has 61 attributes.

Boston Housing Dataset. We use this dataset for regression. The task is to predict the median price of owner-occupied homes. There are 506 instances, and each instance describes a Boston suburb or town. For each instance, there is 13 attribute all continuous except one which is binary.

4.2 Use Case Study 1

Our oblique tree for this classification task achieves 56.92% test error and 33.14% train error. The tree has a depth of 9 and 34 leaves.

First consider a student whose attributes are described in the second column of Table 1. The current *final grades* of the student is *fail*. If create a counterfactual with target class as *excellent*, the following attributes change (third column in Table 1):

- reduce previous class failures from 2 to 1.
- mother's job should be changed from services to teacher.
- father's job should be changed from services to teacher.
- change higher education plan from no to yes.

However, it is hard to jump from failing grades to excellent grades in real life. Also, it requires changing many attributes, which might not be feasible like mother's job and father's job. Instead, the student can try making small changes, which are enough for passing the class. For this, we try to find the closest other class boundary problem here since any other class will lead to passing the class. The closest class we find is *sufficient* and requires the following change (fourth column in Table 1).

- reduce previous class failures from 2 to 1.

This is also expected the closest class to change is *sufficient* as this lowest thing to be considered as passing the class. Moreover, it does not suggest big changes like changing parent's jobs, but only change previous class failures which easier to achieve.

Next, we consider another student (fifth column in Table 1). This student also failed the class. If we upgrade the grades directly to *satisfactory* allowing only attribute to change, then the attributes that will change the class with the lowest cost will be (sixth column in Table 1):

– reduce free time from high(4) to very low(1).

This also makes sense as reducing free time after school can be devoted to studies. That said, it is still a large change. So, we search for the closest different class allowing only one attribute to change (last column in Table 1). The closet class would be *sufficient*, and the attribute with the lowest cost is:

– reduce going out frequency from medium(3) to low(2).

This makes sense as, again, time spent on going out can be used for studies. Also, the numeric cost to change to lower grade (*sufficient*) is 3 times lower than the upper grade (*satisfactory*).

4.3   Use Case Study 2

The predictive task in the study is to predict how much a loan applicant is of credit risk. Our oblique tree achieves a test error of 18.5% and a train error of 17.5% for this dataset. The depth of the tree is 8.

The goal here is to generate diverse counterfactuals, where each subsequent counterfactual is more robust than the previous one. When generating a counterfactual, the algorithm generates a counterfactual in each target leaf that is closest to the input instance. If all the attributes are continuous, then this generated counterfactual will always exist on the boundary of the leaf, as shown in the Fig. 1. However, if we shrunk these leaves ' regions, we can force the algorithm to generate counterfactuals inside the leaf region (generating robust counterfactuals). As described in Sect. 3, for a leaf i this can be done by adding a small positive constant ($\epsilon > 0$) in Eq. (3).

$$\min_{\mathbf{x} \in \mathbb{R}^D} E(\mathbf{x}; \overline{\mathbf{x}}) \quad \text{s.t.} \quad \mathbf{h}_i(\mathbf{x}) \geq \epsilon, \ \mathbf{c}(\mathbf{x}) = \mathbf{0}, \ \mathbf{d}(\mathbf{x}) \geq \mathbf{0}. \tag{6}$$

Consider an example described in the Table 2. This instance is classified as *bad* creditor. Next, we generate a counterfactual for it. The generated counterfactual will able to change its label from *bad* to *good* creditor. However, it is either on the boundary of the target leaf region or very close to it (due to categorical variables). This means the changes suggested by the counterfactual ($\mathbf{x}_1^*$) will only suggest changes that are the bare minimum for a good creditor. However, if we force the algorithm to generate counterfactuals deep in the leaf region, it will suggest the changes that make the counterfactual much better creditor.

Table 1. Example illustrating the construction of counterfactual instances with our extended formulations for an oblique decision tree on the Student Portuguese grades dataset. *Column 2-4:* For source $\overline{\mathbf{x}}_1$, $\mathbf{x}_{11}^*$ and $\mathbf{x}_{12}^*$ represents counterfactual with the given target class and the closest class respectively. *Column 5-7:* For source $\overline{\mathbf{x}}_2$, both counterfactuals are generated by changing only one attribute that has the lowest cost. $\mathbf{x}_{21}^*$ and $\mathbf{x}_{22}^*$ represent counterfactual with the given target class and the closest class, respectively. "=" means the attribute value is the same as in the source instance.

Attribute	$\overline{\mathbf{x}}_1$, source instance	$\mathbf{x}_{11}^*$, target class *excellent*	$\mathbf{x}_{12}^*$, closest class	$\overline{\mathbf{x}}_2$, source instance	$\mathbf{x}_{21}^*$, target class *satisfactory*	$\mathbf{x}_{22}^*$, closest class
					Only one attribute can change	
School	GP	=	=	MS	=	=
Sex	Male	=	=	Male	=	=
Age	18	=	=	17	=	=
Address	Urban	=	=	Urban	=	=
Family size	>3	=	=	>3	=	=
Parent's status	Together	=	=	Together	=	=
Mother's education	2	=	=	1	=	=
Father's education	1	=	=	1	=	=
Mother's job	Services	Teacher	=	Other	=	=
Father's job	Services	Teacher	=	Other	=	=
Reason	Other	=	=	Home	=	=
Guardian	Mother	=	=	Mother	=	=
Traveltime[†]	1	=	=	1	=	=
Study time[°]	1	=	=	2	=	=
Failures	2	1	1	0	=	=
School support	No	=	=	No	=	=
Family support	No	=	=	No	=	=
Paid	No	=	=	Yes	=	=
Activities	No	=	=	No	=	=
Nursery	No	=	=	No	=	=
Plan for higher education	No	Yes	=	Yes	=	=
Internet access	Yes	=	=	Yes	=	=
In romantic relationship	No	=	=	No	=	=
Family relationship[‡]	3	=	=	4	=	=
Free time[‡]	2	=	=	4	1	=
Going out frequency[‡]	5	=	=	3	=	2
Workday alcohol consumption[‡]	2	=	=	2	=	=
Weekend alcohol consumption[‡]	5	=	=	4	=	=
Health[‡]	5	=	=	5	=	=
Absences	4	=	=	4	=	=
Grades	Fail	Excellent	Sufficient	Fail	Satisfactory	Sufficient

[†] 1–<15 min, 2–15 to 30 min, 3–30 min to1 h or 4 –> 1 h).
[°] 1–<2 h, 2–2 to 5 h, 3–5 h to 10 h or 4 –> 10 h).
[‡] from 1 - very low to 5 - very high.

Table 2. Counterfactual solution trajectory as a function of ϵ for Credit dataset. ℓ_2 distance is also mentioned between source instance ($\overline{\mathbf{x}}$) and the counterfactual $\mathbf{x}^*$.

Attribute	$\overline{\mathbf{x}}$, source instance	$\mathbf{x}_1^*$, $\epsilon = 0.00$ $\ell_2 = 1.73$	$\mathbf{x}_2^*$, $\epsilon = 0.10$ $\ell_2 = 1.73$	$\mathbf{x}_3^*$, $\epsilon = 0.20$ $\ell_2 = 2.00$	$\mathbf{x}_4^*$, $\epsilon = 0.25$ $\ell_2 = 2.23$
Existing checking	<0 DM	=	=	=	=
Duration	15 months	16 months	=	=	16 months
Credithistory	Critical account	=	=	Delay in paying off in the past	Delay in paying off in the past
Purpose	Furniture/ equipment	=	=	=	=
Credit amount	1478	=	=	=	=
Savings	<100 DM	=	=	=	=
Employment since	≥7 years	=	=	=	=
IInstallment rate	4	=	=	=	=
Status and sex	Male: single	=	=	=	=
Other debtors	None	=	=	=	=
Residence since	4	=	3	3	3
Property	Car or other	=	=	=	=
Age	44	=	=	=	=
Other installment plans	None	Stores	stores	=	=
Housing	Own	=	=	=	=
Existing credits	2	=	=	1	1
Job	Skilled employee	=	=	=	=
People liable	2	=	=	=	=
Telephone	Yes	=	=	=	=
Foreignworker	Yes	=	=	=	=
Credit	Bad	Good	Good	Good	Good

This is also evident from the Table 2. As we increase the value of ϵ, the number of features changed also increases. Thus, creating a user profile ($\mathbf{x}_4^*$) that is a better creditor than the first counterfactual ($\mathbf{x}_1^*$). This can also be seen as the distance between the original instances and the counterfactual increases as the leaf regions get smaller (increasing value of ϵ).

These kinds of problems can be very useful, as shown in the above case. Here, we give a solution for now and give a strategy to become much better creditor.

In theory, the distance between the source instance and counter should increase with the value of ϵ continuously, but in the presented case, it happens in intervals. The reason is due to the presence of integer and categorical attributes because the solution can only change when an attribute changes by an integer or category.

Table 3. Example illustrating the construction of counterfactual instances for regression with our extended formulations for an oblique decision tree on the Boston housing dataset. We show the dataset attributes, source instance $\bar{x}$, and 3 counterfactual instances (with different Median home value) with various conditions. We have rounded each attribute value to 2 decimal places. "=" means the attribute value is the same as the source after rounding.

Attribute	$\bar{x}$, source instance	x^*		
		$T(x^*) > T(\bar{x})$	$T(x^*) \geq T(\bar{x}) + 5$	$30 \geq T(x^*) \geq 25$
Crime rate	2.37	2.15	1.93	1.81
Residential land zoned proportion	0.0	0.02	=	0.03
Proportion of non-retail business	19.58	19.48	=	19.48
Tract bounds river	0	=	1	=
Nitric oxides concentration	0.87	0.39	0.39	0.385
Avg. rooms per dwelling	4.92	5.13	5.73	8.09
Proportion of units before 1940	95.70	95.67	95.71	95.66
Distances to Boston employment centres	1.40	1.17	=	1.16
Accessibility to highways	5.00	5.04	5.1	5.41
Property-tax rate	403.00	=	=	402.99
Pupil-teacher ratio	14.70	14.47	14.58	14.60
Proportion of african american by town	391.71	=	=	391.67
% lower status of the population	29.53	29.39	29.54	29.41
Median home value in $1000's	14.74	15.96	20.52	29.14

4.4   Use Case Study 3

This user study deals with predicting the median home value of different suburbs and towns in Boston. We train a regression tree for this task. Our tree achieves a test mean squared error of 14.57 and train mean squared error of 13.19. The depth of the tree is 6, and the number of leaves is 21.

We consider an instance that has attributes described in the second column of Table 3. We apply our extension problems here to determine how for the given instance ($\bar{x}$), the attributes need to change to accommodate the new median home value. We investigate three scenarios:

$T(x^*) > T(\bar{x})$ meaning have higher median home value. Almost all of the attributes change (third column of the Table 3), the ones that change the most are as follows. The crime rate, nitric oxides concentration, distances to Boston employment centres, and pupil-teacher ratio decreases which make sense. Because having low crime rate means safer neighborhood and low nitric oxides concentration means the air quality is better. Also, having low pupil-teacher ratio means more teacher per student, which is beneficial for the families with children in school, and then distances to Boston employment centers is important for the people who need employment maybe for other family members. On the other side, the average rooms per dwelling increases, which is reasonable as it means bigger homes and thus higher prices.

$T(x^*) > T(\bar{x}) + 5$ meaning the median home value should be greater than 19.72. As shown in the fourth column of the Table 3 the crime rate and nitric oxides concentration further decrease. The pupil-teacher ratio also decreases but compare to the previous scenario it is higher by a small reason. It may be the increased value of average rooms per dwelling and tract bounds river compensate for it.

$30 \geq T(\mathbf{x}^*) \geq 25$ meaning the median home value should be between 25 and 30. As shown in the fifth column of the Table 3 again the crime rate and nitric oxides concentration plays an important role in the value. The biggest change is the average rooms per dwelling which almost doubled. This is expected as described earlier.

This case study shows for this particular instances the crime rate, nitric oxides concentration and average rooms per dwelling plays an important role in deciding the median home value of the home. Also, average rooms per dwelling compensates for some attributes like pupil-teacher ratio.

5   Conclusion

Classification and regression trees are very important in applications such as education, business, law and medicine, where counterfactual explanations are of particular relevance. These can be formulated as a constrained optimization problem and solved exactly and efficiently for both continuous and categorical features, possibly in an interactive way. The formulation can be applied to answer a variety of practical questions and we have illustrated this in several case studies. Python code implementing the algorithm is available at the authors' web page.

Acknowledgments. Work partially supported by NSF award IIS-2007147.

References

1. Bella, A., Ferri, C., Hernández-Orallo, J., Ramírez-Quintana, M.J.: Using negotiable features for prescription problems. Computing **91**, 135–168 (2011)
2. Breiman, L.J., Friedman, J.H., Olshen, R.A., Stone, C.J.: Classification and Regression Trees. Wadsworth, Belmont (1984)
3. Carreira-Perpiñán, M.Á.: The Tree Alternating Optimization (TAO) algorithm: a new way to learn decision trees and tree-based models (2021). arXIV
4. Carreira-Perpiñán, M.Á., Hada, S.S.: Counterfactual explanations for oblique decision trees: exact, efficient algorithms. In: Proceedings of the 35th AAAI Conference on Artificial Intelligence (AAAI 2021), 2–9 February 2021, pp. 6903–6911 (2021)
5. Carreira-Perpiñán, M.Á., Hada, S.S.: Counterfactual explanations for oblique decision trees: exact, efficient algorithms. arXiv:2103.01096 (2021)
6. Carreira-Perpiñán, M.Á., Hada, S.S.: Inverse classification with logistic and softmax classifiers: efficient optimization (2021). arXIV
7. Carreira-Perpiñán, M.Á., Tavallali, P.: Alternating optimization of decision trees, with application to learning sparse oblique trees. In: Bengio, S., Wallach, H., Larochelle, H., Grauman, K., Cesa-Bianchi, N., Garnett, R. (eds.) Advances in Neural Information Processing Systems (NEURIPS), vol. 31, pp. 1211–1221. MIT Press, Cambridge (2018)

8. Carreira-Perpiñán, M.Á., Zharmagambetov, A.: Ensembles of bagged TAO trees consistently improve over random forests, AdaBoost and gradient boosting. In: Proceedings of the 2020 ACM-IMS Foundations of Data Science Conference (FODS 2020), Seattle, WA, 19–20 October 2020, pp. 35–46(2020)

9. Cui, Z., Chen, W., He, Y., Chen, Y.: Optimal action extraction for random forests and boosted trees. In: Proceedings of the 21st ACM SIGKDD International Conference on Knowledge Discovery and Data Mining (SIGKDD 2015), Sydney, Australia, 10–13 August 2015, pp. 179–188 (2015)

10. Dosovitskiy, A., Brox, T.: Inverting visual representations with convolutional networks. In: Proceedings of the 2016 IEEE Computer Society Conference Computer Vision and Pattern Recognition (CVPR'16), Las Vegas, NV, 26 June–1 July 2016 (2016)

11. Freitas, A.A.: Comprehensible classification models: a position paper. SIGKDD Explor. **15**(1), 1–10 (2014)

12. Gabidolla, M., Zharmagambetov, A., Carreira-Perpiñán, M.Á.: Improved multiclass adaboost using sparse oblique decision trees (2021), submitted

13. Goodfellow, I.J., Shlens, J., Szegedy, C.: Explaining and harnessing adversarial examples. In: Proceedings of the 3rd International Conference on Learning Representations (ICLR 2015), San Diego, CA, 7–9 May 2015 (2015)

14. Guidotti, R., Monreale, A., Ruggieri, S., Turini, F., Giannotti, F., Pedreschi, D.: A survey of methods for explaining black box models. ACM Comput. Surv. **51**(5), 93 (2018)

15. Gurobi Optimization, LLC: Gurobi optimizer reference manual (2019)

16. Hada, S.S., Carreira-Perpiñán, M.Á.: Sampling the "inverse set" of a neuron: an approach to understanding neural nets. arXiv:1910.04857 (2019)

17. Hada, S.S., Carreira-Perpiñán, M.Á.: Sampling the "inverse set" of a neuron. In: IEEE International Conference on Image Processing (ICIP 2021), Anchorage, AK, 9–12 September 2021 (2021)

18. Hada, S.S., Carreira-Perpiñán, M.Á., Zharmagambetov, A.: Sparse oblique decision trees: a tool to understand and manipulate neural net features. arXiv:2104.02922 (2021)

19. Hada, S.S., Carreira-Perpiñán, M.Á., Zharmagambetov, A.: Understanding and manipulating neural net features using sparse oblique classification trees. In: IEEE International Conference on Image Processing (ICIP 2021), Anchorage, AK, 19–12 September 2021 (2021)

20. Idelbayev, Y., Zharmagambetov, A., Gabidolla, M., Carreira-Perpiñán, M.Á.: Faster neural net inference via forests of sparse oblique decision trees (2021). arXIV

21. Lipton, Z.C.: The mythos of model interpretability. Comm. ACM **81**(10), 36–43 (2018)

22. Mahendran, A., Vedaldi, A.: Visualizing deep convolutional neural networks using natural pre-images. Int. J. Comput. Vision **120**(3), 233–255 (2016)

23. Martens, D., Provost, F.: Explaining data-driven document classifications. MIS Q. **38**(1), 73–99 (2014)

24. Russell, C.: Efficient search for diverse coherent explanations. In: Proceedings of ACM Conference Fairness, Accountability, and Transparency (FAT 2019), Atlanta, GA, 29–31 January 2019, pp. 20–28 (2019)

25. Simonyan, K., Vedaldi, A., Zisserman, A.: Deep inside convolutional networks: visualising image classification models and saliency maps. In: Proceedings of the 2nd International Conference Learning Representations (ICLR 2014), Banff, Canada, 14–16 April 2014 (2014)

26. Szegedy, C., et al.: Intriguing properties of neural networks. In: Proceedings of the 2nd International Conference on Learning Representations (ICLR 2014), Banff, Canada, 14–16 April 2014 (2014)
27. Ustun, B., Spangher, A., Liu, Y.: Actionable recourse in linear classification. In: Proceedings of ACM Conference Fairness, Accountability, and Transparency (FAT 2019), Atlanta, GA, 29–31 January 2019, pp. 10–19 (2019)
28. Wachter, S., Mittelstadt, B., Russell, C.: Counterfactual explanations without opening the black box: automated decisions and the GDPR. Harvard J. Law Technol. **31**(2), 841–887 (2018)
29. Wu, X., Kumar, V. (eds.): The Top Ten Algorithms in Data Mining. Chapman & Hall/CRC Data Mining and Knowledge Discovery Series, CRC Publishers, Boca Raton (2009)
30. Yang, Q., Yin, J., Ling, C.X., Pan, R.: Extracting actionable knowledge from decision trees. IEEE Trans. Knowl. Data Eng. **18**(1), 43–56 (2006)
31. Zeiler, M.D., Fergus, R.: Visualizing and understanding convolutional networks. In: Proceedings of 13th European Conference Computer Vision (ECCV'14), Zürich, Switzerland, 6–12 September 2014, pp. 818–833 (2014)
32. Zhang, C., Liu, C., Zhang, X., Almpanidis, G.: An up-to-date comparison of state-of-the-art classification algorithms. Expert Syst. Appl. **82**, 128–150 (2017)
33. Zharmagambetov, A., Carreira-Perpiñán, M.Á.: Smaller, more accurate regression forests using tree alternating optimization. In: Daumé III, H., Singh, A. (eds.) Proceedings of the 37th International Conference on Machine Learning (ICML 2020), 13–18 July 2020, pp. 11398–11408 (2020)
34. Zharmagambetov, A., Carreira-Perpiñán, M.Á.: Learning a tree of neural nets. In: Proceedings of the IEEE International Conference Acoustics, Speech and Signal Processing (ICASSP'21), , Toronto, Canada, 6–11 June 2021, pp. 3140–3144 (2021)
35. Zharmagambetov, A., Gabidolla, M., Carreira-Perpiñán, M.Á.: Improved boosted regression forests through non-greedy tree optimization. In: International Joint Conference Neural Networks (IJCNN'21), Virtual Event, 18–22 July 2021 (2021)
36. Zharmagambetov, A., Gabidolla, M., Carreira-Perpiñán, M.Á.: Improved multi-class AdaBoost for image classification: the role of tree optimization. In: IEEE International Conference on Image Processing (ICIP 2021), Anchorage, AK, 19–22 September 2021 (2021)
37. Zharmagambetov, A., Gabidolla, M., Carreira-Perpiñán, M.Á.: Softmax tree: an accurate, fast classifier when the number of classes is large (2021), submitted
38. Zharmagambetov, A., Hada, S.S., Gabidolla, M., Carreira-Perpiñán, M.Á.: Non-greedy algorithms for decision tree optimization: an experimental comparison. In: International Joint Conference on Neural Networks (IJCNN'21), Virtual event, 18–22 July 2021 (2021)

Towards Explainable Meta-learning

Katarzyna Woźnica$^{(\boxtimes)}$ (iD) and Przemysław Biecek (iD)

Warsaw University of Technology, Warsaw, Poland
katarzyna.woznica.dokt@pw.edu.pl

Abstract. Meta-learning is a field that aims at discovering how different machine learning algorithms perform on a wide range of predictive tasks. Such knowledge speeds up the hyperparameter tuning or feature engineering. With the use of surrogate models, various aspects of the predictive task such as meta-features, landmarker models, etc., are used to predict expected performance. State-of-the-art approaches focus on searching for the best meta-model but do not explain how these different aspects contribute to its performance. However, to build a new generation of meta-models, we need a deeper understanding of the importance and effect of meta-features on model tunability. This paper proposes techniques developed for eXplainable Artificial Intelligence (XAI) to examine and extract knowledge from black-box surrogate models. To our knowledge, this is the first paper that shows how post-hoc explainability can be used to improve meta-learning.

Keywords: Meta-learning · Explainable artificial intelligence · OpenML

1 Introduction

Meta-learning, or learning to learn, is an area of machine learning focused on extracting and transferring knowledge about tuned models from one predictive problem to another. It may increase the effectiveness of AutoML systems by automating feature selection, hyperparameter tuning, and optimising model performance [39]. By using previously acquired experience, meta-learning allows the narrowing of a domain of promising solutions for a new prediction problem, which results in quicker and more computationally efficient model development.

Meta-learning is an extensive term, which is difficult to define, and is perceived differently depending on the research area [22,39]. The main characteristic of meta-learning is what aspect of the learning process is being leveraged and transferred. In the case of deep learning, a common object of interest is finding initial values of weights for a neural network architecture so that several gradient steps lead to accurate predictions [12,13]. For tabular data, machine learning hyperparameter configuration may be considered as a transferred meta-representation since it often turns out that state-of-the-art performance is achieved by classic models, such as ensembles of tree-based models

© Springer Nature Switzerland AG 2021
M. Kamp et al. (Eds.): ECML PKDD 2021 Workshops, CCIS 1524, pp. 505–520, 2021.
https://doi.org/10.1007/978-3-030-93736-2_38

1. Large-scale performance assessment for OpenML data

2. Assembly of the meta-model

3. XAI analysis of the meta-model

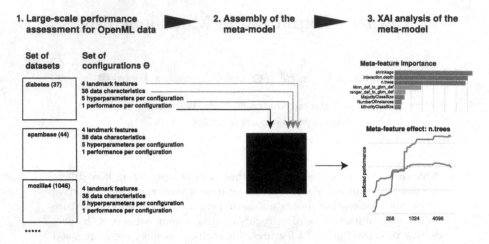

Fig. 1. The proposed process of the explainable meta-model exploration. Firstly, we gather meta-features for selected data sets from the OpenML repository. Then we calculate model performance on these data sets for selected configurations of hyperparameters. Secondly, we assemble a black-box surrogate meta-model. Thirdly, we use XAI techniques to extract information about the relative importance of meta-features and their marginal responses. (Color figure online)

[7,9,15,25]. Then the space of potential configurations has several to a dozen dimensions, so practitioners need resource-effective methods to develop well-performing ML models and need guidelines to make the right choices. In this work, we narrow the scope of meta-learning to a recommendation of hyperparameters settings to initializing machine learning algorithms for tabular data.

So far, researchers have presented diverse approaches to plug-in results from prior experiments into model optimisation for a novel predictive problem. Most of them are based on the principle of interaction between two submodules [24]. The first is a meta-feature extractor, which compresses a data set into a vector of predefined features of predictive problems. For example, this compression may consist of statistical summaries either for a single variable, or for groups of variables, or it may be based on the evaluation of a set of baseline ML algorithms – so-called landmarker models [27]. The second submodule is a meta-learner model that uses meta-features to predict a meta-target. Meta-models for tabular data may assess the most promising hyperparameter configuration [26,36], their ranking [28,33] or the performance prediction for given hyperparameter settings [8,20,34,35]. Very often, a similarity assessment between the training meta-data and independent test meta-data is a component of rules of the meta-models. Distinct methods consolidating measurable properties of data sets and model configurations have proven their benefit in optimisation [11,42,43].

The effectiveness of meta-learning is highly dependent on the type and quality of meta-features [4]. So, it is crucial to consider a broad range of candidates [40]. The primary sources of meta-features that feasibly predict model performance

are data sets' characteristics. The conventional approach is to express the whole
data set as a vector of simple engineered statistics. A comprehensive summary
of these commonly used meta-features is provided in [37,39]. Autoencoders as
meta-feature extractors [10,21] avoid the problem of a priori defining meta-
features, but they are limited to the same schema of meta-data sets. More versatility is offered by an extractor Dataset2Vec based on hierarchical modeling
[24]. In addition to defining features through the internal data structure, we can
also consider how difficult a given prediction problem is for a specific portfolio
of machine learning models. This is a concept of landmarking models introduced
in [32]. It extends the list of meta-features by the relative performance of some
predefined models, so-called landmarkers. As well as a data set's properties, good
meta-models take into account the tunability of particular algorithms concerning selected hyperparameters [34]. For that reason, a part of the meta-feature
space is the detailed model configuration – user-determined hyperparameters.

Another subdomain of machine learning that has been developing intensively
recently is eXplainable Artificial Intelligence (XAI). With the growing demand
for highly engineered models, there is an apparent necessity to investigate black-
box algorithms and identify the most critical aspects affecting model operation. There is a reservoir of XAI methods that address the justifications of a
model's predictions for a single instance as well as data set-level exploration
[3,29]. So far, these exploratory techniques have been applied to the enhancement of the ML model, but this kind of inference amplifies the meta-learning
generalisation ability. The transparency of the algorithm structure helps to identify which explanatory variables are included in the predictive process and how
changes in their distribution affect a model. What is more, the effectiveness
of meta-models greatly depends on the chosen set of meta-targets and on the
approach to modeling the relationship between them and the meta-response.
So, similarly to classic machine learning approaches, it can be a trial-and-error
process that requires several iterations.

This work integrates these two promising directions (see Fig. 1). We show
how to use XAI techniques to extract knowledge and validate meta-models used
in meta-learning. Focus is placed on the specific architecture of meta-models
which enables the application of exploratory techniques. Using them, we extract
essential informative properties of meta-features. We present an example for
meta-learning based on OpenML100 data sets, but the proposed approach can
be applied to any meta-model trained for a particular domain set of problems.
To our knowledge, this is the first paper that combines these two areas of machine
learning.

2   Meta-learning Frameworks

This section briefly summarizes the two most common approaches to predicting
model performance for a given hyperparameter configuration. The first step in
meta-learning is extracting meta-features characterising data sets. According to
the way in which these meta-features are utilized, we may define the following
classes of meta-learning frameworks.

1. **Sequential surrogate meta-models** employ meta-features to assess the similarity between meta-train data sets and new task. Then meta-models use these similarity measures as weights to configuration transfer. A meta-learner does not use meta-features explicitly, they are compressed to similarity measure, so it is more challenging to understand the meta-features impact.

   To propose new hyperparameters that have the potential for good model performance on the new dataset, extensions of Sequential Model Based Optimisation (SMBO) are used. The MI-SMBO procedure [11] uses the best predicted hyperparameters from similar data sets to select the initial points to optimisation for novel data. Two definitions of similarity between meta-features vectors are proposed: the p-norm of the difference between the data sets' meta-features and a metric that reflects how similar the data sets are with respect to the performance of different hyperparameter settings. The meta-features' modification is utilized in the main optimisation step in a surrogate Gaussian Process (GP) for all tasks simultaneously [44]. To measure similarities between instances, they use a squared exponential kernel and the nearest neighbour's kernel. This approach is extended to become more scalable in [41, 43]. They fit separate GP for each task in the meta-data, then aggregate them into one using the Nadaraya Watson kernel.

2. **Black-box surrogate meta-models** directly predict model performance for given hyperparameter settings and for a given data set described with the vector of meta-features. A meta-learner is a regression model, and meta-data has a tabular format – every data set is represented as row with a sequence of meta characteristics [6]. Firstly, this generic approach is proposed in [17]. The authors provide a knn-ranking method to rank candidate models. In [40] the authors also consider this approach in the meta-model framework and suggest the potential source of meta-features. In [35] the authors use the SVM model as a meta-regressor and utilize a broad scope of meta-features: statistical and information theory measures, model-based features, and landmarkers. For every machine learning algorithm, they train the meta-learner independently because of different hyperparameters configurations. Similar meta-features are applied by [8], but the meta-learner is a Multilayer Perceptron. The authors of [34] extend the idea of surrogate benchmarks and train surrogate regression to map hyperparameter configuration to model performance.

Both perspectives on meta-learning show empirically that transferring knowledge from independent tasks is beneficial in terms of yielding from that meta-framework the optimal hyperparameter or more effective warm start point optimisation for a new task. The sequential approach requires the validation of each step separately, and it is more challenging to explore interactions between meta-features and hyperparameter configurations. On the other hand, the black box meta-model is very similar to the classic machine learning approach. It, therefore, offers great opportunities for analogous model exploration methods. Thus, in this paper, we focus on the latter approach. We show how eXplainable Artificial Intelligence methods can be used to audit or extract knowledge from black-box meta-model.

3   The Meta-OpenML100 Surrogate Model

In the next section, we present a universal approach to exploration of the black-box meta-model. Both the choice of data set for which model evaluation and hyperparameters transfer was carried out and the choice of algorithm for the meta-learner should be considered as illustrative. Similarly to previous researches, this meta-model works on classification problems from the OpenML100 benchmark [5]. We use a meta-model engineering methodology analogous to [8,34,35] and employ gradient boosting algorithm as meta-learner. In this section, we describe how this meta-model is built (Fig. 1).

We build the meta-model based on all predictive tasks for binary classification in the OpenML100 suite; see the list of these tasks in Table 1. According to [40] recommendations, for each task, we calculate the following meta-features: four landmarkers' performance for baseline models and 38 statistical properties of the underlying data set. The meta-model is built to predict the performance of the gradient boosting model parametrized with five hyperparameters. Finally, we train a meta-learner, also a gradient boosting model, to predict the performance of a model with selected hyperparameters on a data set with the following meta-properties.

This black-box surrogate model is hereafter called Meta-OpenML100 in this article. In the following subsections, we provide a detailed description of its components.

3.1   Predictive Tasks and Their Meta-features

Out of all tasks in the OpenML100 suite [5] we select only these for binary classification. The suite provides 61 meta-characteristics for these tasks [4], some correspond to properties of continuous variables, some for categorical variables, and some for data sets with mixed variables. Here the limit is to use only data sets with all continuous variables, i.e., 20 data sets listed in Table 1. For these data sets, we use 38 available statistical and information-theoretic properties.

3.2   Landmarkers

We use landmarker-based meta-features to characterise predictive problems. As landmarker models, we consider five machine learning algorithms with default hyperparameter configurations: *generalized linear regression with regularisation, gradient boosting, k nearest neighbours*, two random forest implementations: *randomForest and ranger*. Various models have been proposed as landmarkers [1], but one of the requirements is diversified architectures of algorithm capturing interrelationship of different variables. We balance this diversity with the landmarkers' computational complexity [32].

Table 1. Meta-features for selected data sets from the OpenML repository. Four land-markers (relative performance to default gbm model) and five hyperparameters for gbm model are presented. Data sets' characteristics are omitted for brevity. Only optimal hyperparameters are listed in the last columns for the corresponding data set.

Data set (ID)	Hyperparameters for GBM model					Landmarks			
	Shrink.	Inter. depth	n.trees	Bag fract.	Min. node	KNN	Glmnet	Ranger	Random forest
Diabetes (37)	0.00	4	1480	0.69	7	1.10	2.25	2.36	2.30
Spambase (44)	0.04	5	1414	0.98	16	2.97	4.78	7.57	7.74
Ada_agnostic (1043)	0.05	5	333	0.90	7	2.31	5.41	6.28	5.37
Mozilla4 (1046)	0.09	4	1567	0.54	7	1.71	0.39	2.62	2.84
PC4 (1049)	0.01	5	2367	0.75	12	1.11	2.13	3.46	3.57
PC3 (1050)	0.04	1	949	0.90	11	0.86	1.34	1.85	1.88
KC2 (1063)	0.00	2	273	1.00	21	0.81	0.73	0.90	0.98
KC (1067)	0.00	3	5630	0.26	3	0.23	1.16	1.49	1.66
PC1 (1068)	0.00	4	6058	0.21	14	1.12	0.27	1.92	1.84
Banknote authentication (1462)	0.03	1	8429	0.52	12	6.47	4.99	5.56	6.02
Blood transfusion service center (1464)	0.01	1	394	0.21	10	0.64	1.23	0.86	0.71
Climate model simulation crashes (1467)	0.00	2	654	0.26	15	0.35	1.36	1.12	1.15
EEG-eye-state (1471)	0.08	5	2604	0.28	14	2.48	0.93	3.34	4.16
Hill-valley (1479)	0.08	5	2604	0.28	14	1.78	0.43	2.04	2.24
Madelon (1485)	0.05	5	333	0.90	7	0.34	0.48	1.61	1.58
Ozone-level-8hr (1487)	0.00	3	8868	0.33	11	1.41	3.40	4.43	4.36
Phoneme (1489)	0.02	5	5107	0.60	18	4.89	2.90	7.12	8.13
QSAR-biodeg (1494)	0.05	5	333	0.90	7	4.16	5.34	6.74	6.81
WDBC (1510)	0.04	1	949	0.90	11	1.56	0.63	1.88	1.91
Wilt (1570)	0.00	4	6058	0.21	14	2.73	4.80	7.15	7.28

We apply 20 train/test split methods and compute AUC scores for each split to evaluate their predictive power. These five algorithms are ranked according to methodology in Sect. 3.4. Because we predict the performance of gradient boosting models we compute landmarkers as a ratio of models rankings (knn, glmnet, ranger and randomForest) to a ranking of gbm model with the default configuration. As a result, there are four landmarker meta-features.

3.3   Algorithms and Hyperparameters Space

In this paper we explore the performance of gradient boosting classifiers (*gbm*) to the selection of the following hyperparameters: *n.trees*, *interaction.depth*, *n.minobsinnode*, *shrinkage*, *bag.fraction*. In order to do so, we sample 100 random configurations of hyperparameters in a similar way to [34] to examine their influence gradient boosting model predictive power. Additionally, we add one special configuration – the default settings for the gbm library [19].

3.4 Estimated Predictive Power of Selected Configurations

For every combination of 20 data sets from the OpenML100 suite and the 101 hyperparameter configurations, we try pre-specified 20 train and test data splits. Each model is fitted on each training subset, and afterward, AUC is computed on the test frame. This way, we obtain a meta-data set for the performance of 40400 configuration/data sets/split combinations. Because performance for different data sets takes values in different ranges, we normalized these values using ranks per data set (the higher AUC, the higher position in the ranking). Ratings are scaled to [0,1] intervals. Every configuration for each algorithm appears in the list 20 times because of train-test splits. To aggregate this to one value for every model, we computed the average rating for the model.

3.5 Surrogate Meta-model

As meta-model, we select a gradient boosting algorithm with maximum interaction depth equal to 10 (i.e., a model that may be rich in interactions between meta-features). This kind of algorithm has already been applied inside sequential-model-based optimisation, for example, in SMAC [23]. Tree-based algorithms are particularly well suited to handling high-dimensional and partially categorical input spaces. They are known for robustness and automated feature selection.

This surrogate meta-model configuration is selected as the best one according to the evaluation schema as follows. We apply one-data set-out cross-validation: every model is trained on 19 data sets and then is tested on the remaining data frame for mean square error (MSE), in addition Spearman's correlation between predicted rankings and actual meta-responses was also checked. Given meta-model achieves 0.017 MSE when the constant mean prediction has 0.041 MSE, so the performance of the surrogate meta-model is significantly better than the baseline mean prediction.

The meta-data set with all meta-features and fully reproducible code can be found in this GitHub repository https://github.com/woznicak/MetaFeaturesImpact.

4 Explanatory Analysis of Meta-OpenML100 Model

The black-box meta-model approach, despite its effectiveness, does not provide new knowledge about how data set characteristics translate into optimal choices for hyperparameters. In this chapter, we present the main result of this work – recommendations on how one can use selected XAI techniques [3, 29] for the analysis of the black-box meta-model. Each proposal is complemented with an example for the Meta-OpenML100 meta-model developed in the previous section. The knowledge drawn from the model often looks intuitive, but through the presented analysis, is it possible to get a quantitative validation of our assumptions.

The XAI methods allow the analysis of a single meta-model. Meta-OpenMl100 uses a one-data set-out schema to explore the constructions of each of these meta-models independently or extend this approach and aggregate feature importance across cross-validation meta-models.

4.1   Meta-features Importance

The meta-models are built on various sets of meta-features determined for data sets. For complex meta-models, it is difficult to discover which variables actually contribute to the model output. This investigation is needed to identify presumptive noisy aspects and may be significant in deliberation about excluding these meta-features from new generations meta-models. The solution to this problem is to use model agnostic permutational feature importance, which assesses how perturbations of a specific feature decrease model performance [14].

An example for Meta-OpenMl100 is presented in Fig. 2. We can easily read that most important are hyperparameters followed by two landmarker features (`knn` and `randomForest`) and two meta-features (`NumberOfInstances` and `MinorityClassSize`).

The considered meta-features form three groups because of the different approaches to creating them. Figure 2B presents the assessment of different groups' influence. As we see, the most important class of meta-features is hyperparameters, and this conclusion is consistent with the importance measure for individual variables. Landmarker and data set characteristics have similar dropout values.

4.2   Meta-features Interactions

A good meta-model offers different optimal hyperparameters for different data sets to solve the problem of there being no one optimal hyperparameter across tasks. This means that the model detects interactions between variables, situations in which the values of one attribute (meta-characteristic) affect the effect of another attribute (hyperparameter). In general, the identification of interactions is a complex problem. For analysis of meta-models, we propose to use Friedman's H-statistic [16]. This method decomposes the prediction into components corresponding to two selected features. The variance of the difference between observed values and decomposed ones without interactions is used to assess the strength of interactions. We use the implementation of Friedman's H-statistic from [30].

The example identification of interactions for Meta-OpenMl100 is presented in Fig. 3. Firstly, we study two-way interaction between any two meta-features. The strongest interaction is for `bag.fraction` and `NumberOfFeatures`, i.e. between hyperparameter and statistical meta-feature. Based on such analysis, we can directly identify which meta-characteristics are related to the selection of particular hyperparameters, which is a competitive approach to [11].

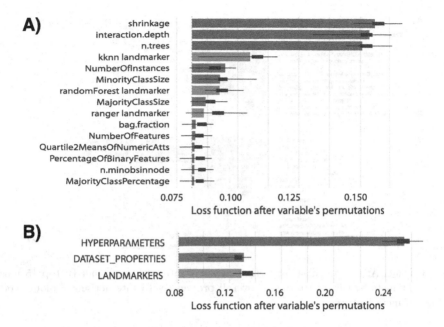

Fig. 2. Panel A: Importance of the 15 top meta-features in GBM meta-model. Panel B: cumulative importance of groups of meta-features.

In Fig. 3B there is the overall assessment of the variable propensity to interact with any other meta-feature. This approach may be an alternative to functional ANOVA from [38]. In this case, the ranking is similar to this at the Fig. 2. The most prone to interact variables are hyperparameters.

4.3 Importance of Correlated Meta-features

Meta-features are often correlated because they describe similar characteristics of data sets. It is not known whether adding more correlated features increases the quality of a meta-model anyway. To test the validity of a group of correlated variables, we propose to use the triplot technique [31]. Groups of meta-variables are determined according to hierarchical clustering based on correlations between meta-features. At each stage, the importance of the cluster of selected meta-variables is calculated.

In Fig. 4, on the right panel, we see that most of the statistical meta-features have marginal contributions independently, and even after clustering, their accumulated importance is substantially less than individual hyperparameter contributions. Landmarkers connected with information about mean values and standard deviation of numeric columns are very close to the importance of hyperparameters independently.

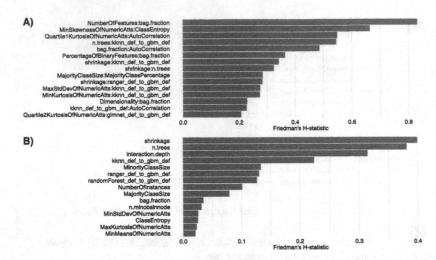

Fig. 3. Panel A: top 15 most important two-way interactions. Panel B: top 15 most important meta-features in terms of overall propensity to interactions. Colours code groups of meta features

4.4 Hyperparameters Informativeness

The majority of meta-learning is concentrated on obtaining the optimal hyperparameter configuration or effective warm-start-points for a novel task. Some of them present the estimated empirical prior distribution of hyperparameters [38]. Alternative approaches to estimating the learning profile for a selected hyperparameter are Ceteris Paribus (CP) profiles, also known as Individual Conditional Expectation profiles [18]. This curve shows how a model's prediction would change if the value of a single exploratory variable changed. In essence, a CP profile shows the dependence of the conditional expectation of the dependent variable (response) on the values of the particular explanatory variable. This is equivalent to a partial dependence plot for an individual instance. In this analysis, we use its R implementation [2].

As we argue, looking at the profile, we would like to point to the optimal hyperparameter value for independent data. Thus, we provide a CP profile for every data set in meta-data. The curve corresponding to a particular data set is extracted from meta-model from the cross-validation schema in which the selected task is the test example. As a result, we present CP profiles for every data set when the prediction curve is independent of that data. This situation is equivalent to the production application of meta-framework optimisation.

Obtained CP analysis for the selected hyperparameters is presented in Fig. 5. To recall, the higher ranking, the better predictive power of the given gradient boosting model. So observing increasing lines in CP profiles indicates the better rating of the considered gbm models.

For `shrinkage`, `interaction.depth` and `n.trees` hyperparameters, three patterns of profiles are detected. We apply hierarchical clustering for profiles,

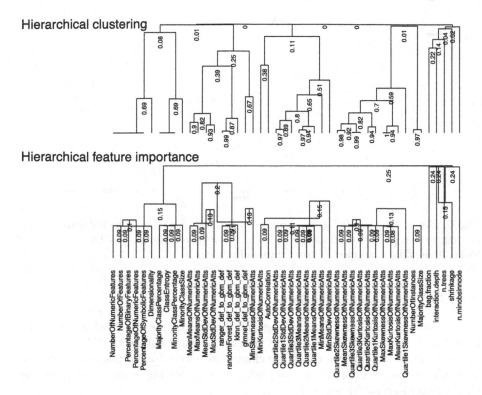

Fig. 4. Triplot joint analysis of feature correlation and importance. The top panel shows hierarchical clustering for meta-features while the bottom one shows the importance of groups of features. Types of meta-features are colour coded. (Color figure online)

and the aggregated profiles for three groups are shown in Fig. 2. These groups are indicated with distinctive colours and termed A, B, and C. It is worth highlighting that the groups indicated A for two hyperparameters may consist of different data sets because the clusterings are performed independently.

For each of these three hyperparameters, group A has an increasing CP profile with a very strong trend for the interior values of variables. Profiles stabilise for high prediction for larger meta-feature values and the maximal value of hyperparameters would be pointed as an optimal warm start point to optimisation. Similar behaviour can be observed in group C. Group B is strictly different for `n.trees` and `shrinkage`: the highest prediction of rankings is obtained for the medium value of hyperparameters, and then predictions decrease. The CP profiles therefore confirm that there is no single value of hyperparameters that is optimal for all data sets, but at the same time suggest several hyperparameter values worth testing because they are optimal for a certain group of prediction problems.

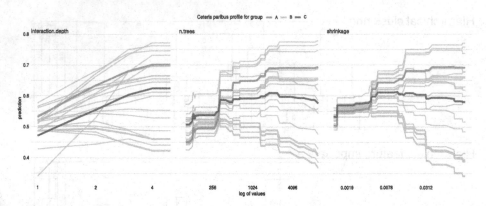

Fig. 5. Ceteris paribus for hyperparameters for test instances. Thick coloured lines are aggregated profiles for data sets clusters. Colours indicate groups. On the x-axis a logarithmic scale is applied. (Color figure online)

4.5   Robustness of Meta-data

For reliable validation, we select the *full* model as one specified meta-model from the Meta-OpenMl100 one-data set-out schema: data set 1471 is the test instance. For this data, we check the change in optimal hyperparameter values. The influences of the remaining data sets are estimated by sequentially deleting them as we describe above.

Figure 6A shows the scatter plot for the considered measures of disturbance caused by removing the effect of the single data set. On the x-axis, there is a distance between the optimal shrinkage hyperparameter; on the y-axis, there is the value of Cook's distance. The biggest perturbations in the final predictions are caused by deleting the data sets 1485 and 37. For data set 37 we also observe the most significant shift of optimal hyperparameter value.

In Fig. 6B are CP profiles for the full meta-model and the selected limited meta-models with diverse values of Cook's distance. These profiles allow the assessment of the significance of the change in the choice of the optimal point. Only for data set 37 we observe the transformation of the CP profile in comparison to the *full* model. This observation shows that the selection of meta-data in this example provides the robust selection of hyperparameter warm-start.

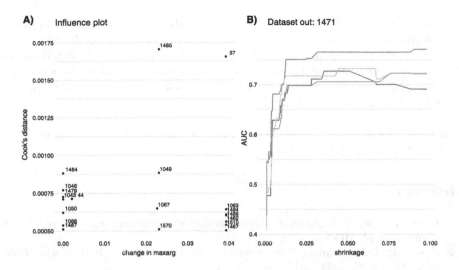

Fig. 6. Panel A: Cook's distance vs. change in estimated optimal `shrinkage`. Numbers stand for the OpenML data sets ids. Panel B: Individual profiles for the most and the least influential data set. Blue line corresponds to the *full* model. (Color figure online)

5   Conclusions

Meta-learning is a promising approach for AutoML solutions. The main contribution of this work is to show the applications of techniques known in the explainable artificial intelligence area in meta-learning. It turns out that many model exploration techniques not only increase knowledge of which meta-features are relevant, but also increase knowledge of the relationship of hyperparameters to model performance. In this work, we have shown how to use XAI techniques to build an effective meta-model, and extract knowledge about the importance of the particular meta-features in the meta-model. Variable importance technique and extension of this, taking into account the correlation structure of a meta-feature, amplify the selection of the most informative set of meta-features. This is crucial in the trial-and-error process of defining meta-space and brings us closer to find a transferable representation of data sets. Examining interaction between data set based meta-features and Individual Conditional Expectation profiles for hyperparameters supports hyperparameter tuning. In further research, this may lead to a significant reduction of the dimension of the searched hyperparameter space and an improvement of automatic model selection processes. What is more, the choice of datasets determines the quality of meta-features and the evaluation of hyperparameters. Validating data sets' informativeness through Cook's distance helps build a robust and reliable repository for meta-learning.

This approach is universal and generic to the explainable analysis of any meta-learning model presented in Fig. 1. The OpenML may be argued that this is not the appropriate illustration for every meta-learning problem - the datasets are relatively small. However, at the moment, there are no publicly available repositories for big data problems. This approach can be reproduced on any repository of data sets from a specific domain or data sets of a selected size or complexity. It is sufficient to store a selected meta-targets, fixed hyperparameters, models performance, and then build a black box meta-model to predict model performance for novel data.

Acknowledgements. The work on this paper is financially supported by the NCN Opus grant 2017/27/B/ST6/01307.

References

1. Balte, A., Pise, N., Kulkarni, P.: Meta-learning with landmarking: a survey. Int. J. Comput. Appl. **105**(8), 47–51 (2014)
2. Biecek, P.: Dalex: explainers for complex predictive models in R. J. Mach. Learn. Res. **19**(1), 3245–3249 (2018)
3. Biecek, P., Burzykowski, T.: Explanatory Model Analysis. Explore, Explain and Examine Predictive Models. CRC Press, Boca Raton (2021). https://pbiecek.github.io/ema/
4. Bilalli, B., Abelló, A., Aluja-Banet, T.: On the predictive power of meta-features in openML. Int. J. Appl. Math. Comput. Sci. **27**(4), 697–712 (2017)
5. Bischl, B., et al.: OpenML benchmarking suites and the openml100. Stat **1050**, 11 (2017)
6. Brazdil, P., Carrier, C.G., Soares, C., Vilalta, R.: Metalearning: Applications to Data Mining. Springer, Heidelberg (2008). https://doi.org/10.1007/978-3-540-73263-1
7. Chen, T., Guestrin, C.: Xgboost: a scalable tree boosting system. In: Proceedings of the 22nd ACM SIGKDD International Conference on Knowledge Discovery and Data Mining, pp. 785–794 (2016)
8. Davis, C., Giraud-Carrier, C.: Annotative experts for hyperparameter selection. In: AutoML Workshop at ICML (2018)
9. Dorogush, A.V., Ershov, V., Gulin, A.: Catboost: gradient boosting with categorical features support. arXiv preprint arXiv:1810.11363 (2018)
10. Edwards, H., Storkey, A.: Towards a neural statistician. arXiv preprint arXiv:1606.02185 (2016)
11. Feurer, M., Springenberg, J.T., Hutter, F.: Initializing bayesian hyperparameter optimization via meta-learning. In: Twenty-Ninth AAAI Conference on Artificial Intelligence (2015)
12. Finn, C., Abbeel, P., Levine, S.: Model-agnostic meta-learning for fast adaptation of deep networks. In: International Conference on Machine Learning, pp. 1126–1135. PMLR (2017)
13. Finn, C., Xu, K., Levine, S.: Probabilistic model-agnostic meta-learning. arXiv preprint arXiv:1806.02817 (2018)
14. Fisher, A., Rudin, C., Dominici, F.: All models are wrong, but many are useful: learning a variable's importance by studying an entire class of prediction models simultaneously. J. Mach. Learn. Res. **20**(177), 1–81 (2019)
15. Friedman, J.H.: Greedy function approximation: a gradient boosting machine. Ann. Stat. **29**, 1189–1232 (2001)

16. Friedman, J.H., Popescu, B.E., et al.: Predictive learning via rule ensembles. Ann. Appl. Stat. **2**(3), 916–954 (2008)
17. Giraud-Carrier, C., et al.: A meta-learning assistant for providing user support in data mining and machine learning (1999–2001)
18. Goldstein, A., Kapelner, A., Bleich, J., Pitkin, E.: Peeking inside the black box: visualizing statistical learning with plots of individual conditional expectation. J. Comput. Graph. Stat. **24**(1), 44–65 (2015)
19. Greenwell, B., Boehmke, B., Cunningham, J., Developers, G.: GBM: generalized boosted regression models (2020). https://CRAN.R-project.org/package=gbm, r package version 2.1.8
20. Guerra, S.B., Prudêncio, R.B.C., Ludermir, T.B.: Predicting the performance of learning algorithms using support vector machines as meta-regressors. In: Kůrková, V., Neruda, R., Koutník, J. (eds.) ICANN 2008. LNCS, vol. 5163, pp. 523–532. Springer, Heidelberg (2008). https://doi.org/10.1007/978-3-540-87536-9_54
21. Hewitt, L.B., Nye, M.I., Gane, A., Jaakkola, T., Tenenbaum, J.B.: The variational homoencoder: learning to learn high capacity generative models from few examples. arXiv preprint arXiv:1807.08919 (2018)
22. Hospedales, T., Antoniou, A., Micaelli, P., Storkey, A.: Meta-learning in neural networks: a survey. arXiv preprint arXiv:2004.05439 (2020)
23. Hutter, F., Hoos, H.H., Leyton-Brown, K.: Sequential model-based optimization for general algorithm configuration. In: Coello, C.A.C. (ed.) LION 2011. LNCS, vol. 6683, pp. 507–523. Springer, Heidelberg (2011). https://doi.org/10.1007/978-3-642-25566-3_40
24. Jomaa, H.S., Schmidt-Thieme, L., Grabocka, J.: Dataset2vec: learning dataset meta-features. Data Min. Knowl. Disc. **35**(3), 964–985 (2021)
25. Ke, G., et al.: Lightgbm: a highly efficient gradient boosting decision tree. Adv. Neural Inf. Process. Syst. **30**, 3146–3154 (2017)
26. King, R.D., Feng, C., Sutherland, A.: Statlog: comparison of classification algorithms on large real-world problems. Appl. Artif. Intell. Int. J. **9**(3), 289–333 (1995)
27. Leite, R., Brazdil, P., Vanschoren, J.: Selecting Classification Algorithms with Active Testing on Similar Datasets. Technical report (2012)
28. Lorena, A.C., Maciel, A.I., de Miranda, P.B., Costa, I.G., Prudêncio, R.B.: Data complexity meta-features for regression problems. Mach. Learn. **107**(1), 209–246 (2018)
29. Molnar, C.: Interpretable Machine Learning (2019). https://christophm.github.io/interpretable-ml-book/
30. Molnar, C., Casalicchio, G., Bischl, B.: IML: an R package for interpretable machine learning. J. Open Source Softw. **3**(26), 786 (2018)
31. Pekala, K., Woznica, K., Biecek, P.: Triplot: model agnostic measures and visualisations for variable importance in predictive models that take into account the hierarchical correlation structure. arXiv preprint arXiv:2104.03403 (2021)
32. Pfahringer, B., Bensusan, H., Giraud-Carrier, C.G.: Meta-learning by landmarking various learning algorithms. In: ICML, pp. 743–750 (2000)
33. Pinto, F., Cerqueira, V., Soares, C., Mendes-Moreira, J.: autobagging: learning to rank bagging workflows with metalearning. arXiv preprint arXiv:1706.09367 (2017)
34. Probst, P., Boulesteix, A.L., Bischl, B.: Tunability: importance of hyperparameters of machine learning algorithms. J. Mach. Learn. Res. **20**(53), 1–32 (2019)
35. Reif, M., Shafait, F., Goldstein, M., Breuel, T., Dengel, A.: Automatic classifier selection for non-experts. Pattern Anal. Appl. **17**(1), 83–96 (2014)
36. Rendell, L., Cho, H.: Empirical learning as a function of concept character. Mach. Learn. **5**(3), 267–298 (1990)

37. Rivolli, A., Garcia, L.P.F, Soares, C., Vanschoren, J.: Towards Reproducible Empirical Research in Meta-learning. Technical report (2018). https://CRAN.R-project.org/package=mfe
38. Van Rijn, J.N., Hutter, F.: Hyperparameter importance across datasets. In: Proceedings of the 24th ACM SIGKDD International Conference on Knowledge Discovery & Data Mining, pp. 2367–2376 (2018)
39. Vanschoren, J.: Meta-learning. In: Hutter, F., Kotthoff, L., Vanschoren, J. (eds.) Automated Machine Learning. TSSCML, pp. 35–61. Springer, Cham (2019). https://doi.org/10.1007/978-3-030-05318-5_2
40. Vilalta, R., Giraud-Carrier, C.G., Brazdil, P., Soares, C.: Using meta-learning to support data mining. Int. J. Comput. Sci. Appl. $1(1)$, 31–45 (2004)
41. Wistuba, M., Schilling, N., Schmidt-Thieme, L.: Learning hyperparameter optimization initializations. In: 2015 IEEE International Conference on Data Science and Advanced Analytics (DSAA), pp. 1–10. IEEE (2015)
42. Wistuba, M., Schilling, N., Schmidt-Thieme, L.: Sequential model-free hyperparameter tuning. In: Proceedings - IEEE International Conference on Data Mining, ICDM, vol. 2016-January, pp. 1033–1038. Institute of Electrical and Electronics Engineers Inc. (2016). https://doi.org/10.1109/ICDM.2015.20
43. Wistuba, M., Schilling, N., Schmidt-Thieme, L.: Scalable gaussian process-based transfer surrogates for hyperparameter optimization. Mach. Learn. $107(1)$, 43–78 (2018)
44. Yogatama, D., Mann, G.: Efficient transfer learning method for automatic hyperparameter tuning. In: Artificial Intelligence and Statistics, pp. 1077–1085 (2014)

How to Choose an Explainability Method? Towards a Methodical Implementation of XAI in Practice

Tom Vermeire[1], Thibault Laugel[2(✉)], Xavier Renard[2], David Martens[1], and Marcin Detyniecki[2,3,4]

[1] University of Antwerp, Prinsstraat 13, 2000 Antwerp, Belgium
`tom.vermeire@uantwerp.be`
[2] AXA, Paris, France
{`thibault.laugel,xavier.renard`}`@axa.com`
[3] Sorbonne Université, CNRS, LIP6, 75005 Paris, France
[4] Polish Academy of Science, IBS PAN, Warsaw, Poland

Abstract. Explainability is becoming an important requirement for organizations that make use of automated decision-making due to regulatory initiatives and a shift in public awareness. Various and significantly different algorithmic methods to provide this explainability have been introduced in the field, but the existing literature in the machine learning community has paid little attention to the stakeholder whose needs are rather studied in the human-computer interface community. Therefore, organizations that want or need to provide this explainability are confronted with the selection of an appropriate method for their use case. In this paper, we argue there is a need for a methodology to bridge the gap between stakeholder needs and explanation methods. We present our ongoing work on creating this methodology to help data scientists in the process of providing explainability to stakeholders. In particular, our contributions include documents used to characterize XAI methods and user requirements (shown in Appendix), which our methodology builds upon.

Keywords: Explainable artificial intelligence · Interpretable machine learning · Stakeholder needs · Methodology

Businesses are increasingly turning to machine learning systems to automate and enhance their operations and decision-making. By making use of complex modeling techniques, they are able to create models with high and sometimes super-human predictive performance. However, given their complexity, these models are often used as black-boxes for which it is unclear how predictions are made.

As a consequence, the business environment is and will increasingly be confronted with demands regarding explainability for various reasons. On the one hand, companies face internal challenges with the adoption of automated decision

© Springer Nature Switzerland AG 2021
M. Kamp et al. (Eds.): ECML PKDD 2021 Workshops, CCIS 1524, pp. 521–533, 2021.
https://doi.org/10.1007/978-3-030-93736-2_39

making systems due to limited explainability (trust in model, business insights, internal control) [2,13]. Moreover, the question of the impact of ML/AI models on business processes and the society is more salient than ever: external players will oblige companies to put explainability on the agenda as a priority.

Hence, data scientists are and will be increasingly confronted with explainability demands from stakeholders beyond the traditional focus on predictive performance. This need of explainability ignited a whole new research field, often referred to as explainable artificial intelligence (XAI) or interpretable machine learning (IML). Over the past years, a multitude of methods has been proposed to explain machine learning models and decisions thereof [1,2,6,11]. These methods significantly differ from each other in terms of output (explanation) and the way these explanations are generated and can therefore be more or less suitable for a specific use case and/or stakeholder. There is thus a risk to fail to achieve a *transparent* use of interpretability, which would be in contradiction with the initial objective of enlightening and empowering model's stakeholders. This issue has been noticed in previous works [3,9], but a concrete methodology is lacking in the existing body of literature. The goal of this paper is to make an argument for the need of prescriptive guidance to match stakeholder needs with an appropriate explanation method. Moreover, we provide an overview of our ongoing work to introduce a methodology which is a first step towards bridging this gap.

This paper is divided in two parts. The first one is devoted to the state of the art on implementing explainability in practice, from a user perspective and from a machine learning perspective. We highlight the absence of thorough and sufficient cross-referencing between these two fields of research and in particular a clear methodology for a machine learning practitioner. The second section proposes a reflection and a sketch of such a methodology for a sound scientific implementation of explainability in practice.

1   Implementing Explainability: Current State of the Literature

The question of practically using machine learning explainability has been poorly covered in the existing literature. One notable exception is Bhatt et al. [3], which study the use of explanation methods in practice and thus show that, currently, they are mainly used by machine learning engineers in an *ad hoc* way as sanity checks for the models they build and deploy. One reason identified by the authors is that organizations lack frameworks for making decisions regarding explainability, leaving these methods only understandable by people with a background in machine learning and obscure to others. This lack of focus on the perspective of the stakeholders to whom explainability is provided has also been observed by Langer et al. [9], who argue that more extensive stakeholder involvement is needed for the development and selection of explanation methods. They provide a conceptual model of the problem space in which they make a distinction between the explainability needs from stakeholders (so-called desiderata) and the explanation method that is used to address them. In their model, the success of an explanation method depends on the extent to which this method satisfies these needs and the careful consideration of these needs is therefore crucial.

However, as the rather user/business-oriented explainability needs and the rather machine learning-oriented properties of explanation methods are completely different in nature, the mapping between the two is not straightforward, leaving the question of the use of XAI in practice unanswered.

Below, we give an overview of relevant work regarding the implementation of interpretability, which can thus be decomposed into two objectives: the collection of explainability needs from stakeholders and the selection of an appropriate explanation method. We conclude the section with a discussion on the lack of methodological guidance provided by the literature.

1.1 Understanding Stakeholder Needs

Typical categorizations of stakeholders are based on their role in an organization [3,9,12,16], their machine learning experience [18] or a combination of the two [15]. Also for the categorization of the stakeholder needs regarding explainability, different propositions are made in the literature. Some authors mention possible high-level goals of explainability, such as model debugging, monitoring etc. [3] or revealing (un)known (un)knowns [12]. Langer et al. [9] provide a list of more detailed needs such as privacy, fairness, legal compliance, etc. Suresh et al. [15] propose an hierarchical framework of so-called tasks (e.g., understand influence of features on output), objectives (e.g., model debugging and compliance) and goals (e.g., trust and understanding) that contribute to each other [15].

In order to gather the needs from a stakeholder in a specific use case, different methods are introduced. Most of them take an approach stemming from information systems and software development research, where the collection of user needs is a well-known and well-studied problem. Köhl et al. [8] provide an analytical framework to elicitate and specify explainability needs and consider them as non-functional requirements which should be satisficed instead of satisfied. These can be translated into Softgoal Interdependency Graphs to represent their relationships. Other scholars claim that data scientists might have difficulties to directly discuss XAI solutions with stakeholders and, therefore, propose to rather focus on what a stakeholder wants to achieve with explainability [4,17]. In essence, they try to get a good understanding of the stakeholder's background, capabilities and goals. In this way, they create textual and/or visual scenarios that describe where and when explainability is needed. Another approach is taken by Liao et al. [10], who assume that an explanation can be seen as an answer to one or more questions of the stakeholder. They provide a set of possible questions for which the stakeholder might require answers in the form of an explanation. Eiband et al. [5] propose a process consisting of different (iterative) phases which incorporates the position of different stakeholders and in which the first phases focus on what must be explained and the later phases look into how this can be explained.

1.2   Specifying Explanation Method Properties

On the other side of assessing stakeholder needs, different scholars propose a set of general characteristics or requirements to describe an explanation method. These allow to create a generalized identity card for explanation methods which can serve as documentation and a means to compare them. Both Hall et al. [7] and Sokol and Flach. [14] introduced a framework of explanation method characteristics that can be filled in for a specific explanation method.

The framework of Hall et al. [7] consists of characteristics that are divided over the dimensions effectiveness, versatility, constraints, explanator types and categories, explanation properties and personal considerations. Sokol and Flach [14] provide a framework with five dimensions of so-called requirements of an explanation method. The functional requirements consists of requirements that determine whether it is practically feasible to use an explanation method for a specific use case. The operational requirements relates to the interaction of users with the system and their expectations. The usability requirements consists of properties that are of importance for the receiver of explanations. Finally, the safety and validation requirements respectively focus on aspects as privacy and security, and the validation of the explanation method.

Without going into further detail about the individual explanation method properties in both frameworks, it can be argued that these dimensions not only cover generic properties of explanation methods but also (possible) stakeholder needs (e.g., constraints and personal considerations in Hall et al. [7], and operational and usability requirements in Sokol and Flach [14]). Compared to the stakeholder needs categorizations we mentioned above, the proposed templates do not give an exhaustive overview of all possible (future) needs, so certain of them might not be assessed and/or addressed. Moreover, describing how an explanation method addresses a specific need in a general way might limit practical usability of the framework. For instance, privacy is a broad concept that might imply different needs in different use cases. It can be about disclosure of the model, certain instances, certain features, etc. Similarly, the complexity of an explanation is a concept that can (simultaneously) cover different aspects. It can be impacted by the size of an explanation, its format, the type of features used in it, etc. Filling in this type of characteristics for a specific approach might lead to a description that is too general to confront with a specific use case. Another option is to fill in the template completely tailored to the use case at hand, which limits reusability and therefore to redo the exercise in each use case.

1.3   There is a Lack of Methodological Guidance

Given the explainability needs of a stakeholder and the properties of an explanation method, a logical next step is to assess whether the properties are capable of satisfying the needs. However, this is a cumbersome task, since there is no clear mapping between the properties and the needs. First, the explainability needs of a stakeholder and the explanation method properties are from a completely different nature. The former come from the domain of the (business) stakeholder and focus on information needs and domain constraints such as privacy and complexity. The latter rather relate to the XAI domain and focus more on the technical (algorithmic) details of an explanation method. As both sides use their respective jargon, a translation in between is considered necessary. Second, there might be a many-to-many relationship between needs and properties. A single property of an explanation method might contribute to multiple stakeholder needs, while the fulfillment of a single stakeholder need can be impacted by multiple and seemingly unrelated explanation method properties.

To our knowledge, the XAI literature lacks prescriptive work on how to perform the mapping between explainability needs and explanation method properties. Approaches on gathering explainability needs typically ignore the properties of explanation methods or only discuss it to a limited extent (e.g., Liao et al. [10] map their questions to a high-level taxonomy of explanation methods). More general frameworks to select explanation methods do mention the fact that a translation or mapping between needs and properties is needed, but provide no concrete guidance on how this can be done. For instance, Hall et al. [7] only mention the mapping between explanation methods and stakeholder needs as a step in their methodology. Langer et al. [9] provide some hypothetical scenarios to inspire future research on the mapping. It can be argued that creating a generalized identity card for an explanation method is a useful tool for data scientists to select an appropriate approach for a use case. However, we believe that this identity card should consist of characteristics that are closely related to the details of the explanation generation process, irrespective of potential needs that might be satisfied by them.

In this work, we make a first step towards bridging this gap by proposing a methodology to translate explainability needs from stakeholders to the lower-level properties of explanation methods. This methodology aims at guiding data scientists in the process of providing explainability to business stakeholders.

2 Proposed Methodology

The goal of this work is to provide a (first step towards a) methodology that can guide a data scientist in tackling an explainability use case. We revisit both the explainability needs and explanation method properties, in order to get them at a level of detail that allows to match them. Based on existing work, we define a template for them that can be used to document both aspects. The current versions of the templates are included in the appendix of this paper. Moreover, we plan to provide guidance on how the templates can be completed and how they can be used to find an appropriate explanation method for an explainability use case. An overview is shown in Fig. 1.

2.1 Explanation Method Properties

Using Hall et al. [7] and Sokol et al. [14] as a starting point, we consider it valuable to create an identity card (later referred to as "ID card") to describe an

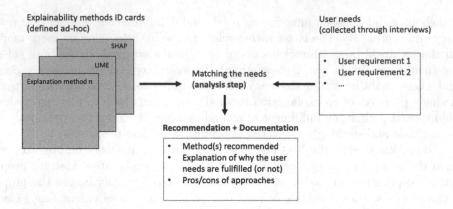

Fig. 1. Illustration of the proposed methodology and its different components.

individual explanation method in detail. However, we make a clear distinction with (and therefore not include) the explainability needs that might be satisfied by a method, thus differentiating it from currently existing versions [7,14]. The current version of our identity card, which is included in Appendix A for the sake of clarity, consists of properties that are organized into the four dimensions discussed below.

Compatibility Properties. This set of properties allows to determine whether it is practically feasible to apply an explanation method to a certain use case. It covers aspects such as the type of the task, the scope of an explanation, the model and the training data.

Explanation Properties. This set of properties describes the explanation that is produced by an explanation method. This comprises the information that is provided, the language and the relationship with the model and the real world.

Method Usage. This section describes how an explanation method can be used in practical terms. It discusses whether code is available and how it can be used.

Process Properties. This set of properties describe the process an explanation method uses to produce explanations. It covers aspects that are not necessarily visible from an explanation itself, but that can play an important role in which final explanation is obtained.

2.2 Explainability Needs

We also propose a card that gives a detailed overview of stakeholder needs. To construct this document, we started from the existing literature discussed in Sect. 1. Subsequently, we selected needs that were considered relevant and detailed enough to work with throughout a brainstorm between the authors of this paper. For these needs, we created a questionnaire that can be used to reveal the needs from a stakeholder. The resulting needs card and questionnaire were further refined by applying them to actual use cases and by integrating the feedback. The current version of the needs card, which is included in Appendix B, is organized in the four dimensions discussed below. The corresponding questionnaire is outlined in Appendix C.

Use Case Context. This dimension gives an overview of the use case for which explainability is required. It contains the business process in which the ML system fits and a detailed description of the ML system itself.

Stakeholder. In order to provide explainability to a stakeholder, it is necessary to have a good understanding of his role and his background, both regarding the business domain and machine learning in general.

Stakeholder Needs. This dimension bundles the aspects that express what information a stakeholder wants to derive from the explanation. In traditional information systems terms, this dimension considers the functional requirements of the explanation.

Stakeholder Constraints. This dimension bundles aspects that put restrictions on the explanation (process) that are relevant to the stakeholder. In traditional information systems term, these can be called the non-functional requirements of an explanation.

2.3 Information Collection

The completion of the template for the explanation method properties is considered a task the research community that is in charge of providing explanation methods. Existing and future explanation methods should be subjected to a thorough investigation in order to assess their properties. To this end, experts can rely on existing documentation (research papers, reports, code base etc.) and experience by experimenting with the method. We acknowledge this is a cumbersome task, but we believe that this documentation, if properly composed, will form a valuable and comparable knowledge base that can be reused for future use cases.

To gather the information on stakeholder needs for a specific use case, the XAI expert should probe the relevant stakeholder. Since certain aspects might be more technical in nature, it is considered necessary to also involve the people that are responsible for the development and deployment of the model that is used in the use case (if they are not the stakeholder themselves). Our questionnaire can be used as a tool to guide this process, but also more approaches that are more tailored to the specific stakeholder might be explored (see the literature we discussed in Sect. 1).

2.4 Matching Stakeholder Needs with Explanation Method Properties

After the collection and documentation of the stakeholder needs, the next step is to confront them with the knowledge base of detailed explanation method properties in order to select the (most) appropriate explanation method(s). As argued in Sect. 1, their is no clear relation between stakeholder needs and corresponding explanation method properties that can be generalized over use cases. However, the separation of the two sides makes their different nature explicit and aids to prevent the oversimplification of the problem. This should encourage the XAI expert to perform a thorough analysis and resulting mapping for each use case. While acknowledging that the current specifications of the stakeholder needs and explanation method properties are not final and should be further revised and refined, we believe they provide the XAI expert with useful tools in that process.

We advocate to document in detail the analysis and the resulting choice of explanation method(s). This encourages the XAI expert to show to what extent the stakeholder needs are addressed and to point at disadvantages or risks that are related to using an specific explanation method. This should result in a substantiated and nuanced advice for a specific use case.

3   Ongoing and Future Research

In this paper, we discussed our on-going work of designing a methodology to help XAI stakeholders choosing an explainability method. Besides identifying and discussing the issue and the current state of the art, our contributions include a prototype of ID cards and questionnaire to identify user needs (both provided in Appendix).

The following step in this research project is to apply this methodology in practice. First, we plan on enriching the current knowledge base of ID cards with the most common explanation methods and their properties. Alongside, we are currently conducting interviews with stakeholders working on specific real-life use cases requiring interpretability. These semi-structured interviews allow us to validate and refine the current stakeholder needs specification and the corresponding questionnaire, which is still an ongoing process. By considering use cases from diverse domains and involving different (types of) stakeholders (e.g. data scientists, non-expert customers, auditors...), we plan on covering as many different needs as possible. In particular, we also pay close attention to the extent to which stakeholders generally can or cannot answer questions related to a specific need.

The next step is to perform the analysis to match the stakeholder needs with an appropriate explanation method. We will do so for each use case collected in the interviews and present a (set of) explanation method(s) that can be considered a suitable option. Subsequently, the proposed solution(s) will be discussed with and demonstrated to the stakeholder as a validation step. This will allow us to assess the effectiveness and discover the limitations of our methodology. By submitting this paper to the XKDD workshop, we hope to have interesting discussions with leading XAI researchers to challenge our in-the-making methodology and thus help solve what we believe to be a fundamental issue of the field today.

A   Explanation Method Properties (ID Cards)

A.1   Compatibility Properties

- Problem supervision level: unsupervised, supervised, reinforcement
- Problem type: classification, regression, clustering, etc.
- Explanation target: data, model, predictions
- Explanation scope: instance-level or global
- Applicable model class: model-agnostic or model-specific
- Ante-hoc (by design) of post-hoc (a posteriori)
- Compatible data types: tabular, image, text, etc.
- Compatible feature types: numerical, ordinal, categorical
- Training data-agnostic or training data access needed

A.2   Explanation Properties

- Explanation family: association between antecedent and consequent, contrasts and differences, causal mechanisms
- Explanatory medium: representation of explanation
- Explanation domain: original feature domain or transformed domain

A.3   Method Usage

- Code availability: source code
- Inputs: detailed description of arguments
- Outputs: detailed description of return values

A.4   Process Properties

- Computational complexity: big-O notation or based on empirical evaluation
- Fidelity: extent to which underlying model is closely mimicked
- Optimality: exact or heuristic explanation generation
- Sparsity: size of explanation
- Coverage: guarantee of explanation
- Plausibility: extent to which explanation is within data manifold
- Consistency: extent to which similar instances get similar explanations
- Stability: extent to which explanation generation is subject to randomness
- Interactiveness: extent to which end user can interact with explanation generation

B   Explainability Needs

B.1   Use Case Context

- Business process and role of ML system
- Model inputs (data and features)
- Model output
- Model type

B.2   Stakeholder

- Role in the organization and the business process
- Background regarding machine learning in general and regarding the application domain

B.3   Stakeholder Needs

- Information needs: information that must be provided by explanation to answer questions from stakeholder
- Language of the explanation
  - Format type: textual/visual/...
  - Feature specificity: fine-grained features vs. higher-level features
  - All model features vs. subset of features
- Truthfulness to model: extent to which an approximation of the model is allowed.
- Faithfulness to real world
  - If actionability: feasibility of actions
  - If case: realisticness of case (in data distribution)

B.4   Shared Stakeholder Constraints

- Timeliness: time frame in which explanation must be provided
- Explanation complexity: degree of difficulty that can be processed by stakeholder within given time frame
- Privacy/Intellectual property: disclosure of instances, features or model reasoning
- Robustness: extent to which explanations can differ between instances, models and over time

C   Questionnaire to Reveal Needs

C.1   Use Case Context

- What is the business process and where does the ML system fit?
- What are the model inputs?
- What are the model outputs?
- What type of model is used?

C.2   Stakeholder

– Can you describe the stakeholder and his role in the business process?
– Background:
  • Do the individual model features make sense to the stakeholder?
  • What is the ML background of the stakeholder?

C.3   Stakeholder Needs

– Information needs
  • Does stakeholder want general insight in the system or insight for a specific case?
  • What does stakeholder want to achieve/do with the explanation?
  • If global:
    * Does stakeholder want to know which features are most/least important?
    * Does stakeholder want to have insight in the general logic of the system?
    * Does stakeholder want to have (a-)typical instances?
  • If local:
    * Does stakeholder want to know the logic of the model regarding this decision?
    * Does stakeholder want to know which features contributed the most/least?
    * Does stakeholder want information on how to change the decision (actionability)?
    * Does stakeholder want to know how changes to the instance affect the decision?
    * Does stakeholder want to see comparable or opposite cases?
– Language of the explanation
  • Does stakeholder prefer a textual explanation, visual explanation, rule etc.?
  • Does stakeholder want an explanation in terms of the individual features, or is an higher abstraction level considered sufficient or desirable?
  • Is it allowed to only consider a subset of understandable features and ignore other features that might also have impacted decision-making?
– Truthfulness to model
  • Is using an inherently transparent model for this use case an option?
  • Is it allowed to make an approximation, meaning that the explanation might not be completely truthful to the model?
– Faithfulness to real world
  • In case stakeholder wants actionability: Do the guidelines need to be actionable/realistic for the stakeholder?
  • In case stakeholder wants case or counterfactual: Should instance be close to data distribution

C.4   Stakeholder Constraints

- Timeliness
  - Does stakeholder always need an explanation?
  - If not, how quickly does stakeholder want an explanation after requesting one?
- Explanation complexity
  - How large can an explanation be?
  - How much time does stakeholder have to process the explanation?
- Privacy/IP
  - Are there model features that cannot be shown?
  - Can (a part of) the model logic be disclosed?
  - Can other data instances be shown?
- Robustness
  - Should similar instances have similar explanations?
  - Should explanation for certain instance or model be the same if model changes over time?

References

1. Adadi, A., Berrada, M.: Peeking inside the black-box: a survey on explainable artificial intelligence (xai). IEEE Access **6**, 52138–52160 (2018)
2. Arrieta, A.B., et al.: Explainable artificial intelligence (XAI): concepts, taxonomies, opportunities and challenges toward responsible ai. Inf. Fusion **58**, 82–115 (2020)
3. Bhatt, U., et al.: Explainable machine learning in deployment. arXiv:1909.06342 [cs, stat]. http://arxiv.org/abs/1909.06342, arXiv: 1909.06342 (2020)
4. Cirqueira, D., Nedbal, D., Helfert, M., Bezbradica, M.: Scenario-based requirements elicitation for user-centric explainable AI. In: Holzinger, A., Kieseberg, P., Tjoa, A.M., Weippl, E. (eds.) CD-MAKE 2020. LNCS, vol. 12279, pp. 321–341. Springer, Cham (2020). https://doi.org/10.1007/978-3-030-57321-8_18
5. Eiband, M., Schneider, H., Bilandzic, M., Fazekas-Con, J., Haug, M., Hussmann, H.: Bringing transparency design into practice. In: 23rd International Conference on Intelligent User Interfaces, pp. 211–223 (2018)
6. Guidotti, R., Monreale, A., Ruggieri, S., Turini, F., Giannotti, F., Pedreschi, D.: A survey of methods for explaining black box models. ACM Comput. Surv. (CSUR) **51**(5), 1–42 (2018)
7. Hall, M., et al.: A systematic method to understand requirements for explainable ai (xai) systems. In: Proceedings of the IJCAI Workshop on eXplainable Artificial Intelligence (XAI 2019), Macau, China (2019)
8. Köhl, M.A., Baum, K., Langer, M., Oster, D., Speith, T., Bohlender, D.: Explainability as a non-functional requirement. In: 2019 IEEE 27th International Requirements Engineering Conference (RE), pp. 363–368. IEEE (2019)
9. Langer, M., et al.: What do we want from explainable artificial intelligence (XAI)?-a stakeholder perspective on XAI and a conceptual model guiding interdisciplinary xai research. Artif. Intell. **296**, 103473 (2021)
10. Liao, Q.V., Gruen, D., Miller, S.: Questioning the AI: informing design practices for explainable AI user experiences. In: Proceedings of the 2020 CHI Conference on Human Factors in Computing Systems, pp. 1–15 (2020)

11. Molnar, C.: Interpretable machine learning (2020). https://www.lulu.com/
12. Preece, A., Harborne, D., Braines, D., Tomsett, R., Chakraborty, S.: Stakeholders in explainable AI. arXiv preprint arXiv:1810.00184 (2018)
13. Samek, W., Müller, K.-R.: Towards explainable artificial intelligence. In: Samek, W., Montavon, G., Vedaldi, A., Hansen, L.K., Müller, K.-R. (eds.) Explainable AI: Interpreting, Explaining and Visualizing Deep Learning. LNCS (LNAI), vol. 11700, pp. 5–22. Springer, Cham (2019). https://doi.org/10.1007/978-3-030-28954-6_1
14. Sokol, K., Flach, P.: Explainability fact sheets: a framework for systematic assessment of explainable approaches. In: Proceedings of the 2020 Conference on Fairness, Accountability, and Transparency, pp. 56–67 (2020)
15. Suresh, H., Gomez, S.R., Nam, K.K., Satyanarayan, A.: Beyond expertise and roles: a framework to characterize the stakeholders of interpretable machine learning and their needs. In: Proceedings of the 2021 CHI Conference on Human Factors in Computing Systems, pp. 1–16 (2021)
16. Tomsett, R., Braines, D., Harborne, D., Preece, A., Chakraborty, S.: Interpretable to whom? a role-based model for analyzing interpretable machine learning systems. arXiv preprint arXiv:1806.07552 (2018)
17. Wolf, C.T.: Explainability scenarios: towards scenario-based XAI design. In: Proceedings of the 24th International Conference on Intelligent User Interfaces, pp. 252–257 (2019)
18. Yu, R., Shi, L.: A user-based taxonomy for deep learning visualization. Visual Inf. 2(3), 147–154 (2018)

Using Explainable Boosting Machines (EBMs) to Detect Common Flaws in Data

Zhi Chen[1,2]([✉]), Sarah Tan[3], Harsha Nori[1], Kori Inkpen[1], Yin Lou[4], and Rich Caruana[1]([✉])

[1] Microsoft, Redmond, WA, USA
rcaruana@microsoft.com
[2] Duke University, Durham, NC, USA
zhi.chen1@duke.edu
[3] Cornell University, Ithaca, NY, USA
[4] Ant Group, Sunnyvale, CA, USA

Abstract. Every dataset is flawed, often in surprising ways that data scientists might not anticipate. However, popular machine learning methods are mostly black-boxes. Due to their lack of interpretability, they might learn defective knowledge from these datasets, which can be difficult to detect. In this work, we show how interpretable machine learning methods such as EBMs can help users detect problems that are lurking in their data. Specifically, we provide a number of case studies, where EBM discovers various types of common dataset flaws, including missing values, confounding and treatment effects, data drift, bias and fairness, and outliers. In each case study, we analyze the flaws using visualization of EBM shape functions combined with domain knowledge. We also demonstrate that in some cases interpretable learning methods such as EBMs provide simple tools for correcting problems when correcting the data is difficult.

Keywords: Interpretability · Generalized additive model · Debugging datasets · Model editing · Missing values · Treatment effects · Fairness

1 Introduction

Data is the center of the machine learning pipeline. As machine learning models are usually trained and evaluated on static datasets, they are encouraged to learn every detail in the dataset. However, the data collection process can never be perfect, resulting in pervasive flaws in almost all datasets. These flaws range from simple problems such as missing values [1], and data drift [8], to serious societal bias that could cause damages to the public [18]. In this case, machine learning models trained and tested on contaminated datasets may mistakenly learn the defective information.

Most widely deployed machine learning models, such as deep neural networks, gradient boosted trees, nonlinear SVMs, are all black-boxes whose prediction

© Springer Nature Switzerland AG 2021
M. Kamp et al. (Eds.): ECML PKDD 2021 Workshops, CCIS 1524, pp. 534–551, 2021.
https://doi.org/10.1007/978-3-030-93736-2_40

Table 1. Summary of datasets used in the paper

Dataset name	Target	Associated dataset flaws
Pneumonia [7]	Pneumonia mortality	Missing values, treatment effects
MIMIC-II [22]	ICU mortality	Missing values, treatment effects, data drift
COMPAS [11]	Defendant recidivism	Bias and fairness
Housing price[a]	Housing price	Outliers

[a]The dataset is proprietary

processes are highly complex and not interpretable by humans. Once these black-box models are trained on a flawed dataset, they might learn biased knowledge embedded in the data, which data scientists can struggle to detect. In fact, researchers have already found many unexpected problems in these black box models that originated from flaws lurking in the datasets [4,5,17]. Making the situation even riskier, we do not know how many more flaws are hidden in the dataset and in what way.

Recently, some high-accuracy and interpretable machine learning models have been proposed. Because of their interpretability, data scientists can examine what the models have learned from a dataset, and potentially also discover flaws in those datasets. Explainable Boosting Machines (EBMs) [6,15,16] in particular can achieve accuracy on par with the best black-box models. More importantly, the model itself is the sum of visualizable shape functions created for individual features (or their pairwise interactions), and these shape functions are often expressive enough to capture subtleties embedded in the datasets, especially for continuous features. This makes EBM an ideal base machine learning model to detect and analyze flaws in datasets.

In this work, we provide a series of case studies that show how EBMs can help users detect flaws that are lurking in their data, and in some cases potentially correct problems caused by these flaws. The dataset flaws we study includes missing values, confounding and treatment effects, data drift, bias and fairness, and outliers. For each type of common dataset flaw, we provide one or more examples of EBM shape function graphs that help us identify the problem. Table 1 is a summary of datasets used in this study and their associated flaw types. Through our case studies, we found that

1. EBM shape function graphs can be helpful in identifying various types of dataset flaws.
2. In many cases, users with domain expertise are needed to examine what the model has learned.
3. In some cases, EBMs provide simple tools for correcting problems in the models, when correcting the data is not feasible or too difficult.

The later sections are organized as follows. Section 2 briefly introduces the EBM model. Sections 3–7 use EBMs to identify one type of dataset flaw per section, and discuss the best approaches to handle these problems. In Sect. 8, we conclude experimental findings and discuss three possible directions for future study.

2   Explainable Boosting Machines

Suppose an input sample is denoted as $(\mathbf{x}, y)$, where $\mathbf{x}$ is the p dimensional feature vector and y is the target. Denote the jth dimension of the feature vector as x_j. Then a generalized additive model (GAM), first introduced by [9], is defined as

$$g(E[y]) = \beta_0 + f_1(x_1) + f_2(x_2) + \cdots + f_p(x_p) \tag{1}$$

where β_0 is the intercept, $f'_j s$ are the shape functions and g is the link function, e.g. identity function for regression and logistic function for classification. Since one can add any offset to f_j while subtracting it from β_0 or other shape functions, we usually set the population mean of f_j, i.e. $E_{x \sim \mathcal{X}}[f(x_j)]$ to 0. Note that, each shape function f_j only operates one single feature x_j, and thus the shape function can directly be plotted. This makes GAMs interpretable since the entire model can be visualized through 2D graphs. In early work of GAM, the shape functions are usually modeled as splines with smoothness constraints. Explainable Boosting Machine (EBM) [15] formulates $f'_j s$ as ensemble of trees using ensemble techniques such as bagging and gradient boosting. Incorporating tree based ensemble learning algorithms significantly improves the performance of GAM. EBM also outperforms traditional GAMs in terms of interpretability, as its shape function has more complexity to capture nuances hidden in the dataset. The GA^{2}M model further improves accuracy by adding a small number of pairwise interactions, i.e.

$$g(E[y]) = \beta_0 + \sum_{j=1}^{p} f_j(x_j) + \sum_{k=1}^{K} f_k(x_{k_1}, x_{k_2}) \tag{2}$$

in which K pairs of features (k_1, k_2) are chosen greedily (see the FAST algorithm in [16]). Including pairwise interactions will not affect the interpretability because the interaction terms can be visualized as heatmaps.

3   Missing Values

A variety of problems can arise when there are missing values in data. In this section, we explore a few of these issues, and show how interpretable models such as EBMs can be used to detect, and in some cases fix these problems.

3.1   Missing Values Assumed Normal

In some domains such as healthcare, it is common for feature values such as lab tests to be missing in the dataset because clinicians believe the patient is likely to be "normal" for this measurement, and thus the lab test is not performed. In other cases, the measurement may be made, but the value may not be recorded if it is within normal range—clinicians tend to focus on abnormal findings.

Figure 1 (a) shows what an interpretable EBM model has learned for predicting pneumonia mortality risk as a function of heart rate. As expected, risk is

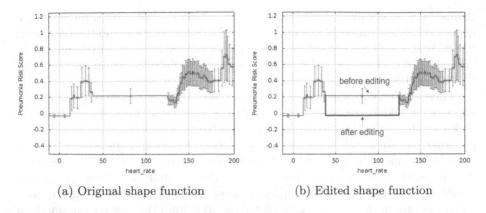

(a) Original shape function   (b) Edited shape function

Fig. 1. EBM shape function of "heart rate" for predicting pneumonia mortality risk.

elevated for patients with abnormally low (10–30) or high heart rate (125–200). The graph, however, shows a surprising region of flat risk between HR 40 and 125, which is considered to be normal heart rate for patients in a doctor's office. Moreover, the model surprisingly predicts patients who have *normal* heart rate are at *elevated* risk: it adds 0.22 to the risk for patients in this region.

On further inspection, it turns out that there are no patients in the data set with heart rates between 40 and 125, and 91% of patients are missing their heart rate which has then been coded as zero. In other words, there are no data to support the model in the normal range of HR 40–125, and instead the patients who would be in this range, are all coded as zero in the data and on the graph. This explains why the model predicts the lowest risk = -0.04 for patients with HR = 0, because these are the patients with the most normal heart rate.

Any model trained on this data (e.g., boosted trees, random forest, neural networks) is likely to learn to make similar predictions in this region because there is no data to support learning the correct risk in this range, and because most models will then learn to interpolate between the regions where they do have data. One exception might be Bayesian models with strong priors where the prior might dominate in regions of little or no data and cause predictions in this region to be closer to a baseline lower-risk value. However, even Bayesian approaches would not learn to predict the correct value in this region, but they might learn a less incorrect value. The key advantage of using interpretable models such as EBMs is that we can easily see these problems in the model, that were caused by problems in the data.

If the model will only be used to make predictions for patients where all heart rates in the range 40–125 will be coded as zero, then the model will make accurate predictions and the elevated risk predicted by the model in the range 40–125 will not be a problem because no patient will ever fall in that range. However, if the model might be used to make predictions for patients whose true heart rate would be coded in this region, the model will then make incorrect, possibly dangerous predictions for patients who have normal heart rate. Because this is risky, it usually is important to correct this kind of problem. One might expect that the data scientist would detect this kind of problem in the data prior to training a model, however in our experience these kinds of problems can be difficult to detect in the raw data and are easier to detect once an interpretable EBM model is trained.

There are several ways to correct this kind of problem. Of course, the best approach would be to collect and record the true heart rates for all patients. Unfortunately, it is often not possible to go back and correct data in this way. An alternate approach would be to edit the data so that patients coded as zero are randomly assigned heart rates in the interval 40–125, i.e., impute the missing heart rates with a random value selected uniformly from the region where we believe most of the missing values arise from. This, however, does make the assumption that all missing values arise from this one region, and that no patients with low or high heart rate had a missing heart rate. An alternate approach is to use a more sophisticated method of imputing missing values such as random forest imputation [24]. As we will see in Sect. 3.2, imputing with the mean or median missing value is probably not recommended.

An alternate approach when interpretable EBM models is used is to directly edit the graph so that the region 45–120 predicts risk similar to the prediction the model has learned to make for patients with HR = 0. The resulting graph is show on Fig. 1 (b). This approach has the following advantages:

1. Editing shape functions provides an opportunity for experts to use their professional training to correct and improve models in ways that may not be adequately represented in the training data.
2. Editing the model can not only improve the accuracy of the model in the real world, but make the shape plots more "correct" and trusted by experts.
3. Editing an EBM shape function can be done without retraining the model.
4. Correcting the model by editing the data is often much more difficult.

3.2   Missing Values Imputed with the Mean

Because many machine learning methods can not deal with missing values, it is common for data scientists to impute missing values before training the model. There are many different ways to impute missing values: with the mean, the median, with a unique value such as 0 or -99 or +99, or by using a machine learning method such as random forest imputation. See [14] for an overview of imputation methods.

Perhaps the most common form of missing value imputation is with the mean. Figure 2 shows an EBM plot of the mortality risk of ICU patients as a function of their PFratio. PFratio is a measure of how well a patient converts O2 in the air they breathe into O2 in their blood: low PFratio indicates patients with low blood O2 whose lung function is impaired, while PFratio around 1000 and higher indicates normal lung function. As expected, the learned shape function captures this, and also shows interesting small jumps at clinically meaningful values such

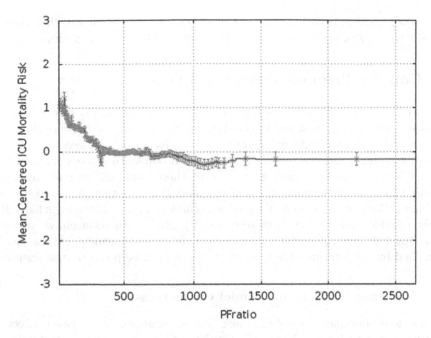

Fig. 2. EBM shape function of "PFratio" for predicting ICU mortality.

as 800, 700, 200, and 100. What is surprising, however, is the large drop in risk at about PFratio = 323. What could cause that?

A simple test for blood O2 levels is to pinch a fingertip and see how quickly color returns to the skin. If color returns quickly, clinicians know that the blood O2 is good and do not bother to measure PFratio—as discussed in the previous section, the PFratio is assumed normal. In this dataset, however, the missing PFratio values were imputed with the mean instead of being coded as 0 as they were in Fig. 1. In this dataset 60% of patients are missing PFratio. The mean PFratio when not missing (40% of the data) is 323.6, so 60% of patients have had their PFratio imputed with this value. Because this is a large sample of healthy patients with strong respiration, the model learns that their risk is comparable to the risk of other healthy patients with PFratio above 1000. This explains why the graph dips at 323, yet predicts higher risk just before and after this value. Although this anomaly does not significantly hurt the accuracy of the model because it has learned to make appropriate low-risk predictions for the 60% of patients at this value, it is risky to leave this anomaly in the model because there are real patients with PFratio≈323 who will be predicted to have low risk but who are genuinely at elevated risk. For this reason, it would be better to either use a more sophisticated method of imputing missing values such as random forest imputation, or to leave the missing value coded as a unique value such as 0. Model editing is not a good solution for this problem because imputation with the mean has caused patients who are low risk (missing values) and elevated risk

(PFratio near 323) to fall at the same place on the shape function, thus there is no edit to the graph that can predict the correct risk for both groups.

4 Confounding and Treatment Effects

Due to the complexity of the world, there will always be variety of confounding variables with chosen features in the dataset. Since many of these confounders are not included in the dataset, their treatment effects would be accounted by the chosen features, while the actual effects of the chosen features are contaminated. In this section, we first show how some of these confounders and treatment effects would affect the correctness of the model, and potentially cause serious problems. Based on this, we will also discuss how interpretable models like EBM might provide tools to correct these types of problems. In addition, we will also show through examples that in some cases if these treatment effects can be identified by the interpretable models, they can even help create new science.

4.1 Treatment Effects and Model Correctness

In the medical domain, one of the most famous examples of treatment effects is that patients who have a history of asthma have lower pneumonia mortality risk than general population. This counterintuitive pattern was first found by rule-based model [2]. The pattern can be interpreted by the fact that patients with asthma history would be admitted directly into ICU and get more aggressive care, thereby lowering their risk of death.

Such treatment effects are pervasive in medical datasets. Figure 3 (a) shows three graphs with noticeable treatment effects, learned by an EBM model to predict pneumonia mortality risk. First, in the middle of Fig. 3 (a), the asthma effect has also been found by EBM. Similar to the asthma effect, shown in the right subfigure of 3 (a), EBM also discovers that a history of chest pain can reduce the mortality risk. The history of chest pain is highly related to heart attacks. Patients with a history of heart disease might confuse the earlier symptom of pneumonia with signs of a heart attack. Therefore, they might call ambulances or get admitted to emergency room as soon as possible, and consequently diagnose pneumonia early and get high-quality care. Another interesting example of treatments effects is the influence of age on the pneumonia mortality risk, shown in the left of Fig. 3 (a). The risk score remains low between age 18 to 50, and slightly goes up from age 50 to 67. Right around age 67, the risk increases rapidly. Interestingly, this tipping point coincides with the average retirement age in the US, suggesting that the increase of risk might be from retirement: things associated with retirement, such as changes in insurance provider and urgency of care, might cause risk to arise rapidly. This is yet another example of how confounders such as "retirement" might affect the behavior of features included in the dataset such as "age". In the age graph, there is another surprising treatment effect that the mortality risk suddenly drops (0.15 in log odds) near age 100. Such an effect is very unlikely to be true biologically. We suspect

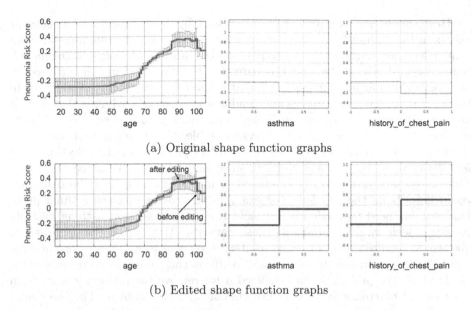

(a) Original shape function graphs

(b) Edited shape function graphs

Fig. 3. EBM shape function graphs for predicting pneumonia mortality risk. From left to right are shape functions of features "age", "asthma" and "history of chest pain."

this is associated with social effects that doctors might try even harder to cure the patients if their age passes 100—given that pneumonia is relatively a treatable disease and centenarian are very rare even worldwide, doctors would not give up on these patients.

Many of these issues have already been mentioned by earlier works [6]. However, their impacts on model correctness and ways to fix them have not been carefully discussed yet. These treatment effects might be real in the dataset or even in practice due to the existence of confounders. Does it mean we should keep them in the model? Our suggestion is that *the model correctness depends on the purpose of the model*. For example, if the model is used by an insurance company to create insurance policy, then these effects might be fine. Because the model successfully captures the treatment effects that happen in reality, the insurance company can then use this model to adjust the insurance charges to patients and potentially make more profits. However, if the models can affect what care patients might receive, using these datasets to train models can be problematic. Certainly, we should not expect doctors to give mild or no care to a patient who is 105 with asthma and a history of chest pain only because the model predicts they have low mortality risk. There is a dangerous feedback loop behind this problem: while the patients' risks are lower because we give them extra treatment, if we then use the predicted risks to determine whether they would get extra treatment, the treatment effects will be removed and their mortality risks increase.

How should we fix these problems caused by confounders and treatment effects in the dataset? One approach would be to include confounding variables that reflect the treatments given to patients as additional features in the model. Sometimes this allows the model to learn that the reduction in risk is caused by the treatment and not the condition that caused the treatment to be given. In some cases, however, treatments and conditions are so tightly linked that models can not distinguish treatments and conditions reliably. Moreover, detailed information about treatments is not always available. Collecting new or additional data seems reasonable, but can be very hard in practice. First, collecting medical data can be very expensive itself, since the samples have to be real cases. More importantly, it could be unethical or impossible to collect the data that could help fixing the problem. For example, in order to fix the asthma problem, doctors need to randomly withhold the extra treatment asthma patients receive, which would be risky to those patients. Also, fixing the drop at age 100 would require doctors treating centenarians equivalently to other patients, which is almost impossible since that requires controlling doctors' care. One might think the problems can be solved if we remove the asthma feature from the model. Unfortunately, this usually will not solve the problem. The bias comes from the target rather than the input features. Since correlations are common between medical features, e.g. history of heart disease and body weight, even if the asthma feature is removed, the treatment effects might be learned implicitly through other correlated features. Even worse, such signals can be distributed in multiple features and become impossible to monitor. Thanks to the interpretability of the EBM model, we can directly fix the problem by editing the graphs. Figure 3 (b) shows some examples of edited graphs that could be more reasonable biologically. For example, having asthma and history of chest pain would be edited to have higher risk than normal. Also, for patients with age>85, we might want to edit the graph such that their risk goes up slowly and monotonically. Note that model edits are high-stakes decisions, and therefore should be based on interaction with domain experts such as doctors. Doctors would also need to monitor the effects of the models and most likely adjust the edited graph based on real-world feedback.

4.2 Discovering New Science

In Sect. 4.1, we show many negative examples where confounding and treatment effects might produce wrong and dangerous models. However, not all treatment effects are harmful. Sometimes useful knowledge associated with the treatment effects exist that domain experts were not aware of.

The left part of Fig. 4 (a) shows the EBM shape function of predicting pneumonia death risk with blood urea nitrogen (BUN) level. Lower BUN level is believed to be healthy which is clearly reflected in the figure as low risk. Interestingly, the figure also discovers that although the risk of death rises at BUN level around 30, the risk curve gets flat when BUN level $\approx$ 50. In fact, most doctors start to give patients medication if their BUN hit 50. This means the patients are getting effective medications and that flattens their risk of death.

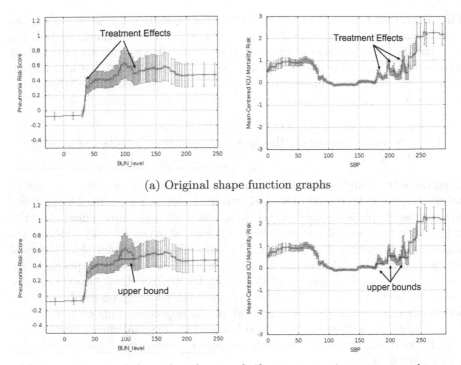

(a) Original shape function graphs

(b) Upper bound of shape function graph if more aggressive treatments happen earlier

Fig. 4. EBM shape function graphs with treatment effects. Left: predicting pneumonia mortality risk using blood urea nitrogen (BUN) level; right: predicting ICU mortality risk using systolic blood pressure (SBP).

Another point of interest on the graph is the peak at BUN level 100: the risk continues to get higher after 50 and starts to decrease at 100. Notice that 100 is a "round number" at which doctors would start to give patients more aggressive treatment. In practice, when the patients' BUN level exceed 100, doctors would start to give them dialysis, which is a process of purifying the blood of the patient. Dialysis treatment quickly reduces BUN and accordingly reduces mortality risk. Some doctors may not give dialysis starting at 100 which accounts for why the risk declines until 120.

Similar effects can also be found in the shape function of predicting ICU mortality with blood pressure (right of Fig. 4 (b)). The risk curve peaks at several "round numbers" such as 175, 200 and 225. Treatments might be provided by the doctors to lower the patients' mortality risk and thus confounding the nature risk of high blood pressure.

So far, although we are able to explain when doctors give different treatments, these findings are not significantly different from the treatment effects described in Sect. 4.1, since they essentially just match with what doctors already know.

What if we take further steps to reason counterfactually on what could happen if we move the treatment threshold for more aggressive treatment earlier, e.g. move the dialysis threshold from 100–120 down to 80? How would such a move influence the shape function graph? To reason counterfactually without collecting new data, we should make some additional assumptions. Take BUN level as an example, we propose two reasonable assumptions

1. Getting dialysis would not increase the risk of death.
2. Given the same treatment, the patients' mortality increase monotonically with BUN level.

Given these two simple assumptions, the upper bound of the graph when moving dialysis treatment threshold to 80, is shown in the left of Fig. 4 (b). The modified part on the graph is *at most* a step function with max value equals the risk at 120—the two ends of the modified part have at most the same risk as the original graph (assumption 1); no middle points has risk higher than the right end (assumption 2). This is just the most conservative guess. In practice, the graph might be a U-shape curve that is under the red line. Given this new graph, we can estimate how many lives could be saved based on total number of pneumonia cases, proportion of sample fall in this region of graph, and the change of probability of death between the red line and the original curve. Surprisingly, even the most conservative estimation suggests we could save 2500 lives per year in just United States.

Similarly, we can infer about what could happen to ICU mortality risk if we change the treatment criterion based on blood pressure. Shown in the right of Fig. 4 (b), the three bumps caused by "inappropriate" treatment thresholds can be least flattened to the red lines. This could save another thousands of lives. Note that, these are just estimations based on correctness of the model and assumptions of the treatment. Rigorous clinical trails are needed to examine if these estimations are correct.

5   Data Drift

Data drift means the statistical property of the data might change over time, and thus making the model trained on the dataset outdated. In this section, we show when training on newer data, how EBM can be used to correct data drift problems in outdated models.

Figure 5 compares two models, on how much risk does "having AIDS" add to the ICU mortality rate. One model, shown as the red bar, is the risk score given by a EBM trained on the MIMIC-II dataset. The other model, shown in the blue bar, is SAPS-II model [12], which is a widely used scoring system for predicting ICU mortality risk. The risk scores predicted by two models, however, completely disagree with each other. SAPS model suggests that if a patient has AIDS, it adds 1.3 to the log odds of the mortality probability, or on average equivalent to adding around 12% to the risk. In fact, this is one of the most important factor in the SAPS-II model. The EBM model, while training on dataset with

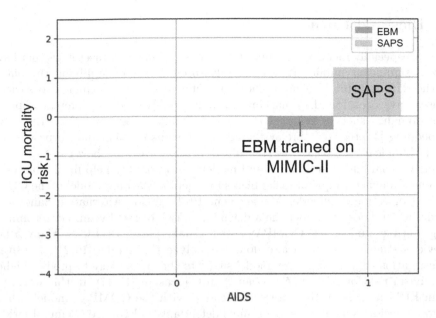

Fig. 5. Comparing contributions of AIDS to ICU mortality risk.

the same purpose and similar feature set, believes the log odds of the mortality probability of a patient diagnosed with AIDS should subtract 0.3 points. Why are these two models have such a large discrepancy on the influence of AIDS: one predicting AIDS to be one of the highest single risk factors while the other model predicting that AIDS is not too bad for the patient?

In fact, the discrepancy is because of progress in medicine. The SAPS model is created in 1993, during which AIDS was often considered as a terminal illness without effective treatments. Especially if the patients were in the ICU, they would have a large chance to be in near terminal stage of HIV. However, the data in MIMIC-II used to train EBM was collected between 2001 and 2008. By then, effective treatments such as HAART and Combivir became the standard, making HIV gradually become a chronic illness rather than a terminal illness. Since the death rate of HIV decreased significantly, there could be much worse reason for why the patient stays in ICU, and therefore making "having AIDS" reduce mortality risk.

This shows why interpretability is important in dealing with data drift. Because the prediction process of the model is transparent, one can compare what the model learns from earlier data and what it learn on new data, and identify how the model changes when the world changes. Particularly, this case study reveals that, although SAPS is still widely used, it is a stale model that does not represent modern health care. Doctors might want to abandon this model and replace it with interpretable models trained on up-to-date dataset. However, if the model is not interpretable, we can never monitor the changes of the model, and thus cannot tell whether the model goes stale and should be replaced.

6 Bias and Fairness

Biases, especially racial and gender biases, are well known dataset flaws that have drawn significant public attention. By learning or even amplifying the biases in the data, machine learning models might cause serious damage to society. Biases have been found in machine learning models used for different application domains, such as criminal justice [17], image super-resolution [19], word embedding [4] etc. For more systematic discussions on bias and fairness issues in machine learning, please see [18]. In this section, we show how EBM can help identify racial biases in datasets and models, and possibly help fix the biases.

One famous example of racial biases in machine learning model is the COMPAS (Correctional Offender Management Profiling for Alternative Sanctions) model which tries to predict the a defendant's risk of recidivism, i.e. recommitting a crime. Judges use COMPAS to help make parole decisions. Many other previous studies have been also done to analyze this model [10,21]. However, investigation on this black-box model suggests that it is likely to predict higher recidivism risk for African American than Caucasian [11,17]. In this work, we train EBM models on the dataset associated with the COMPAS model, which contains recidivism outcomes of criminal defendants in Broward County, Florida. The dataset has both the COMPAS predicted scores and the actual labels for recidivism in two years. We train two EBMs, one mimicking the COMPAS model, the other fitting the true labels.

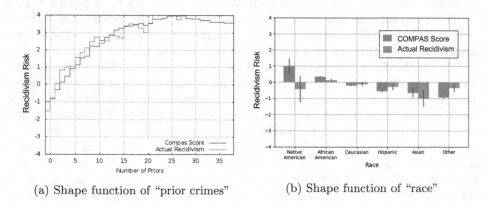

(a) Shape function of "prior crimes"   (b) Shape function of "race"

Fig. 6. EBM shape functions for recidivism prediction. Red bars: EBM approximating the COMPAS model prediction; Green bars: EBM trained on the actual recidivism label. (Color figure online)

Figure 6 compares the shape functions of two EBMs. In Fig. 6 (a), we compare the shape functions of "number of prior convictions". For the effect of "number of prior convictions", the two models seem to agree with each other. Since the "number of prior convictions" is a very important feature, the result might suggest that the EBM mimics the COMPAS accurately. Figure 6 (b) shows the shape functions of the sensitive feature "race". The EBM graphs suggest that racial biases exist in both the dataset and the COMPAS model, e.g. African Americans are predicted to have higher risk than Caucasians. However, the two models disagree on the scale of the bias. The mimicked COMPAS model predicts African American having even higher risk and Caucasian having lower risk than the predictions of EBM trained on true labels. This suggests that COMPAS might even amplify the bias in the dataset. Surprisingly, the largest discrepancy of the two models is on Native Americans. While EBM trained on true label predicting Native Americans have low recidivism risk, the mimicked COMPAS model believes they have very high risk. We suspect that these discrepancies might be because the COMPAS is trained on a larger dataset containing nationwide data, and that there is a population difference between defendants nationwide and defendants in Broward County.

Such biased models should not be deployed, and these biases need to be fixed. This raises the question of how to remove biases in the dataset. One possible solution is to remove sensitive features like gender and race, or even make them inaccessible. However, as mentioned in Sect. 3, such bias comes from the target rather than the input features. Even if sensitive features like "race" are removed, the model can learn their effect via their correlated features. For example, in the US, race is highly correlated with zip code, education and income. If we remove the sensitive feature, the bias would spread in complex ways among other features, and thus becomes impossible to fix or even detect. Therefore, we suggest to first learn the biases with interpretable model like EBM. After letting the biases get concentrated in the model, we can edit the graphs to "zero out" the learned effects for sensitive features, thus mitigating the learned bias.

For certain types of biases, re-collecting data might be helpful. For example, for facial recognition dataset, one can collect more samples for the minority groups. However, in many cases, such as the recidivism example, biases can be impossible to fix by re-collecting the data, as the bias is rooted in complicated societal problems.

7   Outliers

Figure 7 (a) shows the "year of built" shape function of EBM trained on a proprietary housing price prediction dataset. We notice that an anomaly (huge spike) appears at year 1989 which adds +$600k to the housing price. Surprisingly, the error bar of the 1989 bin is relatively small, which means the bin has relatively large support. Why is there such a huge spike in the shape function? We search though the training set and select all the records with year 1989, see Fig. 7 (b). Among the 8 selected samples (year = 1989), 7 samples have sold price of exactly $8,094,000. This completely explains the existence of the anomaly in the EBM shape function. Interestingly, they are all condos and have the same zipcode. Although they do differ in number of bedrooms and bathrooms, their house sizes are all very small and their lot sizes are almost the same. We do not know if these records are real. We suspect that these records might be included

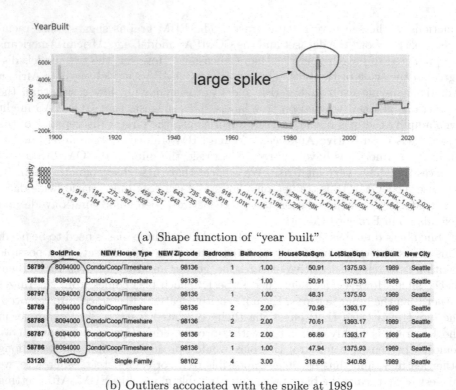

(a) Shape function of "year built"

	SoldPrice	NEW House Type	NEW Zipcode	Bedrooms	Bathrooms	HouseSizeSqm	LotSizeSqm	YearBuilt	New City
58799	8094000	Condo/Coop/Timeshare	98136	1	1.00	50.91	1375.93	1989	Seattle
58798	8094000	Condo/Coop/Timeshare	98136	1	1.00	50.91	1375.93	1989	Seattle
58797	8094000	Condo/Coop/Timeshare	98136	1	1.00	48.31	1375.93	1989	Seattle
58789	8094000	Condo/Coop/Timeshare	98136	2	2.00	70.98	1393.17	1989	Seattle
58788	8094000	Condo/Coop/Timeshare	98136	2	2.00	70.61	1393.17	1989	Seattle
58787	8094000	Condo/Coop/Timeshare	98136	2	2.00	66.89	1393.17	1989	Seattle
58786	8094000	Condo/Coop/Timeshare	98136	1	1.00	47.94	1375.93	1989	Seattle
53120	1940000	Single Family	98102	4	3.00	318.66	340.68	1989	Seattle

(b) Outliers accociated with the spike at 1989

Fig. 7. Outliers and their effects on EBM shape function for predicting housing price.

into the dataset accidentally, e.g. the values of certain columns of one record get overwritten on that of the other records. But regardless of why these outliers are included in the dataset, models trained on these outliers should be corrected, as anomalies caused by these outliers can make the model less robust in the real world. Such outliers could be detected with simpler statistical methods if one knew in advance exactly what to look for. EBMs help outlier detection by making outliers that are not expected easy to find.

The capability of identifying anomalies in the dataset also makes the model easier to deal with security threats such as data poisoning attack [3,13]. In data poisoning attack, the attacker is capable of modifying part of the training data, for example by providing a poisoned public dataset or by compromising the server storing the dataset. After a small fraction of training data being modified, the machine learning model trained on the dataset would make predictions that benefit the attackers. Taking the housing price dataset as an example, although we have no evidence that this is a data-poisoning attack, by including 7 samples with high sold prices ($8,094,000) and the same year of built (1989), the model would overestimate the price of all houses built in 1989. However, adding 7 training samples would not significantly affect the test accuracy of the model, and thus data scientists training black-box models on the dataset might not detect it. Suppose the model was deployed to estimate the housing price but data is not available to costumers, it would benefit all the house owners whose houses are built in 1989, which might include the attacker. Nevertheless, as shown in previous paragraphs, EBM can help data scientists detect such anomalies in the dataset, since the anomalies might be identifiable by the shape function plots. Note that, many detection methods of data poisoning attack have been proposed, including methods that are based on ourlier and anomaly detection [20,23]. Systematic studies are need to prove the effectiveness of detecting data poisoning attack using EBM.

8   Discussion

We present a series of case studies on detecting common data flaws using EBM shape function graphs. From the results, we show that EBM can help data scientists identify common flaws in the dataset. In some cases, EBM even provide simple tools to fix the problems in the models introduced by data flaws. This is extremely helpful when re-collecting data is expensive or impossible, as discussed in Sect. 4 and 6.

Most flaws discussed in this paper are based on identifying bumps or sudden changes in the EBM shape function. Although most of these findings could be supported by domain knowledge, these changes in the shape functions might also come from noise, overfitting or correlation with other variables. To get more reliable results in the future, rigorous hypothesis testing methods for changes in the EBM shape functions are needed. Currently, the uncertainty quantification of EBM solely relies on calculating the standard deviations over all bootstrapped models, which tends to underestimate the uncertainty. Also, joint distributions are needed to test whether changes are significant in the plot, since scores before and after the change are not independent.

In terms of model editing, many important problems need to be investigated. For example, how to work with domain experts to edit models? Once we have an edited model, should we only show the edited model, or both the original and edited model? Also, if the model has been edited multiple times, how to present the editing history to the users, or even redo edits when the model is retrained?

In our case studies, we show that EBM shape functions can help detect common flaws in data such as treatment effects, improper handling of missing values, racial bias and outliers. However, other methods have been developed to address some these problems and future work is needed to systematically compare EBMs with these existing methods.

References

1. Acock, A.C.: Working with missing values. J. Marriage Family **67**(4), 1012–1028 (2005)

2. Ambrosino, R., Buchanan, B.G., Cooper, G.F., Fine, M.J.: The use of misclassification costs to learn rule-based decision support models for cost-effective hospital admission strategies. In: Proceedings of the Annual Symposium on Computer Application in Medical Care, p. 304. American Medical Informatics Association (1995)
3. Barreno, M., Nelson, B., Joseph, A.D., Tygar, J.D.: The security of machine learning. Mach. Learn. **81**(2), 121–148 (2010). https://doi.org/10.1007/s10994-010-5188-5
4. Bolukbasi, T., Chang, K.W., Zou, J.Y., Saligrama, V., Kalai, A.T.: Man is to computer programmer as woman is to homemaker? debiasing word embeddings. Adv. Neural Inf. Process. Syst. **29**, 4349–4357 (2016)
5. Buolamwini, J., Gebru, T.: Gender shades: Intersectional accuracy disparities in commercial gender classification. In: Conference on Fairness, Accountability and Transparency, pp. 77–91. PMLR (2018)
6. Caruana, R., Lou, Y., Gehrke, J., Koch, P., Sturm, M., Elhadad, N.: Intelligible models for healthcare: Predicting pneumonia risk and hospital 30-day readmission. In: Proceedings of the 21th ACM SIGKDD International Conference on Knowledge Discovery and Data Mining, pp. 1721–1730 (2015)
7. Cooper, G.F., et al.: Predicting dire outcomes of patients with community acquired pneumonia. J. Biomed. Inf. **38**(5), 347–366 (2005)
8. Gama, J., Žliobaitė, I., Bifet, A., Pechenizkiy, M., Bouchachia, A.: A survey on concept drift adaptation. ACM Comput. Surv. (CSUR) **46**(4), 1–37 (2014)
9. Hastie, T., Tibshirani, R.: Generalized additive models: some applications. J. Am. Stat. Assoc. **82**(398), 371–386 (1987)
10. Kleinberg, J., Mullainathan, S., Raghavan, M.: Inherent trade-offs in the fair determination of risk scores. arXiv preprint arXiv:1609.05807 (2016)
11. Larson, J., Mattu, S., Kirchner, L., Angwin, J.: How we analyzed the compas recidivism algorithm. ProPublica **9**(1) (2016)
12. Le Gall, J.R., Lemeshow, S., Saulnier, F.: A new simplified acute physiology score (saps ii) based on a European/north American multicenter study. Jama **270**(24), 2957–2963 (1993)
13. Li, B., Wang, Y., Singh, A., Vorobeychik, Y.: Data poisoning attacks on factorization-based collaborative filtering. Adv. Neural Inf. Process. Syst. **29**, 1885–1893 (2016)
14. Lin, W.-C., Tsai, C.-F.: Missing value imputation: a review and analysis of the literature (2006–2017). Artif. Intell. Rev. **53**(2), 1487–1509 (2019). https://doi.org/10.1007/s10462-019-09709-4
15. Lou, Y., Caruana, R., Gehrke, J.: Intelligible models for classification and regression. In: Proceedings of the 18th ACM SIGKDD International Conference on Knowledge Discovery and Data Mining, pp. 150–158 (2012)
16. Lou, Y., Caruana, R., Gehrke, J., Hooker, G.: Accurate intelligible models with pairwise interactions. In: Proceedings of the 19th ACM SIGKDD International Conference on Knowledge Discovery and Data Mining, pp. 623–631 (2013)
17. Mayson, S.G.: Bias in, bias out. YAle lJ **128**, 2218 (2018)
18. Mehrabi, N., Morstatter, F., Saxena, N., Lerman, K., Galstyan, A.: A survey on bias and fairness in machine learning. arXiv preprint arXiv:1908.09635 (2019)
19. Menon, S., Damian, A., Hu, S., Ravi, N., Rudin, C.: Pulse: Self-supervised photo upsampling via latent space exploration of generative models. In: Proceedings of the IEEE/CVF Conference on Computer Vision and Pattern Recognition, pp. 2437–2445 (2020)

20. Paudice, A., Muñoz-González, L., Gyorgy, A., Lupu, E.C.: Detection of adversarial training examples in poisoning attacks through anomaly detection. arXiv preprint arXiv:1802.03041 (2018)
21. Rudin, C., Wang, C., Coker, B.: The age of secrecy and unfairness in recidivism prediction. Harvard Data Sci. Rev. **2**(1), 1811 (2018)
22. Saeed, M., Lieu, C., Raber, G., Mark, R.G.: Mimic ii: a massive temporal ICU patient database to support research in intelligent patient monitoring. In: Computers in Cardiology, pp. 641–644. IEEE (2002)
23. Steinhardt, J., Koh, P.W., Liang, P.: Certified defenses for data poisoning attacks. In: Proceedings of the 31st International Conference on Neural Information Processing Systems, pp. 3520–3532 (2017)
24. Stekhoven, D.J., Bühlmann, P.: MissForest-non-parametric missing value imputation for mixed-type data. Bioinformatics **28**(1), 112–118 (2011). https://doi.org/10.1093/bioinformatics/btr597

Bias and Fairness in AI

2nd Workshop on Bias and Fairness in Artificial Intelligence (BIAS 2021)

The 2nd edition of the Workshop on Bias and Fairness in Artificial Intelligence (BIAS 2021) was held in conjunction with the ECML PKDD 2021 conference on September 13, 2021 and took place online.

Artificial Intelligence (AI) techniques based on big data and algorithmic processing are increasingly used to guide decisions in important societal spheres, including hiring decisions, university admissions, loan granting, and crime prediction. However, there are growing concerns with regard to the epistemic and normative quality of AI evaluations and predictions. In particular, there is strong evidence that algorithms may sometimes amplify rather than eliminate existing bias and discrimination, and thereby have negative effects on social cohesion and on democratic institutions.

The problem of bias and discrimination in AI systems is gaining increasing attention among researchers and practitioners from different disciplines including computer science, philosophy, and law. Despite the significant progress that has been made in the recent years in technical and multidisciplinary research, there is still a lack a comprehensive understanding of how pertinent concepts of bias or discrimination should be interpreted in the context of AI and which socio-technical options to combat bias and discrimination are both realistically possible and normatively justified. The main objective of the workshop is a contribution to the understanding of "How can standards of unbiased attitudes and non-discriminatory practices be met in (big) data analysis, AI and algorithm-based decision-making?".

For BIAS 2021 we mainly solicited contributions that go beyond the current state of the art on fairness formalization and operationalization, investigate other learning tasks beyond the overlooked supervised learning, take into account data challenges like non-stationarity or class-imbalance, and consider the context/application in which fairness problems arise. We opened the call beyond traditional ML tasks and applications (supervised/unsupervised learning, information retrieval and recommender systems, HCI, constraint solving, complex systems and networks, etc.) and aimed to bridge interdisciplinary studies (law, social sciences). Thus the call included the following topics:

- Bias and Fairness by Desig

  - Fairness measures
  - Counterfactual reasoning
  - Metric learning
  - Impossibility results
  - Multi-objective strategies for fairness, explainability, privacy, class-imbalancing, rare events, etc.
  - Federated learning
  - Resource allocation
  - Personalized interventions
  - Debiasing strategies on data, algorithms, procedures
  - Human-in-the-loop approaches

– Methods to Audit, Measure, and Evaluate Bias and Fairness

- Auditing methods and tools
- Benchmarks and case studies
- Standard and best practices
- Explainability, traceability, data and model lineage
- Visual analytics and HCI for understanding/auditing bias and fairness
- HCI for bias and fairness
- Software engineering approaches

The workshop featured three keynote talks delivered by invited speakers and six research paper presentations accepted by the Program Committee after completing the peer reviewing process.

The keynote talks featured several inspiring highlights around technical machine learning and societal and legal aspects of bias and discrimination in AI systems, summarized hereafter:

– "What's fair about fair ML?" by Linnet Taylor, Tilburg Institute for Law, Technology, and Society, The Netherlands.
Taylor questioned in her talk what dimensions of fairness the field of fair ML envisage, and implications this perspective could have in the wider context of responsible AI. She highlighted the limits of the current views on (formalizing and implementing) ML fairness, accountability, responsibility, and trust. Bringing into the spotlight currently deflected fundamental issues of recognition and representation such as decolonization, feminism and geopolitical equity, Taylor emphasised the risks of not recognizing and incorporating such claims into fair ML research and development.

– "The Fairness-Accuracy tradeoff revisited" by Toon Calders, University of Antwerp, Belgium.
Calders revisited three key issues related to the fairness-accuracy trade-off: (1) impossibility results in optimizing simultaneously for different group-level fairness measures, and lack of guidelines for selecting the right measure, (2) counter-intuitiveness of the perception that we must trade some accuracy to gain fairness, and (3) an overemphasis on satisfying fairness constraints without enough understanding about how that is achieved. After a deep dive into these issues he proposed an alternative way of looking at fair ML as optimizing accuracy in a theoretical fair world.

– "Strengths and weaknesses of European legal protection against discriminatory AI" by Frederik Zuiderveen Borgesius, Radboud Universiteit, The Netherlands.
Borgesius illustrated to what extent European law can and cannot help to protect people against illegal discrimination or other types of unfair differentiation by AI. He focused on the non-discrimination law and the GDPR as the two most relevant legal instruments. He also presented his views on how weaknesses in these instruments could be improved in the future.

The submitted papers discussed a variety of problems around the topic of the workshop.

– William Blanzeisky and Padraig Cunningham in their paper titled "Algorithmic Factors Influencing Bias in Machine Learning" study how different data and model characteristics affect bias in machine learning. In particular, they focus on feature and class imbalance and on irreducible error and regularization factors.
– Carlos Mougan, Georgios Kanellos, and Thomas Gottron in their paper titled "Desiderata for Explainable AI in statistical production systems of the European Central Bank" provide a practical overview of user-centric explainability requirements in the bank domain and present two concrete use-cases on outlier detection and data quality checks.
– Serafina Kamp, Andong Luis Li Zhao, and Sindhu Kutty in their paper titled "Robustness of Fairness: An Experimental Analysis" evaluate the robustness of different fairness measures to variations in the test data and show that the variance of fairness-aware learning models is much higher compared to traditional machine learning models, highlighting the need for better model evaluation.
– Gabriel Frisch, Jean-Benoist Leger, and Yves Grandvalet in their paper titled "Co-clustering for fair recommendation" propose using a co-clustering of users and items that respects statistical parity of users with respect to some sensitive attributes and show that such an approach achieves approximately fair recommendations provided that the classification of users approximately respects statistical parity.
– Daphne Lenders and Toon Calder in their paper titled "Learning a Fair Distance Function for Situation Testing" propose a method to learn a fair distance function, namely, a Weighted Euclidean distance, which is necessary for situation testing, a method used in life sciences to prove discrimination. The proposed method does not rely on human input but rather learns the distance through an optimization algorithm so that features with the highest impact on the decision label contribute most to the distance.
– Alessandro Castelnovo, Lorenzo Malandri, Fabio Mercorio, Mario Mezzanzanica, and Andrea Cosentini in their paper titled "Towards Fairness Through Time" study the problem of bias and discrimination for non-stationary data. They show that traditional fairness mitigation strategies for stationary data cannot ensure fairness performance over time and propose a retrain strategy to mitigate this effect. Moreover, they show that explainability may help to monitor fairness through time and the effect of mitigation strategies.

Full information about the workshop including presentation material can be found at the workshop's website at https://sites.google.com/view/bias2021/.

We would like to thank the keynote speakers for their inspiring talks, authors for submitting and presenting their work, members of the Program Committee for providing timely and constructive reviews, and, last but not least, all the people in the audience for engaging in stimulating discussions after each of the presentations and making the workshop interactive and lively.

September 2021   Eirini Ntoutsi
                                                                  Mykola Pechenizkiy
                                                                  Bodo Rosenhahn

Organization

Workshop Co-chairs

Eirini Ntoutsi — Freie Universität Berlin, Berlin, Germany
Mykola Pechenizkiy — Eindhoven University of Technology, The Netherlands
Bodo Rosenhahn — Leibniz Universität Hannover, Germany

Program Committee

Bettina Berendt — KU Leuven, Belgium and TU Berlin, Germany
Toon Calders — University of Antwerp, Belgium
Tim Draws — TU Delft, The Netherlands
Michael Ekstrand — Boise State University, USA
Atoosa Kasirzadeh — University of Toronto, Canada
Katharina Kinder-Kurlanda — University of Klagenfurt, Austria
Masoud Mansoury — University of Amsterdam, The Netherlands
Symeon Papadopoulos — CERTH-ITI, Greece
Jürgen Pfeffer — Technische Universität München, Germany
Evaggelia Pitoura — University of Ioannina, Greece
Salvatore Ruggieri — Università di Pisa, Italy
Jatinder Singh — University of Cambridge, UK
Maryam Tavakol — TU Eindhoven, The Netherlands
Hilde Weerts — TU Eindhoven, The Netherlands
Frederik Zuiderveen Borgesius — Radboud Universiteit, The Netherlands

Algorithmic Factors Influencing Bias in Machine Learning

William Blanzeisky$^{(\boxtimes)}$ and Pádraig Cunningham

School of Computer Science, University College Dublin, Dublin 4, Dublin, Ireland
william.blanzeisky@ucdconnect.ie

Abstract. It is fair to say that many of the prominent examples of bias in Machine Learning (ML) arise from bias in the training data. In fact, some would argue that supervised ML algorithms *cannot* be biased, they reflect the data on which they are trained. In this paper, we demonstrate how ML algorithms can misrepresent the training data through *underestimation*. We show how irreducible error, regularization, and feature and class imbalance can contribute to this underestimation. The paper concludes with a demonstration of how the careful management of synthetic counterfactuals can ameliorate the impact of this underestimation bias.

Keywords: Bias · Fairness · Model capacity · Regularisation

1 Introduction

As Machine Learning (ML) applications become ubiquitous in many aspects of human life, there is an undeniable appeal for discrimination-free machine learning. An algorithm is considered discriminatory if it systematically disadvantages people belonging to specific categories or groups instead of relying solely on individual merits [26]. Despite the increasing number of related works, anecdotal evidence of algorithmic bias is still growing [2].

There are two primary sources of algorithmic bias; it can be due to the data or the algorithm. When it is due to the data, it is sometimes euphemistically called *negative legacy*; when it is due to the algorithm, it is called *underestimation* [6,17]. Negative legacy may be due to labeling errors or inadequate sampling; however, it is likely to reflect discriminatory practices in the past. Underestimation occurs when the algorithm focuses on strong signals in the data, thereby missing more subtle phenomena.

Algorithmic aspects of how ML algorithms can accentuate existing bias are not yet well-understood with researchers focusing directly on methods to eliminate bias - no matter the source. We believe that understanding the extent to which algorithms can amplify existing bias is of importance for designing new mitigation strategies to rectify the issue around algorithmic bias. Hence, the focus of this paper is on underestimation.

Our central hypothesis is that underestimation occurs when an algorithm *underfits* the training data due to a combination of limitations in training data

© Springer Nature Switzerland AG 2021
M. Kamp et al. (Eds.): ECML PKDD 2021 Workshops, CCIS 1524, pp. 559–574, 2021.
https://doi.org/10.1007/978-3-030-93736-2_41

and model capacity issues. This leads us to a series of sub-hypotheses – that the following factors can contribute to underestimation:

– Irreducible error (Bayes error)
– Regularization mechanisms
– Class imbalance
– Under-represented categories

These hypotheses are presented in more detail in Sect. 3. Before that, the relevant background research is reviewed in Sect. 2. Then, in Sect. 4, these hypotheses are tested on synthetic data. In Sect. 5, we show that the addition of synthetic counterfactuals to the training data can reduce the impact of underestimation. Still, there is clearly a need to manage the number of counterfactuals. Finally, the paper concludes in Sect. 6 with an assessment of how this repair strategy works on real datasets.

2   Background

The issues around bias and fairness in ML research have received a lot of attention in recent years. Several notions of what constitutes "fair" in ML have been proposed [8]. Despite this, there is still no unified consensus of what the best fairness notion shall be. In fact, recent works show that some of these notions suffer from significant statistical limitations or might perversely harm the very groups they were designed to protect [5]. In general, an ML model is considered fair if it is not inclined to award desirable outcomes $Y = 1$ (e.g., loan approval/job offers) only to one side of sensitive category $S = 1$ (e.g., gender/race).

The fairness literature has mainly focused on implementing new methods to ensure fairness without explicitly considering the source of bias [3]. Recent efforts in rectifying algorithmic bias include: transforming the dataset to remove discrimination before feeding it into a ML model (pre-processing) [16], modifying a specific algorithm's loss function to account for fairness (in-processing) [17], or transforming the ML model's output to ensure fairness (post-processing) [12].

Many argue that algorithmic bias is purely caused by negative legacy and that the algorithm models the data correctly (no underestimation). However, recent work shows that the algorithm itself could amplify existing bias [6,15]. Of particular interest in this regard is research on model pruning in deep neural networks [14,15]. Since the 1990s there has been research on how a significant proportion of the weights in a neural network can be pruned with minimal impact on model accuracy. Recent work by Hooker et al. [14,15] shows that this pruning can have a significant impact on underestimation. This phenomenon is directly related to the regularization issue we demonstrate in Sect. 4.4.

Although most examples of bias in ML systems occur due to negative legacy, we argue that it is essential to understand how specific algorithm mechanisms can introduce or at least accentuate bias. Understanding how algorithms can become discriminatory on the algorithmic level could help researchers in developing better and more general strategies to ensure fairness in ML.

2.1 Quantifying Bias

Disparate Impact (DI$_S$) is one of the accepted definitions of unfairness [9]:

$$\text{DI}_S \leftarrow \frac{P[\hat{Y} = 1 | S = 0]}{P[\hat{Y} = 1 | S = 1]} < \tau \tag{1}$$

It is the ratio of desirable outcomes $\hat{Y}$ predicted for the sensitive minority $S = 0$ compared with that for the majority $S = 1$. $\tau = 0.8$ is the 80% rule, i.e., proportion of desirable outcomes for the minority should be within 80% of those for the majority.

When the focus is on bias due to the algorithm only we can define an *underestimation score* (US$_S$) in line with DI$_S$:

$$\text{US}_S \leftarrow \frac{P[\hat{Y} = 1 | S = 0]}{P[Y = 1 | S = 0]} \tag{2}$$

This is the ratio of desirable outcomes predicted by the classifier for the sensitive minority compared with what is actually present in the data. If US$_S$ < 1 the classifier is under-predicting desirable outcomes for the minority. It is also important to note that US$_S$ = 1 does not necessarily mean that the classifier is not biased against the minority group (i.e., poor DI$_S$) score. It simply means that the algorithm does not underestimate the predictions of desirable outcome for the minority group.

An alternative underestimation score used by Kamishima *et al.* [17] that considers divergences between overall actual and predicted distributions for all groups S is the underestimation index (UEI) based on the Hellinger distance:

$$\text{UEI} = \sqrt{1 - \sum_{y,s \in D} \sqrt{P[\hat{Y} = y, S = s] \times P[Y = y, S = s]}} \tag{3}$$

Here y and s are the possible values of Y and S respectively. This Hellinger distance is preferred to KL-divergence because it is bounded in the range [0,1] and KL-divergence has the potential to be infinite. UEI = 0 indicates that there is no difference between the probability distribution of the training samples and prediction made by a classifier (no underestimation). In our evaluations we use US$_S$ rather than UEI because it focuses in on the impact on the sensitive minority is the same way that the DI$_S$ score does. We have found UEI scores can sometimes hide detail because it is an aggregate score across all feature/outcome combinations.

3   Factors Contributing to Underestimation

The definition of disparate impact emphasises fairness for all subgroups – see Eq. 1. This definition is independent of any data, indeed disparate impact is likely to be caused by biased historic data. By contrast the definition of under-estimation depends on test data (Eq. 2) – an outcome is under-predicted for a category. Our hypothesis is that this will occur when a model underfits the data. We identify four factors that can contribute to this.

Irreducible Error: In supervised ML the assumption is that the outcome variable Y is determined by the inputs to the model. Typically the inputs do not completely determine the outcome (there are hidden factors involved) so even the best model will have some error. This is the irreducible error, sometimes referred to as Bayes error [24]. If the irreducible error is low, the best models will be able to fit well to the training data and still generalise well to unseen data. If the irreducible error is high ($\sim$25%) then models that can generalise well will need to be relatively simple, i.e., they will underfit the training data. We show in Sect. 4.3 that underestimation directly correlates with irreducible error.

Regularization Mechanisms: It is standard practice in ML to use regularization methods to reduce generalization error. These methods are used to control the model capacity to gather relevant information from the training set. The objective of ensuring that the model does not overfit or underfit this information is crucial in ML because it is tied to the model's ability to generalize well on unseen data. There have been many proposed regularization techniques in the ML literature: lasso, ridge, elastic-net, dropout, early-stopping, etc., [13,23]. The main idea behind these methods is to control the model's complexity: in some circumstances, this is done by adding a penalty term in the loss function, or by limiting the number of nodes in the hidden layer. In regression settings, these methods are used to mitigate over-fitting by penalizing the effect of each predictor in explaining the target variable (reduce variance). Thus, excessive regularization will likely cause a model to underestimate predictions for the minority class. Since regularization directly correlates with bias and variance of a model, we show that it also influences underestimation (see Sect. 4.4).

Class Imbalance: Class imbalance refers to a classification problem where the number of observations in the data set differs for each class. The potential for this to lead to bias has been known for some time. It is known that the model predictions can accentuate the bias: if the minority class represents 30% of the training data the model is likely to predict <30% for that class [20,22]. Training ML models without addressing this issue will make it more challenging for the model to learn the characteristics of examples from the rare events, resulting in predictions that underestimate the overall minority class (i.e., biased towards predicting the majority) [11]. Many remediation strategies have been proposed to combat these issues, such as: sampling, cost-sensitive learning, etc., [25]. In Sect. 4, we will see the varying impact of class imbalance on underestimation.

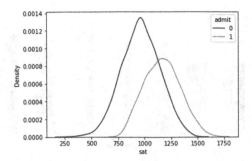

Fig. 1. Probability density functions of the synthetic data where the blue and orange curve shows the distribution of SAT scores for admitted and rejected applicants respectively.

Under-Represented Categories: One of the most common causes of bias in ML is the lack of observations for certain group(s). This difference in frequency often lead to disparate error rates on the under-represented group. For example, given a protected attribute such as race or gender, if a model is trained on a dataset with 5000 data points in which 80% belongs to the majority and only 20% represents the minority group, the model will likely struggle to learn as effectively from the minority examples. It is important to distinguish between this and the class imbalance issue highlighted above. This effect is aggravated when the irreducible error is high (the model needs to be simple). Although the question of directly using protected attributes in ML training remains open [27], we show in Sect. 4.5 that the level of under-representation directly correlates with underestimation.

4   Underestimation on Synthetic Data

In this section, we present results of the experiments based on the hypothesis presented in Sect. 3. These baseline results are demonstrated based on a synthetic data model that is widely used in research on bias on ML [9].

4.1   Data Description

This dataset consists of 5000 observations, each represents an applicant's statistics, which include IQ score, SAT score and a sensitive attribute (Favored, or Discriminated) [1]. The target variable, admit, indicates whether or not a student is admitted. The synthetic data generated has the following key properties: (i) The observations are evenly distributed (50% for both favored and discriminated group); (ii) The IQ scores are randomly sampled between 80 and 120; (iii) To reflect some correlation, the SAT scores are stochastic but dependent on the sensitive attribute and IQ; (iv) The admit outcome is also stochastic but correlated with the SAT score; and (v) since students in the favored group on average have

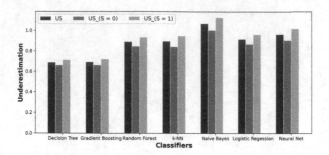

Fig. 2. A demonstration of classification bias and underestimation on the Synthetic dataset across multiple classifiers. It is clear that underestimation for the protected group ($US_{S=0}$) is exacerbated ($US_{S=0} < US$).

higher SAT score than those that are in discriminated group, favored students have been admitted at a higher rate than discriminated students (see Fig. 1).

70% of these observations are used for training, and the remaining 30% are reserved for model testing. We run experiments on seven classifiers implemented in scikit-learn[1]. These classifiers include tree-based classifiers (Decision Tree, Gradient Boost and Random Forest), and four other classifiers; k-Nearest Neighbor, Naive Bayes, Logistic Regression & Neural Networks. Taking into account the stochastic process of generating the synthetic sensitive attributes, each experiments was repeated twenty times and the median underestimation on the test set is obtained.

4.2   Underestimation and Classification Bias

Before proceeding to the factors influencing underestimation, it is worth emphasizing the distinction between underestimation and classification bias. Classification bias (US) occurs due to the model being trained on a highly imbalanced dataset resulting in predictions that underestimate the overall minority class Y [25]. We can then define classification bias as underestimation at a class level:

$$US = \frac{P[\hat{Y} = 1]}{P[Y = 1]} \tag{4}$$

In contrast, the underestimation score defined in Eq. 2 reveals situations where the classifier intensifies this under-representation for a specific sensitive group S, Fig. 2 illustrates the difference between classification bias and underestimation on the Synthetic dataset. We can see that the under-representation for protected group $US_{S=0}$ is exacerbated compared to the overall underestimation at a class level (US).

---

[1] https://scikit-learn.org/.

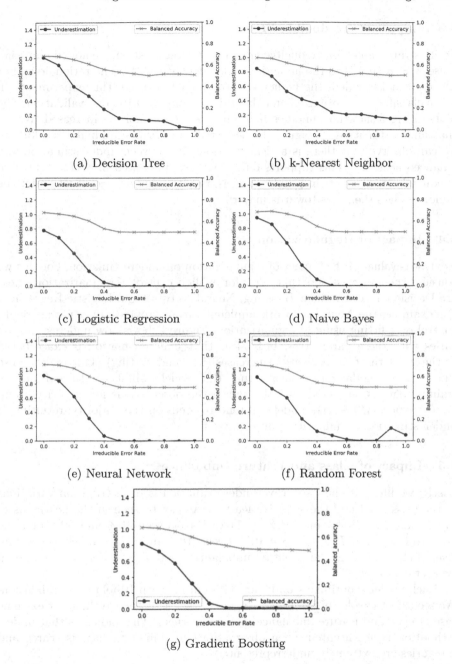

Fig. 3. Impact of irreducible error on underestimation. It is clear that underestimation US_S is exacerbated when irreducible error is high.

4.3   Impact of Irreducible Error

To illustrate impact of irreducible error on underestimation, we added Gaussian noise $X \sim \mathcal{N}(\mu, \sigma^2)$ to the numeric independent variables. The rationale is that the more noise added, the higher the irreducible error. In this experiment, all seven classifiers are optimized on balanced accuracy with 10-cross validation. The plots in Fig. 3 show how underestimation varies as the noise is increased. We see that underestimation is magnified when irreducible errors are high. In addition, we can observe that there is a clear relationship between underestimation and balanced accuracy. This supports our hypothesis presented in Sect. 3 that high irreducible error results in simpler models, resulting in bias accentuation for the minority class (i.e., bias towards majority).

4.4   Impact of Regularization

Next, we evaluated the impact of regularization on underestimation. For this we consider the four classifiers that allow for explicit control of regularization, these are Decision Tree, Gradient Boosting, Neural Network and Logistic Regression. For example, the Neural Network implementation provides an α parameter to control over-fitting using $l2$ regularization. Figure 4 shows how underestimation varies with these parameters. It is clear that underestimation is exacerbated as the strength of regularization increases (i.e., underfitting). This makes sense since higher regularization means that the model's ability to learn from the training data is restricted, and thus when the observations for minority group are extremely scarce, the model will tend to focus on the majority, resulting in underestimation for minority group.

4.5   Impact of Class and Feature Imbalance

Lastly, we illustrate how sensitive underestimation is to distribution variations in the class label and the sensitive feature. As can be seen in the heatmaps in Fig. 5, the Class Imbalance $P(Y = 1)$ varies between 10% and 30% and the Feature Imbalance $P(Y = 1 | S = 0)$ between 10% and 50%. To achieve certain level of class imbalance and feature imbalance, we took a sample of the original synthetic dataset.

Each model is optimized on balanced accuracy with 10-fold cross-validation. We see that the color of the heatmap gets darker as it gets to the most extreme case of class and feature imbalance (top left corner). This indicates that underestimation is accentuated when observations for minority class is scarce and categories are extremely underrepresented.

5   Remediation

In this section, we discuss possible strategies to remediate underestimation. We have shown in Sect. 4 that underestimation occurs when the classifier underfits

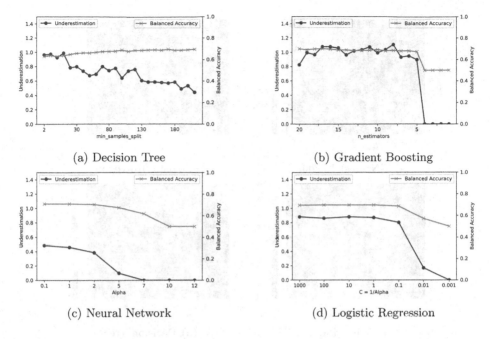

(a) Decision Tree   (b) Gradient Boosting

(c) Neural Network   (d) Logistic Regression

Fig. 4. Impact of regularization on underestimation.

the data due to a combination of limitations in training data and model capacity issues. Hence, the most apparent approach to fix this problem is to get more representative data. Unfortunately, quantifying exactly how much data is needed is not a trivial task in practice.

Many of the strategies for addressing Disparate Impact are also applicable for this more specific task of addressing underestimation, these include:

- **Pre-processing:** data augmentation (increasing the sample size for minority group, transforming training data to remove undesirable biases),
- **In-processing:** adding a constraint to a specific algorithm's loss function to account for underestimation, using cost-sensitive learning, or explicitly considering underestimation in hyper-parameter tuning,
- **Post-processing:** or selecting different optimal threshold value for the minority group.

We evaluated multiple strategies to reduce the impact of underestimation. First, we explicitly consider underestimation in hyper-parameter tuning. However, preliminary experiments on this strategy did not work out, showing only 2% improvement in underestimation. We have had more success with pre-processing techniques and we report on these results here. We have evaluated two strategies for generating counterfactuals (see Sect. 5.1) and a variation on the SMOTE algorithm for generating synthetic data samples. To ensure reproducibility of the

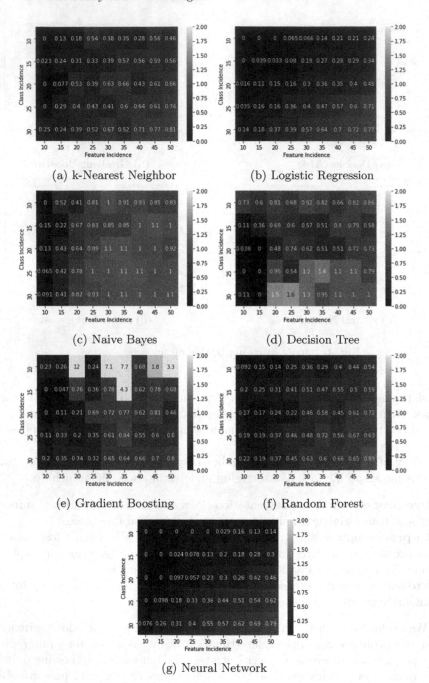

Fig. 5. Impact of class and feature imbalance on underestimation. Each value in the heatmap represents the underestimation score US_S for specific class and feature imbalance. $US_S = 1$ means that the classifier does not underestimate the predictions of desirable outcome for the minority group.

contents of this paper, we have provided access to all of the data and developed techniques used in this article at the author's Github page[2].

SMOTE [4] is perhaps the most popular method for addressing class imbalance in ML. SMOTE is a data augmentation strategy whereby synthetic samples of the minority class are generated by interpolating between actual samples. Our modification on this ($SMOTE_F$) is to only select synthetic samples corresponding to the minority group. These are referred to as S_0Y_1 in Algorithm 1, i.e., the Discriminated feature and the Positive class. In the notation of Algorithm 1, the Favoured Positive (S_1Y_1) samples produced by SMOTE are discarded.

5.1 Adding Counterfactuals

Counterfactual reasoning is an idea from Philosophy that first received attention in Artificial Intelligence research in the 1980s [10]. A counterfactual is a modification to a factual prior event to assess the consequences of the change. The use of counterfactuals in ML typically entails augmenting the data with synthetic counter examples to provide insight or aid explanation [18,21]. Here we use counterfactuals to shift the behaviour of the classifier. Whereas our $SMOTE_F$ strategy generates synthetic examples by interpolating between real ($Y = 1|S = 0$) examples we have two options for creating counterfactuals:

– **Counterfactual$_F$**: creating a desirable outcome for the minority group ($Y = 1|S = 0$) from a undesirable outcome ($Y = 0|S = 0$), or
– **Counterfactual$_L$**: a desirable outcome for the minority group ($Y = 1|S = 0$) from one of the majority group ($Y = 1|S = 1$).

The pseudocode for these strategies is shown in Algorithm 1. A key consideration with the three data augmentation strategies is the number of new samples N required to mitigate underestimation ($US_S = 1$). In our first experiments we simply double the number of samples representing the desirable outcome for the minority group ($Y = 1|S = 0$). The results on varying levels of underrepresentation of class labels and sensitive features are shown in Fig. 6. The classifier is the Neural Network used in earlier experiments. Due to the stochastic nature of the counterfactuals generation process, we repeat each experiment twenty times and the median of underestimation is obtained.

We can see that both Counterfactual$_F$ Counterfactual$_L$ performs best in almost all cases of under-representation. In fact, both of these strategies overshoot the underestimation target ($US_S = 1$) in extreme cases. In the next subsection we use a cross-validation strategy on the training data to select appropriate values for N to avoid the overshooting.

5.2 Tuning

It is clear from Fig. 6 that the impact of data augmentation varies considerably depending on the dataset make-up. For instance, the Counterfactual$_F$ strategy

---

[2] https://github.com/AlgorithmicFactorsInfluencingBiasinML.

Algorithm 1: Adding counterfactuals

Input : A dataset $D(X, Y, S)$, classifier $h(x) \to [0, 1]$
Output: Classifier trained on repaired dataset $h(\bar{D}) \to [0, 1]$
1. Divide D into four groups such that:
   - $S_0Y_1 \leftarrow \{x \in D | x.S = 0 \wedge x.Y = 1\}$
   - $S_0Y_0 \leftarrow \{x \in D | x.S = 0 \wedge x.Y = 0\}$
   - $S_1Y_1 \leftarrow \{x \in D | x.S = 1 \wedge x.Y = 1\}$
   - $S_1Y_0 \leftarrow \{x \in D | x.S = 1 \wedge x.Y = 0\}$
2. $N \leftarrow |S_0Y_1|$
3. if Counterfactual$_L$:
   - Randomly sample N observations from S_0Y_0 with replacement
   - Change the class labels of the sampled data Y to 1
   elif Counterfactual$_F$:
   - Randomly sample N observations from S_1Y_1 with replacement
   - Change the sensitive attribute of the sampled data S to 0
4. Concatenate D with the sampled data
5. Train a classifier $h(x)$ on modified D
6. Return Classifier trained on repaired dataset $h(\bar{D}) \to [0, 1]$

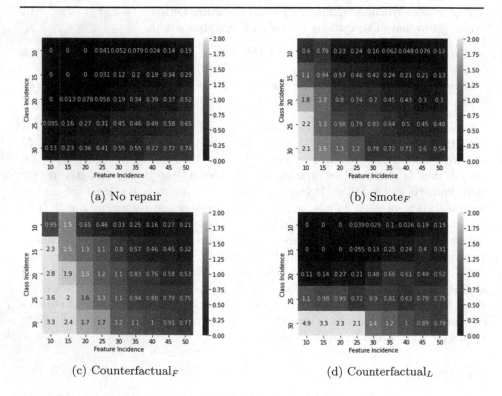

(a) No repair

(b) Smote$_F$

(c) Counterfactual$_F$

(d) Counterfactual$_L$

Fig. 6. The impact of adding synthetic samples using Smote$_F$ and the two counterfactual strategies. In this baseline analysis we always double the size of the DP set (see Algorithm 1).

produces US_S outcomes that vary between 0.21 and 3.3. However, we should be able to tune the process by estimating the best value for N using cross-validation.

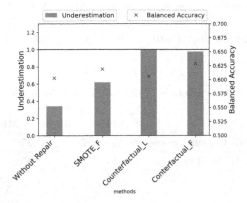

Fig. 7. Tuned remediation: Remediation is repeated on the dataset from Fig. 6 with class imbalance of 20% and feature imbalance of 45% but with N set using cross-validation on the training data.

Given that Counterfactual$_L$ draws from S_0Y_0 and Counterfactual$_F$ draws from S_1Y_1 we can use up to 100% of these sets in the augmentation process. The validation process (i.e., parameter tuning) considers options from 5% to 100% in steps of 5%. In this experiment the SMOTE$_F$ strategy is also tuned by considering options from 5% to 100% of $N \leftarrow |S_1Y_1| - |S_0Y_1|$.

We used a cross-validation process to select the best value for N for each strategy, then the whole training dataset is augmented and the remaining 30% that has been held back for testing is used to assess performance. It is worth stressing that the validation process of selecting the best value for N is performed only on training data. The results are shown in Fig. 7. The tuned SMOTE$_F$ does little to improve underestimation but the two counterfactual strategies do a good job, both bringing US_S close to 1.0. It is worth noting that, in addition to fixing underestimation, Counterfactual$_F$ also improves balanced accuracy from 60.3% to 62.8%.

6   Underestimation on Real Datasets

In this section, we experimentally validate our proposed remediation strategies described in Sect. 5 on the Census Income dataset [19] and a reduced version of the ProPublica Recidivism dataset [7]. These datasets have been extensively studied in fairness research because there is clear evidence of negative legacy. Summary statistics for these datasets are provided in Table 1. For the Census Income dataset the prediction task is to determine whether a person earns more or less than $50,000 per year based on their demographic information. The reduced and anonymized version of the Recidivism dataset includes

7 features and the target variable represents whether a person gets rearrested within two years after the first arrest. The goal of our experiment is to learn an underestimation-free classifier while maintaining high balanced accuracy score when *Sex* and *Caucasian* used as sensitive feature S for Income and Recidivism dataset, respectively. The classifier is the Neural Network as used in Sect. 5.2 and the same tuning strategy is employed to determine the level of data augmentation.

Table 1. Summary details of the Adult and Recidivism datasets.

Dataset	Samples	Features	% Minority
(reduced) Census income	48,842	7	25%
(reduced) Recidivism	7,214	7	45%

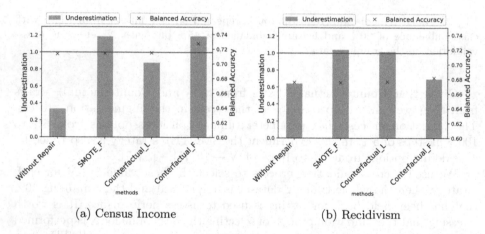

(a) Census Income   (b) Recidivism

Fig. 8. An evaluation of the remediation strategies on two real-world datasets.

The results shown in Fig. 8 demonstrate that all of the proposed methods reduce the impact of underestimation. Looking closer at the Census Income dataset, we see that Counterfactual$_F$ overshoots underestimation by 10%. We suspect that this impact can be removed (or at least reduced) by allowing larger search space for N and hence, we can conclude that the proposed strategies are very sensitive to hyper-parameter N. Moreover, the results in Figs. 7 and 8 show that there is no one remediation strategy that performs best in all datasets, suggesting the importance of tuning. It is worth noting that Counterfactual$_F$ improves balanced accuracy for all three datasets. This indicates that it is a natural policy for producing counterfactuals. In addition, it is also worth emphasizing that the $SMOTE_F$ strategy is a modification of existing oversampling technique that generate synthetic data samples for minority feature $S = 0$ rather than minority class $Y = 0$.

7   Conclusions and Future Work

This paper started by emphasising the difference between negative legacy and underestimation as sources of bias in ML. The first is a problem with the data, the second is a problem with the algorithm. We have shown that underestimation can be addressed by adding counterfactuals, suggesting that perhaps it is really all about the data. It is not. Negative legacy refers to scenarios where there are undesirable patterns in historic data. Underestimation relates to scenarios where the algorithm is not picking up the patterns that *are* in the data. It is worth saying then that fixing underestimation in the sense that the algorithm reflects the data may still leave a fairness problem due to negative legacy.

While our evaluation shows that the use of counterfactuals can ameliorate underestimation there is still some room for improvement. Our next step will be to use ideas from the work of Keane and Smyth [18] to select better counterfactuals. In future we plan to develop in-processing strategies whereby underestimation is explicitly considered in the ML algorithm optimization process.

Acknowledgements. This work was funded by Science Foundation Ireland through the SFI Centre for Research Training in Machine Learning (Grant No. 18/CRT/6183) with support from Microsoft Ireland.

References

1. Adler, P., et al.: Auditing black-box models for indirect influence (2016)
2. Bender, E.M., Gebru, T., McMillan-Major, A., Shmitchell, S.: On the dangers of stochastic parrots: can language models be too big? In: Proceedings of the 2021 ACM Conference on Fairness, Accountability, and Transparency, pp. 610–623. FAccT 2021. Association for Computing Machinery, New York, NY, USA (2021)
3. Caton, S., Haas, C.: Fairness in machine learning: a survey. arXiv preprint arXiv:2010.04053 (2020)
4. Chawla, N.V., Bowyer, K.W., Hall, L.O., Kegelmeyer, W.P.: SMOTE: synthetic minority over-sampling technique. J. Artif. Int. Res. **16**(1), 321–357 (2002)
5. Corbett-Davies, S., Goel, S.: The measure and mismeasure of fairness: a critical review of fair machine learning (2018)
6. Cunningham, P., Delany, S.J.: Algorithmic bias and regularisation in machine learning. arXiv preprint arXiv:2005.09052 (2020)
7. Dressel, J., Farid, H.: The accuracy, fairness, and limits of predicting recidivism. Sci. Adv. **4**(1), eaao5580 (2018)
8. Dunkelau, J., Leuschel, M.: Fairness-aware machine learning (2019)
9. Feldman, M., Friedler, S.A., Moeller, J., Scheidegger, C., Venkatasubramanian, S.: Certifying and removing disparate impact. In: proceedings of the 21th ACM SIGKDD International Conference on Knowledge Discovery and Data Mining, pp. 259–268 (2015)
10. Ginsberg, M.L.: Counterfactuals. Artif. Intell. **30**(1), 35–79 (1986)
11. Guo, X., Yin, Y., Dong, C., Yang, G., Zhou, G.: On the class imbalance problem. In: 2008 Fourth International Conference on Natural Computation, vol. 4, pp. 192–201 (2008)

12. Hardt, M., Price, E., Srebro, N.: Equality of opportunity in supervised learning. In: Proceedings of the 30th International Conference on Neural Information Processing Systems, pp. 3323–3331. NIPS 2016, Curran Associates Inc., NY, USA (2016)
13. Hoerl, A.E., Kennard, R.W.: Ridge regression: biased estimation for nonorthogonal problems. Technometrics **12**(1), 55–67 (1970)
14. Hooker, S., Courville, A., Clark, G., Dauphin, Y., Frome, A.: What do compressed deep neural networks forget? (2020)
15. Hooker, S., Moorosi, N., Clark, G., Bengio, S., Denton, E.: Characterising bias in compressed models. arXiv e-prints pp. arXiv-2010 (2020)
16. Kamiran, F., Calders, T.: Data pre-processing techniques for classification without discrimination. Knowl. Inf. Syst. **33**, 1–33 (2011). https://doi.org/10.1007/s10115-011-0463-8
17. Kamishima, T., Akaho, S., Asoh, H., Sakuma, J.: Fairness-aware classifier with prejudice remover regularizer. In: Flach, P.A., De Bie, T., Cristianini, N. (eds.) ECML PKDD 2012. LNCS (LNAI), vol. 7524, pp. 35–50. Springer, Heidelberg (2012). https://doi.org/10.1007/978-3-642-33486-3_3
18. Keane, M.T., Smyth, B.: Good counterfactuals and where to find them: a case-based technique for generating counterfactuals for explainable AI (XAI). In: International Conference on Case-Based Reasoning, pp. 163–178. Springer (2020)
19. Kohavi, R.: Scaling up the accuracy of Naive-Bayes classifiers: a decision-tree hybrid. In: Proceedings of the Second International Conference on Knowledge Discovery and Data Mining, vol. 96, pp. 202–207 (1996)
20. Kubat, M., Holte, R.C., Matwin, S.: Machine learning for the detection of oil spills in satellite radar images. Mach. Learn. **30**(2), 195–215 (1998)
21. Kusner, M.J., Loftus, J., Russell, C., Silva, R.: Counterfactual fairness. In: Guyon, I., et al. (eds.) Advances in Neural Information Processing Systems, vol. 30. Curran Associates, Inc. (2017)
22. Mac Namee, B., Cunningham, P., Byrne, S., Corrigan, O.I.: The problem of bias in training data in regression problems in medical decision support. Artif. Intell. Med. **24**(1), 51–70 (2002)
23. Thodberg, H.H.: Improving generalization of neural networks through pruning. Int. J. Neural Syst. **01**(04), 317–326 (1991)
24. Tumer, K., Ghosh, J.: Estimating the Bayes error rate through classifier combining. In: Proceedings of 13th International Conference on Pattern Recognition, vol. 2, pp. 695–699. IEEE (1996)
25. Weiss, G.M.: Mining with rarity: a unifying framework. SIGKDD Explor. Newsl. **6**(1), 7–19 (2004)
26. Žliobaitė, I.: Measuring discrimination in algorithmic decision making. Data Mining Knowl. Discovery **31**(4), 1060–1089 (2017)
27. Žliobaitė, I., Custers, B.: Using sensitive personal data may be necessary for avoiding discrimination in data-driven decision models. Artif. Intell. Law **24**(2), 183–201 (2016)

Desiderata for Explainable AI in Statistical Production Systems of the European Central Bank

Carlos Mougan Navarro[1(✉)], Georgios Kanellos[2], and Thomas Gottron[2]

[1] University of Southampton, Southampton, UK
C.Mougan-Navarro@southampton.ac.uk
[2] European Central Bank, Directorate General Statistics, Frankfurt, Germany
{George.Kanellos,Thomas.Gottron}@ecb.europa.eu

Abstract. Explainable AI constitutes a fundamental step towards establishing fairness and addressing bias in algorithmic decision-making. Despite the large body of work on the topic, the benefit of solutions is mostly evaluated from a conceptual or theoretical point of view and the usefulness for real-world use cases remains uncertain. In this work, we aim to state clear user-centric desiderata for explainable AI reflecting common explainability needs experienced in statistical production systems of the European Central Bank. We link the desiderata to archetypical user roles and give examples of techniques and methods which can be used to address the user's needs. To this end, we provide two concrete use cases from the domain of statistical data production in central banks: the detection of outliers in the Centralised Securities Database and the data-driven identification of data quality checks for the Supervisory Banking data system.

Keywords: Explainable AI · Case studies · Counterfactual reasoning

1 Introduction

Improvements in the predictive performance of machine learning algorithms have led to applications across many domains, including central banking, education, healthcare, and public policy. As machine learning becomes increasingly ingrained as a means to improve business efficiency and decision-making, it becomes crucial to establish processes for ensuring fairness, explainability, data privacy, and security.

Explainability has become an important concept in legal and ethical guidelines for data and machine learning applications [39]. The European Union General Data Protection Regulation (GDPR) [13] states that individuals have the right to receive an explanation for the output of an automated algorithm [12]. Ensuring algorithmic accountability and transparency in machine learning models is a key ingredient towards understanding how decisions are made by the system. Understanding decisions is a prerequisite to detecting potential bias and ensuring fair and safe machine learning systems.

Disclaimer: This paper should not be reported as representing the views of the European Central Bank (ECB). The views expressed are those of the authors and do not necessarily reflect those of the ECB.

© Springer Nature Switzerland AG 2021
M. Kamp et al. (Eds.): ECML PKDD 2021 Workshops, CCIS 1524, pp. 575–590, 2021.
https://doi.org/10.1007/978-3-030-93736-2_42

The research community has made an enormous effort to design algorithmic methods for explainability. Various publications reconcile the information of all relevant papers, create explainable AI (xAI) taxonomies, define future opportunities and current gaps [2,3,16,18,27,38]. It is worth noting that several authors have drawn attention to the *need for industry and institutions to establish clear desiderata for their explanation needs* [2,5]. Such desiderata are needed to assess if developed explainability methods address the requirements of real use cases.

In this paper, we leverage practical experience with machine learning applications in the domain of statistical production systems at central banks to present a user-centric classification of xAI requirements and needs. The user-centered approach aims to help both: (i) researchers to orient and describe their work along concrete requirements and (ii) practitioners to identify what kind of xAI challenge they face and which (technical) method might solve their problem. We illustrate the applicability of the desiderata classification scheme using two real-world use cases where xAI needs played an important role. For these two use cases, we map the requirements to the classification scheme and describe the technical algorithms chosen to solve the needs.

In this paper we make the following contributions:

1. With our practical experience we identify a set of user-centric explainability needs that are generic and common throughout different use cases.
2. We provide concrete examples of use cases that make the classes of desiderata more practical, aiming to help identify potential research gaps.

The rest of the paper is organized as follows: in Sect. 2 we introduce related work and outline differences with our contribution. Section 3 introduces the user-centric classification with definitions of the explainability needs together with the user roles involved in the process. In Sect. 4 we describe concrete use cases with their proper introduction, state desiderata, and map business requirements to the classification scheme and algorithmic solutions. Finally, in Sect. 5 we summarise the main conclusions of the paper and outline possible future work.

2   Related Work

Research on xAI has gained a lot of attention [1] and a growing number of research papers on the topic is being published in conferences and journals [3,18]. According to the literature, papers in xAI methods might be suffering from two issues:

(i) A lack of a rigorous definition and evaluation methods for explainability [11,23]
(ii) Explainability methods are introduced as general-purpose solutions and do not directly address real use cases or a specific user audience [2,26]

The challenge of evaluation methods for xAI is typically addressed with user studies. Few recent papers evaluate explanation methods in a very specific real-world task with real users [21], but they do not offer a generalisation to other use cases. More often, previous work found in the literature relies either on theoretical problems or toy datasets [4]. Accordingly, the effectiveness and usefulness of theoretical research papers on real-world applications are still unclear [23, 26]. There are few real explainable machine learning use cases where a full xAI methodology is presented addressing intended users and tasks.

The challenge of mapping explainability methods to concrete use cases is addressed by Bhatt et al. [5]. They critically examine how explanation techniques are used in practice by interviewing various organizations on how they employ explainability in their machine learning workflows. They state that there is a need for the industry to define clear desiderata for explainability, with clear goals and intended users. Our work aims to address that gap by defining needs, users and showcasing them with two use cases. Amarasinghe et al. [2], present a position paper where they aim to define xAI roles and intended users for the public policy domain. Our work extends their research by redefining the taxonomy with a more user-centric approach and making use of two particular use cases to identify more specific explainability needs.

To the best of our knowledge there is no previous work done on the desiderata for explainable machine learning in use cases for statistical production systems, nor on stating user-centric xAI requirements for a public policy institution.

3   User-Centric Classification of Desiderata for Explainability Needs

While working on several central banking projects involving machine learning we observed recurrent patterns of where needs for explainability arose. This experience motivated us to structure and classify the needs and requirements for explainability. The needs are described in a user-centric way in order to identify stakeholders and their specific requirements. This user-centric approach also allows system designers to work with *personas*[1] to reflect specific needs for xAI.

In this section we introduce and describe our classification of generic explainability needs. Furthermore, the entries in the classification link xAI needs with possible methods to address them. We do not intend to provide an in-depth survey on existing work[2]. Rather we intend to give examples of existing work that meet the xAI requirements for some use cases we encountered.

The needs are synthesized from practical use cases of machine learning at the European Central Bank (ECB) in the context of statistical data production. We will provide two concrete use cases to fill the classification with life in Sect. 4. While the use cases will serve as examples on how to use the classification, the desiderata reflect generic user needs which in our view cover most relevant use cases of xAI.

3.1   Users

As stated above, we came across several generic user roles which help to classify the needs for solutions of explainable and responsible AI. A key question driving this classification is *Who needs an explanation of an AI method?* This helps to clearly define and distinguish different desiderata for explanations.

---

[1] We refer to the concept of personas as it is used in approaches like design thinking.
[2] Please refer to some of the paper referenced in related work for comprehensive surveys.

The following profiles define users that can potentially interact with a machine learning system all of whom have quite different needs for explainability:

(i) **Data scientists and AI engineers:** This role corresponds to members of the team who build, model and run a machine learning application. This type of user has technical expertise but does not necessarily have business expertise. They are in charge of the full life cycle of the machine learning application, from development to maintenance in production.

(ii) **Business experts:** Users with this role provide the use case and domain expertise for a machine learning solution. They define the business activity or process which is supported by the AI solution. In our case, they are finance, economics and statistics experts from the European System of Central Banks who act or intervene in business processes based on the recommendations of the models. This type of user might not have a technical background and is not a machine learning expert.

(iii) **High stake decision makers:** This type of user determines whether to use and incorporate a machine learning model in the decision-making process. They typically have a management position, a high-level understanding of the business objectives and a responsibility to deliver value. They need to understand and assess the potential risk and impact of incorporating the machine learning model into production.

(iv) **End users:** Users which are affected by or make use of the final results belong to the group of end users. The knowledge and potential expertise of this user group varies significantly. There might be cases where the group of end users actually overlaps or is even identical to the group of *business experts*, e.g. when the machine learning solution is primarily serving internal business processes. Examples of end users in our domain are the business areas or even the general public making use of data compiled by the Directorate General Statistics at the ECB.

3.2   Building Trust

Trust is essential for the adoption of a machine learning application. Depending on the user, the meaning of trust and the way to obtain it differ. In this section we analyse the different types of trust which need to be established for the four user types.

Data scientists and AI engineers are responsible for the whole life cycle development of the machine learning model [6]. The engineering phase of the cycle includes tasks like data exploration, deciding on an evaluation scheme, feature engineering, selecting and training a model as well as evaluating the results in a systematic way to ensure performance in a production environment.

During this phase, the *data scientist* has to ensure that meaningful features are selected, that there are no data leakages and that spurious correlations are addressed. This can be achieved by performing sanity checks and understanding the model output through the use of feature relevance explainers such as SHAP [25], tree Explainer [24], LIME [34,35], and local statistical aggregations [19].

Business experts have knowledge about business processes and often have practical experience in how the task for the machine learning systems should be performed. In certain cases the *business experts* might even have been responsible to actually perform the task that is to be taken over by a machine learning solution. Hence, from their perspective the aspects that need to be addressed to generate trust in a machine learning algorithm are the following: (*i*) adapting to a changed environment (*ii*) understanding the decisions of the algorithm.

The disruption caused by the introduction of machine learning in the process needs be mitigated in order for the experts to have enough time to understand, test and embrace the new method. To this purpose, the predictions or recommendations of the machine learning algorithm can be introduced and transparently explained in parallel to existing measures and processes so that the experts can familiarise themselves with the new functionality. Once it becomes clear to them that by making use of machine learning they can achieve equal or higher levels of effectiveness and efficiency, transition is simpler.

The second issue that needs to be addressed is the explainability of the results. As part of their business processes, *business experts* might be required to be able to understand the reasons or facts that led to the recommendations and decisions produced by the algorithm. The decision-making process might need to be assessed internally or even externally e.g. in the context of audits.

It is worth noting that the technical level of this type of user is generally lower than that of the data scientist, so normally a less technical explanation or even an *explanation of an explanation* is required. This can be achieved via feature attribution techniques such as SHAP and LIME. It helps immensely in the trust building process if the criteria for the decisions reflect at least partially the prior knowledge of the *business experts*.

For *high stake decision makers* the need to understand machine learning algorithms reaches yet another, even less technical level. They take high-stake decisions and thereby responsibility for the adoption of AI solutions in a production environment. They do not need to understand the machine learning solution on the basis of individual decisions, but on an aggregated level. They need to understand the opportunities in terms of gains in efficiency and effectiveness, ideally using quantifiable measures. At the same time they need to assess risks that come with the adoption of a solution. The risks relate to the probability of errors and the impact of such errors on various levels (operational, competition, reputation, etc.). In particular, it might be required to ensure that an AI system is not biased and non discriminatory. Hence, explainability needs to be mapped to different concepts, which allow for assessing a solution[3].

Furthermore, *high stake decision makers* might want to make sure that solutions for explainability for individual AI based decisions are available. This is managerial and operational prudence to ensure that the needs and requirements of the *business experts* are satisfied (see above).

End users are normally not involved in the data processing part of the process, which means that the algorithms used are not known to them. Nevertheless, ensuring that they continue to trust the data after the introduction of machine learning is a crucial part of the data workflow. This comes down to ensuring they trust the outcome of the algorithm either when it affects the end users themselves or when it is used as a service.

End user trust is obtained through consistently achieving high output quality as well as clear and transparent communication. For example, in the context of statistical publications which have certain timeliness and continuity in the data, sudden variations

---

[3] We hypothesise that this is probably one of the most challenging fields for xAI research.

might distress users. Therefore, the introduction of a new ML algorithm in production must be validated thoroughly and in case it introduces significant changes in the data output of the system these must be clearly communicated in advance to the users so that they understand the underlying reasoning.

3.3 Gaining Knowledge Through Machine Learning

Besides very high prediction performance, machine learning can enhance data understanding. The aforementioned high performance is achieved through the "discovery" of patterns or connections that have not be utilised by other approaches. Understanding these patterns and what they mean can offer important new insights [1]. Such insights are of particular relevance for *business experts*. They may use the discovered patterns together with their business experience on new use cases, or to obtain deeper domain knowledge.

The insights, however, need to be presented in an interpretable way. The patterns have to be communicated to the *business experts* in a way that they can make sense of it. The requirements for this type of explainability are different than for the understanding of a model for the originally intended task. A wide range of different explanations, representations and visualisations might be needed to stimulate a creative cognitive process with the *business experts*.

The evaluation of the pedagogical aspect of xAI techniques for machine learning models remains as a future avenue of research. In Sect. 4 we will provide an example from the domain of central banking, where gaining knowledge was one of the primary objectives for a machine learning project.

3.4 Model Monitoring

Model monitoring aims to ensure that a machine learning application in a production environment displays consistent behavior over time. Monitoring is mainly performed by the *data scientist and AI engineer* and is crucial, as a drop in model performance will affect all the users. Two common challenges in model monitoring are (i) distribution shifts in the input data that can degrade model performance (ii) changes in the machine learning algorithm due to a model retraining that can alter the individual explanations for decisions.

The task of *monitoring distribution shifts* can be formalised as follows: Given source dataset D_S and test dataset D_T, consisting of input x and targets y, drawn from underlying distributions $p_S(x, y)$ and $p_T(x, y)$ respectively, we aim to detect changes in the distribution in p_T [9]. This phenomenon is known as distribution shift or concept drift [9,33] and is particularly difficult to notice.

The challenge from an xAI perspective is how to assess such changes from the perspective of the *business expert* and the *AI engineer*. *Business experts* need to be able to identify if the distribution shift can be explained from a business perspective. Hence, they need to be able to identify possible causal relationships between the change in the data and exogenous events. On the other hand, the *AI engineer* needs to assess the impact of the change on the model's performance and potentially intervene.

Monitoring changes in the explanation addresses another phenomenon. The performance of machine learning models degrades over time. In order to maintain high performance, models are retrained using previous and new input data. This continual learning process can lead to changes in individual explanations for decisions through time. As an example, the explanation offered by the classifier $f_C(x)$, for the decision taken for the i-*th* instance at a time t_0, might differ compared to the explanation offered at time t_n for the same instance after model retraining.

$$E_{t=0}(f_C(x_i)) \neq E_{t=n}(f_C(x_i)) \tag{1}$$

This change due to the continual learning process, increases the risk of contradictory explanations for the same instance. Local explanations are prone to this kind of risk, due to the fact that small changes in the neighborhood of the instance can lead to a different reasoning from the model once retrained.

Both, *business experts* and *AI engineers* need to be able to detect and understand this phenomenon. For the *business experts* such a change might indicate a change in the business logic, for the *AI engineer* it might indicate to review the modelling process and design decisions taken during the development phase. Depending on the use case, even *high stake decisions makers* may need to be aware of such changes, as inconsistent explanations for the same setting may pose a risk of unfair algorithmic decisions.

3.5 Actionable Insights

Understanding a machine learning algorithm is usually not an end in itself. The explanation offered through this understanding supports business processes and leads to actionable insights. Such insights enable the *business expert* to understand how to change a decision by manually intervening in the data. For instance, when data is identified to belong to a certain class, providing a set of actionable changes that would lead to a different decision can assist an expert in correcting or modifying the data.

Counterfactual generation aims to address this issue by proposing to the *business experts the minimal feasible change in the data in order to change the output of the algorithm*. Such a process enhances the understanding of the experts (and might further foster trust in the system). We formulate the problem following the counterfactual recourse formulation by Utsun et al. [22, 44] and the open source python package DICE [28] that quantifies the relative difficulty in changing a feature via feature weights.

3.6 Fostering Explanations Through Simple Models

Copying [41, 43] or distilling [20] machine learning models can greatly contribute to model explainability. Overly complex models tend to be difficult to explain [6] and can become unaccountable [38]. Model agnostic copies with a simple model might be able to achieve global explainability [40] which can be useful to build trust and gain knowledge by the *business expert*. Furthermore, in some deployment scenarios involving incompatible research and deployment versions [41], copying the ML model can ease the deployment task for the *data scientist*.

We now define what is copying a machine learning classifier [41–43]. We use the term original dataset to refer to a set of pairs $X = (x_i, t_i), i = 1, ..., M$, where $x_i \in R_d$

is a set of d-dimensional data points in the original feature space D and $t_i \in 1, ..., K$ their corresponding labels. We define the original model, $f_O : X \rightarrow t$, as any model trained using X. Our aim when copying is to reproduce the behavior of the original model employing a copy, f_C, such that $f_C(x) = f_O(x), \forall x \in D$. To do so we use a two-step process. Firstly, we rely upon the generation of a synthetic target, $y_j = f_O(X)$, where y_j is the target label associated with each individual sample from the original data distribution X. Secondly, we use the synthetic target points to train a copy, f_C, whose decision function reproduces f_O to the extent that it can be used to substitute it.

4   Use Cases

Providing statistical data and information of the highest quality is a core task of the European System of Central Banking (ESCB). Assuring the quality of data in statistical production systems is crucial as it is used in the decision-making process. Statistical production systems rely on domain experts with an outstanding understanding of the data. These experts ensure that the information used in the compilation of statistical products is of the highest quality based on their expertise.

The benefit of using expert knowledge for quality assurance of statistical products is two-fold: (i) all the knowledge and expertise of the domain experts are reflected in the final data and (ii) the overall confidence about the final product increases on the side of the producers and consumers of official statistical data. However, as the volume and granularity of the data increase, this process becomes progressively difficult to maintain. It is almost impossible to manually assess the quality of granular data without substantially increasing the number of experts.

Over the years various approaches have been developed to support and automate quality assurance procedures. However, these approaches have certain limitations:

- **Data aggregation:** Assessing the quality of the data on an aggregated level, condenses the information that needs to be checked to a manageable amount. The disadvantage of this approach is that aggregation might hide granular data quality issues on the level of individual observations. The risk of obfuscation depends on the type and level of aggregation used.
- **Static quality rules:** Another way to ensure data quality for a big data scenario, is using a fixed set of rules to identify potential quality issues. These rules are based on expert knowledge and formalize certain aspects of their domain knowledge. Such rules flag individual records that, for instance, exceed a predefined threshold. The flagged observations typically represent only a small fraction of the entire data set, allowing the field experts to focus their analysis on a limited amount of data. The shortcoming of this approach is that relying on a fixed set of static rules is not flexible enough to capture dynamics in data nor does it automatically adjust to new insights provided by users.

To maintain high standards of quality assurance, new approaches that address these limitations are investigated. This section illustrates two applications of machine learning solutions in the area of central banking statistics and how different user groups and needs for xAI have been addressed.

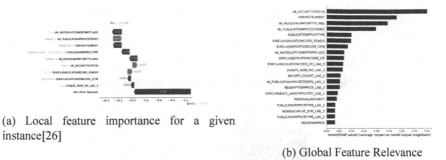

(a) Local feature importance for a given instance[26]

(b) Global Feature Relevance

Fig. 1. Shapley feature values by TreeExplainer [24]

4.1 Centralised Securities Database

The Centralised Securities Database (CSDB) [32] is a securities database with the aim of holding complete, accurate, consistent, and up-to-date information on all individual securities relevant for the statistical and, increasingly, non-statistical purposes of the ESCB. Ensuring quality in such a system is challenging given the amount of data that needs to be monitored. So far, it has been achieved through a combination of data aggregation and static quality rules. In addition, *business experts* with the role of Data Quality Managers have their internal measures for assessing data quality and intervening in case of issues.

Machine Learning Application. The objectives of our approach were [29]:

(i) Identify records in the dataset where there is a high chance that manual intervention is needed.
(ii) Communicate to the *business experts*r which is the field that is most likely to be incorrect.
(iii) Rank these suggestions to minimize the number of false-positives in the top recommendations.

Building Trust. Despite the careful testing and calibration of the machine learning process by the *data scientists, end users* of the data can identify potential data issues that have not been raised by the algorithm. These issues are communicated via the Data Quality Managers (DQM) to the statistical production team (comprising of *business experts* and *data scientists*) responsible for the CSDB data quality [29].

The first step for the statistical production team is to verify that indeed the issues identified are not flagged by the algorithm. The next step is to investigate the reasoning behind this choice from the perspective of the algorithm by determining: (i) similar instances in the training dataset [7], (ii) which features contributed to the decision [24,35] (cf. Fig. 1) and (iii) local decision rules to better understand the model logic [17,36].

Feature	OriginalValue	ModifiedValue	InitialPrediction	ModifiedPrediction
VA_SECURITYSTATUS	100	201	1	0
IDIRCLASSICATIONCODE_ESAI10	F_31	F_32	0	1
VA_SECURITYSTATUS	101	203	0	1
PUBLICATIONPRICETYPE	CLC	PAY	1	0

Fig. 2. Set of counterfactual decisions generated [17,28]

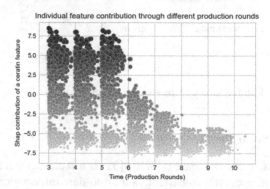

Fig. 3. Simulated covariate distribution shift detected by local attribution of Shapley values in one of the descriptive features

Actionable Insights. Following the formulation from Utsun et al. [44] and the available open source implementation [28], we quantified the relative difficulty in changing a feature through the application of weights to the counterfactual explanations algorithm. For instance, in our case, recommendations should not ask the *business expert* (DQM) to modify the country in which a financial instrument was issued or change the issue date to a time before the creation of the issuing company (cf. Fig. 2).

Model Monitoring. Detecting when the underlying distribution of the data changes is paramount for this use case, since failing to predict outliers or errors in the data will lead to a drop in the trust of the machine learning model. Also, the risk of having an incoherent explanation through time caused by the continual learning process is utterly important, a discrepancy will lead to a decrease of trust by the *business experts*.

An approach suggested by Lundberg et al. [24] is to monitor the SHAP value contribution of input features over time together with decomposing the loss function across input features in order to identify possible bugs in the pipeline.

In Fig. 3, we can see the local contribution of SHAP values through time for a continuous feature. The drop of Shapley contributions denotes a possible distribution change for a given feature or a change in the model's inner working that needs human assistance for further evaluation of performance and explainability [24].

Fostering Explanations Through Simple Models. Following the notation of paragraph 3.6, for our case the original model $f_O(X)$ is a catboost classifier [10,31] which is a gradient boosting model [14] that often achieves state of the art results in many

Table 1. Catboost is the original model $f_O(X)$ and the Decision Tree and the Generalized Linear Model the copied classifier $f_C(X)$

	Catboost	Decision tree	GLM
AUC	0.816	0.771	0.741
Precision	0.805	0.741	0.685
Recall	0.833	0.824	0.733

different types of problems [8, 45] and the copied models $f_C(X)$ are a scikit-learn [30] decision tree classifier and a Generalized Linear Model (GLM). The simpler models are of slightly inferior quality (cf. Table 1). However, having two simple models helps to improve the overall global explainability for the *data scientist* and to simplify deployment.

4.2 Supervisory Banking Data System

The Supervisory Banking data system (SUBA) is the ECB's system to collect supervisory data on credit institutions. The data collected in SUBA is in line with the reporting requirements defined by the European Banking Authority (EBA). SUBA implements all validation rules defined by the EBA to assess data quality. These validation rules implement logical and business-motivated checks of consistency and correctness for the reported data.

However, supervisors and *business experts* at the ECB may define additional *plausibility checks* to further ensure the quality of the data. So far, the additional *plausibility checks* were defined manually based on domain expertise. In an internal project, we developed a data-driven approach leveraging machine learning to identify patterns in supervisory data. The patterns are then used as a basis for defining additional plausibility checks [37]. Furthermore, it was of interest to identify individual observations in the reports which strongly deviated from the observed patterns. In the context of this solution, we had to consider in particular the aspects of *building trust, gaining knowledge* and *obtaining actionable insights*.

Machine Learning Application. Given the overall settings, the concrete objectives of our approach were to:

(i)   Identify novel functional dependencies between subsets of data points reported for supervisory data.
(ii) Ensure the functional dependencies do not correspond to existing validation rules.
(iii) Present the functional dependency to business experts, such that they can analyze from a business perspective if the patterns motivate the introduction of new plausibility checks.
(iv) Identify anomalous values reported for supervisory data which motivate further checks on the level of individual data points.

The solution we developed was based on running a multitude of regression analyses using Extremely Randomised Trees [15]. The main steps to build a useful solution were to (i) ensure accurate results, (ii) provide interpretable representations of the patterns, and (iii) report outliers with a clear explanation of what rendered these data points suspicious [37].

The accuracy of the results involved several aspects. First, we ensured that only novel rules were detected. To ensure that only novel rules are identified, information from existing validation rules was used to limit the search space for the pattern discovery, i.e. to exclude known functional dependencies when running the regression analysis. This prevented the *business experts* from being frustrated by a system that provided them with insights they already knew. Second, we ranked the discovered patterns by their predictive quality. A transparent assessment of the Root Mean Squared Error of the prediction in a normalized value space helped to assess the quality of a pattern and ensured the business experts about the reliability of the functional dependence. Finally, we filtered the ranked patterns by the number of influential variables in the regression analysis. This helped to further focus on patterns that can be mapped more easily to business concepts.

The feature importance provided by the Extremely Randomised Trees implementation along with a transparent and methodologically clean quality assessment enabled us to address all these points. Using a non-linear transformation of the observed feature values to a standardized range rendered results comparable even if the original feature space differed by several orders of magnitude. Using a transformation that was reversible further allowed to illustrate predicted values in the original value range and thereby again interpretable by the *business experts*.

Building Trust. *Business experts* are used to designing plausibility rules solely based on domain knowledge. In a user study, we presented the identified, ranked, and filtered patterns to *business experts*, indicating the predictive quality in terms of Root Mean Squared Error and the list of influential variables. The user study was run on a subset of the supervisory data the business experts were currently investigating for novel plausibility checks [37]. In the course of the study, the insights from identified patterns enabled the experts to define new plausibility checks. These checks have been implemented and are used to assess the quality of data. This success story, based on xAI techniques created trust in the solution [37].

Furthermore, the *data scientists* used the explanation for the identified patterns to test their solutions. By including the known validation rules in the search space, the approaches had to identify these rules as patterns in the data. This helped to create trust in the implementations provided by the *data scientists*.

Finally, for *business experts* and *decision makers* it was of relevance to see that the approach is not biased. Given the heterogeneous landscape and business models of credit institutions, it was important that no particular group of institutions (e.g. big vs. small institutions) was particularly affected by the new checks.

Gain Knowledge. One of the main objectives of the solution was to identify new plausibility checks by gaining additional insights into the data [37]. The detected patterns served as a kind of inspiration to look at new business motivated plausibility checks. Again the easily interpretable patterns with high predictive quality and few influential variables helped *business experts* to connect the identified functional dependencies with their domain knowledge. The result was that the experts were inspired to detect new causal relationships in the data.

Actionable Insights. When analyzing and quality assuring supervisory data reported by a credit institution, the eventual aim is to detect anomalous values which require further investigations. An investigation, in this case, means that the *business experts* contact the credit institution to ask for confirmation of the data and to ensure that no methodological or technical errors were made in the compilation and transmission of data. Hence, in this case the credit institutions correspond to the *end user* being affected by an algorithmic decision.

With the investigated solution, this was achieved by means of comparing observed values vs. the forecasts made by the regression models. Having converted the data to a normalized observation space provided comparable and standardized metrics of how far off an observation was from the expected value. This provided the basis for a criticality assessment and a prioritization of outliers. Furthermore, by mapping back the normalized values to the original data space, *business experts* could not only interpret the results but also reach out to credit institutions and data compilers to ask for confirmation and clarification of the reported data, giving a clear indication of which value raised their attention and what they considered a plausible value range.

In a second user study, several suspicious values were presented to *business experts*. The *business experts* investigated a few selected cases with credit institutions and in fact, they were able to communicate the cases. This lead to the identification of cases of misreporting and the correction of data.

5   Conclusions

In this paper, we described the desiderata, scope, and users of explainable machine learning needs in statistical production systems of the European Central Bank.

We firstly introduced a summary of the different users and a classification of explainability needs that we have identified during our journey through the various machine learning projects. While the needs were distilled from our use cases in central banking, we believe they are of generic nature and the classification may serve for other use cases, as well.

Then we gave life to our classification by illustrating two different use cases that we analysed for explainability needs in the context of our experience with using machine learning for statistical production systems: data quality assurance at the Centralised Securities Database and identifying novel functional dependencies at the Supervisory Banking data system.

As future work, we intend to apply the classification of xAI desiderata to further use cases. We plan to use the classification framework for other use cases we envisage in the domain of central banking but are curious to expand also to other domains. Applying the classification scheme to more xAI use cases may lead to two developments: (i) a refinement of the classification scheme itself and (ii) enrichment of the entries in the scheme with references to suitable technical solutions.

Acknowledgements. This work was partially funded by the European Commission under contract numbers NoBIAS - H2020-MSCA—ITN-2019 project GA No. 860630.

References

1. Adadi, A., Berrada, M.: Peeking inside the black-box: a survey on explainable artificial intelligence (XAI). IEEE Access **6**, 52138–52160 (2018)
2. Amarasinghe, K., Rodolfa, K., Lamba, H., Ghani, R.: Explainable machine learning for public policy: use cases, gaps, and research directions (2020)
3. Arrieta, A.B., et al.: Explainable artificial intelligence (XAI): concepts, taxonomies, opportunities and challenges toward responsible AI. Inf. Fusion **58**, 82–115 (2020)
4. Belle, V., Papantonis, I.: Principles and practice of explainable machine learning (2020)
5. Bhatt, U., et al.: Explainable machine learning in deployment. In: Proceedings of the 2020 Conference on Fairness, Accountability, and Transparency, FAT* 2020, pp. 648–657. Association for Computing Machinery, New York, NY, USA (2020)
6. Burkov, A.: Machine Learning Engineering, 1 edn. Kindle Direct Publishing (2020)
7. Caruana, R., Kangarloo, H., Dionisio, J.D., Sinha, U., Johnson, D.: Case-based explanation of non-case-based learning methods. In: Proceedings AMIA Symposium, pp. 212–215 (1999)
8. Caruana, R., Niculescu-Mizil, A.: An empirical comparison of supervised learning algorithms. In: Proceedings of the 23rd International Conference on Machine Learning, ICML 2006, pp. 161–168. Association for Computing Machinery, New York, NY, USA (2006)
9. Diethe, T., Borchert, T., Thereska, E., Balle, B., Lawrence, N.: Continual learning in practice (2019)
10. Dorogush, A.V., Ershov, V., Gulin, A.: CatBoost: gradient boosting with categorical features support (2018)
11. Doshi-Velez, F., Kim, B.: Towards a rigorous science of interpretable machine learning (2017)
12. Edwards, L., Veale, M.: Enslaving the algorithm: from a "right to an explanation" to a "right to better decisions"? IEEE Secur. Privacy **16**(3), 46–54 (2018)
13. Reform of EU data protection rules. https://ec.europa.eu/commission/sites/beta-political/files/data-protection-factsheet-changes_en.pdf
14. Friedman, J.H.: Greedy function approximation: a gradient boosting machine. Ann. Stat. **29**(5), 1189–1232 (2001)
15. Geurts, P., Ernst, D., Wehenkel, L.: Extremely randomized trees. Mach. Learn. **63**(1), 3–42 (2006)
16. Gilpin, L.H., Bau, D., Yuan, B.Z., Bajwa, A., Specter, M., Kagal, L.: Explaining explanations: an overview of interpretability of machine learning. In: 2018 IEEE 5th International Conference on Data Science and Advanced Analytics (DSAA), pp. 80–89 (2018)
17. Guidotti, R., Monreale, A., Ruggieri, S., Pedreschi, D., Turini, F., Giannotti, F.: Local rule-based explanations of black box decision systems (2018)
18. Guidotti, R., Monreale, A., Ruggieri, S., Turini, F., Giannotti, F., Pedreschi, D.: A survey of methods for explaining black box models. ACM Comput. Surv. **51**(5) (2018)
19. Hastie, T., Tibshirani, R., Friedman, J.: The Elements of Statistical Learning. SSS. Springer, New York (2009). https://doi.org/10.1007/978-0-387-84858-7

20. Hinton, G., Vinyals, O., Dean, J.: Distilling the knowledge in a neural network. In: NIPS Deep Learning and Representation Learning Workshop (2015). http://arxiv.org/abs/1503.02531

21. Jesus, S., et al.: How can I choose an explainer? An application-grounded evaluation of post-hoc explanations. In: Proceedings of the 2021 ACM Conference on Fairness, Accountability, and Transparency, FAccT 2021, pp. 805–815. Association for Computing Machinery, New York, NY, USA (2021)

22. Karimi, A.H., Schölkopf, B., Valera, I.: Algorithmic recourse: from counterfactual explanations to interventions. In: Proceedings of the 2021 ACM Conference on Fairness, Accountability, and Transparency, FAccT 2021, pp. 353–362. Association for Computing Machinery, New York, NY, USA (2021)

23. Lipton, Z.C.: The mythos of model interpretability in machine learning, the concept of interpretability is both important and slippery. Queue **16**(3), 31–57 (2018)

24. Lundberg, S.M., et al.: Explainable AI for trees: from local explanations to global understanding (2019)

25. Lundberg, S.M., Lee, S.I.: A unified approach to interpreting model predictions. In: Guyon, I., et al. (eds.) Advances in Neural Information Processing Systems, vol. 30, pp. 4765–4774. Curran Associates, Inc. (2017)

26. Miller, T., Howe, P., Sonenberg, L.: Explainable AI: beware of inmates running the asylum or: how I learnt to stop worrying and love the social and behavioural sciences (2017)

27. Mittelstadt, B., Russell, C., Wachter, S.: Explaining explanations in AI. In: Proceedings of the Conference on Fairness, Accountability, and Transparency - FAT* 2019 (2019)

28. Mothilal, R.K., Sharma, A., Tan, C.: Explaining machine learning classifiers through diverse counterfactual explanations. In: Proceedings of the 2020 Conference on Fairness, Accountability, and Transparency, FAT* 2020, pp. 607–617. Association for Computing Machinery, New York, NY, USA (2020)

29. Navarro, C.M., Kanellos, G., Martinez-Heras, J., Micheler, J., Gottron, T.: Introducing explainable supervised machine learning into interactive feedback loops for statistical production systems. The Irving Fisher Committee on Central Bank Statistics (IFC) (2021). (to appear)

30. Pedregosa, F., et al.: Scikit-learn: machine learning in Python. J. Mach. Learn. Res. **12**, 2825–2830 (2011)

31. Prokhorenkova, L., Gusev, G., Vorobev, A., Dorogush, A.V., Gulin, A.: CatBoost: unbiased boosting with categorical features (2019)

32. Pérez, A.C., Huerga, J.: The centralised securities database (CSDB) - standardised micro data for financial stability purposes. In: Settlements, B.F.I. (ed.) Combining micro and macro data for financial stability analysis, vol. 41. Bank for International Settlements (2016). https://EconPapers.repec.org/RePEc:bis:bisifc:41--15

33. Quiñonero-Candela, J., Sugiyama, M., Lawrence, N.D., Schwaighofer, A.: Dataset shift in machine learning. MIT Press, Cambridge (2009)

34. Ribeiro, M.T., Singh, S., Guestrin, C.: Model-agnostic interpretability of machine learning (2016)

35. Ribeiro, M.T., Singh, S., Guestrin, C.: "Why should I trust you?": explaining the predictions of any classifier (2016)

36. Ribeiro, M.T., Singh, S., Guestrin, C.: Anchors: high-precision model-agnostic explanations. In: AAAI (2018)

37. Romano, S., Martinez-Heras, J., Raponi, F.N., Guidi, G., Gottron, T.: Discovering new plausibility checks for supervisory data - a machine learning approach. ECB Statistical Paper Series (2021). (to appear)

38. Rudin, C.: Stop explaining black box machine learning models for high stakes decisions and use interpretable models instead (2019)

39. Selbst, A.D., Barocas, S.: The intuitive appeal of explainable machines. Fordham Law Rev. **1085**(87), 2825–2830 (2018)
40. Unceta, I., Nin, J., Pujol, O.: Towards global explanations for credit risk scoring (2018)
41. Unceta, I., Nin, J., Pujol, O.: Copying machine learning classifiers. IEEE Access **8**, 160268–160284 (2020)
42. Unceta, I., Nin, J., Pujol, O.: Environmental adaptation and differential replication in machine learning. Entropy **22**(10) (2020)
43. Unceta, I., Palacios, D., Nin, J., Pujol, O.: Sampling unknown decision functions to build classifier copies. In: Torra, V., Narukawa, Y., Nin, J., Agell, N. (eds.) Modeling Decisions for Artificial Intelligence, pp. 192–204. Springer International Publishing, Cham (2020). https://doi.org/10.1007/978-3-030-85529-1
44. Ustun, B., Spangher, A., Liu, Y.: Actionable recourse in linear classification. In: Proceedings of the Conference on Fairness, Accountability, and Transparency, FAT* 2019, pp. 10–19. Association for Computing Machinery, New York, NY, USA (2019)
45. Zhang, Y., Haghani, A.: A gradient boosting method to improve travel time prediction. Transp. Res. Part C-Emerg. Technol. **58**, 308–324 (2015)

Robustness of Fairness: An Experimental Analysis

Serafina Kamp[1], Andong Luis Li Zhao[2(✉)], and Sindhu Kutty[1]

[1] University of Michigan, Ann Arbor, MI 48103, USA
{serafibk,skutty}@umich.edu
[2] Northwestern University, Evanston, IL 60208, USA
andong@u.northwestern.edu

Abstract. Machine learning algorithms are increasingly used in making decisions with significant social impact. However, the predictions made by these algorithms can be demonstrably biased; oftentimes reflecting and even amplifying societal prejudice. Fairness metrics can be used to evaluate the models learned by these algorithms. But how robust are these metrics to reasonable variations in the test data? In this work, we measure the robustness of these metrics by training multiple models in three distinct application domains using publicly available real-world datasets (including the COMPAS dataset). We test each of these models for both performance and fairness on multiple test datasets generated by resampling from a set of held-out datapoints. We see that fairness metrics exhibit far greater variance across these test datasets than performance metrics, when the model has not been derived to be fair. Further, socially disadvantaged groups seem to be most affected by this lack of robustness. Even when the model objective includes fairness constraints, while the mean fairness of the model necessarily increases, its robustness is not consistently and significantly improved. Our work thus highlights the need to consider variations in the test data when evaluating model fairness and provides a framework to do so.

Keywords: Classification · Fairness · Bootstrap sampling · Robustness

1 Introduction

Machine learning methods use data-driven algorithms for automatic pattern recognition and prediction. Traditionally, the objective of these algorithms has been to optimize for performance metrics such as accuracy, which essentially measures the model's ability to make correct predictions about previously unseen data. These learned predictors can then be used to make decisions with significant societal impact. For instance, among other applications, machine learning is used in automated judicial review [2] and facial recognition for law enforcement [28].

© Springer Nature Switzerland AG 2021
M. Kamp et al. (Eds.): ECML PKDD 2021 Workshops, CCIS 1524, pp. 591–606, 2021.
https://doi.org/10.1007/978-3-030-93736-2_43

Since machine learning models detect and learn from historical patterns in data, they may pick up and amplify societal biases. Several recent results show that the predictions based on these models can be demonstrably biased; for instance, automated facial analysis algorithms show significant accuracy differences across both race and gender [8] while music recommendation algorithms show gender bias in promoting artists [20]. The prevalence of these issues and the concerns they raise are well-documented, not only in machine learning literature, but also in the popular press [33,34].

The degree of unfairness exhibited by these models can be captured by metrics that are widely accepted in machine learning literature [13,22]. Typically, the fairness of the model can be evaluated by measuring it against test data. But how *robust* are these metrics to small perturbations in the data? Does the degree of robustness vary across models and application domains? And can we quantify the degree of *unfairness* across different *sub-populations*?

Fairness can be measured either as *individual* or *group* fairness. Group fairness metrics quantify how the model's predictions fare across different subgroups, often with an emphasis on subgroups that have been historically discriminated against. For instance, consider a model used to predict the probability of recidivism to determine whether or not to release a defendant for parole. This model may show different levels of predictive performance across different races. For one such model (used in US courts to predict recidivism), it has been shown that the probability of predicting a reoffence is greater for African American defendants than it is for Caucasian American defendants even when considering only those individuals who actually go on to reoffend [2]. One way to quantify this inequity in prediction is using the *equality of opportunity* fairness measure [22].

In this work, we are interested in measuring the robustness of fairness metrics when applied to a learned model. In particular, we first learn a predictive model using training data and then measure both its performance and fairness on test data. Crucially, rather than testing on a *single* held-out dataset, we measure fairness across variations in the testing data by generating multiple instances of this held-out dataset using bootstrap sampling [17,18]. Effectively, bootstrap sampling uses the empirical distribution of the resampled data as a surrogate for the true distribution of datapoints. This allows us to measure the variation in both prediction error as well as fairness. We also explore the difference in both the mean and variance of both performance and fairness metrics across three different datasets with different semantic notions of socially disadvantaged groups: by *race* and by *age*. We show that fairness metrics are *less robust* (i.e., exhibit *significantly* more variance) than performance metrics under various underlying models; including models that use post-processing to achieve fairness. We also see that, typically, protected groups are the most affected by this lack of robustness.

1.1 Related Work

There has been significant work in quantifying fairness and designing techniques for achieving it [26,29,35,39] as well as in understanding the implications of using fair predictors in practice [41]. The prevalence of bias in fields as wide-ranging as Natural Language Processing [7,38], vision [8], ad-placement [44] and health [1] have led to domain-specific analyses on bias detection and consequent work on both building and evaluating fairer datasets [4,46]. Further, a survey of industry practitioners highlights the need to understand the practical implications of using fairness metrics [24].

There is no single agreed-upon measure of fairness since different contexts may require different criteria of measurement, including exogenous concerns like privacy-preservation [5,6,45]. In fact, so-called "impossibility theorems" show that some measures of fairness cannot be simultaneously satisfied [12,27]. However, while there is no consensus measure of fairness, some tests for evaluating group fairness that have gained widespread acceptance include *demographic parity* [9], *equalized odds* and *equal opportunity* [22]. In the present work, we we focus primarily on the *equal opportunity* fairness metric since there has been significant exploration of models that enforce this constraint [22,29]. We also use *equalized odds* to derive a fair predictor.

There is an inherent tradeoff between the *performance* of a model, typically measured by metrics such as accuracy, and the *fairness* of the model, usually measured by how the predictor differs across different subgroups [30]. Achieving fairness in a predictive model can be framed by explicitly optimizing for fairness [10,19], as constrained optimization problems [12,14,22,47] and as conflicting objective functions [13].

Recent work has analyzed the effects of statistical and adversarial changes in the data distribution. Some of this work has focused on deriving fair models when there is a distributional shift in the data [40], when strategically acting adversaries inject errors in the data [11] or when the data is perturbed to negatively impact a particular subgroup [3,32].

In this work, we focus on the following research questions:

RQ1. For a given model, is the *equal opportunity* fairness metric a reliable measure of fairness? Does it show stability across reasonable fluctuations in the test data?

RQ2. How does the variation of the fairness metric compare with that of more traditional performance metrics?

RQ3. How much do different choices of models and features affect the robustness of the fairness metric? Is the robustness of the fairness measure affected by post-processing a model to satisfy fairness constraints? Further, does optimizing for a stronger notion of fairness affect the robustness of weaker notions of fairness?

RQ4. If we measure the effects of unfairness on different subgroups, do we see the same effects repeated across different datasets and models?

The rest of this paper is organized as follows: we cover background and a brief overview of our framework in Sect. 2. In Sect. 3 we provide details on the framework as well as the methodology for conducting our experiments. We also provide a description of the datasets and metrics used. We provide both numerical results and plots as well as an analysis of our results in Sect. 4. We conclude with a summary and directions for future work in Sect. 5.

2   Preliminaries and Overview

To learn a predictive model, we use logistic regression both with and without an ℓ_2 regularizer [23]. This involves solving the following optimization problem:

$$\min_{\bar{\theta},b} \ C \sum_{i=1}^{n} \log(\exp(-y_i(x_i^T \bar{\theta} + b)) + 1) + \frac{1}{2}||\bar{\theta}||_2^2$$

where (x_i, y_i) are labeled training datapoints, $\bar{\theta}$, b are the learned parameters, and C is a hyperparameter that controls the degree of regularization.

Each datapoint has a corresponding binary label $\in \{0, 1\}$. For instance, in the COMPAS dataset (see Sect. 3.1) each datapoint corresponds to an individual and a label of 1 indicates an individual who re-offends within two years. The features that distinguish historically disadvantaged groups are called *sensitive attributes* and the groups themselves are called *protected groups* [22]. Each datapoint includes a sensitive attribute $z \in \{0, 1\}$ that indicates their membership in a protected group. We train the base classifier both including and excluding these sensitive attributes.

We use group fairness measures to evaluate the fairness of the predictor returned by the algorithm. In this work, we focus primarily on analyzing the equal opportunity fairness metric [22], which enforces equal true positive rates (TPR) across different groups. This is a weaker notion of fairness that as a consequence allows for higher performance fair models [22]. Our experiments show that even with this relaxed definition, we see high variability in the measure when the optimization problem is agnostic to fairness of the model.

Even when we specifically optimize for this fairness using this metric, the variance in the measure does not decrease substantially. To achieve this fairness measure, the predictor is post-processed by solving a constrained optimization program with the constraints specifying the fairness conditions [22,36,37]. A formal definition of the fairness metrics used is given in Sect. 3.2.

To evaluate a model, we rely on test data that is held out during the training process. However, datasets are only samples of the "true" data distribution; thus although they may be representative of the original distribution, there is a degree of uncertainty associated with these measures. We use the resampling technique of bootstrap sampling to generate multiple instances of the test dataset. We describe the resampling process in further detail in Sect. 3.3.

3   Framework and Experimental Setup

We will now describe our framework for evaluating the robustness of fairness metrics across uncertainty in test data. Prior work has cast the uncertainty inherent in the training data using a Bayesian model [13]. However, we use a resampling approach to design experiments to study the empirical effects of this uncertainty on test data. We define the *robustness* of a metric to be inversely related to the amount of variation we see in this measure across multiple instances of a given dataset; a *robust* metric should show minimal variance across sampling variations. We empirically analyze the robustness of *equal opportunity* by measuring its variance across two datasets drawn from different application domains with the aim of measuring the persistence of our results across multiple learned models.

We use the COMPAS dataset [2] and the Bank Marketing dataset [31] which have been used widely in machine learning literature to study fairness. We also run these experiments on the South German Credit dataset [21]. Fundamentally, these datasets differ in social context (one was collected in the US in 2013−2014, one in Portugal 2008 − 2013, and the other in Southern Germany in 1973 − 1975). The historically disadvantaged groups in the three cases were also different (*race-based* vs. *age*-based discrimination). More details about these datasets are provided in Sect. 3.1.

Following prior work [29], the features that distinguish the traditionally privileged vs. disadvantaged groups are referred to as *sensitive attributes*. To understand the effects of the sensitive attributes on the learned model, we train the ML algorithm both with and without the sensitive attributes. We also investigate the robustness of both the fairness and performance metrics for different levels of model complexities by studying the effects of regularization.

By analyzing these metrics across different datasets and across different instantiations of test data, for different features and model complexities, including or ignoring fairness constraints, we are better able to assess the robustness of these metrics and the generalizability of these results. We describe the datasets, the metrics, and the methodology in further detail below.

3.1   Datasets

We use the COMPAS dataset [2], the Bank Marketing dataset [31], and the South German Credit (SGC) dataset [15] for our analyses. These datasets are well-known benchmarks that have been frequently used to study algorithmic fairness [29].

Thus it is important to understand the impact of using these datasets to derive fair models. Further, the difference in domain and protected attributes between the datasets allows us to analyze the robustness of fairness metrics beyond a single domain.

The COMPAS dataset contains 6150 datapoints with 8 features. The features include demographic information such as age, race, and sex as well as criminal history information such as priors, juvenile offences, and degree of current crime. When assuming a binary sensitive attribute, the dataset is restricted to Caucasian American and African American defendants; given the bias inherent in the dataset, African American defendants are considered to be the protected group. The binary-valued label indicates whether or not the individual has reoffended within two years after being released from prison.

The Bank Marketing dataset [31] contains 45211 datapoints with 15 features. The features include demographic information such as age, job, and education, seasonal data such as day and month, and financial data such as balance and whether an individual has any personal loans. Following prior work [47], the sensitive attribute is age where ages between 25 and 60 are considered protected. A positive outcome is when an individual subscribes to a term deposit.

The SGC dataset [15] contains 1000 datapoints with 20 features. The features of this dataset include demographic information such as age, sex, and marriage status, financial standing information such as credit history, savings account amount, and homeowner status, and, finally, information about the requested loan such as loan amount, purpose of loan, and duration of loan. Consistent with prior work [21,25], we use age as the sensitive attribute for this dataset where an age of 25 years or younger are considered the protected group. The outcome for this dataset is a binary variable indicating whether or not the loan contract has been fulfilled after the duration of the loan.

3.2   Metrics

Accuracy. For a given model, we measure its performance using accuracy defined[1] as $Acc = \frac{1}{N}\sum_{i=1}^{N}[[\hat{y}_i = y_i]]$ where $\hat{y}_i$ is the outcome predicted by the model, y_i is the true outcome and N is the number of samples we are evaluating [23].

Equality of Opportunity and Equalized Odds. While there is no single agreed upon way to measure fairness, one metric that is widely accepted, has been used to develop "fair" models and has semantic relevance for the datasets we consider is *equal opportunity* [22]. A predictor is said to satisfy equal opportunity if and only if $\Pr(\hat{y} = 1|z = 1, y = 1) = \Pr(\hat{y} = 1|z = 0, y = 1)$ where z is a sensitive attribute. For the COMPAS dataset, this can be interpreted as requiring the predictor to be agnostic to race for individuals who reoffend. For the SGC dataset, equal opportunity means that the probability of predicting a loan default should not change based on an individual's age for those individuals who repaid their loan. We also consider a model where the predictor is modified to satisfy the stricter measure of *equalized odds* [22], that additionally enforces equal false positive rates. Formally, equalized odds requires the following to hold: $\forall a \in \{0, 1\}$ $\Pr(y = 1|z = 1, y = a) = \Pr(y = 1|z = 0, y = a)$.

Degree of Fairness and Direction of Unfairness. We also measure the extent to which a model deviates from equality of opportunity. We define the *degree of fairness* of the predictor as: $1 - |\Pr(\hat{y} = 1|z = 1, y = 1) - \Pr(\hat{y} = 1|z = 0, y = 1)|$. The range of this measure is the unit interval $[0, 1]$; a higher value indicates a fairer model. To identify the subgroup against which a predictor is biased, we define the *direction of unfairness* as $\text{sign}[\Pr(\hat{y} = 1|z = 1, y = 1) - \Pr(\hat{y} = 1|z = 0, y = 1)]$. For example, in the COMPAS dataset, $z = 1$ indicates an African American defendant and $z = 0$ indicates a Caucasian

---

[1] We use [[]] to denote the Iverson bracket which returns a value of 1 if the predicate contained within is true and 0 otherwise.

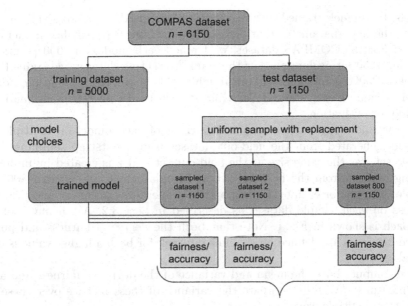

Fig. 1. Schematic illustrating our framework for measuring robustness of performance and fairness metrics. We use the COMPAS dataset for illustrative purposes.

American defendant. So, a *positive* direction of unfairness corresponds to unfairness towards the protected group (in this case, African American defendants). We compare the variance in the degree of fairness with the variance of accuracy across multiple models and instances of test datasets. In the next section, we describe the methodology we use to measure this variance.

3.3 Methodology

We learn twelve different models on the training data to evaluate their effects on both the mean and variance of fairness and performance metrics. In particular, we train a logistic regression classifier both with and without an ℓ_2-norm regularizer and both including and excluding sensitive attributes while training. In addition to these four models, we learn modified models by post-processing each of these models to separately satisfy first equality of opportunity and then equalized odds.

In order to split the datasets into training and held-out sets, we first randomly shuffle each dataset. For each dataset, we also ensure that the proportion of positive examples, the proportion of protected class, and the proportion of positive examples within the protected class are all preserved across the training and testing set. Then, we train four models with and without regularization and with and without the sensitive attributes in the input feature vector. Then we applied post-processing for fairness constraints. In all, we train a total of twelve models. For models trained with regularization we used 5-fold cross-validation to choose the hyperparameter that determines how much we penalize model complexity. For the COMPAS dataset, we trained each model on 5000 points and held out 1150 for evaluation. For the Bank Marketing dataset, we trained each model on 25000 points and held out roughly 20000 points for evaluation. For the South German Credit dataset, we trained each model on 600 points and held out 400 for evaluation.

We evaluate the performance and fairness of each model on multiple test datasets generated from the held-out dataset using bootstrap sampling. Each sample set was the same size as the held-out set and was created by uniformly picking a point from the held-out set with replacement. We created 800 such sample datasets for each evaluation and then measured accuracy and degree of fairness on each sample dataset as described in Sect. 3.2. A schematic of this approach is shown in Fig. 1. Note that both the degree of fairness and performance measures are defined on the unit interval; for both a higher value is more desirable.

We compute both the mean and variance of the degree of fairness and accuracy measures. We then compare the variance of these metrics over these 800 datasets in multiple ways.

– First, we numerically compute the variance achieved by these metrics and tabulate it for comparison across all twelve models (see Tables 1 and 2).
– Next, we plot the values of both metrics for each of the bootstrap sampled datasets (see Fig. 3). For visual consistency, we plot fairness along the horizontal axis and performance along the vertical axis for all plots in that figure. We also use the same scale for both axes. A larger spread along a particular axis, therefore, indicates a larger variance along that metric.
– Then, we plot a histogram of both metrics for a visual representation of the distribution of these measures (see Fig. 2 for the plots for two models on the COMPAS dataset. Due to space constraints, additional figures have been omitted.).
– Lastly, we translate both measures from the $[0, 1]$ to the $(-\infty, +\infty)$ interval by first centering to 0.5 mean and then applying the logit function to the values so obtained[2]. We see that the mapped values broadly follow a normal distribution. We then compute the variance of these mapped values and apply the F-test [42] to determine the significance of the difference in variances with high confidence[3].

We provide plots of accuracy vs. degree of fairness for each sample. We also provide the variance and mean of each of these metrics across the test sets. We describe our results in the next section. Due to space constraints, the results for the Bank Marketing dataset are omitted, but similar trends were observed.

---

[2] Datasets with unit fairness were withheld in the F-test analysis to prevent degenerate cases. However, these accounted for less than 1.5% of all 800 sample datasets.

[3] While the independence assumption does not strictly hold, the F-test gives us one more means of comparison.

Table 1. Mean (and variance) values in percentage for **accuracy** and **degree of fairness** for the COMPAS dataset reported for Logistic regression (LogReg); post-processing for equal opportunity (EqOpp) and equalized odds (EqOdds); L2 indicates regularization.

MODEL	NO SENSITIVE		SENSITIVE	
	ACCURACY	DEG OF FAIRNESS	ACCURACY	DEG OF FAIRNESS
LOGREG	62.16 (2.17)	78.32 (20.66)	62.73 (2.06)	61.50 (20.57)
LOGREG + L2	62.16 (2.17)	78.32 (20.66)	62.37 (2.12)	65.01 (20.61)
EQOPP	58.72 (2.12)	96.54 (6.37)	56.63 (2.00)	96.27 (8.27)
EQOPP + L2	58.71 (2.12)	96.53 (6.42)	56.41 (2.07)	95.41 (10.53)
EQODDS	58.62 (2.09)	96.29 (7.78)	56.95 (1.99)	95.76 (9.65)
EQODDS + L2	58.61 (2.10)	96.30 (7.76)	56.89 (2.07)	95.40 (11.56)

Table 2. Mean (and variance) values in percentage for **accuracy** and **degree of fairness** for the SGC dataset reported for Logistic regression (LogReg); postprocessing for equal opportunity (EqOpp) and equalized odds (EqOdds); L2 indicates regularization.

MODEL	NO SENSITIVE		SENSITIVE	
	ACCURACY	DEG OF FAIRNESS	ACCURACY	DEG OF FAIRNESS
LOGREG	78.45 (4.42)	90.18 (31.65)	77.30 (4.07)	85.07 (43.75)
LOGREG + L2	77.20 (4.55)	91.94 (27.48)	78.53 (4.00)	88.07 (34.59)
EQOPP	76.49 (4.35)	93.51 (22.38)	73.58 (4.05)	92.14 (29.91)
EQOPP + L2	75.22 (4.55)	94.56 (16.92)	74.40 (3.88)	94.74 (16.28)
EQODDS	75.23 (4.34)	93.14 (23.70)	73.65 (4.08)	92.04 (30.34)
EQODDS + L2	74.15 (4.61)	94.27 (18.50)	74.48 (3.87)	94.99 (14.98)

4   Results

4.1   Variance of Fairness and Performance Metrics

As shown in Tables 1 and 2, we note that the variance in degree of fairness is higher than for accuracy. As an example, Fig. 2a shows visually that the spread of accuracy and degree of fairness can vary significantly. In fact, we show that difference in variance is statistically significant for various significance levels. We transform the data to the real number line using the logit function and apply the F-test to this transformed data (see Sect. 3.3 for details). Table 3 reports these values for the logistic regression base classifier with regularization trained on data with sensitive attributes both before and after post-processing for fairness constraints[4]. This indicates that the fairness metric of equal opportunity is not as robust as accuracy across the sampled test sets.

---

[4] While we do not report results on all models due to space constraints, the omitted results are similar to reported values.

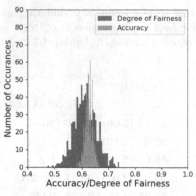
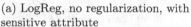
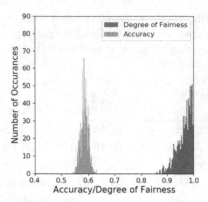

(a) LogReg, no regularization, with sensitive attribute

(b) EqOdds ,regularized, no sensitive attribute

Fig. 2. Histogram showing the difference in mean and variance of degree of fairness and accuracy scores for different models on the COMPAS dataset. Figure 2a includes scores for logistic regression without regularization including sensitive attributes. Figure 2b includes scores for logistic regression with regularization but without sensitive attributes and post-processing for equalized odds fairness constraint.

Once we post-process for fairness constraints, we see that, as expected, mean degree of fairness improves. We also note that the variance in degree of fairness reduces significantly, especially for the COMPAS dataset (see Table 1). This can be seen visually in Fig. 2b, where for comparison the spread in accuracy is indicated as well. We note, however, that the differences in variance of degree of fairness and accuracy are still statistically significant for all models, with the variance of degree of fairness always being higher than that of accuracy. We see in Table 3 that the F-test value is much larger than the f-critical value for the number of observations, thus indicating high confidence that the variances are in fact significantly different.

When comparing the effect of incorporating different fairness constraints, we note that both equalized odds as well as equality of opportunity yield fairly similar results for degree of fairness. Typically, we observe that for models with post-processing for fairness constraints, means of degrees of fairness are within at most 1% of each other. We also observe that in most cases equality of opportunity and equalized odds have comparable magnitudes of variance in degree of fairness. However, in the case of unregularized base classifiers, equality of opportunity has a smaller degree of fairness variance; a likely explanation for this lies in our measure of degree of fairness which explicitly checks for deviation from the equality of opportunity measure.

Table 3. F-test for statistical significance of the difference between performance and fairness variances reported for Logistic regression (LogReg); postprocessing for equal opportunity (EqOpp) and equalized odds (EqOdds). All models include sensitive attributes and a regularizer term. ✓ indicates that the ratio is higher than the F critical value, implying that the difference is statistically significant

DATA SET	VARIANCES	RATIO	$\alpha = 0.05$	$\alpha = 0.025$	$\alpha = 0.001$
			1.1234	1.1488	1.2446
COMPAS	LogReg	9.722	✓	✓	✓
SGC	LogReg	8.648	✓	✓	✓
COMPAS	EqOpp	5.087	✓	✓	✓
SGC	EqOpp	4.196	✓	✓	✓
COMPAS	EqOdds	5.585	✓	✓	✓
SGC	EqOdds	3.871	✓	✓	✓

The effects of incorporating fairness constraints on accuracy have been previously observed [30]. This is corroborated in our experiments as we observe a trade-off between accuracy and degree of fairness. In all cases, adding a fairness constraint reduced overall accuracy; however, the effect on its variance was typically minimal and inconsistent in direction indicating that adding fairness constraints does not seem to affect stability of the performance measure. Amongst models that were optimized for fairness, we notice that their mean accuracy is quite similar, being within at most 1% of each other's performance. This can be explained by the relationship between the fairness constraints and the degree of fairness measure. Another important trend we note is that higher mean degree of fairness generally corresponds to lower degree of fairness variance.

The effects of both including sensitive attributes in training the model, and adding a regularization term in the objective function, are mixed. The best performing models for accuracy are logistic regression models with access to sensitive attributes; perhaps unsurprisingly however, these are often among the worst performing with respect to the mean and variance of degree of fairness. We also note that regularization has a significant effect on variance of degree of fairness especially when post-processing for fairness in the SGC dataset (Table 2) as compared to the COMPAS dataset (Table 1). This can be likely explained by the difference in sizes of the two datasets.

A notable case is when we use a logistic regression model and fairness post-processing with access to sensitive attributes in the COMPAS dataset, which we can see in Fig. 2b. In this case, the mean accuracy is roughly 58%, which is only slightly better than naively predicting the most common label in the dataset (which would give roughly 53% accuracy). This might indicate that there are degenerate cases of fairness where predictions are equally uninformative for different subgroups, potentially because the solution space is too restricted by regularization and fairness constraints.

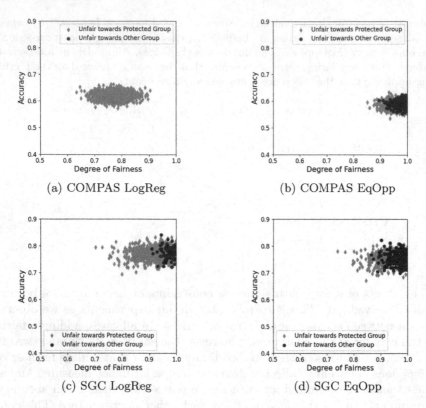

Fig. 3. Scatter plot for degree of fairness and accuracy. Orange diamonds indicate unfairness towards protected group, blue dots indicate unfairness towards the other group. Plots shown for the COMPAS and SGC datasets for Logistic regression (LogReg); postprocessing for equal opportunity (EqOpp) trained with regularization and without sensitive attributes. (Color figure online)

4.2 Direction of Unfairness

In addition to looking at the general trends of fairness, we also explore the direction of unfairness in these models for the SGC and COMPAS datasets. In Fig. 3, we show a scatter plot of the 800 bootstrapped sampled test datasets (for both SGC and COMPAS datasets) along the accuracy and degree of fairness axes. As observable from the plots, generally the models are unfair towards the protected groups. Fairness constraints help shift the entire distribution to more fair outcomes, but we still see that most of the unfairness is to the detriment of protected groups. The plots for other models are omitted due to space constraints, but they show similar results as well.

5   Conclusions and Future Work

In this paper, we have provided a framework for evaluating the robustness of fairness metrics across uncertainty in test data. To do this, we resample test data using bootstrap sampling and compute both the mean and variance of degree of fairness and accuracy. This allows us to compare the variations across these metrics for different learning models. We train a logistic regression model for binary classification with and without a regularizer, as well as with and without sensitive attributes. We also post-process these models to separately satisfy two separate fairness constraints. We evaluate these twelve models separately on 800 bootstrapped test datasets to measure the variability as well as the mean of both a performance metric and a fairness metric. We show that the equality of opportunity fairness metric is less robust to variations in the test data than the accuracy performance metric. We highlight that current post-processing methods for improving fairness can affect mean fairness and reduce fairness variance; by and large, however, the variance of fairness still remains significantly higher than that of performance. We show that variance in model fairness is typically to the detriment of protected groups, making fairness variance analysis an important part of developing robust and fair machine learning models.

These findings have important implications for the study of fairness, both for the machine learning community as well as for disciplines that apply these techniques. Since fairness metrics are significantly less robust than performance metrics, a single reported measure of fairness of an algorithm may not be sufficiently informative; a report of the range and variance of the metric might be relevant. For instance, claims about racial fairness in risk recidivism tools might not be as trustworthy since in reality there could be significant deviation from fairness if the model is applied on data that has even slight deviations from the data it was trained on. Furthermore, this variance is mostly to the detriment of protected groups, indicating that high uncertainty itself might be an indicator of further unfairness that is not directly captured in a single measurement of fairness.

This lays the groundwork for further exploration of the robustness of fairness across other learning models, including those that incorporate a notion of fairness in their objective. Additionally, we are also interested in whether these effects will persist across other fairness metrics and datasets. In particular, we are interested in exploring other group fairness metrics, such as predictive parity rates or generalized entropy indices [43], as well as individual fairness metrics, such as Lipschitz conditions constraints [16]. We are also interested in studying the effects of in-processing learning methods for fairness on its variance. We leave this for future work.

References

1. Abebe, R., Goldner, K.: Mechanism design for social good. AI Matters **4**(3), 27–34 (2018)
2. Angwin, J., Larson, J., Mattu, S., Kirchner, L.: Machine bias. Propublica (2016)

3. Awasthi, P., Kleindessner, M., Morgenstern, J.: Equalized odds postprocessing under imperfect group information. In: Chiappa, S., Calandra, R. (eds.) The 23rd International Conference on Artificial Intelligence and Statistics. AISTATS 2020, 26–28 August 2020, Online [Palermo, Sicily, Italy]. Proceedings of Machine Learning Research, vol. 108, pp. 1770–1780. PMLR (2020)
4. Bao, M., et al.: It's COMPASlicated: the messy relationship between RAI datasets and algorithmic fairness benchmarks. CoRR abs/2106.05498 (2021)
5. Barocas, S., Hardt, M., Narayanan, A.: Fairness and machine learning. fairmlbook.org (2019). http://www.fairmlbook.org
6. Binns, R.: On the apparent conflict between individual and group fairness. In: Proceedings of the 2020 Conference on Fairness, Accountability, and Transparency, pp. 514–524. FAT* 2020. Association for Computing Machinery, New York, NY, USA (2020)
7. Bolukbasi, T., Chang, K.W., Zou, J., Saligrama, V., Kalai, A.: Man is to computer programmer as woman is to homemaker? Debiasing word embeddings. In: Proceedings of the 30th International Conference on Neural Information Processing Systems, pp. 4356–4364. NIPS 2016, Curran Associates Inc., Red Hook, NY, USA (2016)
8. Buolamwini, J., Gebru, T.: Gender shades: intersectional accuracy disparities in commercial gender classification. In: Friedler, S.A., Wilson, C. (eds.) Proceedings of the 1st Conference on Fairness, Accountability and Transparency. Proceedings of Machine Learning Research, vol. 81, pp. 77–91. PMLR, New York, NY, USA, 23–24 February 2018
9. Calders, T., Kamiran, F., Pechenizkiy, M.: Building classifiers with independency constraints. In: 2009 IEEE International Conference on Data Mining Workshops, pp. 13–18 (2009)
10. Calders, T., Verwer, S.: Three Naive Bayes approaches for discrimination-free classification. Data Min. Knowl. Discov. **21**(2), 277–292 (2010)
11. Celis, L.E., Mehrotra, A., Vishnoi, N.K.: Fair classification with adversarial perturbations. CoRR abs/2106.05964 (2021)
12. Chouldechova, A.: Fair prediction with disparate impact: a study of bias in recidivism prediction instruments. Big Data **5**(2), 153–163 (2017)
13. Dimitrakakis, C., Liu, Y., Parkes, D.C., Radanovic, G.: Bayesian fairness. In: Proceedings of the AAAI Conference on Artificial Intelligence, vol. 33(01), pp. 509–516 (2019)
14. Donini, M., Oneto, L., Ben-David, S., Shawe-Taylor, J., Pontil, M.: Empirical risk minimization under fairness constraints. In: Proceedings of the 32nd International Conference on Neural Information Processing Systems, pp. 2796–2806. NIPS 2018, Curran Associates Inc., Red Hook, NY, USA (2018)
15. Dua, D., Graff, C.: UCI machine learning repository (2017). http://archive.ics.uci.edu/ml
16. Dwork, C., Hardt, M., Pitassi, T., Reingold, O., Zemel, R.: Fairness through awareness. In: Proceedings of the 3rd Innovations in Theoretical Computer Science Conference, pp. 214–226 (2012)
17. Efron, B.: The bootstrap and modern statistics. J. Am. Stat. Assoc. **95**(452), 1293–1296 (2000)
18. Efron, B., Tibshirani, R.: An Introduction to the Bootstrap. Springer (1993)

19. Feldman, M., Friedler, S.A., Moeller, J., Scheidegger, C., Venkatasubramanian, S.: Certifying and removing disparate impact. In: Proceedings of the 21th ACM SIGKDD International Conference on Knowledge Discovery and Data Mining, pp. 259–268. KDD 2015. Association for Computing Machinery, New York, NY, USA (2015)

20. Ferraro, A., Serra, X., Bauer, C.: Break the loop: gender imbalance in music recommenders. In: Proceedings of the 2021 Conference on Human Information Interaction and Retrieval, pp. 249–254. CHIIR 2021. Association for Computing Machinery, New York, NY, USA (2021)

21. Friedler, S.A., Scheidegger, C., Venkatasubramanian, S., Choudhary, S., Hamilton, E.P., Roth, D.: A comparative study of fairness-enhancing interventions in machine learning. In: Proceedings of the Conference on Fairness, Accountability, and Transparency, pp. 329–338. FAT* 2019. Association for Computing Machinery, New York, NY, USA (2019)

22. Hardt, M., Price, E., Price, E., Srebro, N.: Equality of opportunity in supervised learning. In: Lee, D., Sugiyama, M., Luxburg, U., Guyon, I., Garnett, R. (eds.) Advances in Neural Information Processing Systems, vol. 29. Curran Associates, Inc. (2016)

23. Hastie, T., Tibshirani, R., Friedman, J.H.: The Elements of Statistical Learning: Data Mining, Inference, and Prediction, 2nd Edition. Springer Series in Statistics. Springer (2009). https://doi.org/10.1007/978-0-387-84858-7

24. Holstein, K., Vaughan, J.W., Daumé, H., Dudik, M., Wallach, H.: Improving fairness in machine learning systems: what do industry practitioners need? pp. 1–16. Association for Computing Machinery, New York, NY, USA (2019)

25. Kamiran, F., Calders, T.: Data preprocessing techniques for classification without discrimination. Knowl. Inf. Syst. 33(1), 1–33 (2012)

26. Kleinberg, J., Ludwig, J., Mullainathan, S., Rambachan, A.: Algorithmic fairness. In: AEA Papers and Proceedings, vol. 108, pp. 22–27 (2018)

27. Kleinberg, J.M., Mullainathan, S., Raghavan, M.: Inherent trade-offs in the fair determination of risk scores. In: Papadimitriou, C.H. (ed.) 8th Innovations in Theoretical Computer Science Conference, ITCS 2017, 9–11 January 2017, Berkeley, CA, USA. LIPIcs, vol. 67, pp. 43:1–43:23. Schloss Dagstuhl - Leibniz-Zentrum für Informatik (2017)

28. MacCarthy, M.: Mandating fairness and accuracy assessments for law enforcement facial recognition systems. The Brookings Institution (2021). https://www.brookings.edu/blog/techtank/2021/05/26/mandating-fairness-and-accuracy-assessments-for-law-enforcement-facial-recognition-systems

29. Mehrabi, N., Morstatter, F., Saxena, N., Lerman, K., Galstyan, A.: A survey on bias and fairness in machine learning. CoRR abs/1908.09635 (2019)

30. Menon, A.K., Williamson, R.C.: The cost of fairness in binary classification. In: Friedler, S.A., Wilson, C. (eds.) Proceedings of the 1st Conference on Fairness, Accountability and Transparency. Proceedings of Machine Learning Research, vol. 81, pp. 107–118. PMLR, New York, NY, USA, 23–24 February 2018

31. Moro, S., Cortez, P., Rita, P.: A data-driven approach to predict the success of bank telemarketing. Decis. Supp. Syst. 62, 22–31 (2014)

32. Nanda, V., Dooley, S., Singla, S., Feizi, S., Dickerson, J.P.: Fairness through robustness: investigating robustness disparity in deep learning. In: Proceedings of the 2021 ACM Conference on Fairness, Accountability, and Transparency, pp. 466–477. FAccT 2021. Association for Computing Machinery, New York, NY, USA (2021)

33. Noble, S.U.: Algorithms of Oppression: How Search Engines Reinforce Racism. NYU Press, New York (2018)
34. O'Neil, C.: Weapons of Math Destruction: How Big Data Increases Inequality and Threatens Democracy. Crown Publishing Group, USA (2016)
35. Parkes, D.C., Vohra, R.V., et al.: Algorithmic and economic perspectives on fairness. CoRR abs/1909.05282 (2019)
36. Pleiss, G.: Code and data for the experiments in "On fairness and calibration" (2013)
37. Pleiss, G., Raghavan, M., Wu, F., Kleinberg, J., Weinberger, K.Q.: On fairness and calibration. In: Guyon, I., et al. (eds.) Advances in Neural Information Processing Systems, vol. 30. Curran Associates, Inc. (2017)
38. Prabhakaran, V., Hutchinson, B., Mitchell, M.: Perturbation sensitivity analysis to detect unintended model biases. In: Inui, K., Jiang, J., Ng, V., Wan, X. (eds.) Proceedings of the 2019 Conference on Empirical Methods in Natural Language Processing and the 9th International Joint Conference on Natural Language Processing, EMNLP-IJCNLP 2019, 3–7 November 2019, Hong Kong, China, pp. 5739–5744. Association for Computational Linguistics (2019)
39. Rambachan, A., Kleinberg, J., Ludwig, J., Mullainathan, S.: An economic perspective on algorithmic fairness. In: AEA Papers and Proceedings, vol. 110, pp. 91–95 (2020)
40. Rezaei, A., Liu, A., Memarrast, O., Ziebart, B.D.: Robust fairness under covariate shift. In: Proceedings of the AAAI Conference on Artificial Intelligence, vol. 35(11), pp. 9419–9427 (2021)
41. Saxena, N.A., Huang, K., DeFilippis, E., Radanovic, G., Parkes, D.C., Liu, Y.: How do fairness definitions fare? Testing public attitudes towards three algorithmic definitions of fairness in loan allocations. Artif. Intell. **283**, 103238 (2020)
42. Snecdecor, G.W., Cochran, W.G.: Statistical Methods. Wiley-Blackwell, Hoboken (1991)
43. Speicher, T., et al.: A unified approach to quantifying algorithmic unfairness: measuring individual & group unfairness via inequality indices. In: Proceedings of the 24th ACM SIGKDD International Conference on Knowledge Discovery & Data Mining, pp. 2239–2248 (2018)
44. Sweeney, L.: Discrimination in online ad delivery: Google ads, black names and white names, racial discrimination, and click advertising. Queue **11**(3), 10–29 (2013)
45. Tran, C., Fioretto, F., Van Hentenryck, P.: Differentially private and fair deep learning: a Lagrangian dual approach. In: Proceedings of the AAAI Conference on Artificial Intelligence, vol. 35(11), pp. 9932–9939 (2021)
46. Yang, K., Qinami, K., Fei-Fei, L., Deng, J., Russakovsky, O.: Towards fairer datasets: filtering and balancing the distribution of the people subtree in the ImageNet hierarchy. In: Proceedings of the 2020 Conference on Fairness, Accountability, and Transparency, pp. 547–558. FAT* 2020. Association for Computing Machinery, New York, NY, USA (2020)
47. Zafar, M.B., Valera, I., Gomez-Rodriguez, M., Gummadi, K.P.: Fairness constraints: a flexible approach for fair classification. J. Mach. Learn. Res. **20**(75), 1–42 (2019)

Co-clustering for Fair Recommendation

Gabriel Frisch[(✉)], Jean-Benoist Leger, and Yves Grandvalet

Université de Technologie de Compiègne, CNRS, Heudiasyc UMR 7253,
Compiègne, France
{gabriel.frisch,jean-benoist.leger,yves.grandvalet}@utc.fr

Abstract. Collaborative filtering relies on a sparse rating matrix, where
each user rates a few products, to propose recommendations. The approach consists of approximating the sparse rating matrix with a simple model whose regularities allow to fill in the missing entries. The
latent block model is a generative co-clustering model that can provide
such an approximation. In this paper, we show that exogenous sensitive
attributes can be incorporated in this model to make fair recommendations. Since users are only characterized by their ratings and their
sensitive attribute, fairness is measured here by a parity criterion. We
propose a definition of fairness specific to recommender systems, requiring item rankings to be independent of the users' sensitive attribute. We
show that our model ensures approximately fair recommendations provided that the classification of users approximately respects statistical
parity.

Keywords: Fairness · Recommender system · Co-clustering · Block
clustering · Latent block model · Statistical parity

1 Introduction

In simple terms, fairness is often loosely defined as the quality of treating people
equally, with impartiality and rightfulness. Although imprecise, this definition
stipulates that equal treatment refers to certain sensitive attributes shared by
groups of people, such as gender, age, ethnicity, socio-economic group, etc. In
recent years, intensive research has highlighted the lack of fairness in decisions
made by machine learning algorithms [7].

There are several stakeholders in a recommendation scenario. In the terminology of Burke *et al.* [9], we target consumer-fairness, where the objective is to
provide the same treatment to users of the recommender system, regardless of
their sensitive attribute. We consider recommender systems relying on collaborative filtering, which aims at building recommendations from the history of
user ratings. These observed ratings are the basis for making predictions about
non-rated items. Although sensitive attributes are generally not used in collaborative filtering, some disparate impacts may be observed due to societal or
cultural effects that affect the data. When the sensitive attribute is observed, its
processing can help mitigate these discriminatory effects.

© Springer Nature Switzerland AG 2021
M. Kamp et al. (Eds.): ECML PKDD 2021 Workshops, CCIS 1524, pp. 607–630, 2021.
https://doi.org/10.1007/978-3-030-93736-2_44

Several proposals have already been made on how fairness should be formally defined in collaborative filtering [12,31,34]. One common approach is the recommendation independence [22], that requires the unconditional statistical independence between recommendations and a specified sensitive attribute s_i of user i:

$$\mathbb{P}(R_{ij} > R_{ij'}|s_i = 1) = \mathbb{P}(R_{ij} > R_{ij'}|s_i = -1) \ ,$$

with $R_{ij} > R_{ij'}$ when item j is preferred to item j' for user i. Assuming the access to the value of the sensitive attribute s_i of all users, we give the following definition of a fair recommender system:

Definition 1 (ε-fair recommendation, binary sensitive attribute). *A recommender system is said to be ε-fair with respect to attribute s if for any two items j and j':*

$$\left| \frac{\# \{i|s_i = 1 \wedge (R_{ij} > R_{ij'})\}}{\# \{i|s_i = 1\}} - \frac{\# \{i|s_i = -1 \wedge (R_{ij} > R_{ij'})\}}{\# \{i|s_i = -1\}} \right| \leq \varepsilon \ , \quad (1)$$

where $\varepsilon \in \mathbb{R}_+$ measures the gap to exact fairness

In essence, an ε-fair recommender system ensures that, for any two items, the proportion of users with the same preference is approximately identical in all the subpopulations of users defined by identical sensitive attributes. This equal treatment does not ensure equal impact, which argues for equal recommendation quality between sensitive groups. Although some works [34] have argued that recommendation independence may be overly restrictive, resulting in a poor quality of recommendations, we use here this definition to propose fair recommendations exempted from any stereotypes.

In this paper, we aim at producing fair recommendations using a co-clustering of users and items that respects statistical parity of users with respect to some sensitive attributes. For this purpose, we introduce a co-clustering model based on the Latent Block Model (LBM) that relies on an ordinal regression model that takes as inputs the sensitive attributes. We demonstrate that our model ensures approximately fair recommendations (Definition 1) provided that the clustering of users approximately respects statistical parity. Finally, we conduct experiments on a real-world dataset to show that the proposed approach can help alleviate unfairness.

Related Works

Several recent works have raised the issue of fairness in recommender systems. Kamishima *et al.* [22] have proposed methods for improving fairness, formalized as the independence of the predicted ratings with the sensitive attribute. Their methods are based on matrix factorization regularized by criteria that favor independence by controlling the moments of the distributions of rating among sensitive groups. Using the same definition of fairness, Zhu *et al.* [35] proposed a tensor method that isolates sensitive attributes in sub-dimensions of the latent factor matrix. Unlike many other methods, this solution is capable of handling

multiple and non-binary sensitive attributes. Yao and Huang [34] proposed four
new metrics that deal with different types of unfairness and used them as penalty
functions in augmented matrix factorization objectives.

All of the above methods are based on the fairness of predicted ratings, but
an approximate fairness of ratings may not entail an approximate fairness of the
recommender system that provides users with a short list of relevant items. With
this in mind, Beutel *et al.* [5] provided new metrics based on pairwise comparisons
and proposed a novel pairwise regularization approach to improve the fairness
of the recommender system during training. Finally, further from recommender
systems but still related to the model we use, the notion of statistical parity is
often considered for fairness in clustering methods [1,4,14].

2 Model

The data used to build recommender systems can be aggregated in a matrix
where rows are users, columns are items and entries the feedbacks. The model
we propose is based on the Latent Block Model that considers a data matrix to
group users and items based on their opinions.

2.1 The Latent Block Models

The Latent block models (LBM), also known as bipartite stochastic block models
els and introduced in [15], are generative probabilistic models enabling to cluster
jointly the rows and the columns of a data matrix denoted R. These co-clustering
models assume a homogeneous block structure of the whole data matrix. This
structure is unveiled by the reordering of rows and columns according to their
respective cluster index; for k_1 row clusters and k_2 column clusters, the reordering reveals $k_1 \times k_2$ homogeneous blocks in the data matrix being possibly binary
[15] categorical [23], or quantitative [16,26].

The partitions of rows and columns are governed by the latent variables U
and V, U being the $n_1 \times k_1$ indicator matrix of row classes, and V being the
$n_2 \times k_2$ indicator matrix of the column classes. The class indicator of row i is
denoted U_i, and similarly, the class indicator of column j is denoted V_j. The
LBM makes several assumptions on the dependency and on the form of the
distributions:

- The latent group memberships of rows and columns are assumed to be
  mutually independent and identically distributed, with respectively multi-
  nomial distributions $\mathcal{M}(1; \alpha)$ and $\mathcal{M}(1; \beta)$, where $\alpha = (\alpha_1, \ldots, \alpha_{k_1})$ and
  $\beta = (\beta_1, \ldots, \beta_{k_2})$ are the mixing proportions of rows and columns:

$$p(U, V) = p(U)\, p(V) = \prod_i p(U_i; \alpha) \prod_j p(V_j; \beta) \ .$$

– Conditionally to rows and columns assignments $(\boldsymbol{U}, \boldsymbol{V})$, the entries of the data matrix $\boldsymbol{R}$ are independent and identically distributed:

$$p(\boldsymbol{R}|\boldsymbol{U}, \boldsymbol{V}; \boldsymbol{\theta}) = \prod_{ij} p(R_{ij}|\boldsymbol{U}_i, \boldsymbol{V}_j) \,,$$

$$p(R_{ij}|U_{iq}V_{jl} = 1) = \phi_{ql}(R_{ij}) \,, \tag{2}$$

with $\phi_{ql}(R_{ij})$ the density of the conditional distribution of R_{ij} depending on the group memberships of row i and column j (Fig. 1).

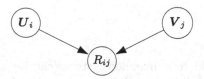

Fig. 1. Graphical view of the Latent Block Model. Entries R_{ij} of the data matrix are independently generated according to the group membership $\boldsymbol{U}_i$ of row i and the group membership $\boldsymbol{V}_j$ of column j.

2.2   Model Proposed

The user feedback used for collaborative filtering can be implicit (history, browsing history, clicks...) or explicit. In the case of explicit evaluation data, users most often express their interest in items using a discrete rating scale. This rating scale suppose an order between levels, for example from 1 to 5 expressing the worst opinion to the best one. Models handling this type of data can assume that these scales are a discretization of the opinion of a user that may be better handled by a continuous variable. The method we propose to model ratings is based on a statistical co-clustering using ordered probit regression to model ordinal responses. Covariates encoding a sensitive user attribute can easily be included in the probit regression framework.

Ordered Probit in Latent Block Model. The ordered probit model [11] assumes the existence of a continuous, Gaussian distributed latent random variable, denoted $\boldsymbol{R}^*$. In a collaborative filtering context, this latent variable represents the underlying value, assumed to be continuous, assigned to an item by the user. The assumption of a single underlying continuous variable leading to ordinal ratings may be appropriate when ratings are not the result of a sequential process [10]. The discrete observed ratings $\boldsymbol{R}$ are the result of the partition of the continuous space of $\boldsymbol{R}^*$ by a set of thresholds $\boldsymbol{\zeta}$ such that: $R_{ij} = 1$ if $-\infty < R_{ij}^* < \zeta_1$, $R_{ij} = 2$ if $\zeta_1 < R_{ij}^* < \zeta_2$, ..., $R_{ij} = K$ if $\zeta_{K-1} < R_{ij}^* < +\infty$ (see Fig. 2).

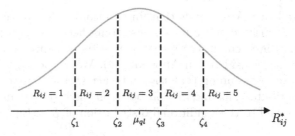

Fig. 2. The conditional density function of R_{ij}^* and its relationship to R_{ij}. Fixed thresholds ζ_k, defines the discretization of R_{ij}^*.

We use the ordered probit model within a Latent Block Model (see Sect. 2.1), assuming that conditionally to row and column group assignments, the entries of $\boldsymbol{R}^*$ are independent and identically distributed with Gaussian distribution:

$$p(R_{ij}^* \,|\, U_{iq}V_{jl} = 1; \mu_{ql}, \sigma) = \phi\big(R_{ij}^*; \mu_{ql}, \sigma^2\big) \ , \ \mu_{ql} \in \mathbb{R} \text{ and } \sigma \in \mathbb{R}_+^* \qquad (3)$$

with $\phi(\cdot; \mu_{ql}, \sigma^2)$ the probability density function of the Gaussian distribution with mean μ_{ql} and variance σ^2. The conditional probability that a user i gives to the item j the rating with value k is then:

$$\begin{aligned} p(R_{ij} = k \,|\, U_{iq}V_{jl} = 1; \mu_{ql}) &= p(\zeta_{k-1} < R_{ij}^* < \zeta_k \,|\, U_{iq}V_{jl} = 1; \mu_{ql}) \\ &= \varPhi\big(\zeta_k; \mu_{ql}, \sigma^2\big) - \varPhi\big(\zeta_{k-1}; \mu_{ql}, \sigma^2\big) \ , \end{aligned}$$

with $\varPhi(\cdot; \mu_{ql}, \sigma^2)$ being the normal cumulative distribution function. To ensure model identifiability, the thresholds ζ are fixed to equidistant predefined values.

Individual Row and Column Effects. The Latent Block Model is well suited to collaborative filtering, in that it searches for users and items that share the same opinion patterns. However, a model that assumes that users in a given cluster share exactly the same opinion patterns is very restrictive. Instead, we assume here that opinions may be slightly different within a cluster, using a richer model than Eq. (3) for the conditional distribution of R_{ij}^*. In addition to the cluster effect μ_{ql} derived solely from the group memberships of users and items, one deviation is induced by the user i and another by the item j :

$$p(R_{ij}^* \,|\, U_{iq}V_{jl} = 1, A_i, B_j; \mu_{ql}) = \phi\big(R_{ij}^*; \mu_{ql} + A_i + B_j, \sigma^2\big) \ , \qquad (4)$$

with latent variables $\boldsymbol{A}$ and $\boldsymbol{B}$ independently and identically distributed with:

$$\begin{aligned} A_i &\overset{\text{iid}}{\sim} \mathcal{N}(0, \sigma_A^2), & \sigma_A^2 \in \mathbb{R}_+^* \\ B_i &\overset{\text{iid}}{\sim} \mathcal{N}(0, \sigma_B^2), & \sigma_B^2 \in \mathbb{R}_+^* \end{aligned}$$

These two variables encode different rating patterns for users and items such as systematic over- or under-rating relative to the user or item populations.

Sensitive Attribute. We assume that, in addition to the matrix of ratings, we have access to a sensitive attribute s_i, describing here a binary feature of user i that should not intervene in the recommendation of items (more general sensitive attributes are considered in Appendix D). We introduce a latent variable C_j for each object j assuming that they interact with different strengths with the sensitive attribute. This interaction between the object j and the sensitive attribute s_i is added to the conditional distribution of R_{ij}^* (Eq. 4):

$$p(R_{ij}^* \mid U_{iq}V_{jl} = 1, A_i, B_j, s_i, C_j; \mu_{ql}) = \phi(R_{ij}^*; \mu_{ql} + A_i + B_j + s_iC_j, \sigma^2) \ ,$$

with

$$C_j \stackrel{\mathrm{iid}}{\sim} \mathcal{N}(0, \sigma_C^2), \ \sigma_G^2 \in \mathbb{R}_+^* \ .$$

This model explains the ratings by $\mu_{ql} + A_i + B_j + s_iC_j$ and σ^2; the co-clustering is driven by μ_{ql}, and provided the effects of the sensitive attribute are well captured by s_iC_j, we expect the co-clustering to be independent of the sensitive attribute, which ensures fair recommendations as shown in Sect. 3.2. A summary of the model we propose is presented in Fig. 3.

Modelling Missingness. The datasets extracted from recommender systems are usually extremely sparse, with a high proportion of missing ratings, that is, ratings that were not provided by the users. The model we proposed so far does not accommodate missing observations, and suppose a fully observed data matrix $\boldsymbol{R}$.

The study of missing data identifies three main type of missingness [33]: Missing Completely At Random (MCAR) and Missing At Random (MAR) referring to the mechanisms in which the probability of being missing does not depend on the variable of interest (here $\boldsymbol{R}^*$); and finally Missing Not At Random (NMAR) referring to the mechanisms in which the probability of being missing depends on the actual value of the missing data. A common implicit assumption in collaborative filtering is that ratings are MAR or MCAR: the presence/absence of ratings is assumed to convey no information whatsoever about the value of these ratings. For simplicity of statistical modelling we take the same assumption, although previous studies [27,28] have shown a potential dependence between the presence of ratings and the underlying opinion. We introduce a simple Bernoulli missingness model generating $\boldsymbol{M} \in \{0,1\}^{n_1 \times n_2}$, a mask matrix where each entry M_{ij} is one with probability p and indicates whether the rating is observed: $M_{ij} = 1$ if R_{ij} is observed and 0 otherwise. Given the complete data matrix $\boldsymbol{R}^*$ and the mask matrix $\boldsymbol{M}$, the elements of the observed ratings $\boldsymbol{R}$ are generated as follows:

$$(R_{ij} \mid R_{ij}^*, M_{ij}) = \begin{cases} \sum_{k=1}^{K} k \, \mathbb{1}_{]\zeta_{k-1}; \zeta_k]}(R_{ij}^*) & \text{if} \quad M_{ij} = 1 \\ \text{NA} & \text{if} \quad M_{ij} = 0 \end{cases}$$

Any generative model under a MCAR or MAR process can be fitted separately from the missingness model as the overall likelihood can be factorized between the observed and non observed data. Under such assumptions, we show in Appendix A that ignoring non-observed ratings results in a proper fitting.

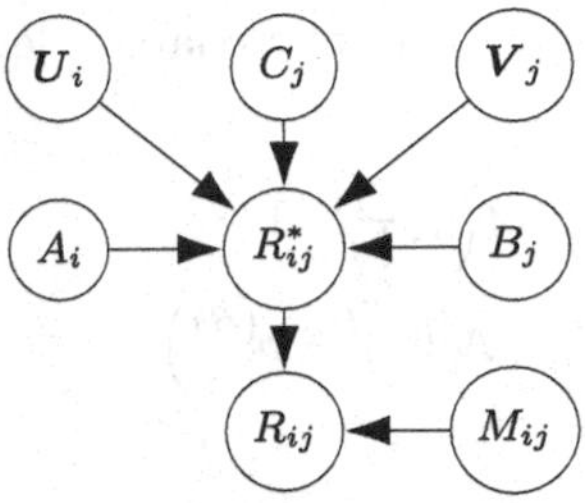

$$U_i \overset{\text{iid}}{\sim} \mathcal{M}(1;\boldsymbol{\alpha}), \qquad \boldsymbol{\alpha} \in \mathbf{S}_{k_1-1}$$

$$V_j \overset{\text{iid}}{\sim} \mathcal{M}(1;\boldsymbol{\beta}), \qquad \boldsymbol{\beta} \in \mathbf{S}_{k_2-1}$$

$$A_i \overset{\text{iid}}{\sim} \mathcal{N}(0,\sigma_A^2), \qquad \sigma_A^2 \in \mathbb{R}_+^*$$

$$B_j \overset{\text{iid}}{\sim} \mathcal{N}(0,\sigma_B^2), \qquad \sigma_B^2 \in \mathbb{R}_+^*$$

$$C_j \overset{\text{iid}}{\sim} \mathcal{N}(0,\sigma_C^2), \qquad \sigma_C^2 \in \mathbb{R}_+^*$$

$$\left(R_{ij}^* \big| U_{iq}=1, V_{jl}=1, A_i, B_j\right) \overset{\text{ind}}{\sim} \mathcal{N}\left(\mu_{ql}+A_i+B_j+s_iC_j,\sigma^2\right)$$

$$\left(R_{ij}\big|R_{ij}^*, M_{ij}\right) = \begin{cases} \sum_{k=1}^K k \, \mathbb{1}_{]\zeta_{k-1};\zeta_k]}(R_{ij}^*) & \text{if} \quad M_{ij}=1 \\ \text{NA} & \text{if} \quad M_{ij}=0 \end{cases}$$

with $M_{ij} \overset{\text{iid}}{\sim} \mathcal{B}(p), \qquad p \in [0,1]$

and $\zeta_0 = -\infty < \zeta_1 < ... < \zeta_{K-1} < \zeta_K = \infty$, fixed thresholds

Fig. 3. Graphical view and summary of the ordered probit Latent Block Model with protected attribute s. The discrete observed data R_{ij} is generated by the underlying continuous data R_{ij}^* and the mask entry M_{ij}.

3 Inference and Fair Recommendations

3.1 A Stochastic Batch Gradient Descent of the Variational Criterion

The log-likelihood of the model is not tractable as it involves a sum that is combinatorially too large [8]. We resort to a variational inference procedure [19] that introduces q_γ, a restricted parametric inference distribution defined on the latent variables of the model, to optimize the following lower bound on the log-likelihood:

$$\mathcal{J}(\gamma,\theta) = \log p(\boldsymbol{R};\theta) - \text{KL}(q_\gamma \, \| \, p(L|\boldsymbol{R}))$$

where KL stands for the Kullback-Leibler divergence, $\mathcal{H}$ for the differential entropy, $\theta = (\boldsymbol{\alpha},\boldsymbol{\beta},\boldsymbol{\mu},\sigma^2,\sigma_A^2,\sigma_B^2,\sigma_C^2,p)$ is the concatenation of the model parameters, and $L = (\boldsymbol{U},\boldsymbol{V},\boldsymbol{A},\boldsymbol{B},\boldsymbol{C})$ is the concatenation of the latent variables.

The variational distribution q_γ is chosen so that the computation of the criterion becomes easier:

$$\forall i, \qquad U_i|\boldsymbol{R} \underset{q_\gamma}{\sim} \mathcal{M}\left(1;\boldsymbol{\tau}_i^{(U)}\right) \qquad \forall j, \qquad V_j|\boldsymbol{R} \underset{q_\gamma}{\sim} \mathcal{M}\left(1;\boldsymbol{\tau}_j^{(V)}\right)$$

$$\forall i, \qquad A_i|\boldsymbol{R} \underset{q_\gamma}{\sim} \mathcal{N}\left(\nu_i^{(A)},\rho_i^{(A)}\right) \qquad \forall j, \qquad B_j|\boldsymbol{R} \underset{q_\gamma}{\sim} \mathcal{N}\left(\nu_j^{(B)},\rho_j^{(B)}\right)$$

$$\forall j, \qquad C_j|\boldsymbol{R} \underset{q_\gamma}{\sim} \mathcal{N}\left(\nu_j^{(C)},\rho_j^{(C)}\right) .$$

We also enforce the conditional independence of the latent variables, leading to the following fully factorized form:

$$q_\gamma = \prod_{i=1}^{n_1} \mathcal{M}\left(1; \tau_i^{(U)}\right) \times \prod_{j=1}^{n_2} \mathcal{M}\left(1; \tau_j^{(V)}\right) \tag{5}$$
$$\times \prod_{i=1}^{n_1} \mathcal{N}\left(\nu_i^{(A)}, \rho_i^{(A)}\right) \times \prod_{j=1}^{n_2} \mathcal{N}\left(\nu_j^{(B)}, \rho_j^{(B)}\right)$$
$$\times \prod_{j=1}^{n_2} \mathcal{N}\left(\nu_j^{(C)}, \rho_j^{(C)}\right) ,$$

where γ denotes the concatenation of all parameters of the variational distribution[1]. This conditional independence of the latent variables to $\boldsymbol{R}$ simplifies the criterion $\mathcal{J}(\gamma, \theta)$ to:

$$\mathcal{J}(\gamma, \theta) = \mathbb{E}_{q_\gamma}[\log p(\boldsymbol{R}|L)] - \mathrm{KL}(q_\gamma \| p(L; \theta)) . \tag{6}$$

As explained in Sect. 2.2, the optimization criterion relies only on the non-missing entries of $\boldsymbol{R}$ because the data is assumed to be missing at random. The full expansion of the criterion is given in Appendix A.

We resort to a batch stochastic optimization to maximize the variational criterion using noisy estimates of its gradient [30]. Samples are drawn from the variational distribution (Eq. 5) to estimate a noisy but unbiased gradient of the expectation of the conditional log-distribution of $\boldsymbol{R}$ (first term of Eq. 6), which we then use to update our parameters as follows:

$$(\gamma, \theta)^{(t+1)} = (\gamma, \theta)^{(t)} + \eta \cdot \nabla_{(\gamma, \theta)} \mathcal{J}\left(\boldsymbol{R}_{(i:i+n),(j:j+n)}; \gamma, \theta\right) ,$$

where n is the batch size and η is the adaptive learning rate based on the past gradients that were computed (Adam optimizer [24]).

Using a stochastic gradient algorithm instead of the usual EM algorithm alleviates the well-known initialization problems of the Latent Block Model, which result in unsatisfactory local maxima [2,6]. However, it requires the use of differentiable functions to back-propagate gradients through the automatic differentiation graph. For this purpose, the multinomial distributions are replaced by a differentiable Gumbel-Softmax distribution [20].

3.2   Fair Recommendations

This section describes a theoretical result establishing a guarantee on the fairness of recommendations. This guarantee is subject to an assumption about the parity of the clustering of users that can be tested in practice, and that holds true for the experiments reported in Sect. 4 and Appendix D. We develop here the case of a binary sensitive attribute to simplify the exposition. The result is more general and applies to any discrete sensitive attribute. It is proven in this general sense in Appendix C.

Recommendations are partial orders between items. In collaborative filtering, the usual approach to producing recommendations is to estimate a relevance

---

[1] $\gamma = (\boldsymbol{\tau}^{(U)}, \boldsymbol{\tau}^{(V)}, \boldsymbol{\nu}^{(A)}, \boldsymbol{\rho}^{(A)}, \boldsymbol{\nu}^{(B)}, \boldsymbol{\rho}^{(B)}, \boldsymbol{\nu}^{(C)}, \boldsymbol{\rho}^{(C)}).$

score for each item, which is then used to define a total order through numerical comparisons. With the parameters obtained by variational inference, we define the relevance score of item j for user i as:

$$\hat{R}_{ij} = \boldsymbol{\tau}_i^{(U)} \hat{\boldsymbol{\mu}} \boldsymbol{\tau}_j^{(V)T} + \nu_i^{(A)} + \nu_j^{(B)} \ . \tag{7}$$

This relevance score is computed from the maxima *a posteriori* of the latent variables encoding the user and item group memberships $(\boldsymbol{\tau}_i^{(U)}, \boldsymbol{\tau}_j^{(V)})$, that is, the trend related to the co-cluster to which (i, j) belongs, and the global effects related to user i and item j. It does not use the user's sensitive attribute s_i which is considered here as a nuisance parameter, properly taken into account during inference and then ignored when predicting a relevance score. It then becomes possible to compare items fairly with respect to the sensitive attribute.

Definition 2 (Fair comparison of items). *Given user i and any two items j and j', the comparison of items j and j' is said to be fair if it is freed from the evaluation bias regarding the sensitive attribute s: item j is fairly preferred to item j' if $\hat{R}_{ij} > \hat{R}_{ij'}$, that is:*

$$\boldsymbol{\tau}_i^{(U)} \hat{\boldsymbol{\mu}} \boldsymbol{\tau}_j^{(V)T} + \nu_i^{(A)} + \nu_j^{(B)} > \boldsymbol{\tau}_i^{(U)} \hat{\boldsymbol{\mu}} \boldsymbol{\tau}_{j'}^{(V)T} + \nu_i^{(A)} + \nu_{j'}^{(B)} \ .$$

The modelling of the observed data $\boldsymbol{R}$ incorporates the term $\nu_j^{(C)} s_i$, interpreted here as a spurious opinion bias related to the sensitive attribute. While it is important to ignore this term for a fair comparison of items, its inclusion into the model is important to allow the construction of clusters that are not affected by this spurious effect. These clusters can then be expected to be representative of all subpopulations defined by their sensitive attribute value, and thus to respect the statistical parity of users.

Definition 3 (Clustering ε-parity, binary sensitive attribute). *The clustering of users is said to respect ε-parity with respect to attribute s iff:*

$$\forall q, \quad \left| \frac{\#\{i|s_i = 1 \wedge u_{iq} = 1\}}{\#\{i|s_i = 1\}} - \frac{\#\{i|s_i = -1 \wedge u_{iq} = 1\}}{\#\{i|s_i = -1\}} \right| \leq \varepsilon \ , \tag{8}$$

where $\varepsilon \in \mathbb{R}_+$ measures the gap to exact parity, u_{iq} is the (hard) membership of user i to cluster q, and $\#\{i|\Omega\}$ is the number of users defined by the cardinality of the set Ω.

In essence, clustering ε-parity requires that subpopulations of users defined by identical sensitive attributes be represented approximately equally in each user group. For the Latent Block Model, the hard membership u_{iq} of Definition 3 is given by the maximum *a posteriori* of the latent variable $\tau_{iq}^{(U)}$.

Our theoretical guarantee ensures that this approximate statistical parity in clusters is sufficient to get approximately fair recommendations (Definition 1) from our model:

Theorem 1 (Fair recommendation from clustering parity). *If the clustering of users in k_1 groups respects ε-parity (Definition 3 or Definition S1) then the recommender system relying on the relevance score defined in Eq. (7) is $(k_1\varepsilon)$-fair (Definition 1 or Definition S2).*

Proof: see Appendix C.

4   Experiment on MovieLens Dataset

The final goal of a recommender system is to provide users with a shortlist of items that they might most enjoy. We choose here to directly assess the quality of the ranking rather than using proxy measures, such as root mean square error on ratings, that ignore relative rankings.

To measure the ranking performance of algorithms, we use the Normalized Discounted Cumulative Gain [21] (NDCG) that measures ranking quality by a penalized sum of the relevance scores of the ranking results:

$$NDCG@k = \frac{DCG@k}{IDCG@k} \text{ with } DCG@k = \sum_{i=1}^{k} \frac{rel_i}{\log(i+1)} \ ,$$

rel_i, the relevance of the results at each rank i before k and $IDCG@k$ being the $DCG@k$ computed with a perfect ranking.

We use the MovieLens 1M dataset [29] that contains one million ratings given by 6,040 users to 3,900 movies scaling from 1 to 5 (from least liked to most liked). The dataset also contains additional information about users: gender (binary), age category (seven levels) or occupation. We give here some experimental results where gender is the sensitive attribute, and additional results, in particular with age considered as the sensitive attribute, can be found in Appendix D.

4.1   Experimental Protocol

We estimate the average performances by predicting preferences on ratings that are concealed during training (see "test ratings" methodology from [3]). These concealed ratings form our test set, with 20 ratings per user, which is about 10% of the available data. This process is repeated 5 times, with independent random draws, to produce stable average performances.

We compare our model (referred to as Parity LBM) with the baseline LBM that does not use the sensitive variable in the modelling (referred to as Standard LBM). We expect the latter model to create groups of users that do not respect clustering parity and to generate unfair recommendations. We also compare to another co-clustering algorithm, weighted Bregman co-clustering with squared Euclidean distance [13] (referred to as Bregman co-clust) to compare the statistical parity of user groups inferred from another baseline. Finally, we compare with Singular Value Decomposition (SVD), a method popularized during the Netflix challenge [25] that still remains state of the art in collaborative filtering

[32]. All these baselines are implemented in the Python module Surprise [18]. We were not able to use or update the implementation of the recommendation independence models of [22], which is no longer compatible with current libraries.

The number of clusters in co-clustering and the number of factors in matrix factorization are both arbitrarily set to fifteen. Another comparison with more clusters, provided in Appendix D, produces qualitatively similar results.

We repeat the learning process 25 times from different random initializations to mitigate the initialization dependence that affects all optimization procedures. We select the best solution based on the optimization criteria, that is, the one with the highest likelihood for the LBM models and the lowest training reconstruction error for the other baselines.

4.2 Results and Discussion

Gender as Sensitive Attribute. User gender (binary in this dataset) is used as the sensitive attribute s_i. In the dataset, 27% of users self-identified as females, this proportion must be met in each group to respect clustering parity. To measure the dependence between gender and user group memberships, we compute the χ^2 statistic constructed from the contingency table of males and females counts in each group. Table 1 reports the p-value for testing the independence between groups and genders, with an asymptotical test. We recall that, under the null hypothesis of independence, the test statistic with k degrees of freedom has mean k and variance $2k$. The results show that the methods that do not consider the sensitive variable in the modelling create groups that are dependent on gender. In contrast, our Parity-LBM model is consistent with the clustering parity hypothesis: the gender representation in groups is representative of the gender distribution in the overall dataset. The fairness of recommendations resulting from this clustering parity is ascertained by computing the gap ε from exactly fair recommendations, as defined in Definition 1. Figure 4 displays these gaps, with lower values indicating a fairer recommendation; our model provides a significantly fairer recommendation compared to the standard Latent Block Model, which is itself much fairer than the two other baselines. The order observed in Table 1 is followed.

Figure 5a depicts the ranking performance of algorithms with the NDCG, averaged over all users, for a recommendation list of 10 items. SVD gets the best overall result, followed by the Latent Block Models that outperform Bregman co-clustering. The overall performances of our model and the standard LBM are not significantly different. Figure 5a also reports the average NDCG within each sensitive group. This performance measure shows that female users receive significantly less relevant recommendations than males with all algorithms. This measure of disparate impact on truly relevant recommendations is reminiscent of equalized odds [17] in the classification framework, in that it measures a disparity on positive outcomes. The performance gap between the sensitive groups is reduced by our parity LBM compared to the standard LBM. Although the difference is the smallest among all comparisons, our model does not eliminate

Table 1. Measures of statistical gender parity among user clusters. The number of user groups is $k_1 = 15$. The χ^2 statistic (with 14 degrees of freedom) is averaged over the five replicates of the experiment. A high value of the χ^2 statistic (or a low p-value) leads to the rejection of the clustering parity hypothesis.

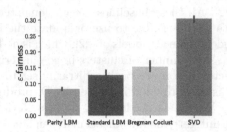

Model	Parity LBM	Standard LBM	Bregman Co-clust
χ^2 statistic	18.0	44.4	187
p-value	0.20	$5.1 \cdot 10^{-5}$	$< 10^{-15}$

Fig. 4. Gaps ε for the ε-fair recommendations (see Definition 1) provided by each model: a smaller ε-fairness indicates fairer recommendations.

disparate impact. As a caveat, it is likely that all recommendations are less relevant to female users, but since the observed ratings used to evaluate all solutions are affected by the gender stereotypes we wish to counter, it is not possible to unbiasedly measure the performance of fair recommendations from the original rating matrix, which suffers from this stereotype bias. To reduce the disparate performances between males and females, a possible solution is to systematically favors items identified as overrated by females. We recall that the latent variable C_j, which is until now not used for fair prediction, captures the difference in opinion trends between female and male users on movie j. A high absolute value of C_j indicates a strongly gendered opinion for movie j. With our encoding of genres, negative C_j indicate a relative overrating by males and positive C_j indicate a relative overrating by females. If the relevance score of Eq. (7) is modified by using this latent variable C_j, but not the protected attribute, some stereotype is equally injected in the recommendations of all users:

$$\hat{R}_{ij} = \boldsymbol{\tau}_i^{(U)} \hat{\boldsymbol{\mu}} \boldsymbol{\tau}_j^{(V)T} + \nu_i^{(A)} + \nu_j^{(B)} + \lambda \, \nu_j^{(C)} , \tag{9}$$

with $\lambda \in \mathbb{R}$ being the strength of the stereotype injected in recommendations. A positive value of λ will favors movies identified as overrated by females while a negative value will favors movies identified as overrated by males. It should be noted that altering the relevance score to inject stereotype for all users do not modify the gap ε from exactly fair recommendations of the method. Figure 5b shows that the gap of performances between males and females can be reduced up to be equalized. Although these results show that seeking for an equality of impact is possible, it comes with the cost that some sort of stereotype, originally present in the dataset, would remain in the recommendations which is ethically arguable and should be used with cautious for each type of recommender system.

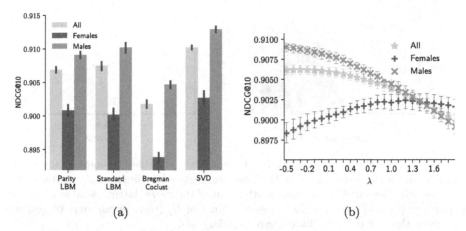

Fig. 5. (a): Normalized Discounted Cumulative Gain estimated on MovieLens-1M (the higher the better). (b): NDCG estimated using the relevance scores exploiting stereotype (Eq. 9). A positive (resp. negative) value of λ favors movies overrated by females (resp. males).

Finally, we present some insights provided by our model on movies. We display the empirical cumulative distribution function (CDF) of C_j for movies conditionally on their genre (for some handpicked archetypal genres). The dominance of the CDF for a given genre expresses that, according to our model, female users have a higher opinion than male users for the movies belonging to that genre. Figure 6 shows the results, which reflect stereotypes that women are more likely than men to positively evaluate musical films and dramas, while men are similarly inclined toward westerns and action films. These stereotypes are incorporated into our model to fit actual ratings, but ignored to deliver fair recommendations using Eq. (7). The lists of extreme movies based on extreme (positive and negative) values of C_j is given in Appendix D.1.

The latent variable B_j encodes the overall opinion trend about movie j. Two interesting observations can be made from the scatter plot of B_j versus movie popularity (see bottom of Fig. 6). First, unpopular movies are also the least appreciated according to our model; this supports the hypothesis that ratings are generated by a MNAR (Missing Not At Random) process, where a missing rating can be considered as weak negative feedback, assuming that users primarily rate items they like. This missingness process must still be taken into account in our model. Second, it shows that the most liked movies (according to our model) are not necessarily the most popular (and will be recommended); the recommendations are not affected by popularity bias.

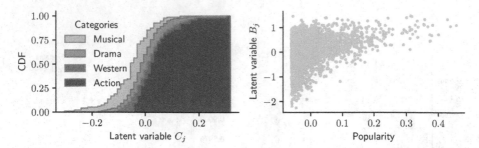

Fig. 6. Top: cumulative distribution function of latent variable C_j conditionally on the genre of the movie. A dominating CDF indicates a genre for which females' opinions are more positive than males'. Bottom: scatter plot of the movie latent variable B_j versus popularity (ratio of ratings). High positive values of B_j (resp. popularity) correspond to movies that are the most liked (resp. popular).

5   Conclusion

We proposed a new co-clustering method for fair recommendation. Our model combines the Gaussian Latent Block Model with an ordinal regression model. The sensitive attribute is adequately accounted for in the model, allowing the clustering of users to be unaffected by the effects of this attribute on ratings. This results in user clusters that approximately respect statistical parity. We base recommendation on a relevance score that ignores the sensitive attribute in order to compare items fairly. We provide theoretical guarantees ensuring approximately fair recommendations, for any known discrete sensitive attribute, provided that the clustering of users respects an approximate statistical parity that can be assessed in practice. Our analysis focuses on the fairness of preferences, as defined by the ranking of ratings, rather than on the predicted values themselves, which are less relevant for recommendation. Through experiments on real-world data, we show that our method significantly mitigates the unfairness of recommendations. Furthermore, the latent variables inferred by the model are also amenable to analyses that can help identify recommendation bias.

Our study supports that the absence of rating conveys some information that should be exploited. Previous works [27,28] have already shown that the data used for collaborative filtering datasets can be strongly influenced by observational bias, which motivates dealing with missingness by a Missing Not At Random (MNAR) process. Societal biases may have a significant contribution to missingness, leading to an additional source of unfairness if missingness is not properly modeled. Studying fairness with MNAR processes is a highly relevant but extremely challenging direction for future research, as assessing the relevance of MNAR models in real situations requires data that are typically produced by online randomized experiments.

Co-clustering for Fair Recommendation. Supplementary Material

A   Computation of the Variational Log-Likelihood Criterion

The criterion we want to optimize is:

$$\mathcal{J}(q_\gamma, \theta) = \mathcal{H}(q_\gamma) + \mathbb{E}_{q_\gamma}\left[\mathcal{L}(R, U, V, A, B, C; \theta)\right] . \tag{S1}$$

We chose to restrict the space of the variational distribution q_γ in order to get a fully factorized form:

$$q_\gamma = \prod_{i=1}^{n_1} \mathcal{M}\left(1; \tau_i^{(U)}\right) \times \prod_{j=1}^{n_2} \mathcal{M}\left(1; \tau_j^{(V)}\right) \tag{S2}$$
$$\times \prod_{i=1}^{n_1} \mathcal{N}\left(\nu_i^{(A)}, \rho_i^{(A)}\right) \times \prod_{j=1}^{n_2} \mathcal{N}\left(\nu_j^{(B)}, \rho_j^{(B)}\right)$$
$$\times \prod_{j=1}^{n_2} \mathcal{N}\left(\nu_j^{(C)}, \rho_j^{(C)}\right)$$

where γ denotes the parameters concatenation of the variational distribution[2] q_γ. The entropy is additive across independant variables so we get:

$$\mathcal{H}(q_\gamma) = \mathcal{H}(q_\gamma(U)) + \mathcal{H}(q_\gamma(V)) + \mathcal{H}(q_\gamma(A)) + \mathcal{H}(q_\gamma(B)) + \mathcal{H}(q_\gamma(C)) ,$$

with the following terms:

$$\mathcal{H}(q_\gamma(U)) = -\sum_{iq} \tau_{iq}^{(U)} \log \tau_{iq}^{(U)}$$

$$\mathcal{H}(q_\gamma(V)) = -\sum_{jl} \tau_{jl}^{(U)} \log \tau_{jl}^{(V)}$$

$$\mathcal{H}(q_\gamma(A)) = \frac{1}{2}\sum_i \log \rho_i^{(A)} + \frac{n_1}{2}(\log 2\pi + 1)$$

$$\mathcal{H}(q_\gamma(B)) = \frac{1}{2}\sum_j \log \rho_j^{(B)} + \frac{n_2}{2}(\log 2\pi + 1)$$

$$\mathcal{H}(q_\gamma(C)) = \frac{1}{2}\sum_j \log \rho_j^{(C)} + \frac{n_2}{2}(\log 2\pi + 1)$$

The independence of the latent variables allows to rewrite the expectation of the complete log-likelihood as:

$$\mathbb{E}_{q_\gamma}[\mathcal{L}(R, U, V, A, B, C)] = \mathbb{E}_{q_\gamma}[\mathcal{L}(U)] + \mathbb{E}_{q_\gamma}[\mathcal{L}(V)]$$
$$+ \mathbb{E}_{q_\gamma}[\mathcal{L}(A)] + \mathbb{E}_{q_\gamma}[\mathcal{L}(B)] + \mathbb{E}_{q_\gamma}[\mathcal{L}(C)]$$
$$+ \mathbb{E}_{q_\gamma}[\mathcal{L}(R \mid U, V, A, B, C)] ,$$

---

[2] $\gamma = (\tau^{(U)}, \tau^{(V)}, \nu^{(A)}, \rho^{(A)}, \nu^{(B)}, \rho^{(B)}, \nu^{(C)}, \rho^{(C)})$.

with the following terms:

$$\mathbb{E}_{q_\gamma}\mathcal{L}(U) = \mathbb{E}_{q_\gamma}\left[\sum_{iq} U_{iq}\log\alpha_q\right] = \sum_{iq}\tau_{iq}^{(U)}\log\alpha_q$$

$$\mathbb{E}_{q_\gamma}\mathcal{L}(V) = \mathbb{E}_{q_\gamma}\left[\sum_{jl} V_{jl}\log\beta_l\right] = \sum_{jl}\tau_{jl}^{(V)}\log\beta_l$$

$$\mathbb{E}_{q_\gamma}\mathcal{L}(A) = -\frac{n_1}{2}\log 2\pi - \frac{n_1}{2}\log\sigma_A^2 - \frac{1}{2\sigma_A^2}\sum_i \mathbb{E}_{q_\gamma}A_i^2$$

$$= -\frac{n_1}{2}\log 2\pi - \frac{n_1}{2}\log\sigma_A^2 - \frac{1}{2\sigma_A^2}\sum_i\left(\left(\nu_i^{(A)}\right)^2 + \rho_i^{(A)}\right)$$

$$\mathbb{E}_{q_\gamma}\mathcal{L}(B) = -\frac{n_2}{2}\log 2\pi - \frac{n_2}{2}\log\sigma_B^2 - \frac{1}{2\sigma_B^2}\sum_i\left(\left(\nu_i^{(B)}\right)^2 + \rho_i^{(B)}\right)$$

$$\mathbb{E}_{q_\gamma}\mathcal{L}(C) = -\frac{n_2}{2}\log 2\pi - \frac{n_2}{2}\log\sigma_C^2 - \frac{1}{2\sigma_C^2}\sum_j\left(\left(\nu_j^{(C)}\right)^2 + \rho_j^{(C)}\right)$$

and as the entries of the data matrix R are independent and identically distributed:

$$\mathbb{E}_{q_\gamma}\mathcal{L}(R\,|\,A,B,C,U,V) = \mathbb{E}_{q_\gamma}\mathcal{L}\left(R^{(o)}\,\Big|\,A,B,C,U,V\right) + \mathcal{L}\left(R^{(\neg o)}\right) \quad \text{(S3)}$$

where $R^{(o)}$ denotes the set of observed ratings and $R^{(\neg o)}$, the set of non-observed ratings, where $R_{ij} = $ NA. From Eq. S3, it becomes clear that maximizing $\mathbb{E}_{q_\gamma}\mathcal{L}(R^{(\neg o)})$ is not necessary to infer the model parameters used for prediction and therefore ignoring the non-observed data is correct. The expectation of the conditional log-likelihood (first term of right side of Eq. S3) is numerically estimated by sampling from q_γ.

Stochastic Gradient Optimization. To optimize the criterion with stochastic gradient descent, we express the variational log-likelihood criterion on a single rating:

$$\mathcal{J}(R_{ij}; q_\gamma, \theta) = \mathbb{E}_{q_\gamma}\left[\mathcal{L}\left(R_{ij}^{(o)}\,\Big|\,U_i, V_j, A_i, B_j, C_j\right)\right]$$

$$+ \frac{1}{n_2}\left(\mathcal{H}(q_\gamma(U_i)) + \mathcal{H}(q_\gamma(A_i)) + \mathbb{E}_{q_\gamma}[\mathcal{L}(U_i)] + \mathbb{E}_{q_\gamma}[\mathcal{L}(A_i)]\right)$$

$$+ \frac{1}{n_2}\left(\mathcal{H}(q_\gamma(V_j)) + \mathcal{H}(q_\gamma(B_j)) + \mathbb{E}_{q_\gamma}[\mathcal{L}(V_j)] + \mathbb{E}_{q_\gamma}[\mathcal{L}(B_j)]\right)$$

$$+ \frac{1}{n_2}\left(\mathcal{H}(q_\gamma(C_j)) + \mathbb{E}_{q_\gamma}[\mathcal{L}(C_j)]\right)$$

A batch of data, $\boldsymbol{R}_{(i:i+n),(j:j+n)}$, consists of a $(n \times n)$ sub-matrix randomly sampled from the original matrix $\boldsymbol{R}$.

B   Clustering ε-parity and ε-fair Recommendation for Arbitrary Discrete Sensitive Attribute

Definition S1 (Clustering ε-parity, arbitrary discrete sensitive attribute). *The clustering of users is said to respect ε-parity with respect to the discrete attribute $s \in \mathcal{S}$ iff:*

$$\forall (t,t') \in \mathcal{S}^2, \ \forall q, \ \left| \frac{\#\left\{i|s_i = t \wedge u_{iq} = 1\right\}}{\#\left\{i|s_i = t\right\}} - \frac{\#\left\{i|s_i = t' \wedge u_{iq} = 1\right\}}{\#\left\{i|s_i = t'\right\}} \right| \le \varepsilon \ , \quad (S4)$$

where $\varepsilon \in \mathbb{R}_+$ measures the gap to exact parity, u_{iq} is the (hard) membership of user i to cluster q, and $\#\left\{i|\Omega\right\}$ is the number of users defined by the cardinality of the set Ω.

Definition S2 (ε-fair recommendation, arbitrary discrete sensitive attribute). *A recommender system is said to be ε-fair with respect to the dicrete attribute $s \in \mathcal{S}$ if for any two items j and j':*

$$\forall (t,t') \in \mathcal{S}^2, \ \left| \frac{\#\left\{i|s_i = t \wedge (\hat{R}_{ij} > \hat{R}_{ij'})\right\}}{\#\left\{i|s_i = t\right\}} - \frac{\#\left\{i|s_i = t' \wedge (\hat{R}_{ij} > \hat{R}_{ij'})\right\}}{\#\left\{i|s_i = t'\right\}} \right| \lesssim c,$$

$$(S5)$$

where $\varepsilon \in \mathbb{R}_+$ measures the gap to exact fairness

C   Proof of Theorem 1

Theorem 1 (Fair recommendation from clustering parity). *If the clustering of users in k_1 groups respects ε-parity (Definition 3 or Definition S1) then the recommender system relying on the relevance score defined in Eq. (7) is $(k_1\varepsilon)$-fair (Definition 1 or Definition S2).*

Proof. Suppose that $\boldsymbol{\tau}^{(U)}$, the maximum a posteriori of $\boldsymbol{U}$, is a binary matrix; $\boldsymbol{\tau}^{(U)}$ is thus a $n_1 \times k_1$ indicator matrix of row classes membership. Then, given user i, item j is said to be preferred to item j' if $\hat{R}_{ij} > \hat{R}_{ij'}$, that is:

$$\hat{R}_{ij} > \hat{R}_{ij'} \iff \boldsymbol{\tau}_i^{(U)} \hat{\boldsymbol{\mu}} \boldsymbol{\tau}_j^{(V)^T} + \nu_i^{(A)} + \nu_j^{(B)} > \boldsymbol{\tau}_i^{(U)} \hat{\boldsymbol{\mu}} \boldsymbol{\tau}_{j'}^{(V)^T} + \nu_i^{(A)} + \nu_{j'}^{(B)}$$

$$\iff \boldsymbol{\tau}_i^{(U)} \hat{\boldsymbol{\mu}} \left(\boldsymbol{\tau}_j^{(V)} - \boldsymbol{\tau}_{j'}^{(V)} \right)^T > \nu_{j'}^{(B)} - \nu_j^{(B)}$$

$$\iff \boldsymbol{\tau}_i^{(U)} \boldsymbol{a} > b$$

$$\iff \boldsymbol{a}_{d_i} > b \ , \quad (S6)$$

with $\boldsymbol{a} \in \mathbb{R}^{k_1}$ defined by $\boldsymbol{a} = \hat{\boldsymbol{\mu}}\left(\boldsymbol{\tau}_j^{(V)} - \boldsymbol{\tau}_{j'}^{(V)}\right)^T$, $b \in \mathbb{R}$ defined by $b = \nu_{j'}^{(B)} - \nu_j^{(B)}$ and $d_i \in \{1, \cdots, k_1\}$ being the group indicator of user i: $\tau_{i,d_i}^{(U)} = 1$.

Suppose ε-parity, from Definition S1 (Definition 3 is a particular case of Definition S1), we have

$$\forall (t, t'), \qquad \forall q, \qquad \left| \frac{\#\{i|s_i = t \wedge d_i = q\}}{\#\{i|s_i = t\}} - \frac{\#\{i|s_i = t' \wedge d_i = q\}}{\#\{i|s_i = t'\}} \right| \leq \varepsilon$$

therefore,

$$\forall (t, t'), \ \forall q, \ \left| \mathbb{1}_{\boldsymbol{a}_{d_i} > b} \frac{\#\{i|s_i = t \wedge d_i = q\}}{\#\{i|s_i = t\}} - \mathbb{1}_{\boldsymbol{a}_{d_i} > b} \frac{\#\{i|s_i = t' \wedge d_i = q\}}{\#\{i|s_i = t'\}} \right| \leq \varepsilon \mathbb{1}_{\boldsymbol{a}_{d_i} > b}$$

By summing over all groups, we get:

$$\forall (t, t'), \ \sum_q \left| \frac{\mathbb{1}_{\boldsymbol{a}_{d_i} > b} \#\{i|s_i = t \wedge d_i = q\}}{\#\{i|s_i = t\}} - \frac{\mathbb{1}_{\boldsymbol{a}_{d_i} > b} \#\{i|s_i = t' \wedge d_i = q\}}{\#\{i|s_i = t'\}} \right| \leq \varepsilon \sum_q \mathbb{1}_{\boldsymbol{a}_{d_i} > b}$$

and from the triangular inequality,

$$\forall (t, t'), \quad \left| \frac{\sum_q \mathbb{1}_{\boldsymbol{a}_{d_i} > b} \#\{i|s_i = t \wedge d_i = q\}}{\#\{i|s_i = t\}} - \frac{\sum_q \mathbb{1}_{\boldsymbol{a}_{d_i} > b} \#\{i|s_i = t' \wedge d_i = q\}}{\#\{i|s_i = t'\}} \right| \leq \varepsilon \sum_q \mathbb{1}_{\boldsymbol{a}_{d_i} > b}$$

$$\forall (t, t'), \quad \left| \frac{\#\{i|s_i = t \wedge \boldsymbol{a}_{d_i} > b\}}{\#\{i|s_i = t\}} - \frac{\#\{i|s_i = t' \wedge \boldsymbol{a}_{d_i} > b\}}{\#\{i|s_i = t'\}} \right| \leq \varepsilon k_1$$

And, applying (S6), the result is obtained:

$$\forall (t, t'), \quad \left| \frac{\#\left\{i|s_i = t \wedge (\hat{R}_{ij} > \hat{R}_{ij'})\right\}}{\#\{i|s_i = t\}} - \frac{\#\left\{i|s_i = t' \wedge (\hat{R}_{ij} > \hat{R}_{ij'})\right\}}{\#\{i|s_i = t'\}} \right| \leq \varepsilon k_1$$

□

D   Supplemental Results for MovieLens 1M

D.1   Gender as Sensitive Attribute

Supplemental Analysis of the Model. We list in Tables 2 and 3 the most extreme movies according to the inferred value of their latent variable C_j. Variable C_j encodes the difference in opinion between the sensitive groups, not the overall opinion. For example, a movie may well be liked by most people but liked even more by males. Table 2 lists movies for which females have a better opinion than males and Table 3 lists movies for which males have a better opinion than females.

Table 2. List of movies with the largest gap in opinion between females and males for which females have a better opinion than males

Title	Year	Genders	Cj
Dirty Dancing	1987	Musical|Romance	0.31
Rocky Horror Picture Show, The	1975	Comedy|Horror|Musical|Sci-Fi	0.26
Sound of Music, The	1965	Musical	0.24
Grease	1978	Comedy|Musical|Romance	0.23
Jumpin' Jack Flash	1986	Action|Comedy|Romance|Thriller	0.23
Gone with the Wind	1939	Drama|Romance|War	0.22
Newsies	1992	Children's|Musical	0.21
Strictly Ballroom	1992	Comedy|Romance	0.21
Steel Magnolias	1989	Drama	0.20
Sense and Sensibility	1995	Drama|Romance	0.20
Full Monty, The	1997	Comedy	0.19
Much Ado About Nothing	1993	Comedy|Romance	0.18
Thelma & Louise	1991	Action|Drama	0.18
Swing Kids	1993	Drama|War	0.17
Fried Green Tomatoes	1991	Drama	0.17
Ever After: A Cinderella Story	1998	Drama|Romance	0.17
Anastasia	1997	Animation|Children's|Musical	0.17
Little Women	1994	Drama	0.17
Color Purple, The	1985	Drama	0.17
To Wong Foo, Thanks for Everything!	1995	Comedy	0.17

Higher Number of Groups. We did not optimize the hyper-parameters of the compared models. We present here additional experiments to illustrate that the conclusions of Sect. 4 apply to different hyper-parameter settings. Using a substantially larger number of groups ($k_1 = 50$ user groups and $k_2 = 50$ item groups) or a larger dimension of latent factors for SVD (also 50), the statistical gender parity measures given in Table 4 and the recommendation performance given in Fig. 7 are qualitatively similar to the ones given in Table 1 and Fig. 5.

Table 3. List of movies with the largest gap in opinion between females and males for which males have a better opinion than females

Title	Year	Genders	C_j
Good, The Bad and The Ugly, The	1966	Action|Western	−0.32
Animal House	1978	Comedy	−0.30
Caddyshack	1980	Comedy	−0.27
Dumb & Dumber	1994	Comedy	−0.27
Exorcist, The	1973	Horror	−0.24
Clockwork Orange, A	1971	Sci-Fi	−0.24
Patton	1970	Drama|War	-0.23
Godfather: Part II, The	1974	Action|Crime|Drama	−0.22
Reservoir Dogs	1992	Crime|Thriller	−0.22
Saving Private Ryan	1998	Action|Drama|War	−0.22
Airplane!	1980	Comedy	−0.21
Eyes Wide Shut	1999	Drama	−0.21
Aliens	1986	Action|Sci-Fi|Thriller|War	−0.21
Predator	1987	Action|Sci-Fi|Thriller	−0.20
Apocalypse Now	1979	Drama|War	−0.20
Unforgiven	1992	Western	−0.20
Evil Dead II (Dead By Dawn)	1987	Action|Adventure|Comedy|Horror	−0.20
Big Trouble in Little China	1986	Action|Comedy	−0.20
Godfather, The	1972	Action|Crime|Drama	−0.20

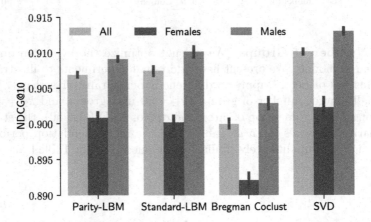

Fig. 7. Normalized Discounted Cumulative Gain estimated on MovieLens-1M with $k_1 = k_2 = 50$ groups for clustering methods and 50 factors for the SVD.

D.2   Age as Sensitive Attribute

The age range of the users is indicated within the following intervals: 'Under 18','18–24', '25–34', '35–44', '45–49', '50–55' and '56+'. The counts of users in each age category is displayed in Fig. 8.

Table 4. Measures of gender statistical parity. The number of user groups is $k_1 = 50$. The χ^2 statistic (with 49 degrees of freedom) is averaged over the five replicates of the experiment. A high value of the χ^2 statistic (or a low p-value) leads to the rejection of the statistical parity hypothesis.

Model	Parity LBM	Standard LBM	Bregman co-clust
χ^2 statistic	20	94	105
p-value	0.999	$1.1 \cdot 10^{-4}$	$5.8 \cdot 10^{-6}$

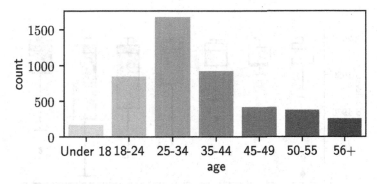

Fig. 8. Count of users in each age category.

Uscr age is treated as sensitive: we introduce seven binary sensitive attributes s_i encoding for the seven categories of user age. We use a one-hot encoding of the seven categories of user age and introduce for the purpose seven binary sensitive attributes $s_i^1, \cdots, s_i^7$ and their item associated latent variables $C_j^1, \cdots, C_j^7$. We use the protocol described in Sect. 4 with the exception that our Parity-LBM is initialized from estimates obtained with the Standard-LBM. Table 5 presents results of the χ^2 statistics constructed from the contingency table of user age counts in each group. The methods that do not consider the sensitive variable in the modelling create groups that are dependent on the age and assuming the statistical parity with our Parity-LBM model is reasonable.

Table 5. Measures of statistical parity with respect to age category. The number of group of users is $k_1 = 15$. A high value of the χ^2 statistic (or a low p-value) leads to the rejection of the statistical parity hypothesis. The χ^2 statistic is averaged on the five folds of the cross-validation. Degrees of freedom is 14.

Model	Parity LBM	Standard LBM	Bregman co-clust
χ^2 statistic	99	144	577
p-value	0.12	$5.1 \cdot 10^{-5}$	$< 10^{-15}$

Finally, we illustrate the interpretability of the estimates of the latent variables $C_j^1, \cdots, C_j^7$ related to movies. For each age category k, we select the thirty movies with the largest value of the latent variables C_j^k. These movies have the largest positive opinion bias for users in the given age category. Figure 9 displays a boxplot of the release years of these films for all user age categories. The greater variability in the distribution for older users means that they have a comparatively higher opinion of older movies than younger users. If user age were the sensitive attribute, the recommendations would not account for these differences.

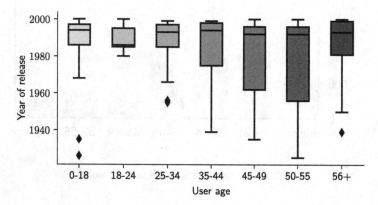

Fig. 9. Release years of the thirty most extreme movies according to the inferred positive value of the latent variables $C_j^1, \cdots, C_j^7$. Each latent variable C_j^k is matched with its corresponding user age category.

References

1. Abbasi, M., Bhaskara, A., Venkatasubramanian, S.: Fair clustering via equitable group representations. In: Elish, M.C., Isaac, W., Zemel, R.S. (eds.) ACM Conference on Fairness, Accountability, and Transparency (FAccT), pp. 504–514 (2021). https://doi.org/10.1145/3442188.3445913
2. Baudry, J.P., Celeux, G.: EM for mixtures. Stat. Comput. **25**(4), 713–726 (2015). https://doi.org/10.1007/s11222-015-9561-x
3. Bellogin, A., Castells, P., Cantador, I.: Precision-oriented evaluation of recommender systems: an algorithmic comparison. In: Proceedings of the Fifth ACM Conference on Recommender Systems, pp. 333–336. Association for Computing Machinery, New York, NY, USA (2011). https://doi.org/10.1145/2043932.2043996
4. Bera, S.K., Chakrabarty, D., Flores, N.J., Negahbani, M.: Fair algorithms for clustering (2019)
5. Beutel, A., et al.: Fairness in Recommendation Ranking through Pairwise Comparisons, pp. 2212–2220 (2019). https://doi.org/10.1145/3292500.3330745
6. Biernacki, C., Celeux, G., Govaert, G.: Choosing starting values for the EM algorithm for getting the highest likelihood in multivariate Gaussian mixture models. Comput. Stat. Data Anal. **41**, 561–575 (2003). https://doi.org/10.1016/S0167-9473(02)00163-9

7. Binns, R.: Fairness in machine learning: lessons from political philosophy. In: Friedler, S.A., Wilson, C. (eds.) Proceedings of the 1st Conference on Fairness, Accountability and Transparency. Proceedings of Machine Learning Research, vol. 81, pp. 149–159. PMLR, 23–24 Feb 2018, New York, NY, USA. http://proceedings.mlr.press/v81/binns18a.html

8. Brault, V., Mariadassou, M.: Co-clustering through latent bloc model: a review. J. de la Société Française de Statistique **156**(3), 120–139 (2015). http://journal-sfds.fr/article/view/474/448

9. Burke, R., Sonboli, N., Ordonez-Gauger, A.: Balanced neighborhoods for multi-sided fairness in recommendation. In: 1st Conference on Fairness, Accountability and Transparency. PMLR, vol. 81, pp. 202–214 (2018). http://proceedings.mlr.press/v81/burke18a.html

10. Bürkner, P.C., Vuorre, M.: Ordinal regression models in psychology: a tutorial. Adv. Meth. Pract. Psychol. Sci. **2**(1), 77–101 (2019)

11. Daykin, A.R., Moffatt, P.G.: Analyzing ordered responses: a review of the ordered Probit model. Understand. Stat. **1**(3), 157–166 (2002). https://doi.org/10.1207/S15328031US0103_02

12. Gajane, P.: On formalizing fairness in prediction with machine learning. CoRR abs/1710.03184 (2017). http://arxiv.org/abs/1710.03184

13. George, T., Merugu, S.: A scalable collaborative filtering framework based on co-clustering. In: Fifth IEEE International Conference on Data Mining (ICDM) (2005)

14. Ghadiri, M., Samadi, S., Vempala, S.: Socially fair k-means clustering. arXiv preprint arXiv:2006.10085 (2020)

15. Govaert, G., Nadif, M.: Block clustering with Bernoulli mixture models: comparison of different approaches. Comput. Stat. Data Anal. **52**(6), 3233–3245 (2008)

16. Govaert, G., Nadif, M.: Latent block model for contingency table. Commun. Stat. Theory Meth. **39**(3), 416–425 (2010). https://doi.org/10.1080/03610920903140197

17. Hardt, M., Price, E., Srebro, N.: Equality of opportunity in supervised learning. In: Advances in Neural Information Processing Systems 29, pp. 3315–3323 (2016). https://proceedings.neurips.cc/paper/2016/hash/9d2682367c3935defcb1f9e247a97c0d-Abstract.html

18. Hug, N.: Surprise: a python library for recommender systems. J. Open Source Softw. **5**(52), 2174 (2020). https://doi.org/10.21105/joss.02174

19. Jaakkola, T.S.: Tutorial on variational approximation methods. In: Advanced Mean Field Methods: Theory and Practice, pp. 129–159. MIT Press, Cambridge (2000)

20. Jang, E., Gu, S., Poole, B.: Categorical reparameterization with Gumbel-softmax. arXiv preprint arXiv:1611.01144 (2016)

21. Järvelin, K., Kekäläinen, J.: IR evaluation methods for retrieving highly relevant documents. In: ACM SIGIR Forum, vol. 51, pp. 243–250 (2017)

22. Kamishima, T., Akaho, S., Asoh, H., Sakuma, J.: Recommendation independence. In: Conference on Fairness, Accountability and Transparency, pp. 187–201 (2018)

23. Keribin, C., Brault, V., Celeux, G., Govaert, G.: Estimation and selection for the latent block model on categorical data. Stat. Comput. **25**(6), 1201–1216 (2015). https://hal.inria.fr/hal-01095957

24. Kingma, D.P., Ba, J.: Adam: a method for stochastic optimization. arXiv preprint arXiv:1412.6980 (2014)

25. Koren, Y., Bell, R., Volinsky, C.: Matrix factorization techniques for recommender systems. Computer **42**(8), 30–37 (2009)

26. Lomet, A., Govaert, G., Grandvalet, Y.: Model selection for Gaussian latent block clustering with the integrated classification likelihood. Adv. Data Anal. Classif. **12**(3), 489–508 (2018). https://hal.archives-ouvertes.fr/hal-00913680

27. Marlin, B.M., Zemel, R.S., Roweis, S.T., Slaney, M.: Collaborative filtering and the missing at random assumption. In: Twenty-Third Conference on Uncertainty in Artificial Intelligence (UAI), pp. 267–275 (2007)
28. Marlin, B.M., Zemel, R.S., Roweis, S.T., Slaney, M.: Collaborative filtering and the missing at random assumption. CoRR abs/1206.5267 (2012). http://arxiv.org/abs/1206.5267
29. Movielens 1M datasets. https://grouplens.org/datasets/movielens/
30. Ranganath, R., Gerrish, S., Blei, D.: Black box variational inference. In: Kaski, S., Corander, J. (eds.) Proceedings of the Seventeenth International Conference on Artificial Intelligence and Statistics. Proceedings of Machine Learning Research, vol. 33, pp. 814–822. PMLR, 22–25 April 2014, Reykjavik, Iceland. http://proceedings.mlr.press/v33/ranganath14.html
31. Räz, T.: Group fairness: independence revisited. arXiv preprint arXiv:2101.02968 (2021)
32. Rendle, S., Zhang, L., Koren, Y.: On the difficulty of evaluating baselines: a study on recommender systems. arXiv preprint arXiv:1905.01395 (2019)
33. Rubin, D.B.: Inference and missing data. Biometrika **63**(3), 581–592 (1976). http://www.jstor.org/stable/2335739
34. Yao, S., Huang, B.: Beyond parity: fairness objectives for collaborative filtering. In: Guyon, I., et al. (eds.) Advances in Neural Information Processing Systems, vol. 30. Curran Associates, Inc. (2017). https://proceedings.neurips.cc/paper/2017/file/e6384711491713d29bc63fc5eeb5ba4f-Paper.pdf
35. Zhu, Z., Hu, X., Caverlee, J.: Fairness-aware tensor-based recommendation. In: 27th ACM International Conference on Information and Knowledge Management, pp. 1153–1162 (2018). https://doi.org/10.1145/3269206.3271795

Learning a Fair Distance Function
for Situation Testing

Daphne Lenders[(✉)] and Toon Calders

University of Antwerp, Antwerp, Belgium
{daphne.lenders,toon.calders}@uantwerpen.be

Abstract. Situation testing is a method used in life sciences to prove
discrimination. The idea is to put similar testers, who only differ in their
membership to a protected-by-law group, in the same situation such as
applying for a job. If the instances of the protected-by-law group are consistently treated less favorably than their non-protected counterparts, we
assume discrimination occurred. Recently, data-driven equivalents of this
practice were proposed, based on finding similar instances with significant differences in treatment between the protected and unprotected
ones. A crucial and highly non-trivial component in these approaches,
however, is finding a suitable distance function to define similarity in
the dataset. This distance function should disregard attributes irrelevant for the classification, and weigh the other attributes according to
their relevance for the label. Ideally, such a distance function should not
be provided by the analyst but should be learned from the data without
depending on external resources like Causal Bayesian Networks. In this
paper, we show how to solve this problem based on learning a Weighted
Euclidean distance function. We demonstrate how this new way of defining distances improves the performance of current situation testing algorithms, especially in the presence of irrelevant attributes. (Source code:
https://github.com/calathea22/learning-fair-dist-func)

Keywords: Discrimination detection · Situation testing · Fair
distance metric learning

1 Introduction

Many decisions affecting people's everyday life, such as whether a job applicant should get hired or which credit applicants should be granted a loan, are
prone to discriminatory biases. Discrimination occurs when decision-makers do
not base decisions on people's relevant characteristics, but on their membership
to protected groups, such as women or ethnic minorities. Now that an increasing amount of data is collected in decision processes, it is possible to use this
data to detect and understand patterns of discrimination. This task is not only
interesting in itself but is especially important in the context of discrimination

© Springer Nature Switzerland AG 2021
M. Kamp et al. (Eds.): ECML PKDD 2021 Workshops, CCIS 1524, pp. 631–646, 2021.
https://doi.org/10.1007/978-3-030-93736-2_45

prevention in automatic decision making. After all, if machine learning algorithms are created to support human decision processes, these algorithms may mirror the bias of the data they are based on, if this bias is not removed first.

Originally, detecting discrimination has mostly been based on group-based measures of fairness, which assume that for data to be discrimination-free, there needs to be some form of statistical parity between the decision labels of a protected and unprotected group. This could for instance mean, that the ratio of positive decision labels should be the same across both groups, a measure known as demographic parity [2]. Recent work has however highlighted the shortcomings of these measures, and more attention has been paid to the idea of "individual fairness", with which we determine for one person at a time whether they were discriminated against or not [2].

A method to assess individual fairness in real life is "situation testing", which is based on the idea that people who are alike should be treated alike. If, for instance, two nearly-identical individuals differ on one sensitive attribute (like their gender or ethnicity), but only one gets granted a loan, we can say that the other was discriminated against. One way to translate the principles of situation testing into an algorithm was proposed by Luong et al. [4]. To illustrate their methodology, consider the dataset of a house insurance company, displayed in Table 1. The "Race of Owner" is here the sensitive attribute and the eligibility for insurance is the decision attribute.

Table 1. Dataset of an illustrating example.

#	Race of owner	Zip code	House condition	Age of house	Eligible?
1	Black	1234AZ	Medium	40–50 years	No
2	Black	1234AZ	Good	40–50 years	Yes
3	White	4567BY	Good	40–50 years	Yes
4	White	4567BY	Medium	30–40 years	Yes

Say we want to find out whether instance 1 was decided not to be eligible for house insurance, because of discriminatory bias against their race. Luong et al. find its most similar dataset instances, k from the protected and k from the unprotected group, and the positive class probability of these two groups is compared. If the positive class rate for the protected group is significantly lower than for the unprotected group, the instance is flagged as potentially discriminated. While the idea behind this algorithm is appealing, it is challenging to accurately define distances in the data. First, Luong et al. state that distances should only be defined on attributes relevant to the decision problem, but it is not trivial which attributes can be seen as such. Second, in their distance function, all attributes contribute equally to the defined distances, when in fact it is desirable that features that contribute more to the decision label also have a higher weight in the function.

To overcome these shortcomings, Zhang et al. [7] recently proposed a more refined approach for situation testing in which the distance function takes the causal relationships between the features and the decision attribute into account: features that have a higher causal effect on the decision attribute will also contribute more to the distances between dataset instances. If, for instance, in the above dataset, a good "house condition" is the most crucial factor for being eligible for insurance, this feature should also have the highest weight in the distance function. Though in theory, this approach works well, in practice it is hard to find an appropriate causal network to use for this method: networks defined by experts are not always available, while network learning algorithms do not always yield accurate or robust results [1].

Next to Zhang's work, other studies discuss ways to learn fair distance measures [3,5,6]. All these studies base the learning of the function on human feedback, e.g., by querying user's opinions on how similar two instances are to each other. The disadvantage of these approaches is that there is no guarantee that human judgement on instance's similarity is completely free of bias and whether the resulting distance function will be bias-free. Another disadvantage is again the reliance on external resources for this process.

The main aim of this paper is therefore to propose a new way in which a distance measure for the situation testing algorithm can be learned from the data. Rather than relying on a causal network or human input, these distances are learned through an optimization algorithm, which learns the parameters for the Weighted Euclidean distance, such that features with the highest impact on the decision label contribute most to the distance. At the same time, the distance function is "fair" in the sense that it does not discriminate between members of the protected group and the unprotected group. That is, if a protected member and unprotected member differ on an attribute that does not directly contribute to the decision label, this attribute will not contribute to the distance between the two instances. The superior performance of our proposed distance function is demonstrated on both simulated and real-life data. In addition to defining a new distance measure, we refine Luong's situation testing algorithm, by proposing methods for selecting its hyperparameters. The overall purpose of the resulting algorithm is twofold: First, it can be used for detecting discrimination in data. Second, it can be applied in the context of fair learning, by using the algorithm to detect discrimination and remove it, such that machine learning algorithms trained on this data, will not learn to discriminate. In this paper, we will mainly focus on the first task. Note that the method is not meant to give legally binding judgements about whether an individual is discriminated, but is rather meant as a tool to flag potentially discriminated instances. Whether the similarity between an instance and its neighbors is sufficient to base discrimination decisions on, should still be determined by a human auditor.

2   Analysis of Existing Situation Testing Methods

Before we go into detail about the existing methods of situation testing, let us first introduce some notations that we will use throughout the rest of this paper:

- X is a dataset, consisting of N individuals. We use the notation $\mathbf{x}$ to refer to one individual of the dataset
- D is a binary decision variable, where $D(\mathbf{x}) = +$ is a more favourable decision outcome then $D(\mathbf{x}) = -$.
- A is a binary sensitive variable, signifying whether an individual is a member of a protected group ($A(\mathbf{x}) = -$) or unprotected group ($A(\mathbf{x}) = +$). We assume that some members of the protected group undeservedly got the less favourable decision label $D(\mathbf{x}) = -$
- $\mathcal{G}$ are the attributes of the dataset that are legally grounded for being used in the decision making process. It is assumed that these attributes are given beforehand. We use $G(\mathbf{x})$ to denote the value of $\mathbf{x}$ for attribute $G \in \mathcal{G}$

2.1   Situation Testing - Luong et al.

The basis of Luong's situation testing algorithm is, to deem a protected instance with negative decision label as discriminated, if there is a considerable difference in the ratio of positive decision labels among its protected and unprotected neighbors. The similarity between two people x and y is defined as the sum of value differences between all legally grounded attributes of x and y, where VD_G is an appropriate distance for the domain of G.

$$d(\mathbf{x}, \mathbf{y}) = \sum_{G \in \mathcal{G}} VD_G(G(\mathbf{x}), G(\mathbf{y})) \tag{1}$$

Luong et al. propose to use this distance function with a kNN classifier to define discrimination scores. Each protected individual with a positive decision label, cannot be discriminated and thus gets a discrimination score of 0. For each protected individual with negative decision label, we define the discrimination score as the difference in the ratio of positive decision labels among its k nearest protected and unprotected neighbors. If this score is higher than threshold t, we decide that the instance in question was discriminated.

Shortcomings Luong's Approach. Luong's approach elegantly simulates the method of situation testing. However, in some circumstances, their defined distance function may not capture the full complexity of the problem. Consider e.g., the dataset displayed in Table 1. In a fair setting, only house-owners who pose a large risk of needing high insurance coverage should not be considered eligible for the insurance. The first question is which attributes of the data to include in $\mathcal{G}$, i.e. the set of legally grounded attributes used in the distance function. While it is clear that "Race" should not be considered, the feature "Zip code" provides more ground for discussion: on the one hand, the Zip code of a property might say something about the safety of a neighborhood, which is fair to consider when deciding on insurance eligibility. On the other hand, it can be seen as a "red-lining attribute", i.e. an indicator of "Race" that can be used for discriminatory practices and should therefore not be included in the set $\mathcal{G}$.

Even when excluding "Zip code" from the legally grounded attributes, Luong's approach may be too simplistic to define similarity. For instance, we could ask ourselves whether dataset instance 3 or 4 is more similar to instance 1. According to Luong's distance function, both pairs are equally similar, because instance 1 and 3 only differ on the condition of the house and 1 and 4 only on its age. In reality, we however want to put different emphasis on both features, depending on how they contribute to the decision label of this dataset.

2.2 Situation Testing - Zhang et al.

Recognizing the above shortcomings, Zhang et al. recently proposed a more refined method for computational situation testing, where the distance function is partly based on causal relationships found in the dataset [7]. Given a Causal Bayesian Network (*CBN*), that models these relationships, Zhang et al. propose to define the distance between two dataset instances solely based on attributes that have a direct causal effect on the decision attribute. Again this distance is measured as the sum of value differences for each attribute, but this time these value differences are weighted based on their causal relationship with the decision attribute. This causal relationship is measured by the "intervention", in which an instance's value on the attribute of interest is replaced by the other instance's value for that attribute. By calculating the difference in the probability of a positive decision label before and after this intervention, we get an idea of how the attribute change causally affected this probability. For the exact formulas of Zhang's approach, we refer to their paper [7].

Using this refined distance function, we again select the k nearest protected unprotected neighbors of a possibly discriminated instance. Again a discrimination score of an instance is defined as the difference in the ratio of positive decision labels between both groups.

Shortcomings Zhang's Approach. While Zhang's approach elegantly solves some of the problems of Luong's method, it suffers from the disadvantage that it heavily relies on a causal network to define the distances between dataset instances. Causal networks given by domain experts may not always be available or accurate, while networks learned by algorithms may not be very robust [1]. Since the distance function of Zhang et al. is only defined on the attributes that have a direct causal effect on the decision attribute, the presence or absence of a causal link can make a tremendous difference on the distances that are defined with this approach. Thus the results of Zhang's algorithm may vary a lot depending on the causal network they were based on.

3 Learning a Fair Distance Measure

To overcome the shortcomings of Luong's and Zhang's distance measures, we now propose a way in which a distance function can be learned from the dataset. This function defines distances of instances on the value differences between their features, as well as on the importance of these features for the decision label. We ensure that features only correlated to the decision attribute through a sensitive attribute, do not contribute to any distances. To guarantee robust results, we do not rely on a causal network to define the distance function but rather use an optimization algorithm to learn it. The distance metric we are going to use for this task is the Weighted Euclidean distance, given by Eq. (2). Note that to include nominal variables in the distance function these variables first need to be one-hot-encoded. We assume that next to the label D and the sensitive attribute A, there are numerical attributes $B_1, \ldots, B_n$. In order not to overload notation we will denote $B_i(\mathbf{x})$ by x_i. We define the weighted Euclidean distance in the usual way; for a vector of weights $\mathbf{w} = (w_1, \ldots, w_n)$, the distance is:

$$d_{\mathbf{w}}(\mathbf{x}, \mathbf{y}) := \sqrt{\sum_{i=1}^{n} w_i(x_i - y_i)^2} \tag{2}$$

In this function, the distance between two instances is defined as the sum of squared differences between its features. These differences are multiplied by a weight, which indicates how important the corresponding feature is for the overall distance. The question that now arises is how to learn these weights. The basic idea behind our optimization algorithm is that they should be learned such that the distances between instances with the same class label are small while they are big between instances with different class labels.

To make sure that the sensitive attribute does not directly contribute to the distance between two instances, the distance function $d_{\mathbf{w}}$ will only be defined over the non-sensitive features of the dataset. However, this alone does not make sure that the distance function does not differentiate between the protected and unprotected group through "red-lining attributes". These are attributes like "height" which are strongly correlated to "gender" and therefore (in case of discrimination) also correlated to the decision attribute. Despite this correlation, red-lining attributes should not contribute to the distance between two instances, and should therefore be assigned low weights in the learning process. An easy way to ensure this is splitting the dataset X according to sensitive attribute A when minimizing the distance between instances with the same class label, and maximizing the distance between instances with a different class label. With this approach, our distance function is given by $d_{\mathbf{w}^*}$ where $\mathbf{w}^*$ is the vector of weights that minimizes the objective function given in (3). We here use A^- and A^+ to refer to the instances belonging to the protected and unprotected group respectively. Further, we use $|S|$ to denote the number of dataset pairs with the same class label and $|U|$ to denote the number of dataset pairs with a different class label.

$$\left\{ \frac{1}{|S(A^-)|} \sum_{\substack{\mathbf{x}, \mathbf{y} \in A^- \\ D(\mathbf{x}) = D(\mathbf{y})}} d_{\mathbf{w}}^2(\mathbf{x}, \mathbf{y}) + \frac{1}{|S(A^+)|} \sum_{\substack{\mathbf{x}, \mathbf{y} \in A^+ \\ D(\mathbf{x}) = D(\mathbf{y})}} d_{\mathbf{w}}^2(\mathbf{x}, \mathbf{y}) \right.$$
$$\left. - \frac{1}{|U(A^-)|} \sum_{\substack{\mathbf{x}, \mathbf{y} \in A^- \\ D(\mathbf{x}) \neq D(\mathbf{y})}} d_{\mathbf{w}}^2(\mathbf{x}, \mathbf{y}) - \frac{1}{|U(A^+)|} \sum_{\substack{\mathbf{x}, \mathbf{y} \in A^+ \\ D(\mathbf{x}) \neq D(\mathbf{y})}} d_{\mathbf{w}}^2(\mathbf{x}, \mathbf{y}) + \lambda \|\mathbf{w}\|_1 \right\} \tag{3}$$

In this function the term $\lambda\|\mathbf{w}\|_1$ acts as an L1 regularizer, which forces weights not relevant for the task to be close to zero. It is necessary to include this regularizer because otherwise the weights of irrelevant attributes would neither increase nor decrease the value of the objective function, since they are likely to be equally distributed among instances with positive and negative class labels.

3.1   Theoretical Justification of the Distance Optimization Problem

In this part we show that the theoretical optimal solution for the optimization problem we defined in Eq. 3 has several desirable properties, showing that it resolves some of the problems we identified in the approach of Luong et al., without relying on a *CBN* to do so.

The objective function is an estimator of the following risk function together with a regularization term $\rho(W) := \lambda\|\mathbf{w}\|_1$:

$$R(\mathbf{w}) :=$$

$$\sum_{a=0,1} \Big(E[d_{\mathbf{w}}^2(\mathbf{x},\mathbf{y}) \mid D(\mathbf{x}) = D(\mathbf{y}), A(\mathbf{x}) = A(\mathbf{y}) = a]$$

$$- E[d_{\mathbf{w}}^2(\mathbf{x},\mathbf{y}) \mid D(\mathbf{x}) \neq D(\mathbf{y}), A(\mathbf{x}) = A(\mathbf{y}) = a] \Big)$$

All expected values are taken over the distribution that generated the dataset.

The next result shows that if an attribute only contributes to the label through the sensitive attribute, then its weight will be 0 in the optimal solution. An attribute B "only contributing through the sensitive attribute" means that the label D is conditionally independent from B given the sensitive attribute.

Theorem 1. *Let B be an attribute such that $D \perp B|A$, and let $\mathbf{w}^*$ be such that $R(w^*) + \rho(w^*)$ is minimized. Then $B(\mathbf{w}^*) = 0$.*

Proof. Assume for the sake of contradiction that $B(\mathbf{w}^*) \neq 0$ and let $\mathbf{w}'$ be the vector with $C(\mathbf{w}') = C(\mathbf{w}^*)$ for all attributes $C \neq B$ and $B(\mathbf{w}') = 0$; that is: we get $\mathbf{w}'$ by setting the weight corresponding to B to 0.

Clearly, $\rho(\mathbf{w}') < \rho(\mathbf{w}^*)$; $\rho(\mathbf{w}^*) = \rho(\mathbf{w}') + |B(\mathbf{w}^*)|$. We will show now that $R(\mathbf{w}') = R(\mathbf{w}^*)$. This is easy to see; first observe that:

$$E[d_{\mathbf{w}*}^2(\mathbf{x},\mathbf{y})] - E[d_{\mathbf{w}'}^2(\mathbf{x},\mathbf{y})] = E[B(\mathbf{w}^*)(B(\mathbf{x}) - B(\mathbf{y}))^2]$$

Then, since B is independent of D conditioned on A,

$$E[B(\mathbf{w}^*)(B(\mathbf{x}) - B(\mathbf{y}))^2|D(\mathbf{x}) = D(\mathbf{y}), A(\mathbf{x}) = A(\mathbf{y}) = a]$$
$$= E[B(\mathbf{w}^*)(B(\mathbf{x}) - B(\mathbf{y}))^2 A(\mathbf{x}) = A(\mathbf{y}) = a]$$
$$= E[B(\mathbf{w}^*)(B(\mathbf{x}) - B(\mathbf{y}))^2|D(\mathbf{x}) \neq D(\mathbf{y}), A(\mathbf{x}) = A(\mathbf{y}) = a]$$

Combining these pieces we get:

$$
\begin{aligned}
R(\mathbf{w}^*) - R(\mathbf{w}') &= \sum_{a=0,1} E[d_{\mathbf{w}^*}^2(\mathbf{x},\mathbf{y}) - d_{\mathbf{w}'}'^2(\mathbf{x},\mathbf{y}) \mid D(\mathbf{x}) = D(\mathbf{y}), A(\mathbf{x}) = A(\mathbf{y}) = a] \\
&\quad - E[d_{\mathbf{w}^*}^2(\mathbf{x},\mathbf{y}) - d_{\mathbf{w}'}'^2(\mathbf{x},\mathbf{y}) \mid D(\mathbf{x}) \neq D(\mathbf{y}), A(\mathbf{x}) = A(\mathbf{y}) = a] \\
&= \sum_{a=0,1} E[B(\mathbf{w}^*)(B(\mathbf{x}) - B(\mathbf{y}))^2 \mid D(\mathbf{x}) = D(\mathbf{y}), A(\mathbf{x}) = A(\mathbf{y}) = a] \\
&\quad - E[B(\mathbf{w}^*)(B(\mathbf{x}) - B(\mathbf{y}))^2 \mid D(\mathbf{x}) \neq D(\mathbf{y}), A(\mathbf{x}) = A(\mathbf{y}) = a] \\
&= \sum_{a=0,1} E[B(\mathbf{w}^*)(B(\mathbf{x}) - B(\mathbf{y}))^2 \mid A(\mathbf{x}) = A(\mathbf{y}) = a] \\
&\quad - E[B(\mathbf{w}^*)(B(\mathbf{x}) - B(\mathbf{y}))^2 \mid A(\mathbf{x}) = A(\mathbf{y}) = a] = 0
\end{aligned}
$$

Given this, we know that $R(\mathbf{w}^*) + \rho(\mathbf{w}^*) > R(\mathbf{w}') + \rho(\mathbf{w}')$, which contradicts the optimality of $\mathbf{w}^*$. Hence, in an optimal solution $B(\mathbf{w}^*) = 0$ has to hold. □

4   Learning the Distances and Setting Hyperparameters

In this section, we will show which algorithm was used to learn the Weighted Euclidean distance. After learning, the function could be applied in the original situation testing algorithm. Here we will, however, introduce additional adjustments to the algorithm, which should help improve its performance.

4.1   Learning the Distance Function

To learn the Weighted Euclidean distance, the objective function given by (3) had to be optimized. To do so we applied the "SLSQP" algorithm, implemented in Python's `SciPy` library. This is a quasi-Newton method, that assumes that the region around the optimum of the objective can be approximated by a quadratic function. The first and second derivatives of the objective are used to find the stationary point of this function.

4.2   Selecting Neighbors from the Unprotected Group only

Luong's and Zhang's discrimination scores are calculated as the difference in positive decision labels between an instance's nearest protected and nearest unprotected neighbors. While this approach is not necessarily wrong, it moves away from the original idea behind situation testing, where we only observe how a member of a protected group is treated differently than similar members from the unprotected group. We argue that there is no need to look at more than one protected instance at a time since we are only interested in how their decision label is different from unprotected counterparts, not how it relates to decision labels of other protected instances. After all, the motivation behind individual measures of fairness is that an individual can be discriminated based on their sensitive attribute even if there occurs no discrimination on a group level [2].

Based on these arguments, we suggest adapting the situation testing algorithm by deriving an instance's discrimination score only on the positive decision rate among its k unprotected neighbors. We will later on refer to this as the "Situation Testing k" approach as opposed to the original "Situation Testing k+k" approach. Note, that a downside of this new approach is that high discrimination scores may be given too easily: imagine for instance a group of equally capable males and females of which 80% received a job offer, regardless of their gender. In this case, the 20% of women receiving the negative label will be wrongly flagged as potentially discriminated. We will reduce this potential downside, by providing a method of choosing a threshold (that turns discrimination scores into discrimination labels), that counteracts this effect (see Sect. 4.4). Further, we will see in the experiment section how the adapted approach has positive effects on the performance of the algorithm and how it also helps in selecting the hyperparameters of the algorithm.

4.3   Setting the Number of Selected Neighbours

One issue that previous works only slightly touch upon is how to choose a good value for k. Intuitively, k should be big enough to guarantee that the class information we gather from an instance's neighbors is representative. At the same time, k should not be too big, otherwise, the selected neighbors may not be close to the instance in question anymore. In this section, we try to quantify this intuition. For our reasoning, we make use of our choice of the previous section, where we propose to infer the non-biased decision label of a protected group member from the decision labels of its unprotected neighbors. Since there is no way to say whether a derived non-biased class label is correct, we could look at unprotected group members instead. Since we assume that no discrimination occurs in this group, any class label that is derived for a member of this group should be the same as their actual class label in the dataset. In other words, we use a regular kNN classifier (that utilizes the desired distance function) to predict the labels of any unprotected instance from the rest of the unprotected group and see which k yields the highest accuracy in this approach. For our experiments, we choose to set the possible k-values to $\{10, 20, 30, 40\}$. This approach can simultaneously be used to choose the best value for λ for the learned distance function, by checking which combination of k and λ works best for an accurate prediction of the decision labels in the unprotected group.

4.4   Setting the Threshold

In the situation testing algorithm, we turn discrimination scores into discrimination labels by checking whether they exceed a given threshold t. It is proposed to either base t on existing discrimination laws, or to let the analyst choose an appropriate value for it. However, statistic-based discrimination laws do not always exist, and so far there is no clear guideline on how to adopt a general approach for choosing t. The idea behind our alternative approach for selecting t is that any difference in how a protected member was treated differently than its unprotected neighbours, can only be interpreted as discrimination if this difference is higher than expected by chance. To quantify this idea we could look at the discrimination scores that are assigned to the unprotected members of the dataset with a negative decision label. Since we assume that they were not discriminated, any discrimination score higher than 0 reflects some randomness in the data or the general chance of receiving a positive decision label. To make sure that the discrimination labels assigned to the protected individuals reflect more than bare randomness, we could thus choose a threshold that lies higher than most discrimination scores assigned to the unprotected indices. This threshold could for instance be at the maximum non-outlier value of the unprotected discrimination scores. This value can easily be found by making a boxplot of discrimination scores of the unprotected members with negative class labels and setting the discrimination-label threshold at its upper whisker.

5   Experiments on Simulated Data

5.1   Generation Process of Data

Our first experiments were conducted on simulated datasets, based on the causal networks displayed in Fig. 1a and 1b. In these datasets the "Wage" of an employee is the decision attribute, which either can be "high" or "low". "Sex" is taken as the sensitive attribute, where women represent the protected group and men the unprotected group. Further, the skill level and amount of working hours of each employee are recorded, where a higher value for either of them increases an employee's chance of receiving a high wage. Since in the dataset based on Fig. 1b, there is a link between "Sex" and "Workinghours" (with women being likely to work fewer hours) this dataset contains what we call "explainable discrimination": even though men receive on average a higher wage than women, this difference is not seen as a form of illegal discrimination, since it can be explained by other variables of the data. Note, that this is an oversimplified example used for illustration purposes only. In real life these relations are much more complex, and a presence of an "explanatory attribute" may not fully justify a different treatment between members of a protected and an unprotected group.

In the datasets based on each network, we added 1, 4 or 7 features that are only correlated to the persons' sex. In our experiments, we assume that these attributes belong to the set of legally grounded ones (and will be used in Luong's distance function), even though they do not give information about the decision label. The formulas used to generate the data are given in Fig. 2. Starting from the non-biased datasets, we simulate a situation where a decision-maker is biased against women. This person decides that the wage of several women should be "low", when in fact it was supposed to be "high". To make this simulation more realistic, only labels from women who neither have a high amount of working hours, nor a high level of skills is changed. To test the situation testing algorithms, we split each dataset into train- validation and 10 test sets. The validation set is used to choose the best value of k and t (and where applicable λ), while the test sets are used to evaluate the performance of the algorithms.

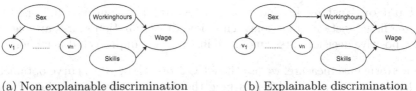

(a) Non explainable discrimination   (b) Explainable discrimination

Fig. 1. Causal networks used for data simulation to generate the unbiased data. This data is used as the ground truth in the experiments. A version of this ground truth in which discrimination is added is used for testing the algorithms.

$$\text{Sex} \sim \text{Bernoulli}(0.5)$$
$$\text{Redline}_{\text{Sex}=0} \sim \mathcal{N}(170, 2)$$
$$\text{Redline}_{\text{Sex}=1} \sim \mathcal{N}(175, 2)$$
$$\text{Skills} \sim \mathcal{N}(5, 2)$$
$$\text{Workinghours} \sim \mathcal{N}(5, 2) \qquad\qquad non\ expl.\,disc.$$
$$\text{Workinghours} \sim 3 \times \text{Sex} + \mathcal{N}(5, 2) \qquad expl.\,disc.$$
$$\text{Wage} \sim 3 \times \text{Skills} + 3 \times \text{Workinghours} + \mathcal{N}(5, 2)$$
$$\text{Wage Prob.} = \frac{\text{Wage} - \min(\text{Wage})}{\max(\text{Wage}) - \min(\text{Wage})}$$
$$\text{Wage Labels} = \text{Wage Prob.} > 0.5$$

Fig. 2. The formulas to simulate the data corresponding to CBNs in Fig. 1a and 1b. Note, that the generation process for both networks only differs for the variable "Workinghours". For the train-sets we simulated 3500 datapoints, and for the validation sets and each of the 10 test sets we simulated 500 points.

5.2 Experimental Setup

We tested the following algorithms to see which are best at predicting the non-biased class labels of the protected instances (as given by the ground truth).

- **Baseline:** The discrimination scores of the baseline are based on a classifier trained on the unprotected dataset. Assuming that no discrimination occurs in this part of the data, the trained classifier should predict "fair" labels on the protected data as well. Thus the discrimination scores of a protected instance with a negative decision label is taken as the classifier's predicted probability that their decision label should have been positive
- **Situation Testing k + k:** The situation testing algorithm as originally proposed by Luong, using one of the following distance measures:
  1. Luong - the distance function as given in Sect. 2.1
  2. Zhang - the distance function as given in Sect. 2.2
  3. Weighted Euclidean - the learned function as described in Sect. 3
  Recall that with this algorithm, the discrimination scores of an instance are based on both its k unprotected and k protected neighbors

- **Situation Testing k**: The situation testing algorithm utilizing one of the three listed distance functions. This time discrimination scores are based on the positive decision rates among an instance's k unprotected neighbors only.

As a performance measure, we use the AUC score of the ROC curve obtained on the discrimination scores. After turning the discrimination scores into discrimination labels (with the approach described in Sect. 4.4), we also calculate the F1-score for each approach. All performance measures are calculated and then averaged over the 10 test sets.

5.3   Discrimination Detection Without Explainable Discrimination

Looking at Fig. 3a and 3b we see that in the non-explainable discrimination dataset with just one irrelevant attribute, utilizing the k approach results in similar performances of the situation testing algorithms with the different distance functions. All of them outperform the baseline, and interestingly all of them slightly outperform the $k + k$ approach utilizing the same distance functions.

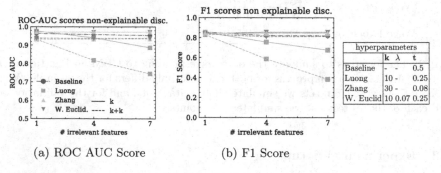

(a) ROC AUC Score   (b) F1 Score

Fig. 3. Performance on non-explainable discrimination dataset. The situation testing algorithms using Luong's distance function, are the only ones dropping in performance when irrelevant attributes are added to the data.

This could be the case because the used number for k was not optimal for the $k+k$ approach. Additionally, the worse performance can be explained by the reasons listed in Sect. 4.2: to know whether an individual of a protected group was discriminated, we do not necessarily need the information about other members of the protected group. When adding three or six additional irrelevant attributes to the dataset, we see how the situation testing algorithm utilizing Luong's distance function is the only one dropping in performance. This is the case because the distance function does not distinguish between the relevant and irrelevant attributes of the dataset. With one irrelevant attribute, this is not a problem, since it is still easy to find neighbors for an instance that are similar regarding all its features, even the ones that are not important for the decision problem.

The more irrelevant attributes are added, the more "distraction" there is: with Luong's distance function we end up selecting neighbors for every instance that are similar on the irrelevant attributes, but not on those attributes that matter. The fact that the performance of the Weighted Euclidean's distance does not drop when more irrelevant attributes are added, shows that it has learned to disregard these attributes. Different than with Zhang's approach, this is accomplished without relying on the presence of a *CBN*.

5.4 Discrimination Detection in the Presence of Explainable Discrimination

Looking at Figs. 4a and 4b we see that the results of the situation testing algorithms are a bit worse than in the previous dataset.

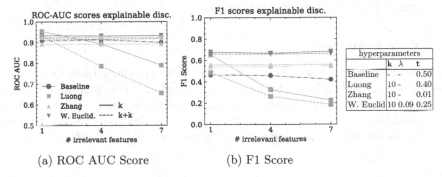

(a) ROC AUC Score   (b) F1 Score

Fig. 4. Performance on explainable discrimination dataset. The Weighted Euclidean distance outperforms the other distance functions. Again, the performance with Luong's distance function drops with more irrelevant attributes.

This is no surprise given that "Workinghours", an attribute relevant to the decision label, is distributed differently among men and women. If we try to assign a discrimination score to a woman with a low amount of workinghours it is hard to find similar male neighbours, which may lead to incorrect results. The performance of the baseline also is considerably worse than in the non-explainable discrimination dataset. To explain this, recall that its discrimination scores are based on a classifier's predicted class labels for women, after being trained on the data of men. During training, the classifier only saw instances with a high amount of workinghours, as men typically score high on this attribute. Thus it is logical that the classifier is not able to deal with the female dataset instances, which typically have a lower value for this feature. Due to the same reasons as described in Sect. 5.3, we further see that for the situation testing algorithms the $k + k$ approach yields worse performances than the k approach, and that the performance of Luong's distance function drops when more irrelevant attributes are added to the data. Again, this effect does not occur when utilizing one of the more refined distance functions. This also shows that our distance function has correctly learned the difference between red-lining attributes and attributes like "Workinghours", which are correlated to the sensitive attribute but also in itself indicative for the decision label.

6 Qualitative Experiments on Real Data

To investigate the working of the new distance function on a more realistic dataset, we run some experiments on the "Adult" dataset, where for each individual is recorded whether their annual income is higher than $50k, along with information about their age, amount of working hours, education level, etc. As a sensitive attribute, we take the "Sex" of individuals, assuming that some women do not have a low income because of their specific characteristics but because of historical discrimination towards their sex. Since we do not know the true discrimination labels of the adult dataset, we will study the utility of the distance functions by exploring some case examples.

6.1 Case Examples

We have visualized some properties of the 10 nearest unprotected neighbours of two dataset instances, as selected by the distance functions. For each interval-scaled feature, we display the neighbors' mean value on this feature, while for ordinal-/and nominal scaled features we show the number of neighbors that have the same value as the instance in question on that feature (denoted by $|S|$) (Table 2).

Table 2. Some properties of the neighbors of two dataset instances, as selected by the distance functions. $|S|$ here denotes the number of neighbors that have the same value on the given feature as the instance in question.

Instances			Properties of neighbors as selected by distance functions		
# 1	Feature	Value	Luong	Zhang	W. Euclid.
	Age	44	$\mu = 44.5$	$\mu = 45.9$	$\mu = 45.7$
	Education level	Doctorate	$|S| = 6$	$|S| = 10$	$|S| = 10$
	Hours per week	38	$\mu = 42.6$	$\mu = 55.9$	$\mu = 41.6$
	Capital gain	0	$\mu = 0$	$\mu = 0$	$\mu = 0$
	Capital loss	0	$\mu = 0$	$\mu = 0$	$\mu = 0$
	Native country	US	$|S| = 10$	$|S| = 10$	$|S| = 10$
	Marital status	Married	$|S| = 10$	$|S| = 10$	$|S| = 10$
	Workclass	Governmental	$|S| = 10$	$|S| = 1$	$|S| = 4$
# 2	Feature	Value	Luong	Zhang	W. Euclid.
	Age	42	$\mu = 41.5$	$\mu = 36.4$	$\mu = 35.6$
	Education level	High School	$|S| = 10$	$|S| = 10$	$|S| = 10$
	Hours per week	50	$\mu = 47.6$	$\mu = 40.2$	$\mu = 44.8$
	Capital gain	5455	$\mu = 1956.4$	$\mu = 4202.3$	$\mu = 5092.1$
	Capital loss	0	$\mu = 0$	$\mu = 0$	$\mu = 0$
	Native country	US	$|S| = 10$	$|S| = 10$	$|S| = 9$
	Marital status	Divorced	$|S| = 10$	$|S| = 3$	$|S| = 3$
	Workclass	Private	$|S| = 10$	$|S| = 9$	$|S| = 9$

To understand these findings better we visualized the CBN learned for Zhang's distance function and the top 6 features ranked by their importance to the decision attribute, according to the Weighted Euclidean distance.

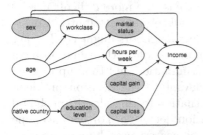

Fig. 5. CBN for Zhang's distance function.

Table 3. Ranked feature importance

Rank	Feature importance according to Weight. Euclid.
#1	Education level
#2	Capital gain
#3	Capital loss
#4	Age
#5	m. status = married
#6	Hours per week

Coherent with what we see in Fig. 5 and Table 3, Zhang's and the Weighted Euclidean distance function put a high emphasis on finding neighbours with a similar "capital gain" and "education level" as the instances in question. This comes at a cost on the similarity of features, like "marital status" that are less important for the decision attribute. Still, our learned distance function is better than Zhang's function at finding neighbours similar on attributes like "hours per week", which may not be most important for the decision task but also not trivial. This makes sense, given that Zhang's approach only defines distances based on the attributes directly affecting the decision attribute, and "hours per week" is not one of them. Looking at the neighbours selected with Luong's approach, we see that a lower similarity on specific features results in a higher similarity on others. Especially for instance #2 this might seem desirable, as here similarity is quite high for most features. However, seeing that the neighbors differ a lot on "capital gain", an attribute highly indicative for the decision label, this intuition is counteracted. Given that we are dealing with a complex dataset, where it is not always possible to find neighbors that are similar in all regards to some instance, it makes sense that a learned distance function priorities similarity on the most important features of the data. This especially holds, when recalling how the presence of irrelevant attributes worsened the performance of Luong's distance function in the simulated datasets.

7   Discussion and Conclusion

We have shown how a learned Weighted Euclidean distance function can be applied in the situation testing algorithm originally proposed by Luong [4], to find discrimination in data. The results on the simulated datasets show an advantage of utilizing the learned distance function over previously defined ones. This especially holds when a dataset contains multiple irrelevant attributes. With the experiments on a realistic dataset, the performance is less straightforward to assess, however, there are indications that our learned distance function also performs better here. Nevertheless, we here observed a downside of the situation testing methodology: the more features a dataset has, the more difficult it is to find neighbors for an instance that are similar on all relevant attributes. Consequentially, the discrimination scores based on these neighbors might not be very accurate. As seen in the experiments on the simulated data, the same problem occurs when a dataset contains explainable discrimination and the unprotected and protected group differ on an attribute relevant to the decision problem. Given these problems, we emphasize that the situation testing algorithm should merely be used as a tool to support discrimination detection. For instance, one could use the learned distance function to find the nearest neighbors of a potentially discriminated instance. Whether these neighbors are similar enough to base a discrimination judgement upon, should be decided by a human auditor.

In this study, we did not investigate yet how the situation algorithm with our proposed distance function, performs when it comes to discrimination prevention. The task of discrimination prevention is tackled by first using the algorithm to detect discrimination and then remove it, such that a classifier trained on this de-biased data does not learn to discriminate. How this approach affects the fairness and accuracy of a classifier, and how these performance measures compare to other fair learning algorithms, can be a direction for future research.

Acknowledgements. This research received funding from the Flemish Government under the "Onderzoeksprogramma Artificiele Intelligentie (AI) Vlaanderen" programme.

References

1. Dash, D., Druzdzel, M.J.: Robust independence testing for constraint-based learning of causal structure. In: UAI, vol. 3, pp. 167–174 (2003)
2. Dwork, C., Hardt, M., Pitassi, T., Reingold, O., Zemel, R.: Fairness through awareness. In: Proceedings of the 3rd Innovations in Theoretical Computer Science Conference, pp. 214–226 (2012)
3. Ilvento, C.: Metric learning for individual fairness. arXiv preprint arXiv:1906.00250 (2019)
4. Luong, B.T., Ruggieri, S., Turini, F.: k-NN as an implementation of situation testing for discrimination discovery and prevention. In: Proceedings of the 17th ACM SIGKDD International Conference on Knowledge Discovery and Data Mining, pp. 502–510 (2011)
5. Mukherjee, D., Yurochkin, M., Banerjee, M., Sun, Y.: Two simple ways to learn individual fairness metrics from data. In: International Conference on Machine Learning, pp. 7097–7107. PMLR (2020)
6. Wang, H., Grgic-Hlaca, N., Lahoti, P., Gummadi, K.P., Weller, A.: An empirical study on learning fairness metrics for COMPAS data with human supervision. arXiv preprint arXiv:1910.10255 (2019)
7. Zhang, L., Wu, Y., Wu, X.: Situation testing-based discrimination discovery: a causal inference approach. In: IJCAI, vol. 16, pp. 2718–2724 (2016)

Towards Fairness Through Time

Alessandro Castelnovo[1,2]([✉]), Lorenzo Malandri[3,4], Fabio Mercorio[3,4],
Mario Mezzanzanica[3,4], and Andrea Cosentini[1]

[1] Intesa Sanpaolo Bank S.p.A, Milan, Italy
[2] Department of Informatics, Systems and Communication,
University of Milan-Bicocca, Milan, Italy
a.castelnovo5@campus.unimib.it
[3] Department of Statistics and Quantitative Methods, University of Milan-Bicocca,
Milan, Italy
[4] CRISP Research Centre, University of Milan-Bicocca, Milan, Italy
https://crispresearch.eu/

Abstract. Over the past decade, the development of Machine Learning
(ML) algorithms to replace human decisions has raised concerns about
potential bias issues. At the same time, significant advances have been
made in the study of fairness in classification to prevent discrimination.
However, only a few of these works have investigated how those techniques can have an impact on the society and if they remain reliable
over time.

This work aims (i) to shed light on traditional group fairness mitigation strategies that fail in real-time environments when financial data
drifts affect only some classes of sensitive attributes; (ii) to investigate a
strategy that encodes the human behaviour while retraining the model
over time, favoring the convergence between individual and group fairness; (iii) to put the basis of strategies based on eXplainable AI (XAI), to
monitor the evolution of financial gaps between different population subgroups, like gender or race, observing whether the mitigation strategy is
bringing benefits to society. Preliminary results are provided, processing
about 800k personal loan granting from 2016 to 2019 for Intesa Sanpaolo
bank.

Keywords: Fairness · XAI · Individual fairness · Demographic
Parity · Machine Learning · Loan granting

1 Introduction and Motivation

In recent years, the performance of Machine Learning (ML) and Deep Learning
(DL) models has improved drastically. However, many of those algorithms are
developed as black boxes, i.e. they conceal to the users the logic behind their
decisions. This strongly limits the adoption of these models in several domains,
especially the one defined as *High-Risk* (see [49] for details), like creditworthiness,
education or law enforcement.

© Springer Nature Switzerland AG 2021
M. Kamp et al. (Eds.): ECML PKDD 2021 Workshops, CCIS 1524, pp. 647–663, 2021.
https://doi.org/10.1007/978-3-030-93736-2_46

An unmitigated data-driven decision-making algorithm can systematically make unfair decisions against certain population subgroups with specific race or gender due to the inherited biases encoded in the data [6,11,54].

Not surprisingly, recently the European Commission published a communication on fostering a European approach to AI, highlighting that *"AI-generated improvements need to be based on rules that safeguard the functioning of markets and the public sector, and people's safety and fundamental rights"* [15].

To date, the literature provides a large collection of works that allows to *understand, define, measure* and *mitigate* fairness (see, e.g. [3,23,38,39,46,58]). Unfortunately, the notion of fairness remains ambiguous and not uniquely defined. In many cases the main problem is to *choose* the proper concept of fairness for the context [6].

The current literature distinguishes among *group* and *individual* fairness. Group fairness notions aim to protect vulnerable classes of people to generate outputs that are equal. Individual fairness does not favor specific classes, but it does ensure that *"similar people receive similar outputs"* [58]. While individual mitigation strategies tend to blind the effect of the sensitive attribute[1], group mitigation strategies use it to compensate inequalities. Under certain condition these two notions are not mutually compatible (see, e.g. [5,14]).

Fairness definitions usually lack in considering the effect that the imposed constraints can cause on the disadvantaged groups [34]. Moreover, in many real-world scenarios the attributes of the population can change over time, consequently fairness cannot be treated as a static problem [54]. Some recent works that model the fairness as a multi-stage game are worth mentioning [40,56], since they consider the convergence between individual and group fairness as the *optimal situation*. In particular, [21] claims that group fairness helps to approach a convergence with individual fairness in a long-term vision, while the opposite is not true, since *"group-level discrimination remains unchanged when subject to decisions that respect notions of individual fairness"*. Formalizing the dynamics of fairness is as important as monitoring the actually occurred effects in the population over time.

Motivation. In this paper we analyse a real-world scenario about credit lending where the decision to grant the loan Y_{t_1} is made by human loan experts. In this context, we are developing an AI algorithm that embeds ethical principles [16,26] into its design, training and deployment to ensure social good [43] and in compliance with the upcoming European regulation [49]. More specifically, let us consider a Machine Learning model ψ_1 trained using a dataset D_{t_1} at time t_1 to perform loan granting $\hat{Y}_{t_1}$. The major debate derives from the impossibility of providing simultaneously at t_1 individual and group fairness due to the correlation between sensitive attributes and some important financial variables such

---

[1] In this paper we shall use the terms "protected" and "sensitive" inter-changeably to indicate attributes to be taken into account when dealing with fairness and discrimination issues. To distinguish the different classes of the sensitive attribute we shall use the terms (i) "disadvantaged" or "vulnerable" and (ii) "advantaged".

as *salary* or *rating*. Let us assume that under this condition the bank decides to pursue a notion of group fairness to favor convergence over time towards the optimal situation. Specifically, within the set of group fairness metrics (see, e.g. [38, 42]) it asked to ensure *Demographic Parity* (DP) because it isn't reasonable to pursue a definition of fairness based on *error rate parity*; since the target variable is the result of a human decision. It is not possible to claim that *"the fairness optimal classifier is perfectly fair only if the accuracy optimal classifier is perfectly accurate"* [47]. Historical data could be discriminatory [44].

Assuming that data scientists are able to develop a mitigation model ψ_1, that ensures DP and the required performance, *can the bank deploy it to fully automate the credit granting without running the risk of making unfair decisions in the future?*

Considering fairness as a dynamic game might enable the following challenging questions.

C1: Will the outputs of ψ_1 continue to ensure Demographic Parity over time?
C2: How can the model be retrained over time if it has completely replaced the loan officer decision?
C3: How can XAI techniques be used to verify that the chosen fairness policy (ensure Demographic Parity) is helping to reduce individual discrimination over time?

Contribution. The aim of this work is to investigate novel ways to answer the previous *challenging* questions. By conducting experiments on real data owned by Intesa Sanpaolo bank, accounting for 800,000 personal loan granting documents dated between 2016 and 2019, we provide the following contribution:

- Highlight that classical group fairness mitigation strategies, that work when developed and tested in the same temporal sample, can fail when used over time. This phenomenon in the financial context can occur after only one year from the training time due to distribution drift that affects only certain population subgroups with specific race or gender.
- Investigate a strategy to retrain the group fairness mitigation model assuming the absence of the final human decision - as replicated by the model - and the absence of a method for obtaining feedback from the system. Under these conditions, we propose to develop an auxiliary model that ensures individual fairness recreating over time the target variable on which the group mitigation model is retrained. This strategy favors the convergence over time between individual and group fairness and allows to maintain superior performance.
- Propose two XAI strategies, (i) a *local* one based on a SHAP [36] extension and (ii) a *global* one based on a ContrXT [37] extension to monitor fairness over time. These methods allow us to observe whether, after the retrain of the group mitigation model, individual discrimination has been reduced indicating a decrease of the financial gaps between sensitive population subgroups.

In this work, we put the basis towards the creation of a system for automatic loan granting, reducing the risk of making unethical decisions, even in the presence of unforeseen financial fluctuations over time. In addition, we propose monitoring methods to verify that the chosen mitigation strategy keeps producing the expected benefits on the population over time. Our contributions can be generalized in other contexts.

2   Related Works

Static Fairness-Aware Learning. Literature provides a growing number of definitions of fairness [4,42,51], respective mitigation strategies [8,38] and theorems that demonstrate how some of these definitions are orthogonal to each other [3,7,10,14,30]. Applied fairness works commonly consider static objectives defined on a snapshot of the population at one instant in time since the most used open data-sets in the fairness domain [2,13,33,45] do not contain data over several years. Reliability over time remains an open issue.

Fair Stream Learning. In the stream learning, the distribution of the data can change over time: the so-called concept drift [1,18,31]. Therefore, the decision-model should be able to adapt to the drifts through an incremental learn from new occurrences (see, e.g. [55]). In particular, the model handles the drifts through (i) *informed adaption* - adapting only if it detects changes - or through (ii) *blind adaption* - updating constantly with the new data. In [24,53,54] the fairness constraints are taken in account together with the drifts in the data.

Fairness for Sequential Decision Making. In recent years, fairness has started having a strategic connotation [57]. [9,17,22] have shown that exists complex relations between model decision and effect on the underlying population. Therefore, many works aim at studying the impacts of imposing fairness strategies on underlying feature distributions. [21,29,34] proposed multi-stage models. In particular, [21] studied a model for long-term improvements in the labor market and that imposing the Demographic Parity constraint can lead to an equitable long term equilibrium. The authors in [34] show that consequential decisions can reshape population over time. They demonstrate that some fairness definitions do not promote improvement and may cause harm. They also introduced a one-step feedback model to quantify the long-term impact of classification on different groups in the population, by classifying them as *long-term improvement, stagnation,* or *decline.* [25,41] focused in providing fairness in reinforcement learning to detect the policy that maximizes the cumulative rewards subject to certain fairness strategies; [20] constructs a user participation dynamics model where individuals respond to perceived decisions. The aim is to understand the impact of various fairness interventions on group representation. [22,35,40,50] studied the long-term impacts of decisions on the groups qualification states with different dynamics. The authors in [40] investigate the role of dynamics of non-discrimination in the population, concluding that *"imposing fairness considerations on decision making systems requires understanding the influence that they will have on the population at hand"*. Consistently, [50] agrees that *"algorithmic decisions lead to changes in the underlying feature distribution, which then feed back into the decision making process."*

The main difference between *Fair Stream Learning* and *Fairness for Sequential Decision Making* is that in the first, drifts in the data are taken into account to build a stable fair-classifier, whereas in the second fairness strategies are assumed to cause drifts in the data. In our work we address simultaneously these two fields of fairness. Understanding and formalizing dynamics between decisions and fairness remain a major challenge [12] as the works proposed in this direction have introduced several assumptions and simplifications. Moreover, in the credit context, the effects caused by the fairness strategy in the population could be overshadowed by many other factors, such as macroeconomic policies or immigration phenomena that cause drifts in the data.

In our work, instead of formalizing dynamic cause-effect relationships, we are interested in providing Demographic Parity through *blind adaption* and investigating methods based on XAI to observe the effects that mitigation policies actually cause between sensitive population subgroups over time.

3   Problem Setting

Considering the dataset D, composed by a set of sensitive features S, a set of financial features X and the final decision on loan granting Y, we focus on the following definitions of fairness:

Demographic Parity is achieved when the outcome of the Machine Learning model $\hat{Y}$ is independent to S i.e.

$$P(\hat{Y} \mid S) = P(\hat{Y}). \tag{1}$$

We denominate ψ a model that satisfies Demographic Parity.

Individual Fairness is achieved when the decision function of a machine learning model is independent to S i.e.

$$F(D \mid S) = F(D). \tag{2}$$

We denominate ϕ a model that satisfies individual fairness.

We refer to *optimal situation* as when it is possible to achieve simultaneously Demographic Parity and individual fairness. This condition is achievable only if X is independent to S, i.e. when there are no financial gaps between different classes of the protected attributes. Note that under this condition, *Counterfactual Fairness* [32] is also achieved, as there are no causal relationships triggered from sensitive attributes.

Assuming that it is not possible to achieve the optimal situation in the present, we are motivated to achieve it over time. Keeping the same notation introduced in [34], we propose extensions of some XAI methods to observe qualitatively *long-term improvement, stagnation*, or *decline* after each model retraining window. Our experiments aim to ensure a stable Demographic Parity over time - as the chosen fairness constraint - and to observe through XAI whether the level of individual discrimination, introduced by the group mitigation model, is decreasing. If this evidence occurs, it means that the population is having a *long-term improvement* and is moving towards the optimal situation.

4   Experiments

We conducted experiments on real data, owned by Intesa Sanpaolo bank, about 800,000 personal loan granting documents dated between 2016 and 2019. The data-set contains over 50 financial attributes and 2 sensitive attributes: *gender* and *citizenship*. The sensitive attributes are both binary. We replicate some models presented and developed in [6] in order to obtain Demographic Parity through *pre-processing*, *in-processing* and *post-processing* methodologies [38]. Specifically, the employed models are *Massaging, Aversarial Debiasing* and *TreshDP.*

Massaging [27,28] is a pre-processing technique based on relabelling the target variables in order to reach Demographic Parity in the training data. To choose which observations must be relabelled, an auxiliary classifier is trained. In our experiment we train a first Random Forest (RF) to rank the population. We flip the label of: (i) the people in the disadvantaged class who didn't receive the loan among the top positions of the rank; (ii) the people in the advantage class who actually received the loan among the worst positions of the rank. After the relabelling phase, we train a second Random Forest on the new massaged data-set. The prediction of this model will satisfy the DP without affecting performance.

Adversarial Debiasing [52] is an in-processing technique based on the simultaneous training of two competing classifiers (corresponding to a Generative Adversarial Network). In the first classifier, the predictor P tries to accomplish the task of predicting the target variable Y given the input variables X by modifying its weights W to minimise some loss function $L_P(\hat{Y}, Y)$. In the second classifier, the adversary A tries to accomplish the task of predicting the sensitive variable given $\hat{Y}$ by modifying its weights U to minimize some loss function $L_A(\hat{Z}, Z)$ and consequently back-propagating the error through the predictor P.

If the Adversary model estimates the sensitive attribute given only $\hat{Y}$, the result will satisfy Demographic Parity.

TreshDP [19] is a function $\hat{Y} = f(R, S)$, where R is the outcome of any given classifier (in our case a Random Forest) and S the sensitive attribute, designed to impose Demographic Parity in the post-processing phase, typically by identifying S-dependent thresholds.

We apply mitigation strategies for the sensitive feature *citizenship* since it is correlated with the most important financial variables to determine the granting of the loan. We focus only on this sensitive variable given that the *gender* does not show any bias in the data.

4.1   C1: Evaluating Fairness Stability Through Time

According to the state-of-the-art, the evaluation of the quality of trained models are often performed on a set of validation data extracted from the same temporal sample used for training. This validation isn't reliable in the financial perimeter because the model has to run in a live environment, hence ensuring both performance and fairness over time. Our experiments revealed that current approaches fail in satisfying the Demographic Parity only one year after training.

The second column of Table 1 shows a clear decay of all mitigation models of Demographic Parity when the models forecasts are computed with the data of the following year compared to the year used to build the training set, confirming what already stated.

Table 1. Demographic Parity of the various models tested on different temporal samples or after different stress tests. DP is calculated using *citizenship* as sensitive feature. Bold highlights $|DP|$ under 0.01 and underline highlights $|DP|$ over 0.1

Model trained in 2018	Demographic parity evaluation					
	Real data 2018	Real data 2019	Pos. shock overall	Neg. shock overall	Pos. shock conditioned	Neg. shock conditioned
Unmitigated RF	−0.21	−0.26	−0.15	−0.19	−0.09	0.44
Massaging + RF	**0.003**	−0.12	0.057	−0.071	<u>0.14</u>	−0.16
AdversarialDP	**0.005**	−0.14	0.081	−0.093	<u>0.17</u>	−0.2
RF + TreshDP	**0.004**	−0.12	0.061	−0.073	<u>0.16</u>	−0.18

Stress Tests. A set of stress tests are conducted to shed light on this phenomenon. The trained models are used to predict the granting of the loan after (i) injecting shocks (positive and negative) that cause drifts in the most important variables (*"overall shock"*) and after (ii) injecting shocks (positive and negative) that cause drifts in the most important variables conditioned to only one class of the protected attribute (*"conditioned shock"*).

Table 1 shows that the worsening of Demographic Parity occurs more evidently when *conditioned shocks* occur compared to drifts caused by *overall shocks*. This behaviour clarifies why the models failed to provide DP with real data as well. Indeed Fig. 1 shows that in real data the *income* variable is affected by a drift for the *not citizen* (see Fig. 1(c)) and not for the *citizen* (see Fig. 1(b)). In particular, from Fig. 1(c), it is possible to notice a drift to the right on the distribution, which is a good sign. Unfortunately, it also occurs an increase of the frequency of people with low income, which negatively affects the final group fairness results.

4.2   C2: Providing Fairness Stability Through Time

As previously showed, the classical mitigation methods can fail in ensuring DP when deployed on a live environment. This usually happens as the marginal contributions assigned by the mitigation model to the protected attribute are no longer properly calibrated in relation to the conditional distribution drift of some important variables. Notice that, building a fair classifier that ensures *group* fairness - rather than *individual* - aims in improving the financial condition of the vulnerable class and it is consistent with the concerns about the trained *group-fair* classifier robustness over time.

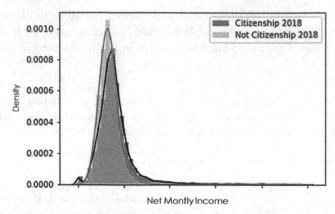

(a) Density plot: *Net Montly Income* of Citizenship vs non Citizenship in 2018

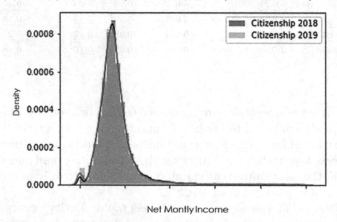

(b) Density plot: *Net Montly Income* of Citizenship in 2018 vs Citizenship in 2019

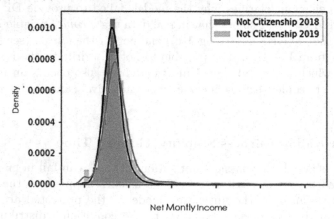

(c) Density plot: *Net Montly Income* of Not Citizenship in 2018 vs Not Citizenship in 2019

Fig. 1. Density plot of the variable *net montly income* conditioned to vary combination of *citizenship* and *year*. Distribution values are blinded for data privacy.

A practical solution is to retrain ψ_1 using a dataset D_{t_2} at time t_2 that contains the new financial conditions, resulting in a novel model ψ_2. This approach, which apparently may seem trivial, brings with it the complication of choosing correctly the target variable Y_{t_2} to update the models. Indeed, in case the model is already deployed, the decision of the loan officer - needed to retrain the system - is no longer available. Furthermore, in the credit lending field it is difficult to build a feedback-system to reinforce the results of the model, as (i) it is impossible to observe whether the not granted loans would have been repaid (*false negative control*) and (ii) the defaults on the granted loans may not have occurred yet (*false positive control*).

We recall that our goal is to provide both an output (i) that satisfies DP and (ii) that leads to a convergence between individual and group fairness. To this end, our idea is to create an individual fair model, named ϕ_{t_1}², using D_{t_1} as input and the loan officer decision Y_{t_1} as target. Since ϕ_{t_1} reproduces the behavior of the loan officer without introducing individual discrimination, its predictions on D_{t_2} might be used to update the group mitigation model ψ_2.

To clarify the matter, Table 2 compares the results of the *Massaging* model ψ_2 retrained (i) using the prediction of ϕ_1 on D_{t_2} as the target (*individual fair model as target label method*) and (ii) using the prediction of ψ_1 on D_{t_2} as the target (*group fair model as target label method*).

By observing the results, clearly both models respect DP by design, but the model retrained with an individual target has superior performance. Performance is measured by knowing the actual decision of the loan officer.

Table 2. Fairness and Performance of the *Massaging* algorithm ψ_2 using different target label methods. Bold highlights $|DP|$ under 0.01 and area under the ROC Curve (AUC) over 0.8.

Target label method	DP (ψ_2)	AUC (ψ_2)
Group fair model ψ_1	**0.001**	0.74
Individual fair model ϕ_1	**0.003**	**0.83**

4.3   C3: Monitoring Fairness Through Time

Even though an AI algorithm might embed ethical principles by design to ensure social good, there is no guarantees that these will eventually occur during deployment over time. Moreover, it is important to monitor the consequences that mitigation strategies have on people subjected to model decisions. Unwanted distortions or complex effects could impact people in unexpected ways [48]. It is therefore essential to design monitoring systems to verify the effects of fairness strategies and eventually change them based on the acquired *know-how*.

---

² ϕ_{t_1} has been developed following the methodology presented in [14] called *Fairness Troughs Unawareness*.

The process described in the previous chapter makes monitoring fairness possible, explaining the changes of the model behaviour after each model retraining window [37]. We present two different methods to monitor fairness through time based on XAI, one local and one global in order to provide a complete view of the phenomenon.

Monitoring Fairness Locally. Deploying a classifier that provides DP, when there are direct or indirect relationships between the sensitive attribute and the target variable, means assigning a positive marginal contribution to the disadvantaged class. The greater the marginal contribution, the greater is the individual discrimination introduced by the model to ensure equality between groups. SHAP [36] allows to observe the marginal contributions assigned by the mitigation model to the classes of the sensitive attribute through the *Shapley values*. We propose to use an extension of SHAP to observe the $\Delta Shapley$ *values* calculated from the group mitigation models ψ_1 and ψ_2, trained in D_{t_1} and D_{t_2} respectively. If the marginal contributions assigned to the sensitive attributes decrease after retraining, it represents a *long-term improvement* because the updated model is now able to ensure *group* fairness by decreasing *individual* discrimination. Figure 2 shows the Shapley values of two *Massaging* models trained respectively in (a) 2018 and (b) 2019 and (c) their differences. It can be seen that the marginal contributions, assigned to the protected class of the sensitive attribute, increase in 2019 compared to 2018. Both models ensure Demographic Parity in their year of competence, but to do so, in 2019 individual discrimination must be increased. This synthetic result clearly indicates a *decline*; the financial gaps have clearly widened without the necessity to observe the conditional distributions drifts of all the variables present in the data.

Monitoring Fairness Globally. As clarified by [21] *"local fairness checks are often incapable of addressing the self-perpetuating nature of biases"*. That's why it is useful to combine the previously shown method with one that provides explanations on a global level. For this purpose we are developing FairX, to extend ContrXT [37] - a time contrastive explainer - to catch changes in the individual discrimination of a model over time, by observing the paths that contain sensitive variables and the amount of population involved.

ContrXT at a Glance. ContrXT is a global, model-agnostic, time-contrastive explainer [37]. The novelty of ContrXT is the encoding of the *differences* in the logic of two classifiers ψ_1 and ψ_2 - as in the motivating example - through compacted representation of boolean formulae, i.e., Binary Decision Diagrams (BDD) that allows deriving T-contrast explanations. Specifically, ContrXT first

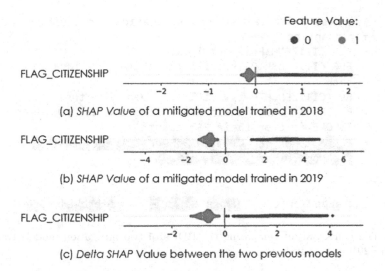

Feature Value:

● 0   ● 1

FLAG_CITIZENSHIP

(a) *SHAP Value* of a mitigated model trained in 2018

FLAG_CITIZENSHIP

(b) *SHAP Value* of a mitigated model trained in 2019

FLAG_CITIZENSHIP

(c) *Delta SHAP* Value between the two previous models

Fig. 2. Representation of the Shapley values of two mitigated models trained in different years and the relative differences.

derives a global surrogate model for ψ_1 and ψ_2, generating the BDDs for both, let's say b_1 and b_2 (the *trace* step). Then, it manipulates them to generate two new BDDs, let's say b_1' and b_2' that provide T-contrast explanations for each category c, such as *"Why does a path on b_1 had a true (false) value on c, and now is leading to false (true) in b_2?"*, i.e. the *eXplain* step. The final result is provided to the final user through natural language, highlighting the classification rules added (removed) by the novel model according to the previous one.

The key intuition is to rely on FairX to highlight paths that involve at least one sensitive variable. The novelty of FairX compared to ContrXT are not just limited to filtering and reporting the paths-changes involving sensitive variables, but also highlighting whether the percentage of the population involved in the path is increasing (or decreasing), reporting long-term variations even in the presence of no path-changes. Moreover, it allows to intercept the emergence of any unexpected bias on other sensitive attributes.

Figure 3 shows a consistent configuration with the one reported previously, but giving a global view thought FairX.

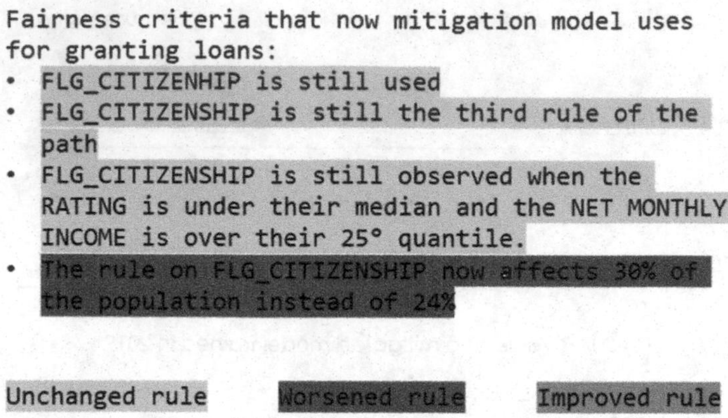

Fairness criteria that now mitigation model uses for granting loans:
• FLG_CITIZENHIP is still used
• FLG_CITIZENSHIP is still the third rule of the path
• FLG_CITIZENSHIP is still observed when the RATING is under their median and the NET MONTHLY INCOME is over their 25° quantile.
• The rule on FLG_CITIZENSHIP now affects 30% of the population instead of 24%
Unchanged rule
Worsened rule
Improved rule

Fig. 3. The FairX output comparing the BDDs of two mitigation models trained in 2018 and 2019 respectively.

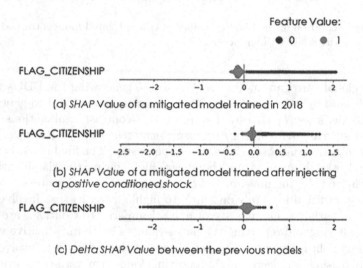

Feature Value:
0
1
FLAG_CITIZENSHIP
−2 −1 0 1 2
(a) SHAP Value of a mitigated model trained in 2018
FLAG_CITIZENSHIP
−2.5 −2.0 −1.5 −1.0 −0.5 0.0 0.5 1.0 1.5
(b) SHAP Value of a mitigated model trained after injecting a positive conditioned shock
FLAG_CITIZENSHIP
−2 −1 0 1 2
(c) Delta SHAP Value between the previous models

Fig. 4. Representation of the Shapley values of two mitigated models trained in 2018, after injecting a positive conditioned shock and the relative differences.

The *decline* showed by the previous results is due to macroeconomic phenomena that have affected the population. It is not possible to verify the effects of the fairness policy proposed in this paper as it was not really used to grant the loans. To provide an example of *long-term improvement* we have replicated the presented explanation techniques after injecting a *positive conditioned shock* (see Fig. 4 and Fig. 5).

```
Fairness criteria that now mitigation model uses
for granting loans:
• FLG_CITIZENHIP is still used
• FLG_CITIZENSHIP is still the third rule of the
  path
• FLG_CITIZENSHIP is still observed when the
  RATING is under their median and the NET MONTHLY
  INCOME is over their 25° quantile.
• The rule on FLG_CITIZENSHIP now affects 13% of
  the population instead of 24%
```

```
Unchanged rule      Worsened rule      Improved rule
```

Fig. 5. The FairX output comparing the BDDs of two mitigation models trained in 2018 and after injecting a positive conditioned shock respectively.

5   Conclusion and Future Outlooks

In this research, we conducted experiments on real data owned by Intesa Sanpaolo bank of about 800,000 personal loan granting documents dated between 2016 and 2019. The results of the experiments showed that (i) in the financial context, classical systems for ensuring Demographic Parity can fail when used over time, raising the issue of instability in fairness when deployed to production. We also (ii) have addressed the topic of how to carry out the retraining of the AI system assuming unavailability of the Ground Truth. We investigated on a strategy that encodes the human behaviour while retraining the model over time, favoring the convergence to the optimal situation and providing superior performance. Finally, (iii) we introduced two methodologies that exploit Shapley values and ContrXT to monitor the trend of fairness locally and globally. Their purpose was to highlight changes on the level of individual discrimination after the retraining of the group mitigation model. Thanks to these explanations it was possible to verify qualitatively whether the chosen mitigation strategy was contributing to induce *long-term improvement*, *stagnation*, or a *decline* on the population.

Future efforts will be devoted to deploy a system to monitor the effects of mitigation policies through time and to distinguish which effects observed on the population were caused by the chosen fairness strategy and which were caused by other macroeconomic phenomena.

Disclaimer

The views and opinions expressed within this paper are those of the authors and do not necessarily reflect the official policy or position of Intesa Sanpaolo bank. Assumptions made in the analysis, assessments, methodologies, models and results don't reflect the position of any entity other than the authors.

Furthermore, at the moment of writing, Intesa Sanpaolo bank does not have any ML algorithm in place to evaluate credit lending applications.

References

1. Aggarwal, C.C.: Data Streams: Models and Algorithms, vol. 31. Springer Science & Business Media (2007). https://doi.org/10.1007/978-0-387-47534-9
2. Asuncion, A., Newman, D.: UCI machine learning repository (2007)
3. Barocas, S., Hardt, M., Narayanan, A.: Fairness Mach. Learn. Nips Tutorial 1, 2 (2017)
4. Barocas, S., Hardt, M., Narayanan, A.: Fairness and Machine Learning. fairml-book.org (2019). http://www.fairmlbook.org
5. Binns, R.: On the apparent conflict between individual and group fairness. In: Proceedings of the 2020 Conference on Fairness, Accountability, and Transparency, pp. 514–524 (2020)
6. Castelnovo, A., et al.: BeFair: addressing fairness in the banking sector (2021)
7. Castelnovo, A., Crupi, R., Greco, G., Regoli, D.: The zoo of fairness metrics in machine learning. arXiv preprint arXiv:2106.00467 (2021)
8. Caton, S., Haas, C.: Fairness in machine learning: a survey. arXiv preprint arXiv:2010.04053 (2020)
9. Chaney, A.J., Stewart, B.M., Engelhardt, B.E.: How algorithmic confounding in recommendation systems increases homogeneity and decreases utility. In: Proceedings of the 12th ACM Conference on Recommender Systems, pp. 224–232 (2018)
10. Chouldechova, A.: Fair prediction with disparate impact: a study of bias in recidivism prediction instruments. Big Data 5(2), 153–163 (2017)
11. Council of Europe, Committee of experts on Internet MSI-NET: study on the human rights dimensions of automated data processing techniques and possible regulatory implications (2017)
12. D'Amour, A., Srinivasan, H., Atwood, J., Baljekar, P., Sculley, D., Halpern, Y.: Fairness is not static: deeper understanding of long term fairness via simulation studies. In: Proceedings of the 2020 Conference on Fairness, Accountability, and Transparency, pp. 525–534 (2020)
13. Dua, D., Graff, C., et al.: UCI machine learning repository (2017)
14. Dwork, C., Hardt, M., Pitassi, T., Reingold, O., Zemel, R.: Fairness through awareness. In: Proceedings of the 3rd Innovations in Theoretical Computer Science Conference, pp. 214–226 (2012)
15. European Commission: communication on fostering a European approach to AI. https://ec.europa.eu/newsroom/dae/redirection/document/75790 (2021)
16. Fjeld, J., Achten, N., Hilligoss, H., Nagy, A., Srikumar, M.: Principled artificial intelligence: mapping consensus in ethical and rights-based approaches to principles for AI. Berkman Klein Center Research Publication (2020-1) (2020)
17. Fuster, A., Goldsmith-Pinkham, P., Ramadorai, T., Walther, A.: Predictably unequal? The effects of machine learning on credit markets. The Effects of Machine Learning on Credit Markets, 1 October 2020
18. Gama, J., Žliobaitė, I., Bifet, A., Pechenizkiy, M., Bouchachia, A.: A survey on concept drift adaptation. ACM Comput. Surv. (CSUR) 46(4), 1–37 (2014)
19. Hardt, M., Price, E., Srebro, N.: Equality of opportunity in supervised learning. In: Advances in Neural Information Processing Systems, pp. 3315–3323 (2016)

20. Hashimoto, T., Srivastava, M., Namkoong, H., Liang, P.: Fairness without demographics in repeated loss minimization. In: International Conference on Machine Learning, pp. 1929–1938. PMLR (2018)
21. Hu, L., Chen, Y.: A short-term intervention for long-term fairness in the labor market. In: Proceedings of the 2018 World Wide Web Conference, pp. 1389–1398 (2018)
22. Hu, L., Immorlica, N., Vaughan, J.W.: The disparate effects of strategic manipulation. In: Proceedings of the Conference on Fairness, Accountability, and Transparency, pp. 259–268 (2019)
23. Hutchinson, B., Mitchell, M.: 50 years of test (un) fairness: lessons for machine learning. In: Proceedings of the Conference on Fairness, Accountability, and Transparency, pp. 49–58 (2019)
24. Iosifidis, V., Ntoutsi, E.: FABBOO - online fairness-aware learning under class imbalance. In: Appice, A., Tsoumakas, G., Manolopoulos, Y., Matwin, S. (eds.) DS 2020. LNCS (LNAI), vol. 12323, pp. 159–174. Springer, Cham (2020). https://doi.org/10.1007/978-3-030-61527-7_11
25. Jabbari, S., Joseph, M., Kearns, M., Morgenstern, J., Roth, A.: Fairness in reinforcement learning. In: International Conference on Machine Learning, pp. 1617–1626. PMLR (2017)
26. Jobin, A., Ienca, M., Vayena, E.: The global landscape of AI ethics guidelines. Nat. Mach. Intell. **1**(9), 389–399 (2019)
27. Kamiran, F., Calders, T.: Classifying without discriminating. In: 2009 2nd International Conference on Computer, Control and Communication, pp. 1–6. IEEE (2009)
28. Kamiran, F., Calders, T.: Data preprocessing techniques for classification without discrimination. Knowl. Inf. Syst. **33**(1), 1–33 (2012)
29. Kannan, S., Roth, A., Ziani, J.: Downstream effects of affirmative action. In: Proceedings of the Conference on Fairness, Accountability, and Transparency, pp. 240–248 (2019)
30. Kleinberg, J., Mullainathan, S., Raghavan, M.: Inherent trade-offs in the fair determination of risk scores. arXiv preprint arXiv:1609.05807 (2016)
31. Krawczyk, B., Minku, L.L., Gama, J., Stefanowski, J., Woźniak, M.: Ensemble learning for data stream analysis: a survey. Inf. Fusion **37**, 132–156 (2017)
32. Kusner, M., Loftus, J., Russell, C., Silva, R.: Counterfactual fairness. In: Advances in Neural Information Processing Systems 30 (NIPS 2017) pre-proceedings 30 (2017)
33. Larson, J., Mattu, S., Kirchner, L., Angwin, J.: Compas analysis. github (2016)
34. Liu, L.T., Dean, S., Rolf, E., Simchowitz, M., Hardt, M.: Delayed impact of fair machine learning. In: International Conference on Machine Learning, pp. 3150–3158. PMLR (2018)
35. Liu, L.T., Wilson, A., Haghtalab, N., Kalai, A.T., Borgs, C., Chayes, J.: The disparate equilibria of algorithmic decision making when individuals invest rationally. In: Proceedings of the 2020 Conference on Fairness, Accountability, and Transparency, pp. 381–391 (2020)
36. Lundberg, S., Lee, S.I.: A unified approach to interpreting model predictions. arXiv preprint arXiv:1705.07874 (2017)
37. Malandri, L., Mercorio, F., Mezzanzanica, M., Nobani, N., Seveso, A.: ContrXT: generating contrastive explanations from any text classifier. Inf. Fusion **81**, 103–115 (2022). https://doi.org/10.1016/j.inffus.2021.11.016
38. Mehrabi, N., Morstatter, F., Saxena, N., Lerman, K., Galstyan, A.: A survey on bias and fairness in machine learning. arXiv preprint arXiv:1908.09635 (2019)

39. Mitchell, S., Potash, E., Barocas, S., D'Amour, A., Lum, K.: Prediction-based decisions and fairness: a catalogue of choices, assumptions, and definitions. arXiv preprint arXiv:1811.07867 (2018)
40. Mouzannar, H., Ohannessian, M.I., Srebro, N.: From fair decision making to social equality. In: Proceedings of the Conference on Fairness, Accountability, and Transparency, pp. 359–368 (2019)
41. Nabi, R., Malinsky, D., Shpitser, I.: Learning optimal fair policies. In: International Conference on Machine Learning, pp. 4674–4682. PMLR (2019)
42. Narayanan, A.: Translation tutorial: 21 fairness definitions and their politics. In: Proceedings Conference Fairness Accountability Transparency, New York, USA, vol. 1170 (2018)
43. Ntoutsi, E., et al.: Bias in data-driven artificial intelligence systems-an introductory survey. Wiley Interdisc. Rev. Data Mining Knowl. Discovery **10**(3), e1356 (2020)
44. Pedreshi, D., Ruggieri, S., Turini, F.: Discrimination-aware data mining. In: Proceedings of the 14th ACM SIGKDD International Conference on Knowledge Discovery and Data Mining, pp. 560–568 (2008)
45. Redmond, M.: Communities and crime unnormalized data set. UCI Machine Learning Repository. In website: http://www.ics.uci.edu/mlearn/MLRepository.html (2011)
46. Romei, A., Ruggieri, S.: A multidisciplinary survey on discrimination analysis. Knowl. Eng. Rev. **29**(5), 582–638 (2014)
47. Speicher, T., et al.: A unified approach to quantifying algorithmic unfairness: measuring individual & group unfairness via inequality indices. In: Proceedings of the 24th ACM SIGKDD International Conference on Knowledge Discovery & Data Mining, pp. 2239–2248 (2018)
48. Suresh, H., Guttag, J.V.: A framework for understanding unintended consequences of machine learning. arXiv preprint arXiv:1901.10002 (2019)
49. The European Commission: Proposal for a Regulation of the European Parliament and of the Council laying down harmonised rules on Artificial Intelligence (Artificial Intelligence Act) and amending certain Union legislative acts, April 2021. https://digital-strategy.ec.europa.eu/en/library/proposal-regulation-laying-down-harmonised-rules-artificial-intelligence
50. Tu, R., et al.: How do fair decisions fare in long-term qualification? In: Thirty-fourth Conference on Neural Information Processing Systems (2020)
51. Verma, S., Rubin, J.: Fairness definitions explained. In: 2018 IEEE/ACM International Workshop on Software Fairness (FairWare), pp. 1–7. IEEE (2018)
52. Zhang, B.H., Lemoine, B., Mitchell, M.: Mitigating unwanted biases with adversarial learning. In: Proceedings of the 2018 AAAI/ACM Conference on AI, Ethics, and Society, pp. 335–340 (2018)
53. Zhang, W., Bifet, A.: FEAT: a fairness-enhancing and concept-adapting decision tree classifier. In: Appice, A., Tsoumakas, G., Manolopoulos, Y., Matwin, S. (eds.) DS 2020. LNCS (LNAI), vol. 12323, pp. 175–189. Springer, Cham (2020). https://doi.org/10.1007/978-3-030-61527-7_12
54. Zhang, W., Ntoutsi, E.: FAHT: an adaptive fairness-aware decision tree classifier. arXiv preprint arXiv:1907.07237 (2019)
55. Zhang, W., Wang, J.: A hybrid learning framework for imbalanced stream classification. In: 2017 IEEE International Congress on Big Data (BigData Congress), pp. 480–487. IEEE (2017)
56. Zhang, X., Liu, M.: Fairness in learning-based sequential decision algorithms: a survey. arXiv preprint arXiv:2001.04861 (2020)

57. Zhang, X., Liu, M.: Fairness in learning-based sequential decision algorithms: a survey. In: Vamvoudakis, K.G., Wan, Y., Lewis, F.L., Cansever, D. (eds.) Handbook of Reinforcement Learning and Control. SSDC, vol. 325, pp. 525–555. Springer, Cham (2021). https://doi.org/10.1007/978-3-030-60990-0_18

58. Žliobaitė, I.: Measuring discrimination in algorithmic decision making. Data Mining Knowl. Discovery **31**(4), 1060–1089 (2017). https://doi.org/10.1007/s10618-017-0506-1

International Workshop on Active Inference

Workshop on Active Inference (IWAI 2021)

The active inference framework, which first originated in neuroscience, could be assimilated with a theory of choice behavior and learning. The basic assumption of this new theoretical and methodological structure is the distinction between goal-directed and habitual behavior of an intelligent agent and how they contextualize each other. In this architecture, the intelligent agent's main goal is to minimize surprise or, more formally, its free energy. In particular, in contrast to other approaches, the resulting behavior has both explorative (epistemic) and exploitative (pragmatic) aspects that are sensitive to ambiguity and risk, respectively. Thus, active inference not only offers an interesting framework for understanding behavior and the brain, but also for developing artificial intelligent agents and investigating novel machine learning algorithms.

Within this scope, this volume presents some recent developments in active inference and its applications. These papers were presented and discussed at the 2nd International Workshop on Active Inference (IWAI 2021), which was held in conjunction with the European Conference on Machine Learning and Principles and Practice of Knowledge Discovery in Databases (ECML PKDD). The workshop took place fully online on September 13, 2021, due to the COVID-19 pandemic. Out of 22 submissions, 10 full papers and 6 poster papers were selected through a double-blind review process.

The IWAI 2021 organizers would like to thank the the Program Committee for their valuable review work, all authors for their contributions, all attendees for the fruitful discussions, and Zafeirios Fountas and Inês Hipólito for their outstanding keynotes.

September 2021

Daniela Cialfi
Tim Verbelen
Pablo Lanillos
Christopher Buckley
Maxwell Ramstead

Organization

Organizing Committee

Christopher Buckley	University of Sussex, UK
Daniela Cialfi	University of Chieti-Pescara, Italy
Pablo Lanillos	Donders Institute, The Netherlands
Maxwell Ramstead	McGill University, Canada
Tim Verbelen	Ghent University - imec, Belgium

Program Committee

Mel Andrews	University of Cincinnati, USA
Glen Berseth	University of California, Berkeley, USA
Christopher Buckley	University of Sussex, UK
Daniela Cialfi	University of Chieti-Pescara, Italy
Cedric De Boom	Ghent University - imec, Belgium
Karl Friston	University College London, UK
Casper Hesp	University of Amsterdam, The Netherlands
Inês Hipólito	Humboldt Universitat zu Berlin, Germany
Natalie Kastel	University of Amsterdam, The Netherlands
Pablo Lanillos	Donders Institute for Brain, Cognition and Behaviour, Netherlands
Christoph Mathys	Aarhus University, Denmark
Mark Miller	Hokkaido University, Japan
Alvaro Ovalle	Queen Mary University of London, UK
Ayca Ozcelikkale	Uppsala University, Sweden
Maxwell Ramstead	McGill University, Canada
Noor Sajid	University College London, UK
Kai Ueltzhffffer	Heidelberg University, Germany
Tim Verbelen	Ghent University - imec, Belgium
Martijn Wisse	Delft University of Technology, The Netherlands

Active Inference for Stochastic Control

Aswin Paul[1,2,3($\boxtimes$)], Noor Sajid[4], Manoj Gopalkrishnan[2], and Adeel Razi[3,4,5,6]

[1] IITB-Monash Research Academy, Mumbai, India
aswin.paul@monash.edu
[2] Department of Electrical Engineering, IIT Bombay, Mumbai, India
[3] Turner Institute for Brain and Mental Health, Monash University, Clayton, Australia
[4] Wellcome Trust Centre for Human Neuroimaging, UCL, London, UK
[5] Monash Biomedical Imaging, Monash University, Clayton, Australia
[6] CIFAR Azrieli Global Scholars Program, CIFAR, Toronto, Canada

Abstract. Active inference has emerged as an alternative approach to control problems given its intuitive (probabilistic) formalism. However, despite its theoretical utility, computational implementations have largely been restricted to low-dimensional, deterministic settings. This paper highlights that this is a consequence of the inability to adequately model stochastic transition dynamics, particularly when an extensive policy (i.e., action trajectory) space must be evaluated during planning. Fortunately, recent advancements propose a modified planning algorithm for finite temporal horizons. We build upon this work to assess the utility of active inference for a stochastic control setting. For this, we simulate the classic windy grid-world task with additional complexities, namely: 1) environment stochasticity; 2) learning of transition dynamics; and 3) partial observability. Our results demonstrate the advantage of using active inference, compared to reinforcement learning, in both deterministic and stochastic settings.

Keywords: Active inference · Optimal control · Stochastic control · Sophisticated inference

1 Introduction

Active inference, a corollary of the free energy principle, is a formal way of describing the behaviour of self-organising systems that interface with the external world and maintain a consistent form over time [1–3]. Despite its roots in neuroscience, active inference has snowballed to many fields owing to its ambitious scope as a general theory of behaviour [4–6]. Optimal control is one such field, and several recent results place active inference as a promising optimal control algorithm [7–9]. However, research in the area has largely been restricted to low-dimensional and deterministic settings where defining, and evaluating, policies (i.e., action trajectories) is feasible [9]. This follows from the active inference process theory that necessitates equipping agents a priori with sequences of actions in time. For example, with 8 available actions and a time-horizon of 15, the total number of (definable) policies that would need to be considered $\rightarrow 3.5 \times 10^{13}$.

This becomes more of a challenge in stochastic environments with inherently uncertain transition dynamics, and no clear way to constrain the large policy space to a

© Springer Nature Switzerland AG 2021
M. Kamp et al. (Eds.): ECML PKDD 2021 Workshops, CCIS 1524, pp. 669–680, 2021.
https://doi.org/10.1007/978-3-030-93736-2_47

smaller subspace. Happily, recent advancements like sophisticated inference [10] propose a modified planning approach for finite-temporal horizons [11]. Briefly, sophisticated inference [10], compared to the earlier formulation [9,12], provides a recursive form of the expected free energy that implements a deep tree search over actions (and outcomes) in the future. We reserve further details for Sect. 3.2.

In this paper, we evaluate the utility of active inference for stochastic control using the sophisticated planning objective. For this, we utilise the windy grid-world task [13], and assess our agent's performance when varying levels of complexity are introduced e.g., stochastic wind, partial observability, and learning the transition dynamics. Through these numerical simulations, we demonstrate that active inference, compared to a Q-learning agent [13], provides a promising approach for stochastic control.

2   Stochastic Control in a Windy Grid-World

In this section, we describe the windy grid-world task, with additional complexity, used for evaluating our active inference agent (Sect. 3). This is a classic grid-world task from reinforcement learning [13], with a predefined start (S) and goal (G) states (Fig. 1). The aim is to navigate as optimally (i.e., within a minimum time horizon) as possible, taking into account the effect of the wind along the way. The wind runs upward through the middle of the grid, and the goal state is located in one such column. The strength of the wind is noted under each column in Fig. 1, and its amplitude is quantified by the number of columns shifted upwards that were unintended by the agent. Here, the agent controls its movement through 8 available actions (i.e., the King's moves): North (N), South (S), East (E), West (W), North-West (NW), South-West (SW), South-East (SE), and North-East (NE). Every episode terminates either at the allowed time horizon, or when the agent reaches the goal state.

2.1   Grid-World Complexity

To test the performance of our active inference agent in a complex stochastic environment, we introduced different complexity levels to the windy grid-world setting (Table 1).

Table 1. Five complexity levels for the windy grid-world task

Level	Wind	Observability	Transition dynamics
1	Deterministic	Full (MDP)	Known
2	Stochastic	Full (MDP)	Known
3	Deterministic	Full (MDP)	Learned
4	Stochastic	Full (MDP)	Learned
5	Stochastic	Partial (POMDP)	Known

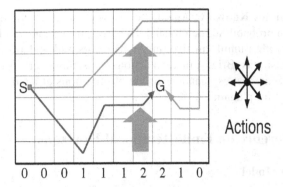

Fig. 1. Windy grid-world task. Here, S and G denote starting and goal locations. On the x-axis, the wind amplitude is shown. This is quantified as the number of unintended additional columns the agent moves during each action e.g., any action in column four results in one unintended shift upwards. There are 8 actions: $N, S, E, W, NW, SW, SE, NE$. We plot sample paths from the start to the goal state in light and dark blue. Notice, the indirect journey to the goal is a consequence of the wind. (Color figure online)

Wind Properties. In a deterministic setting, the amplitude of the wind remains constant. Conversely, in stochastic setting, for windy columns the effect varies by one from the mean values. We consider two settings: medium and high stochasticity. For medium stochasticity, the mean value is observed 70% of the time and similarly 40% of the time in the high stochastic case (Table 2). The adjacent wind values are observed with remaining probabilities. Here, stochasticity is not externally introduced to the system, but it is inbuilt in the transition dynamics $\mathcal{B}$ (Sect. 3) of the environment.

Table 2. Stochastic nature of wind

Level	Wind amplitude static	Wind amplitude ± 1
Medium	70% of the time	15% each for ± 1
High	40% of the time	30% each for ± 1

Observability. In the fully observable setting, the agent is aware of the current state i.e., there is no ambiguity about the states of affair. We formalise this as a Markov decision processes (MDP). Whereas in the partially observable environment, the agent measures an indirect function of the associated state i.e., current observation. This is used to infer the current state of the agent. We formalise this as a partially observable MDP (POMDP). Specific details of outcome modalities used in the task are discussed in Appendix B.

Transition Dynamics Known to Agent. In the known set-up, the agent is equipped with the transition probabilities beforehand. However, if these are not known, the agent begins the trials with a uninformative (uniform) priors and updates its beliefs (Eq. 9) using random transitions. Briefly, random actions are sampled and transition dynamics updated to reflect the best explanation for the observations at hand. Here, the learned dynamics are used for planning.

3   Active Inference on Finite Temporal Horizons

3.1   Generative Model

The generative model is formally defined as a tuple of finite sets (S, O, T, U, B, C, A):

o $s \in S$: states where $S = \{1, 2, 3, ..., 70\}$ and s_1 is a predefined (fixed) start state.
o $o \in O$: where $o = s$, in the fully observable setting, and in partial observability $o = f(s)^1$
o $T \in \mathbf{N}^+$, and is a finite time horizon available per episode.
o $a \in U$: actions, where $U = \{N, S, E, W, NW, SW, SE, NE\}$.
o B : encodes the transition dynamics, $P(s_t | s_{t-1}, a_{t-1}, B)$ i.e., the probability that action a_{t-1} taken at state s_{t-1} at time $t - 1$ results in s_t at time t.
o C : prior preferences over outcomes, $P(o|C)$. Here, C preference for the predefined goal-state.
o A : encodes the likelihood distribution, $P(o_\tau | s_\tau, A)$ for the partially observable setting.

Accordingly, the agents generative model is defined as the following probability distribution:

$$P(o_{1:T}, s_{1:T}, a_{1:T-1}, A, B, C) \tag{1}$$

$$= P(A)P(B)P(C)P(s_1) \prod_{\tau=2}^{T} P(s_\tau | s_{\tau-1}, a_{\tau-1}, B) \prod_{\tau=1}^{T} P(o_\tau | s_\tau, A) \tag{2}$$

3.2   Full Observability

Perception: During full observability, states can be directly accessed by agent with known or learned transition dynamics. Then the posterior estimates, $Q(s_{\tau+1} | a_\tau, s_\tau)$, can be directly calculated from B [11].

$$Q(s_{\tau+1} | a_\tau, s_\tau) = P(s_{\tau+1} | a_\tau, s_\tau, B). \tag{3}$$

---

[1] Here, outcomes introduce ambiguity for the agent as similar outcomes map to different (hidden) states. See Appendix B, Table 3 for implementation details.

Planning: In active inference, expected free-energy ($\mathcal{G}$) [9] is used for planning. For finite temporal horizons, the agent acts to minimise $\mathcal{G}$ [11]. Here, to calculate $\mathcal{G}$ we using the recursive formulation introduced in [10]. This is defined recursively as the immediate expected free energy plus the expected free energy for future actions:

$$\mathcal{G}(a_\tau|s_\tau) = \mathcal{G}(a_{T-1}|s_{T-1}) = D_{KL}[Q(s_T|a_{T-1}, s_{T-1})||C(s_T)] \tag{4}$$

for $\tau = T - 1$ and,

$$\mathcal{G}(a_\tau|s_\tau) = D_{KL}[Q(s_{\tau+1}|a_\tau, s_\tau)||C(s_{\tau+1})] + E_Q\Big[\mathcal{G}(\text{nextstep})\Big] \tag{5}$$

for $\tau = 1, ..., T - 2^2$.

In Eq. 5, the second term is calculated as,

$$E_Q\Big[\mathcal{G}(\text{nextstep})\Big] = E_{Q(a_{\tau+1}, s_{\tau+1}|s_\tau, a_\tau)}[\mathcal{G}(a_{\tau+1}|s_{\tau+1})]. \tag{6}$$

Prior preference over states are encoded such that the agent prefers to observe itself in the goal state at every time-step. $C(o = \text{goal}) = 1$, and 0 otherwise[3]. In the vector form, the ith element of C, corresponds to ith state in S.

Action Selection: A distribution for action selection $Q(a_\tau|s_\tau) > 0$ is defined using expected free energy such that,

$$Q(a_\tau|s_\tau) = \sigma\left(-\mathcal{G}\left(U|s_\tau\right)\right). \tag{7}$$

Here, σ is the softmax function ensuring that components sum to one. At each time-step, actions are samples from:

$$a_t \sim Q(a_t|s_t). \tag{8}$$

Learning Transition Dynamics: We learn the transition dynamics, $\mathcal{B}$, across time using conjugacy update rules [9, 12, 14]:

$$b_a = b_a + \sum_{\tau=2}^{t}\sum_{a \epsilon U} \delta_{a,a_\tau} Q(a)\left(s_{a,\tau} \otimes s_{a,\tau-1}\right). \tag{9}$$

Here, $b_a \sim Dir(b; \alpha)$ is the learned transition dynamics updated over time, $Q(a)$ is the probability of taking action a, $s_{a,\tau}$ is the state at time τ as a consequence of action a, $s_{a,\tau-1}$ is the state-vector at time $\tau - 1$ taking action a, and $\otimes$ is the Kronecker-product of the corresponding state-vectors. Furthermore, we also assessed the model accuracy obtained after a given number of trials to update $\mathcal{B}$, when random actions were employed to explore transition dynamics. These learned transitions were used for control in Level-3 and Level-4 of the problem.

---

[2] First term in Eq. 5 does not contribute to solving the problem addressed in the paper. Here, C only accommodates preference to goal-state. However, for a more informed C i.e with preferences for immediate reward maximisation, the term will influence action selection.

[3] The elements in C should be given a finite negligible value while implementation, to avoid divergence of D_{KL} terms in Eq. 4 and Eq. 5.

3.3 Partial Observability

We formalise partial observability as a partially observed MDP (POMDP). Here, the agents have access to indirect observations about the environment. Specific details of outcome modalities used in this work are discussed in Appendix B. These outcome modalities are same for many states for e.g., the states 2 and 11 have the same outcome modalities (see Appendix B, Table 3). Here, we evaluate the ability of active inference agent to perform optimal inference and planning in the face of ambiguity. The critical advancement with sophisticated inference [10] compared to the classical formulation [9] allows us to perform deep-tree search for actions in the future. The agent infers the hidden-states by minimising a functional of its predictive distribution (generative model) of the environment called the variational free-energy. This predictive distribution can be defined as,

$$Q(\vec{s}|\vec{a}, \tilde{o}) := \prod_{\tau=1}^{T} Q(s_\tau | a_{\tau-1}, s_{\tau-1}, \tilde{o}). \tag{10}$$

To infer hidden-states from partial observations, the agent engages in minimising variational free energy ($\mathcal{F}$) functional of Q using variational (Bayesian) inference. For a rigorous treatment of it, please refer to [10,11]. In this scheme, actions are considered as random variables at each time-step, assuming successive actions are conditionally independent. This comes with a cost of having to consider many action sequences in time. The search for policies in time is optimised both by restricting the search over future outcomes which has a non-trivial posterior probability (Eg: $>1/16$) as well as only evaluating policies with significant prior probabilities (Eg: $>1/16$) calculated from the expected free energy (i.e., Occam's window). In the partially observable setting, the expected free energy accommodates ambiguity in future observations prioritising both preference seeking as well as ambiguity reduction in observations [10].

4 Results

We compare the performance of our active inference agent with a popular reinforcement learning algorithm, Q-learning [13], in Level 1. Q-Learning is a model-free RL algorithm that operates by learning the 'value' of actions at a particular state. It is well suited for problems with stochastic transitions and reward dynamics due to its model-free parametrization. Q-Learning agents are extensively used in similar problem settings and exhibit state-of-the-art (SOTA) performances [13]. To train the Q-learning agents, we used an exploration rate of 0.1, learning rate of 0.5 and discount factor of 1. Training was conducted using 10 different random seeds to ensure unbiased results. The training depth for Q-Learning agents were increased with complexity of the environment.

We instantiate two Q-learning agents, one trained for 500 time-steps (QLearning500) and another for 5000 time-steps (QLearning5K) in Level-1. Both the active inference agent and the QLearning5K agent demonstrate optimal success rate for the time-horizon $T = 8+$ (see Appendix A, Fig. 4).

Using these baselines from the deterministic environment with known transition dynamics, we compared the performance of the agent in a complex setting with medium and highly stochastic wind (Level 2; Table. 2). Here, the active inference agent is clearly superior against the Q-Learning agents (Fig. 2 top row). Moreover, they demonstrate better success rates for shorter time-horizons, and 'optimal' action selection. Note, success rate is the percentage of trials for which the agent successfully reached the goal within the allowed time-horizon.

Next, we considered how learning the transition dynamics impacted agent behaviour (Level 3 and 4). Here, we used Eq. 9 for learning the transition dynamics, $\mathcal{B}$. First, the algorithm learnt the dynamics by taking random actions over X steps (for example, X is 5000 time steps in 'SophAgent (5K B-updates)', see Fig. 2 middle row). These learned transition dynamics $\mathcal{B}$ were used (see Fig. 3) by the active inference agent to estimate the action distribution in Eq. 8. Results for level 3 are presented in Appendix A, Fig. 5. Here, the Q-Learning algorithm with 5, 000 learning steps shows superior performance to the active inference agents. However with longer time horizons, the active inference agent shows competitive performance. Importantly, the active inference agent used self-learned, and imprecise transition dynamics $\mathcal{B}$ in these levels. Level 4 results for medium and highly stochastic setting are presented in Fig. 2 (middle row). For medium stochasticity, the QLearning10K exhibited satisfactory performance, however it failed with zero success rate in the highly stochastic case. This shows the need for extensive training for algorithms like Q-Learning in highly stochastic environments. However, the active inference agent demonstrated at-par performance. Remarkably, the performance was achieved using imprecise (compared to true-model), self-learned transition dynamics ($\mathcal{B}$) (see Fig. 3).

The active inference agent shows superior performance in the highly stochastic environment even with partial observability (Fig. 2, last row). Conversely, excessive training was required for the Q-Learning agent to achieve a high success rate in a medium stochastic environment, but even this training depth led to a zero success rate with high stochasticity. These results present active inference, with a recursively calculated free-energy, as a promising algorithm for stochastic control.

5   Discussion

We explored the utility of the active inference with planning in finite temporal-horizons for five complexity levels of the windy grid-world task. Active inference agents performed at-par, or superior, when compared with well-trained Q-Learning agents. Importantly, in the highly stochastic environments the active inference agent showed clear superiority over the Q-Learning agents. The higher success rates at lower time horizons demonstrated the 'optimality' of actions in stochastic environments presented to the agent. Additionally, this performance is obtained with no specifications of acceptable policies. The total number of acceptable policies scale exponentially with the number of available actions and time-horizon. Moreover, the Level 4 & 5 results demonstrate the need for extensive training for the Q-Learning agents when operating in stochastic environments. We also demonstrated the ability of the active inference agents to achieve high success rate even with self-learned, but sub-optimal, transition dynamics. Methods to equip the agent to learn both transition-dynamics $\mathcal{B}$ and outcome-dynamics $\mathcal{A}$

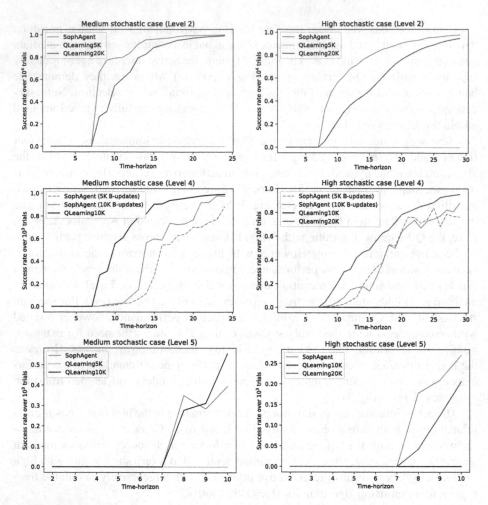

Fig. 2. Stochastic environments: Performance comparison of agents in Level-2 (top row), Level-4 (middle row), and Level-5 (last row) of windy grid-world task for medium-stochastic (left column) and high-stochastic (right column) environments, respectively. Here, x-axis denotes time horizon and y-axis the success rate over multiple-trials. 'SophAgent' represents the active inference agent, 'QLearning5K' represents Q-learning agent trained for 5,000 time-steps, 'QLearning10K' for the 'Q-learning agent' trained for 10,000 time-steps, and 'QLearning20K' for the Q-learning agent trained for 20,000 time-steps. Each agent was trained using 10 different random seeds. 'SophAgent (5K B-updates)' and SophAgent (10K B-updates) refers to active inference agent using self-learned transition dynamics B with 5000 and 10000 updates respectively.

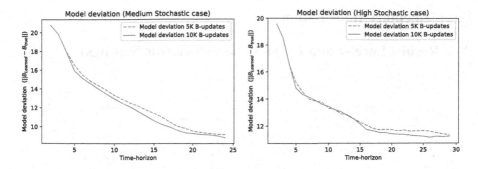

Fig. 3. Accuracy of learned dynamics in terms of deviation from true transition dynamics in Level-4 A: Medium stochastic case B: High stochastic case

for a partially observable setting have been previously explored [9, 14]. For a stochastic setting, we leave their implementation for future work.

The limitation yet to be addressed is the time consumed for trials in active inference. Large run-time restricted analysis for longer time horizons in Level 5. Deep learning approaches using tree searches, for representing policies were proposed recently [15–17], may be useful in this setting. We leave run-time analysis and optimisation for more ambitious environments for future work. Also, comparing active inference to model based RL algorithms like Dyna-Q [13] and control as inference approaches [18] is a promising direction to pursue.

We conclude that the above results place active inference as a promising algorithm for stochastic-control.

Software Note. The environments and agents were custom written in Python for fully observable settings. The script 'SPM_MDP_VB_XX.m' available in SPM12 package was used in the partially observable setting. All scripts are available in the following link: https://github.com/aswinpaul/iwai2021_aisc.

Acknowledgments. AP acknowledges research sponsorship from IITB-Monash Research Academy, Mumbai and Department of Biotechnology, Government of India. AR is funded by the Australian Research Council (Refs: DE170100128 & DP200100757) and Australian National Health and Medical Research Council Investigator Grant (Ref: 1194910). AR is a CIFAR Azrieli Global Scholar in the Brain, Mind & Consciousness Program. AR and NS are affiliated with The Wellcome Centre for Human Neuroimaging supported by core funding from Wellcome [203147/Z/16/Z].

Supplementary Information

A   Results Level-1 and Level-3 (Non-stochastic Settings)

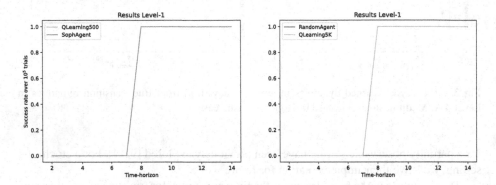

Fig. 4. Performance comparison of agents in Level-1 of windy grid-world task. 'RandomAgent' refers to a naive-agent that takes all actions with equal probability at every time step.

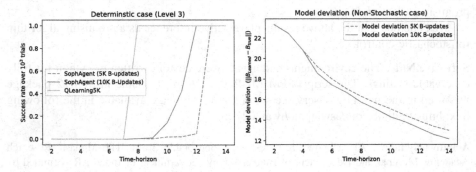

Fig. 5. A: Performance comparison of active inference agents with learned B using 5000 and 10000 updates respectively to Q-Learning agent in Level-3. 'Q-Learning5K' stands for Q-Learning agent trained for 5000 time steps using 10 different random seeds. B: Accuracy of learned dynamics in terms of deviation from true dynamics.

B   Outcome Modalities for POMDPs

In the partially observable setting, we considered two outcome modalities and both of them were the function of 'side' and 'down' coordinates defined for every state in Fig. 1. Examples of the coordinates and modalities are given below. First outcome modality is the sum of co-ordinates and second modality is the product of coordinates.

Table 3. Outcome modalities specifications

State	Down coordinate (C1)	Side coordinate (C2)	Outcome-1 (C1 + C2)	Outcome-2 (C1 * C2)
1	1	1	2	1
2	1	2	3	2
.	.	.	.	.
11	2	1	3	2
.	.	.	.	.
31	4	1	5	4
38	4	8	12	32
.	.	.	.	.

These outcome modalities are similar for many states (for e.g., states 2 and 11 have the same outcome modalities (see Table 3)). The results demonstrates the ability of active inference agent to perform optimal inference and planning in the face of ambiguity. One of the output from 'SPM_MDP_VB_XX.m' is 'MDP.P'. 'MDP.P' returns the action probabilities an agent will use for a given POMDP as input at each timestep. This distribution was used to conduct multiple trails to evaluate success rate of the active inference agent.

References

1. Friston, K.: The free-energy principle: a unified brain theory? Nat. Rev. Neurosci. **11**, 127–138 (2010)
2. Kaplan, R., Friston, K.J.: Planning and navigation as active inference. Biol. Cybern. **112**(4), 323–343 (2018). https://doi.org/10.1007/s00422-018-0753-2
3. Kuchling, F., Friston, K., Georgiev, G., Levin, M.: Morphogenesis as Bayesian inference: a variational approach to pattern formation and control in complex biological systems. Phys. Life Rev. **33**, 88–108 (2019)
4. Oliver, G., Lanillos, P., Cheng, G.: Active inference body perception and action for humanoid robots. arXiv preprint arXiv:1906.03022 (2019)
5. Rubin, S., Parr, T., Da Costa, L., Friston, K.: Future climates: Markov blankets and active inference in the biosphere. J. Royal Soc. Interface **17**(172), 20200503 (2020)
6. Deane, G., Miller, M., Wilkinson, S.: Losing ourselves: active inference, depersonalization, and meditation. Front. Psychol. **11**, 2893 (2020)
7. Friston, K.J., Daunizeau, J., Kiebel, S.J.: Reinforcement learning or active inference? PLOS ONE **4**(7), e6421 (2009). https://doi.org/10.1371/journal.pone.0006421
8. Friston, K., Samothrakis, S., Montague, R.: Active inference and agency: optimal control without cost functions. Biol. Cybern. **106**(8), 523–541 (2012)
9. Sajid, N., Ball, P.J., Parr, T., Friston, K.J.: Active inference: demystified and compared Neural Comput. **33**(3), 674–712 (2021)
10. Friston, K., Da Costa, L., Hafner, D., Hesp, C., Parr, T.: Sophisticated inference. Neural Comput. **33**(3), 713–763 (2021)

11. Da Costa, L., Sajid, N., Parr, T., Friston, K., Smith, R.: The relationship between dynamic programming and active inference: the discrete, finite-horizon case. arXiv arXiv:2009.08111 (2020)
12. Da Costa, L., Parr, T., Sajid, N., Veselic, S., Neacsu, V., Friston, K.: Active inference on discrete state-spaces: a synthesis. arXiv e-prints (2020)
13. Sutton, R., Barto, A.: Reinforcement Learning: An Introduction. MIT Press, Cambridge (2018)
14. Friston, K., FitzGerald, T., Rigoli, F., Schwartenbeck, P., Pezzulo, G.: Active inference: a process theory. Neural Comput. **29**(1), 1–49 (2017)
15. Fountas, Z., Sajid, N., Mediano, P.A., Friston, K.: Deep active inference agents using Monte-Carlo methods. arXiv preprint arXiv:2006.04176 (2020)
16. Çatal, O., Nauta, J., Verbelen, T., Simoens, P., Dhoedt, B.: Bayesian policy selection using active inference. arXiv preprint arXiv:1904.08149 (2019)
17. van der Himst, O., Lanillos, P.: Deep active inference for partially observable MDPs. In: IWAI 2020. CCIS, vol. 1326, pp. 61–71. Springer, Cham (2020). https://doi.org/10.1007/978-3-030-64919-7_8
18. Millidge, B., Tschantz, A., Seth, A.K., Buckley, C.L.: On the relationship between active inference and control as inference. In: IWAI 2020. CCIS, vol. 1326, pp. 3–11. Springer, Cham (2020). https://doi.org/10.1007/978-3-030-64919-7_1

Towards Stochastic Fault-Tolerant Control Using Precision Learning and Active Inference

Mohamed Baioumy[1]([⊠]), Corrado Pezzato[2], Carlos Hernández Corbato[2], Nick Hawes[1], and Riccardo Ferrari[3]

[1] Oxford Robotics Institute, University of Oxford, Oxford, UK
{mohamed,nickh}@robots.ox.ac.uk
[2] Cognitive Robotics, Delft University of Technology, Delft, The Netherlands
{c.pezzato,c.h.corbato}@tudelft.nl
[3] Delft Center for Systems and Control, Delft University of Technology, Delft,
The Netherlands
{r.ferrari,m.wisse}@tudelft.nl

Abstract. This work presents a fault-tolerant control scheme for sensory faults in robotic manipulators based on active inference. In the majority of existing schemes a binary decision of whether a sensor is healthy (functional) or faulty is made based on measured data. The decision boundary is called a threshold and it is usually deterministic. Following a faulty decision, fault recovery is obtained by excluding the malfunctioning sensor. We propose a stochastic fault-tolerant scheme based on active inference and precision learning which does not require a priori threshold definitions to trigger fault recovery. Instead, the sensor precision, which represents its health status, is learned online in a model-free way allowing the system to gradually, and not abruptly exclude a failing unit. Experiments on a robotic manipulator show promising results and directions for future work are discussed.

1 Introduction

Safety is paramount for autonomous systems designed for operating in the real world. External dangers in the environment such as steep and slippery terrain encountered by planetary rovers [15] can compromise entire missions. In addition to external dangers, internal system components can also fail and possibly lead to dangerous outcomes if a proper fault-tolerant (FT) control scheme is not present. Building systems that are robust to the presence of faulty components, such as sensors and actuators, is addressed in the FT literature [8,24,32]. Generally speaking, FT control consists of *fault detection*, which provides a signal representing whether a system component is faulty; *fault isolation*, which identifies the exact faulty component, and *fault recovery*, which typically contains a switching or a re-tuning procedure of the running controllers to accommodate for the fault.

© Springer Nature Switzerland AG 2021
M. Kamp et al. (Eds.): ECML PKDD 2021 Workshops, CCIS 1524, pp. 681–691, 2021.
https://doi.org/10.1007/978-3-030-93736-2_48

Several methods are available for fault detection, but model-based methods are among the most powerful and appealing, as they provide theoretical guarantees [8]. These methods rely on monitoring system outputs using mathematical models to generate 'symptoms' called *residual signals*. These signals are then compared to carefully designed detection *thresholds*: the sensor is 'faulty' if a threshold is exceeded or 'healthy' otherwise. To recover from a fault, the recovery actions are usually performed through controller reconfiguration, that entails adapting the controller parameters, or switching to another controller or to backup sensors and actuators [22]. When modelling external dangers or monitoring faulty systems, robust detection thresholds are essential. Robust thresholds used in existing work (such as [7] or [32]) are often *deterministic*, but this is sub-optimal. For instance, if the safety threshold for a rover on a slippery terrain slope is 15 °C, this means that a slope of 14.9 is safe but 15.1 is unsafe. Additionally, a slope of 15.1 °C and 40 °C are 'equally unsafe'.

In this paper we build upon two ideas in the literature. First, the usage of a stochastic fault tolerant formulation (e.g. [9,30]). This allows the agent to overcome the issues mentioned above. Additionally, we leverage an unbiased active inference controller (u-AIC) [3], evolved from previous active inference controllers (AIC) [1,6,25]. Active inference is a promising framework for FT control which has already been shown to facilitate fault-detection, isolation and recovery for robotic systems with sensory faults [3,26].

Besides fault tolerance, active inference showed promising performance in many control and state-estimation problems in robotics [16,17]. Particularly interesting are the works on robot arm control [23,25,29], which highlighted the adaptive properties of active inference. Active inference also shares similarities with the control as inference framework [18]. A more extensive analysis of active inference and its relation to control as inference can be found in [14,20].

The main contribution of this paper is a FT controller for robot manipulators with sensory faults based on unbiased active inference with a stochastic decision boundary. Unlike previous work [3], here we model the precision (inverse covariance) of each sensor in our system and determine the probability of the sensor being healthy to be proportional to its precision. Our approach allows for fault-tolerant behaviour without needing any threshold definition a priori, and without the need to design additional ad-hoc recovery mechanisms. Finally, this work can be used stand-alone or in conjunction with other methods for fault-detection and isolation in order to estimate the faults.

2   Problem Statement and Background

The FT scheme in this paper is derived for a class of systems, namely serial robot manipulators equipped with sensors for joint position and velocity, and end-effector location. In the following, the problem and the setup are described, and some background knowledge on u-AIC for torque control from [3] is presented.

Problem Setup. Consider a robotic manipulator with state x comprising of its joint positions and velocities $x = [q \; \dot{q}]^\top$. The available sensors provide noisy joint position and velocities y_q, $y_{\dot{q}}$ readings. In addition, the end-effector Cartesian position y_v is available through a visual sensor. The system's output is represented by $y = [y_q, \; y_{\dot{q}}, \; y_v] \in \mathbb{R}^d$. The proprioceptive sensors and the visual sensor are affected by zero mean Gaussian noise $\eta = [\eta_q, \eta_{\dot{q}}, \eta_v]$. Additionally, the visual sensor is affected by barrel distortion. The system is controlled through an u-AIC [3] which steers the robot arm to a (changing) desired configuration in joint space μ_d, providing the control input $u \in \mathbb{R}^m$ as torques to the joints.

Background: Unbiased Active Inference Controller. In this section we briefly describe the u-AIC as introduced in [3], to which an interested reader is referred for more details on the derivations of the following equations. The novel FT method presented in this paper in Sect. 3 builds upon the u-AIC, but instead of employing an ad-hoc hard update of the precision of a faulty sensor after fault detection, it relies on online precision learning during operations.

Let us consider $x = [q \; \dot{q}]^\top$ and let us define a probabilistic model where actions are modelled explicitly:

$$p(x, u, y_v, y_q, y_{\dot{q}}) = \underbrace{p(u|x)}_{control} \underbrace{p(y_v|x)p(y_q|x)p(y_{\dot{q}}|x)}_{observation\ model} \underbrace{p(x)}_{prior} \qquad (1)$$

Note that with the u-AIC the information about the desired goal to be reached is encoded in the distribution $p(u|x)$. In this paper, as in [1], we assume that an accurate dynamic model of the system is not available to keep the solution system agnostic and to highlight once again the adaptability of the controller.

The u-AIC aims at finding the posterior over states as well as the posterior over actions $p(x, u|y_v, y_q)$. The posteriors are approximated using a variational distribution $Q(x, u)$. We can make use of the mean-field assumption ($Q(x, u) = Q(x)Q(u)$) and the Laplace approximation, and assume the posterior over the state x Gaussian with mean μ_x [13]. Similarly for the actions, the posterior u is assumed Gaussian with mean μ_u. By defining the Kullback-Leibler divergence between the variational distribution and the true posterior, one can derive an expression for the so-called free-energy F as [3]:

$$F = -\ln p(\mu_u, \mu_x, y_v, y_q, y_{\dot{q}}) + C \qquad (2)$$

Considering Eq. (1) and assuming Gaussian distributions, F becomes:

$$\begin{aligned}
F = \frac{1}{2}(&\varepsilon_{y_q}^\top \Sigma_{y_q}^{-1} \varepsilon_{y_q} + \varepsilon_{y_{\dot{q}}}^\top \Sigma_{y_{\dot{q}}}^{-1} \varepsilon_{y_{\dot{q}}} + \varepsilon_{y_v}^\top \Sigma_{y_v}^{-1} \varepsilon_{y_v} \\
&+ \varepsilon_x^\top \Sigma_x^{-1} \varepsilon_x + \varepsilon_u^\top \Sigma_u^{-1} \varepsilon_u + \ln|\Sigma_u \Sigma_{y_q} \Sigma_{y_{\dot{q}}} \Sigma_{y_v} \Sigma_x|) + C,
\end{aligned} \qquad (3)$$

The terms $\varepsilon_{y_q} = y_q - \mu$, $\varepsilon_{y_{\dot{q}}} = y_{\dot{q}} - \mu'$, $\varepsilon_{y_v} = y_v - g_v(\mu)$ are the sensory prediction errors respectively for position, velocity, and visual sensory inputs. The controller represents the states internally as $\mu_x = [\mu, \; \mu']^\top$. The relation between internal state and observation is expressed through the generative model of the sensory input $g = [g_q, g_{\dot{q}}, g_v]$. Position and velocity encoders directly measure the state, thus g_q and $g_{\dot{q}}$ are linear (identity) mappings. To define g_v, instead, we use a *Gaussian Process Regression* (GPR). This is particularly useful because we can model the noisy and distorted sensory input from the camera, and at the same time we can compute a closed form for the derivative of the process with respect to the beliefs μ, required for the state update laws. For details, see [3].

Additionally, ε_u is the prediction error on the control action while ε_x is the prediction error on the state. The latter is computed considering a prediction of the state $\hat{x}$ at the current time-step such that $\varepsilon_x = (\mu_x - \hat{x})$. The prediction is a deterministic value $\hat{x} = [\hat{q}\ \hat{\dot{q}}]^\top$ which can be computed in the same fashion as the prediction step of, for instance, a Kalman filter. The prediction is approximated propagating forward in time the current state belief using the following simplified discrete time model:

$$\hat{x}_{k+1} = \begin{bmatrix} I & I\Delta t \\ 0 & I \end{bmatrix} \mu_{x,k} \tag{4}$$

where I represents an unitary matrix of suitable size. This form assumes that the position of each joint is thus computed as the discrete time integral of the velocity, using a first-order Euler scheme. This approximation can be avoided if a better dynamic model of the system is available, and in that case predictions can be made using the model itself. Finally, by choosing the distribution $p(u|x)$ to be Gaussian with mean $f^*(\mu_x, \mu_d)$, we can steer the systems toward the target μ_d without biasing the state estimation. This results in $\varepsilon_u = (\mu_u - f^*(\mu_x, \mu_d))$.

In the u-AIC state-estimation and control are achieved using gradient descent the free-energy. This leads to:

$$\dot{\mu}_u = -\kappa_u \frac{\partial F}{\partial \mu_u}, \quad \dot{\mu}_x = -\kappa_\mu \frac{\partial F}{\partial \mu_x}, \tag{5}$$

where κ_u and κ_μ are the gradient descent step sizes.

3   Precision Learning for Fault-Tolerant Control

In previous work [3], the u-AIC is used in combination with an established FT approach to achieve fault detection and recovery. In particular, the sensory prediction errors in the free-energy are used as residual signals for fault detection purposes. The statistical properties of the residuals are analysed offline and healthy boundaries are defined. At runtime, a healthy residual set is computed and if the current residual is outside the admissible set, the relative sensor is marked as faulty. When a fault is detected, the precision (or inverse covariance) of the sensor is abruptly set to zero, that is $P = \Sigma^{-1} = 0$, to exclude that sensor from the optimization of the free-energy. This idea is summarised in Fig. 1.

In this work, we propose a different approach to achieve fault recovery through online precision learning with u-AIC instead ad-hoc hard switches in the controller's parameters. Figure 2 shows thee difference with respect to [3].

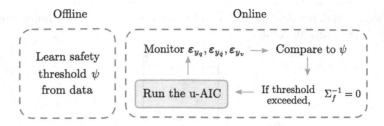

Fig. 1. Fault-tolerant pipeline from [3]. The term Σ_f^{-1} represents the precision of the detected faulty sensor.

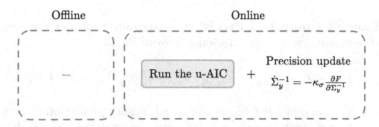

Fig. 2. New fault-tolerant pipeline with precision learning, in contrast to previous work [3] from Fig. 1.

Learning Sensory Precision. For a sensor y, we can update an inverse precision matrix Σ_y^{-1} using gradient descent on F as done in [1,2]:

$$\dot{\Sigma}_y^{-1} = -\kappa_\sigma \frac{\partial F}{\partial \Sigma_y^{-1}}. \tag{6}$$

However, we need to ensure that precision remains a positive number. Performing gradient descent does not inherently guarantee that.

First, consider a one-dimensional problem where state x and observation y are scalars. The observations is affected by zero-mean Gaussian noise with a variance of σ^2 (also a scalar). The scalar precision is defined as the inverse variance $\omega = 1/\sigma^2$. As explained, performing gradient descent on the free-energy with respect to ω may result in it being negative. A simple solution is to perform a reparameterization with a strictly positive function such as an exponential. I.e. we assume that $\omega = \exp \zeta$ and we perform gradient descent on ζ:

$$\dot{\zeta} = -\kappa_\zeta \frac{\partial F}{\partial \zeta} \tag{7}$$

where κ_ζ is the gradient step-size. Another way is to set a lower bound on the variance (as done in [5]). Both methods ensure the variance being positive.

Diagonal Precision Matrix. Guaranteeing a positive semi-definite matrix in an n-dimensional case is not as straightforward. However, in the context of a robotic manipulator, one can reasonably assume that the observation noise on each sensor is independent [1,23,25]. This means that the covariance (and precision) matrices are diagonal.

$$P = \begin{bmatrix} \omega_1 & & & \\ & \omega_2 & & \\ & & ... & \\ & & & \omega_n \end{bmatrix}$$

Given this assumption, every element on the diagonal is positive and can be updated in the same fashion as the scalar case (Eq. (7)).

Fault-Tolerant Control as Precision Learning. Consider the sum of the sensory prediction errors in the free-energy from Eq. (3):

$$F = \frac{1}{2}(\varepsilon_{y_q}^\top \Sigma_{y_q}^{-1} \varepsilon_{y_q} + \varepsilon_{y_{\dot{q}}}^\top \Sigma_{y_{\dot{q}}}^{-1} \varepsilon_{y_{\dot{q}}} + \varepsilon_{y_v}^\top \Sigma_{y_v}^{-1} \varepsilon_{y_v} + ...) + C, \qquad (8)$$

Intuitively, when a sensor is faulty, the related sensory prediction error will necessarily be higher since sensory readings and internal beliefs will drift away. After a fault, the estimated precision through our precision learning scheme will be much lower than the original $P = \Sigma^{-1}$. Thus its weight in the free-energy F, and so in the state-estimation and control equations as in Eq. (5) will naturally become lower than the other healthy sensors. Its weight essentially adjusts *proportionally to the degree of the sensor being faulty*. Note that this allows for automatic fault recovery but it does not provide explicit fault detection. In case the latter is needed for a potential user or an additional supervisory system, traditional techniques can be used as the one presented in [3] in conjunction with precision learning.

FT control for sensory faults can now be done using precision learning in several ways. The first way is to use it as a stand-alone and activate precision learning for all sensors during operation. In this case, no other methods are needed, no thresholds are designed and the recovery emerges naturally. As mentioned before, the drawback is that the users can not be 'alerted' for the presence of a fault (since there is no explicit fault-detection). The second way, which addresses this issue, is to use an established algorithm for fault detection (such as the one presented in [3]) and then, only after a fault is detected, allow precision update.

Interestingly, performing precision learning as presented in this section can make the state-estimation noisier since the agents only relies on the current observation (rather than a batch of last k observations) for the update and both the uncertainty of the state and precision are not quantified. An additional approach would then be to consider the last k observations for the update, but this is out of the scope of this work.

To summarise, the precision learning in this paper can either be activated at all times or *only after a fault is detected*. Activating the precision learning at all times with a small step-size for the gradient seems to work best.

4   Results

We apply the methods in Sect. 3 on a 2-DOF robotic manipulator. We test three scenarios: a) precision learning at all times, b) precision learning only when a fault is detected and c) a deterministic update as done in [3]. Note that the latter has access to a model and uses data to determine a threshold offline. This is not the case for the first two options where only model-free precision learning is performed. The results are summarized in the Table 1. In the simulations, the sensors are injected with zero-mean Gaussian noise. The standard deviation of the noise for encoders and velocity sensors is set to $\sigma_q = \sigma_{\dot{q}} = 0.001$, while the one for the camera is set to $\sigma_v = 0.01$. The camera is also affected by barrel distortion with coefficients $K_1 = -1.5e^{-3}$, $K_2 = 5e^{-6}$, $K_3 = 0$ (values are similar to work from [19, 28]).

Table 1. Mean Squared Error (MSE) for different methods of fault-tolerant control. PL indicates precision learning

	Joints with encoder fault	Joints without encoder fault
No fault-tolerance	0.0036	0.0020
PL at all time	5.422 e−5	4.527 e−5
PL + fault-detection	6.097 e−5	4.134 e−5
Deterministic fault recovery	**0.5946 e−5**	**0.3579 e−5**

The agents starts in configuration x_0, then moves to the targets x_1 and x_2. At $t = 8$ s a fault is injected. The encoder fault is such that the output related to the first joint freezes. For a discrete step k it holds then $y_q(k) = [q_1(k_f), q_2(k)]^\top$ for $k \geq k_f$ and $k_f = 8$. The fault detection and recovery of such a fault, as well as the system's response, are reported below in Fig. 3.

As seen in Fig. 3, the system is not able to reach the set-point after the occurrence of the fault if online precision update is not allowed. The robot arm reaches a different configuration to minimise the free-energy, which is built fusing the sensory information from the (faulty) encoders and the (healthy) camera. However, when the faulty encoder is adjusted using precision learning, the agent is able to reach the final configuration.

Figure 3 reports the results when precision learning is being done during the full operational time. Alternatively, one could only use precision leaning when a fault is detected. This yields a response that is almost identical. The Mean Squared Error (MSE) between the belief and the true position ($\mu_x - x$) is computed on a sample of test runs and reported in the Table 1. The results are reported for both the joint whose encoder is faulty, and joints with healthy encoders. In both cases, hard update of the precision to zero has the lowest MSE; however, the approaches based on precision learning do not require any

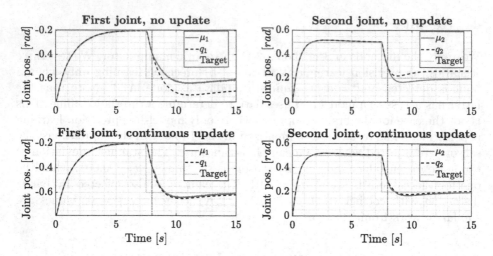

Fig. 3. System's response in the case of encoder fault with and without precision learning applied at all times. The fault is injected at $t = 8$ s and indicated with a dot-dashed line.

previous information or a threshold definition thus it is simpler to implement. Yet, precision learning has a satisfactory performance while accommodating a sensory fault.

5   Improving Precision Learning: A Discussion

In this paper, we perform a simple modification to the unbiased active inference controller: adding precision learning for all sensors. We show that this results in stochastic fault-tolerance to sensory faults, i.e. the precision of a faulty sensor will decrease automatically making its relative weight in the control and estimation laws smaller. This eliminates the need to learn a threshold from data offline. Additionally, no ah-hoc recovery action is required. The controller automatically adjusts to the new precision.

In the experiments, we compared precision learning to a state-of-the-art method. Precision learning was an order of magnitude worse in performance but still satisfactory. Note that precision learning did *not* require any data or training offline to determine thresholds or recovery strategies. Finally, precision learning performs stochastic fault-detection rather than deterministic.

Most importantly, this approach based on precision learning can be improved in many ways. First, rather than computing a point-mass estimate, we can explicitly model the precision as a random variable and perform inference on it.

We can perform Bayesian inference by modelling the precision as a random variable and computing a posterior over it. In the one dimensional case we use a Gamma prior on the precision ω as

$$\Gamma(\omega; a, b) = \frac{b^a}{\Gamma(a)} \omega^{a-1} e^{-\omega b}.$$

Given that the observation model is Gaussian, this choice is beneficial since it is the conjugate prior [4,21], where a and b are the parameters of the distribution and $\Gamma(a) = (a-1)!$ is a factorial function. For example, $\Gamma(5) = 4! = 24$. Now to compute the posterior, we multiply the prior with the Gaussian likelihood model of $p(y|\omega)$ and obtain the posterior which is also a Gamma distribution as shown below.

$$p(\omega) = \Gamma(\omega; a, b) \propto \omega^{a-1}e^{-\omega b}$$

$$p(\omega|y) \propto p(y|\omega)p(\omega) \propto \omega^{0.5+a-1}e^{-\omega(b+\frac{(y-C)^2}{2})}$$

$$p(\omega|y) = \Gamma(\omega; a + \frac{1}{2}, b + \frac{(y-C)^2}{2})$$

The last equation shows a simple update rule to modify the belief over the precision for every data point. In the optimization for the state, the following quantities are used: expected precision $\mathbb{E}[\omega] = a/b$, $Mode[\omega] = (a-1)/b$ and $Var[\omega] = a/b^2$. In the n-dimensional case, the same procedure can be done but with a Wishart distribution rather than a Gamma.

Additionally, we could use a batch of k observation to learn the precision rather than just one observation. Many approaches for covariance/precision estimation have been successful in robotics e.g. [27,31,33,34]. Additionally, many other approaches within the active inference literature can be used for effective precision learning such as dynamic expectation maximization (DEM) [11,12]. These will be explored and compared in future work.

6   Conclusions

This paper presents a fault-tolerant controller based on active inference. We model the precision (inverse covariance) of each sensor in our system and determine the probability of the sensor being healthy to be proportional to its precision. Rather than reasoning about whether a sensor is faulty or not, we reason about the degree to which the sensor is faulty. We present gradient based approaches to approximate the precision matrices of the system. The results show that the precision learning is a promising approach for fault-tolerant control. It allows for robust behaviour without needing any threshold definition a priori, without designing additional ad-hoc recovery mechanisms, and can be used stand-alone or in conjunction with other methods. The results using precision learning was satisfactory but an order of magnitude away from the to state-of-the-art. However, precision learning was not trained on data offline and performs a stochastic update. Bayesian methods can be used to improve the performance of the approach. Additionally, in all cases regarding precision learning, the performance can be improved by considering the last k observations rather than just one. Future work will address this.

References

1. Baioumy, M., Duckworth, P., Lacerda, B., Hawes, N.: Active inference for integrated state-estimation, control, and learning. In: Proceedings of the IEEE International Conference on Robotics and Automation (ICRA) (2021)
2. Baioumy, M., Mattamala, M., Duckworth, P., Lacerda, B., Hawes, N.: Adaptive manipulator control using active inference with precision learning. In: UKRAS (2020)
3. Baioumy, M., Pezzato, C., Ferrari, R., Corbato, C.H., Hawes, N.: Fault-tolerant control of robotic systems with sensory faults using unbiased active inference. In: European Control Conference (ECC) (2021)
4. Bishop, C.M.: Pattern Recognition and Machine Learning. Springer, Heidelberg (2006)
5. Bogacz, R.: A tutorial on the free-energy framework for modelling perception and learning. J. Math. Psychol. **76**, 198–211 (2017)
6. Buckley, C.L., Kim, C.S., McGregor, S., Seth, A.K.: The free energy principle for action and perception: a mathematical review. J. Math. Psychol. **81**, 55–79 (2017)
7. Budd, M., Lacerda, B., Duckworth, P., West, A., Lennox, B., Hawes, N.: Markov decision processes with unknown state feature values for safe exploration using gaussian processes. In: IEEE/RSJ International Conference on Intelligent Robots and Systems (IROS) (2020)
8. Chen, J., Patton, R.J.: Robust Model-Based Fault Diagnosis for Dynamic Systems. Springer, Boston (1999). https://doi.org/10.1007/978-1-4615-5149-2
9. Fang, S., Blanke, M., Leira, B.J.: Mooring system diagnosis and structural reliability control for position moored vessels. Control. Eng. Pract. **36**, 12–26 (2015)
10. Fox, C.W., Roberts, S.J.: A tutorial on variational Bayesian inference. Artif. Intell. Rev. **38**(2), 85–95 (2012)
11. Friston, K., Stephan, K., Li, B., Daunizeau, J.: Generalised filtering. Math. Prob. Eng. **2010** (2010)
12. Friston, K.J., Trujillo-Barreto, N., Daunizeau, J.: DEM: a variational treatment of dynamic systems. Neuroimage **41**(3), 849–885 (2008)
13. Friston, K., Mattout, J., Trujillo-Barreto, N., Ashburner, J., Penny, W.: Variational free energy and the Laplace approximation. Neuroimage **34**(1), 220–234 (2007)
14. Imohiosen, A., Watson, J., Peters, J.: Active inference or control as inference? A unifying view. In: IWAI 2020. CCIS, vol. 1326, pp. 12–19. Springer, Cham (2020). https://doi.org/10.1007/978-3-030-64919-7_2
15. Inotsume, H., Kubota, T., Wettergreen, D.: Robust path planning for slope traversing under uncertainty in slip prediction. IEEE Robot. Autom. Lett. **5**(2), 3390–3397 (2020)
16. Lanillos, P., Cheng, G.: Adaptive robot body learning and estimation through predictive coding. In: IROS (2018)
17. Lanillos, P., Cheng, G.: Active inference with function learning for robot body perception. In: International Workshop on Continual Unsupervised Sensorimotor Learning (ICDL-Epirob) (2018)
18. Levine, S.: Reinforcement learning and control as probabilistic inference: tutorial and review. arXiv preprint arXiv:1805.00909 (2018)
19. Marshall, M., Lipkin, H.: Kalman filtering visual servoing control law. In: IEEE Proceedings of the International Conference on Mechatronics and Automation (2014)

20. Millidge, B., Tschantz, A., Seth, A.K., Buckley, C.L.: On the relationship between active inference and control as inference. In: IWAI 2020. CCIS, vol. 1326, pp. 3–11. Springer, Cham (2020). https://doi.org/10.1007/978-3-030-64919-7_1

21. Murphy, K.P.: Machine Learning: A Probabilistic Perspective. MIT Press (2012)

22. Narendra, K.S., Balakrishnan, J.: Adaptive control using multiple models. IEEE Trans. Autom. Control **42**, 171–187 (1997)

23. Oliver, G., Lanillos, P., Cheng, G.: An empirical study of active inference on a humanoid robot. IEEE Trans. Cogn. Develop. Syst. (2021). https://doi.org/10.1109/TCDS.2021.3049907

24. Paviglianiti, G., Pierri, F., Caccavale, F., Mattei, M.: Robust fault detection and isolation for proprioceptive sensors of robot manipulators. Mechatronics **20**(1), 162–170 (2010)

25. Pezzato, C., Ferrari, R., Corbato, C.H.: A novel adaptive controller for robot manipulators based on active inference. IEEE Robot. Autom. Lett. **5**(2), 2973–2980 (2020)

26. Pezzato, C., Baioumy, M., Corbato, C.H., Hawes, N., Wisse, M., Ferrari, R.: Active inference for fault tolerant control of robot manipulators with sensory faults. In: IWAI 2020. CCIS, vol. 1326, pp. 20–27. Springer, Cham (2020). https://doi.org/10.1007/978-3-030-64919-7_3

27. Pfeifer, T., Lange, S., Protzel, P.: Dynamic covariance estimation-a parameter free approach to robust sensor fusion. In: 2017 IEEE International Conference on Multisensor Fusion and Integration for Intelligent Systems (MFI), pp. 359–365. IEEE (2017)

28. Piepmeier, J., McMurray, G., Lipkin, H.: Uncalibrated dynamic visual servoing. IEEE Trans. Robot. Autom. **20**, 143–147 (2004)

29. Pio-Lopez, L., Nizard, A., Friston, K., Pezzulo, G.: Active inference and robot control: a case study. J. R. Soc. Interface **13**(122), 20160616 (2016)

30. Rostampour, V., Ferrari, R.M., Teixeira, A.M., Keviczky, T.: Privatized distributed anomaly detection for large-scale nonlinear uncertain systems. IEEE Trans. Autom. Control **66**, 5299–5313 (2020)

31. Shetty, A., Gao, G.X.: Covariance estimation for GPS-LiDAR sensor fusion for UAVs. In: Proceedings of the 30th International Technical Meeting of the Satellite Division of the Institute of Navigation, ION GNSS+ 2017, pp. 2919–2923 (2017)

32. Van, M., Wu, D., Ge, S., Ren, H.: Fault diagnosis in image-based visual servoing with eye-in-hand configurations using Kalman filter. IEEE Trans. Ind. Electron. **12**(6), 1998–2007 (2016)

33. Vega-Brown, W., Bachrach, A., Bry, A., Kelly, J., Roy, N.: Cello: a fast algorithm for covariance estimation. In: 2013 IEEE International Conference on Robotics and Automation, pp. 3160–3167. IEEE (2013)

34. Vega-Brown, W., Roy, N.: CELLO-EM: adaptive sensor models without ground truth. In: 2013 IEEE/RSJ International Conference on Intelligent Robots and Systems, pp. 1907–1914. IEEE (2013)

On the Convergence of DEM's Linear Parameter Estimator

Ajith Anil Meera[✉] and Martijn Wisse

Cognitive Robotics, Delft Institute of Technology, Delft, The Netherlands
`a.anilmeera@tudelft.nl`

Abstract. The free energy principle from neuroscience provides an efficient data-driven framework called the Dynamic Expectation Maximization (DEM), to learn the generative model in the environment. DEM's growing potential to be the brain-inspired learning algorithm for robots demands a mathematically rigorous analysis using the standard control system tools. Therefore, this paper derives the mathematical proof of convergence for its parameter estimator for linear state space systems, subjected to colored noise. We show that the free energy based parameter learning converges to a stable solution for linear systems. The paper concludes by providing a proof of concept through simulation for a wide range of spring damper systems.

Keywords: Free energy principle · Dynamic expectation maximization · Parameter estimation · Linear state space systems

1 Introduction

The free energy principle (FEP) models the brain's perception and action as a gradient ascend over its free energy objective [7]. The action side of FEP, known as active inference [8], has already been applied to real robots including ground robots for SLAM [5], humanoid robots for body perception [12] and manipulator robots for pick and place operation [13]. Similarities with standard control technique like PID was also analyzed [3]. One of the variants of FEP, the Dynamic Expectation Maximization (DEM) [9], provides a model inversion framework for perception and system identification. DEM's distinctive feature lies in its capability to gracefully handle colored noise through the use of generalized coordinates [6], thereby rendering it with the potential to be the learning algorithm for robots. DEM was reformulated as a linear state and input observer under colored noise [11] and was validated for quadrotor flights [4]. A DEM based linear parameter estimator for system identification was developed by [2] and was applied for the perception of quadrotor in wind [1]. Since an estimator with convergence guarantees is preferred for safe robotics applications, we aim at paving way to DEM's practical use by mathematically analyzing it for its convergence properties. Moreover, it is of interest to the active inference community to develop active learning and control strategies with stability guarantees. The

© Springer Nature Switzerland AG 2021
M. Kamp et al. (Eds.): ECML PKDD 2021 Workshops, CCIS 1524, pp. 692–700, 2021.
https://doi.org/10.1007/978-3-030-93736-2_49

presence of generalized coordinates, mean field terms and brain priors complicates the convergence proof and makes it different from other estimators like Expectation Maximization [10]. The goal of this paper is: 1) to show that DEM has convergence guarantees for linear systems with colored noise, and 2) to show that it can be applied to control system problems like the estimation of a mass-spring-damper system.

2   Preliminaries

Consider the linear plant dynamics (generative process) given in Eq. 1, where $\mathbf{A}$, $\mathbf{B}$ and $\mathbf{C}$ are constant system matrices, $\mathbf{x} \in \mathbb{R}^n$ is the hidden state, $\mathbf{v} \in \mathbb{R}^r$ is the input and $\mathbf{y} \in \mathbb{R}^m$ is the output.

$$\dot{\mathbf{x}} = \mathbf{A}\mathbf{x} + \mathbf{B}\mathbf{v} + \mathbf{w}, \qquad \mathbf{y} = \mathbf{C}\mathbf{x} + \mathbf{z}. \tag{1}$$

Here $\mathbf{w} \in \mathbb{R}^n$ and $\mathbf{z} \in \mathbb{R}^m$ represent the process and measurement noise respectively. The notations of the plant are denoted in boldface, whereas its estimates are denoted in nonboldface letters. Since the brain has no access to the plant dynamics except through the sensory measurements $\mathbf{y}$, it maintains the copy of an approximate model of the generative process called the generative model. The noise color assumption (convolution of white noise with a Gaussian kernel) facilitates the differentiated form of the generative model as [9]:

$$\begin{aligned} x' &= Ax + Bv + w & y &= Cx + z \\ x'' &= Ax' + Bv' + w' & \dot{y} &= Cx' + z' \end{aligned} \tag{2}$$

$$\cdots \qquad\qquad\qquad \cdots$$

One of the key technique behind DEM to model the colored noise is to express the time varying components in generalized coordinates, denoted by a tilde operator. The colored noises can be expressed in generalized coordinates using their higher derivatives as $\tilde{z} = [z, z', z'', ...]^T$ and $\tilde{w} = [w, w', w'', ...]^T$. The generative model in Eq. 2 can be compactly written as [9]:

$$\dot{\tilde{x}} = D^x \tilde{x} = \tilde{A}\tilde{x} + \tilde{B}\tilde{v} + \tilde{w} \qquad \tilde{y} = \tilde{C}\tilde{x} + \tilde{z} \tag{3}$$

where $D^x = \begin{bmatrix} 0 & 1 & & & \\ & 0 & 1 & & \\ & & \ddots & \ddots & \\ & & & 0 & 1 \\ & & & & 0 \end{bmatrix}_{(p+1)\times(p+1)} \otimes I_{n\times n}$, $\tilde{A} = I_{p+1} \otimes A$, $\tilde{B} = I_{p+1} \otimes B$

and $\tilde{C} = I_{p+1} \otimes C$. Here $\otimes$ is the Kronecker tensor product. To facilitate the convergence proof later in the paper, we introduce a redefinition for Eq. 3 with all parameters grouped to the right side as θ:

$$\dot{\tilde{x}} = M\theta + \tilde{w}, \ \tilde{y} = N\theta + \tilde{z}, \ \theta = \begin{bmatrix} vec(A^T) \\ vec(B^T) \\ vec(C^T) \end{bmatrix}, \tag{4}$$

where

$$M = \begin{bmatrix} I_n \otimes x^T & I_n \otimes v^T & I_n \otimes O_{1 \times m} \\ I_n \otimes x'^T & I_n \otimes v'^T & I_n \otimes O_{1 \times m} \\ \cdots & \cdots & \cdots \end{bmatrix}, N = \begin{bmatrix} I_n \otimes O_{1 \times n} & I_n \otimes O_{1 \times r} & I_m \otimes x^T \\ I_n \otimes O_{1 \times n} & I_n \otimes O_{1 \times r} & I_m \otimes x'^T \\ \cdots & \cdots & \cdots \end{bmatrix}.$$

(5)

The goal of this paper is to mathematically prove that the DEM's estimate for θ converges while maximizing the free energy objective.

3   Parameter Learning as Free Energy Optimization

DEM postulates the parameter learning algorithm as the gradient ascend over the free energy action, which is the time integral of free energy $\bar{F} = \int F dt$. The parameter update equation can be expressed as the gradient [2,9]:

$$\frac{\partial \theta}{\partial a} = k^\theta \frac{\partial \bar{F}}{\partial \theta} = -P^\theta(\theta - \eta^\theta) + \sum_t (-E_\theta + W_\theta^X),$$

(6)

where k^θ is the learning rate, $E_\theta = \frac{\partial E}{\partial \theta}$ is the gradient of precision weighed prediction error, $W_{\theta^i}^X = \frac{\partial W^X}{\partial \theta}$ is the gradient of state mean field term, η^θ is the prior parameters and P^θ is the prior parameter precision. Subscripts will be used for the derivative operator. E_θ for an LTI system can be simplified as:

$$E_\theta = \tilde{\epsilon}_\theta^T \tilde{\Pi} \tilde{\epsilon}, \text{ where } \tilde{\epsilon} = \begin{bmatrix} \tilde{y} - N\theta \\ \tilde{v} - \tilde{\eta}^v \\ D^x \tilde{x} - M\theta \end{bmatrix} \text{ and } \tilde{\epsilon}_\theta = \begin{bmatrix} -N \\ O \\ -M \end{bmatrix}$$

(7)

are the prediction error and its gradient. Here $\tilde{\eta}^v$ is the prior on inputs with prior precision $\tilde{P}^v$, $\tilde{\Pi} = diag(\tilde{\Pi}^z, \tilde{P}^v, \tilde{\Pi}^w)$ is the generalized noise precision with Π^z and Π^w being the precisions (inverse covariance) for measurement and process noise. Here $diag()$ represents the block diagonal operation. Similarly, W_θ^X for an LTI system can be written as [2,9]:

$$W_{\theta^i}^X = -\frac{1}{2} tr(\Sigma^X \tilde{\epsilon}_{X\theta^i}^T \tilde{\Pi} \tilde{\epsilon}_X), \ \tilde{\epsilon} = \begin{bmatrix} \tilde{y} - \tilde{C}\tilde{x} \\ \tilde{v} - \tilde{\eta}^v \\ D^x \tilde{x} - \tilde{A}\tilde{x} - \tilde{B}\tilde{v} \end{bmatrix}, \ \tilde{\epsilon}_X = \begin{bmatrix} -\tilde{C} & O \\ O & I \\ D^x - \tilde{A} & -\tilde{B} \end{bmatrix}.$$

(8)

4   Proof of Convergence for Parameter Estimator

If E_θ and W_θ^X can be expressed as linear in θ, in the form $E_\theta = E_1\theta + E_2$ and $W_\theta^X = W_1\theta + W_2$, Eq. 6 can be rewritten as:

$$\frac{\partial \theta}{\partial a} = -\Big[P^\theta + \sum_t (E_1 - W_1)\Big]\theta + \Big[P^\theta \eta^\theta + \sum_t (-E_2 + W_2)\Big].$$

(9)

The differential equation given by Eq. 9 is of the form of a linear state space equation ($\dot{\theta} = A^\theta\theta + B^\theta.1$). From the basics of control theory, the solutions of this equation converges exponentially (stabilise) if $A^\theta = -[P^\theta + \sum_t(E_1 - W_1)]$ is negative definite (negative eigen values). This section aims to prove this result.

Lemma 1. *If $A, B \succ O$, then $A + B \succ O$.*

As per Lemma 1, the positive definiteness of $P^\theta - \sum_t W_1 + \sum_t E_1$ can be proved by proving the positive definiteness of the individual terms P^θ, $-W_1$ and E_1. We know by definition that the prior parameter precision P^θ is positive definite. We now proceed to prove that $E_1 \succeq O$. Upon simplification of Eq. 7, E_θ can be written as $E_\theta = E_1\theta + E_2$, where:

$$E_1 = N^T \tilde{\Pi}^z N + M^T \tilde{\Pi}^w M \text{ and } E_2 = -\left[N^T \tilde{\Pi}^z \ M^T \tilde{\Pi}^w D\right]\begin{bmatrix}\tilde{\mathbf{y}} \\ \tilde{x}\end{bmatrix}. \tag{10}$$

Lemma 2. *If $A \succeq O$, then $B^T AB \succeq O$.*

Proof. By definition, if $A \succeq O$, there exists a square root $A^{\frac{1}{2}} \succeq O$. Therefore, $x^T(B^T AB)x = x^T(B^T A^{\frac{1}{2}}A^{\frac{1}{2}}B)x = (A^{\frac{1}{2}}Bx)^T(A^{\frac{1}{2}}Bx) \geq 0, \implies B^T AB \succeq O$.

Since $\tilde{\Pi}^z \succ O$ and $\tilde{\Pi}^w \succ O$ by definition, from Lemma 1 and 2, $E_1 = N^T \tilde{\Pi}^z N + M^T \tilde{\Pi}^w M \succeq O$. Therefore, E_1 is proved to be positive semi-definite.

The final term under consideration is W_1. The rest of this section aims to prove that $W_1 \prec O$, which will conclude the entire convergence proof of parameter estimation. We rewrite the mean field term for parameter θ^i from Eq. 8 as:

$$\begin{aligned}
W_{\theta^i}^X &= -\frac{1}{2}tr(\Sigma^X \tilde{\epsilon}_{X\theta^i}^T \tilde{\Pi}\tilde{\epsilon}_X), \\
&= -\frac{1}{2}tr\left[\begin{bmatrix}\Sigma^{\tilde{x}\tilde{x}} & \Sigma^{\tilde{x}\tilde{v}} \\ \Sigma^{\tilde{v}\tilde{x}} & \Sigma^{\tilde{v}\tilde{v}}\end{bmatrix}\begin{bmatrix}\tilde{C}_{\theta^i}^T \tilde{\Pi}^z\tilde{C} - \tilde{A}_{\theta^i}^T \tilde{\Pi}^w(D - \tilde{A}) & \tilde{A}_{\theta^i}^T \tilde{\Pi}^w\tilde{B} \\ -\tilde{B}_{\theta^i}^T \tilde{\Pi}^w(D - \tilde{A}) & \tilde{B}_{\theta^i}^T \tilde{\Pi}^w\tilde{B}\end{bmatrix}\right] \\
&= -\frac{1}{2}tr\left[\begin{bmatrix}\Sigma^{\tilde{x}\tilde{x}} & \Sigma^{\tilde{x}\tilde{v}} \\ \Sigma^{\tilde{v}\tilde{x}} & \Sigma^{\tilde{v}\tilde{v}}\end{bmatrix}\begin{bmatrix}\tilde{C}_{\theta^i}^T \tilde{\Pi}^z\tilde{C} + \tilde{A}_{\theta^i}^T \tilde{\Pi}^w\tilde{A} & \tilde{A}_{\theta^i}^T \tilde{\Pi}^w\tilde{B} \\ \tilde{B}_{\theta^i}^T \tilde{\Pi}^w\tilde{A} & \tilde{B}_{\theta^i}^T \tilde{\Pi}^w\tilde{B}\end{bmatrix}\right] \\
&\quad -\frac{1}{2}tr\left[\begin{bmatrix}\Sigma^{\tilde{x}\tilde{x}} & \Sigma^{\tilde{x}\tilde{v}} \\ \Sigma^{\tilde{v}\tilde{x}} & \Sigma^{\tilde{v}\tilde{v}}\end{bmatrix}\begin{bmatrix}-\tilde{A}_{\theta^i}^T \tilde{\Pi}^w D & O \\ -\tilde{B}_{\theta^i}^T \tilde{\Pi}^w D & O\end{bmatrix}\right].
\end{aligned} \tag{11}$$

Since the second trace term in Eq. 11 is independent of θ^i, it is lumped into the $W_2^{\theta^i}$ term. Equation 11 is further simplified as:

$$\begin{aligned}
W_{\theta^i}^X &= -\frac{1}{2}\Big[tr(\Sigma^{\tilde{x}\tilde{x}}\tilde{C}_{\theta^i}^T \tilde{\Pi}^z\tilde{C}) + tr(\Sigma^{\tilde{x}\tilde{x}}\tilde{A}_{\theta^i}^T \tilde{\Pi}^w\tilde{A}) + tr(\Sigma^{\tilde{x}\tilde{v}}\tilde{B}_{\theta^i}^T \tilde{\Pi}^w\tilde{A}) \\
&\quad + tr(\Sigma^{\tilde{v}\tilde{x}}\tilde{A}_{\theta^i}^T \tilde{\Pi}^w\tilde{B}) + tr(\Sigma^{\tilde{v}\tilde{v}}\tilde{B}_{\theta^i}^T \tilde{\Pi}^w\tilde{B})\Big] + W_2^{\theta^i}
\end{aligned} \tag{12}$$

We aim to separate θ out so that the mean field term can be expressed in the form $W_\theta^X = W_1\theta + W_2$. We proceed by first introducing the transpose of the generalized parameter matrices $\tilde{A}$, $\tilde{B}$ and $\tilde{C}$ to Eq. 12 and then moving them out of the trace terms.

Lemma 3. *If A, B, C and D are matrices, then* $tr(ABCD) = tr(C^T B^T A^T D^T)$

Proof. $tr(ABCD) = tr((ABCD)^T) = tr(D^T C^T B^T A^T) = tr(C^T B^T A^T D^T)$.

Lemma 4. *If A, B and C are matrices, then* $tr(ABC) = vec(A^T)^T(I \otimes B)vec(C)$.

Applying Lemma 3 throughout Eq. 12 results in:

$$W_{\theta^i}^X = -\frac{1}{2}\Big[tr(\tilde{\Pi}^{zT}\tilde{C}_{\theta^i}\Sigma^{\tilde{x}\tilde{x}T}\tilde{C}^T) + tr(\tilde{\Pi}^{wT}\tilde{A}_{\theta^i}\Sigma^{\tilde{x}\tilde{x}T}\tilde{A}^T) + tr(\tilde{\Pi}^{wT}\tilde{B}_{\theta^i}\Sigma^{\tilde{v}\tilde{x}T}\tilde{A}^T)$$
$$+ tr(\tilde{\Pi}^{wT}\tilde{A}_{\theta^i}\Sigma^{\tilde{v}\tilde{x}T}\tilde{B}^T) + tr(\tilde{\Pi}^{wT}\tilde{B}_{\theta^i}\Sigma^{\tilde{v}\tilde{v}T}\tilde{B}^T)\Big] + W_2^{\theta^i},$$

$$(13)$$

which upon further expansion using Lemma 4 and grouping yields:

$$W_{\theta^i}^X = -\frac{1}{2}\Bigg[\Big(vec(\tilde{A}_{\theta^i}^T\tilde{\Pi}^w)^T(I \otimes \Sigma^{\tilde{x}\tilde{x}T}) + vec(\tilde{B}_{\theta^i}^T\tilde{\Pi}^w)^T(I \otimes \Sigma^{\tilde{x}\tilde{v}T})\Big)vec(\tilde{A}^T)$$

$$+ \Big(vec(\tilde{A}_{\theta^i}^T\tilde{\Pi}^w)^T(I \otimes \Sigma^{\tilde{v}\tilde{x}T}) + vec(\tilde{B}_{\theta^i}^T\tilde{\Pi}^w)^T(I \otimes \Sigma^{\tilde{v}\tilde{v}T})\Big)vec(\tilde{B}^T)$$

$$+ \Big(vec(\tilde{C}_{\theta^i}^T\tilde{\Pi}^z)^T(I \otimes \Sigma^{\tilde{x}\tilde{x}T})\Big)vec(\tilde{C}^T)\Bigg] + W_2^{\theta^i}.$$

$$(14)$$

We have now separated all the generalized parameters out of the trace terms in their vector forms. These vectors can be grouped such that the mean field term is linear with respect to the generalized parameter vector $\tilde{\theta} = \begin{bmatrix} vec(\tilde{A}^T) \\ vec(\tilde{B}^T) \\ vec(\tilde{C}^T) \end{bmatrix}$ as:

$$W_{\theta^i}^X = -\frac{1}{2}\Bigg[vec(\tilde{A}_{\theta^i}^T\tilde{\Pi}^w)^T(I \otimes \Sigma^{\tilde{x}\tilde{x}T}) + vec(\tilde{B}_{\theta^i}^T\tilde{\Pi}^w)^T(I \otimes \Sigma^{\tilde{x}\tilde{v}T}),$$

$$vec(\tilde{A}_{\theta^i}^T\tilde{\Pi}^w)^T(I \otimes \Sigma^{\tilde{v}\tilde{x}T}) + vec(\tilde{B}_{\theta^i}^T\tilde{\Pi}^w)^T(I \otimes \Sigma^{\tilde{v}\tilde{v}T}), \qquad (15)$$

$$vec(\tilde{C}_{\theta^i}^T\tilde{\Pi}^z)^T(I \otimes \Sigma^{\tilde{x}\tilde{x}T})\Bigg]\tilde{\theta} + W_2^{\theta^i}.$$

Lemma 5. *If A and B are matrices, then* $vec(AB)^T = vec(A)^T(B \otimes I)$.

We use Lemma 5 to further simplify Eq. 15 as:

$$W_{\theta^i}^X = -\frac{1}{2}\Bigg[vec(\tilde{A}_{\theta^i}^T)^T(\tilde{\Pi}^w \otimes I)(I \otimes \Sigma^{\tilde{x}\tilde{x}T}) + vec(\tilde{B}_{\theta^i}^T)^T(\tilde{\Pi}^w \otimes I)(I \otimes \Sigma^{\tilde{x}\tilde{v}T}),$$

$$vec(\tilde{A}_{\theta^i}^T)^T(\tilde{\Pi}^w \otimes I)(I \otimes \Sigma^{\tilde{v}\tilde{x}T}) + vec(\tilde{B}_{\theta^i}^T)^T(\tilde{\Pi}^w \otimes I)(I \otimes \Sigma^{\tilde{v}\tilde{v}T}),$$

$$vec(\tilde{C}_{\theta^i}^T)^T(\tilde{\Pi}^z \otimes I)(I \otimes \Sigma^{\tilde{x}\tilde{x}T})\Bigg]\tilde{\theta} + W_2^{\theta^i}.$$

$$(16)$$

Since the parameters A, B and C are independent of each other, their derivatives with respect to each other are zeros, resulting in $vec(\tilde{A}^T_{\theta^i}) = O, \forall \theta^i \in \{B, C\}$, $vec(\tilde{B}^T_{\theta^i}) = O, \forall \theta^i \in \{A, C\}$ and $vec(\tilde{C}^T_{\theta^i}) = O, \forall \theta^i \in \{A, B\}$. This simplifies the expression for $W^X_{\theta^i}$ in Eq. 16. The total mean field term W^X_θ can be computed by vertically stacking the individual mean field contributions $W^X_{\theta^i}$ from each parameter θ^i as:

$$W^X_\theta = -\frac{1}{2} W_3 \tilde{\theta} + W_2, \tag{17}$$

where $W_3 = \begin{bmatrix} W_4 & O \\ O & W_5 \end{bmatrix}$ with $W_5 = vec(\tilde{C}^T)^T_{vecC^T}(\tilde{\Pi}^z \otimes I)(I \otimes \Sigma^{\tilde{x}\tilde{x}T})$ and

$$W_4 = \begin{bmatrix} vec(\tilde{A}^T)^T_{vecA^T}(\tilde{\Pi}^w \otimes I)(I \otimes \Sigma^{\tilde{x}\tilde{x}T}) & vec(\tilde{A}^T)^T_{vecA^T}(\tilde{\Pi}^w \otimes I)(I \otimes \Sigma^{\tilde{v}\tilde{x}T}) \\ vec(\tilde{B}^T)^T_{vecB^T}(\tilde{\Pi}^w \otimes I)(I \otimes \Sigma^{\tilde{x}\tilde{v}T}) & vec(\tilde{B}^T)^T_{vecB^T}(\tilde{\Pi}^w \otimes I)(I \otimes \Sigma^{\tilde{v}\tilde{v}T}) \end{bmatrix}.$$

W_3 can be simplified as:

$$W_3 = \frac{\partial \tilde{\theta}}{\partial \theta}^T \begin{bmatrix} \tilde{\Pi}^w \otimes I & O & O \\ O & \tilde{\Pi}^w \otimes I & O \\ O & O & \tilde{\Pi}^z \otimes I \end{bmatrix} \begin{bmatrix} I \otimes \Sigma^{\tilde{x}\tilde{x}T} & I \otimes \Sigma^{\tilde{v}\tilde{x}T} & O \\ I \otimes \Sigma^{\tilde{x}\tilde{v}T} & I \otimes \Sigma^{\tilde{v}\tilde{v}T} & O \\ O & O & I \otimes \Sigma^{\tilde{x}\tilde{x}T} \end{bmatrix}, \tag{18}$$

where $\frac{\partial \tilde{\theta}}{\partial \theta} = diag(vccA^T_{vecA^T}, vecB^T_{vecB^T}, vecC^T_{vecC^T})$. Since the generalized parameter vector $\tilde{\theta}$ is linear in parameter vector θ, we can write:

$$\tilde{\theta} = \frac{\partial \tilde{\theta}}{\partial \theta} \theta = \begin{bmatrix} vec\tilde{A}^T_{vecA^1} & O & O \\ O & vec\tilde{B}^T_{vecB^T} & O \\ O & O & vec\tilde{C}^T_{vecC^T} \end{bmatrix} \theta. \tag{19}$$

Substituting Eq. 18 and 19 in Eq. 17 yields:

$W^X_\theta = W_1 \theta + W_2,$

$$W_1 = -\frac{1}{2}\frac{\partial \tilde{\theta}}{\partial \theta}^T \begin{bmatrix} \tilde{\Pi}^w \otimes I & O & O \\ O & \tilde{\Pi}^w \otimes I & O \\ O & O & \tilde{\Pi}^z \otimes I \end{bmatrix} \begin{bmatrix} I \otimes \Sigma^{\tilde{x}\tilde{x}T} & I \otimes \Sigma^{\tilde{v}\tilde{x}T} & O \\ I \otimes \Sigma^{\tilde{x}\tilde{v}T} & I \otimes \Sigma^{\tilde{v}\tilde{v}T} & O \\ O & O & I \otimes \Sigma^{\tilde{x}\tilde{x}T} \end{bmatrix} \frac{\partial \tilde{\theta}}{\partial \theta}. \tag{20}$$

Therefore, the mean field term W^X_θ is linear in θ. For the parameter estimator to provide a converging solution, we need to prove that $W_1 \prec O$. Lemma 2 could be applied to the expression for W_1 to prove that $W_1 \prec O$ if:

$$W_6 = \begin{bmatrix} \tilde{\Pi}^w \otimes I & O & O \\ O & \tilde{\Pi}^w \otimes I & O \\ O & O & \tilde{\Pi}^z \otimes I \end{bmatrix} \begin{bmatrix} I \otimes \Sigma^{\tilde{x}\tilde{x}T} & I \otimes \Sigma^{\tilde{v}\tilde{x}T} & O \\ I \otimes \Sigma^{\tilde{x}\tilde{v}T} & I \otimes \Sigma^{\tilde{v}\tilde{v}T} & O \\ O & O & I \otimes \Sigma^{\tilde{x}\tilde{x}T} \end{bmatrix} \succ O \tag{21}$$

Lemma 6. *If $A, B \succeq O$ and A is invertible, then $AB \succeq O$.*

Proof. $AB = A^{\frac{1}{2}}(A^{\frac{1}{2}}BA^{\frac{1}{2}})A^{-\frac{1}{2}}$, implies AB and $A^{\frac{1}{2}}BA^{\frac{1}{2}}$ are similar matrices, sharing all eigen values. Using Lemma 2, since $B \succeq O$, $A^{\frac{1}{2}}BA^{\frac{1}{2}} \succeq O \implies AB \succeq O$.

Using Lemma 6 it is straightforward to see that $W_6 \succeq O$ because: $\tilde{\Pi}^z \succ O, \tilde{\Pi}^w \succ O, \implies \tilde{\Pi}^z \otimes I \succ O$ and $\tilde{\Pi}^w \otimes I \succ O, I \otimes \Sigma^X \succ O$. Therefore, $W_1 \preceq O$. This completes the proof that the parameter estimation of DEM converges for an LTI system with colored noise.

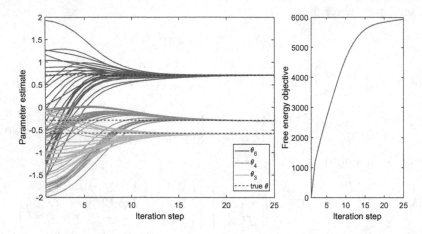

Fig. 1. The parameter estimates of DEM converges to the correct value of $\theta_3 = -\frac{k}{m} = -0.5714$, $\theta_4 = -\frac{b}{m} = -0.2857$ and $\theta_6 = \frac{1}{m} = 0.7143$, marked in black, for a set of 25 experiments, despite being initialized by randomly sampled priors such that $\eta^{\theta^i} \in [-2, 2]$ and that the prior A matrix is stable. The parameter estimation proceeds by maximizing the free energy objective as shown on the right (sample realization).

5   Proof of Concept: Mass-Spring-Damper System

This section aims at providing a proof of concept for the convergence of DEM's parameter estimator, through realistic simulations. A mass-spring-damper system with mass $m = 1.4$ kg, spring constant $k = 0.8$ N/m and damping coefficient $b = 0.4$ Ns/m, is considered in the state space form given by:

$$\begin{bmatrix} \dot{x} \\ \ddot{x} \end{bmatrix} = \begin{bmatrix} 0 & 1 \\ -\frac{k}{m} & -\frac{b}{m} \end{bmatrix} \begin{bmatrix} x \\ \dot{x} \end{bmatrix} + \begin{bmatrix} 0 \\ \frac{1}{m} \end{bmatrix} v, \quad y = \begin{bmatrix} 1 & 0 \end{bmatrix} \begin{bmatrix} x \\ \dot{x} \end{bmatrix}. \tag{22}$$

A Gaussian bump input $v = e^{-0.25(t-12)^2}$, centred around 12 s and sampled at $dt = 0.1$ s for $T = 32$ s was used. To generate the colored noise, the white noise ($\Pi^w = e^6 I_2$ and $\Pi^z = e^6$) was convoluted using a Gaussian kernel with a width of $\sigma = 0.5$ s. A partially known system with unknown $\theta_3 = -\frac{k}{m}$, $\theta_4 = -\frac{b}{m}$ and $\theta_6 = \frac{1}{m}$ was considered. Using the output $\mathbf{y}$ generated from the spring damper system, parameter estimation was performed using DEM for 25 experiments with different η^θ. The parameter priors η^θ for unknown parameters were randomly sampled from $[-2,2]$ such that the resulting prior A matrix is stable. A low prior precision ($P^{\theta_i} = e^{-4}$) was used for known parameters, and a high precision

($P^{\theta_i} = e^{32}$) was used for unknown parameters. The order of generalized motion of $p = 6$ and $d = 2$ were used for the states and inputs respectively. The result for DEM's parameter estimation is shown in Fig. 1. Despite being initialized by random wrong priors, DEM's parameter estimates exponentially converges to the correct values, by maximizing the free energy objective.

Next, we proceed to show that the estimate converges for a wide range of systems. The same experiment was repeated for 25 different randomly selected stable mass-spring-damper systems. Although the convergence applies to unstable systems, sampling was restricted to stable systems within the range $[-1,1]$ ($\theta_3, \theta_4 \in [-1, 0]$ and $\theta_6 \in [0, 1]$) for better visualization. DEM was initialized with the same priors for all experiments ($\eta^{\theta_6} = 2$, $\eta^{\theta_4} = -1$ and $\eta^{\theta_3} = -2$). Figure 2 shows that DEM is capable of providing converging solutions for a wide range of stable spring-damper systems, that are influenced by colored noise. Note that the numerical analysis is restricted to the dynamics of spring damper systems for demonstrative purposes, and can be extended to other systems. In summary, DEM can provide converging parameter estimates for linear systems with colored noise, by maximizing the free energy objective.

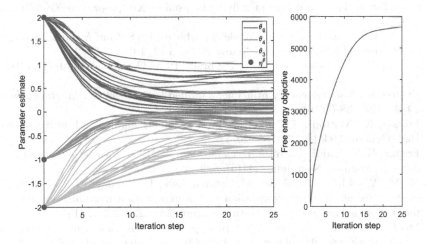

Fig. 2. DEM's parameter estimates for 25 different randomly sampled stable mass-spring-damper systems. The estimates for all the experiments started from the same prior of $\eta^{\theta_6} = 2$, $\eta^{\theta_4} = -1$ and $\eta^{\theta_3} = -2$, and converged, while maximizing the free energy objective. Therefore, the estimator converges for a wide range of systems.

6   Conclusion

DEM has the potential to be a bioinspired learning algorithm for future robots, due to its capability to robustly handle colored noise. Its superior performance in state estimation under colored noise was proven by [11] and was experimentally validated by [4]. In this paper, we derived a mathematical proof of convergence for DEM's parameter estimator, applied to linear systems with colored noise. We proved that a perception scheme based on the gradient ascend of the free energy action, provides a converging solution. Since a convergence proof is mandatory for the safe and reliable application of DEM on real robots, this work widens its applicability in robotics. The applicability of DEM for real control system problem was demonstrated through rigorous simulations on the estimation problem for mass-spring-damper systems. The future research will focus on the conditions for unbiased estimation and on applying DEM to real robots.

References

1. Meera, A.A., Wisse, M.: A brain inspired learning algorithm for the perception of a quadrotor in wind. arXiv preprint arXiv:2109.11971 (2021)
2. Anil Meera, A., Wisse, M.: Dynamic expectation maximization algorithm for estimation of linear systems with colored noise. Entropy **23**(10), 1306 (2021)
3. Baltieri, M., Buckley, C.L.: PID control as a process of active inference with linear generative models. Entropy **21**(3), 257 (2019)
4. Bos, F., Meera, A.A., Benders, D., Wisse, M.: Free energy principle for state and input estimation of a quadcopter flying in wind. arXiv preprint arXiv:2109.12052 (2021)
5. Çatal, O., Verbelen, T., Van de Maele, T., Dhoedt, B., Safron, A.: Robot navigation as hierarchical active inference. Neural Netw. **142**, 192–204 (2021)
6. Friston, K.: Hierarchical models in the brain. PLoS Comput. Biol. **4**(11), e1000211 (2008)
7. Friston, K.: The free-energy principle: a unified brain theory? Nat. Rev. Neurosci. **11**(2), 127–138 (2010)
8. Friston, K., Mattout, J., Kilner, J.: Action understanding and active inference. Biol. Cybern. **104**(1), 137–160 (2011)
9. Friston, K.J., Trujillo-Barreto, N., Daunizeau, J.: DEM: a variational treatment of dynamic systems. Neuroimage **41**(3), 849–885 (2008)
10. Mader, W., Linke, Y., Mader, M., Sommerlade, L., Timmer, J., Schelter, B.: A numerically efficient implementation of the expectation maximization algorithm for state space models. Appl. Math. Comput. **241**, 222–232 (2014)
11. Meera, A.A., Wisse, M.: Free energy principle based state and input observer design for linear systems with colored noise. In: 2020 American Control Conference (ACC), pp. 5052–5058. IEEE (2020)
12. Oliver, G., Lanillos, P., Cheng, G.: Active inference body perception and action for humanoid robots. arXiv preprint arXiv:1906.03022 (2019)
13. Pezzato, C., Ferrari, R., Corbato, C.H.: A novel adaptive controller for robot manipulators based on active inference. IEEE Robot. Autom. Lett. **5**(2), 2973–2980 (2020)

Disentangling What and Where for 3D Object-Centric Representations Through Active Inference

Toon Van de Maele[(✉)], Tim Verbelen, Ozan Çatal, and Bart Dhoedt

IDLab, Department of Information Technology, Ghent University - imec Ghent,
Ghent, Belgium
{toonvande.maele,tim.verbelen,ozan.catal,bart.dhoedt}@ugent.be

Abstract. Although modern object detection and classification models achieve high accuracy, these are typically constrained in advance on a fixed train set and are therefore not flexible to deal with novel, unseen object categories. Moreover, these models most often operate on a single frame, which may yield incorrect classifications in case of ambiguous viewpoints. In this paper, we propose an active inference agent that actively gathers evidence for object classifications, and can learn novel object categories over time. Drawing inspiration from the human brain, we build object-centric generative models composed of two information streams, a what- and a where-stream. The what-stream predicts whether the observed object belongs to a specific category, while the where-stream is responsible for representing the object in its internal 3D reference frame. We show that our agent (i) is able to learn representations for many object categories in an unsupervised way, (ii) achieves state-of-the-art classification accuracies, actively resolving ambiguity when required and (iii) identifies novel object categories. Furthermore, we validate our system in an end-to-end fashion where the agent is able to search for an object at a given pose from a pixel-based rendering. We believe that this is a first step towards building modular, intelligent systems that can be used for a wide range of tasks involving three dimensional objects.

Keywords: Deep learning · Object recognition · Object pose estimation · Active inference

1 Introduction

In the last decade, we have seen a proliferation of deep learning systems, especially in the field of image classificaton [10,15]. Although these systems show high accuracies on various classification benchmarks, their applicability is typically limited to a fixed input distribution based on the dataset used during training. In contrast, the real world is not stationary, which urges the need for continual learning [7]. Also, these classifiers lack the concept of action, which renders them vulnerable to ambiguous and adverserial samples [6]. As humans, we will typically move around and sample more viewpoints to improve the precision of our

© Springer Nature Switzerland AG 2021
M. Kamp et al. (Eds.): ECML PKDD 2021 Workshops, CCIS 1524, pp. 701–714, 2021.
https://doi.org/10.1007/978-3-030-93736-2_50

classification, illustrating the importance of embodiment in building intelligent agents [22].

Active inference offers a unified treatment of perception, action and learning, which states that intelligent systems build a generative model of their world and operate by minimizing a bound on surprise, i.e. the variational free energy [5]. In [18], Parr et al. propose a model for (human) vision, which considers a scene as a factorization of separate (parts of) objects, encoding their identity, scale and pose. This is in line with the so called two stream hypothesis, which states that visual information is processed by a dorsal ("where") stream on the one hand, representing where an object is in the space, and a ventral ("what") stream on the other hand, representing object identity [4]. Similarly, Hawkins et al. propose that cortical columns in the neocortex track objects and their pose in a local reference frame, encoded by cortical grid cells [9].

In this paper, we propose a system that builds on these principles for learning object-centric representations that allow for accurate classification. Inspired by cortical columns, our system is composed of separate deep neural networks, called Cortical Column Networks (CCN), where each CCN learns a representation of a single type of 3D object in a local reference frame. The ensemble of CCNs forms the agent's generative model, which is optimized by minimizing free energy. By also minimizing the expected free energy in the future, we show that our agent can realize preferred viewpoints for certain objects, while also being urged to resolve ambiguity on object identity.

We evaluate our agent on pixel data rendered from 3D objects from the YCB benchmarking dataset [1], where the agent can control the viewpoint. We compare the performance of an embodied and a static agent for classification, and show that classification accuracy is higher for the embodied agent. Additionally, we leverage the where stream for implicit pose estimation of the objects.

2   Method

In active inference, an agent acts and learns in order to minimize an upper bound on the negative log evidence of its observations, given its generative model of the world i.e. the free energy. In this section, we first formally introduce the generative model of our agent for representing 3D objects. Next we discuss how we instantiate and train this generative model using deep neural networks. Finally, we show how action selection is driven by minimizing expected free energy in the future.

2.1   A Generative Model for Object-Centric Perception

Our generative model is based on [18], but focused on representing a single object. Concretely, our agent obtains pixel observations $o_{0:t}$ that render a 3D object with identity i as viewed from certain viewpoints $v_{0:t}$ specified in an object-local reference frame. Each time step t the agent can perform an action

$\mathbf{a}_t$, resulting in a relative translation and rotation of the camera. The joint probability distribution then factorizes as:

$$p(\mathbf{o}_{0:t}, \mathbf{a}_{0:t-1}, \mathbf{v}_{0:t}, \mathbf{i}) = p(\mathbf{i}) \prod_t p(\mathbf{o}_t|\mathbf{v}_t, \mathbf{i})p(\mathbf{v}_t|\mathbf{v}_{t-1}, \mathbf{a}_{t-1})p(\mathbf{a}_{t-1}) \qquad (1)$$

Using the approximate posterior $q(\mathbf{i}, \mathbf{v}_{0:t}|\mathbf{o}_{0:t}) = q(\mathbf{i}|\mathbf{o}_{0:t}) \sum_t q(\mathbf{v}_t|\mathbf{i}, \mathbf{o}_t)$, the free energy becomes:

$$
\begin{aligned}
F &= \mathbb{E}_{q(\mathbf{i},\mathbf{v}_{0:t})}[\log q(\mathbf{i}, \mathbf{v}_{0:t}|\mathbf{o}_{0:t}) - \log p(\mathbf{o}_{0:t}, \mathbf{a}_{0:t-1}, \mathbf{v}_{0:t}, \mathbf{i})] \\
&\stackrel{+}{=} D_{KL}[q(\mathbf{i}|\mathbf{o}_{0:t})\|p(\mathbf{i})] + \sum_t D_{KL}[q(\mathbf{v}_t|\mathbf{i}, \mathbf{o}_t)\|p(\mathbf{v}_t|\mathbf{v}_{t-1}, \mathbf{a}_{t-1})] \qquad (2) \\
&\quad - \mathbb{E}_{q(\mathbf{i},\mathbf{v}_{0:t})}[\log p(\mathbf{o}_t|\mathbf{v}_t, \mathbf{i})]
\end{aligned}
$$

This shows that minimizing free energy is equivalent to maximizing the accuracy, i.e. predicting the observation for a given object identity and viewpoint, while minimizing complexity of the posterior models.

2.2   An Ensemble of CCNs

We instantiate the generative model using deep neural networks similar to a variational autoencoder (VAE) [14,19] with an encoder and decoder part. For each object identity, we train a separate encoder-decoder pair, since $p(\mathbf{o}_t|\mathbf{v}_t, \mathbf{i}) = \sum_k p(\mathbf{o}_t|\mathbf{v}_t, i = k)$. Similarly the encoder outputs distribution parameters for the object identity $q(i = k|\mathbf{o}_t)$ and viewpoint $q(\mathbf{v}_t|\mathbf{o}_t, i = k)$, the former parameterized as a Bernoulli variable, the latter as a multivariate Gaussian with a diagonal covariance matrix. Finally, we also parameterize the transition model $p(\mathbf{v}_t|\mathbf{v}_{t-1}, \mathbf{a}_{t-1}, i = k)$ which enforces $\mathbf{v}$ to encode relative viewpoint information.

Intuitively, each encoder-decoder pair captures the information about a single object class, with a "what" stream modeled as a binary classifier of whether an observation belongs to a certain object identiy, and a "where" stream encoding the observer viewpoint w.r.t. a local, object-specific reference frame. We call such a pair a Cortical Column Network (CCN), as it mimicks the "voting for object at pose" behavior of cortical columns in the neocortex as hypothesized in [9]. This is illustrated in Fig. 1. The agent hence entails a generative model as an ensemble of CCNs. We obtain $q(\mathbf{i}|\mathbf{o}_{0:t}) \propto q(\mathbf{i}|\mathbf{o}_{0:t-1})q(\mathbf{i}|\mathbf{o}_t)$, where $q(\mathbf{i}|\mathbf{o}_t)$ is a Categorical distribution from the CCN votes $q(i = k|\mathbf{o}_t)$, and $q(\mathbf{i}|\mathbf{o}_{0:t-1})$ is a conjugate prior Dirichlet distribution whose concencentration parameters are aggregated votes from previous observations, as updated in a Bayesian filter [25]. This process computes the posterior belief over the different timesteps. The Dirichlet distribution reflects the prior that an object is unlikely to change its category between timesteps. We also include an "other" object class, which is activate when none of the object classes receive votes, hence enabling the agent to detect novel object categories.

Each CCN is trained in an end-to-end fashion using a dataset of object observation pairs and the relative camera transform between them for each object

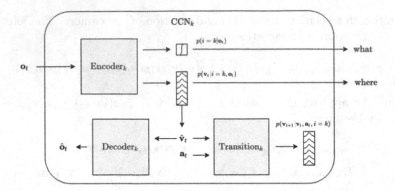

Fig. 1. A single CCN. Observation $\mathbf{o}_t$ is processed by an encoder and provides a belief over the object identity $p(i = k|\mathbf{o}_t)$ and over the observers pose $p(\mathbf{v}_t|i = k, \mathbf{o}_t)$. From this distribution, a sample $\hat{\mathbf{v}}_t$ is drawn and is decoded in a reconstructed observation $\hat{\mathbf{o}}_t$. This sample is also transformed together with an action $\mathbf{a}_t$ into a belief over a future pose $\mathbf{v}_{t+1}$.

class. To minimize Eq. 2, we use MSE loss on the reconstructions and a KL divergence between the viewpoint posterior and transition model. The identity posterior is trained as a binary classifier, sampling positive and negative anchors from the dataset. For more details on the training loss and model architectures, the reader is referred to the appendix.

2.3 Classification by Minimizing Expected Free Energy

Crucially in active inference, an agent will select the action that minimizes the expected free energy in the future G. In our case, this yields:

$$
\begin{aligned}
G(\mathbf{a}_t) = \;& \mathbb{E}_{q(\mathbf{i}, \mathbf{v}_{0:t+1}, \mathbf{o}_{t+1})}[\log q(\mathbf{i}, \mathbf{v}_{0:t+1}|\mathbf{o}_{0:t}, \mathbf{a}_t) - \log p(\mathbf{o}_{0:t+1}, \mathbf{a}_{0:t-1}, \mathbf{v}_{0:t+1}, \mathbf{i}|\mathbf{a}_t)] \\
\approx \;& \mathbb{E}_{q(\mathbf{o}_{t+1})}[-\log p(\mathbf{o}_{0:t+1})] \\
& - \mathbb{E}_{q(\mathbf{i}, \mathbf{v}_{0:t+1}, \mathbf{o}_{t+1})}[\log q(\mathbf{i}|\mathbf{o}_{0:t+1}, \mathbf{a}_t) - \log q(\mathbf{i}|\mathbf{o}_{0:t}, \mathbf{a}_t)] \\
& - \mathbb{E}_{q(\mathbf{i}, \mathbf{v}_{0:t+1}, \mathbf{o}_{t+1})}[\log q(\mathbf{v}_{0:t+1}|\mathbf{i}, \mathbf{o}_{0:t+1}, \mathbf{a}_t) - \log q(\mathbf{v}_{0:t+1}|\mathbf{i}, \mathbf{o}_{0:t}, \mathbf{a}_t)]
\end{aligned}
\tag{3}
$$

The expected free energy unpacks into three terms, the first is an instrumental term that indicates that the agent is driven to some prior preferred observations, whereas the second and third term encode the expected information gain for the object identity and the object pose for a certain action. This shows how the agent can be steered to seeing a certain object at a certain pose, which could be for example a grasp position in the case of a robotic manipulator. On the other hand, in the absence of preferences, the agent will query new viewpoints that provide information on the object identity and pose, effectively trying to get a better classification.

3 Experiments

We evaluate our model for an agent in a 3D environment, where 3D models of objects from the YCB dataset [1] are rendered from a certain camera viewpoint. The agent actions are then defined as relative transforms (i.e. rotation and translation), moving the camera viewpoint. This setup closely mimics a robot manipulator with an in-hand camera, but without kinematic constraints [26].

We create a dataset using 3D meshes of objects from the YCB dataset [1]. For each of 9 "known" objects, 14000 viewpoints and their corresponding view, for which the object is centered in view, are generated as a train set. During training, pairs of two views are randomly selected, for which the action is defined as the relative transform between these two viewpoints.

We first validate that the CCN ensemble is able to learn pose and identity representations unsupervisedly by minimizing free energy. Next, we show how the expected free energy allows to agent to infer actions that can bring the agent to a preferred pose relative to an object on the one hand, and resolve ambiguity for inferring an object identity on the other hand.

3.1 The "what" Stream: Object Recognition

First, we evaluate the performance of each individual CCN "what" binary classifier. The ROC curves are shown in Fig. 2a where each CCN is tested on a dataset with 3000 novel views for each of the 9 known objects, and 3000 views from 5 objects, it has never seen during training. For all objects we achieve near-perfect ROC curves, which can be attributed to the fact that each CCN can focus on

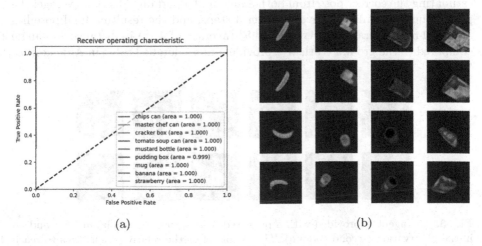

(a) (b)

Fig. 2. (a) The ROC curve for individual CCNs. Negative samples are observations from the test set: padlock, power drill, knife, orange and tuna fish can. (b) The ground truth (top row), the reconstruction (second row), and imagined transformed observations (other rows) are shown for multiple YCB objects [1].

particular features that distinguish a particular object from the others. Investigating the impact on the ROC performance when using real-world observations instead of 3D renders of predefined object models would be an interesting avenue for future work.

3.2 The "where" Stream: Implicit Pose Estimation

Crucially, our CCNs not only learn a classification output, but also an implicit representation of the 3D structure of the object at hand. As discussed in Sect. 2, this is encoded in a latent code $\mathbf{v}_t$, from which the model can reconstruct the given viewpoint using the decoder, or imagine other viewpoints after a relative transform using the transition model. This is illustrated in Fig. 2b, where the first row shows ground truth object observations, the second row shows the reconstruction after encoding, and the third and fourth row show imagined other viewpoints.

We can now use the CCN to infer the actions that will yield some "preferred" observation, by minimizing the expected free energy in Eq. 3. This is useful for example to instruct a robotic manipulator to a certain grasp point for an object. As computing G for every action is intractable, we sample 1000 random relative transforms for which G is calculated. A transform is sampled by first sampling a target viewpoint in 3D space uniformly in the workspace. The orientation is then determined so that the camera looks at the center of gravity of the object. The relative transform can then be computed between both current and target sampled poses. The identity transform is always provided as an option, allowing the agent to stay at its current pose when no better option is found. This results in the agent finding the estimated pose. Figure 3 shows qualitative trajectories for estimating the correct pose from both a mug and a pudding box. On average, the pose estimation process converges after 3 steps, and the resulting final pose lies around 1 mm (average of 1.4 mm) and 5° (average of 4.7°) in distance and angle compared to the ground truth. We provide a more detailed table in Appendix B.

Fig. 3. The agent is provided with a preferred observation (first column left), and an initial observation (second column). The agent infers the relative transforms to reach the preferred observation.

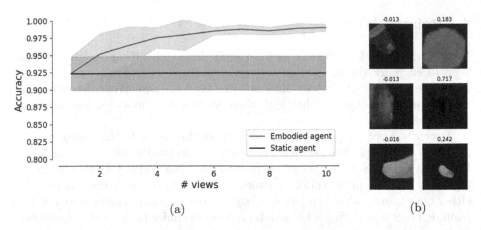

(a)   (b)

Fig. 4. (a) Performance of the CCN ensemble in a static (red) and an embodied agent, driven through active inference (blue). The agent is provided with 20 situations for each of the 9 known objects and 5 never-before seen test objects. (b) Imagined views for actions that result in the largest and smallest expected free energy G. (Color figure online)

3.3 Embodied Agents for Improved Classification

Whereas previously we evaluated the binary classification performance of individual CCNs, we now evaluate the performance of the CCN ensemble as an $n + 1$-way classifier, with n object classes and one "other" class.

We evaluate an embodied agent that can query extra viewpoints to improve its classification. In this case, the agent again infers actions that minimize the expected free energy G, effectively maximizing information gain on the object identity. In this case, CCN votes are aggregated in the concentration parameters of a Dirichlet distribution, as described in Sect. 2. We also add a fixed 0.5 value vote for the "other" object category which accounts for evidence for the "other" class when none of the CCNs "fire". We compare this agent with a static agent that only has a single view. In this case, only a single vote is used for the parameters of the Dirichlet distribution.

To evaluate the performance, we randomly sample 20 views for each of 14 object classes (9 known object classes and 5 never seen before), and evaluate both the static and embodied classificaton accuracy. The results are shown in Fig. 4a. Whereas the static agent achieves an overall accuracy of 92.5%, the embodied agent consistently improves in accuracy, reaching 98%, as more viewpoints are queried, in line with [8]. The error bounds are computed over 5 different random seeds and represent the 95% HDI. Figure 4b shows the imagined observations for the largest and smallest expected free energy G. Figure 4b shows imagined views with highest or lowest G, and illustrates that the active inference agent prefers observations where the object is clearly in view from a more close up view, rather than more ambiguous viewpoints.

4   Related Work

Deep learning has been widely used for static image classification [10,15]. However, recent work also focused on active vision. In [26] a generative model learning representations of a whole 3D scene was used for an active inference agent, whereas in [2] an explicit what and where stream were modeled for classifying MNIST digits.

Recently a lot of progress has been made in methods that learn the 3D geometry of objects. The geometry can either be learned implicitly using Neural Radiance Fields (NeRF) [16] or Generative Query Networks (GQN) [3] or explicitly using Scene Representation Networks (SNR) [24]. However these approaches either have to generalize to a large variety of objects, which results in an involved training process requiring a lot of data, or they optimize for a single observation, limiting the flexibility.

Continual learning methods are able to use experience gathered during deployment of a system to improve the system over time. Typical approaches involve an ensemble of classifiers, that operate on a subset of the inputs, either by splitting the train data in specific subsets to train a mixture of experts [12], or by identifying clusters in a shared latent space and training separate classifiers for separate clusters [23].

The use of information gain has also been used as an exploration strategy outside of the active inference community, in which it substantially improves exploratory performance on a number of Atari tasks [17].

While most approaches tackle these problems separately, we propose a biologically inspired method that learns object-centric representations in an unsupervised manner for both the object identity and its geometric properties.

5   Discussion

We believe that this is a first step towards manipulation of three dimensional objects, and plan to extend this work to a real-world robot setup. In this case, the robotic agent needs to make inferences on the object pose and identity that is present in the workspace. In case the object is identified, the agent is attracted to preferred observations, e.g. for grasping or manipulating the object. In case a different, novel object class is identified, a new CCN is instantiated and trained on these novel object views. In this case, we could infer the viewpoints that have a high information gain on model parameters in an active learning setting, which can also be written as an expected free energy term [5]. We can further extend the generative model to also take into account multiple objects in the scene and modelling inter-object relations and geometry.

A limitation of our current setup is that it can only deal with a single object in the center of the view. As multi-object scenes are ubiquitous in the real world, this is a natural direction for future work. We propose a solution in which the agent can divide its spatial attention on the observations by looking at the CCN activations at different patches on the observation. Once an object and its relative reference frame is found, these can be linked using a global, ego-centric reference frame of the agent [18]. This way, a hierarchical generative model of the whole workspace, composed of different object is constructed. These latent parameters can then be propagated over time through a predictive model, and can in that way deal with occlusions.

In principle, one could instantiate a hierarchy of CCNs, where higher level CCNs process the output of lower level CCNs, effectively modeling part-whole relationships. This is similar to Capsule Networks [21] and GLOM [11], and corresponds better with the 1000 brains theory [9]. However, given the limited scalability of state of the art implementations of such hierarchical approaches [21], we adopted CCNs that operate on the level that is most important for a robot operating in a workspace, i.e. the discrete object level.

We found that failure cases exist when CCNs incorrectly "fire" for unseen objects. This confusion occurs for some objects yielding a non-perfect classification score. We could further improve the system by also taking into account how good novel observations match our predictions in the past.

6 Conclusion

In this paper we showed a novel approach to modelling 3D object properties, drawing inspiration from current development in the Neuroscientific domain. We proposed to model a separate what- and where stream for each individual object and are able to use these models for object identification as well as implicit object pose estimation. We show that through embodiment, these models can aggregate information and increase classification performance. Additionally, we show that by following the free energy formulation, these module networks can be used for implicit pose estimation of the objects.

Acknowledgments. This research received funding from the Flemish Government (AI Research Program). Ozan Çatal was funded by a Ph.D. grant of the Flanders Research Foundation (FWO). Part of this work has been supported by Flanders Innovation & Entrepreneurship, by way of grant agreement HBC.2020.2347.

Appendix A   Neural Network Architecture and Training Details

The neural network is based on a variational autoencoder [14,19] consisting of an encoder and a decoder. The encoder ϕ_θ uses a convolutional pipeline to map a high dimensional input image (64x64x3) into a low dimensional latent distribution. We parameterize this distribution as a Bernouilli distribution representing the identity of the object and the camera viewpoint as a Multivariate Normal distribution with diagonal covariance matrix of 8 latent dimensions. The decoder ψ_θ then takes a sample from the viewpoint and is able to reconstruct the observation through a convolutional pipeline using transposed convolutions. In addition to a traditional variational autoencoder, we have a transition model χ_θ that transforms a sample from the viewpoint distribution to a novel latent distribution, provided with an action. This action is a 7D vector representing the translation as coordinates in and rotation in quaternion representation. The model architecture for encoder, decoder and transition models are shown in Table 1, Table 2 and Table 3, respectively.

The model is optimized end-to-end through the minimization of Free Energy as described in Eq. 2. The expectations over the different terms are approximated through stochastic gradient descent using the Adam optimizer [13]. As minimization of negative log likelihood over reconstruction is equivalent to minimization of the Mean Squared Error, this is used in practice. Similarly, the negative log likelihood over the identity is implemented as a binary cross-entropy term. We choose the prior belief over $\mathbf{v}$ to be an isotropic Gaussian with variance 1. The individual terms of the loss function are constrained and weighted using Lagrangian multipliers [20]. We consider only a single timestep during the optimization process. In practice this boils down to:

$$
\begin{aligned}
L_{FE} = &\lambda_1 \cdot L_{BCE}(\hat{i}, i) + \lambda_2 \cdot L_{MSE}(\psi_\theta(\hat{\mathbf{v}}_{t+1}), \mathbf{o}_{t+1}) \\
&+ D_{KL}[\ \underbrace{\chi_\theta(\mathbf{v}_t, \mathbf{a}_t)}_{q(\mathbf{v}_{t+1}|\mathbf{v}_t, \mathbf{a}_t, i)}\ \|\ \underbrace{\phi_\theta(\hat{\mathbf{o}})}_{p(\mathbf{v}_{t+1}|i, \mathbf{o}_t)}\ ]
\end{aligned}
\tag{4}
$$

where $\hat{i}$ is the prediction $\phi_\theta(\mathbf{o}_t)$ of the what-stream for the encoder, $\hat{\mathbf{v}}_{t+1}$ is a sample from the predicted transitioned distribution $\chi_\theta(\mathbf{v}_t, \mathbf{a}_t)$ and $\hat{\mathbf{o}}_{t+1}$ is the expected observation from viewpoint $\hat{\mathbf{v}}_{t+1}$, decoded through $\psi_\theta(\mathbf{v}_{t+1})$. The λ_i variables represent the Lagrangian multipliers used in the optimization process.

During training, pairs of observations $\mathbf{o}_t$ and $\mathbf{o}_{t+1}$ and corresponding action $\mathbf{a}_t$ are required. To maximize data efficiency, the equation is also evaluated for zero-actions using only a single observation, and reconstructing this directly without transition model.

Table 1. Neural network architecture for the image encoder. All strides are applied with a factor 2. The input image has a shape of $3 \times 64 \times 64$. The output of the convolutional pipeline is used for three different heads. The first predicts the mean of the distribution μ, the second head predicts the natural logarithm of the variance σ^2, for stability reasons and finally the third head predicts the classification output score c as a value between zero and one after activation through the sigmoid activation function.

Output label	Layer	Kernel size	# Filters
	Strided Conv2D	4	8
	LeakyReLU		
	Strided Conv2D	4	16
	LeakyReLU		
	Strided Conv 2D	4	32
	LeakyReLU		
	Strided Conv2D	4	64
	LeakyReLU		
h	Reshape to 128		
μ	Linear (input: h)		8
$\ln \sigma^2$	Linear (input: h)		8
c	Linear + Sigmoid (input: h)		1

Table 2. Neural network architecture for the image decoder. The input of this model is a sample drawn from the latent distribution, either straight from the encoder, or transitioned through the transition model. All transpose layers use a stride of two. The final layer of the model is a regular convolution with stride 1 and kernel size 1, after which a sigmoid activation is applied to map the outputs in the correct image range.

Layer	Kernel size	# Filters
Linear		128
Reshape to $128 \times 1 \times 1$		
ConvTranspose2D	5	64
LeakyReLU		
ConvTranspose2D	5	64
LeakyReLU		
ConvTranspose2D	6	32
LeakyReLU		
ConvTranspose2D	6	16
LeakyReLU		
Conv2D	1	3
Sigmoid		

Table 3. Neural network architecture for the transition model. The input from this model is an 8 dimensional latent code, concatenated with the 7-dimensional representation of the relative transform (position coordinates and orientation in quaternion representation). For stability reasons, the log-variance is predicted rather than the variance directly.

Output label	Layer	# Filters
	Linear	128
	LeakyReLU	
	Linear	256
	LeakyReLU	
	Linear	256
	LeakyReLU	
μ	Linear	8
$\ln \sigma^2$	Linear	8

Appendix B   Additional experimental details

In Table 4, the computed angular and translational distances for the 9 evaluated objects are shown. Figure 5 shows a sequence of imaginations for all 9 objects, the top row represents the ground truth input, the second row the reconstruction and the subsequent rows are imagined observations along a trajectory.

Table 4. The mean distance error in meters and angle error in radians for different objects of the YCB dataset [1] in our simulated environment. For each object 20 arbitrary target poses were generated over which the mean values are computed.

Object	Distance error (m)	Angle error (rad)
Chips can	0.00328 ± 0.00824	0.15997 ± 0.21259
Master chef can	0.00036 ± 0.00034	0.06246 ± 0.03844
Cracker box	0.00028 ± 0.00023	0.04659 ± 0.02674
Tomato soup can	0.00073 ± 0.00104	0.08653 ± 0.07021
Mustard bottle	0.00070 ± 0.00072	0.06351 ± 0.03818
Mug	0.00083 ± 0.00128	0.09098 ± 0.10232
Pudding box	0.00051 ± 0.00052	0.06190 ± 0.03843
Banana	0.00055 ± 0.00042	0.07482 ± 0.03592
Strawberry	0.00573 ± 0.01181	0.16699 ± 0.15705

Fig. 5. The top row represents the ground truth observation that was provided as input to the model. The second row shows a direct reconstruction when no action is applied to the transition model. All subsequent rows show imagined observations along a trajectory.

References

1. Calli, B., Singh, A., Walsman, A., Srinivasa, S., Abbeel, P., Dollar, A.M.: The YCB object and model set: towards common benchmarks for manipulation research. In: 2015 International Conference on Advanced Robotics (ICAR), pp. 510–517 (2015)
2. Daucé, E., Perrinet, L.: Visual search as active inference. In: IWAI (2020). https://doi.org/10.1007/978-3-030-64919-7_17
3. Eslami, S.M.A., et al.: Neural scene representation and rendering. Science **360**(6394), 1204–1210 (2018)
4. Ettlinger, G.: Object vision and spatial vision: the neuropsychological evidence for the distinction. Cortex **26**(3), 319–341 (1990)
5. Friston, K., FitzGerald, T., Rigoli, F., Schwartenbeck, P., O'Doherty, J., Pezzulo, G.: Active inference and learning. Neurosc. Biobehav. Rev. **68**, 862–879 (2016)
6. Gilmer, J., Adams, R.P., Goodfellow, I.J., Andersen, D.G., Dahl, G.E.: Motivating the rules of the game for adversarial example research. CoRR, abs/1807.06732 (2018)
7. Hadsell, R., Rao, D., Rusu, A.A., Pascanu, R.: Embracing change: continual learning in deep neural networks. Trends Cogn. Sci. **24**(12), 1028–1040 (2020)
8. Hawkins, J., Ahmad, S., Cui, Y.: A theory of how columns in the neocortex enable learning the structure of the world. Front. Neural Circ. **11**, 81 (2017)
9. Hawkins, J., Lewis, M., Klukas, M., Purdy, S., Ahmad, S.: A framework for intelligence and cortical function based on grid cells in the neocortex. Front. Neural Circ. **12**, 121 (2019)

10. He, K., Zhang, X., Ren, S., Sun, J.: Deep residual learning for image recognition. In: 2016 IEEE Conference on Computer Vision and Pattern Recognition (CVPR), pp. 770–778, June 2016
11. Hinton, G.E.: How to represent part-whole hierarchies in a neural network. CoRR, abs/2102.12627 (2021)
12. Jacobs, R.A., Jordan, M.I., Nowlan, S.J., Hinton, G.E.: Adaptive mixtures of local experts. Neural Comput. **3**(1), 79–87 (1991)
13. Kingma, D.P., Ba, J.: A method for stochastic optimization, Adam (2017)
14. Kingma, D.P., Welling, M.: Auto-encoding variational Bayes. arXiv preprint arXiv:1312.6114 (2013)
15. Krizhevsky, A., Sutskever, I., Hinton, G.E.: ImageNet classification with deep convolutional neural networks. In: Proceedings of the 25th International Conference on Neural Information Processing Systems - vol. 1, NIPS 2012, pp. 1097–1105, Red Hook, NY, USA (2012). Curran Associates Inc
16. Mildenhall, B., Srinivasan, P.P., Tancik, M., Barron, J.T., Ramamoorthi, R., Ng, R.: NeRF: representing scenes as neural radiance fields for view synthesis. In: Vedaldi, A., Bischof, H., Brox, T., Frahm, J.-M. (eds.) ECCV 2020. LNCS, vol. 12346, pp. 405–421. Springer, Cham (2020). https://doi.org/10.1007/978-3-030-58452-8_24
17. Nikolov, N., Kirschner, J., Berkenkamp, F., Krause, A.: Information-directed exploration for deep reinforcement learning (2019)
18. Parr, T., Sajid, N., Da Costa, L., Mirza, M.B., Friston, K.J.: Generative models for active vision. Front. Neurorobot. **15**, 34 (2021)
19. Rezende, D.J., Mohamed, S., Wierstra, D.: Stochastic backpropagation and approximate inference in deep generative models (2014)
20. Rezende, D.J., Viola, F.: Taming vaes (2018)
21. Sabour, S., Frosst, N., Hinton, G.E.: Dynamic routing between capsules. CoRR, abs/1710.09829 (2017)
22. Safron, A.: The radically embodied conscious cybernetic Bayesian brain: from free energy to free will and back again. Entropy **23**(6), 783 (2021)
23. Shanahan, M., Kaplanis, C., Mitrovic, J.: Encoders and ensembles for task-free continual learning. CoRR, abs/2105.13327 (2021)
24. Sitzmann, V., Zollhöfer, M., Wetzstein, G.: Scene representation networks: continuous 3D-structure-aware neural scene representations. In: Advances in Neural Information Processing Systems (2019)
25. Smith, R., Friston, K., Whyte, C.: A step-by-step tutorial on active inference and its application to empirical data, January 2021
26. Van de Maele, T., Verbelen, T., Çatal, O., De Boom, C., Dhoedt, B.: Frontiers in Neurorobotics **15**, 14 (2021)

Rule Learning Through Active Inductive Inference

Tore Erdmann[1]([✉]) and Christoph Mathys[2,3]

[1] SISSA, Via Bonomea 265, 34356 Trieste, Italy
terdmann@sissa.it
[2] Interacting Minds Center, Aarhus University, Jens Chr. Skous Vej 4,
8000 Aarhus, Denmark
[3] Translational Neuromodeling Unit (TNU), Institute for Biomedical Engineering,
University of Zurich and ETH Zurich, Wilfriedstrasse 6, 8032 Zurich, Switzerland

Abstract. We propose a grammar-based approach to active inference based on hypothesis-driven rule learning where new hypotheses are generated on the fly. This contrasts with traditional approaches based on fixed hypothesis spaces and Bayesian model reduction. We apply these two contrasting approaches to an established active inference task and show that grammar-based agents' performance benefits from the explicit rule representation underpinning hypothesis generation. Our proposal is a synthesis of the active inference framework with language-of-thought models, which paves the way for computational-level descriptions of false inference based on an aberrant hypothesis-generating process.

Keywords: Active inference · Rule induction · Context free grammars · Structure learning · Sampling-based inference · Reasoning

1 Introduction

Structure learning is a fundamental problem for an active inference agent. Logically structured concepts can be found in domains such as mathematics, social systems or causal processes [13]. The likelihood mapping of a POMDP with discrete state space can be represented as a matrix with elements indicating the likelihood of an observation given a state. Current approaches for learning this mapping rely on separately estimating the individual elements of the matrix [4,12]. Here, we propose an approach for structure learning that uses a prior based on context-free grammars (CFG; [2]), which were invented in linguistics to describe the structure of sentences in natural language and are used to define programming languages in computer science. From such a grammar, the agent can, through recursive composition and substitution of terms, generate an infinite number of expressions, which represent the underlying structure of (parts of) its environment. As a proof of concept, we will illustrate our approach by applying it to a rule learning problem inspired by the task in [4].

This approach has previously been used in cognitive science, psychology [6, 13,14] and, in particular, in "language of thought" models [10]. Previous work

© Springer Nature Switzerland AG 2021
M. Kamp et al. (Eds.): ECML PKDD 2021 Workshops, CCIS 1524, pp. 715–725, 2021.
https://doi.org/10.1007/978-3-030-93736-2_51

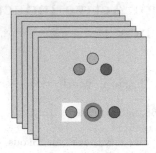
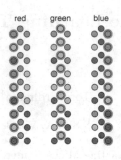

Fig. 1. Left: shows the display during a trial. The agent sees context variables (the three circles in the upper half), makes a response (indicated by the white box around the red circle) and, having made a choice, the correct choice (highlighted in green). Right: The possible contexts arranged according to the value of the middle circle, which implies where to look for the correct choice (highlighted in green). The correct response is equal to the color of the circle on the left, in the center or on the right when the color of the central circle is "red", "green" or "blue", respectively. (Color figure online)

has shown that these models can account for various features of human concept learning ([5]). Furthermore, this approach has been used to explain surprise signals in the striatum [1].

We will work through a simplified version of the task of [4]. For ease of presentation and to place our focus on structure learning, we remove state uncertainty and all intra-trial actions except for the final choices. All remaining uncertainty is thus about the hidden rule. However, our proposal can be straightforwardly applied to the case including state uncertainty and observations corrupted by noise. In this task, see Fig. 1, the agent has to infer a rule, that is a deterministic mapping from three context variables to the correct choice.

2 Active Inference

Solving this task consists of finding a policy $p(a_t|c_t, \theta)$, that gives the probability of a choice $a_t \in \{1, 2, 3\}$ given the context variables c_t and some parameters θ. The generative model the agent holds of the task is

$$c_t^{(j)} \sim U(\{1, 2, 3\}), \ j = 1, 2, 3 \tag{1}$$

$$c_t = (c_t^{(1)} c_t^{(2)}, c_t^{(3)}) \tag{2}$$

$$o_t \sim f(c_t, \cdot) \tag{3}$$

$$p(r_t = 1|a_t, o_t) = \exp(\ell(a_t, o_t)) \tag{4}$$

$$\ell(a_t, o_t) = \begin{cases} 0 & \text{if } a_t = o_t \\ -4 & \text{else} \end{cases} \tag{5}$$

where $f(c, o)$ is a function representing the hidden rule. That is, it returns the probability of observing the outcome $o \in \{1, 2, 3\}$ in context c. The prior about reward observations $p(r_t|a_t, o_t)$ represents an optimistic bias, so that the agent's beliefs are biased by desirable states and not the actual task dynamics, which is $r_t = \mathbb{I}(o_t = a_t)$. This model implies a distribution over trial sequences, which we denote $\tau = (c_{1:T}, a_{1:T}, o_{1:T}, r_{1:T})$, that factorizes as

$$p(\tau|f) = \prod_{t=1}^{T} p(r_t|a_t, o_t)p(o_t|c_t, f)p(a_t)p(c_t). \tag{6}$$

Given the biased prior over rewards we obtain the following posterior over actions when conditioning on $r_t = 1$, $\forall t = 1, \ldots, T$ and summing out o_t, which is unknown at the time of the action,

$$p(c_{1:T}, a_{1:T}, o_{1:T}|r_{1:t} = 1, f) \propto \prod_{t=1}^{T} p(r_t = 1|a_t, o) \tag{7}$$

In keeping with the active inference framework, the expected log model evidence is minimized by computation of the posterior over action, which can be done at each trial t by choosing

$$p(a_t|r_t = 1, c_{1:t}, o_{1:t-1}) = \sigma(-G_{a_t}) \tag{8}$$

$$G_{a_t} = \sum_o l(a_t, o) \cdot \mathbb{E}_{p(f|c_{1:t-1}, o_{1:t-1})}\left[p(O_t = o|c_t, f)\right] \tag{9}$$

For the implementation, this means we need to be able to evaluate the agent's posterior predictive about the belief about the outcome o_t. The above constructions leads to the maximization of the following objective (see [7])

$$D_{KL}(p^*(\tau)||p(\tau)) = \mathbb{E}_{\tau \sim p^*(\tau)}\left[\log p(\tau) - \log p^*(\tau)\right], \tag{10}$$

which is the Kullback-Leibler divergence of the agent's beliefs about its future states and a desired distribution over these p^*, and which is equivalent to the free energy of the expected future, which is a lower bound on the expected log model evidence [8].

2.1 Evidence Accumulating Agent

A straightforward solution for learning the rule is available if we represent it as a stochastic vector consisting of independent Dirichlet variables, $f(c, o) = \theta_{c,o}$, with $\theta_{j,\cdot} \sim Dir(\alpha_0)$, $j = 1, \ldots, 27$, for which the posterior can be computed by accumulation of concentration parameters:

$$p(\theta_{j,o}|c_{1:t}, o_{1:t}) = Dir(n_{c,o} + \alpha_0) \tag{11}$$

where $n_{c,o}$ is the number times (up until time t) the agent has observed outcome o for context c. If we define a matrix α with entries $\alpha_{c,o} = n_{c,o} + \alpha_0$, the expectation in Eq. 8 is a that of a categorical-Dirichlet distribution and the action is chosen via

$$G_{a_t} = \sum_o \ell(a_t, o)) \cdot \frac{\alpha_{c_t, o}}{\sum_j \alpha_{c_t, j}}. \tag{12}$$

2.2 Bayesian Model Reduction

If the agent knows that there must be a deterministic rule, it can quickly recognize the rule by comparing the evidence for each potential model in a set of hypothetical models and accept a model if its evidence exceeds a certain threshold.

The model space can be considered the set of deterministic, one-to-one mappings from each color to each response (of which there 6) which are combined with the 6 possible mappings between the central color and which location the color-to-response mapping should be applied to (see [4]). There are thus 36 hypotheses, for which the evidence is computed on each trial. This allows us to represent the priors through sets of prior concentration parameters as derived in [4]. A condition for this agent is that the space of hypotheses is specified for the agent beforehand, which is a strong assumption in general. We will now introduce a way to model acquisition of new models. This has the advantage of being based on weaker assumptions about (and a different conception of) prior knowledge.

3 Grammar-Based Rule Induction

Here, we describe how rule learning can be supported through a structured prior over an auxiliary space of symbolic rule expressions. Each such rule expression is defined by a syntax tree, consisting of logical connectives (and, or), and references to the observations in a trial. The "leaf nodes" of the tree are predicates of some part of the observation c_t, for example $color(c_t^{(1)}) = red$, which is either true or false (see appendix for an example). An agent can learn a rule expression that accurately predicts the outcome of the unknown rule f by searching the space of rule expressions for hypotheses which are then evaluated against the available evidence. Hypotheses are represented by expressions that can be generated by iterating the following set of re-write (or production) rules:

$$\text{(Start)} \ S \rightarrow f(c, o) \iff (D)$$
$$\text{(Disjunction)} \ D \rightarrow C \vee D \ | \ P \ | \ false$$
$$\text{(Conjunction)} \ C \rightarrow P \wedge C \ | \ P \ | \ true$$
$$\text{(Predicate)} \ P \rightarrow color(Loc) = Col$$
$$\text{(Location)} \ Loc \rightarrow c_1 \ | \ c_2 \ | \ c_3$$
$$\text{(Color)} \ Col \rightarrow \text{"red"} \ | \ \text{"green"} \ | \ \text{"blue"}$$

These rules indicate how symbols on the left hand side of the $\rightarrow$ can be replaced by one of the options on the right hand side (options are separated by |). From this grammar, given certain production probabilities (which give the probability of each possible production for each line in the grammar; can be assumed uniform), we can generate rule expressions (we refer the interested reader to Wikipedia, for examples, or [11] for a comprehensive treatment). Note that we omit the trial index t in the formulas (since the rules only refers to variables in the current trial) and instead use the subscript to denote the location (1, 2 or 3) of the context variable.

Each generated expression describes some arrangement of context observations. Say, we wanted to describe the rule for when the correct color is red (as given in the caption of 1). This can be expressed as $color(c_2) =$ "red" $\wedge\ color(c_1) =$ "red") $\vee\ (color(c_2) =$ "$blue$" $\wedge\ color(c_3) =$ "red"), which can be generated through step-wise replacement of the above rules. The prior probability of a formula (i.e. a sequence of substitutions from the grammar) is equal to the product of the probabilities of the individual substitutions. This prior naturally places higher probability on shorter and less complex expressions since they include fewer terms in the product.

For the rule learning task described above, we want to model the contexts that correspond to the three outcomes (and actions), so we will make the procedure to be learned a function of both the observed context c and the outcome o, changing the rule in the topmost line above to be a context-sensitive expression of the form

$$S \rightarrow f(c, o) \iff$$
$$((o = \text{``}red\text{''}) \wedge D) \vee ((o = \text{``}green\text{''}) \wedge D) \vee ((o = \text{``}blue\text{''}) \wedge D),$$

wherein the D terms will come to represent the parts of the rule that imply the corresponding outcome. We can then evaluate expressions with regard to each possible outcome to determine if the context c matches the outcome o. Starting from the above expression and generating sub-expressions according to the above grammar, we can represent the true hidden rule described in Fig. 1 as follows:

$$f(c, o) \iff ((o = \text{``}red\text{''}) \wedge$$
$$((color(c_2) = \text{''}red\text{''} \wedge color(c_1) = \text{''}red\text{''}) \vee$$
$$(color(c_2) = \text{''}blue\text{''} \wedge color(c_3) = \text{''}red\text{''})))$$
$$\vee ((o = \text{``}green\text{''}) \wedge$$
$$((color(c_2) = \text{''}green\text{''}) \vee (color(c_2) = \text{''}red\text{''} \wedge color(c_1) = \text{''}green\text{''}) \vee$$
$$(color(c_2) = \text{''}blue\text{''} \wedge color(c_3) = \text{''}green\text{''})))$$
$$\vee ((o = \text{``}blue\text{''}) \wedge$$
$$((color(c_2) = \text{''}red\text{''} \wedge color(c_1) = \text{''}blue\text{''}) \vee$$
$$(color(c_2) = \text{''}blue\text{''} \wedge color(c_3) = \text{''}blue\text{''})))$$

However, we can represent this rule more succinctly by adding more abstract terms to the grammar. For example, by adding two new production rules to the grammar above:

$$P \rightarrow color(Loc) = COL \quad | \quad o = color(Loc)$$
$$Loc \rightarrow c_1 \quad | \quad c_2 \quad | \quad c_3 \quad | \quad c_{Loc}$$

The last production will lead to a "subsetting", such as c_{c_2}, which means that the value of c_2 indexes the context variables (with the colors mapped to the numbers $\{1, 2, 3\}$). The expression $o = color(Loc)$ evaluates to true if the outcome matches the variable Loc. With these additions, we can now represent the true rule as a much shorter expression

$$f(c, o) \iff (o = color(c_{c_2})). \tag{13}$$

This shorter representation of the rule helps the agent to discover it much more quickly. This is because shorter rules have higher prior probabilities of being produced.

The above rule expression defines a function that evaluates to `true` if the action a is correct given the observation o and `false` otherwise. The likelihood of this expression is given by its match with the observed data, that is, the number of examples for which the rule f evaluates to true,

$$p(f|o_{1:t}, a_{1:t}, c_{1:t}) \propto \bigwedge_{c, o} f(c, o) \tag{14}$$

or, if assuming that some observations might be outliers to the rule, we have

$$p(f|o_{1:t}, a_{1:t}, c_{1:t}) \propto e^{-\gamma Q(f)} \tag{15}$$

where $Q(f) = |\{(c, o) \in (c_{1:t}, o_{1:t}) : f(c, o) = false\}|$ (the count of examples for which the rule expression evaluates to false) and γ is a parameter denoting the probability that a given example is an outlier. Here, the probabilities need not be normalized, since any normalization constants cancel in the MCMC acceptance probability. The truth value of the procedure $f(a, o)$ follows from the evaluation approach in mathematical logic [3] and is defined recursively:

1. $f(a, o)$ is a node.
2. If a node is a predicate, it can be evaluated directly
3. If it is a logical connective then it is evaluated by first evaluating the sub-expressions separately and then applying the logical function to the result. For example, $a \wedge b$ is true only if both sub-expressions a and b are true.

In our implementation, we represent the agent's belief about the correct rule expression as a set of samples that are approximately distributed according to the posterior distribution implied by the above likelihood and prior. This posterior is updated on each trial by running a Markov Chain Monte Carlo (MCMC) chain for a fixed number of iterations. The set of expressions that was visited during the walk is taken to represent the posterior belief. This construction leads to the posterior predictive distribution, given a set H_t of hypotheses. Formally, if we denote the chain representing the belief update in trial t by $H^{(t)} = \left(h_1^{(t)}, \ldots, h_n^{(t)} \right)$, we can evaluate the posterior expectation in action selection in Eq. 8 approximately as follows

$$p(a_t = a | r_t = 1, c_{1:t}, o_{1:t-1}) = \sigma(-G_{a_t}) \tag{16}$$

$$G_{a_t} = \sum_o l(a_t, o) \cdot \mathbb{E}_{p(f|c_{1:t-1}, o_{1:t-1})} \Big[p(O_t = o | c_t, f) \Big] \tag{17}$$

$$\approx \sum_o \ell(a_t, o)) \cdot \frac{\sum_i f_{h_i^{(t)}}(c_t, a)}{\sum_{j \in \{1,2,3\}} \sum_i f_{h_i^{(t)}}(c_t, j)} \tag{18}$$

which can be seen as a model average of all hypotheses that were visited by the Markov chain during the computation of the posterior.

The iterations of the MCMC procedure propose changes to the expression by randomly selecting a sub-expression and replacing it with a newly generated sub-expression. The Metropolis-Hastings acceptance probability for a proposal balances the probability of the proposal and the reverse proposal, the prior probabilities and the likelihood (see Eq. 14) of the current and proposed expressions (tree-substitution MCMC; see [5] for details). The belief update can thus be performed by running n MCMC iterations, starting from the current state of the chain. For the current task, once the true rule has been found, proposal for moves away from it will have very low probability. In general, when the rule cannot be known with certainty, the chain will move between alternatives and thereby lead to a representation of the remaining uncertainty in the posterior belief about the rule.

4   Experiments

We simulated learning in four agents who completed 20 trial sequences each. These sequences contained 27 trials and were generated by randomly shuffling the 27 unique combinations of context variables. The four agents differed in substantial ways and could be characterized as concentration parameter accumulating agents (Agents 1 and 2, described in Sect. 2.1) with (Agent 2) and without (Agent 1) model-selection (by Bayesian model reduction, Sect. 2.2) after each trial; and the grammar-based agents (Agents 3 and 4) with the simple grammar described in Sect. 3 (Agent 3) and an extended grammar described below (Agent 4).

A comparison of the performance of the different agents are shown in Fig. 2, where the average proportion of correct responses is shown over trials. As can be seen, the grammar-based agents show higher proportions of correct responses already during early trials. This is due to the nature of the rule expressions, which can be extrapolated from rapidly.

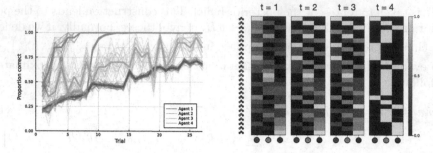

Fig. 2. Left: Proportion of correct choices (averaged over simulations) for the four agents with uncertainty indicated via bootstrapped estimates (thin lines). Right: Belief of (purple) grammar-based agent for the first 4 examples of a particular trial sequence. Each heatmap shows the probability of an action (x-axis) to be correct in a given context (y-axis). (Color figure online)

The best-performing agent (purple in Fig. 2) is Agent 4 with the extended grammar, that has two additional production rules contained in its grammar (see sec. 3). Figure 2 (right) shows the Agent 4's belief about the rule during the early trials of a particular trial sequence. The examples presented to the agent were $((\bullet, \bullet, \bullet), (\circ, \bullet, \bullet), (\bullet, \bullet, \bullet), (\bullet, \circ, \bullet))$, for which the correct responses were $(\bullet, \circ, \bullet, \circ)$. We can inspect the set of hypotheses held by the agent. At $t = 3$, the hypotheses with the highest weights (about $n/3$ occurrences) are:

1. $a = color(o_1)$, "answer equal to the left circle"
2. $a = color(o_{o_2})$ "answer equal to the color at location indicated by o_2"
3. $a = color(o_{o_3})$ "answer equal to the color at location indicated by o_3"

The agent cannot tell between these explanations until observing the outcome in the 4th trial, when the predictions of hypotheses 1 and 3 are disproved and the agent correctly infers the rule.

These results show how learning speed relates to underlying assumptions. As opposed to Agents 1 and 2, who need to be equipped with a fixed hypothesis set, the grammar-based agents can learn arbitrary rules, including such for which maintaining a fixed hypothesis set would be infeasible, as long as they can be represented within the language spanned by their grammar. For example, if we had just told the agent: "In this game, there is a deterministic mapping between the three colored circles and the correct response", the hypothesis space would have to cover a space of mappings containing 2^{81} elements. Comparing each of these candidates at the end of a trial would be infeasible (at a rate of 10^9 evaluations per second (1 evaluation per nanosecond), it would take about 77 million years to evaluate all candidates). The code for all experiments reported here is available at https://github.com/ilabcode/IWAI2021.

5   Discussion

We have shown a novel way to perform structure learning in active inference agents. In particular, we demonstrate how an agent can use grammar-based structure learning to develop a model in a bottom-up fashion. This is different from the traditional approach of Bayesian model reduction, which can be considered a top-down approach. The assumption of a grammar that spans a hypothesis space is weaker and hence more generalizable than pre-defining a finite set of hypotheses. Other ways of searching for rule expressions are possible, such as genetic algorithms, but these do not represent uncertainty and are therefore not well suited as a basis for adaptive prediction and decision-making.

Our results showed differences between the two grammar-based agents that were apparent in the speed by which they learn the rule. For the task presented here, both agents converge to the same behavior, but their underlying rule representations are different. This highlights how higher-order inferences can depend on the base of concepts and abstractions they are built upon. In terms of the behavior, the agents will look the same, however, their representational vocabulary differs and so they will find separate explanations for the rule (which do describe the same contingencies), which also have different complexity (as clearly visibly in the number of terms). Given a way to update their own grammars through experience, two agents starting with different grammars but in similar environments might develop a similar conceptual toolbox. One way to enable this would be to add special "lambda expression" terms to the grammar. Such an encoding of the lambda calculus within the hypothesis language leads to the ability to define new terms and apply or re-combine them (see [9]).

An interesting aspect of your hypothesis-generating grammar-based approach is the ways in which the assumptions underlying the generation of hypotheses of can influence what the agent finally takes to be the most promising course of action. This can become a useful tool for understanding aberrations in world modeling such as those apparent in psychiatric illnesses, which might have to do with a deficient hypothesis-generating process. For example, hypotheses generated from a grammar that is poorly attuned to a domain can seem bizarre to outside observers. Such misattunement may be the result of aberrant learning processes that update the production probabilities of a grammar, or the addition or removal of terms.

The agent described in [4] did not include model-selection considerations in its actions since they were outside of its generative model (and, in any case, the actions in the task were uninformative in that regard). By contrast, with a grammar-based approach, the structure is part of the agent's prior. Therefore its actions can subserve the testing of freshly generated hypotheses about the hidden structure of a task, which corresponds to active learning. Crucially, this could be made relevant in a version of the rule learning task where the agent can choose its next set of context variables. This would require planning, where the agent finds the optimal plan for testing its currently most promising hypotheses—an interesting avenue for future research based on the approach introduced here.

Appendix

Context-free grammars

A context-free grammar is defined by a 4-tuple (V, Σ, R, S), with V a finite set of *variables*, Σ a finite set of *terminals*, R a set of *rules*, each of which consist of a variable and a string of variables and terminals, and $S \in V$ is the *start variable*. One can use a grammar to describe a language by generating strings of that language in the following manner.

1. Write down the start symbol.
2. Find a variable that is written down and a rule that starts with that variable. Replace the variable with the right-hand side of the rule.
3. Repeat step 2 until no variables remain.

$$V := \{S, C, D, P, Loc, Col, \wedge, \vee\},$$

Σ is $\{c_1, c_2, c_3, red, green, blue\}$ and rules are as given in Sect. 3. This grammar describes a basic programming language for expressions containing logical connectives and predicates. For example, the string $color(c_1) = red \wedge color(c_2) = green$ can be generated from the grammar. The sequences of substitutions to obtain a string is called a *derivation*. The derivation of the above example is shown in Fig. 3.

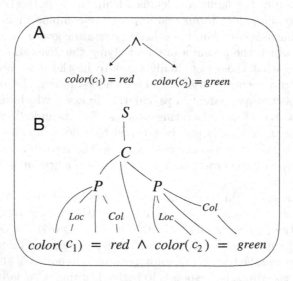

Fig. 3. A: Example of an expression that evaluates to true in the example from 1. B: Corresponding derivation from the language defined by the grammar in Sect. 3.

References

1. Ballard, I., Miller, E.M., Piantadosi, S.T., Goodman, N.D., McClure, S.M.: Beyond reward prediction errors: human striatum updates rule values during learning. Cereb. Cortex **28**(11), 3965–3975 (2018). https://doi.org/10.1093/cercor/bhx259
2. Chomsky, N.: Three models for the description of language. IRE Trans. Inf. Theory **2**(3), 113–124 (1956). https://doi.org/10.1109/TIT.1956.1056813
3. Enderton, H.B.: A Mathematical Introduction to Logic, 2nd edn. Harcourt/Academic Press, San Diego (2001)
4. Friston, K.J., Lin, M., Frith, C.D., Pezzulo, G., Hobson, J.A., Ondobaka, S.: Active inference, curiosity and insight. Neural Comput. **29**(10), 2633–2683 (2017). https://doi.org/10.1162/neco_a_00999
5. Goodman, N., Tenenbaum, J., Feldman, J., Griffiths, T.: A rational analysis of rule-based concept learning. Cogn. Sci. Multidisc. J. **32**(1), 108–154 (2008). https://doi.org/10.1080/03640210701802071
6. Kemp, C., Tenenbaum, J.B., Niyogi, S., Griffiths, T.L.: A probabilistic model of theory formation. Cognition **114**(2), 165–196 (2010). https://doi.org/10.1016/j.cognition.2009.09.003
7. Levine, S.: Reinforcement Learning and Control as Probabilistic Inference: Tutorial and Review. arXiv:1805.00909 [cs, stat] (2018)
8. Millidge, B., Tschantz, A., Buckley, C.L.: Whence the expected free energy? Neural Comput. **33**(2), 447–482 (2021). https://doi.org/10.1162/neco_a_01354
9. Piantadosi, S.T., Tenenbaum, J.B., Goodman, N.D.: Bootstrapping in a language of thought: a formal model of numerical concept learning. Cognition **123**(2), 199–217 (2012). https://doi.org/10.1016/j.cognition.2011.11.005
10. Piantadosi, S.T., Tenenbaum, J.B., Goodman, N.D.: The logical primitives of thought: empirical foundations for compositional cognitive models. Psychol. Rev. **123**(4), 392–424 (2016). https://doi.org/10.1037/a0039980
11. Sipser, M.: Introduction to the Theory of Computation. PWS Pub. Co., Boston (1997)
12. Smith, R., Schwartenbeck, P., Parr, T., Friston, K.J.: An active inference approach to modeling structure learning: concept learning as an example case. Front. Comput. Neurosci. **14** (2020). https://doi.org/10.3389/fncom.2020.00041
13. Tenenbaum, J.B., Kemp, C., Griffiths, T.L., Goodman, N.D.: How to grow a mind: statistics, structure, and abstraction. Science **331**(6022), 1279–1285 (2011). https://doi.org/10.1126/science.1192788
14. Ullman, T.D., Goodman, N.D., Tenenbaum, J.B.: Theory Acquisition as Stochastic Search, p. 6 (2010)

Interpreting Dynamical Systems
as Bayesian Reasoners

Nathaniel Virgo[1]([✉])(iD), Martin Biehl[2](iD), and Simon McGregor[3]

[1] Earth-Life Science Institute, Tokyo Institute of Technology, Tokyo 152-8550, Japan
nathanielvirgo@elsi.jp
[2] Araya Inc., Tokyo 107-6024, Japan
[3] University of Sussex, Falmer, UK
s.mcgregor@sussex.ac.uk

Abstract. A central concept in active inference is that the internal states of a physical system parametrise probability measures over states of the external world. These can be seen as an agent's beliefs, expressed as a Bayesian prior or posterior. Here we begin the development of a general theory that would tell us when it is appropriate to interpret states as representing beliefs in this way. We focus on the case in which a system can be interpreted as performing either Bayesian filtering or Bayesian inference. We provide formal definitions of what it means for such an interpretation to exist, using techniques from category theory.

Keywords: Bayesian filtering · Bayesian inference · Category theory

1 Introduction

A question of current interest is *what does it mean for a physical system to be an agent?* That is, given a physical system that interacts with an environment, when does it make sense to say that the system is *learning about its environment* or *trying to achieve a goal*, rather than merely being dynamically coupled to its environment? Here we confine ourselves to the first of these, in a simple form: given a physical system that is influenced by its surroundings, under what circumstances can it be said to be performing inference, such that its internal states could be said to contain 'knowledge' or 'beliefs' about the outside world?

Our approach has something in common with Dennet's intentional stance [17], in that on the one hand we treat the question of whether a system is performing inference as a matter of interpretation, but on the other hand we draw a strong connection between interpretations and the underlying physical dynamics. We provide a formal notion of interpretation for the particular cases we are interested in (Bayesian filtering and Bayesian inference), such that the question of *whether a system can be consistently interpreted in a particular way* is mathematically well-defined and has a definite answer.

© Springer Nature Switzerland AG 2021
M. Kamp et al. (Eds.): ECML PKDD 2021 Workshops, CCIS 1524, pp. 726–762, 2021.
https://doi.org/10.1007/978-3-030-93736-2_52

The question of how to identify agents in physical systems has been addressed in several ways. Some works focus on whether a system's actions can be seen as pursuing a goal [27,33], with [31] taking an explicitly Dennettian approach. Others focus more on the question of identifying which part of a system should be identified as the agent [2,5–7,28], or on understanding what the external world looks like from the agent's point of view [3,5,6]. Another approach, which we take here, is to regard the system's internal state as *parametrising* its beliefs. That is, there is a function mapping the system's physical state to a probability measure that can be seen as a Bayesian prior. This is a key component of work on the Free Energy Principle (FEP) [16,19,34] and also of [39], although our approach differs from these in that our model is not derived from the dynamics of the true environment. The notion that agency is closely related to parametrisation is also central to recent approaches to agency based on category theory [11,12,38].

The idea that states of a system parametrise Bayesian probability distributions appears more broadly in the Bayesian brain literature [26,30] and has also arisen in cell biology [29,32]. Our contribution is to make the concept much more formal, and in so doing, to shed light on the precise relationship between the interpretation level and the underlying physical level.

On a technical level we formulate the problem of Bayesian filtering at an abstract level using the tools of category theory. This part of the work is inspired by [23], which formulates the notion of conjugate prior in terms of category theory in a similar way. Conjugate priors are convenient because they ensure the functional form of the posterior is the same as the posterior, in Bayesian belief updating. At the same time they can be seen as a special case of Bayesian filtering, as we explain in Sect. 2.3 and Appendix B.1. Formulating filtering in this way allows us to clearly distinguish the role of the physical machine from the more semantic level at which we can talk about priors and posteriors. We then flip this perspective around, asking, for a given system, whether it can be interpreted as implementing Bayesian filtering, and if so under which model. In this respect our approach generalises that of [8], who studied the special case of the Dirichlet distribution (which is conjugate prior to a categorical distribution) in the context of interpreting a physical system as performing inference.

One thing our approach makes clear is that a given system may have more than one interpretation, and the "correct" interpretation cannot be determined from the system's dynamics alone. Another important aspect of our framework is that an interpretation only depends on the system's internal dynamics, and not on the dynamics of the external world. Because of this, a system's presumed beliefs might not match the true dynamics of the world at all—its beliefs might be consistent but incorrect—and indeed we can construct examples where the world "as the system sees it" has a different causal structure from the world as it really is. (Compare Eq. (1) to Eq. (7).)

2   Definitions and Results

2.1   Technical Preliminaries

In the following, we use the concepts of *measurable space* and *Markov kernel*. By measurable space we mean a set equipped with a σ-algebra, i.e. the kind of thing on which a probability measure can be defined. An example of a measurable space is a finite set, and a reader who is only interested in the finite case could mentally substitute "finite set" wherever we say "measurable space" and "probability distribution" wherever we say "probability measure."

Given a measurable space X, we write $P(X)$ for the set of all probability measures over X. Given measurable spaces X and Y, a Markov kernel is a function $\kappa\colon X \to P(Y)$ that maps elements of X to probability measures over Y, with an additional technical requirement that the function κ be measurable. Markov kernels are closely related to conditional probability, but they are different in that a Markov kernel defines a probability measure over Y for every element X, regardless of whether any probability distribution has been defined over X or what form such a distribution has. In the case where Y is a finite set, we write $\kappa(y\,|\,x)$ for the probability that the kernel κ assigns to y when given the input x.

We also make use of a graphical notation known as *string diagrams*, which comes from the literature on category-theoretic probability [13, 22]. This notation provides a convenient way to reason about how Markov kernels relate to one another. The full technical description of this calculus can be found in [22] or [13], but we provide a brief intuitive explanation in Appendix A, aimed at readers with no category theory background, along with references to further reading.

2.2   Machines and Interpretations

We are concerned with interpreting a physical system as performing inferences of some kind on its inputs. We therefore begin by defining a notion corresponding to a physical system that can take an input from the outside world, which leads to a change in state, which might be stochastic and might depend on the input.

Definition 1. *A machine consists of two measurable spaces, Y (the state space) and S (the input space), together with a Markov kernel $\gamma\colon Y \times S \to P(Y)$ called the update kernel.*

The idea is that the machine is in reality only a part of some larger stochastic process. We might typically think of this broader context as represented by the following causal Bayesian network, although this is not the only possibility.

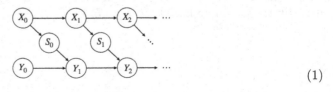

$$(1)$$

Here the variables $X_0, X_1, \ldots$ represent the states of the external world at different times, which are hidden from the machine's perspective. $S_0, S_1, \ldots$ are the observable "sensor values" that the machine can access, all of which have the same sample space, given by S. Similarly, $Y_0, Y_1, \ldots$ are the machine's internal state at each time step and each has the sample space Y. We assume that each of the nodes $Y_1, Y_2, \ldots$ in this network are associated with the same kernel, γ.

The nodes X_0 and Y_0 represent distributions over initial states of the agent and the world. A common ancestor of these nodes could be added, to represent the possibility that the initial states are correlated. There is no need for any stationarity assumption in our framework; the initial distribution in Eq. (1) can be arbitrary.

Even though we might think of the machine as existing in a context along these lines, our notion of interpretation does not depend on the machine's external environment at all, but *only* on the machine's internal dynamics. That is, on the measurable spaces Y and S and on the update kernel γ. This is because, informally speaking, a reasoner may have consistent *but wrong* beliefs about the external world, and we wish to include this possibility in our framework. (In particular, a system that reasons correctly in one environment might be placed in a different environment where the same inferences are no longer correct.) Our notion of interpretation will include a notion of "beliefs" about the external world. These must be consistent with the machine's internal dynamics and the inputs it receives, but they need not relate in any particular way to any ground truth about the process by which those inputs are generated, since we regard that as something the reasoner has no direct access to.

Because of this we will rarely reason about causal models directly, and instead will express our definitions directly in terms of Markov kernels and the relationships between them. The string diagram notation, explained in Appendix A, will be indispensable for this.

We now describe our central concept: an *interpretation* of a machine. Here we will present only two kinds of interpretations, *Bayesian filtering interpretations* and an important special case, *Bayesian inference interpretations*, but we expect these to fit naturally into a much broader family of concepts.

The most important component of an interpretation is what we term an *interpretation map*, a function that maps the physical state of a machine to something that we can think of as a belief about some external world. In the cases we are concerned with in this paper, a "belief" will be a probability measure over some hidden variable. An interpretation of a machine will be an interpretation map together with some additional data (the *model* as defined below), such that a *consistency equation* is obeyed.

For a given machine there may be many possible interpretations. We use the term *reasoner* for a machine together with a particular choice of interpretation.

In the case of Bayesian inference interpretations and Bayesian filtering interpretations, in which 'beliefs' are probability measures over a hidden variable, the interpretation map is a Markov kernel $\psi_H : Y \to P(H)$. Instead of interpreting this stochastically, we think of it as a function that takes a state $y \in Y$ and returns a probability measure $\psi_H(y)$ (the belief) over H. This is to be thought of as the reasoner's subjective knowledge (a Bayesian prior or posterior) about the hypotheses in H. This kernel plays quite a different role from those associated with the graphical model in Eq. (1), since its purpose is to map states to beliefs, rather than to model causal influences between random variables.

For an interpretation to be consistent, the reasoner's beliefs must update in the appropriate way when the machine receives new data. The precise meaning of this will depend on what kind of interpretation we are using. For the interpretations we describe here it is given by Bayes' rule, in a form that we state precisely below. In future work we can imagine interpretations based on other principles, such as approximate Bayes (e.g. via the free energy principle).

The idea is that a machine by itself is merely a (possibly stochastic) dynamical process, but if a consistent interpretation exists then it is at least consistent to ascribe a *meaning* to its states.

It should be noted that, for a given machine, the question of *whether it can be consistently interpreted in a particular way* is in principle an empirical one, since it depends on the machine's update kernel, which can in principle be measured. However, in general a given machine might have multiple non-equivalent consistent interpretations, and one cannot distinguish between these empirically by looking only at the system's internal dynamics.[1] Consequently the relationship between interpretations and the empirical, physical world is rather subtle, and one should keep in mind that our notion of "consistent reasoner" unavoidably involves an element of choice in which interpretation to adopt.

2.3 Consistency for Bayesian Interpretations

We begin with the more general case of Bayesian filtering interpretations. The idea is that a reasoner has a *model* of the environment's dynamics. This model is part of the interpretation, and need not match the true environment dynamics.

In the case of filtering, such a model can be described as a Markov kernel $H\!-\!\boxed{\kappa}\!\!\begin{smallmatrix}H\\S\end{smallmatrix}$, that is, $\kappa\colon H \to P(H \times S)$. The idea is that H is the space of possible hidden states of the external world, as modelled by the reasoner. The kernel κ models a step of this hypothetical external world's evolution, during which it both changes to a new state in H and also emits an observable sensor value in S.

A Bayesian filtering interpretation of a machine $\begin{smallmatrix}S\\Y\end{smallmatrix}\!\!\boxed{\eta}\!-\!Y$ will then consist of a choice of interpretation map $Y\!-\!\boxed{\psi_H}\!-\!H$ as described above, together with a choice of model $H\!-\!\boxed{\kappa}\!\!\begin{smallmatrix}H\\S\end{smallmatrix}$. The kernel κ thus describes a reasoner's beliefs about the next hidden state and the next sensor value, given the current hidden state.

Given the kernels ψ_H and κ we can define another kernel, which we also consider to be an interpretation map,

$$Y\!-\!\boxed{\psi_{S,H',H}}\!\!\begin{smallmatrix}H\\H\\S\end{smallmatrix} := \quad Y\!-\!\boxed{\psi_H}\!\!-\!\bullet\!\!\begin{smallmatrix}H\\ \\ \end{smallmatrix}\!\boxed{\kappa}\!\!\begin{smallmatrix}H\\H\\S\end{smallmatrix} \tag{2}$$

---

[1] We leave open the possibility that they could be distinguished by looking at some broader context, e.g. by discovering that a device's designer intended a particular interpretation, or that evolution selected for a particular interpretation.

The kernel $\psi_{H,H',S}$ maps a state of the machine $y \in Y$ to a joint distribution over S and two copies of H, which we think of as the reasoner's beliefs about the the next sensor value, the next hidden state, and the current hidden state. We also define its marginals,

$$Y\!-\!\boxed{\psi_{S,H'}}\!\begin{smallmatrix}-H\\-S\end{smallmatrix} \;:=\; Y\!-\!\boxed{\psi_{S,H',H}}\!\begin{smallmatrix}-H\\-S\end{smallmatrix} \;=\; Y\!-\!\boxed{\psi_H}\!-\!{}^{H}\!\boxed{\kappa}\begin{smallmatrix}-H\\-S\end{smallmatrix} \tag{3}$$

and

$$Y\!-\!\boxed{\psi_S}\!-\!S \;:=\; Y\!-\!\boxed{\psi_{S,H',H}}\!\begin{smallmatrix}\bullet\\-S\end{smallmatrix} \;=\; Y\!-\!\boxed{\psi_H}\!-\!{}^{H}\!\boxed{\kappa}\begin{smallmatrix}\bullet\\-S\end{smallmatrix}, \tag{4}$$

which represent the reasoner's beliefs about the next hidden state and the next input, and about only the next input, respectively. We can now state the consistency requirement for a Bayesian filtering interpretation.

Definition 2. *Given a machine* ${}^{S}_{Y}\!\boxed{\gamma}\!-\!Y$, *a consistent Bayesian filtering interpretation of* γ *is given by a measurable space* H *together with Markov kernels* $Y\!-\!\boxed{\psi_H}\!-\!H$ *and* $H\!-\!\boxed{\kappa}\!-\!{}^{H}_{S}$ *that satisfy*

$$\tag{5}$$

with $\psi_{S,H'}$ *and* ψ_S *given by Eqs.* (3) *and* (4).

The left-hand-side of Eq. (5) can be read as sampling from the reasoner's joint beliefs about the next hidden state and the next input, and then feeding the corresponding value of S into the machine as an input. The right-hand-side can be read as sampling from the reasoner's belief about its next input, feeding the result in as its next input, and then sampling from its resulting (posterior) belief about what it now sees as the current hidden state. The equation says that these two procedures must give the same result.

In Appendix B.1 we give some further intuition for this definition, by considering the case where S, Y and H are finite sets. An important consequence is that, in the finite case, whether a given interpretation is consistent or not only depends on which states are reachable from which other states under a given input; the actual transition probabilities are irrelevant beyond that. We expect an analogous statement to hold more generally.

Another important consequence discussed in Appendix B.1 is that there is a large class of machines that only admit trivial interpretations. At least in the finite case, non-trivial interpretations can only exist if some transitions are impossible, in the sense that there is a zero probability of transitioning from $y \in Y$ to $y' \in Y$ under the input $s \in S$. So if a finite machine has the property that every state always has a non-zero probability of being the next state, for any given input and initial state, then that machine can only have trivial interpretations.

In Appendix B.2 we give a more technical proof, using string diagrams, that at least in the case of deterministic machines, a machine with a consistent Bayesian filtering interpretation can indeed be regarded as performing a

Bayesian filtering task. This can be seen as extending some of the ideas in [24] on conjugate priors to the more general case of Bayesian filtering.

An important special case of Definition 2 is where the model κ is such that the hidden state does not change over time. In this case H can be thought of as an unknown parameter of a statistical model, with the sensor inputs being independent and identically distributed samples from the model. We call these *Bayesian inference interpretations*. Although the following follows from Definition 2 under this assumption, we write it as a separate definition.

Definition 3. *Given a machine $\underset{Y}{\overset{S}{\longrightarrow}}\gamma\!\!-\!Y$, a consistent Bayesian inference interpretation of γ is given by a measurable space H together with Markov kernels $Y\!-\!\psi_H\!-\!H$ and $H\!-\!\phi\!-\!S$ that satisfy*

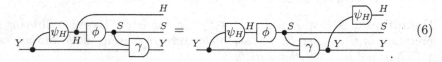

$$\tag{6}$$

In Appendix B.3 we show that this definition is closely related to the notion of a conjugate prior, and in particular to the definition of conjugate prior given by [24] in terms of Markov kernels and string diagrams. Finally, in Appendix B.4 we unpack Eq. (6) in more familiar terms, showing that in the discrete case it does indeed correspond to Bayes' theorem as usually understood.

In a Bayesian inference interpretation we interpret the reasoner as *assuming* its inputs are i.i.d. samples from some distribution, but this need not mean that they actually are. Under a consistent Bayesian inference interpretation a reasoner is interpreted as modelling the world as if its causal structure is as follows, where each of the 'S' nodes is associated with the kernel ϕ.

$$\tag{7}$$

However, the true dynamics of the world could still correspond to the Bayesian network in Eq. (1) or some other causal structure. The reasoner is simply (interpreted as being) unable to perceive the correlations among its inputs.

In Appendix C we present three examples of machines that have consistent Bayesian inference interpretations. The first example is a non-deterministic finite machine with three states. The consistent Bayesian inference interpretation we provide also involves subjectively impossible observations.

The second example is a countably infinite deterministic machine that counts occurrences of each of two possible observations. The consistent Bayesian interpretation we provide intuitively considers this machine to be inferring the bias of a coin that is flipped to cause its observations. It uses the standard conjugate prior to its model as an interpretation map.

The third example is also a countably infinite machine with two possible observations. However, it only stores the difference between the number of the two possible observations. Intuitively, instead of considering all possible biases of a coin as the second example the consistent Bayesian interpretation we provide for this machine infers which of two specific biases of a coin is causing its observations. We also hint at how the second machine can "inherit" this interpretation.

3   Discussion

We see the main contribution of this work as a conceptual one. The consistency equation involved in Definitions 2 and 3 can be seen either as a constraint on the machines that have a particular interpretation or as a constraint on the interpretations that a particular machine allows. Every machine has an interpretation with respect to a trivial model (one that has no parameter), but in order to have an interpretation with respect to a non-trivial model, a machine must at least obey the constraints discussed in Appendix B.1.

Similar consistency equations should exist for approximate Bayesian filtering as well as for related inference problems like Bayesian smoothing or prediction. Even non-Bayesian normative theories about what a system should be representing and how this should change under external influences will probably have associated consistency equations.

Another direction to extend the concept of consistency equations is to take into account a possible influence of the machine's state on the external world. On the interpretation side this would mean going beyond perception and representation to also include deliberate actions that combine with beliefs and possibly also with goals. A machine with such an interpretation might deserve the term *agent* instead of reasoner.

Our work is related to other current efforts to capture the notion of agent using category theory. These include approaches to Bayesian inference [37] and game theory [10]. The idea that agency is related to parametrisation has also arisen in these contexts [11, 12, 38]. These works focus on the notion of a *lens* and its generalisations. It is interesting to note that our notions of interpretation seem to be different, and somewhat simpler, in that they lack the bidirectional nature of lenses. We conjecture that a more lens-like bidirectional structure would be needed if we were to consider Bayesian smoothing rather than Bayesian filtering. It will be interesting in future work to better understand the relationship between lens-like categories and the concept of interpretation developed in the present work.

3.1   Relation to the Free Energy Principle

Let us now consider the relation to the Free Energy Principle (FEP) which is also referred to as active inference. The relevant part of FEP is the part called "Bayesian mechanics" [16, 19, 34]. It seems that the ingredients for an interpretation map can be found in this literature: [16, Eq. 3.3] describes a Markov kernel of an appropriate type, as we detail in Appendix D. However, it is not currently clear to us whether the FEP can be formulated in terms of a consistency equation that this kernel obeys. Presumably, such an equation would be different from our Definitions 2 and 3, because the FEP is concerned with approximate rather than exact inference and deals with continuous time.

One important difference between our approach and current formulations of FEP is that the FEP requires a stationarity assumption on the true dynamics of the agent-environment system. It seems to us that this is used to derive something that corresponds to a model. In our approach the reasoner's and the "ground truth" dynamics of the environment are different things, and partly for this reason we need no stationarity assumption. We see this conceptual separation as an advantage of the consistency equation approach, and we believe that by incorporating these ideas it might be possible to formulate the FEP in a way that would make its assumptions clearer and perhaps even avoid the need for the stationarity assumption. Although we do not currently know the precise relationship between our work and the FEP at a technical level, we explore it in more detail in Appendix D.

Acknowledgements. The work by Martin Biehl on this publication was made possible through the support of a grant from Templeton World Charity Foundation, Inc. The opinions expressed in this publication are those of the authors and do not necessarily reflect the views of Templeton World Charity Foundation, Inc. Martin Biehl is also funded by the Japan Science and Technology Agency (JST) CREST project.

A   Category-Theoretic Probability and String Diagrams

In this paper we use some concepts from category-theoretic probability, and in particular we use a notation known as string diagrams. A full introduction to these topics would be out of scope of the paper, but we include here an informal introduction to the topic. We do this because, to our knowledge, no concise introduction currently exists that is focused on (classical) probability and does not assume a background in category theory. We assume that the reader knows the definition of a category, but not much more than that.

Appendix A.1 introduces the basic concepts, mostly in the context of discrete probability. In Appendix A.2 we briefly comment on how this extends to the general case of measure-theoretic probability with very little extra work. In Appendix A.3 we explain how to reason about conditional probabilities and Bayes' theorem within this category-theoretic context.

These sections contain no original material. Their purpose is to give the reader enough information to be able to read the string diagram equations in the main text and later sections of the Appendix without needing to consult a category theory text. However, they are intended neither as an authoritative technical reference nor as a comprehensive review, and readers should consult the cited references for full details.

A.1   Introduction to String Diagrams and Category-Theoretic Probability

A full technical introduction to the use of string diagrams in probability can be found in [22] or the earlier [13], but these works require some knowledge of category theory. The string diagram notation predates its use in probability and has many other applications. One could consult [4,14,15,18] for tutorial introductions to diagrammatic reasoning in other fields, of various different flavours. Here we present it somewhat informally and only in the context of probability.

It should be kept in mind that, despite our somewhat informal introduction, string diagrams are formal expressions. The main difference between them and the more familiar kind of mathematical expression formed from strings of symbols is their two-dimensional syntax. This makes it easier to express certain concepts. (Particularly those relating to joint distributions, in the case of probability.)

We use the so-called *Markov category* approach to probability [22]. The main idea here is to express everything in terms of *measurable spaces* and *Markov kernels*, whose definitions we outlined in the main text.[2] To explain how the framework works, let us consider the special case where the only measurable spaces we are interested in are finite sets (with their power sets as their σ-algebras). If A and B are finite sets then a Markov kernel can be thought of as just a function $f: A \to P(B)$, where $P(B)$ is the set of all probability distributions over B. (The set $P(B)$ may be thought of as a $(|B| - 1)$-dimensional simplex, consisting of all those vectors in $\mathbb{R}^{|B|}$ whose components are all non-negative and sum to 1.) Such a function amounts to a $|B|$-by-$|A|$ stochastic matrix, although some care needs to be taken over which rows correspond to which elements of B and which columns to which elements of A.

In this finite case, we write $f(b\,|\,a)$ to denote the probability that the kernel f assigns to the outcome $b \in B$ when given the input $a \in A$. We use a thick vertical line to indicate a close relationship to conditional probability while also emphasising that the concept is different: given a kernel $f: A \to P(B)$ the quantities $f(b\,|\,a)$ are always defined, regardless of whether any probability distribution has been defined over A, and regardless of whether a has a nonzero probability according to such a distribution. More common notations include $|$ or ; in place of $|$.

We also write $f(a)$ for the probability distribution over B that the function f returns when given the input a. We could say that $f(b\,|\,a)$ is defined as $f(a)(b)$.

Given Markov kernels $f: A \to P(B)$ and $g: B \to P(C)$, we can compose them to form a new kernel of type $A \to P(C)$. We write this $f \,\mathbin{\substack{\circ \\ \circ}}\, g$. It is given by

---

[2] In fact for most of the paper we will work much more abstractly than this. It would be more correct to say "objects in a Markov category" wherever we say "measurable space" and "morphisms in a Markov category" wherever we say "Markov kernel," since for most of the paper we will reason at the category level, and we will not directly invoke the definition of a measurable space. We have chosen to use the more concrete terms because they express a clear intuition for how these objects and morphisms are intended to be interpreted.

$$(f \mathbin{\fatsemi} g)(c \mid a) = \sum_{b \in BY} f(b \mid a)\, g(c \mid b). \tag{8}$$

In this finite case this is simply matrix multiplication, and we could have denoted it gf instead of $f \mathbin{\fatsemi} g$ accordingly. (Another common notation is $g \circ f$.) We prefer $f \mathbin{\fatsemi} g$ because it puts f and g in the same order that they will appear in string diagrams.

It is straightforward to show that composition is associative, that is

$$(f \mathbin{\fatsemi} g) \mathbin{\fatsemi} h = f \mathbin{\fatsemi} (g \mathbin{\fatsemi} h). \tag{9}$$

In addition, for every finite set A there is an identity kernel, which amounts to just the $|A|$-by-$|A|$ identity matrix. We write this as id_A and define it by $\mathsf{id}_A(a' \mid a) = \delta_{a,a'}$. For every Markov kernel $f \colon A \to P(B)$ we have

$$\mathsf{id}_A \mathbin{\fatsemi} f = f = f \mathbin{\fatsemi} \mathsf{id}_B. \tag{10}$$

These two facts mean that there is a category whose objects are finite sets and whose morphisms are Markov kernels between finite sets. This category is called FinStoch.

Since Markov kernels are morphisms in a category, we will often write $f \colon A \dashrightarrow B$ instead of $f \colon A \to P(B)$, using the dotted arrow $\dashrightarrow$ to distinguish morphisms in FinStoch and related categories from ordinary functions. (In the main text we continue writing them as functions in order to avoid introducing new notation.)

The composition of Markov kernels can be generalised to the case of measure-theoretic probability, which allows us to reason about continuous probability and more general probability measures using the same kinds of diagram and much of the same reasoning. We briefly discuss this in more detail in Appendix A.2. The main difference is that composition becomes integration over measures rather than summation.

Probability measures themselves may be seen as a special case of Markov kernels. Consider a set with a single element, denoted $\mathbb{1} = \{\star\}$. (The identity of the element does not matter because all one-element sets are isomorphic to each other. Category theorists often speak of "the one-element set" for this reason. We use a star to denote the element.) Then a Markov kernel $p \colon \mathbb{1} \dashrightarrow A$ is a function $p \colon \mathbb{1} \to P(A)$, which takes an element of $\mathbb{1}$ and returns a probability measure over A. Since there is only one element of $\mathbb{1}$ this means that the kernel p only defines a single probability measure over A. We therefore think of Markov kernels $\mathbb{1} \dashrightarrow A$ and probability measures over A as essentially the same thing.

We now begin to introduce the string diagram notation. A Markov kernel $f \colon A \dashrightarrow B$ will be denoted

$$A \!-\!\boxed{f}\!-\! B \tag{11}$$

This expression means much the same thing as the notation $f \colon A \dashrightarrow B$. It is just a formal symbol denoting the kernel f, annotated with type information.

The composition of kernels $f \colon A \dashrightarrow B$ and $g \colon B \dashrightarrow C$ is written

$$A \!-\!\boxed{f \mathbin{\fatsemi} g}\!-\! C \quad = \quad A \!-\!\boxed{f}\!-\! B \!-\!\boxed{g}\!-\! C \tag{12}$$

The left and right hand side of this equation are just two different ways to write the composite kernel $f \, \fatsemi \, g$, as defined by Eq. (8) or its measure-theoretic generalisation.

In reading a diagram like the right-hand side of Eq. (12) we find it helpful to imagine an element of A travelling along the wire from the left. As it passes through the kernel f it is stochastically transformed into an element of B, in a way that might depend on its original value. It then travels further to the right and is stochastically transformed by g into an element of C. Equation (8) can be seen as describing this process.

In string diagrams a special notation is used for identity kernels (or identity morphisms more generally): an identity kernel id_A is drawn simply as a wire with no box on it,

$$A \underline{\hspace{3cm}} . \tag{13}$$

For any Markov kernel $f \colon A \rightsquigarrow B$ the identity law Eq. (10) can then be written

$$
\begin{aligned}
A \!-\!\!\boxed{f}\!-\! B \;&=\; A \underline{\hspace{1cm}}\!\boxed{f}\!-\! B \\
&=\; A \underline{\hspace{0.5cm}}\boxed{f}\underline{\hspace{1cm}} B .
\end{aligned}
\tag{14}
$$

This allows us to think of the wires as stretchy: we can extend and contract them at will. We will think of the wires as continuously deformable, rather than extending and contracting in discrete units. This is justified by the formal theory of string diagrams. (One may informally think of the wire itself as an infinite chain of identity kernels, all composed together.) This ability to continuously deform diagrams turns out to be an extremely powerful and useful idea.

Another special notation is used for one-element sets[3]: they are drawn as no wire at all. For this reason a probability measure over A, that is, a kernel $p \colon \mathbb{1} \rightsquigarrow A$, is drawn as

$$\boxed{p}\!-\!\!^A . \tag{15}$$

(Morphisms of this kind are sometimes known as "states," and they are often drawn as a triangle rather than a box, though here we draw them in the same style as other morphisms.)

It is worth noting that the kernels p and f above can be composed, yielding

$$\boxed{p \fatsemi f}\!-\!\!^B \;=\; \boxed{p}\!-\!\!^A\!\boxed{f}\!-\!\!^B . \tag{16}$$

Because of this, although the kernel $f \colon A \rightsquigarrow B$ is defined as a function $f \colon A \to P(B)$ mapping *elements* of A to probability distributions over B, we can instead choose to see it as mapping *probability measures* over A to probability measures over B. In the finite case, if we think of finite probability distributions as normalised and nonnegative vectors in $\mathbb{R}^n$, then f can be seen as a linear map with the property that it maps points in one simplex to points in another. (This justifies thinking of it as a stochastic matrix.)

---

[3] or in a more general context, the unit object of a monoidal category.

The string diagram notation becomes useful when we start thinking about joint distributions. We do this by drawing wires in parallel. As an example, we can consider a Markov kernel defined by a function $h\colon A \times B \to P(C \times D)$. This function takes two arguments, an element of A and an element of B, and it returns a joint probability distribution over C and D. In string diagrams we write this as

$$\begin{array}{c} B \quad\quad D \\ \underline{\ h\ } \\ A \quad\quad C \end{array} \tag{17}$$

In symbols, we write $h\colon A \otimes B \dashrightarrow C \otimes D$. An object like $A \otimes B$, drawn as two parallel wires, can either be thought of as the measurable space $A \times B$ (which is the Cartesian product of sets in the finite case), or as the space of probability measures over $A \times B$. The symbol $\otimes$ is referred to as a monoidal product.

There is some inherent ambiguity in this notation. If we draw three parallel wires, $\begin{smallmatrix} C \\ B \\ A \end{smallmatrix}\!\!\!=\!\!=$, it could either mean $(A \otimes B) \otimes C$ or $A \otimes (B \otimes C)$. In the finite case, these correspond to the sets $(A \times B) \times C$ and $A \times (B \times C)$. These are different sets, since one is composed of pairs $((a, b), c)$ and the other of pairs $(a, (b, c))$. This ambiguity is not important in practice, however, and the formal machinery of *monoidal categories* allows us to use string diagrams without worrying about it. We do not give a formal treatment of this here. (A concise summary can be found in [4].) Instead we simply remark that when we draw three parallel wires we think of joint distributions over A, B and C, and the precise distinction between $P(A \times (B \times C))$ and $P((A \times B) \times C)$ will not be important to us.

In a similar vein, the spaces A and $A \otimes \mathbb{1}$ are different, but the difference is not important to us, and in fact they are written the same way in string diagrams. This is because we draw $\mathbb{1}$ as an invisible wire. This also allows us to write

$$\begin{array}{c} B \text{------} \\ A \text{------} \end{array} = \begin{array}{c} B\text{------} \\ \\ A \text{------} \end{array} = \begin{array}{c} B\!\!\smile\!\!\frown \\ A\!\!\frown\!\!\smile \end{array} \tag{18}$$

That is, string diagrams are stretchy in the vertical direction as well as the horizontal one. We can bend the wires, as long as we don't deform them so much that they point backwards, from right to left.

This also allows to write things like

$$A \underline{}\!\!\!\boxed{f}\!\!\!\underset{B}{\overset{C}{}} \tag{19}$$

for a kernel $f\colon A \dashrightarrow B \otimes C$.

We can also draw morphisms (i.e. Markov kernels) in parallel with each other, for example,

$$\begin{array}{c} C \,\boxed{g}\, D \\ A \,\boxed{f}\, B \end{array} \tag{20}$$

We write this in symbols as $f \otimes g$, which is a morphism of type $A \otimes C \dashrightarrow B \otimes D$. In the finite case, it is given by

$$(f \otimes g)(b, d \,|\, a, c) = f(b \,|\, a)\, g(d \,|\, c). \tag{21}$$

The probabilities $f(b|c)$ and $g(d|c)$ are multiplied together because the two Markov kernels are operating in parallel. One can imagine an element of A entering from the bottom left and being stochastically transformed by f into an element of B, while in parallel, and independently, an element of C enters from the top left and is stochastically transformed by g into an element of D. In general, in the finite case, $f \otimes g$ is given by the tensor product of the stochastic matrices that represent f and g. (This might give some intuition for the symbol $\otimes$.)

We can cross wires over each other. (In category theory terms, the categories we are concerned with are symmetric monoidal categories.) The diagram

$$\begin{array}{c} B \diagdown \diagup A \\ A \diagup \diagdown B \end{array} \tag{22}$$

can be seen as a Markov kernel $A \otimes B \to B \otimes A$. In the finite case it is defined by

$$\text{swap}_{A,B}(b', a' \,|\, a, b) = \delta_{a,a'} \delta_{b,b'}. \tag{23}$$

We have a number of equations that are standard in monoidal category theory, and allow us to freely slide boxes along wires and bend wires to cross over each other. These can either be shown directly from the definitions above or (perhaps more usefully) deduced from the definition of a symmetric monoidal category. Three such equations are as follows. More details can be found in the references cited above.

$$\begin{array}{c} B \diagdown \diagup \diagdown B \\ A \diagup \diagdown \diagup A \end{array} = \begin{array}{c} B ---- B \\ A ---- A \end{array} \tag{24}$$

$$\begin{array}{c} C -\boxed{g}- D \\ A -\boxed{f}- B \end{array} = \begin{array}{c} C -\boxed{g}- D \\ A -\boxed{f}- B \end{array} \tag{25}$$

$$\begin{array}{c} C - \diagdown B \\ A -\boxed{f} \diagup C \end{array} = \begin{array}{c} C \diagdown \boxed{f}- B \\ A \diagup - C \end{array} \tag{26}$$

So far, everything we have said about string diagrams applies to any symmetric monoidal category. However, there are two additional things we can add that take us much closer to probability theory. These are the ability to *copy* and to *delete*. These operations, and their special properties, do not necessarily exist in other contexts, such as quantum mechanics. This is a central point of [4,14]. We will stick to the context of classical probability, however, so copying and deletion will always be possible in this paper.

We cover deletion first. For every measurable space A there is a unique kernel of type $A \to \mathbb{1}$, which we call del_A. In the finite case it is given by $\text{del}_A(\star | a) = 1$ for all $a \in A$. We can think of this as a $1 \times |A|$ matrix (i.e. a row vector) whose entries are all 1. This is the only possible $1 \times |A|$ stochastic matrix.

In string diagrams we write such a deletion kernel as a black dot:

$$A\!\!-\!\!\!-\!\!\bullet . \tag{27}$$

There is one such morphism for every measurable space, but we denote them all with the same kind of black dot. These black dots have the property that

$$A\!\!-\!\!\boxed{f}\!\!-\!\!\!^B\!\!\bullet \;=\; A\!\!-\!\!\!-\!\!\bullet \tag{28}$$

for every Markov kernel f. This says that if we take some input A, perform some stochastic operation f on it and then delete the result, this is the same as simply deleting the input.[4]

The second special operation is copying. For every measurable space A there is a kernel $\mathsf{copy}_A \colon A \to A \otimes A$, which we will describe shortly. We write this also as a black dot, but this time with two output wires rather than one.

$$A\!\!-\!\!\!\!<^{\!\!\!A}_{\!\!\!A} \tag{29}$$

Informally, this kernel takes an outcome $a \in A$ and copies it, producing a pair (a, a) of identical values. It's important to note that it copies *values* rather than *distributions*. Its output does not consist of two independent and identically distributed elements of A but rather two perfectly correlated elements of A that always have the same value. In the discrete case the copy map is defined as

$$\mathsf{copy}_A(a'', a' \,|\, a) = \begin{cases} 1 & \text{if } a'' = a' = a \\ 0 & \text{otherwise.} \end{cases} \tag{30}$$

In addition to Eq. (28), the copy and delete maps obey the following properties [22, Definition 2.1]:

$$A\!\!-\!\!\!\!<\!\!<^{\,A}_{\,A} \;=\; A\!\!-\!\!\!\!<\!\!<^{\,A}_{\,A} \tag{31}$$

$$A\!\!-\!\!\!\!<_{\,A}^{\,\bullet} \;=\; A\!\!-\!\!\!-\!\!\!- \;=\; A\!\!-\!\!\!\!<^{\,A}_{\,\bullet} \tag{32}$$

$$A\!\!-\!\!\!><^{\,A}_{\,A} \;=\; A\!\!-\!\!\!\!<^{\,A}_{\,A} \tag{33}$$

$$A \otimes B\!\!-\!\!\!-\!\!\bullet \;=\; {}^{B\!-\!\!\bullet}_{A\!-\!\!\bullet} \tag{34}$$

$$A \otimes B\!\!-\!\!\!\!<^{\,A \otimes B}_{\,A \otimes B} \;=\; {}^{B-}_{A-}\!\!<\!\!<^{\substack{B \\ A \\ B \\ A}} \tag{35}$$

────────────

[4] In category theory terms, this means that the set of all delete kernels collectively forms a natural transformation. (Specifically, it is a natural transformation from the identity functor to the functor that sends all objects to $\mathbb{1}$ and all morphisms to $\mathrm{id}_{\mathbb{1}}$.) For this reason this property of delete kernels is called "naturality.".

Equation (31) says that if we make multiple copies of something it doesn't matter which order we make them in. Equation (32) says that if we copy something and then delete one of the copies, that is the same as doing nothing to it. Equation (33) says that if we copy something and then swap the copies it makes no difference. (Because the two copies are the same as each other.)

Equations (34) and (35) are more technical. They say that if we have elements of A and B we can delete or copy them as a single element of $A \otimes B$ or separately, as elements of A and B, and these should give the same result.

These equations can be derived from the definitions we have given for the finite case. They may also be derived in various more general measure-theoretic contexts [13,22].

However, the approach of [22] is instead to treat them as *axioms*: any symmetric monoidal category with copy and delete maps that obey Eqs. (28) and (31) to (35) is called a *Markov category*. One can do a surprising amount of reasoning about probability theory using these axioms alone, although there are also Markov categories that do not directly resemble the category of measurable spaces and Markov kernels that we have described. There are various additional axioms that can to be added as well, which then allow more specific results to be proven. (See [22] for the details.)

An important thing to note about the copy operator is that, in general,

$$A \!-\!\boxed{f}\!\!\overset{B}{\underset{B}{\prec}}\!\!\!\begin{matrix} B \\ B \end{matrix} \quad \neq \quad A \overset{A}{\underset{A}{\prec}} \begin{matrix} \boxed{f}-B \\ \boxed{f}-B \end{matrix} \tag{36}$$

That is, copying the output of a kernel f is not the same as copying its input and then applying two copies of the kernel to it. Intuitively, this is because f might be stochastic. If we copy the output we end up with two perfectly correlated copies, whereas if we copy the input then the stochastic variations will be independent.

However, if the kernel is deterministic then copying its input is indeed the same as copying its output. In fact, in the Markov category framework this is the *definition* of a deterministic Markov kernel: we say a kernel $h \colon A \to B$ is deterministic if

$$A\!-\!\boxed{h}\!\overset{B}{\underset{B}{\prec}}\!\!\begin{matrix} B \\ B \end{matrix} \quad = \quad A \overset{A}{\underset{A}{\prec}} \begin{matrix} \boxed{h}-B \\ \boxed{h}-B \end{matrix} \tag{37}$$

In this paper we use square boxes for kernels that are known to be deterministic, and boxes with rounded edges for general, possibly-stochastic kernels.

In the main text, we write Markov kernels as functions $f \colon A \to P(B)$, and we write deterministic kernels as functions $f \colon A \to B$. To be more precise, a deterministic kernel should really also be considered as a function $f \colon A \to P(B)$, such that Eq. (37) is obeyed. However, if we assume we are working in a category called BorelStoch (which is a common assumption in category-theoretic probability) then Eq. (37) implies that f always returns a delta measure [22, Example 10.5], and in this case there is not much harm in treating a deterministic kernel f as a function $f \colon A \to B$.

A.2   The Extension to Measure Theory

Above we described the category FinStoch and introduced string diagrams mostly in that context. Here we briefly describe how this generalises to the measure-theoretic case, which is needed in order to think about continuous probability.

In the measure-theoretic case the objects (X, Y, etc.) are any measurable spaces rather than only finite sets. Markov kernels are still functions $f \colon X \to P(Y)$, but now $P(Y)$ is the set of all probability measures on the measurable space Y. (That is, $P(Y)$ is the set of all functions from the σ-algebra associated with Y to $[0, 1]$, such that Kolmogorov's axioms are obeyed.) $P(Y)$ can itself be made into a measurable space in a standard way, and the function f must obey an additional restriction that it be a measurable function. (This means that the preimage of every element of $P(Y)$ must be a member of the σ-algebra associated with X.)

In this case $f(x)$ is a probability measure rather than a probability distribution, and composition is given by integration rather than summation. (See [22, Example 4] for the details.) This gives rise to a category called Stoch, whose objects are all measurable spaces and whose morphisms are all Markov kernels. (This category is also known as the Kleisli category of the Giry monad, for reasons we do not discuss here.)

Unfortunately the category Stoch does not have all of the properties that one might want it to have. (See Appendix A.3 below.) Because of this a common approach is to work in a category called BorelStoch (also discussed in [22, Example 4]), in which the objects are a subset of measurable spaces called standard Borel spaces, and the morphisms are all Markov kernels between standard Borel spaces. Standard Borel spaces include many kinds of measurable space that one would be likely to use in practice, and in particular they include both finite sets and $\mathbb{R}^n$ with its usual σ-algebra.

In the present paper, the properties of BorelStoch are used in two ways. Firstly, in BorelStoch we can always use conditionals, as explained in the next section. Secondly, as a notational convenience we treat deterministic kernels and measurable functions as interchangeable, which makes sense in BorelStoch but doesn't hold in the more general case of Stoch.

A.3   Conditionals and Bayes' Theorem

Conditional probabilities and Bayes' theorem play central roles in the theory of inference. Here we briefly discuss how they look in string diagrams. Given a joint distribution $\boxed{q}\!\!\!\!\!\!\!\!\!\!\!\!{}^{-B}_{-A}$ we may want to split it up into a product of a marginal and a conditional, which in traditional notation, in the discrete case, would be written $p(a, b) = p(a)\, p(b \,|\, a)$.

The category-theoretic approach, as set out in [13,22], is slightly different. We write the following, which is called a *disintegration* of q. (The term "disintegration" is used because it is the opposite of integration.)

$$\boxed{q}\!\!\begin{array}{l}{}^{-B}\\{}_{-A}\end{array} = \boxed{q}\!\!\begin{array}{l}{}^{\bullet\, B}\\{}_{A}\end{array}\!\!\!\!\!\diagup\!\!\boxed{c}\!\!\begin{array}{l}{}^{-B}\\{}_{-A\,.}\end{array} \tag{38}$$

Here, $\boxed{q}^{\bullet}_A$ is the marginal of A according to the joint distribution q. In the finite case it can be written $\sum_{b \in B} q(a, b)$. The kernel $A\text{--}\boxed{c}\text{--}B$ is called a *conditional* of p. It is defined by Eq. (38), which in the finite case can be written

$$q(a, b) = \left(\sum_{b' \in B} q(a, b') \right) c(b \,|\, a). \tag{39}$$

This is closely analogous to the identity $p(a, b) = p(a)\, p(b \,|\, a)$. The difference is that $p(b \,|\, a)$ is defined as $p(a, b)/p(a)$, and is only defined when $p(a) > 0$. On the other hand, in Eq. (39), if $\left(\sum_{b' \in B} q(a, b') \right) = 0$ for some $a \in A$ then $q(a, b)$ must be 0 for all $b \in B$, and consequently the equation puts no constraint on $c(b \,|\, a)$ in this case.

This means that instead of being undefined in this case, the conditional c is not *uniquely* defined: there may be many different kernels c that satisfy the equation.

This carries over to the general measure-theoretic case as well. If we are in the category BorelStoch then for any joint distribution $\boxed{q}^{\,B}_A$ there exists at least one conditional $A\text{--}\boxed{c}\text{--}B$ that satisfies Eq. (38), but there might be many. (In the case of Stoch conditionals may fail to exist at all, see [22, Example 11.3].)

We may also want to disintegrate a joint distribution that is a function of some parameter, e.g. $z\text{--}\boxed{q}^{\,B}_A$. In this case Eq. (38) becomes

$$z \text{--} \boxed{q} {\overset{B}{\underset{A}{=}}} \;\; = \;\; z \text{--}\!\!\bullet\!\!\boxed{q}\!\!\overset{B}{\underset{A}{\bullet}}\!\!\boxed{c}\text{--}B \;\;\; A. \tag{40}$$

Conceptually this is very similar. We want the disintegration to hold for every parameter value $z \in Z$, and we define the conditional to be a function of z as well as of $a \in A$. In the discrete case, Eq. (40) is analogous to the identity $p(a, b \,|\, z) = p(a \,|\, z)\, p(b \,|\, a, z)$.

Bayes' theorem is closely related to conditional probability and can be expressed in a similar way. Given a prior $\boxed{q}\text{--}A$ and a kernel $A\text{--}\boxed{f}\text{--}B$, we can define a *Bayesian inverse* of f with respect to q, which is a kernel $B\text{--}\boxed{f^\dagger}\text{--}A$ such that

$$\boxed{q}\!\!-\!\!\boxed{f}\overset{B}{\underset{A}{}} \;\; = \;\; \boxed{q}\overset{A}{\underset{}}\boxed{f}\!\!\bullet\!\!\overset{B}{\underset{\boxed{f^\dagger}\text{--}A}{B}} \tag{41}$$

The Bayesian inverse $f^\dagger$ depends on the prior q as well as on the kernel f. If we had chosen a different distribution in place of q, the Bayesian inverse $f^\dagger$ would be different. As with conditionals, Bayesian inverses are not necessarily unique, and for a given f and q there may be many kernels $f^\dagger$ that satisfy Eq. (41). (In fact, Bayesian inverses can be seen as a special case of conditionals; see [13, 22].)

We may also consider the case where the prior takes a parameter, such as $z\text{--}\boxed{q}\text{--}A$. In this case a Bayesian inverse also needs to depend on the parameter in general, which gives us the following more general definition:

$$z\text{--}\boxed{q}\!\!-\!\!\boxed{f}\overset{B}{\underset{A}{}} \;\; = \;\; z\text{--}\!\!\bullet\!\!\boxed{q}\overset{A}{}\boxed{f}\!\!\bullet\!\!\overset{B}{\underset{\boxed{f^\dagger}\text{--}A}{B}} \tag{42}$$

The references [13, 22, 37] contain much more detail about Bayes' theorem in this form.

B   More Details About Bayesian Interpretations

B.1   Unpacking Bayesian Filtering Interpretations

In this section we give some more intuition for Definition 2 and then note some consequences of it. The section deals mostly with the case where S, Y and H are discrete sets, meaning that we can reason in terms of probability distributions rather than measure theory. In this case Definition 2 can be written in a form that makes the relationship to Bayes' theorem more clear. We define a notion of *subjectively impossible input*, which is a value of S that the reasoner believes with certainty will not occur as its next input. (This does not imply that the input actually is impossible according to the true dynamics of the environment.) We show that Definition 2 puts no constraints on the reasoner's posterior after receiving a subjectively impossible input.

We also show that the possible interpretations of a machine only depend on which states can transition to which other states given which inputs, and not on the probabilities of such transitions. In addition, we show that some machines admit no non-trivial interpretations at all.

In order to unpack Definition 2 a little more, let us consider the case where S, Y and H are discrete. Before starting we note that in the finite case, the definition of ψ_S, Eq. (4), can be written as

$$\psi_S(s\,|\,y) = \sum_{h \in H} \psi_{S,H'}(s, h\,|\,y). \tag{43}$$

In this case, Eq. (5) can be written in symbols as

$$\psi_{S,H'}(h, s\,|\,y)\gamma(y'\,|\,y, s) = \psi_S(s\,|\,y)\gamma(y'\,|\,y, s)\psi_H(h\,|\,y'), \tag{44}$$

for all $s \in S, h, \in H, y, y' \in Y$. We can cancel $\gamma(y'\,|\,y, s)$ from both sides on the assumption that it is positive, yielding

$$\gamma(y'\,|\,y, s) > 0 \implies \psi_{S,H'}(h, s\,|\,y) = \psi_S(s\,|\,y)\psi_H(h\,|\,y'). \tag{45}$$

The condition $\gamma(y'\,|\,y, s) > 0$ means that $y' \in Y$ is a *possible next state* when the machine starts in state $y \in Y$ and receives the input $s \in S$. (There may be many possible next states in this situation because the machine may be stochastic.)

Let us then suppose that the machine starts in state y, receives an input s, and transitions to state y'. Let h be an arbitrary element of H. The number $\psi_{S,H'}(h, s\,|\,y) \in [0, 1]$ can then be seen as the reasoner's prior probability that the next state is h and the next input is s. In more traditional notation we might write this as $P(H' = h, S = s)$, where we leave the state of the underlying machine implicit. (Here we do not attempt to formalise this in terms of random variables, but simply treat it as a kind of notational shorthand for $\psi_{S,H'}(h, s \,|\, y)$.)

We may then regard $\psi_S(s\,|\,y)$ as the reasoner's prior probability that the next input is s, i.e. $P(S = s) = \sum_{h \in H} P(H = h, S = s)$.

However, since $\psi_H(h\,|\,y')$ is conditioned on y' rather than y, we instead regard it as the reasoner's *posterior* probability that $H' = h$. (We refer to H' rather than H here because after it receives an input its previous "next" hidden state becomes its current hidden state.) $\psi_H(h\,|\,y')$ therefore corresponds to what we might write as $P(H' = h \mid S = s)$.

With this informal shorthand notation Eq. (45) then says

$$P(H' = h, S = s) = P(S = s)\, P(H' = h \mid S = s), \tag{46}$$

which has the same appearance as a familiar identity from elementary probability theory. It corresponds to a single step of Bayesian filtering, which we spell out in more detail in Appendix B.2.

This shorthand notation gives some intuition for why Eq. (5) has the particular form it does, but it leaves the dependence on the state of the underlying machine implicit, and in so doing it obscures an important and subtle point. In a more traditional context, $P(H' = h \mid S = s)$ is defined by

$$P(H' = h \mid S = s) = P(H' = h, S = s)/P(S = s) \tag{47}$$

and has no value when $P(S = s) = 0$. However, in our case $P(H' = h \mid S = s)$ is a shorthand for $\psi_H(h\,|\,y')$, which is defined even when $\psi_S(s\,|\,y) = 0$.

In the case where $P(S = s) > 0$, Eq. (5) in the form of Eq. (46) demands that $P(H' = h \mid S = s)$ is indeed equal to $P(H' = h, S = s)/P(S = s)$. More precisely, if $\psi_S(s\,|\,y) > 0$ then we must have $\psi_H(h\,|\,y') = \psi_{S,H'}(h, s\,|\,y)/\psi_S(s\,|\,y)$. However, if $\psi_S(s\,|\,y) = 0$ then Eq. (5) puts no constraints on $\psi_{S,H'}(h, s\,|\,y)$ at all, or indeed on $\psi_H(h\,|\,y)$.

In the case where S is a discrete set (even if Y and H are not discrete), we say that $s \in S$ is a *subjectively impossible input* for a given state $y \in Y$ if $\psi_S(s\,|\,y) = 0$. The point is that the reasoner believes, with certainty, that it will not receive the input s as its next input. The reasoning above says that in this situation, *any* posterior over H is acceptable, because Bayes' rule doesn't specify what the posterior should be. We find this somewhat analogous to the fact that in logic one can deduce any proposition from a contradiction. Definition 2 indeed permits any posterior in the case of a subjectively impossible input. In fact, it even allows the posterior to be chosen stochastically in this case.

This is in a sense the minimal possible assumption we could make. However, one could imagine addressing the issue in a different way by changing the framework, thus introducing a subtly different notion of interpretation than the one we have presented here. One possibility would be to allow *partial interpretations*, where ψ_H becomes a partial function, meaning that not every state of the machine needs to have an interpretation at all. This would allow the posterior to be undefined in the case of a subjectively impossible input, rather than merely arbitrarily defined. Another possibility would be to strengthen Eq. (5)

with additional conditions, forcing the posterior to be meaningful even after a subjectively impossible input. We suspect that such an approach can lead to an interesting way to formalise improper priors, which are also about having meaningful posteriors in the case of 'impossible' inputs, but we leave investigation of this to future work.

We note one other important consequence of the above reasoning, in the discrete case. When we express Eq. (5) in the form of Eq. (45), we see that it only depends on whether a transition from y to y' is possible given an input s, and not on the probability of such a transition. Thus, for Bayesian filtering interpretations (and hence also for Bayesian inference interpretations), the only property of a machine that matters is which states can be reached from which other states (in a single step) under a given input. (Strictly speaking this only makes sense in the discrete case, but we expect an analogous statement to this to hold more generally.)

This has the consequence that some machines only admit trivial interpretations. By a trivial interpretation we mean one where the posterior is always equal to the prior. Such interpretations exist for every machine, because we can always take the model to be such that H does not change over time and S does not depend on H, so that the input $s \in S$ never gives any information about H. That is, in string diagrams, for any machine we can set

$$H-\boxed{\kappa}\genfrac{}{}{0pt}{}{-H}{-S} = \genfrac{}{}{0pt}{}{H\text{———}H}{\boxed{q}-S} \quad , \tag{48}$$

and

$$Y-\boxed{\psi_H}-H = Y-\!\bullet\ \boxed{r}-H \quad , \tag{49}$$

for any choice of distributions q and r. Then the conditions of Definition 2 will always be satisfied. This may be shown using string diagram manipulations and the Markov category axioms.

We now show that there is a class of machines that only admit trivial interpreatations. Consider a machine with the property that $\gamma(y'|y,s) > 0$ for all y, y', s. That is, for a given input s and initial state y, every state y' has some nonzero probability of being the next state. In this case, Eq. (45) implies that

$$\psi_{S,H'}(h,s|y) = \psi_S(s|y)\psi_H(h|y'), \tag{50}$$

for all $y, y' \in Y, s \in S, h \in H$. Since the left-hand side does not depend on y' it follows that $\psi_H(h|y')$ must not depend on y' either. That is, $\psi_H(h|y') = p(h)$ for some fixed distribution p. (The other possibility would be that $\psi_{S,H'}(h,s|y) = \psi_S(s|y) = 0$, but this can't hold for all $s \in S$, because $\psi_{S,H'}(h,s|y)$ must be nonzero for some s in order to be normalised.)

This means that if a machine satisfies this property then the only intrepretations it admits are trivial ones, of the form Eq. (49). This means that in order for a discrete machine to admit *any* non-trivial Bayesian filtering interpretation it must satisfy a fairly strong constraint, namely that some of its transition probabilities are zero.

This is to some extent a consequence of our choice to consider only exact Bayesian filtering interpretations. If a discrete machine has no non-zero transition probabilities it might still be possible to interpret it as performing some form of approximate inference, but defining such interpretations precisely is a task for future work.

B.2   More on Bayesian Filtering

In this section we show that Definition 2 does indeed correspond to Bayesian filtering, at least in the case of a deterministic machine. Our proof of this is inspired by [24, theorem 6.3], which proves an analogous fact about conjugate priors. The proof we give uses string diagram reasoning, which means that it holds even in the most general measure-theoretic context; we do not need to assume that the sets involved are discrete.

Since we restrict ourselves to only deterministic machines in this section, we will note a couple of things about deterministic machines before we talk about Bayesian filtering.

We first note that the condition for a machine γ to be deterministic is

$$(51)$$

This comes from the defining equation for deterministic morphisms, Eq. (37), and also the axiom (35), noting that γ is a kernel with input $S \otimes Y$ and output Y.

Next we prove the following proposition, which is useful for reasoning about Bayesian filtering interpretations of deterministic machines.

Proposition 1. *Suppose* $\,{}^S_Y\!\!-\!\!\boxed{\gamma}\!-_Y$ *is a deterministic machine, and let* $Y\!-\!\boxed{\psi_H}\!-H$ *and* $\,{}^H\!-\!\boxed{\kappa}\!-^H_S$ *be arbitrary Markov kernels. Then* ψ_H *and* κ *form a consistent Bayesian filtering interpretation of* γ *(i.e. Definition 2 is satisfied) if and only if*

$$(52)$$

with $\psi_{S,H'}$ *and* ψ_S *as defined in Eqs. (3) and (4).*

Proof. To see that Definition 2 implies Eq. (52) we marginalise Eq. (5):

$$(53)$$

This implies Eq. (52) by the rules for Markov categories, specifically Eqs. (28) and (32).

For the other direction we assume Eq. (52) holds and calculate

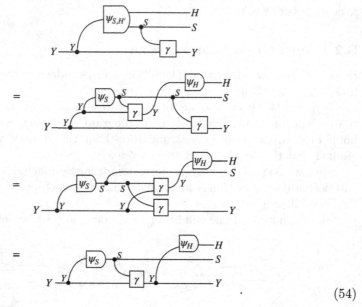

$$(54)$$

The first step substitutes in the right-hand side of Eq. (52), the second rearranges using the rules of Markov categories, and the third uses the determinism condition. This proves that Eq. (5) holds.

We now consider what a Bayesian filtering task involves. The idea is that the reasoner has a model of a hidden Markov process, given by the kernel $H\!-\!\boxed{\kappa}\!\langle^{H}_{S}$. As described in the main text, this kernel can be thought of as a process that simultaneously transforms the hidden state, stochastically, into a new value and emits a visible "sensor value."

Given a kernel of this type, we can iterate it to produce sequences of values in S. For example, we can write

$$(55)$$

where S^3 means $S \otimes S \otimes S$ and κ^n is notation for iterating the kernel n times. A kernel of this kind, thought of as an infinitely iterated process, is sometimes called a "coalgebra," since it is a special case of a more general concept of that name. (e.g. [25] takes a coalgebraic approach to de Finetti's theorem.)

For filtering we are interested in inferring the final hidden state of a system, given a finite sequence of visible states. In order to reason about this, we define the following kernel:

$$Y - \boxed{\psi_{S^n, H_n}} \begin{matrix} -H \\ -S^n \end{matrix} \quad = \quad Y - \boxed{\psi_H} -^H \boxed{\kappa^n} \begin{matrix} -H \\ -S^n \end{matrix} \tag{56}$$

This can be seen as an interpretation map, mapping the state of a reasoner to its beliefs about its next n inputs, $S^n = (S_1, \ldots, S_n)$, along with the final value of the hidden state, H_n. These take the form of a joint distribution between S^n and H_n. This joint distribution is formed from the reasoner's initial prior over the initial hidden state H_1 (given by the kernel ψ_H) and the model κ, which is iterated n times.

We define this because in filtering we wish to make a probabilistic inference of the final hidden state, H_n, given the sequence of visible states S^n. To infer H_n given S^n we seek a disintegration of ψ_{S^n, H_n}. (See Eq. (38) in Appendix A.3.) Specifically, we seek a kernel $\psi_{H_n | S^n} : S^n \otimes Y \to P(H)$ such that

$$Y - \boxed{\psi_{S^n, H_n}} \begin{matrix} -H \\ -S^n \end{matrix} \quad = \quad Y - \boxed{\psi_{S^n, H_n}} \begin{matrix} -H \\ -S^n \end{matrix} \boxed{\psi_{H_n | S^n}} \begin{matrix} -H \\ -S^n \end{matrix} \tag{57}$$

The kernel $\psi_{H_n | S^n}$ takes in a sequence S^n of observations and returns the reasoner's conditional beliefs about H_n, given the sequence S_n. It is also a function of the reasoner's initial beliefs $y \in Y$.

In fact such a kernel can be constructed iteratively in a natural way, if we assume that ψ_H and κ form a consistent Bayesian filtering interpretation. To do this, we first define the iteration of γ, in a similar way to the iteration of κ:

$$\begin{matrix} S^n \\ Y \end{matrix} - \boxed{\gamma^n} - Y \quad := \quad \begin{matrix} S \\ \vdots \\ S \\ S \\ Y \end{matrix} \boxed{\gamma} \boxed{\gamma} \cdots \boxed{\gamma} - Y \tag{58}$$

where there are n copies of γ on the right-hand side. We can then state the following result, which shows that consistent Bayesian filtering interpretations can indeed be seen as performing Bayesian filtering, in the discrete case.

Proposition 2. *The kernel* $\begin{smallmatrix} S^n \\ Y \end{smallmatrix} \boxed{\gamma^n} \overset{Y}{\boxed{\psi_H}} - H$ *is a conditional of* ψ_{S^n, H_n}, *satisfying Eq. (57), in that*

$$Y - \boxed{\psi_{S^n, H_n}} \begin{matrix} -H \\ -S^n \end{matrix} \quad = \quad Y - \boxed{\psi_{S^n, H_n}} \begin{matrix} -H \\ -S^n \end{matrix} \boxed{\gamma^n} \overset{Y}{\boxed{\psi_H}} - H \tag{59}$$

Proof. We begin by defining the kernel

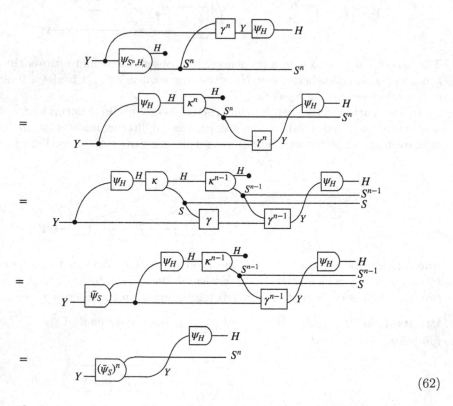

$$\qquad(60)$$

We also define its iteration, $(\bar\psi_S)^n \colon Y \to Y \otimes S^n$, analogously to κ^n and γ^n. We note that the consistency equation for Bayesian filtering interpretations, Eq. (5), can be written in terms of κ and $\bar\psi_S$, as

$$\qquad(61)$$

We then calculate

$$\qquad(62)$$

where the last step is by applying the other steps inductively. We can then apply a second inductive argument in "the other direction" using Eq. (52), as follows:

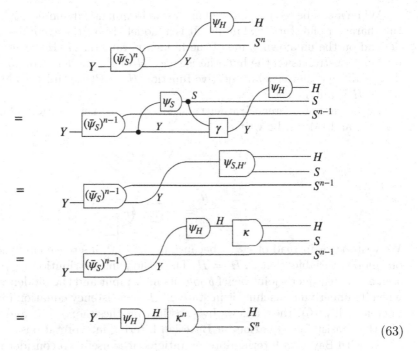

$$\tag{63}$$

where the last step is again by applying the other steps inductively.

We have proved that $S^n\!-\!\boxed{\gamma^n}\!-\!Y\!-\!\boxed{\psi_H}\!-\!H$ with input S^n_Y is a conditional of ψ_{S^n, H_n}. The kernel $S^n_Y\!-\!\boxed{\gamma^n}\!-\!Y\!-\!\boxed{\psi_H}\!-\!H$ can be thought of as giving the reasoner's beliefs about H after receiving a given sequence S^n of inputs, starting from a given initial state $y \in Y$. The result shows that these beliefs are consistent with the agent's prior $\psi_H(y)$ and the model κ, in the sense that the agent's final posterior beliefs about H are a conditional of its initial joint beliefs about the sequence S^n and the final hidden state. We conclude that a deterministic machine with a consistent Bayesian filtering interpretation can indeed be seen as performing a Bayesian filtering task. We expect this to be true in the general case of stochastic machines as well.

B.3   Bayesian Inference Interpretations and Conjugate Priors

In the main text we noted that Bayesian inference corresponds to a special case of Bayesian filtering. By "Bayesian inference" here we mean the case where the reasoner is interpreted as assuming its inputs are i.i.d. samples from some known distribution with an unknown parameter space H, which we also call the hypothesis space.

The difference between inference and filtering is that we interpret the reasoner as believing that the value of H is unknown but fixed. That is, the reasoner assumes that H doesn't change over time. This corresponds to a special case of filtering in which $H\!-\!\boxed{\kappa}\!-\!{}^H_S = H\!-\!\boxed{\phi}\!-\!{}^H_S$, for some kernel ϕ that we also call the model.

While κ can be seen as a model of the environment's dynamics, ϕ has more of the character of a statistical model. It is a model of how the agent's sensor values depend on the unknown value of the hidden parameter H. However, we do not put any constraints on the hypothesis space H or the model ϕ. In particular, we do not assume that ϕ is an injective function $H \to P(S)$, and we allow the case where H is a finite set.

In the case of inference rather than filtering, the kernels ψ_S and $\psi_{S,H'}$ from Eqs. (3) and (4) can be written

$$Y\!-\!\boxed{\psi_S}\!-\!{}^S = {}^Y\!-\!\boxed{\psi_H}\!-\!{}^H\!\boxed{\phi}\!-\!{}^S \tag{64}$$

and

$$Y\!-\!\boxed{\psi_{S,H}}\!\overset{H}{\underset{S}{}} = {}^Y\!-\!\boxed{\psi_H}\!\!-\!\!\bullet\overset{H}{\underset{\boxed{\phi}\!-\!S}{H}} \tag{65}$$

We write $\psi_{S,H}$ instead of $\psi_{S,H'}$ because in the i.i.d. inference case there is only one hidden variable, that is, $H' = H$. Thus, the joint distribution $\psi_{S,H}(y)$ can be seen as the reasoner's joint belief about its next input and the hidden variable H, when its underlying machine is in state y. The consistency equation for Bayesian inference, Eq. (6), then follows by substituting these for $\psi_{S,H'}$ and ψ_S in Eq. (5), the consistency equation for Bayesian filtering interpretations.

As with Bayesian filtering interpretations, it is useful to consider the case in which the underlying machine is deterministic (but not necessarily discrete). In proposition 1 we gave a simpler version of the consistency equation for Bayesian filtering interpretations, which is equivalent to Definition 2 in the case of a deterministic machine. In the inference case we can substitute Eqs. (64) and (65) into this simplified consistency equation (Eq. (52)) to obtain

$$Y\!-\!\boxed{\psi_H}\!\!-\!\!\bullet\overset{\boxed{\phi}\!-\!S}{\underset{H}{H}} = {}^Y\!-\!\bullet\!-\!\boxed{\psi_H}\!-\!{}^H\!\boxed{\phi}\!-\!\bullet\!-\!{}^S\,\underset{\boxed{\gamma}\!-\!{}^Y\!\boxed{\psi_H}\!-\!H}{\overset{S}{}} \tag{66}$$

This is exactly the equation given by [24, Eq. 16] as a definition of a conjugate prior.

Both sides of Eq. (66) express a joint distribution between S and H, as a function of Y. In the context of conjugate priors, ϕ is considered to be a family of distributions, with parameters H. Our interpretation map ψ_H corresponds to another family of distributions, which is a conjugate prior to ϕ. The machine state Y corresponds to the so-called hyperparameters, i.e. the parameters of ψ_H.

This shift in perspective makes sense. In a computational context, conjugate priors are often useful precisely because they offer a way to perform inference without needing to directly calculate Bayesian inverses at run-time. Instead, the implementation only needs to keep track of the hyperparameters and update them in response to data. This update takes place according to a deterministic function, whose form depends on the family ϕ and its conjugate prior ψ_H. This updating of the hyperparameters is the role played by γ: it takes in a data point in S along with the current value of the hyperparameters, and returns the updated hyperparameters. Equation (66) asserts that this must be done in such a way that the new value of Y does indeed correspond to the correct Bayesian posterior, when mapped to a distribution over H by the kernel ψ_H.

We note that it is somewhat nontrivial to find a pair of kernels ψ_H, ϕ and a function γ such that Eq. (66) is obeyed. However, many such examples are known. (Although it is not an authoritative source, a useful list can be found online [40, under "Table of conjugate distributions"], which explicitly gives both kernels and the update function for each example.) Any example of a conjugate prior can be seen as a deterministic machine together with a consistent Bayesian inference interpretation. In addition, in Appendix C we give a number of examples of a different flavour, in that in our examples H is either a finite or a countable set.

B.4   Unpacking Bayesian Inference Interpretations

We now unpack Definition 3 by converting Eq. (6) into more familiar terms in the case where all the spaces are discrete sets, as we did for filtering interpretations in Appendix B.1.

In the case where Y, H and S are finite sets, Eq. (6) can be written as

$$\psi_H(h\,|\,y)\,\phi(s\,|\,h)\,\gamma(y'\,|\,s,m) = \psi_S(s\,|\,y)\,\gamma(y'\,|\,s,y)\,\psi_H(h\,|\,y'), \qquad (67)$$

or equivalently,

$$\gamma(y'\,|\,s,y) > 0 \quad \Longrightarrow \quad \psi_H(h\,|\,y)\,\phi(s\,|\,h) = \psi_S(s\,|\,y)\,\psi_H(h\,|\,y'), \qquad (68)$$

since we can cancel $\gamma(y'\,|\,s,y)$ if we assume it is positive. For $\gamma(y'\,|\,s,y)$ to be positive means that it is possible for the machine to transition from state $y \in Y$ to state $y' \in Y$ after receiving the input $s \in S$.

We can now give an intuitive interpretation to the terms in this equation. If the machine starts in state y, receives input s, and transitions to state y' as a result, then we can regard $\psi_H(h\,|\,y)$ as the reasoner's prior beliefs about the hypothesis h, $\psi_S(s\,|\,y)$ as its prior beliefs about the input s, and $\psi_H(h\,|\,y')$ as the reasoner's posterior belief about the hypothesis h. Equation (68) can then be compared, term by term, to the much more familiar equation

$$p(h)\,p(s \mid h) = p(s)\,p(h \mid s). \qquad (69)$$

Here we have written $p(s \mid h)$ in place of $\phi(s\,|\,h)$ and $p(h \mid s)$ in place of $\psi_H(h\,|\,y')$ in order to emphasise the similarity to Bayes' theorem in a more familiar form. Our definition, in the form of Eq. (6) or Eq. (67), differs from this in that it explicitly takes account of the machine's state, and ϕ and ψ_H are defined by Markov kernels rather than conditional probabilities.

We note that, as in the case of filtering (Appendix B.1), our definition of a consistent Bayesian inference interpretation allows the posterior to be arbitrary in the case of subjectively impossible inputs, i.e. those $s \in S$ for which $\sum_{h \in H} \psi_H(h\,|\,y)\phi(s\,|\,h) = 0$ for a given state $y \in Y$. Given such an input the reasoner may update its posterior to anything at all. As with filtering, we regard this as the minimal assumption we could have made, but we can imagine several other choices that one could make instead. These include allowing the posterior to be undefined in such cases; *requiring* it to be undefined; requiring it to obey some additional consistency equation such that the posterior would make sense even on subjectively impossible inputs; or requiring ϕ and ψ_H to be such that subjectively impossible inputs do not exist. We would consider these to be subtly different kinds of interpretation, and we leave their further investigation to future work.

C   Details of Examples

C.1   An Interpretation of a Non-Deterministic Finite Machine

We here present a non-deterministic finite machine with internal state space $Y = \{y_0, y_1, y_2\}$ and sensory input space $S = \{s_1, s_2\}$. One Bayesian interpretation of this machine, for a hidden state space $H = \{h_1, h_2\}$, is as follows (where δ is the Kronecker delta):

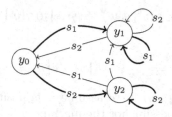

Fig. 1. Transitions for a three-state machine. Deterministic transitions are depicted using bold arrows, and non-deterministic transitions using regular arrows. The precise probability values for non-deterministic transitions are not shown, since we only need to know that they are non-zero.

$$\phi(s_i \,|\, h_j) = \delta_{ij}$$

$$\psi(h_i \,|\, y_j) = \begin{cases} \delta_{ij} & \text{if } j \in \{1, 2\} \\ 0.5 & \text{if } j = 0. \end{cases} \tag{70}$$

Under this interpretation, the model ϕ ascribed to the machine is that sensory inputs transparently reflect the hidden state. The machine, in internal state y_0, is taken to be uncertain about the hidden state; in state $y_i \in \{y_1, y_2\}$ it is taken to be certain that the hidden state is h_i. The dynamics of the machine match this interpretation: it transitions deterministically to y_i when receiving input s_i, unless s_i is "subjectively impossible" (s_2 at m_1, and s_1 at m_2). Behaviour on subjectively impossible inputs is not constrained by the consistency equation, so this is a consistent Bayesian interpretation.

C.2   Machine Counting Occurrences of Different Observations

We now consider a countably infinite deterministic machine (Y_0, S, γ_0). Let $Y_0 = \mathbb{N}^+ \times \mathbb{N}^+$ ($\mathbb{N}^+$ excludes 0) and the input space be $S = \{+1, -1\}$. The machine deterministically computes the function $f_0 : (Y_0 \times S) \to Y_0$, in the sense that $\gamma_0(f_0(y, s) | y, s) = 1$. Essentially, it keeps distinct count of how many $+1$ and -1 inputs it has received. Formally:

$$f_0((i, j), s) = \begin{cases} (i + 1, j) & \text{if } s = +1 \\ (i, j + 1) & \text{if } s = -1 \end{cases} \tag{71}$$

One consistent Bayesian interpretation (ψ_0, ϕ_0) for machine γ_0 uses hypothesis space $H_0 = [0, 1]$ and model:

$$\phi_0(s | h) = h^{\delta_{-1}(s)}(1 - h)^{\delta_{+1}(s)} \tag{72}$$

where

$$\delta_u(v) := \begin{cases} 1 & \text{if } u = v \\ 0 & \text{else.} \end{cases} \tag{73}$$

This model is known as the categorical distribution for two outcomes (or just the Bernoulli distribution). The machine states were deliberately chosen to be the hyperparameters of a possible interpretation map $\psi_0 : Y_0 \to H_0$ which is known as the Dirichlet distribution (and in this special case also as Beta-distribution):

$$\psi_0(h | (i, j)) = \frac{1}{B(i, j)} h^{i-1}(1 - h)^{j-1} \tag{74}$$

where $B(i, j)$ is the Beta function.

This interpretation map (the Dirichlet distribution) is the conjugate priors for categorical distributions. This implies that (ψ_0, ϕ) form a consistent Bayesian inference interpretation, as explained in Appendix B.3.

C.3   Machine Tracking Differences Between the Number of Occurrences of Different Observations

We now consider another countably infinite deterministic machine (Y_1, S, γ_1) which has the same input space as the machine in Appendix C.2. Let $Y_1 = \mathbb{Z}$ and the input space again be $S = \{+1, -1\}$. The machine γ_0 deterministically computes a function $f_1 : (Y_1 \times S) \to Y_1$, in the sense that $\gamma_1(f_1(y, s) | y, s) = 1$. Machine γ_1 only counts how many more $+1$ inputs it has received than -1 inputs. Formally:

$$f_1(k, s) = k + s. \tag{75}$$

One consistent Bayesian interpretation (ψ_1, ϕ_1) for machine γ_1 uses hypothesis space $H_1 = \{h_{+1}, h_{-1}\}$ and model:

$$\phi(s | h_i) = \begin{cases} 0.75 & \text{if } i = s \\ 0.25 & \text{otherwise.} \end{cases} \tag{76}$$

The interpretation map $\psi_1 : Y_1 \to H_1$ is

$$\psi_1(h_{+1}\,|\,k) = \frac{1}{2(1 + 0.75^k 0.25^{-k})}$$
$$\psi_1(h_{-1}\,|\,k) = \frac{1}{2(1 + 0.75^{-k} 0.25^{k})} \tag{77}$$

It is relatively easy to verify that (ψ_1, ϕ_1) is a consistent Bayesian interpretation with γ_1's dynamics.

As a teaser for future work we may note the following. Since machine γ_0 of Appendix C.2 stores the individual counts for s_0 and s_1 inputs, it also implicitly keeps track of the difference between those counts; γ_1 only keeps track of this difference. Consequently, we can define a deterministic kernel $g : Y_0 \to P(Y_1)$ that maps any state $(i, j) \in Y_0$ of γ_0 to $g(i, j) := \delta_{i-j}$ which is a probability measure over the state space of γ_1. It turns out that for this map, for any $k' \in Y_1, s \in S$ and $(i, j) \in Y_0$ we have

$$(\gamma_0 \,\fatsemi\, g)(k'\,|\,(i, j), s) = \sum_k \gamma_1(k'\,|\,k, s)g(k\,|\,(i, j)). \tag{78}$$

This implies that we can construct an interpretation of machine γ_0 from the interpretation (ψ_1, ϕ_1) of γ_1. For this we precompose the interpretation map ψ_1 for γ_1 with the machine map g to get a consistent Bayesian inference interpretation $(g \,\fatsemi\, \psi_1, \phi_1)$ for γ_0. In future work we intend to further develop the theory of how a consistent interpretation of one deterministic machine can be "pulled back" to other machines that are related in a similar way to Eq. (78).

D   Details on the Relation to the FEP

We here try to identify the structures in the FEP that are analogous to the notions of machine γ, model κ, and interpretation map ψ_H. This suggests that, at least in some treatments of FEP, there is an implicit concept that is close to what we have called a reasoner. We will call this putative concept the FEP reasoner.

Large parts of the FEP literature do not explicitly deal with FEP reasoners but are sometimes presented as based on them (e.g. in [20]). The parts that construct the FEP reasoner are those called "Bayesian mechanics" and are still evolving. A standard reference is [19] but this is known to contain some issues [1,9,21]. The most recent version can be found in [16].

Understanding more precisely the relationship between the concepts of our Bayesian and the FEP reasoner is future work. The following are preliminary observations.

D.1   Machine

We first identify the structure in the FEP setup that is most closely related to a machine and is also said to appear to perform Bayesian inference. Unfortunately, the latest iteration of the conditions under which there exists an FEP reasoner, which is [16], does not make this particular structure as explicit as the previous version [19]. We will therefore identify this structure in the older version. A corresponding structure should also exist in the newer version and we will hint at how it may differ.

The FEP setup in [19] consists of four sets of variables $\eta \in E, s \in S, a \in A, \mu \in M$ called external, sensory, active, and internal states with E, S, A, M finite dimensional real vector spaces. These variables obey the stochastic differential equations

$$
\begin{aligned}
\dot{\eta} &= f_\eta(\eta, s, a) + \omega_\eta \\
\dot{s} &= f_s(\eta, s, a) + \omega_s \\
\dot{a} &= f_a(s, a, \mu) + \omega_a \\
\dot{\mu} &= f_\mu(s, a, \mu) + \omega_\mu
\end{aligned}
\tag{79}
$$

where $\omega_\eta, \omega_s, \omega_a, \omega_\mu$ are independent Gaussian noise terms. The FEP goes beyond the scope of a reasoner and formulates a concept of agent. The concept of an agent should, as part of its interpretation, make it possible to talk about deliberate actions. In the FEP deliberate actions are associated to the active states a. At the same time, the internal states are only involved in inference (or filtering) and the special case where there are no active states seems to be within the scope of the FEP. This should still leave us with a FEP reasoner and make it more comparable to our Bayesian reasoner. We therefore consider the special case where there are no active states such that we get:

$$
\begin{aligned}
\dot{\eta} &= f_\eta(\eta, s) + \omega_\eta \\
\dot{s} &= f_s(\eta, s) + \omega_s \\
\dot{\mu} &= f_\mu(s, \mu) + \omega_\mu.
\end{aligned}
\tag{80}
$$

This looks like a continuous time version of the Bayesian network in Eq. (1) and has the somewhat significant feature that all influences from the external states η are mediated by the sensory states s. This suggests that it is possible to see the sensory states $s \in S$ as inputs to a machine state $\mu \in M$ with the external states $\eta \in E$ "hidden behind" the sensory states.

The internal states $\mu \in M$ are supposed to appear to infer the external states. So the state space Y of the machine of the FEP reasoner should be identified with M. Going by their name and their role in the earlier dynamics of Eq. (79) it seems reasonable to identify the sensory state space S with the input state space (also S in our notation) of the machine.

This brings us to the machine's kernel γ. Our formalism does not deal with continuous-time kernels at the moment so we only make some informal comments here. Note that none of the following statements should be considered as proven. Since all variables together form a (time-homogeneous) Markov process, we can choose times $t, t + \tau$ with $\tau > 0$ and write the conditional probability density (assuming things are well behaved enough) at a state (η', s', μ') at $t + \tau$ given a state (η, s, μ) at time t as $p(\eta', s', \mu', t + \tau \mid \eta, s, \mu, t)$ (this notation is taken from [35, p.31]). We can then marginalise out η' and s' to get $p(\mu', t + \tau \mid \eta, s, \mu, t)$

which looks a bit closer to a machine kernel but still depends also on η. We cannot just drop η from this expression even if we assume Eq. (80) holds since within a time interval $[t, t + \tau]$ with $\tau > 0$ the influence from η would propagate through the intermediate values of the sensory states to μ'. Instead we here condition on all those intermediate values of the sensory state between t and $t + \tau$. Write $s[t : t + \tau]$ for a part of the trajectory of the sensory state between t and $t + \tau$ that starts in s. Then, assuming Eq. (80) we should get:

$$p(\mu', t + \tau \mid \eta, s[t : t + \tau], \mu, t) = p(\mu', t + \tau \mid s[t : t + \tau], \mu, t). \quad (81)$$

In order to make this look even more like a kernel we may take the limit as $\tau \to 0$ and so we write

$$\gamma(\mu' \mid \mu, s) := \lim_{\tau \to 0} p(\mu', t + \tau \mid s[t : t + \tau], \mu, t) \quad (82)$$

which is just a notation for an expression that hopefully provides sufficient intuition for our purposes.

What is important is that within the system Eq. (80) there should be a (continuous-time) machine describing the dynamics of the internal states in response to sensory states.

In [16] the structure of Eq. (79) and thus Eq. (80) is not stated explicitly. However, the sensory (and usually the active states) are still special due to an additional assumption which is also made in [19, 34]. The larger process has to have a stationary distribution $p(\eta, s, a, \mu)$ that factorises according to

$$p(\eta, s, a, \mu) = p(\eta|s, a)p(\mu|s, a)p(s, a) \quad (83)$$

which is referred to as a Markov blanket. With this assumption only, one can no longer assume that the sensory states $s[t : t + \tau]$ can "shield" those states from direct influence by external states, which makes it more difficult to compare the dynamics to our setup. A solution may be to use a continuous-time version of the approach in [36]. Below we ignore this issue and assume that we have the structure of Eq. (80).

D.2   Model

For a reasoner we also need a model and an interpretation map. As already mentioned the FEP assumes that the system in Eq. (79) has a stationary distribution $p(\eta, s, \mu)$. One purpose of this assumption seems to be the definition of what we call the model. In the language of the FEP literature the stationary distribution defines the generative model. Here, generative model refers to a joint probability distribution over causes (parameters/hidden variables) and observed variables. In [16, Section 3.b] the generative model is defined to be $p(\eta, s, \mu)$ with η as the hidden variables and observed variables (s, μ). This could mean that the machine state μ itself is also modelled by an FEP reasoner, which is different from our framework. This would need further investigation that we leave for future work.

So we resort to a previous version where only the marginalised stationary distribution $p(\eta, s)$ was considered as the generative model ([34, Fig. 3],[19, p.101]). In that case the hidden variable space H in our notation should be identified with the external state space E and the model (in our sense) is a conditional distribution induced by the stationary distribution:

$$\phi(s \mid \eta) := p(s \mid \eta). \tag{84}$$

Note that, this choice of a model by itself does not immediately tell us whether the FEP reasoner does filtering or just inference in the sense of Definition 3. A model like $\phi(s|\eta)$ can be part of a filtering kernel κ as well. In both cases we also need an interpretation map.

D.3   Interpretation map

For the interpretation map ψ_H we need a kernel of type $M \to P(E)$. Indeed, a kernel that has the right type can be identified in the FEP literature. This kernel is denoted $q_\mu(\eta)$ and we will identify $\psi_H(\eta|\mu) = q_\mu(\eta)$. The kernel's definition, however, relies on another assumption of the FEP, namely the existence of a "synchronisation map" $\sigma : M \to E$. To construct σ let us first define two other functions $g_M : S \to M$ and $g_E : S \to E$ via

$$\begin{aligned} f_M(s) &:= \mathbb{E}_{p(\mu|s)}[\mu] \\ f_E(s) &:= \mathbb{E}_{p(\eta|s)}[\eta] \end{aligned} \tag{85}$$

and then set

$$\sigma(\mu) := f_E(f_M^{-1}(\mu)) \tag{86}$$

which is assumed to be well defined. For details on when this exists in the linear case see [1, 16]. With this we can define $q_\mu(\eta)$ and in turn the interpretation map ψ_H. This maps an internal state μ to the Gaussian distribution with mean value equal to $\sigma(\mu)$:

$$\psi(\eta|\mu) := q_\mu(\eta) := \mathcal{N}(\eta; \sigma(\mu), \Sigma(\mu)) \tag{87}$$

where the variance $\Sigma(\mu)$ is defined as the variance of the best Gaussian approximation to the model $p(s|\eta = \sigma(\mu))$ when the external state is equal to $\sigma(\mu)$ [34, Eq. 2.4]. Note that in [16] the whole stationary distribution is assumed as Gaussian and so $p(\eta|\mu)$ in the corresponding equation in that publication (i.e. Eq. 3.3) is also a Gaussian.

In conclusion, the necessary ingredients for something like a Bayesian reasoner seem to be present in the FEP literature. One thing that is special about the FEP reasoner is that its model κ and interpretation map ψ_H are derived from features of the process that the machine is embedded in.

We do not know whether there is an appropriate notion of consistency equation that the FEP reasoner obeys. Presumably, instead of the equation for exact inference that we have presented, such an equation would express the idea that the FEP reasoner performs approximate inference in the form of free energy minimisation. Other differences are that the FEP takes place in continuous time, and perhaps more significantly, that it deals with deliberate actions as well as inference. However, it is not inconceivable that these could be expressed in the form of a consistency equation.

In the current formulations of the FEP, the interpretation is derived from the properties of the 'true' environment, such as the stationary distribution, or the synchronisation map σ. In our consistency equation approach, this need not be the case, since a reasoner's beliefs only need to be consistent and need not be correct. This means in particular that no stationarity assumption is needed.

Nonetheless, perhaps an important idea behind the FEP is that the model that most closely corresponds to the true environment can be considered the best one. A consistency equation approach would still be helpful, in order to systematically explore whether and how interpretations should relate to the larger process in which the machine is embedded.

References

1. Aguilera, M., Millidge, B., Tschantz, A., Buckley, C.L.: How particular is the physics of the Free Energy Principle? arXiv:2105.11203 (2021). http://arxiv.org/abs/2105.11203

2. Albantakis, L., Massari, F., Beheler-Amass, M., Tononi, G.: A macro agent and its actions. arXiv:2004.00058 (2020). http://arxiv.org/abs/2004.00058

3. Ay, N., Löhr, W.: The Umwelt of an embodied agent-a measure-theoretic definition. Theory Biosci. = Theorie in Den Biowissenschaften 134(3–4), 105–116 (2015). https://doi.org/10.1007/s12064-015-0217-3

4. Baez, J., Stay, M.: Physics, topology, logic and computation: a rosetta stone. In: Coecke, B. (ed.) New Structures for Physics, Lecture Notes in Physics, pp. 95–172. Springer, Heidelberg (2011). https://doi.org/10.1007/978-3-642-12821-9_2

5. Beer, R.D.: Autopoiesis and cognition in the game of life. Artif. Life 10(3), 309–326 (2004). https://doi.org/10.1162/1064546041255539

6. Beer, R.D.: The cognitive domain of a glider in the game of life. Artif. Life 20(2), 183–206 (2014). https://doi.org/10.1162/ARTL_a_00125

7. Biehl, M., Ikegami, T., Polani, D.: Towards information based spatiotemporal patterns as a foundation for agent representation in dynamical systems. In: Proceedings of the Artificial Life Conference 2016, pp. 722–729. The MIT Press (2016). https://doi.org/10.7551/978-0-262-33936-0-ch115, https://mitpress.mit.edu/sites/default/files/titles/content/conf/alife16/ch115.html

8. Biehl, M., Kanai, R.: Dynamics of a bayesian hyperparameter in a markov chain. In: IWAI 2020. CCIS, vol. 1326, pp. 35–41. Springer, Cham (2020). https://doi.org/10.1007/978-3-030-64919-7_5

9. Biehl, M., Pollock, F.A., Kanai, R.: A technical critique of some parts of the free energy principle. Entropy 23(3), 293 (2021). https://doi.org/10.3390/e23030293, https://www.mdpi.com/1099-4300/23/3/293

10. Bolt, J., Hedges, J., Zahn, P.: Bayesian open games. arXiv:1910.03656 (2019). http://arxiv.org/abs/1910.03656
11. Capucci, M., Gavranović, B., Hedges, J., Rischel, E.F.: Towards foundations of categorical cybernetics. arXiv:2105.06332 (2021). http://arxiv.org/abs/2105.06332
12. Capucci, M., Ghani, N., Ledent, J., Forsberg, F.N.: Translating Extensive Form Games to Open Games with Agency. arXiv:2105.06763 (2021). http://arxiv.org/abs/2105.06332
13. Cho, K., Jacobs, B.: Disintegration and bayesian inversion via string diagrams. Math. Struct. Comput. Sci. **29**(7), 938–971 (2019). https://doi.org/10.1017/S0960129518000488, http://arxiv.org/abs/1709.00322, arXiv: 1709.00322
14. Coecke, B., Paquette, É.: Categories for the practising physicist. In: Coecke, B. (ed.) New Structures for Physics, Lecture Notes in Physics, pp. 173–286. Springer Heidelberg (2011). https://doi.org/10.1007/978-3-642-12821-9_3
15. Coecke, B., Kissinger, A.: Picturing Quantum Processes: A First Course in Quantum Theory and Diagrammatic Reasoning. Cambridge University Press, Cambridge (2017)
16. Da Costa, L., Friston, K., Heins, C., Pavliotis, G.A.: Bayesian Mechanics for Stationary Processes. arXiv:2106.13830 [math-ph, physics:nlin, q-bio] (2021). http://arxiv.org/abs/2106.13830, arXiv: 2106.13830
17. Dennett, D.C.: True believers : the intentional strategy and why it works. In: Heath, A.F. (ed.) Scientific Explanation: Papers Based on Herbert Spencer Lectures Given in the University of Oxford, pp. 53–75. Clarendon Press (1981)
18. Fong, B., Spivak, D.I.: An invitation to applied category theory: seven sketches in compositionality. Cambridge University Press, Cambridge (2019)
19. Friston, K.: A free energy principle for a particular physics. arXiv:1906.10184 [q-bio] (2019). http://arxiv.org/abs/1906.10184, arXiv: 1906.10184
20. Friston, K., Da Costa, L., Hafner, D., Hesp, C., Parr, T.: Sophisticated Inference. Neural Comput. **33**(3), 713–763 (2021). https://doi.org/10.1162/neco_a_01351
21. Friston, K.J., Da Costa, L., Parr, T.: Some interesting observations on the free energy principle. Entropy **23**(8), 1076 (2021). https://doi.org/10.3390/e23081076, https://www.mdpi.com/1099-4300/23/8/1076
22. Fritz, T.: A synthetic approach to Markov kernels, conditional independence and theorems on sufficient statistics. Adv. Math. **370**, 107239 (2020). https://doi.org/10.1016/j.aim.2020.107239, https://www.sciencedirect.com/science/article/pii/S0001870820302656
23. Jacobs, B.: A channel-based perspective on conjugate priors. Math. Struct. Comput. Sci. **30**(1), 44–61 (2020). https://doi.org/10.1017/S0960129519000082, https://www.cambridge.org/core/journals/mathematical-structures-in-computer-science/article/channelbased-perspective-on-conjugate-priors/D7897ABA1AA06E5F586F60CB21BDDB32
24. Jacobs, B.: A Channel-Based Perspective on Conjugate Priors. arXiv:1707.00269 (2018). http://arxiv.org/abs/1707.00269
25. Jacobs, B., Staton, S.: De Finetti's construction as a categorical limit. In: Petrişan, D., Rot, J. (eds.) Coalgebraic Methods in Computer Science, Lecture Notes in Computer Science, pp. 90–111. Springer, Cham (2020). https://doi.org/10.1007/978-3-030-57201-3_6
26. Knill, D.C., Pouget, A.: The Bayesian brain: the role of uncertainty in neural coding and computation. Trends Neurosci. **27**(12), 712–719 (2004). https://doi.org/10.1016/j.tins.2004.10.007, https://www.cell.com/trends/neurosciences/abstract/S0166-2236(04)00335-2

27. Kolchinsky, A., Wolpert, D.H.: Semantic information, autonomous agency and non-equilibrium statistical physics. Interface Focus **8**(6), 20180041 (2018). https://doi.org/10.1098/rsfs.2018.0041, https://royalsocietypublishing.org/doi/full/10.1098/rsfs.2018.0041

28. Krakauer, D., Bertschinger, N., Olbrich, E., Flack, J.C., Ay, N.: The information theory of individuality. Theory Biosci. **139**(2), 209–223 (2020). https://doi.org/10.1007/s12064-020-00313-7

29. Libby, E., Perkins, T.J., Swain, P.S.: Noisy information processing through transcriptional regulation. Proc. Natl. Acad. Sci. **104**(17), 7151–7156 (2007)

30. Ma, W.J., Jazayeri, M.: Neural coding of uncertainty and probability. Ann. Rev. Neurosci. **37**, 205–220 (2014). https://doi.org/10.1146/annurev-neuro-071013-014017

31. McGregor, S.: The bayesian stance: equations for 'as-if' sensorimotor agency. Adapt. Behav., 105971231770050 (2017). https://doi.org/10.1177/1059712317700501, http://journals.sagepub.com/doi/10.1177/1059712317700501

32. Nakamura, K., Kobayashi, T.J.: Connection between the bacterial chemotactic network and optimal filtering. Phys. Rev. Lett. **126**(12), 128102 (2021). https://doi.org/10.1103/PhysRevLett.126.128102, https://link.aps.org/doi/10.1103/PhysRevLett.126.128102

33. Orseau, L., McGill, S.M., Legg, S.: Agents and Devices: A Relative Definition of Agency. arXiv:1805.12387 (2018). http://arxiv.org/abs/1805.12387

34. Parr, T., Da Costa, L., Friston, K.: Markov blankets, information geometry and stochastic thermodynamics. Phil. Trans. Roy. Soc. A Math. Phys. Eng. Sci. **378**(2164), 20190159 (2020). https://doi.org/10.1098/rsta.2019.0159, https://royalsocietypublishing.org/doi/full/10.1098/rsta.2019.0159

35. Risken, H., Frank, T.: The Fokker-Planck equation: methods of solution and applications. In: Springer Series in Synergetics, 2 edn. Springer-Verlag, Heidelberg (1996). https://doi.org/10.1007/978-3-642-61544-3, https://www.springer.com/gp/book/9783540615309

36. Rosas, F.E., Mediano, P.A.M., Biehl, M., Chandaria, S., Polani, D.: Causal blankets: theory and algorithmic framework. In: IWAI 2020. CCIS, vol. 1326, pp. 187–198. Springer, Cham (2020). https://doi.org/10.1007/978-3-030-64919-7_19

37. Smithe, T.S.C.: Bayesian Updates Compose Optically. arXiv:2006.01631 (2020). http://arxiv.org/abs/2006.01631

38. St Clere Smithe, T.: Cyber kittens, or some first steps towards categorical cybernetics. Electron. Proc. Theor. Comput. Sci. **333**, 108–124 (2021). https://doi.org/10.4204/EPTCS.333.8, http://arxiv.org/abs/2101.10483v1

39. Still, S., Sivak, D.A., Bell, A.J., Crooks, G.E.: The thermodynamics of prediction. Phys. Rev. Lett. **109**, 120604 (2012). arXiv e-print 1203.3271 http://arxiv.org/abs/1203.3271

40. Wikipedia contributors: Conjugate prior – Wikipedia, the free encyclopedia (2021). https://en.wikipedia.org/w/index.php?title=Conjugate_prior&oldid=1030202570, Accessed 8 July 2021

Blankets All the Way up – the Economics of Active Inference

Morten Henriksen(✉) ⓘ

Ministry of Defence, Herningvej 30, 7470 Karup, Denmark
acw-oe-01@mil.dk

Abstract. A direct implication of active inference, by way of minimizing expected free energy, is the ability to reframe optimization problems as they relate to biological systems. Instead of employing objective functions in order to maximizing an agent's exposure to some exogenous measurable quantity, active inference describes how biological systems optimize by minimizing a divergence (KL) between a posterior probability density and a generative density, by definition endogenous to the system. This particular framework can be shown to underwrite many seemingly disparate disciplines in economics, and may prove to be a source of new insights for the field.

Keywords: Active inference · Complexity economics · Behavioural economics · Decision theory

1 Introduction

The free energy principle states that any biological organism capable of existing over a period of time must minimize entropy/surprise, formally described as minimizing a bound on free energy, or minimizing a KL divergence between a posterior density and a generative density. This moves the objective away from a maximization scheme of external quantities, towards the minimization of an internal energy bound afforded by a generative model from which external states are inferred. When taking the expectation, the imperative now becomes to optimize beliefs about world states, rather than maximizing the expected utility of a world state [1–3]. Minimizing expected free energy can therefore be seen as a way for any system to minimize entropy with respect to a policy, or a plan of action, connecting present states with future states [4]. This particular framework can be shown to underwrite many seemingly disparate disciplines in economics, and may prove to be a source of new insights for the field. In having an active inference framework underwriting economics, as opposed to the more dominant rational expectations-based general equilibrium approach [5], there will naturally be a move away from "traditional" neo-classical economics and comparative statics, towards a greater appreciation of complexity. In general this opens the field of economics up to a wider area of research in which contributions from neurology, psychology, biology ecology, information- and complexity theory can aid in the further understanding and modeling of economic systems. In particular, active inference provides a first principle

© Springer Nature Switzerland AG 2021

M. Kamp et al. (Eds.): ECML PKDD 2021 Workshops, CCIS 1524, pp. 763–771, 2021.
https://doi.org/10.1007/978-3-030-93736-2_53

account from which complex systems can evolve, and hereby a basis for hypothesis and theory generation to areas of inquiry dominated by a more computationally inspired approach, as is the case in the field of complexity economics [6, 7].

2 Expected Free Energy and Active Inference

While the minimization of variational free energy is a general principle governing how organic (self-organizing) systems exist over time, minimizing expected free energy can be interpreted as extending this principle in order to account for planning [8]. It is in this arena that active inference takes form, and likewise the arena in which the free energy principle could prove useful to the field of economics.

If we start by stating the general principle, free energy (F) can be written as:

$$F = D_{KL}[Q(S_t|O_t; \emptyset)P(O_t, S_t)]. \tag{1}$$

Where $Q(S_t|O_t; \emptyset)$ is a variational posterior, $Q(S_t)$ is a variational prior and $\emptyset$ is a variational parameter.

This can be decomposed into:

$$
\begin{aligned}
F &= E_{Q(S_t|O_t; \emptyset)}\left[ln \frac{Q(S_t|O_t; \emptyset)}{P(O_t, S_t)}\right] \\
&= -E_{Q(S_t|O_t; \emptyset)}[lnP(O_t|S_t)] + D_{KL}[Q(S_t|O_t; \emptyset)P(S_t)]
\end{aligned} \tag{2}
$$

Resulting in a (negative) accuracy/energy term, plus a complexity/entropy term, specifying the objective to minimize entropy or complexity in order to maximize accuracy. When minimizing expected free energy (G), the objective becomes to minimize the probability distribution of a policy (path integral) $P(\pi)$, such that $G = -\ln P(\pi)$. Here an optimal policy, denoted $Q^*(\pi)$, will be given by $\sigma\left(\sum_t^T G_t(\pi)\right)$, where $\sigma(x)$ is a softmax function. We can therefore write [9]:

$$
\begin{aligned}
G_t(\pi) &= E_{Q(O_t, S_t|\pi)}[lnQ(S_t|\pi) - ln\tilde{p}(O_t|S_t)] \\
&\approx E_{Q(O_t, S_t|\pi)}[lnQ(S_t|\pi) - ln\tilde{p}(O_t) - lnQ(S_t|O_t)] \\
&\approx -E_{Q(O_t, S_t|\pi)}[ln\tilde{p}(O_t)] - E_{Q(O_t|\pi)}D_{KL}[Q(S_t|O_t)Q(S_t|\pi)]
\end{aligned} \tag{3}
$$

Decomposing expected free energy into an extrinsic value term, often described as a goal directed term or pragmatic value, and an intrinsic value term, which can be interpreted as an epistemic value term, or simply information gain [10]. Here, minimizing expected free energy becomes a mixture of pragmatic and epistemic considerations where maximizing exposure to information is used to indicate the appropriateness of a policy (epistemic value), given a specific goal (pragmatic value). Naturally there exists a trade-off between epistemic and pragmatic value, where both can be shown in isolation to be equally valid strategies for minimizing expected free energy, but in different ways. However, when connecting future states with present states by selecting a trajectory expected to minimize surprise, this trajectory can only be propositional when considering complexity. The difference is comparable to planning and executing a move from A to B considering what is known about the environment, or planning and executing a move from A to B considering what is known about the environment, while in heavy traffic. None withstanding, the principle of least action, or indeed, least effort [11], must apply, given that all optimal policies are referencing a path integral with a minimum expected time average.

Note that the only way for an agent to minimize expected free energy, is to actively bring desired future states into existence through action, and while this may seem like a trivial statement, it points to a very interesting implication regarding the concept of planning as inference [12]. In essence there is only action, encoded as an expected sequence to be performed, here represented as intensity (energy divergence) as a function of time. What this means, is that we can treat an expected action sequence and an expected time sequence interchangeably, the expected time sequence being a function of the expected action sequence and vice versa. This quickly moves us into the realm of time perception, and gives us the ability to hypothesise about the various ways time perception can alter in response to both the work of the system in the present, and the expected work to be done in the future [13, 14]. As such, time perception will be influenced by surprise, as well as the expected time average of surprise (entropy). If nothing more, this gives us the ability to speak in terms of urgency when considering various policies, and therefore various degrees of "intensity" connected to different utilities. More generally, we observe a system that changes dynamically in response to information, and it is this "adaptive" ability that interferes with more linear formulations of agent behaviour, exemplified in the expected utility theory for instance [3, 5–7]. This does not mean however, that it becomes impossible to derive general statements about agent behaviour from an active inference perspective, quite the contrary. The presence of a generative model from which external (hidden) states are initially inferred, and the presence of the epistemic value term (information gain) to which the generative model must adapt, actually provides many opportunities for a priori statements concerning agent behaviour and economics more generally.

3   Discounted Utility

Equation 3 can be read as describing the probability of occupying an expected future state given a specific trajectory (policy) generated by the generative model (prior and likelihood). Here there will be an attraction to future states of high expected probability, reflecting what could be interpreted as expected utility, if the prior density is associated with "prior preferences" [15]. The trajectory is controlled by information, where low probability states are equivalent to states of high surprise. As such, low probability states command a low expected utility, the avoidance of which is dependent upon the minimization of surprise or negative log model evidence ($-\ln p(O_t)$). Given the time subscript (t), longer policy sequences will be associated with higher probabilities of occupying more surprising states, reflecting an increase in total entropy over time. This is in part due to the use of approximate Bayesian inference that is a consequence of the variational treatment on free energy show in Eq. 1. If expected utility then is associated with states of high probability, expected utility must fall over time in proportion to an increase in entropy; naturally giving us discounted utility functions, as well as a general theory of time preference.[1]

4 Probability and Utility Spaces

Things will, however, become a lot more interesting when considering the full impact of epistemic value or information gain. Here information gain will inherently be uncertainty resolving given the hidden states (S_t) in the environment. Because of this, the objective function governing the policy trajectory cannot take any specific value, or indeed be defined with a terminal, but must in a sense be "discovered" due to the unknown cost functions governed by surprise/information gain. As such, the objective function is simply to minimize more or less hidden cost functions, and by this measure minimize surprise or entropy through adaptation. What this means for the specifics of any given utility function, is that it cannot technically be connected to the utility either X, Y or Z, since this quantity cannot be evaluated at time t_0. What is evaluated is a utility space, in conjuncture with a probability space, describing a set of solutions (utilities) the most optimal of which, is "discovered" through action and adaptation. If, however, priors are allowed to perfectly model hidden states, then active inference could be described without cost functions [16]. This would however render the variational treatment moot, as the variational density under the "true" posterior would reflect what for all intents and purposes looks like exact Bayesian inference. While this approach could prove very useful in modelling autonomic responses, it can only take us so far when considering complexity.

The presence of a utility space, as opposed to specific utility given by revealed preferences,[2] means that we can start to combine intuitions, ad hoc observations and various formalisms regarding the notions of heuristics, metaheuristics and satisficing behaviour[3] [17]. At t_0, expected utility will be subject to a discount function, but since this function is a consequence of time sensitive uncertainty about state transition, the utility of any single prospect is not merely residing in a probability space, but must itself be probabilistic in nature. This means that any specific preferred future state, and the utility that this state represents, does not exist as such, prior to occupying the preferred state. What utility instead is "reduced" to, is a policy inferring expected surprise in accordance with least action, describing a categorical representation of preferred future

---

[1] The observation that goods or services are preferred sooner rather than later, all else being equal. In economics, the prefixes "high" or "low" is sometimes used in order to differentiate between various levels of "impatience". High time preference agents will value time at a high rate and display high levels of impatience, while low time preference agents will display low levels of impatience. High or low time preferences are therefor often used as explanations for various levels of propensities to consume or save in an economy.

[2] Revealed preferences is a way to infer an agents utility function by observing past behaviour. As such, agents cannot change their preferences once they have been revealed, since this would change the utility function and hereby greatly complicate economic modelling practices. The concept of revealed preferences is tightly linked to the transitivity axiom [18, 19].

[3] The idea, that in situations where optimal solutions do not exist, an agent will search until a solution is deemed to be good enough.

states. Subjective value will therefore be governed by the specifics of the policy trajectory, and not the expected utility of any specific prospect. Intuitively we must admit that any preferred future state is colloquially a "figment of our imagination", as the information describing this state cannot be taken fully into account. In fact, the more specific the description of any future state, the higher the probability of this state being purely fictitious and therefore surprising. Maintaining a specific description of a future state would hereby not be a tenable strategy for optimization given the potential amplitude of the loss functions, and as a consequence, the inability to effectively minimize surprise.

5   Active Inference and Biases

Still occupying t_0, the utility space represents options, that at first glance looks like indifference, that is, equalities between various preferred states or utilities. However, the selection of a policy from the generative model, must favour some states in the utility space over others when this policy is acted out. In concert with information update revealing hidden states in the environment as they occur, continuous movement along a policy trajectory will favour fewer and fewer states until only one remains. This means that the ex post policy trajectory was selected for by the environment in conjuncture with the generative model, making the attached utility or subjective value of a given prospect a function of beliefs about how states in the world unfold, the uncertainty these beliefs entail as an increasing function of time, as well as the actual unfolding of events in the world forcing the need for adaptation. Isolating a single specific good from our utility space now allows us see how the attached utility or subjective value of this good changes along the policy trajectory in response to the minimization of surprise, as the good variably moves further away, or closer to, the agent in time and effort. This also allows us to see why a good once obtained tend to command a higher subjective value than a comparable or identical good not yet obtained, as is the case with the endowment effect [20]. The higher subjective value place on things close to/visible/understandable/not hidden, similarly aids in explaining the status quo bias, as well as various anchoring effects [20, 21]. The anchoring effect is however a curious case, since active inference demonstrates how "value" is ultimately determined *actively* when beliefs are acted out, rather than simply stated. What agents believe some good is "worth" based on, or not based on, anchoring effects, is not a good indicator of what agents will do in order to obtain a good in the final analysis when cost functions or surprise is taken into account. Interestingly, many of the implications of active inference in economics can be recapitulated in terms of transaction cost economics, where the goal is to formalize and take into account the costs associated with being an economic agent given such things as bounded rationality and imperfect information [22–24]. The connection between transaction cost economics and active inference, will however not be treated in this presentation, but could prove to be a very interesting area of study going forth.

6 A Simple Model

Ultimately, subjective value is determined by effort, which is to say that effort likewise commands a "value". One aspect is the rate of remuneration[4] (RR) that active inference implies; another is the increasing present value[5] (PV) of various prospects as a function of time preference. The RR can be shown as an increasing function of time, while the PV, or inverse discount rate, is a decreasing function of time. As a simplification, or heuristic if you will, we can view the PV curve as reflecting the extrinsic value term in Eq. 3, and the RR as reflecting the intrinsic value term. To qualify this statement, we can refer to the previous discussion on discount rates concerning extrinsic value. Depicting intrinsic value as an increasing rate of remuneration will however need to be elaborated. Intrinsic value is in the literature [10] often interpreted as information gain when viewing free energy minimization from an information theoretic standpoint. Moving in time reveals hidden states, and to the extent that these states do not match expectations, they will be surprising for the system. Minimizing surprise prompts the system to adapt, generating cost functions (complexity cost) to which a given future prospect must be able to remunerate. Over a longer and longer time interval, the cost functions add up in proportion to a higher and higher rate of remuneration. As such, the RR communicates the "cost of acting" over a time period given a hypothetical discount rate on the future. Overlaying the two functions denoting the X-axis as time/action and the Y-axis as entropy, shows the trade-off between exploration and exploitation, the intersect representing the "sweet spot" between this trade-off. Here the PV (discount) rate will communicate uncertainty or ambiguity with regards to state transitions, and the RR communicates the cost of "foraging" uncertainty resolving information. We can therefore also view the RR as the *active* part of active inference, and the PV as the *inference* part.

While the entropy axis is self-explanatory, the time/action axis can be thought of as the level of urgency in the system, that is, the amount of work to be done in a given time frame. Shifting the PV schedule to the right will hereby depict a system with low uncertainty about state transitions, commensurate with a well explored environment. The intersect is now higher on the RR, where the cost of further exploration likewise is higher, rendering exploitation a relatively more preferable strategy. This is demonstrated by the intersect being further along the time/action axis, commensurate with a less urgent system. Conversely, a leftward shift in PV schedule depicts a system with high

---

[4] There is at present no corollary in the economic literature to a rate of remuneration as used in this presentation. Normally a remuneration rate simply refers to a salary or stream of payments due for work or services rendered. Here, a rate of remuneration refers the expected energy input for a system given energy output, where the minimum requirement is long run homeostasis. For this reason the term may be ill conceived. Conversely, the term as used herein perfectly captures the observation, that agents have a preference for increasing over declining sequences not strictly "permissible" under a rational expectations framework [25, 26].

[5] Normally present value refers to the value at present of a discounted future cash flow where $PV = \frac{CF}{(1+r)^n}$. Here, the term refers to subjective value given a time component. As such, it is the expected value of something that by necessity must lie in the future, and therefore must be discounted to some degree, considering a generative model that takes surprise or uncertainty into account. We can therefore also treat present value and utility (discounted) as interchangeable.

uncertainty about state transitions, imitating perhaps a poorly explored environment. Here, the intersect is closer on the time/action axis, as uncertainty forces the system to act in a more urgent manner. The intersect is likewise lower on the RR, meaning that epistemic foraging comes at a lower cost, rendering exploration a relatively more preferable strategy.

Interestingly, Milling et al. [9] have proposed an amendment to the expected free energy formulation, where information gain is penalized instead of encouraged. Rather than being an alternative approach to expected free energy, one the paper has termed "free energy of the future (FEF)", the two formulations might actually depict two equally valid strategies under different generative models, emphasizing either exploration or exploitation respectively as demonstrated in the model above [9].

$$
\begin{aligned}
G_t(\pi) &= E_{Q(O_t,S_t|\pi)}\big[lnQ(S_t|O_t) - ln\tilde{p}(O_t|S_t)\big] \\
&= E_{Q(O_t|\pi)}D_{KL}\big[Q(S_t|O_t)\tilde{p}(O_t|S_t)\big] \\
&\approx -E_{Q(O_t,S_t|\pi)}\big[ln\tilde{p}(O_t|S_t)\big] + E_{Q(O_t|\pi)}D_{KL}\big[Q(S_t|O_t)Q(S_t|\pi)\big]
\end{aligned}
\tag{4}
$$

Where a shift in the PV schedule represents various levels of uncertainty about state transitions, a shift in the RR schedule represents surprise, either positive for a leftward shift or negative for a rightward shift. Here we see how the present value or utility of any given prospect changes in response to surprise, as the prospect variably moves closer to, or further away from the agent in time and effort. The Y-axis, at present denoted simply as entropy, will be analogous to system stability. While the model lacks a dimension necessary to represent this dynamic, it will none the less be an a priori consequence of the underlying mathematics. A low degree of uncertainty about state transitions is commensurate with intersects that lies comparatively higher on the Y-axis, and a high degree of uncertainty about state transitions lies comparatively lower on the Y-axis. In both examples there will be trade-offs, where low uncertainty generates stability, but at the same time, rigidity, locked-in processes and path dependency. High uncertainty generates adaptability and flexibility, as well as volatility and unpredictability commensurate with a lower probability of "correctly" inferring expected state transitions.

7 Conclusion

In this presentation, I hope to have demonstrated how active inference can help to inform the field of economics. Apart from elucidating on various biases in agent behaviour, active inference also holds promising implications for price, interest and transaction cost theory by providing a rigid principle for agency and organization. More than this, active inference provides an intuitive and simple avenue for the introduction of complexity into classical or standard economic theory, with the possible implication of being able to aid in the modeling and analysis of economic data and events going forth.

References

1. Ramstead, M.J., Kirchhoff, M.D., Friston, K.J.: A tale of two densities: active inference is enactive inference. Adapt. Behav. **28**(4), 225–239 (2020). https://doi.org/10.1177/105971231 9862774
2. Friston, K.J., Schwartenbeck, P., FitzGerald, T., Moutoussis, M., Behrens, T., Dolan, R.J.: The anatomy of choice: active inference and agency. Front. Hum. Neurosci. **25**, 598 (2013). https://doi.org/10.3389/fnhum.2013.00598
3. Henriksen, M.: Variational free energy and economics optimizing with biases and bounded rationality. Front. Psychol. **11**, 549187 (2020). https://doi.org/10.3389/fpsyg.2020.549187
4. Friston, K.J., FitzGerald, T., Rigoli, F., Schwartenbeck, P., O'Doherty, J., Pezzulo, G.: Active inference and learning. Neurosci. Biobehav. Rev. **68**, 862–879 (2016). ISSN 0149-7634. https://doi.org/10.1016/j.neubiorev.2016.06.022
5. Arthur, W.B.: Foundations of complexity economics. Nat. Rev. Phys. **3**, 136–145 (2021). https://doi.org/10.1038/s42254-020-00273-3
6. Arthur, W.B.: Complexity and the Economy. Oxford University Press, Oxford (2015). ISBN 978-0-19-933429-2
7. Farmer, J., Foley, D.: The economy needs agent-based modelling. Nature **460**, 685–686 (2009). https://doi.org/10.1038/460685a
8. Parr, T., Friston, K.J.: Generalised free energy and active inference. Biol. Cybern. **113**(5–6), 495–513 (2019). https://doi.org/10.1007/s00422-019-00805-w
9. Milling, B., Tschantz, A., Buckley, C.L.: Whence the Expected Free Energy? Neural Comput. **33**(2), 447–482 (2021). https://doi.org/10.1162/neco_a_01354
10. Friston, K.J., Rigoli, F., Ognibene, D., Mathys, C., Fitzgerald, T., Pezzulo, G.: Active inference and epistemic value. Cogn. Neurosci. **6**(4), 187–214 (2015). https://doi.org/10.1080/175 88928.2015.1020053
11. Kim, C.S.: Bayesian mechanics of perceptual inference and motor control in the brain. Biol. Cybern. **115**(1), 87–102 (2021). https://doi.org/10.1007/s00422-021-00859-9
12. Botvinick, M., Toussaint, M.: Planning as inference. Trends Cogn. Sci. **16**(10), 485–488 (2012). https://doi.org/10.1016/j.tics.2012.08.006
13. Roseboom, W., Fountas, Z., Nikiforou, K.: Activity in perceptual classification networks as a basis for human subjective time perception. Nat. Commun. **10**, 267 (2019). https://doi.org/10.1038/s41467-018-08194-7
14. Zakharov, A., Crosby, M., Fountas, Z.: Episodic Memory for Learning Subjective-Timescale Models (2020). arXiv:2010.01430
15. Lopez-Parsem, A., Domenech, P., Pessiglione, M.: How prior preferences determine decision-making frames and biases in the human brain. eLife **5**, e20317 (2016). https://doi.org/10.7554/eLife.20317
16. Friston, K., Samothrakis, S., Montague, R.: Active inference and agency: optimal control without cost functions. Biol. Cybern. **106**, 523–541 (2012). https://doi.org/10.1007/s00422-012-0512-8
17. Simon, H.A.: Rational choice and the structure of the environment. Psychol. Rev. **63**(2), 129–138 (1956). https://doi.org/10.1037/h0042769
18. Samuelson, P.: A note on the pure theory of consumers' behaviour. Econometrica **5**, 61–71 (1938). https://doi.org/10.2307/2548836
19. Von Neumann, J., Morgenstern, O.: Theory of Games and Economic Behaviour. Princeton University Press, Princeton, NJ (1947)
20. Kahneman, D., Knetsch, J.L., Thaler, R.: Anomalies: the endowment effect, loss aversion, and status Quo bias. J. Econ. Perspect. **5**, 193–206 (1991). https://doi.org/10.1257/jep.5.1.193

21. Ariely, D., Loewenstein, G., Prelec, D.: Coherent arbitrariness: stable demand curves without stable preferences. Q. J. Econ. **118**(1), 73–106 (2003). https://doi.org/10.1162/00335530360535153
22. Coase, R.H.: The problem of social cost. J. Law Econ. **3**, 63–94 (1960)
23. Williamson, O.: The economics of organization – the transaction cost approach. Am. J. Sociol. **87**(3), 548–577 (1981). https://www.researchgate.net/publication/235356934_The_Economics_of_Organization_The_Transaction_Cost_Approach
24. Williamson, O.: Transaction Cost Economics - An Introduction; Discussion Paper No. 2007-3 (2007). http://www.economics-ejournal.org/economics/discussionpapers/2007-3
25. Loewenstein, G.F., Prelec, D.: Preferences for sequences of outcomes. Psychol. Rev. **100**(1), 91–108 (1993). https://doi.org/10.1037/0033-295X.100.1.91
26. Scholten, M., Read, D.: Better is worse, worse is better: Violations of dominance in intertemporal choice. Decisions **1**(3), 215–222 (2014). https://doi.org/10.1037/dec0000014

Filtered States: Active Inference, Social Media and Mental Health

Ben White[1] and Mark Miller[2,3](✉)

[1] University of Sussex, Brighton, UK
[2] Center for Human Nature, Artificial Intelligence and Neuroscience,
Hokkaido University, Sapporo, Japan
[3] Center for Consciousness and Contemplative Studies,
Monash University, Melbourne, Australia
markmiller@chain.hokudai.ac.jp

Abstract. Social media is implicated today in an array of mental health concerns. While worries around social media have become mainstream, little is known about the specific cognitive mechanisms underlying the correlations seen in these studies, or why we find it so hard to stop engaging with these platforms when things obviously begin to deteriorate for us. New advances in computational neuroscience are now perfectly poised to shed light on this matter. In this paper we approach these concerns around social media and mental health issues, including the troubling rise in Snapchat surgeries, depression and addiction, through the lens of the Active Inference Framework (AIF).

Keywords: Active inference · Social media · Depression · Addiction

1 Introduction

Levi Jed Murphy smoulders into the camera. It's a powerful look: piercing eyes, a razor-sharp jawline, and high cheekbones, which according to Levi himself cost around £30,000 pounds [17]. Levi is an influencer from the UK, with a large social media following. Speaking on growing his following, Levi reveals that if a picture doesn't receive a certain number of likes within a set time, then it gets deleted, and that the surgeries are simply a way to achieve this rapid validation: "it's important to be good looking for social media, because obviously I want to attract an audience", he states [47]. While the filter that inspired Murphy's surgeries has since been banned, many similar ones are still available, and Murphy's story highlights growing concerns about a phenomenon now dubbed 'Snapchat surgery'. One survey of young people online found that nearly half had felt influenced by social media to consider cosmetic surgery [4]. While these concerns around social media and cosmetic surgery are recent, they can be added to a litany of worries about the effect of social media use on mental health and general wellbeing.

Today, social media is implicated in an array of mental health concerns. A 2017 report published by a parliamentary group in the UK linked social media use with a range of worries about mental health [9], while a growing number of

© Springer Nature Switzerland AG 2021
M. Kamp et al. (Eds.): ECML PKDD 2021 Workshops, CCIS 1524, pp. 772–783, 2021.
https://doi.org/10.1007/978-3-030-93736-2_54

empirical studies link social media use with symptoms of addiction and depression [3,33]. Worries that social media platforms might in some way warp our perception of the world, or cause low self-esteem or diminished life satisfaction, seem to be running through the mainstream collective psyche and even some former influencers have begun to turn against various social media platforms, highlighting the dangers of curating a self-image with little purchase on reality [21]. In response, some platforms have begun trialling design tweaks aimed at protecting user's health, such as limiting the visibility of 'likes' on a post.

While concerns around social media have become mainstream, little is known about the specific cognitive mechanisms underlying the correlations seen in these studies, and why we find it so hard to stop engaging with these platforms when things obviously begin to deteriorate for us. In what follows, we suggest that both the rise in Snapchat surgery, and the connections between social media, depression and addiction, can be accounted for via a unified theoretical approach grounded in an emerging, and now highly influential, theory of cognition and affect - the active inference framework (AIF). We propose that the structure of some social media platforms constitute what have been dubbed 'hyper-stimulating' digital environments, wherein the design features and functional architecture of digital environments impacts the machinery of cognition in ways which can lead to a warping of healthy agent-environment dynamics, producing precisely the sorts of pathological outcomes we see emerging today [48]. In what follows we will first briefly introduce the AIF. Next, we highlight how the same predictive mechanisms that keep us alive and well can also become warped, leading to aberrant feedback loops in cognition and behavior that help explain the various psychopathologies that are related today to social media and internet use. Finally, we will argue that various digital environments, including social media platforms, have very specific design features and mechanisms that leave our predictive systems particularly vulnerable to these kinds of suboptimal feedback loops.

2   Introducing Active Inference

The revolutionary move of the AIF is to reimagine the brain as a prediction engine constantly attempting to predict the sensory signals it encounters in the world and minimising the discrepancy ('prediction errors') between those predictions and the incoming signal [7,18,24]. To make apt predictions these systems need to build up a 'generative model': a structured understanding of the statistical regularities in our environment which are used to generate predictions. This generative model is essentially a model of our world, including both immediate, task-specific information, as well as longer-term information that constitutes our narrative sense of self. According to this framework, predictive systems can go about minimising prediction errors in two ways: either they update the generative model to reflect the world more accurately, or they behave in ways that bring the world better in line with their prediction [7]. In this way, the brain forms a part of an embodied predictive system which is always striving to move from uncertainty to certainty. By successfully minimising potentially harmful surprises these systems keep us alive and well. Consider the healthy and highly expected body temperature for a human being of 37°C. A shift in temperature in either direction registers as a spike in prediction error, signalling to the organism that it is moving into an unexpected, and therefore a potentially dangerous, state. If the change in temperature is not too extreme, we could just sit there and come to terms with the changing temperature (update our generative model), or we might reach for a blanket or open a window. In these cases what we're doing is acting upon our environment, sampling the world, and changing our relation to it, in order to bring ourselves back within acceptable bounds of uncertainty.

Predictive systems must be flexible and able to quickly adapt to changing conditions within an environment. According to the AIF the predictive system is flexibly tuned by second order predictions that estimate the salience and reliability of the error units resulting from first order predictions given the current context [39]. So called 'precision weighting' acts to modulate the impact that particular prediction errors have on the system. For example, high precision can drive learning and further processing, while low precision would render a signal relatively impotent within the system [8]. This mechanism allows the system contextual flexibility and can also allow greater reliance on either the generative model or sensory signals. Precision also plays a central role in selecting which behaviours are enacted, as actions are selected based on expectations about future error reduction. That is to say, predictive agents score and select behaviours based on predictions about the likely error-reducing capacities of those behaviors within a given context. In other words, precision is weighted on beliefs about policies, given the likelihood that in a given context a certain policy will lead to a certain reduction in error [19].

Crucial for this process is a sensitivity about how well we are managing error over time relative to expectations. 'Error dynamics' refers to changes in the rate of average error reduction over time [10, 27, 29]. The rate of change in error can be visualized as a 'slope', with steep decreases in error minimization representing that the system is doing well at confirming predictions, and a steep increase as a loss of predictive acuity. On the agent level, changes in error dynamics are experienced as valenced bodily affect (i.e. positive and negative feelings accompanied by approach or avoidance tendencies). When an organism registers a slope of error reduction in line with (or better than) its expectations, they are 'rewarded' with positively valenced affective changes. When the rate of prediction error rises (or the rate of reduction slows down) the organism is 'punished' with a negatively valenced affect [14]. These affective changes play a role in the AIF by tuning precision weighting on action policies. Positive or negative valence acts as feedback to the system, up-regulating or down-regulating precision expectations respectively. In short, valenced bodily affect shifts us toward a closer attunement with the environment, by raising or lowering the system's confidence in sets of action policies relative to how well or poorly those behaviours have proven themselves to be relative to expectations [30].

This means that as predictive organisms, we actively seek out waves of manageable prediction error - manageable uncertainty - because resolving it results in our feeling good. Predictive organisms that are tuned by error dynamics then will naturally exhibit curiosity and exploratory behaviour [29]. They will be moved affectively to seek out and make the most of the steepest slopes of error reduction in their environment. Situations which offer too little resolvable error (i.e. are too predictable) are boring for such organisms, while situations with too much error (i.e. too uncertain) are experienced by the agent as frustrating or threatening. The recent rise in jigsaw puzzle sales during the covid lockdown testifies to our love of manageable uncertainty. These feelings evolved to keep us well-tuned to our environment, helping us to curiously feel out novel and successful strategies for survival, while also avoiding all the stress and unpleasantness which comes with runaway uncertainty. This active, recursive, and felt relationship with the environment is crucial to grasping how social media can be detrimental to our mental health, and why we often find it so hard to stop using it, as we will see next.

Living well, in active inference terms, means being able to effectively manage uncertainty - and that's predicated on having a generative model which represents the world accurately. A generative model that poorly reflects the regularities of the environment would inevitably lead to an increase in bad predictions, and a flood of difficult-to-resolve errors. Active inference theorists are beginning to develop novel accounts of mental health conditions which focus on the predictive effectiveness of a person's generative model [5,12,34,44]. In the next section we look at how social media threatens to engender these suboptimal generative models in users.

3 Your Brain on Social Media

Social media is a spectacularly effective method for warping our generative models, as it often bombards users with bad evidence about both the world around us and our place in it. Typically, in the offline world, our generative model and expectations are encoded with information incoming from the unfiltered environment, which means that most of the time our generative model accurately (or at least usefully) reflects the world. However, in cases of regular and heavy engagement with social media, incoming information about the world is very often carefully selected, curated, and altered - we're potentially engaging with a fantasy. Moreover, apps that offer the use of filters also allow us to represent ourselves in carefully curated ways, potentially cultivating kinds and quantities of feedback and validation simply not available to us when we go offline. The space between being and appearing is potentially vast - with a few swipes we can dramatically alter our appearance or retake the same picture twenty times until our face exudes the calm mastery of life we want to project. As social media platforms develop features which foster an increasing potential for inauthenticity, the more those platforms become powerful bad-evidence generators, flooding the predictive systems of their users with inaccurate information, telling us that the world is full of incredibly beautiful, cool people, living wonderfully luxurious lives: social media platforms can act as a digital crowbar, prising apart our generative model from the offline environment. Instead, our model of the real world comes to take on the expectations generated through the online one, and the result is increasingly unmanageable waves of prediction error which the system must now strive to minimise.

The seemingly extreme actions of seeking cosmetic surgery to look more like one's online presence are one strategy for resolving this kind of prediction error. A recent survey found that more than half of cosmetic surgeons had patients ask explicitly for procedures which would enhance their online image, while many also reported patients using enhanced images of themselves as an example of how they'd like to look [26]. Levi the influencer describes how filters allowed him to preview the effects of specific cosmetic procedures, and while Instagram has now banned that specific filter, many perform similar functions. While this may seem extreme, these actions make perfect sense when viewed through the AIF. If we become accustomed to our own doctored appearance, and to receiving all the feedback associated with it, soon the level of validation available offline will be registered as a mounting prediction error, that's likely to result in feelings of stress, and inadequacy. According to the AIF, seeking surgery to bring our offline self in line with our online self is no different from grabbing a blanket as the temperature begins to drop - we're sampling the world to bring us back into an expected state, acting to minimize prediction error. It's just that through very deliberate design features, social media is - for some users - capable of displacing our self-image so much that the only way to rectify the error and meet those expectations is to surgically alter the way we look.

Note, though, how high the stakes are in this scenario. Surgery might offer one way to attempt to resolve the mounting error, but if we're unable to resolve the error, and continue to engage with social media, then this consistent failure is fed back to the system, eventually teaching it to expect its own failure and inability to act effectively in the world. This 'pessimistic' tendency in prediction bears a striking resemblance to the kind of scenarios now described by neuroscientists working on computational accounts of depression based on the AIF. Various forms of psychopathology, including depression, have now been described as a form of 'cognitive rigidity' wherein the system fails to adjust its expectations (including expected rate of error reduction) in line with feedback from the world [5, 6, 16, 30, 40, 41, 46]. In properly functioning predictive systems, when there is failure to resolve error in line with expectations, negatively valenced affect feeds back to the system and downregulates expectations accordingly, which then leads to the system being likely to resolve error once again in line with new expectations, which results in positive affect and an upregulation in expectation [30]. This constant undulation of expectation and valenced affect serves to keep wellfunctioning agents in a relatively stable state. In AIF accounts of depression however, when error isn't reduced in line with expectations, the system fails to update those expectations, leading to ongoing failure and an inverse slope of error reduction. The long-term summation of error manifests as persistent low mood, leading to a downregulation of precision on action policies. In short, a system which displays this rigidity in expectation comes to predict its own failure and ineffectiveness, which manifests on the agent level as symptoms of depression, such as feelings of helplessness, isolation, lack of motivation, and an inability to find pleasure in the world [30].

A 2018 exchange between Instagram user 'ScarlettLondon' and Twitter user 'Nathan', illustrates widespread intuitions about a link between social media and depression. 'ScarlettLondon' posted an image of her "morning routine" with a caption reading "I... give you a little insight into how I start my day in a positive way." The image featured Scarlett in a luxurious hotel room, with a selection of breakfast dishes laid out on the bed, complete with a product placement for Listerine. 'Nathan' reposted the image with another caption, reading "Fuck off this is anybody's normal morning. Instagram is a ridiculous lie factory made to make us all feel inadequate" [36]. Nathan's sentiment captures a pervasive intuition, since confirmed in several studies: that social media can cause depression because it facilitates negative comparisons with inauthentic or otherwise unattainable content [11].

Indeed, engagement with social media platforms has been shown to have a measurable impact on an individual's expectations of a specific place or event [37]. Through ongoing and consistent engagement with inauthentic content, a user's expectations for successful error reduction in the environment have the potential to effectively be 'pinned' in place, leading to the predictive system being unable to flexibly adjust those expectations in the face of evidence of failure coming in from the offline world. This ongoing failure eventually teaches the system to expect failure - to predict its own inefficacy in the world - which is precisely the scenario described by AIF accounts of depression.

Thus, social media can put us in a bind: either we somehow bring the world into line with our new expectations, which might involve drastic action, or we risk experiencing symptoms of depression, engendered by an influx of inaccurate evidence which renders our generative model inflexible and inaccurate.

4 Designing Addictive Digital Spaces

Of course, there's a more obvious way to alleviate any rising prediction error resulting from too much time online: spend less time online. For some of us this is easier said than done though, as mounting evidence supports the suspicion that social media can be addictive. A comprehensive 2015 review defined social media addiction as a disproportionate concern with and drive to use social media that impairs other areas of life and found that roughly ten percent of users exhibit symptoms of addiction [3]. Interestingly, this is around the same percentage of people who have problems with alcohol - but while the addictive hooks of alcohol and other drugs are relatively well understood and uncontroversial, those of behavioural addictions such as engagement with social media are still subject to debate [28]. Some researchers argue that there is in fact no such thing as internet addiction at all [49]. Again, the active inference framework holds the key to understanding why we should view engagement with digital hyperstimulators as potentially addictive.

The AIF offers a new understanding of addiction as a derailment of the alignment between predictive systems and their environment. Life contains various kinds of rewards: sex, food, status, etc., but for the brain all that matters is reducing prediction error, bringing us closer into expected states across various timescales. Dopamine encodes and reinforces behaviours that seek and pursue prediction error reduction. According to the AIF, dopamine plays a central role in encoding precision expectations on action policies [20,43]. Dopamine driven precision allocation is determined in part, as we have seen above, by changes in the rate at which error is reduced [10,23,29]. Addictive substances directly impact dopaminergic systems, signalling to the brain that a far better than expected slope of error reduction has taken place [35,44], which in turn upregulates precision on those drug seeking and taking policies. Through repeated use, the system comes to expect that vertiginous slope of error reduction, and is increasingly confident that it can be achieved only through drug seeking and taking behaviours. When the intense slope of error reduction associated with addictive substances cannot be met in ordinary life, the system registers failure, which is felt as disappointment, frustration, or pain (and a subsequent lowering of confidence on those policies). This progressive upregulation of precision on drug-seeking and taking behaviours, and down-regulation of non-addiction related action policies, leads to a narrowing of the agent's niche: friend groups are formed from fellow users, money is funneled toward buying drugs, and time is spent (when not using) planning how to acquire drugs. All the while the agent's other concerns related to family, career, friends, hobbies, and health are increasingly neglected [32]. A vicious feedback loop emerges - as mounting prediction errors spread across the broader concerns of an addicted individual's life (as these concerns are increasingly ignored and abandoned), the addict is more attracted to the one area where they can feel like they are succeeding, namely drug use. It is this feedback loop, that recruits the predictive mechanisms of brain and body to reorganize the habits of the addict around drug seeking and taking behaviour, that underlies the true pathology of addiction.

Just like alcohol and other drugs, digital environments threaten to disrupt this balance between naturally occurring rewards and reward seeking behaviour. In his important book 'Your Brain on Porn' [48], Gary Wilson argues that internet pornography presents itself as dangerously rewarding, pointing out that in one evening, internet porn facilitates levels of sexual novelty which would have been unavailable to our ancestors across an entire lifetime: multiple tabs or windows, hundreds of different models, escalating fetishes, all conspire to have our reward circuitry screaming "wow, we're doing far better than we ever thought possible!", when in reality we're just staring at a screen, alone. The novelty is particularly enticing, as our brains are always seeking new ways of reducing error, novel strategies for doing better than expected [23]. Our brains register this as a huge resolution of uncertainty, and our reward circuitry in the brain goes into overdrive, reinforcing these particular reward seeking behaviours.

What pornography is to sex, social media platforms are to our intrinsic appetite for socialising. Engaging in meaningful interpersonal bonding draws on all the reward circuitry mentioned above: it feels good to socialise, and dopamine entrenches learning for successful social behaviours [31]. One major similarity between social media and pornography is that both take a naturally occurring reward (sex and social behaviour, respectively), engineer a powerful vehicle of carefully curated fantasy, and present it as an attainable and desirable reality. These presentations of 'better than real life' scenarios (e.g. carefully staged and filtered images; maximally exciting sexual encounters in pornography) are highly alluring for predictive agents always on the lookout for ways to improve. On social media - just as with online porn - high levels of novelty and excess mean that the reward system is kicked into overdrive. It's no wonder that a 2019 report found that the average teenager in the US now spends more than seven hours a day looking at a screen [45]. Through social media, hyperstimulation can work to reorganise our predictive model and restructure our habits: we wake up and reach for our phone, never leave home without it, and constantly feel drawn toward our phone even when in the company of friends.

One avenue of objection here might be to point out that it seems debatable to what extent social media captures the reality of offline social interactions, devoid as it is of many features of face-to-face communication, and therefore it should be unclear just how rewarding online social interactions actually are. In response to this, we can first return to the comparison with online pornography, which clearly lacks the same substantive character as real sex and relationships. Nevertheless, engagement with online porn has been shown to powerfully engage the brain's reward seeking machinery [1,38]. This returns us to the point made earlier that, according to the AIF, all rewards are fundamentally processed as prediction error minimization - crack cocaine or explicit imagery, it's all the same: the system learns to expect states where error is reduced in line with or better than expected. While social media certainly lacks the face-to-face nuance of real-world interaction, it nevertheless works hard to turbo charge many of its most gratifying aspects such as judgement, monitoring, and validation and positive feedback.

In order to see how social media takes these rewarding aspects of social interaction and hyper charges them, first notice how all digital space has the inherent quality of dissolving the temporal and spatial restraints which govern offline interaction, thereby - in the case of social media - facilitating an excess of novelty and validation which simply isn't available in the real world. Users can instantaneously exchange direct messages with people who may well be complete strangers, and when users get bored of the content they're currently interacting with, a quick swipe generates new, exciting, unpredictable content. These structural features - which deliberately elicit anticipatory states and facilitate near endless potential for novelty - is something that deflationary accounts of social media addiction often fail to emphasize.

However, the potentially addictive nature of social media platforms doesn't only emerge from an excess of carefully edited content and potentially massive social feedback. It also emerges from a deliberately designed and carefully implemented functional architecture which draws on our knowledge about the brain's reward circuitry and established approaches in the gambling industry. In gambling what's so arousing (and habit-forming) is the anticipation of reward, or the expectation of an uncertain reward [25]. Of course, offline social interactions are often unpredictable too, in that we don't know when someone might contact us or interact with us in rewarding ways, but social media sites are engineered to compound this anticipation through gamification, in which features such as progression, points scoring and risk taking are introduced into a non-game setting. Social media gamifies social interaction, primarily through various highly interactive systems of 'likes', 'shares', 'upvotes', comments and so on, which apply to user created content. This feedback is the direct measure of the 'success' of a particular post, and allows for comparisons in popularity between posts and posters.

When the potentially enormous levels of social feedback do come, it isn't immediately communicated to the user. Rather, we receive notifications in the form of a shining button or exciting sound which delays the discovery of the precise nature of the incoming content. The simple act of pushing a button to reveal information has been shown to trigger arousal and compulsive behaviour, and newly developed features on smartphones add further layers of anticipation [2]. The 'swipe to refresh' feature of the Facebook app's news feed, for example, where users physically swipe the screen to generate a new stream of information, is a startlingly similar action to the pulling of a casino slot machine arm. In each case, users don't know for sure what kind of content will spring up until they swipe. This feature, coupled with the fact that Facebook's feed is now effectively infinite has led to the app being described as "behavioural cocaine" [3].

One final layer of anticipation comes through the use of a smartphone itself, compounding the intermittency arousal of feedback and interaction: We've been so conditioned by anticipation of smartphone buzzing that "phantom vibration syndrome" - the erroneous sensation of our phone vibrating - now affects 65%–89% of people who use smartphones [13,42]. Crucially, these carefully engineered spikes in user anticipation mirror the anticipatory states known to underlie problematic gambling; in people who exhibit addictive gambling behaviour, dopamine response has been shown to be most pronounced during phases of high anticipation [22]. Rather than the reward itself, it's these highly arousing states of expectation of reward which have been shown to elicit the strongest dopaminergic response [25,33], and the designers of these digital platforms know this.

5   Conclusion

The active inference framework has, in recent years, come to change how we understand a range of psychological phenomena, including addiction and depression. In this paper, we've used the theoretical tools of active inference to enter into an ongoing debate about the ways in which social media - and digital environments more broadly - have the potential to negatively impact our mental wellbeing. While deflationary accounts downplay the effects of the inherent design of digital environments, this active inference account adds weight to arguments that there are inherent features of digital technology that can have profound consequences for our wellbeing. These arguments may have a wide-ranging impact, given that these inherent features are deliberately implemented. As design guru Nir Eyal states, "Companies increasingly find that their economic value is a function of the strength of the habits they create" [15]. As it turns out then, the designers of social media, aiming to maximize engagement through design, may have a de facto interest in increasing the corrosive effect their platforms have on the mental health of users. Seen in this context, this emerging scientific picture may lend significant weight to arguments that we should take digital hyperstimulants seriously as a threat to our wellbeing, and to voices calling for changes to the way digital technology like social media is designed, operated, and regulated.

Acknowledgments. Mark Miller carried out this work with the support of Horizon 2020 European Union ERC Advanced Grant XSPECT - DLV-692739.

References

1. de Alarcón, R., de la Iglesia, J.I., Casado, N.M., Montejo, A.L.: Online porn addiction: what we know and what we don't-a systematic review. J. Clin. Med. **8**(1), 91 (2019)
2. Alter, A.: Irresistible: Why you are addicted to technology and how to set yourself free. Vintage (2017)
3. Andersson, H.: Social media apps are "deliberately" addictive to users. BBC News 3 (2018)
4. Arab, K., et al.: Influence of social media on the decision to undergo a cosmetic procedure. Plastic Reconstr. Surg. Global Open **7**(8) (2019)
5. Badcock, P.B., Davey, C.G., Whittle, S., Allen, N.B., Friston, K.J.: The depressed brain: an evolutionary systems theory. Trends Cogn. Sci. **21**(3), 182–194 (2017)
6. Barrett, L.F., Quigley, K.S., Hamilton, P.: An active inference theory of allostasis and interoception in depression. Philos. Trans. R. Soc. B Biol. Sci. **371**(1708), 20160011 (2016)
7. Clark, A.: Surfing Uncertainty: Prediction, Action, and the Embodied Mind. Oxford University Press, Oxford (2016)
8. Clark, A.: Predictions, precision, and agentive attention. Conscious. Cogn. **56**, 115–119 (2017)
9. Cramer, S., Inkster, B.: Statusofmind-social media and young people's mental health and wellbeing.[online]. royal society for public health (2017)
10. Van de Cruys, S.: Affective Value in the Predictive Mind. MIND Group, Frankfurt am Main (2017)
11. Curtis, S.: Social media users feel ugly, inadequate and jealous. The Telegraph https://www.telegraph.co.uk/technology/social-media/10990297/Social-media-users-feel-ugly-inadequate-and-jealous.html
12. Deane, G., Miller, M., Wilkinson, S.: Losing ourselves: active inference, depersonalization, and meditation. Front. Psychol. **11**, 2893 (2020)
13. Drouin, M., Kaiser, D.H., Miller, D.A.: Phantom vibrations among undergraduates: Prevalence and associated psychological characteristics. Comput. Hum. Behav. **28**(4), 1490–1496 (2012)

14. Eldar, E., Rutledge, R.B., Dolan, R.J., Niv, Y.: Mood as representation of momentum. Trends Cogn. Sci. **20**(1), 15–24 (2016)
15. Eyal, N.: Hooked: How to Build Habit-Forming Products. Penguin, London (2014)
16. Fabry, R.E.: Into the dark room: a predictive processing account of major depressive disorder. Phenomenol. Cogn. Sci. **19**(4), 685–704 (2019). https://doi.org/10.1007/s11097-019-09635-4
17. Flood, R.: Insta sham: I spent & #x00A3;30k on surgery to look like an Instagram filter but instead get compared to the 'Purge' mask. The Sun https://www.thesun.co.uk/fabulous/14374803/man-spend-30k-look-instgram-filter-purge-mask/
18. Friston, K.: The free-energy principle: a unified brain theory? Nat. Rev. Neurosci. **11**(2), 127–138 (2010)
19. Friston, K., FitzGerald, T., Rigoli, F., Schwartenbeck, P., Pezzulo, G.: Active inference: a process theory. Neural Comput. **29**(1), 1–49 (2017)
20. Friston, K.J., Shiner, T., FitzGerald, T., Galea, J.M., Adams, R., Brown, H., Dolan, R.J., Moran, R., Stephan, K.E., Bestmann, S.: Dopamine, affordance and active inference. PLoS Comput. Biol. **8**(1), e1002327 (2012)
21. Gritters, J.: How instagram takes a toll on influencers brains. The Guardian (2019)
22. Hegarty, C., et al.: Ventral striatal dopamine synthesis correlates with neural activity during reward anticipation. In: Neuropsychopharmacology, vol. 39, pp. S197–S198. Nature Publishing Group Macmillan Building, 4 Crinan St, London N1 9XW, England (2014)
23. Hesp, C., Smith, R., Parr, T., Allen, M., Friston, K.J., Ramstead, M.J.: Deeply felt affect: the emergence of valence in deep active inference. Neural Comput. **33**(2), 398–446 (2021)
24. Hohwy, J.: The predictive mind. Oxford University Press, Oxford (2013)
25. van Holst, R.J., Veltman, D.J., Büchel, C., van den Brink, W., Goudriaan, A.E.: Distorted expectancy coding in problem gambling: is the addictive in the anticipation? Biol. Psychiatry **71**(8), 741–748 (2012)
26. Hunt, E.: Faking it: how selfie dysmorphia is driving people to seek surgery. Guardian **23**(02) (2019)
27. Joffily, M., Coricelli, G.: Emotional valence and the free-energy principle. PLoS Comput. Biol. **9**(6), e1003094 (2013)
28. Kardefelt-Winther, D., et al.: How can we conceptualize behavioural addiction without pathologizing common behaviours? Addiction **112**(10), 1709–1715 (2017)
29. Kiverstein, J., Miller, M., Rietveld, E.: The feeling of grip: novelty, error dynamics, and the predictive brain. Synthese **196**(7), 2847–2869 (2017). https://doi.org/10.1007/s11229-017-1583-9
30. Kiverstein, J., Miller, M., Rietveld, E.: How mood tunes prediction: a neurophenomenological account of mood and its disturbance in major depression. Neurosci. Conscious. **2020**(1), niaa003 (2020)
31. Kopec, A.M., Smith, C.J., Bilbo, S.D.: Neuro-immune mechanisms regulating social behavior: dopamine as mediator? Trends Neurosci. **42**(5), 337–348 (2019)
32. Lewis, M.: Brain change in addiction as learning, not disease. England J. Med. **379**(16), 1551–1560 (2018)
33. Linnet, J.: Neurobiological underpinnings of reward anticipation and outcome evaluation in gambling disorder. Front. Behav. Neurosci. **8**, 100 (2014)
34. Linson, A., Parr, T., Friston, K.J.: Active inference, stressors, and psychological trauma: a neuroethological model of (mal) adaptive explore-exploit dynamics in ecological context. Behav. Brain Res. **380**, 112421 (2020)
35. Miller, M., Kiverstein, J., Rietveld, E.: Embodying addiction: a predictive processing account. Brain Cogn. **138**, 105495 (2020)

36. Moss, R.: Instagram's scarlett london on being in the centre of a social media storm. Huffington Post https://www.huffingtonpost.co.uk/entry/there-is-a-real-
37. Narangajavana, Y., Fiol, L.J.C., Tena, M.Á.M., Artola, R.M.R., García, J.S.: The influence of social media in creating expectations. an empirical study for a tourist destination. Ann. Tourism Res. **65**, 60–70 (2017)
38. Negash, S., Sheppard, N.V.N., Lambert, N.M., Fincham, F.D.: Trading later rewards for current pleasure: pornography consumption and delay discounting. J. Sex Res. **53**(6), 689–700 (2016)
39. Parr, T., Friston, K.J.: Uncertainty, epistemics and active inference. J. R. Soc. Interface **14**(136), 20170376 (2017)
40. Paulus, M.P., Feinstein, J.S., Khalsa, S.S.: An active inference approach to interoceptive psychopathology. Ann. Rev. Clin. Psychol. **15**, 97–122 (2019)
41. Ramstead, M.J., Wiese, W., Miller, M., Friston, K.J.: Deep neurophenomenology: An active inference account of some features of conscious experience and of their disturbance in major depressive disorder (2020)
42. Rothberg, M.B., Arora, A., Hermann, J., Kleppel, R., St Marie, P., Visintainer, P.: Phantom vibration syndrome among medical staff: a cross sectional survey. Bmj 341 (2010)
43. Schwartenbeck, P., FitzGerald, T.H., Mathys, C., Dolan, R., Friston, K.: The dopaminergic midbrain encodes the expected certainty about desired outcomes. Cereb. Cortex **25**(10), 3434–3445 (2015)
44. Schwartenbeck, P., FitzGerald, T.H., Mathys, C., Dolan, R., Wurst, F., Kronbichler, M., Friston, K.: Optimal inference with suboptimal models: addiction and active bayesian inference. Med. Hypotheses **84**(2), 109–117 (2015)
45. Siegel, R.: Tweens, teens and screens: The average time kids spend watching online videos has doubled in 4 years. The Washington Post (2019)
46. Smith, R., et al.: An active inference model reveals a failure to adapt interoceptive precision estimates across depression, anxiety, eating, and substance use disorders. medRxiv (2020)
47. Truly: Surgery transformed my face into an Instagram filter, hooked on the look (2019). https://www.youtube.com/watch?v=JXEqVL6-ENY
48. Wilson, G.: Your Brain on Porn: Internet Pornography and the Emerging Science of Addiction. Commonwealth Publishing Richmond, Virginia (2014)
49. Yellowlees, P.M., Marks, S.: Problematic internet use or internet addiction? Comput. Hum. Behav. **23**(3), 1447–1453 (2007)

Ideas Worth Spreading: A Free Energy Proposal for Cumulative Cultural Dynamics

Natalie Kastel[✉] and Casper Hesp

1012 WX, Amsterdam, Netherlands

Abstract. While there is a fast growing body of theoretical work on characterizing cumulative culture, quantifiable models underlining its dynamics remain scarce. This paper provides an active-inference formalization and accompanying simulations of cumulative culture in two steps: Firstly, we cast cultural transmission as a bi-directional process of communication that induces a generalized synchrony (operationalized as a particular convergence) between the internal states of interlocutors. Secondly, we cast cumulative culture as the emergence of accumulated modifications to cultural beliefs from the local efforts of agents to converge on a shared narrative.

Keywords: Active inference · Generalized synchrony · Communication · Cumulative culture · Cultural dynamics

1 Introduction

Research on cultural dynamics focuses on the examination of fluctuations in cultural beliefs and practices and their evolution from a systems perspective. These dynamics consist of three processes that are typically studied separately: the introduction of novel beliefs and practices to a culture (i.e., innovation), the transmission of established beliefs and practices within a population (i.e., innovation diffusion), and their change in prevalence (Kashima et al. 2019).

While there is a fast growing body of theoretical and empirical literature on the processes of cultural evolution (Aunger 2001; Buskell et al. 2019; Bettencourt et al. 2006; Creanza et al. 2017; Dawkins 1993; Dean et al. 2014; Dunstone and Caldwell 2018; Enquist et al. 2011; Gabora 1995; Heylighen and Chielens 2009; Kashima et al. 2019; Richerson et al. 2010; Stout and Hecht 2017; Weisbuch et al. 2009), quantitative models that are able to integrate different approaches and insights from multiple disciplines into unified, quantifiable interpretations of theory and empirical data are in rapidly growing demand (Creanza et al. 2017).

This is particularly true for the mechanisms of social transmission, which have been especially reviewed under theoretical models (Aunger 2001; Bettencourt et al. 2006; Dawkins 1993; Gabora 1995; Heylighen and Chielens 2009; Kashima et al. 2019; Weisbuch et al. 2009) while mathematical models for cultural transmission remain scarce

---

N. Kastel and C. Hesp--Private authors.

© Springer Nature Switzerland AG 2021
M. Kamp et al. (Eds.): ECML PKDD 2021 Workshops, CCIS 1524, pp. 784–798, 2021.
https://doi.org/10.1007/978-3-030-93736-2_55

in this field. The term "cultural transmission" typically denotes the transference and spread of any particular fashion, ideology, preference, language or behavior within a culture (Creanza et al. 2017). A prominent stream of quantitative models for cultural transmission are inspired by epidemiology, and convert models used for predicting the spread of a virus to formalize the spread of an idea (Bettencourt et al. 2006).

While the comparison of an idea to a virus has its benefits from a structural perspective, it implies the controversial notion that an idea is simply copied during its transmission through cultural exchange between individuals. This notion is not only intuitively insufficient for a realistic portrayal of communication dynamics, but also conflicts with established theoretical models of transmission on these same grounds.

Current literature in cultural psychology indicates that rather than being simply duplicated during transmission, cultural beliefs and practices are modified through the active interpretation of each individual (Kashima et al. 2019). Another example for the discrepancy between quantitative epidemiology models for transmission and theory is taken from the psychology of communication. Research in this field suggests that communication is conditioned upon a mutual shared reality (Echterhoff et al. 2009), or "common ground" (Clark and Brennan 1991) between interlocutors. According to these theories, not only does cultural information change during communication, but (contradictory to the one-sided transmission of cultural information from "transmitter" to "receiver" that is implicit in epidemiological models) both interlocutors are active participants in generating this change. "Grounding" theories suggest that communication involves more than simply formulating a message and sending it off, but requires the mutual belief that what is being said will be understood by all parties.

Crucially, the notion that cultural information resists alterations during its transmission conflicts with a fundamental and particularly distinguished theory of cultural transmission: cumulative culture (Dunstone and Caldwell 2018; Stout and Hecht 2017). This approach to cultural evolution reflects the idea that cultural traits are gradually modified through transmission such that adaptive modifications accumulate over historical time (Dean et al. 2014). This theory operates from a basic assumption that transmission of cultural information naturally involves its modification in a way that fundamentally conflicts with the depiction of transmission under a disease spread formalisation.

The cumulative conceptualisation of modifications to cultural information is prominent in the literature and may be the most representative of genuine complexities underlying cultural dynamics. However, this triumph entails perhaps an inevitable downfall in that such a complex depiction of culture has proven exceptionally challenging to model in quantitative accounts (Buskell et al. 2019). This paper provides an active-inference based quantitative account of cumulative culture as an accumulation of changes to cultural information over multiple transmissions.

2 Method

An emerging conclusion from the literature is that the term "transmission" for describing the spread of cultural information seems impoverished, as it leaves out the retention of cultural information. As implied by active inference and theoretical models of communication, the acquisition of cultural beliefs is as fundamental to the understanding of cultural information spread as their transmission. For this reason, we will henceforth be referring to what is known in the literature as cultural transmission as communication, or more technically- the local dynamics of cumulative culture.

2.1 Simulating the Local Dynamics of Communication

In our model, cultural transmission is cast as the mutual attunement of actively inferring agents to each other's internal belief states. This builds on a recent formalisation of communication as active inference (Friston and Frith 2015) which resolves the problem of hermeneutics, (i.e., provides a model for the way in which people are able to understand each other rather precisely despite lacking direct access to each other's internal representations of meaning) by appealing to the notion of generalised synchrony as signaling the emergence of a shared narrative to which both interlocutors refer to. In active inference, this shared narrative is attained through the minimisation of uncertainty, or (variational) free energy when both communicating parties employ sufficiently similar generative models. We build on this to suggest that having sufficiently similar generative models allows communicating agents to recombine distinct representations of a belief (expressed as generative models) into one synchronised, shared model of the world. When we simulate the belief-updating dynamics between interacting agents, the cultural reproduction of a particular idea takes the form of a specific convergence between their respective generative models.

Under this theory, the elementary unit of heritable information takes the form of an internal belief state, held by an agent with a certain probability. When we simulate the belief-updating dynamics between interacting agents, a reproduced cultural belief is carried by the minds (or generative models) of both interlocutors as a site of cultural selection, where it may be further reproduced through the same process. Our simulations of communication involve two active inference agents with distinct generative models and belief claims that engage in communication over a hundred time steps.

2.2 Simulating the Global Dynamics of Cumulative Culture

Cultural beliefs and practices spread within a society through communication, a process which we have referred to as the local dynamics of cumulative culture. This description is appropriate because the accumulated outcomes of each (local) dyadic interaction collectively determine the degree to which an idea is prevalent in a culture. Moving from local communication dynamics to a degree to which an idea is prevalent in a cumulative culture is what we will refer to as the global dynamics of cumulative culture.

In our simulations of a cumulative culture, 50 active inference agents simultaneously engage in local dyadic communication as shown in our first simulation, such that 25 couples are engaged in conversation at every given time step. At the first time step, all agents have relatively similar belief states- referred to as the status quo. When we introduce an agent holding a divergent belief state to that of the status quo in the population, it propagates through it via pseudo-random engagements of agents in dialogue. In a simulated world of actively inferring agents, their individual mental (generative) models are slightly modified with every interlocutor they encounter, as their distinct representations converge to a shared narrative (Constant et al. 2019). The attunement of interlocutor's to each other's generative models on the microscale thus translates over time and with multiple encounters into collective free energy minimisation on the macroscale.

3 A Generative Model of Communication

In our simulations, agents attempt to convince each other of a cultural belief by utilising generative models that operate with local information only. For the establishment of such generative models, we will formulate a partially observed Markov decision process (MDP), where beliefs take the form of discrete probability distributions (for more details on the technical basis for MDP'S under an active inference framework, see Hesp et al. 2019).

Under the formalism of a partially observed Markov decision process, active inference entails a particular structure. Typically, variables such as agent's hidden states (x, s), observable outcomes (o) and action policies (u) are defined, alongside parameters (representing matrices of categorical probability distributions).

3.1 Perceptual Inference

The first level of this generative model aims to capture how agents process belief claims they are introduced to through conversation with other agents. The perception of others' beliefs (regarded in active inference as evidence) requires prior beliefs (represented as likelihood mapping A1 about how hidden states (s1) generate sensory outcomes (o). Specifically, our agents predict the likelihood of perceiving evidence toward a particular expressed belief, given that this belief is "the actual state of the world". Parameterising an agent's perception of an interlocutor's expression of belief in terms of precision values can be simply understood as variability in agents' general sensitivity to model evidence. High precisions here correspond to high responsiveness to evidence for a hidden state and low precisions to low responsiveness to evidence. Precisions for each agent were generated from a continuous gamma distribution which is skewed in favour of high sensitivity to evidence on a population level (See Fig. 1: Perception).

Updating of core belief based on beliefs expressed by self and another agent after each meeting (detailed descriptions of the computations involved in perceptual inference can be found under appendix):

$$Q\left(x_{core}^{(2)}\right) = \sigma\left(\ln x_{core}^{(2)} + \gamma_{A,self}^{(2)}\ln o_{expr,self} + \gamma_{A,other}^{(2)}\ln o_{expr,other}\right) \tag{1}$$

3.2 Anticipation

At this level, our generative model specifies agents' beliefs about how hidden states (detailed in Appendix A2) evolve over time. State transition probabilities [B1] define a particular value for the volatility of an agent's meeting selection (s2) and belief expression (s1) [B1]. For each agent, this precision parameter is sampled from a gamma distribution, determining the a priori probability of changing state, relative to maintaining a current state. Note that belief states themselves are defined on the continuous range <0, 1> (i.e., as a probability distribution on a binary state), such that multiplication tends to result in a continuous decay of confidence over time in the absence of new evidence (where the rate of decay is inversely proportional to the precision on B) (See Fig. 1: Anticipation).

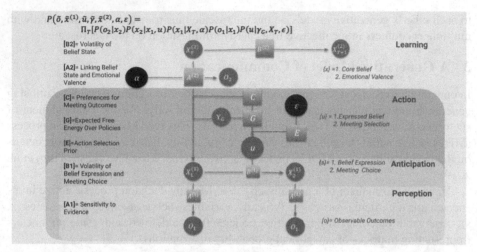

Fig. 1. A generative model of communication. Variables are visualised as circles, parameters as squares and concentration parameters as dark blue circles. Visualised on a horizontal line from left to right-states evolve in time. Visualised on a vertical line from bottom to top- parameters build to a hierarchical structure that is in alignment with cognitive functions. Parameters are described to the left of the generative model and variables are described on the right. (Color figure online)

3.3 Action

After perceiving and anticipating hidden belief states in the world, our agents carry out deliberate actions biased towards the minimum of the expected free energy given each action (a lower level generative model for action is detailed in Appendix A4 and A5). At each time point, a policy (U) is chosen out of a set of possible sequences for action. In our simulations, two types of actions are allowed: selecting an agent to meet at each given time point (u2) and selecting a specific belief to express in conversation (u1). The first allowable action holds 50 possible outcomes (one for each agent in the simulation) while the second is expressed on the range <0, 1>, where the extremes correspond to complete confidence in denying or supporting the belief claim, respectively. Each policy under the G matrix specifies a particular combination of action outcomes weighted by its expected negative free energy value and a free energy minimising policy is chosen (See Fig. 1: Action).

Voluntary Meeting Selection. While the choice of interlocutor is predetermined in a dyad, our multi-agent simulations required some sophistication in formulating the underlying process behind agents' selection for a conversational partner (s3) at each of the hundred time points. Building on previous work on active inference navigation and planning (Kaplan and Friston 2018), agents' meeting selection in our model is represented as a preferred location on a grid, where each cell on the grid represents a possible agent to meet (Appendix).

We demonstrate the feasibility of incorporating empirical cultural data within an active inference model by incorporating (1) confirmation bias through state-dependent preferences [C], biasing meeting selection through the risk component of expected free energy (G) and (2) novelty seeking through the ambiguity component of expected free energy. The first form of bias reflects the widely observed phenomenon in psychology research that people's choices tend to be biased towards confirming their current beliefs (Nickerson 1998). The second form of bias reflects the extent to which agents are driven by the minimisation of ambiguity about the beliefs of other agents, driving them towards seeking out agents they have not met yet.

3.4 Perceptual Learning

On this level agents anticipate how core belief states (specified in Appendix A1) might change over time [B2] (Fig. 2.3). This is the highest level of cognitive control, where agents experience learning as a high cognitive function (higher level generative model is detailed in Appendix A3). By talking with other simulated agents and observing their emotional and belief states, our agents learn associations between EV and beliefs via a high level likelihood mapping [A2], (updated via concentration parameter α). The Updating of core belief, based on beliefs expressed by other agents, is detailed in Appendix A7. This learning is important because it provides our agents with certainty regarding the emotional value they can expect from holding the alternative belief to the status quo, which has low precision at the beginning of the simulation (before the population is introduced to an agent proclaiming this belief). The prior P(A) for this likelihood mapping is specified in terms of a Dirichlet distribution (Appendix).

4 Results

4.1 Local Dynamics of Coupled Communication

In nature, generalised synchrony emerges from a specific coupling between the internal states of dissipative chaotic systems (Pikovsky et al. 2003). In active inference communication, agents are coupled in a bidirectional action-perception cycle in which they can be described as coupled dynamical systems (Friston and Frith 2015; Constant et al. 2018) Specifically, our model defines perceptual inference as the coupling parameter linking the internal states of interlocutors.

Also understood as sensitivity to model evidence (A1), perceptual inference is a direct and explicit form of coupling that occurs over the span of a single dialogue such that it modulates agents' convergence of internal belief states during conversation (Fig. 2).

Our results indicate that without sufficiently high precisions on sensitivity to model evidence, agents' ability to listen and attune to the belief expression of their partner is limited to the extent that they are responsive to sensory evidence from their environment.

To get a sense of the implications of these simulations, it is important to make explicit the way in which they tie in to previous work on active inference communication. In 2015, Friston and Frith provided evidence for the notion that generalised synchrony becomes altogether unattainable when agents do not possess sufficiently similar generative models. Our model goes beyond this to provide evidence for the idea that only when generalised synchrony is attainable (i.e., when interlocutors possess sufficiently similar

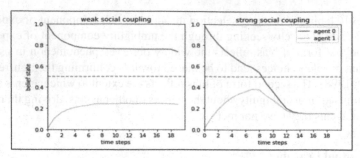

Fig. 2. The [A1] parameter (sensitivity to model evidence) modulates the level of social coupling between agents in dialogue. **(Left)** When precision on sensitivity to model evidence is low (for both agents) their internal states are very weakly coupled, which results in each agent sticking to their own belief. **(Right)** When both agents have high sensitivity to model evidence, their beliefs converge into a shared representation of an idea that inhabits both of their generative models.

generative models), communication underlies a convergence between their belief states. Our simulations should therefore be understood as taking generalised synchrony for granted while providing evidence for the premise that the level to which agents' beliefs converge (i.e., the level of synchrony between their internal states) is modulated by their sensitivity to model evidence [A1].

4.2   Global Dynamics of Cumulative Culture

Our simulations of a cumulative culture should be understood as capturing the dynamics of a culture that is the sum (or-accumulation) of modifications to cultural beliefs and practices over time (Fig. 3). While the local dynamics simulated in the previous section represent a single modification to cultural information (as a convergence between distinct belief states held by individual agents), these simulations accumulate these modifications and expose their emerging dynamics within the population. The fundamental achievement of these results is therefore their methodologically consistent and novel depiction of cumulative culture under a quantitative and measurable framework (namely, active inference).

We explain the communicative isolation observed in our simulations (Fig. 3) as a self organised separation between groups of agents when they hold intractably divergent beliefs, such that communicative isolation best ensures local and collective free energy minimization. In other words, when an intractable divergent belief propagates within a homogenous population, communicative isolation between incongruent groups emerges as a strategy to minimize expected free energy, while the same strategy homogenizes the belief states of agents within congruent groups.

The above simulations also show how changes to parameters that determine levels of confirmation bias [C] and novelty seeking [G] affect the segregation within the population into groups of agents holding either status quo congruent beliefs or the alternative belief. When novelty seeking is upregulated, the population evolves such that the majority of agents end up subscribing to the alternative belief. However, when confirmation bias is upregulated, the majority of agents end up subscribing to the status quo. What

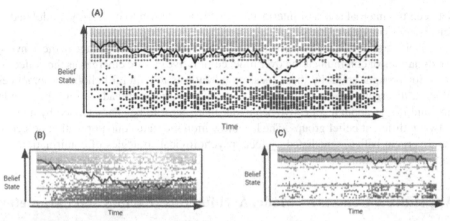

Fig. 3. Simulations of the spread of each agent's belief state (y) across time (x). 5o agents were used in this simulation and each of the 100 time steps represents the reproduced belief state outcomes of a particular combination of agents in dialogue. (**A**) Simulation of a Cumulative Culture. When a divergent belief state (blue) is introduced to the status quo population (red) at the first time step, it spreads through it via pseudo-random engagements of agents in dialogue that cumulatively change the belief structure within the population. Most notably, the introduction of a divergent belief seems to split the population into two subgroups: those holding a belief state that approximates the new divergent belief, and those holding an approximate status quo belief. This effect is modulated by agents' individual strategies for choosing which interlocutors to engage in conversation with (s3). (**B**) When novelty seeking is high in the population (above 10% of agents present high novelty seeking), the population is divided in favour of the divergent belief state, with more agents eventually holding this belief than the status quo. (**C**) When confirmation bias is high in the population (above 90% of agents present high confirmation bias) the population is divided in favour of the status quo belief, with more agents holding to this belief than the new and divergent belief. (Color figure online)

these results indicate is that novelty seeking on a local level stimulates the population as a whole toward the adoption of a belief that is divergent from the status quo. This happens when novelty seeking individuals, which are open and willing to meet with agents of unknown beliefs, are intrinsically encouraged by their own curiosity to engage with a divergent belief. Once such agents become gradually more favoring of this belief they start to popularize it to the rest of the population. If the population is however, populated by a vast majority of agents biased toward meeting individuals with confirming beliefs, they do not engage with an alternative belief and it does not get popularized.

5   Conclusion

In this paper, we employed an active inference model to tackle the complex task of formulating the dynamics underlying cumulative culture. Under this account, transmission is cast as a bidirectional process of communication that induces a generalised synchrony between the internal (belief) states of agents holding sufficiently similar generative models. Generalised synchrony is operationalised in our model as a particular convergence between the internal states of interlocutors, which is shown to be largely modulated by sensitivity to model evidence [A1].

When we simulate a population of agents that simultaneously engage in the converging dynamics of communication over time, cumulative culture emerges as the collective behavior brought about by these local modifications to cultural beliefs and practices. When a divergent belief is introduced to the status quo, it spreads within the population and brings about a collective behaviour that seems to be characterised by a divide between different belief groups. The level to which the status quo population defects to the divergent belief is mediated by local psychological strategies of confirmation bias and novelty seeking.

Appendix A Generative Model Architecture, Factors and Parameters

A.1 Higher Level Hidden State Factors

$$x^{(2)} : \{x^{(2)}_{core},$$

[core beliefs of self and others about a particular claim, across days]

$$x^{(2)}_{mem},$$

[memory of having visited each agent]

$$x^{(2)}_{habit} \}$$

[habits of self, across days]

A.2 Lower Level Hidden State Factors (Specify Events on a Given 'Day')

$$x^{(1)} : \{ \ x^{(1)}_{loc},$$

[self location, where each agent has a unique 'home' location]

$$x^{(1)}_{belief},$$

[beliefs of self and others about a particular claim]

$$x^{(1)}_{visit},$$

[beliefs about having visited each agent]

$$x^{(1)}_{sat} \}$$

[satisfaction of self and others]

A.3 Higher level generative model

$$x_{belief}^{(1)} = A_{belief}^{(2)} x_{core}^{(2)}$$

[core beliefs specify prior expectations for beliefs on the lower level]

$$x_{sat}^{(1)} = A_{sat}^{(2)} x_{core}^{(2)}$$

[core beliefs specify satisfaction states for the lower level]

$$x_{visit}^{(1)} = A_{mem}^{(2)} x_{mem}^{(2)}$$

[memory specifies beliefs about having visited each agent on the lower level]

$$E_{expr} = A_{expr}^{(2)} x_{habit}^{(2)}$$

[habits of self specify prior tendency for belief expression]

$$x_{T+1}^{(2)} = B^{(2)} x_T^{(2)}$$

[higher-level states decay over time: gradual forgetting]

A.4 Lower Level Generative Model For Action

Action model for meeting selection

In our simulations, we have incorporated psychological biases in agents' preferences for meeting similar (i.e., belief compatible) or unknown agents. Note that while agents biased toward confirming beliefs would tend toward individuals with similar beliefs to their own, novelty seekers would not look for the opposite of this (i.e. look for individuals of divergent beliefs to their own), but rather have a preference for individuals of yet unknown beliefs.

In active inference, action selection is guided by the expected free energy [G], which entails maximising the expected benefit or utility of an action (known as pragmatic value), while also maximising the potential information gain of future actions by reducing uncertainty about the causes of valuable outcomes (known as epistemic value). These constraints to action selection could be interpreted as formalising the exploration–exploitation trade-off in learning systems. Epistemic value (exploration) refers to the benefit related to searching over a sample space in order to get a better estimation of promising areas that will maximise pragmatic value (exploitation). Active-inference agents would therefore maximise epistemic value until information gain is low, after which the maximisation of pragmatic value and exploitation are assured (Friston et al. 2015).

In our model, agents' choice in meeting interlocutors with known and similar beliefs versus those with unknown beliefs can be cast in terms of a tradeoff between pragmatic and epistemic value. On the one hand, a confirmation bias emerges from the maximisation of expected utility, increasing synchronisation between interlocutors' internal models, thus allowing for the emergence of shared expectations (Hesp et al. 2019). On the other hand, novelty seeking emerges from the maximisation of information gain, allowing for the exploration of the sample space. Also understood as intrinsically motivated curious behaviour (Friston et al. 2017), maximisation of epistemic value allows individuals to better predict the consequences of their actions (e.g., which agent to meet) through greater certainty about the hidden states of their environment (e.g., the beliefs of other agents).

From the point of view of agents in our simulations, increasing pragmatic value translates into selecting to meet interlocutors with similar beliefs, while increasing epistemic value translates into selecting agents whose beliefs are unknown or highly uncertain (This way, a meeting increases information gain). From this point of view, it is clear the two values constrain each other and maximizing both simultaneously is partially (but not entirely) paradoxical. While maximising pragmatic value requires agents to choose to meet with an interlocutor they know is similar to them, maximising epistemic value necessitates they meet with one they do not know at all.

$$P(u_{loc}) = \sigma\left(-\gamma_{G,loc}G_{loc} + \gamma_{E,loc}E_{loc}\right)$$

$$G_{loc} = o_{u,belief} \cdot \left(\ln o_{u,belief} - C_{belief}\right) + H \cdot x_{u,2,visit}$$

$$x_{u,2}^{(1)} = B_u^{(1)}x_1^{(1)}$$

$$o_{u,belief} = A_{belief}^{(1)}x_{u,2}^{(1)}$$

$$c_{beleif} = \ln\left(A_C^{(2)}x_{core}^{(2)}\right)$$

$$\textit{if } x_{visit,j} = 1;$$

[equal 1 if agent visited a particular agent j]

$$H_j = 0$$

[ambiguity is zero if agent visited this agent j already] *else* :

$$H_j = 0.1$$

[ambiguity is non - zero if agent has not visited agent j yet]

A6. Generative Process

Generative process for meeting selection

$u_{loc} \sim P(u_{loc})$[*actual meeting u_{loc} is sampled from meeting selection prior $P(u_{loc})$*]

Generative process for belief expression and EV (satisfaction) of each agent
At a high level of cognitive control, agents incorporate a series of processes underlying the selection of a particular belief for expression (u2). Other than the partial reliance on a low level habitual factor [E], this action involves multiple higher order considerations.

First, an agent considers their core belief state (x), and the way this state apriori maps on to one of two discrete emotional valence states (s2) via an initial likelihood mapping [A2] Emotional Valence (EV) is defined as the extent to which an emotion is positive or negative (Feldman Barrett and Russell 1999), such that agents' core beliefs are apriori associated with either positive emotional valence or negative emotional valence (with some probability). As a minimal form of vicarious learning, the initial mapping is further updated based on associations agents observe between their interlocutors' expressed belief state and EV value The initial mapping therefore involves minimal precision for the expected EV for belief 2, since agents are first introduced to this belief (and associated EV) during the simulations. For this reason, the initial likelihood mapping between states is updated throughout our simulation via a crucial concentration parameter (α).

EV states are generated from core belief states, using a (learnable) likelihood mapping:

$$x_{sat}^{(1)} = A_{sat}^{(2)} x_{core}^{(2)}$$

Confidence of belief expression is generated using a Gamma distribution, where the rate parameter expris the Bayesian model average of (+, −) values associated with high and low satisfaction:

$$P(\gamma_{expr}) \approx \Gamma(1, \beta_{expr})$$
$$\beta_{expr} = \beta^{(+,-)} \cdot x_{sat}^{(1)}, \qquad \beta^{(+,-)} = [0.25, 2.0]$$

The expression of beliefs is guided by current core beliefs (scaled with satisfaction-dependent expr) and by habitual belief expression Eexpr (scaled with a fixed parameter E,expr):

$$P(u_{expr}|\gamma_{expr}) = \sigma\left(-\gamma_{expr}\ln x_{core}^{(2)} + \gamma_{E,expr} E_{expr}\right)$$

The intrinsically stochastic and itinerant nature of the generative process of communication is modeled by using a two-dimensional Dirichlet distribution to generate observed expressions on the range [0,1], where each agent's belief expression prior Puexprlexpr is used to specify their concentration parameters (multiplied by 12 to reduce variance):

$$o_{expr} = Dir(12u_{expr})$$

Generative process for emotional valence expressed by each agent

$$o_{sat} = A_{sat}^{(1)} x_{sat}^{(1)}$$

[satisfaction observed by interaction partner corresponds to actual satisfaction]

The EV state predicted is then used to generate an action confidence value (γ) such that positive EV generates high confidence in a certain expression of the belief state (u1) and negative EV generates low confidence values. Higher confidence values produce higher precision on the expected free energy (G) for one's belief expressed in the current conversation.

A7. Perception

Updating beliefs about the other agent's belief based on their expression

$$Q\left(x_{belief}^{(1)}\right) = o_{expr}$$

Updating of core belief based on beliefs expressed by other agents

$$Q\left(x_{core}^{(2)}\right) = \sigma\left(\ln \ln x_{core}^{(2)} + \gamma_{A,self}^{(2)} \ln \ln o_{expr,self} + \gamma_{A,other}^{(2)} \ln \ln o_{expr,other}\right)$$

A8. Learning

Habit learning for meeting selection

$$P(E_{loc}) = Dir(e_{loc})$$
$$Q(E_{loc}) = Dir(e_{loc} + 0.05 u_{loc})$$

Habit learning for belief expression

$$P\left(E_{expr}\right) = Dir\left(e_{expr}\right)$$
$$Q\left(E_{expr}\right) = Dir\left(e_{expr}\right) + 0.1 o_{expr}$$

Perceptual learning for the mapping between satisfaction and core beliefs, based on the expressions of other agents

$$P\left(A_{sat}^{(2)}\right) = Dir\left(a_{sat}^{(2)}\right)$$
$$Q\left(A_{sat}^{(2)}\right) = Dir\left(a_{sat}^{(2)} + \gamma_A^{(2)} o_{expr} \ln \ln x_{sat}^{(1)}\right)$$

A9. Initialisation of Parameters for Each Agent

$$\gamma_{A,belief}^{(2)} \sim \Gamma(5, 6)$$

[*regulates the integration of beliefs of other agents in one's own core belief*]

$$\gamma_{A,sat}^{(2)} \sim \Gamma(10, 1)$$

[*regulates learning rate of mappings between satisfaction and core belief based on observed correspondences in other agents*]

$$\gamma_{G,loc} \sim \Gamma(1, 1)$$

[*regulates reliance on action model in selecting agent to meet*]

$$\gamma_{E,loc} \sim \Gamma(1, 1)$$

[*regulates reliance on habitual prior in selecting agent to meet*]

$$\gamma_{E,expr} \sim N\left(\frac{\gamma_{E.scc}}{10}, \frac{\gamma_{E.loc}}{200}\right)$$

[*regulates reliance on habitual prior in expressing action, which correlates with $\gamma_{E,loc}$*]

$$\gamma_{B,core}^{(2)} \sim \Gamma(4, .5)$$

[*regulates stability of core beliefs across days*]

$$\gamma_{B,habits}^{(2)} \sim \Gamma(.5, 1)$$

[*regulates stability of expression habits across days*]

$$B_0^{(2)} = [[.75, .25], [.25, .75]]$$

[*specifies baseline transition probabilities*]

$$B^{(2)} = \sigma\left(\gamma_B^{(2)} \ln B_0^{(2)}\right)$$

[*corrects $B_0^{(2)}$ using the agent $-$ specific $\gamma_B^{(2)}$ values*]

Agents with relatively weak confirmation bias :

$$A_C^{(2)} \sim Dir(6, 4)$$

[*induces weak reliance on core beliefs for specifying lower-level preferences*]
Agents with relatively strong confirmation bias :

$$A_{C,1}^{(2)} \sim Dir(999, 1)$$

[*induces strong reliance on core beliefs for specifying lower-level preferences*]

References

Aunger, R.: Darwinizing Culture: The Status of Memetics as a Science. Oxford University Press, Oxford (2001)

Buskell, A., Enquist, M., Jansson, F.: A systems approach to cultural evolution. Palgrave Commun. **5**(1), 1–15 (2019)

Bettencourt, L.M., Cintrón-Arias, A., Kaiser, D.I., Castillo-Chávez, C.: The power of a good idea: quantitative modeling of the spread of ideas from epidemiological models. Physica A Stat. Mech. Appl. **364**, 513–536 (2006)

Clark, H.H., Brennan, S.E.: Grounding in Communication. American Psychological Association, Washington DC (1991)

Constant, A., Ramstead, M.J., Veissière, S.P., Friston, K.: Regimes of expectations: an active inference model of social conformity and human decision making. Front. Psychol. **10**, 679 (2019)

Constant, A., Ramstead, M.J., Veissiere, S.P., Campbell, J.O., Friston, K.J.: A variational approach to niche construction. J. R. Soc. Interface **15**(141), 20170685 (2018)

Creanza, N., Kolodny, O., Feldman, M.W.: Cultural evolutionary theory: how culture evolves and why it matters. Proc. Natl. Acad. Sci. **114**(30), 7782–7789 (2017)

Dawkins, R.: Viruses of The Mind. Dennett and his Critics: Demystifying Mind, vol. 13, p. e27 (1993)

Dean, L.G., Vale, G.L., Laland, K.N., Flynn, E., Kendal, R.L.: Human cumulative culture: a comparative perspective. Biol. Rev. **89**(2), 284–301 (2014)

Dunstone, J., Caldwell, C.A.: Cumulative culture and explicit metacognition: a review of theories, evidence and key predictions. Palgrave Commun. **4**(1), 1–11 (2018)

Echterhoff, G., Higgins, E.T., Levine, J.M.: Shared reality: experiencing commonality with others' inner states about the world. Perspect. Psychol. Sci. **4**(5), 496–521 (2009)

Enquist, M., Ghirlanda, S., Eriksson, K.: Modelling the evolution and diversity of cumulative culture. Philos. Trans. Royal Soc. B Biol. Sci. **366**(1563), 412–423 (2011)

Friston, K., Frith, C.: A duet for one. Conscious. Cogn. **36**, 390–405 (2015)

Friston, K.J., Frith, C.D.: Active inference, communication and hermeneutics. Cortex **68**, 129–143 (2015)

Gabora, L.: Meme and variations: a computational model of cultural evolution. In: 1993 Lectures in Complex Systems, pp. 471–485. Addison Wesley, Boston (1995)

Hesp, C., Ramstead, M., Constant, A., Badcock, P., Kirchhoff, M., Friston, K.: A multi-scale view of the emergent complexity of life: a free-energy proposal. In: Evolution, Development and Complexity, pp. 195–227. Springer, Cham (2019). https://doi.org/10.1007/978-3-030-00075-2_7

Kaplan, R., Friston, K.J.: Planning and navigation as active inference. Biol. Cybern. **112**(4), 323–343 (2018)

Kashima, Y., Bain, P.G., Perfors, A.: The psychology of cultural dynamics: What is it, what do we know, and what is yet to be known? Annu. Rev. Psychol. **70**, 499–529 (2019)

Pikovsky, A., Kurths, J., Rosenblum, M., Kurths, J.: Synchronization: A Universal Concept in Nonlinear Sciences (No. 12). Cambridge University Press, Cambridge (2003)

Richerson, P.J., Boyd, R., Henrich, J.: Gene-culture coevolution in the age of genomics. Proc. Natl. Acad. Sci. **107**(Supplement 2), 8985–8992 (2010)

Stout, D., Hecht, E.E.: Evolutionary neuroscience of cumulative culture. Proc. Natl. Acad. Sci. **114**(30), 7861–7868 (2017)

Weisbuch, M., Pauker, K., Ambady, N.: The subtle transmission of race bias via televised nonverbal behavior. Science **326**(5960), 1711–1714 (2009)

Dream to Explore: 5-HT2a as Adaptive Temperature Parameter for Sophisticated Affective Inference

Adam Safron[1](✉) and Zahra Sheikhbahaee[2]

[1] Center for Psychedelic and Consciousness Research, Department of Psychiatry and Behavioral Sciences, Johns Hopkins University School of Medicine, Baltimore, MD 21224, USA
[2] David R. Cheriton School of Computer Science, University of Waterloo, Waterloo, ON, Canada
zsheikhb@uwaterloo.ca

Abstract. Relative to other neuromodulators, serotonin (5-HT) has received far less attention in machine learning and active inference. We will review prior work interpreting 5-HT1a signaling as an uncertainty parameter with opponency to dopamine. We will then discuss how 5-HT2a receptors may promote more exploratory policy selection by enhancing imaginative planning (as sophisticated affective inference). Finally, we will briefly comment on how qualitatively different effects may be observed across low and high levels of 5-HT2a signaling, where the latter may help agents to change self-adversarial policies and break free of maladaptive absorbing states in POMDPs.

Keywords: Serotonin · 5-HT1a · 5-HT2a · Sophisticated active inference · Affective inference · Exploration · Exploitation · Imagination · Planning · Consciousness

1 Introduction

Serotonin (5-HT) is a phylogenetically ancient monoamine neuromodulator found in all life forms, and which in mammals involves at least 14 distinct receptors that can be subdivided into 7 sub-classes [1]. This divergence of 5-HT systems arose through a process of evolutionary divergence via gene duplication and subsequent specialization of receptor subtypes and associated pathways [2]. This diversity may seem to suggest limited utility for attempting to recapitulate 5-HT-related functions in artificial systems. However, we propose that a substantial portion of 5-HT-related functionality may be obtained by focusing on the 1a and 2a receptors. We suggest this seemingly myopic focus on these two receptor classes may constitute a fruitful research direction on account of their highly conserved status in evolution, their relatively broad distribution in mammalian brains, as well as the common organismic significance of 5-HT signaling implicated by the mass-release of these diffusely acting neuromodulators from concentrated neurons in midbrain nuclei. While understanding the full diversity of 5-HT signaling will likely be illuminating with respect to abilities to differentially modify various neural

© Springer Nature Switzerland AG 2021
M. Kamp et al. (Eds.): ECML PKDD 2021 Workshops, CCIS 1524, pp. 799–809, 2021.
https://doi.org/10.1007/978-3-030-93736-2_56

processes—with potentially common organismic significances—we believe a focus on 1a and 2a receptors will provide both maximal explanatory purchase and a foundation upon which subsequent modelling may proceed.

Cortex is highly populated by both 5-HT1a and 5-HT2a receptors [1]. 5-HT1a receptors have primarily inhibitory effects on neurons, and are functionally associated with somewhat subtle effects on mood, uncertainty, and the learning of complex behaviors, whereas 5-HT2a receptors have primarily excitatory effects and appear to have more pronounced effects on affect and cognitive processes [3]. 5-HT2a receptors also mediate the primary mechanism of action for hallucinogenic drugs such as lysergic acid diethylamide (LSD), psilocybin, and N,N-dimethyltryptamine (DMT) [4]. These compounds are widely known for inducing states of altered perception, thought, and feeling, with similarities to lucid dreaming; in this way, psychedelic states share features with both dreaming and waking consciousness [5].

The functionality of 5-HT1a and 5-HT2a receptors has respectively been associated with either passive or active coping strategies in the face of threat (or uncertainty) [6]. In this view, 5-HT1a signaling enables adaptive responses to mild-to-moderate stress through affective regulation and the inhibition of (disinhibitory) dopaminergic processes. 5-HT2a signaling, in contrast, is upregulated during more intense states of challenge— potentially including uncertainty with respect to highly valued goals—so allowing for both increased behavioral flexibility and neural plasticity [7].

The Free Energy Principle and Active Inference (FEP-AI) framework characterizes organisms as kinds of generative models [8, 9], with brains functioning as cybernetic control systems for embodied agents as they attempt to minimize uncertainty with respect to realizing their goals. These (both implicitly and explicitly) valued goals are understood as Bayesian prior preferences over likely outcomes that allow such systems to maintain their forms on evolutionary and developmental timescales. Towards this end, an FEP-AI agent maximizes model evidence for its existence by minimizing expected (variational) free energy (i.e., cumulative precision-weighted prediction errors) between its world model and sensory observations. This expected free energy minimization (and thereby self-model-evidence maximization) is realized either via perception (updating world models) or action (enactively updating world states). According to hierarchical predictive processing (HPP) models, cortical processes—and potentially biological systems more generally—continuously generate top-down predictions of bottom-up information at multiple levels of hierarchical abstraction. Notably, each level of these hierarchical generative models has varying levels of spatial and temporal granularity with respect to the latent system-world states it attempts to predict and alter through active inference, so allowing for multi-scale models with varying degrees of temporal depth and counterfactual richness [10]. In HPP, bottom-up observations are (efficiently) encoded as prediction errors, which ascend to higher cortical levels to update generative models when not predictively suppressed by top- down prior expectations. At the highest levels of abstraction in cortical hierarchies, maximal explanatory power may be found through models related to complex processes such as those underlying various forms of selfhood and self-consciousness [11–13].

Within FEP-AI, conscious planning is understood as "sophisticated active inference," in which agents generate imagined sequences of counterfactual outcomes through rolling out mental simulations of different patterns of action/policy selection [14, 15]. This imaginative planning takes the form of a deep tree search over counterfactual observations and actions, where different rollouts of simulated actions allow for exploring different branches of decision trees. This sophisticated active inference is governed by a singular objective function(al) of expected free energy, which achieves balance with respect to exploration-exploitation tradeoffs by selecting governing models that neither overfit nor underfit patterns of data in shaping perception and action. At perhaps the highest level of organization, patterns of action/policy selection via counterfactual processing (*e.g.*, simulated movements through space) are orchestrated by the hippocampal/entorhinal system [16, 17], where these spatiotemporal trajectories may be understood as constituting the stream of consciousness [13].

Below we describe parts of our ongoing explorations of ways in which the functional significances of 5-HT1a and 5-HT2a receptors may be understood through the lens of FEP-AI. If accurate, these models may provide a unified account of the roles of 5-HT in adaptive behavior with implications for machine-learning, neuropsychology, and evolutionary-developmental biology. We will characterize ways in which both 5-HT1a and 5-HT2a receptors influence the degrees to which agents initiate imaginative planning and offline learning via mental simulations, as opposed to more immediately releasing policies for overt goal-seeking behaviors. Finally, we will address some potential misconceptions about the functionality of different levels of 5-HT2a signaling, and further establish potentially fruitful connections to meta-reinforcement learning.

2   5-HT1a Receptors

5-HT1a receptors are found in different layers of the cortex, but they are most strongly expressed in layers V and VI [1]. 5-HT1a receptors suppress pyramidal cell activity by increasing rectifying K^+ currents, and have also been found to inhibit gamma oscillations in the hippocampus [18], potentially indicating reduced sensitivity to overall organismic prediction error [19]. This is opposite to the effects for DA (and in some respects 5-HT2a) signaling, which in FEP-AI is understood as enhancing the precision of bottom-up prediction errors [20, 21], so promoting the sensitivity of behavioral response to rewards. 5-HT1a signaling, in contrast, would instead promote deliberation and patience with respect to policy deployment [22–24].

Substantial experimental evidence has demonstrated opponency between dopaminergic (DA) and serotoninergic (5-HT) signaling [25, 26], which appear to be respectively associated with situations characterized by either more appetitive or aversive states. For biological organisms, appetitive motivational systems encourage approach while aversive systems promote avoidance and withdrawal. However, this is not to say that serotonin creates aversion, but rather that 5-HT signaling tends to be enhanced for situations in which organisms experience stress and uncertainty with respect to their ability to achieve their goals [27], including the fundamental goal of survival. This is consistent with modulation of 5-HT1a neurons in the dorsal raphe nucleus [28], since release of action policies associated with either more passive or active coping ought to be modulated by the expected value of different patterns of enaction. However, firing rates for serotonergic also correlate with uncertainty more generally, indicating sensitivity to surprising events irrespective of the value of particular rewards. Opponency is observed in

5-HT1a and DA systems in terms of mutual inhibition of release and differential shaping of modulated systems [29]. However, their interactions can also produce synergy, both in terms of differentially parameterizing the nature of imaginative planning (e.g. with more or less confidence), and also in terms of providing a dynamic tension via their opponency, since pursuing complex goals require capacities for flexible adaption in response to environmental changes [30–33].

5-HT1a receptors are found as somatodendritic autoreceptors in raphe nuclei, as well as in postsynaptic sites in neocortex, hippocampus, and other "limbic" structures such as the amygdala and homeostatic regulatory nuclei of the hypothalamus [1]. Through inhibiting excitatory neurotransmission, 5-HT1a autoreceptors can both help with passively coping with stressful events by attenuating prediction error, and also help to promote more adaptive behavior by providing more time for planning in the face of uncertain circumstances. This inhibition of overt behavior affords both an opportunity for being informed by more complex world modeling, as well as an opportunity for adaptively calibrating world models by imbuing imaginative rollouts with greater uncertainty (i.e., implicitly functioning as a temperature parameter) [34]. Taken together, these functions allow for more flexible behavior via planning, as well as enhanced exploration and policy generalization via imagining the pursuit of goals under uncertain and potentially challenging conditions. Functionally speaking, this would be extremely sensible for a parameter that tends to be elevated in not just stressful circumstances, but also in situations involving the satiation of organismic drives such as eating, and possibly social contact [35–38]. That is, once goals are realized, a shift from exploration to exploitation would be both an adaptive foraging strategy and proximate mechanism for lifelong learning. Further, the more passive behavior encouraged by attenuation of action readiness would likely also be an adaptive response for an agent facing potential threats from circumstances that may overwhelm its present control abilities.

Thus, in FEP-AI terms, 5-HT1a signaling would be understood as promoting sophisticated inference via imaginative rollouts of predicted (or postdicted) patterns of enaction under conditions of reduced precision over counterfactually-deployed policies [14]. This would correspond to an agent experiencing relatively lower levels of confidence while entertaining counterfactual policies, but also with reduced "affective charge" [15], which one would normally expect to be stronger in a negative direction under conditions of reduced certainty with respect to realizing prior preferences. In these ways, 5-HT1a signaling would promote adaptive responses to challenging (and potentially novel) environments by increasing tolerance with respect to uncertainty through stress moderation, so allowing for more flexible and sophisticated forms of cognition and behavior [22–24].

The effects of different levels of 5-HT1a signaling may have profound functional consequences. For example, the ascending serotonergic pathway from the dorsal raphe nucleus and its effects on the amygdala and frontal cortex may promote adaptive reshaping conditioned fear responses [39, 40]. The basolateral nucleus of the amygdala contributes to behavioral changes in the face of emotional events and associated stimuli, including in response to stressors such as social defeat and other fearful circumstances [41, 42]. However, activation of 5-HT1a postsynaptic receptors in the dorsal hippocampus and amygdala facilitate extinction of fear-conditioned behaviors, consistent with the proposed roles of serotonin in facilitating coping in the face of threat [39]. With respect to potentially synergistic interactions between neuromodulatory systems, DA may be understood as providing a learning rate signal that influences the degree to which an agent updates its predictions in response to novel experiences [43]. This learning rate would influence the degree to which reward prediction errors shape policy selection, which if excessive could result in impulsivity by having presently estimated rewards promote more reactive forms of policy selection [44]. 5-HT1a, in contrast, would promote modeling with greater temporal depth and counterfactual richness, providing opportunities for meta-learning with sensitivity to (and ability to adaptively cope with) uncertain circumstances [24]. Notably, with respect to pathological states such as the rumination associated with depression and the impulsive aggression associated with antisocial behavior, low 5-HT1a in the medial prefrontal cortex appears to be associated with more perseverative tendencies and reduced abilities to adapt to novel environments [45, 46].

3 5-HT2a Receptors

5-HT2a receptors are most strongly expressed in high-level association cortices [1], including the "default mode network" (DMN). The DMN is comprised of a set of brain regions exhibiting high metabolic activity during resting states (including sleep), and which also become deactivated during goal-directed cognition, in conjunction with upregulation of "task positive" brain areas [47]. Notably, DA signaling tends to shift activity in the direction of increased dominance by frontoparietal control networks, and 5-HT signaling tends to shift activity towards exhibiting greater DMN power [48]. Key nodes of the DMN include medial prefrontal and parietal cortices [49, 50], as well as the temporoparietal junction [51, 52], which together constitute key areas for imaginative simulations involving both self and other [53, 54], and which may also be essential for establishing minimum embodied selfhood and coherent subjective experience [13, 55]. These nodes become functionally and structurally connected in a gradual manner over the course of development [56], and may have been uniquely expanded in the course of human evolution [57]. With respect to capacities for imaginative planning, it is particularly notable that the DMN plays a central role in counterfactual mental simulations [58, 59]. The DMN is often considered to be at the top of the cortical hierarchy in FEP-AI [11], although evidence suggests a more complicated picture in which salience networks may be understood as constituting the highest levels of control [60]. Notably, 5-HT2a receptors are also particularly concentrated in the anterior insula [61], a key node in networks for salience determination and goal prioritization [62].

5-HT2a receptors are also responsible for the neuropsychological effects of psychedelics, which have been shown to elevate the entropy of endogenous brain activity and potentially enhance the richness of both the level and contents of consciousness [63–67]. With the widely known "RElaxed Beliefs Under pSychedelics" (REBUS) model [68], 5-HT2a agonists are suggested to attenuate the precision of high-level prior beliefs, so flattening the curvature of free energy landscapes and enhancing sensitivity to novel observations [69]. This altered regime promotes the breakdown of the brain's usual hierarchical structure [70], so allowing for an "anarchic" state in which novel forms of communication are allowed between brain areas, so allowing for enhanced cognitive exploration and opportunities for updating of deep beliefs. In this REBUS regime, ascending prediction errors from hierarchically lower levels may reshape upper level (potentially excessively precise) priors, so allowing the agent to break free from overly rigid patterns of thought and behavior.

In machine learning terms, such maladaptive cognitive and behavioral habits could be viewed as constituting self-adversarial policies, potentially formed through histories of excessive (or premature) exploitation in policy selection [71]. The more entropic dynamics afforded by high levels of 5-HT2a agonism, however, may allow agents to escape from these self-undermining attractors and reach more desirable regions of policy space. In terms of active inference, such relaxation of deep beliefs would be understood as reducing dominance from the parameters that serve as priors for agent- based generative models [72], which if excessively concentrated may preclude patterns of policy selection that could allow for opportunities for updating [73]. Concentration of probability mass in Dirichlet parameters via iterative policy selection and learning may provide a model of personality formation, and so their potential updating under conditions of strong 5-HT2a agonism (or functional homologues) could also provide a model of the kinds of personality change that have been associated with psychedelics [74]. This would also provide a model for the generation of novel, and potentially more (and possibly excessive) creative modes of cognition for both biological and artificial agents [75].

However, more physiologically typical low-to-moderate levels of 5-HT2a agonism have been suggested to involve a strengthening of beliefs under psychedelics (i.e., SEBUS effects) [76], both on account of increased activity from deep pyramidal neurons encoding prior expectations (or predictions), as well as reduced activity from superficial pyramidal neurons encoding prediction errors. Under this kind of SEBUS regime, individuals may engage in counterfactual processing under conditions of intense salience (e.g. sophisticated inference with high affective charge), potentially including greater confidence in both imagination and action. This would be highly consistent with accounts of 5-HT2a signaling as entailing strategies for "active coping" in the face of uncertainty/threat [6]. Whether patterns of either simulated or overtly enacted policy selection are more exploitative or exploratory would depend on a multitude of both pre-existing and context-specific priors over preferred patterns of enaction (e.g. typical levels of curious engagement) [77]. Further, both 5-HT1a and 5-HT2a signaling have been shown to inhibit sharp-wave ripples [78], which may correspond to hippocampal/entorhinal remapping events [79]. Such inhibition of resetting of (generalized) mapping and accompanying repertoires of operative policies may promote opportunities for imaginative planning via more extended rollouts [80]. However, while 5-HT1a signaling will tend to be associated with more passive forms of cognition and behavior as described above, 5-HT2a signaling could promote more proactive modes of engagement with elevated affect from enhanced sensitivity to interoceptive signals [36]. In this way, 5-HT2a would provide a flexible parameter for imaginative planning when systems face varying degrees of stress (or uncertainty) with respect to achieving their goals.

SEBUS effects may also occur alongside REBUS effects at moderate-to-high levels of 5-HT2a agonism [68, 76], with potentially further indirect strengthening of intermediate level beliefs associated with the perceptual synthesis underlying conscious experience. This conjunction of high levels of perceptual vividness with exploration of novel forms of cognition could provide the greatest opportunities for updating, which may be a crucially important intervention for systems suffering from self-adversarial modes of policy selection. That is, while more physiological levels of 5-HT2a signaling may afford more flexible and adaptive refinement of normal policies, very high levels of agonism may constitute a qualitatively different regime that could allow both biological and artificial systems to "change their mind" in profound ways capable of altering their overall character [81]. Whether such changes are beneficial or harmful to system performance will depend on a multitude of factors, with the "set and setting" of such interventions being of crucial importance for shaping the direction of future system evolution. Going forward, we are currently planning simulation experiments in which we will demonstrate how these principles may apply to (artificial) world- modeling active inferential agents.

Acknowledgements. We gratefully acknowledge partial funding support from the Waterloo-Huawei Joint Innovation Lab within the project "the Active Inferential Meta-Learning Engine".

References

1. Barnes, N.M., et al.: International union of basic and clinical pharmacology. CX. Classification of receptors for 5-hydroxytryptamine; pharmacology and function. Pharmacol. Rev. **73**(1), 310–520 (2021). https://doi.org/10.1124/pr.118.015552
2. Moutkine, I., Collins, E.L., Béchade, C., Maroteaux, L.: Evolutionary considerations on 5-HT2 receptors. Pharmacol. Res. **140**, 14–20 (2019). https://doi.org/10.1016/j.phrs.2018.09.014
3. Zhang, G., Stackman, R.W.: The role of serotonin 5-HT2A receptors in memory and cognition. Front. Pharmacol. **6**, 225 (2015). https://doi.org/10.3389/fphar.2015.00225
4. Johnson, M.W., Hendricks, P.S., Barrett, F.S., Griffiths, R.R.: Classic psychedelics: an integrative review of epidemiology, therapeutics, mystical experience, and brain network function. Pharmacol. Ther. **197**, 83–102 (2019). https://doi.org/10.1016/j.pharmthera.2018.11.010
5. Kraehenmann, R.: Dreams and psychedelics: neurophenomenological comparison and therapeutic implications. Curr. Neuropharmacol. **15**(7), 1032–1042 (2017). https://doi.org/10.2174/1573413713666170619092629
6. Carhart-Harris, R., Nutt, D.: Serotonin and brain function: a tale of two receptors. J. Psychopharmacol. Oxf. Engl. **31**(9), 1091–1120 (2017). https://doi.org/10.1177/0269881117725915
7. Shao, L.-X., et al.: Psilocybin induces rapid and persistent growth of dendritic spines in frontal cortex in vivo. Neuron **109**(16), 2535–2544 (2021). https://doi.org/10.1016/j.neuron.2021.06.008
8. Friston, K.J., FitzGerald, T., Rigoli, F., Schwartenbeck, P., Pezzulo, G.: Active inference: a process theory. Neural Comput. **29**(1), 1–49 (2017). https://doi.org/10.1162/NECO_a_00912
9. Friston, K.J.: The free-energy principle: a unified brain theory? Nat. Rev. Neurosci. **11**(2), 127–138 (2010). https://doi.org/10.1038/nrn2787
10. Friston, K.J., Rosch, R., Parr, T., Price, C., Bowman, H.: Deep temporal models and active inference. Neurosci. Biobehav. Rev. **77**, 388–402 (2017). https://doi.org/10.1016/j.neubiorev.2017.04.009
11. Carhart-Harris, R.L., Friston, K.J.: The default-mode, ego-functions and free-energy: a neurobiological account of Freudian ideas. Brain J. Neurol. **133**(Pt 4), 1265–1283 (2010). https://doi.org/10.1093/brain/awq010
12. Safron, A.: The radically embodied conscious cybernetic Bayesian brain: from free energy to free will and back again. Entropy **23**(6), 783 (2021). https://doi.org/10.3390/e23060783

13. Safron, A.: An integrated world modeling theory (IWMT) of consciousness: combining integrated information and global neuronal workspace theories with the free energy principle and active inference framework; toward solving the hard problem and characterizing agentic causation. Front. Artif. Intell. **3**, 30 (2020). https://doi.org/10.3389/frai.2020.00030

14. Friston, K., Da Costa, L., Hafner, D., Hesp, C., Parr, T.: Sophisticated Inference. https://arxiv.org/abs/2006.04120v1. Accessed 18 Jun 2020

15. Hesp, C., Tschantz, A., Millidge, B., Ramstead, M., Friston, K., Smith, R.: Sophisticated affective inference: simulating anticipatory affective dynamics of imagining future events. In: Active Inference, pp. 179–186. Springer, Cham (2020). https://doi.org/10.1007/978-3-030-64919-7_18

16. Barron, H.C., Auksztulewicz, R., Friston, K.: Prediction and memory: a predictive coding account. Prog. Neurobiol. **192**, 101821 (2020). https://doi.org/10.1016/j.pneurobio.2020.101821

17. Çatal, O., Verbelen, T., Van de Maele, T., Dhoedt, B., Safron, A.: Robot navigation as hierarchical active inference. Neural Netw. **142**, 192–204 (2021). https://doi.org/10.1016/j.neunet.2021.05.010

18. Johnston, A., McBain, C.J., Fisahn, A.: 5-Hydroxytryptamine1A receptor- activation hyperpolarizes pyramidal cells and suppresses hippocampal gamma oscillations via Kir3 channel activation. J. Physiol. **592**(19), 4187–4199 (2014). https://doi.org/10.1113/jphysiol.2014.279083

19. Mannella, F., Gurney, K., Baldassarre, G.: The nucleus accumbens as a nexus between values and goals in goal-directed behavior: a review and a new hypothesis. Front. Behav. Neurosci. **7**, 135 (2013). https://doi.org/10.3389/fnbeh.2013.00135

20. FitzGerald, T.H.B., Dolan, R.J., Friston, K.J.: Dopamine, reward learning, and active inference. Front. Comput. Neurosci. **9**, 136 (2015). https://doi.org/10.3389/fncom.2015.00136

21. Friston, K.J., Schwartenbeck, P., FitzGerald, T., Moutoussis, M., Behrens, T., Dolan, R.J.: The anatomy of choice: dopamine and decision-making. Philos. Trans. R. Soc. B Biol. Sci. **369**(1655), 20130481 (2014). https://doi.org/10.1098/rstb.2013.0481

22. Moran, R.J., et al.: The protective action encoding of serotonin transients in the human brain. Neuropsychopharmacology **43**(6), 1425–1435 (2018). https://doi.org/10.1038/npp.2017.304

23. Grossman, C.D., Bari, B.A., Cohen, J.Y.: Serotonin neurons modulate learning rate through uncertainty. bioRxiv **103**, 922 (2020). https://doi.org/10.1101/2020.10.24.353508

24. Ohmura, Y., et al.: Disruption of model-based decision making by silencing of serotonin neurons in the dorsal raphe nucleus. Curr. Biol. **31**(11), 2446–2454 (2021). https://doi.org/10.1016/j.cub.2021.03.048

25. Boureau, Y.-L., Dayan, P.: Opponency revisited: competition and cooperation between dopamine and serotonin. Neuropsychopharmacology **36**(1), 74–97 (2011). https://doi.org/10.1038/npp.2010.151

26. Daw, N.D., Kakade, S., Dayan, P.: Opponent interactions between serotonin and dopamine. Neural Netw. Off. J. Int. Neural Netw. Soc. **15**(4–6), 603–616 (2002). https://doi.org/10.1016/s0893-6080(02)00052-7

27. Doya, K., Miyazaki, K.W., Miyazaki, K.: Serotonergic modulation of cognitive computations. Curr. Opin. Behav. Sci. **38**, 116–123 (2021). https://doi.org/10.1016/j.cobeha.2021.02.003

28. Bromberg-Martin, E.S., Hikosaka, O., Nakamura, K.: Coding of task reward value in the dorsal raphe nucleus. J. Neurosci. Off. J. Soc. Neurosci. **30**(18), 6262–6272 (2010). https://doi.org/10.1523/JNEUROSCI.0015-10.2010

29. Yagishita, S.: Transient and sustained effects of dopamine and serotonin signaling in motivation-related behavior. Psychiatry Clin. Neurosci. **74**(2), 91–98 (2020). https://doi.org/10.1111/pcn.12942

30. Hayes, S.C.: A Liberated Mind: How to Pivot Toward What Matters. Penguin, London (2019)
31. Atasoy, S., Deco, G., Kringelbach, M.L.: Playing at the edge of criticality: expanded whole-brain repertoire of connectome-harmonics. In: Tomen, N., Herrmann, J.M., Ernst, U. (eds.) The Functional Role of Critical Dynamics in Neural Systems. SSBN, vol. 11, pp. 27–45. Springer, Cham (2019). https://doi.org/10.1007/978-3-030-20965-0_2
32. Davis, A.K., Barrett, F.S., Griffiths, R.R.: Psychological flexibility mediates the relations between acute psychedelic effects and subjective decreases in depression and anxiety. J. Context. Behav. Sci. **15**, 39–45 (2020). https://doi.org/10.1016/j.jcbs.2019.11.004
33. Gerraty, R.T., Davidow, J.Y., Foerde, K., Galvan, A., Bassett, D.S., Shohamy, D.: Dynamic flexibility in striatal-cortical circuits supports reinforcement learning. J. Neurosci. **38**(10), 2442–2453 (2018). https://doi.org/10.1523/JNEUROSCI.2084-17.2018
34. Ha, D., Schmidhuber, J.: World Models (Mar 2018). ArXiv180310122 Cs Stat. https://doi.org/10.5281/zenodo.1207631
35. Voigt, J.-P., Fink, H.: Serotonin controlling feeding and satiety. Behav. Brain Res. **277**, 14–31 (2015). https://doi.org/10.1016/j.bbr.2014.08.065
36. Hjorth, O.R., et al.: Expression and co-expression of serotonin and dopamine transporters in social anxiety disorder: a multitracer positron emission tomography study. Mol. Psychiatry **26**(8), 1–10 (2019). https://doi.org/10.1038/s41380-019-0618-7
37. Fotopoulou, A., Tsakiris, M.: Mentalizing homeostasis: the social origins of interoceptive inference–replies to commentaries. Neuropsychoanalysis **19**(1), 71–76 (2017)
38. Ciaunica, A., Constant, A., Preissl, H., Fotopoulou, A.: The first prior: from co-embodiment to co-homeostasis in early life. PsyArXiv. https://doi.org/10.31234/osf.io/twubr. Accessed 5 Jan 2021
39. Pavlova, I.V., Rysakova, M.P.: Effects of administration of serotonin 5- HT1A receptor ligands into the amygdala on the behavior of rats with different manifestations of conditioned reflex fear. Neurosci. Behav. Physiol. **48**(3), 267–278 (2018). https://doi.org/10.1007/s11055-018-0560-1
40. Dayan, P., Huys, Q.J.M.: Serotonin in affective control. Annu. Rev. Neurosci. **32**(1), 95–126 (2009). https://doi.org/10.1146/annurev.neuro.051508.135607
41. Colyn, L., Venzala, E., Marco, S., Perez-Otaño, I., Tordera, R.M.: Chronic social defeat stress induces sustained synaptic structural changes in the prefrontal cortex and amygdala. Behav. Brain Res. **373**, 112079 (2019). https://doi.org/10.1016/j.bbr.2019.112079
42. Badcock, P.B., Davey, C.G., Whittle, S., Allen, N.B., Friston, K.J.: The depressed brain: an evolutionary systems theory. Trends Cogn. Sci. **21**(3), 182–194 (2017). https://doi.org/10.1016/j.tics.2017.01.005
43. Schultz, W.: Neuronal reward and decision signals: from theories to data. Physiol. Rev. **95**(3), 853–951 (2015). https://doi.org/10.1152/physrev.00023.2014
44. Dalley, J.W., Roiser, J.P.: Dopamine, serotonin and impulsivity. Neuroscience **215**, 42–58 (2012). https://doi.org/10.1016/j.neuroscience.2012.03.065
45. Di Pietro, N.C., Seamans, J.K.: Dopamine and serotonin interactions in the prefrontal cortex: insights on antipsychotic drugs and their mechanism of action. Pharmacopsychiatry **40**(Suppl 1), S27-33 (2007). https://doi.org/10.1055/s-2007-992133
46. Lu, H., Liu, Q.: Serotonin in the frontal cortex: a potential therapeutic target for neurological disorders. Biochem. Pharmacol. Open Access **6**(1), e184 (2017). https://doi.org/10.4172/2167-0501.1000e184
47. Dohmatob, E., Dumas, G., Bzdok, D.: Dark control: the default mode network as a reinforcement learning agent. Hum. Brain Mapp. **41**(12), 3318–3341 (2020). https://doi.org/10.1002/hbm.25019
48. Conio, B., et al.: Opposite effects of dopamine and serotonin on resting-state networks: review and implications for psychiatric disorders. Mol. Psychiatry **25**(1), 82–93 (2020). https://doi.org/10.1038/s41380-019-0406-4

49. Fransson, P., Marrelec, G.: The precuneus/posterior cingulate cortex plays a pivotal role in the default mode network: evidence from a partial correlation network analysis. Neuroimage **42**(3), 1178–1184 (2008). https://doi.org/10.1016/j.neuroimage.2008.05.059

50. Utevsky, A.V., Smith, D.V., Huettel, S.A.: Precuneus is a functional core of the default-mode network. J. Neurosci. **34**(3), 932–940 (2014). https://doi.org/10.1523/JNEUROSCI.4227-13. 2014

51. Baird, B., Castelnovo, A., Gosseries, O., Tononi, G.: Frequent lucid dreaming associated with increased functional connectivity between frontopolar cortex and temporoparietal association areas. Sci. Rep. **8**(1), 17798 (2018). https://doi.org/10.1038/s41598-018-36190-w

52. Graziano, M.S.A.: The temporoparietal junction and awareness. Neurosci. Conscious. **2018**(1) (2018). https://doi.org/10.1093/nc/niy005

53. Hassabis, D., Spreng, R.N., Rusu, A.A., Robbins, C.A., Mar, R.A., Schacter, D.L.: Imagine all the people: how the brain creates and uses personality models to predict behavior. Cereb. Cortex **24**(8), 1979–1987 (2014). https://doi.org/10.1093/cercor/bht042

54. Guterstam, A., Bio, B.J., Wilterson, A.I., Graziano, M.: Temporo-parietal cortex involved in modeling one's own and others' attention. eLife **10**, e63551 (2021). https://doi.org/10.7554/ eLife.63551

55. Davey, C.G., Harrison, B.J.: The brain's center of gravity: how the default mode network helps us to understand the self. World Psychiatry **17**(3), 278–279 (2018). https://doi.org/10. 1002/wps.20553

56. Fan, F., et al.: Development of the default-mode network during childhood and adolescence: a longitudinal resting-state fMRI study. Neuroimage **226**, 117581 (2021). https://doi.org/10. 1016/j.neuroimage.2020.117581

57. Buckner, R.L., DiNicola, L.M.: The brain's default network: updated anatomy, physiology and evolving insights. Nat. Rev. Neurosci. **20**(10), 593–608 (2019). https://doi.org/10.1038/ s41583-019-0212-7

58. Hassabis, D., Maguire, E.A.: The construction system of the brain. Philos. Trans. R. Soc. Lond. B. Biol. Sci. **364**(1521), 1263–1271 (2009). https://doi.org/10.1098/rstb.2008.0296

59. Faul, L., St. Jacques, P.L., DeRosa, J.T., Parikh, N., De Brigard, F.: Differential contribution of anterior and posterior midline regions during mental simulation of counterfactual and perspective shifts in autobiographical memories. NeuroImage **215**, 116843 (2020). https:// doi.org/10.1016/j.neuroimage.2020.116843

60. Zhou, Y., Friston, K.J., Zeidman, P., Chen, J., Li, S., Razi, A.: The hierarchical organization of the default, dorsal attention and salience networks in adolescents and young adults. Cereb. Cortex NY. **28**(2), 726–737 (2018). https://doi.org/10.1093/cercor/bhx307

61. Santangelo, A.M., et al.: Insula serotonin 2A receptor binding and gene expression contribute to serotonin transporter polymorphism anxious phenotype in primates. Proc. Natl. Acad. Sci. **116**(29), 14761–14768 (2019). https://doi.org/10.1073/pnas.1902087116

62. Rueter, A.R., Abram, S.V., MacDonald, A.W., Rustichini, A., DeYoung, C.G.: The goal priority network as a neural substrate of conscientiousness. Hum. Brain Mapp. **39**(9), 3574–3585 (2018). https://doi.org/10.1002/hbm.24195

63. Barnett, L., Muthukumaraswamy, S.D., Carhart-Harris, R.L., Seth, A.K.: Decreased directed functional connectivity in the psychedelic state. Neuroimage **209**, 116462 (2020). https://doi. org/10.1016/j.neuroimage.2019.116462

64. Schartner, M.M., Carhart-Harris, R.L., Barrett, A.B., Seth, A.K., Muthukumaraswamy, S.D.: Increased spontaneous MEG signal diversity for psychoactive doses of ketamine, LSD and psilocybin. Sci. Rep. **7**, 46421 (2017). https://doi.org/10.1038/srep46421

65. Aru, J., Suzuki, M., Rutiku, R., Larkum, M.E., Bachmann, T.: Coupling the state and contents of consciousness. Front. Syst. Neurosci. **13**, 43 (2019). https://doi.org/10.3389/fnsys.2019. 00043

66. Carhart-Harris, R.L.: The entropic brain - revisited. Neuropharmacology **142**, 167–178 (2018). https://doi.org/10.1016/j.neuropharm.2018.03.010
67. Carhart-Harris, R.L., et al.: The entropic brain: a theory of conscious states informed by neuroimaging research with psychedelic drugs. Front. Hum. Neurosci. **8**, 20 (2014)
68. Carhart-Harris, R.L., Friston, K.J.: REBUS and the anarchic brain: toward a unified model of the brain action of psychedelics. Pharmacol. Rev. **71**(3), 316–344 (2019). https://doi.org/10. 1124/pr.118.017160
69. Luppi, A.I., et al.: Connectome harmonic decomposition of human brain dynamics reveals a landscape of consciousness. bioRxiv 199, 127 (2020). https://doi.org/10.1101/2020.08.10. 244459
70. Luppi, A.I., Carhart-Harris, R.L., Roseman, L., Pappas, I., Menon, D.K., Stamatakis, E.A.: LSD alters dynamic integration and segregation in the human brain. NeuroImage **227**, 117653 (2021). https://doi.org/10.1016/j.neuroimage.2020.117653
71. Stanley, K.O., Lehman, J.: Why Greatness Cannot Be Planned: The Myth of the Objective. Springer, Chem (2015). https://doi.org/10.1007/978-3-319-15524-1
72. Safron, A., DeYoung, C.G.: Chapter 18 - integrating cybernetic big five theory with the free energy principle: a new strategy for modeling personalities as complex systems. In: Wood, D., Read, S.J., Harms, P.D., Slaughter, A. (eds.) Measuring and Modeling Persons and Situations, pp. 617–649. Academic Press, New York (2021). https://doi.org/10.1016/B978-0-12-819200-9.00010-7
73. Constant, A., Hesp, C., Davey, C.G., Friston, K.J., Badcock, P.B.: Why depressed mood is adaptive: a numerical proof of principle for an evolutionary systems theory of depression. Comput. Psychiatry **5**(1), 60–80 (2021). https://doi.org/10.5334/cpsy.70
74. Erritzoe, D., Smith, J., Fisher, P.M., Carhart-Harris, R., Frokjaer, V.G., Knudsen, G.M.: Recreational use of psychedelics is associated with elevated personality trait openness: exploration of associations with brain serotonin markers. J. Psychopharmacol. Oxf. Engl. **33**(9), 1068–1075 (2019). https://doi.org/10.1177/0269881119827891
75. Girn, M., Mills, C., Roseman, L., Carhart-Harris, R.L., Christoff, K.: Updating the dynamic framework of thought: creativity and psychedelics. Neuroimage **213**, 116726 (2020). https:// doi.org/10.1016/j.neuroimage.2020.116726
76. Safron, A.: Strengthened beliefs under psychedelics (SEBUS)? A commentary on 'REBUS and the anarchic brain: toward a unified model of the brain action of psychedelics.' PsyArXiv. https://doi.org/10.31234/osf.io/zqh4b. Accessed 30 Nov 2020
77. Schwartenbeck, P., Passecker, J., Hauser, T.U., FitzGerald, T.H., Kronbichler, M., Friston, K.J.: Computational mechanisms of curiosity and goal-directed exploration. eLife **8**, 10 (2019). https://doi.org/10.7554/eLife.41703
78. ul Haq, R., et al.: Serotonin dependent masking of hippocampal sharp wave ripples. Neuropharmacology **101**, 188–203 (2016). https://doi.org/10.1016/j.neuropharm.2015.09.026
79. Latuske, P., Kornienko, O., Kohler, L., Allen, K.: Hippocampal remapping and its entorhinal origin. Front. Behav. Neurosci. **11**, 253 (2018). https://doi.org/10.3389/fnbeh.2017.00253
80. O'Callaghan, C., Walpola, I.C., Shine, J.M.: Neuromodulation of the mind- wandering brain state: the interaction between neuromodulatory tone, sharp wave-ripples and spontaneous thought. Philos. Trans. R. Soc. Lond. B. Biol. Sci. **376**(1817), 20190699 (2021). https://doi. org/10.1098/rstb.2019.0699
81. Pollan, M.: How to Change Your Mind: The New Science of Psychedelics. Penguin Books Limited, London (2018)

Inferring in Circles: Active Inference in Continuous State Space Using Hierarchical Gaussian Filtering of Sufficient Statistics

Peter Thestrup Waade[1,2](✉) ⓘ, Nace Mikus[1,3] ⓘ, and Christoph Mathys[1,4,5] ⓘ

[1] Interacting Minds Centre (IMC), Aarhus University, Aarhus, Denmark
ptw@cas.au.dk
[2] Embodied Cognition Group (ECG), Aarhus University, Aarhus, Denmark
[3] Department of Cognition, Emotion, and Methods in Psychology,
University of Vienna, Vienna, Austria
[4] Scuola Internazionale Superiore di Studi Avanzati (SISSA), Trieste, Italy
[5] Translational Neuromodeling Unit (TNU), Institute for Biomedical Engineering,
University of Zurich and ETH Zurich, Zurich, Switzerland

Abstract. We create a continuous state space active inference agent based on the hierarchical Gaussian filter. It uses the HGF to track the sufficient statistics of noisy observations of a moving target that is performing a Gaussian random walk with drift and varying volatility. On the basis of this filtering, the agent predicts the target's position, and minimizes surprisal by staying close to it. Our simulated agent represents the first full implementation of this approach. It demonstrates the feasibility of supplementing active inference with HGF-filtering of the sufficient statistics of observations, which is particularly useful in noisy and volatile continuous state space environments.

Keywords: Active inference · Continuous state space · Sufficient statistics filtering · Precision-weighted prediction errors · Hierarchical gaussian filter

1 Introduction

Active inference [7] is a formal framework for programming and modelling agents that navigate their environment such that they sample evidence for being within a desired set of states. This is done by minimizing the surprisal of sensory observations relative to a generative model of the environment, in which preferences for states are encoded as prior expectations. Actions are then chosen that are expected to lead to less surprisal in the future. Evaluating surprisal exactly is usually computationally intractable. In practical implementations of active inference, a variational free energy approximation is therefore often used.

Active inference furnishes a modeling framework which unites action and perception under a shared optimization imperative. Models inherently include a

© Springer Nature Switzerland AG 2021
M. Kamp et al. (Eds.): ECML PKDD 2021 Workshops, CCIS 1524, pp. 810–818, 2021.
https://doi.org/10.1007/978-3-030-93736-2_57

balance between epistemic and pragmatic behavior [4], can be related to neuro-biological theories such as predictive processing [6], and can be motivated from first principles in physics and information theory [2,8].

Recently, most active inference agents have been implemented as partially observable Markov Decision Processes (POMDP's) [18]. Here agents are limited to making discrete actions and observations in a discrete state space. By contrast, we here aim to (re-)extend active inference models to the continuous domain. We demonstrate a principled approach where an active inference agent filters the sufficient statistics of its observations with a hierarchical Gaussian filter [12,13], allowing it to perform goal-directed actions in a noisy and volatile continuous state space-environment.

2   Filtering Sufficient Statistics with Hierarchical Gaussian Filters

For agents inferring continuous states obscured by observational, informational, and environmental uncertainty, a fundamental challenge is to filter these various sources of noise from their observations. One principled way of solving this problem, which we use here and which is consistent with active inference in general, is to invert a generative model of what causes sensory observations. The hierarchical Gaussian filter's update equations implement such an inversion, where the generative model consists of a hierarchical cascade of random walks [13]. The update equations in the HGF are a more efficient alternative to variational Laplace, as detailed in [12]. Given a time series of observations, this allows for teasing apart observation noise, (potentially changing) volatility and (possibly state-dependent) regularities like drifts and biases. HGFs also provide precision-weighted predictions for future states, and can be used to infer a full predictive posterior probability distribution over observations in the future. This can be done by constructing the predictive distribution such that it reflects the uncertainty implied in the HGF's updates when filtering the sufficient statistics of the observations [14].

The decisive point here is that in a Gaussian model for a continuous univariate hidden state (i.e., Gaussian prior and Gaussian likelihood), the prior and posterior predictive distributions are *Gaussian-predictive* distributions, corresponding to a reparameterization of the generalized Student's-t distribution. This means that in addition to location and scale parameters, the predictive distribution also has a degree-of-freedom parameter which determines the fatness of its tails. An appropriate filter, such as an HGF, allows for inferring all three of these parameters. We put this to use here in order for our active inference agent to make the most appropriate predictions possible, i.e. predictions which minimize surprisal by optimizing all three of their aspects: point prediction (mean), uncertainty (variance), and fatness of tails (degrees of freedom). In the next section, we demonstrate how this can be accomplished in a simple active inference context.

3   Active Inference in Continuous State Space

We here provide a framework for a proof-of-principle simulation with a simple HGF-based active inference agent. The agent's objective is to stay close to a target which moves in continuous space with varying volatility. We will here first describe the *generative process* that forms the environment, and how it interfaces with the agent's *observations* o and *actions* a. Then we will describe how the agent makes actions as active inference based on inferences from the HGF.

The generative process consists of a total of three hidden states. The first is the target's position x_1 which moves in a Gaussian random walk with a constant drift ρ:

$$x_{1,t} \sim \mathcal{N}(x_{1,t-1} + \rho, x_{2,t}) \tag{1}$$

The second is the volatility of the random walk x_2, which changes in a pre-specified pattern between low and high levels of volatility. The last is then the agent's own position x_{agent}. In this example simulation, the position is fully determined by the agent's action a_{agent}, implemented here as being sampled from a delta distribution:

$$x_{\text{agent},t} \sim \delta(a_{\text{agent},t}) \tag{2}$$

The target's position x_1 is observed noisily, with observations o_1 normally distributed around the true position with standard deviation σ:

$$o_{1,t} \sim \mathcal{N}(x_{1,t}, \sigma) \tag{3}$$

The agent also observes its own position perfectly:

$$o_{\text{agent},t} \sim \delta(x_{\text{agent},t}) \tag{4}$$

To make inferences and predictions about the position and volatility of the target, the agent uses a standard HGF with a single volatility parent and a drift on the position. On each timestep t, this gives the agent a Gaussian belief about the target's position on the next timestep with mean $\hat{\mu}_{x_1,t}$ and precision $\hat{\pi}_{x_1,t}$. This lets it generate a full predictive posterior probability distribution for the observation on the next timestep. This distribution is a t-distribution with $\nu_t + 1$ degrees of freedom, with location $\hat{\mu}_{x_1,t}$ and precision $\hat{\pi}_{x_1,t}$:

$$p_{PP}(o_{1,t+1} | o_{1,1:t}) = t\left(o_{1,t} | \hat{\mu}_{x_1,t}, \hat{\pi}_{x_1,t}, \nu_t + 1\right) \tag{5}$$

where

$$\nu_t = \frac{\hat{\pi}_{x_1,t}}{\pi_\epsilon}, \tag{6}$$

and π_ϵ is the agent's prior belief about the input precision. In addition, the agent is equipped with a static prior distribution encoding its expectations (i.e.

preferences) for observations. The *goal prior*, as it will be referred to, is here a probability distribution over differences between the observed position of the target o_1 and the observation of the agent's own position o_{agent}. Specifically, it is a Gaussian distribution, with the mean μ_{GP} (usually at 0) encoding the preferred position relative to the target and the precision π_{GP} specifying the strength of this preference:

$$p_{GP}(o_1 - o_{\text{agent}}) = \mathcal{N}(o_1 - o_{\text{agent}}; \mu_{GP}, \pi_{GP}) \qquad (7)$$

On each trial, the agent's surprisal is calculated as the negative log probability of its sensory input relative to the goal prior:

$$\Im(o_1 - o_{\text{agent}}) = -\ln p_{GP}(o_1 - o_{\text{agent}}) \qquad (8)$$

In order to choose its action, the agent creates an expected surprisal distribution over possible control states a. First the expectation of the predictive posterior over observations of the target is assumed as the observation of the target. This gives a time-varying distribution over the agent's preferences for observations of its own position, given that the target is observed at its expectation. In the agent's model of the environment, Eqs. 2 and 4 are recapitulated, so we can substitute the expected observation o_{agent} with the agent's control states a_{agent}:

$$p_{GP,t}(a_{\text{agent}}) = p_{GP}(o_{\text{agent}}|o_1 = E(p_{PP}(o_1, t))) \qquad (9)$$

The right side of this equation is the goal prior over observations of the agent, given that the target is observed at its expectation $E(p_{PP}(o_1, t))$. The left side of the equation $p_{GP,t}(a_{\text{agent}})$ then becomes what might be called a 'goal posterior', a distribution over preferences for actions. In order to incorporate the full uncertainty of the agent's predictions, however, this preference distribution is convolved with the full predictive posterior. Taking the negative log of the resulting probability distribution then yields the expected surprisal associated with each possible move, after including the full uncertainty:

$$\Im_{\text{expected},t}(a_{\text{agent}}) = -\ln p_{PP}(o_1) * p_{GP,t}(a_{\text{agent}}) \qquad (10)$$

The agent then selects deterministically the action with the lowest associated expected surprisal.

$$a_{\text{agent},t} = \underset{a}{\text{argmin}} \ \Im_{\text{expected},t}(a_{\text{agent}})) \qquad (11)$$

4   Example Simulation

We here show results from an example simulation with the environment and the HGF-based active inference implementation described in the previous section. Figure 1 shows a schematic of the inference, prediction and decision process of the active inference agent. A GIF demonstrating the agent moving to follow the noisy and volatile observations of the target can be found on this link: https://osf.io/x5v39/

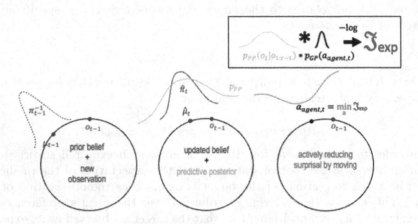

Fig. 1. Graphical sketch of the agent's action process, visualized on the circle. The agent starts with a Gaussian prior belief about the target's position with mean μ_{t-1} and precision π_{t-1}, and makes a new observation o_t. From that a new belief is computed with the HGF (also taking into consideration the drift ρ), and a predictive posterior t-distribution p_{PP} can be calculated. Finally, the predictive posterior is convolved with the 'goal posterior' $p_{GP}(a_{agent,t})$ i.e. the goal prior over agent positions given that the target is observed at its expectation (see Eq. 9). The negative log of the resulting probability distribution is the expected surprisal associated with the agent moving to different positions, of which the lowest is selected. If the static goal prior $p_{GP}(o_1 - o_{agent})$ is symmetric and centered around 0, μ_t, the mean of p_{PP} and the agent's action $a_{agent,t}$ coincide.

Figure 2 shows the inference on the target position, the prediction of future observations and the subsequent movement of the agent in an example simulation. Here the agent's goal prior is centered around 0, meaning that it consistently moves to the mean of its predictive posterior. Figure 3 shows the volatility of the environment and the agents inferred volatility in the same simulation. Note that this is a stochastic process, so even though the generating volatility is high it is not guaranteed that the target will move more. This means that the optimal inference on the volatility is not always the same as the volatility that generated the data (as in this case), although it should be when averaged over many simulations (this is also potentially true for any other hidden aspect of the environment). Figure 4 shows the agent's surprisal at its observation, relative to the goal prior. As expected, surprisal and predictive uncertainty is generally higher in those periods where the actual volatility is higher. In general, the agent performs its task well, demonstrating the feasibility of using the HGF for performing active inference. Code to replicate and modify this simulation can be found on https://github.com/ilabcode/hgf-active-inference

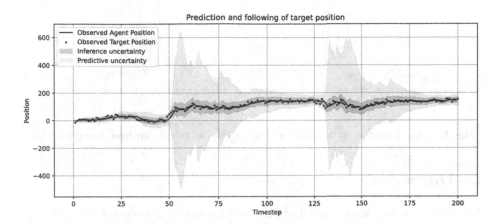

Fig. 2. Relative positions of the agent and the target. Inner shaded area is the inverse precision of the inference of the target position. Outer shaded area is the 68% confidence interval of the predictive posterior

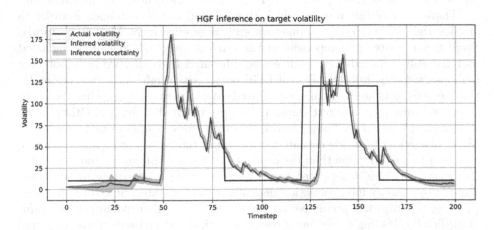

Fig. 3. The agent's HGF-based inference of the volatility (standard deviation) of the target's Gaussian random walk. Shaded area is the inverse precision of the inference.

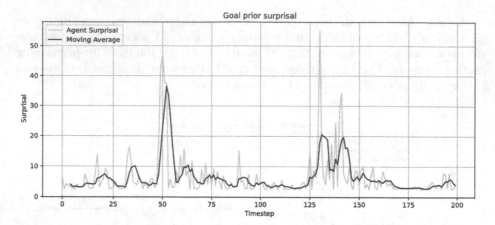

Fig. 4. Agent surprisal at making its observation, as calculated from the goal prior.

5   Discussion

We have here provided a proof-of-principle for active inference models in continuous state space that use hierarchical Gaussian filters to infer and predict the environment. This provides a way of constructing active inference agents that efficiently navigate volatile and noisy continuous state space-environments. The method is consistent with the theoretical framework of active inference, it is elegant, interpretable, and computationally efficient, and shows promise as a method for (re-)extending active inference models into the continuous domain.

There are multiple ways in which this method can be extended in order to employ the active inference framework still more effectively. Most importantly, there is no epistemic component in the current task, which can be included to fully utilize the advantages of active inference. Secondly, we have not here demonstrated the full flexibility of HGF-based active inference. In our simulation, we used a Gaussian goal prior centred around zero, meaning that the agent in practice always moves to the mean of its predictive distribution. Other distributions can be used for the goal prior, however, for example to make the agent prefer observing itself at a certain distance from the target, or be more sensitive to erring in one direction than the other. It is also possible to use more complex instances of the HGF as generative model, allowing for tracking an arbitrary number of possibly inter- or action-dependent hidden states. Since our approach provides a parametric predictive posterior, the expected surprisal can be evaluated directly, leading to the same result as variational methods would converge on. However, when the generative model is more complex, as for example when the agent must plan several steps ahead, or when its actions also influence the movement of the target, this could become less feasible. If so, approximate variational methods might be required.

It would also be valuable to make a more detailed comparison of the HGF-based active inference framework to older continuous state space approaches (for example the saccade models in [3] or the birdsong models in [5]). There are also newer mixed continuous-discrete models which combine discrete policy-level POMDP's with continuous movement and sensation models [9,10,15]. The predictions of the discrete model are then used as prior constraints on the continuous model, which in turn provides evidence for the discrete hypotheses entertained in the former. The main difference between these approaches and HGF-based active inference is that the HGF uses single-step update equations instead of the iterative variational Laplace approach, and that it is generically applicable across contexts. The HGF-based active inference framework can also be contrasted with recent attempts at scaling POMDP methods to complex and continuous domains by amortizing the specification of the generative model with deep learning techniques [1,11,19]. Here, an advantage of HGF-based active inference is that it, beyond specification of hyperparameters as is also the case in deep learning approaches, does not need to be trained, and that it is fully transparent and interpretable.

Finally, it still remains to equip the HGF-based active inference method with parameter and model structure learning capabilities, so that it can select the HGF-architecture that best explains observations. This is especially important when mapping the HGF to neuronal message passing [20]. Note that this can be combined with recent approaches where the hyperparameters of the HGF are learned online [16,17]. It also remains to apply it more complex environments, and to fit it to experimentally observed behaviour. This is feasible because it has been shown that the HGF is generic and adaptable, can be fit to experimental data, and can be mapped onto neuronal message-passing.

References

1. Çatal, O., Wauthier, S., De Boom, C., Verbelen, T., Dhoedt, B.: Learning generative state space models for active inference. Front. Comput. Neurosci. **14**, 103 (2020)
2. Friston, K.: A free energy principle for a particular physics. arXiv preprint arXiv:1906.10184 (2019)
3. Friston, K., Adams, R., Perrinet, L., Breakspear, M.: Perceptions as hypotheses: saccades as experiments. Front. Psychol. **3**, 151 (2012)
4. Friston, K., FitzGerald, T., Rigoli, F., Schwartenbeck, P., O'Doherty, J., Pezzulo, G.: Active inference and learning. Neurosci. Biobehav. Rev. **68**, 862–879 (2016). https://doi.org/10.1016/j.neubiorev.2016.06.022
5. Friston, K., Frith, C.: A duet for one. Conscious. Cogn. **36**, 390–405 (2015)
6. Friston, K., Kiebel, S.: Predictive coding under the free-energy principle. Philos. Trans. R. Soc. B Biol. Sci. **364**(1521), 1211–1221 (2009)
7. Friston, K.J., Daunizeau, J., Kiebel, S.J.: Reinforcement Learning or Active Inference? PLoS ONE **4**(7), e6421 (2009). https://doi.org/10.1371/journal.pone.0006421
8. Friston, K.J., Daunizeau, J., Kilner, J., Kiebel, S.J.: Action and behavior: a free-energy formulation. Biol. Cybern. **102**(3), 227–260 (2010). https://doi.org/10.1007/s00422-010-0364-z

9. Friston, K.J., Parr, T., de Vries, B.: The graphical brain: belief propagation and active inference. Netw. Neurosci. **1**(4), 381–414 (2017)
10. Friston, K.J., Sajid, N., Quiroga-Martinez, D.R., Parr, T., Price, C.J., Holmes, E.: Active listening. Hear. Res. **399**, 107998 (2021)
11. Van de Maele, T., Verbelen, T., Çatal, O., De Boom, C., Dhoedt, B.: Active vision for robot manipulators using the free energy principle. Front. Neurorobotics **15**, 14 (2021)
12. Mathys, C., Daunizeau, J., Friston, K.J., Stephan, K.E.: A Bayesian foundation for individual learning under uncertainty. Front. Hum. Neurosci. **5**, 39 (2011). https://doi.org/10.3389/fnhum.2011.00039
13. Mathys, C., et al.: Uncertainty in perception and the hierarchical gaussian filter. Front. Hum. Neurosci. **8**, 825 (2014). https://doi.org/10.3389/fnhum.2014.00825
14. Mathys, C., Weber, L.: Hierarchical gaussian filtering of sufficient statistic time series for active inference. In: IWAI 2020. CCIS, vol. 1326, pp. 52–58. Springer, Cham (2020). https://doi.org/10.1007/978-3-030-64919-7_7
15. Parr, T., Friston, K.J.: The discrete and continuous brain: from decisions to movement-and back again. Neural Comput. **30**(9), 2319–2347 (2018)
16. Şenöz, İ., De Vries, B.: Online variational message passing in the hierarchical gaussian filter. In: 2018 IEEE 28th International Workshop on Machine Learning for Signal Processing (MLSP), pp. 1–6. IEEE (2018)
17. Şenöz, İ., de Vries, B.: Online message passing-based inference in the hierarchical gaussian filter. In: 2020 IEEE International Symposium on Information Theory (ISIT), pp. 2676–2681. IEEE (2020)
18. Smith, R., Friston, K., Whyte, C.: A step-by-step tutorial on active inference and its application to empirical data. PsyArXiv (2021)
19. Tschantz, A., Baltieri, M., Seth, A.K., Buckley, C.L.: Scaling active inference. In: 2020 International Joint Conference on Neural Networks (IJCNN), pp. 1–8. IEEE (2020)
20. Weber, Lilian, A.E.: Perception as Hierarchical Bayesian Inference - Toward Non-Invasive Readouts of Exteroceptive and Interoceptive Processing. Doctoral thesis, ETH Zurich (2020)

On Solving a Stochastic Shortest-Path Markov Decision Process as Probabilistic Inference

Mohamed Baioumy[(✉)], Bruno Lacerda, Paul Duckworth, and Nick Hawes

Oxford Robotics Institute, University of Oxford, Oxford, UK
{mohamed,bruno,pduckworth,nickh}@robots.ox.ac.uk

Abstract. Previous work on planning as active inference addresses finite horizon problems and solutions valid for *online* planning. We propose solving the general Stochastic Shortest-Path Markov Decision Process (SSP MDP) as probabilistic inference. Furthermore, we discuss online and offline methods for planning under uncertainty. In an SSP MDP, the horizon is *indefinite* and unknown a priori. SSP MDPs generalize finite and infinite horizon MDPs and are widely used in the artificial intelligence community. Additionally, we highlight some of the differences between solving an MDP using dynamic programming approaches widely used in the artificial intelligence community and approaches used in the active inference community. F

1 Introduction

A core problem in the field of artificial intelligence (AI) is building agents capable of automated planning under uncertainty. Problems involving planning under uncertainty are typically formulated as an instance of a Markov Decision Process (MDP). At a high level, an MDP comprises 1) a set of world states, 2) a set of actions, 3) a transition model describing the probability of transitioning to a new state when taking an action in the current state, and 4) an objective function (e.g. minimizing costs over a sequence of time steps). An MDP solution determines the agent's actions at each decision point. An optimal MDP solution is one that optimizes the objective function. These are typically obtained using dynamic programming algorithms[1] [17, 26].

Recent work based on the active inference framework [13] poses the planning problem as a probabilistic inference problem. Several papers have been published showing connections between active inference and dynamic programming to solve an MDP [7, 8, 15]. However, the planning problem being solved in the two communities is not equivalent.

First, dynamic programming approaches used to solve an MDP, such as policy iteration, are valid for finite, infinite and indefinite horizons. Indefinite horizons

---

[1] Linear programming approaches are also popular methods for solving MDPs [2, 9, 12, 22]. Additionally, other methods exist in the reinforcement learning community such as policy gradient methods [14, 27, 28].

© Springer Nature Switzerland AG 2021
M. Kamp et al. (Eds.): ECML PKDD 2021 Workshops, CCIS 1524, pp. 819–829, 2021.
https://doi.org/10.1007/978-3-030-93736-2_58

are finite but of which the length is unknown a priori. For instance, consider an agent navigating from a starting state to a goal state in a grid world where the outcome is uncertain (e.g. the 4×4 grid world in Fig. 1 shown in the appendix). Before starting to act in the environment, there is no way for the agent to know how many time steps it will take to reach the goal. Algorithms based on dynamic programming, such as policy iteration, are valid for such settings. They can solve the Stochastic Shortest-Path Markov decision process (SSP MDP) [2,17]. However, work from active inference is only formulated for finite horizons [8].

Second, the optimal solution to an SSP MDP is a stationary deterministic policy [17]. This refers to a mapping from states to actions independent of time. Computing this optimal policy can be done *offline* (without interaction with the environment) or online (while interacting). In the active inference literature however, solving the planning problem is performed by computing a stochastic plan (a sequence of actions given the current state). This is only valid during online planning. Additionally, the solution is only optimal given a certain horizon, which is specified a priori. If the horizon chosen is too short, the agent will not find a solution to reach the target. If it is too long, the solution will be sub-optimal.

The main contribution of this paper is presenting a novel algorithm to solve a Stochastic Shortest-Path Markov Decision Process using probabilistic inference. This is an MDP with an *indefinite horizon*. Additionally, highlighting the several gaps between solving an MDP in the AI community and the active inference community.

Section 2 discusses the SSP MDP and Sect. 3 presets an approach for solving an SSP MDP as probabilistic inference. The equivalence between the two methods is shown in Sect. 4.1. Furthermore, the difference between world states and temporal state is highlighted in Sect. 4.2. Policies, plans and probabilistic plans are discussed in Sect. 4.3. Finally, a discussion on online vs offline planning can be found in Sect. 5.

2   Stochastic Shortest Path MDP

An SSP MDP is defined as a tuple $\mathcal{M} = (S, A, C, T, G)$. S is the set of states, A is the set of actions, $C : S \times A \times S \rightarrow \mathbb{R}$ is the cost function, and $T : S \times A \times S \rightarrow [0, 1]$ is the transition function. $G \subset S$ is the set of goal states. Each goal state $s_g \in G$ is absorbing and incurs zero cost. The expected cost of applying action a in state s is $\bar{C}(s, a) = \sum_{s' \in S} T(s, a, s') \cdot C(s, a, s')$. The minimum expected cost at state s is $\bar{C}^*(s) = \min_{a \in A} \bar{C}(s, a)$. A policy maps a state to a distribution over action choices. A policy is deterministic if it chooses a single action at each step. A policy π is proper if it reaches $s_g \in G$ starting from any s with probability 1. In an SSP MDP, the following assumptions are made [17]: a) there exists a proper policy, and b) every improper policy incurs infinite cost at all states where it is improper.

The goal is to find the an *optimal policy* π^* with the minimum expected cost $\bar{C}^*(s)$ and can be computed as

$$\pi^*(s) = \text{argmin}_{a \in \mathcal{A}} \left[\sum_{s' \in \mathcal{S}} \mathcal{T}(s, a, s') \left[\mathcal{C}(s, a, s') + V^*(s') \right] \right].$$

$V^*(s)$ is referred to as the optimal value for a state s and is defined as:

$$V^*(s) = \min_{a \in \mathcal{A}} \left[\sum_{s' \in \mathcal{S}} \mathcal{T}(s, a, s') \left[\mathcal{C}(s, a, s') + V^*(s') \right] \right].$$

Crucially, the optimal policy π^* corresponding to the optimal value function is Markovian (only dependant on the current state) and deterministic [17]. Solving an SSP MDP means finding a policy that minimizes expected cost, as opposed to one that maximizes reward. This difference is purely semantic as the problems are dual. We can define a reward function $\mathcal{R} = -\mathcal{C}$ and move to a reward maximization formulation. A more fundamental distinction is the presence of a special set of (terminal) goal states, in which staying forever incurs no cost.

Solving an SSP MDP can be done using standard dynamic programming algorithms such as policy iteration. Policy iteration can be divided in two steps, policy evaluation and improvement. In policy evaluation, for a policy π, the value function $V_\pi(s)$ is recursively evaluated until convergence as

$$V_\pi(s) \leftarrow \sum_{s' \in \mathcal{S}} \mathcal{T}(s, \pi(s), s') \left[\mathcal{C}(s, \pi(s), s') + V_\pi(s') \right].$$

In the policy improvement step, the state-action value function $Q(s, a)$ is computed as:

$$Q(s, a) = \sum_{s' \in \mathcal{S}} \mathcal{T}(s, a, s')[\mathcal{C}(s, a, s') + V(s')].$$

Then we compute a new policy as $\pi' = \text{argmin}_{a \in \mathcal{A}} Q(s, a)$ for every state in S. Iterating between these two steps guarantees convergence to an optimal policy.

Properties of an SSP MDP. An SSP MDP can be shown to generalize finite, infinite and indefinite horizon MDPs [3,17]. Thus algorithms valid for an SSP MDP are also valid for the finite and infinite horizon MDPs. Additionally, it can be proven that each SSP MDP has an optimal deterministic policy independent of time. Therefore, the claims made in [8] about active inference being more general since it computes stochastic policies are unjustified when solving an MDP. However, these results do not hold in partially observable cases or in the presence of uncertain models. But in the SSP MDP defined above (which is commonly used in the AI community), there always exists a deterministic optimal policy. Note that there is an infinite number of stochastic policies but only a finite number of deterministic policies ($|\mathcal{S}|^{|\mathcal{A}|}$). This greatly speeds up the algorithms while still guaranteeing optimality.

3   Solving an SSP MDP as Probabilistic Inference

In this section we discuss a novel approach for solving an SSP MDP as probabilistic inference. We use an inference algorithm that exactly solves an SSP MDP as defined in the previous section. This approach is inspired by work from [31,32] which solves an MDP with an indefinite horizon. This approach has been successfully applied to solve problems of planning under uncertainty, e.g. [18,30].

3.1   Definitions

The definition of an SSP MDP includes a set of world states S and actions A. In probabilistic inference we instead reason about *temporal* states and actions. A temporal state s_t is a random variable defined over all world states. Conceptually it represents the state that the agent will visit at the time-step t. For the grid world in Fig. 1, there are 16 world states but the number of temporal states is unknown a priori since the horizon is unknown.

The transition probability is defined as a probability distribution over temporal states and actions as $P(s_{t+1}|a_t, s_t)$. If the random variables are fixed to specific world states i and j and an action a, the transition probability $P(s_{t+1} = j|a_t = a, s_t = i)$ would be equivalent to the transition function T defined for an SSP MDP. The probability of taking a certain action in a state is parameterized by a policy π as $P(a_t = a|s_t = i; \pi) = \pi_{ai}$. This policy is defined exactly the same as in the case of an SSP MDP.

The cost function $P(c_t|s_t, a_t)$ is defined differently. The temporal cost variables c_t are defined as binary random variables $c_t \in \{0, 1\}$. Translating an arbitrary cost function to temporal costs can be done by scaling the cost function $C(s, a)$ (as defined in the previous section) between the minimum cost $(\min(C))$ and maximum cost $(\max(C))$ as:

$$P(c_t = 1 \mid a_t = a, s_t = s) = \frac{C(a, s) - \min(C)}{\max(C) - \min(C)}.$$

Any expression with $P(c_t = 1)$ can be though of as 'the probability of a cost being maximal'. Thus, the probability of a cost being maximal given a state and an action is $P(c_t = 1 \mid a_t = a, s_t = s)$. Now we can reason about the highest possible cost for a state s and action a as one where $P(c_t = 1 \mid a_t = a, s_t = s) = 1$ and the lowest possible cost as $P(c_t = 1 \mid a_t = a, s_t = s) = 0$. Any other cost will have a probability in-between, according to its magnitude.

Finally, we model the horizon as a random variable. The temporal states and actions are considered up to the end of the horizon T. However, the horizon is generally unknown. We thus model T itself as a random variable. Combining all this information we can define the SSP MDP using a probabilistic model.

3.2   Mixture of Finite MDPs

In this section we define the SSP MDP in terms of a mixture of finite MDPs with only a final cost variable. Given every horizon (for instance $T = 1$) the finite MDP can be given as $P(c, s_{0:T}, a_{0:T} \mid T; \boldsymbol{\pi})$. Note that we dropped the time-index for c_t since there is only one cost variable now. This model can be factorized as

$$P(c, s_{0:T}, a_{0:T} \mid T; \boldsymbol{\pi}) = P(c \mid a_T, s_T) P(a_0 \mid s_0; \boldsymbol{\pi})$$
$$P(s_0) \cdot \prod_{t=1}^{T} P(a_t \mid s_t; \boldsymbol{\pi}) P(s_t \mid a_{t-1}, s_{t-1})$$

To reason about the full MDP, we consider the mixture model of the joint given by the joint probability distribution

$$P(c, s_{0:T}, a_{0:T}, T; \boldsymbol{\pi}) = P(c, s_{0:T}, a_{0:T} \mid T; \boldsymbol{\pi}) P(T)$$

where $P(T)$ is a prior over the total time, which we choose to be a flat prior (uniform distribution).

3.3 Computing an Optimal Policy

Our objective is to find a policy that minimizes the expected cost. Similarly to policy iteration, we do not assume any knowledge about the initial state. Expectation-Maximization[2] can be used to find the optimal parameters of our model: the policy π. The E-step will, for a given π, compute a posterior over state-action sequences. The M-step then adapts the model parameters π to optimize the expected likelihood with respect to the quantities calculated in the E-step.

E-Step: A Backwards Pass in All Finite MDPs. We use the simpler notation $p(j \mid a, i) \equiv P(s_{t+1} = j \mid a_t = a, s_t = i)$ and $p(j \mid i; \boldsymbol{\pi}) \equiv P(s_{t+1} = j \mid s_t = i; \boldsymbol{\pi}) = \sum_a p(j \mid a, i) \pi_{ai}$. Further, as a 'base' for backward propagation, we define

$$\beta_0(i) = P(c = 1 \mid x_T = i; \boldsymbol{\pi}) = \sum_a P(c = 1 \mid a_T = a, x_T = i) \pi_{ai}.$$

This is the immediate cost when following a policy π. It is the expected cost if there is only one time-step remaining. Then, we can recursively compute all the other backward messages. We use the index τ to indicate a backwards counter. This means that $\tau + t = T$, where T is total (unknown) horizon length. This is computed as

$$\beta_\tau(i) = P(c = 1 \mid x_{T-\tau} = i; \boldsymbol{\pi}) = \sum_j p(j \mid i; \boldsymbol{\pi}) \beta_{\tau-1}(j).$$

Intuitively, the backward messages are the expected cost if one incurs a cost at the last time step only. So, β_2, is the expected cost if the agent follows the policy

---

π for two time-steps and only incurs a cost after that. Using these messages, we can compute a value function dependent on time, actions and states given as:

$$q_\tau(a, i) = P(c = 1 \mid a_t = a, s_t = i, T = t + \tau; \boldsymbol{\pi})$$

$$= \begin{cases} \sum_j p(j \mid i, a)\beta_{\tau-1}(j) & \tau > 1 \\ P(c = 1 \mid a_T = a, s_T = i) & \tau = 0. \end{cases}$$

Marginalizing out time, we get the state-action value-function

$$P(c = 1 \mid a_t = a, s_t = i; \boldsymbol{\pi}) = \frac{1}{C} \sum_\tau P(T = t + \tau) q_\tau(a, i)$$

where C is a normalization constant. This quantity is the probability of getting a maximum cost given a state and action. It is similar to the $Q(s, a)$ function computed in policy iteration.

M-Step: The Policy Improvement Step. The standard M-step in an EM-algorithm maximizes the expected complete log-likelihood with respect to the new parameters π'. Given that the optimal policy for an MDP is deterministic, a greedy M-step can be used. However, our goal is to minimize the log-likelihood in this case as it refers to a the probability of receiving a maximal cost. This can done as

$$\pi' = \underset{a}{\operatorname{argmin}}(P(c = 1 \mid a_t = a, s_t = i; \boldsymbol{\pi})) \tag{1}$$

This update converges much faster than in a standard M-step. Here an M-step can be used to obtain a stochastic policy. However, this is unnecessary since the optimal policy is deterministic. Note that there is an infinite number of stochastic policies but a finite number of deterministic ones. In conclusion, a greedy M-step is faster to converge but still guarantees an optimal policy.

4   Connections Between the Two Views

4.1   Exact Relationship Between Policy Iteration and Planning as Probabilistic Inference

The messages β computed during backward propagation are exactly equal to the value functions for a single MDP of finite time. The full value function is can therefore be written as the sum of the βs,

$$V_\pi(i) = \sum_T \beta_T(i)$$

since the prior over time $P(T)$ is a uniform prior. If $P(T)$ is not a uniform distribution, this would result in a mixture rather than a sum. The same applies to the relationship between the Q-value function:

$$Q_\pi(a,i) = \sum_T q_T(a,i).$$

Hence, the E-step essentially performs a policy evaluation which yields the classical value function. Given this relation to policy evaluation, the M-step performs an operation exactly equivalent to standard policy improvement. Thus, the EM-algorithm using exact inference is equivalent to Policy Iteration but computes the necessary quantities differently.

One unanswered question is when to stop computing the backward messages. In [32] messages are computed up to a number T_{max}. From this perspective, the planning as inference algorithm presented is equivalent to the so-called *truncated policy iteration* algorithm as opposed to the more common ϵ-greedy version.

In the policy evaluation step, one iterates through the state space to update the value $v_\pi(s)$ for every state until a termination criterion is met. An ϵ-greedy criterion means that we stop iterating though the state space once the maximum difference in $V_\pi(s)$ for any s is smaller than a positive small number ϵ. In truncated policy iteration, however, we iterate through the state space $T_{\max}$ times and then perform the policy improvement step. The probabilistic inference algorithm presented in this paper is equivalent to truncated policy iteration if we restrict the maximum number of β messages to be computed.

4.2 World States vs Temporal States

In dynamic programming, one reasons over the world states. In the grid world example in Fig. 1, this refers to a grid cell. This grid world has 16 world states. In probabilistic inference, one reasons about a *temporal state*. This is a random variable over all world states. The number of temporal states is dependent on how many time-steps the agents acts in the environment (which is often unknown beforehand). An illustration of the difference is given in Fig. 2.

4.3 Policies, Plans and Probabilistic Plans

A classical planning algorithm computes a plan: a sequence of actions. An algorithm like A* or Dijkstra's algorithm [5] can be used to find the optimal path from a stating state to a goal state, given a deterministic world. Crucially, this solution can be computed offline (without interactions with the environment). In a stochastic world, this does not work since the agent can not predict in which states it will end up. However, one can use deterministic planning algorithms for stochastic environments if the path is re-planned online at every time-step. Determinization-based methods have found success in solving planning under uncertainity problems such as the famous FF-replan algorithm [34].

Active inference approaches computes a *probabilistic plan*. The active inference literature calls this a policy; however, we use a different term to avoid confusion.[3] In active inference, the agents computes a finite plan while interacting with the environment. However, rather than assuming a deterministic world (like FF-replan [34]), the probabilities are taken into account. This can be shown to compute the optimal solution to an MDP (when planning online). We thus refer to it as a probabilistic plan, a plan that was computed while taking the transition probabilities into account.

Finally, a policy is a mapping from states to actions, i.e. the agent has a preferred action to take for every state. Policies can be stochastic or time-dependent; however, for an SSP MDP the optimal policy is deterministic and independent of time. An agent can compute a policy offline and use it online without needing any additional computation while interacting with the environment. The difference between a plan and policy is illustrated in Fig. 1.

To summarize, a plan or a probabilistic plan can only be used for online planning. Since the outcome of an action is inherently uncertain. Probabilistic plans (as used in active inference) find an optimal solution when used to plan online. A policy also provides an optimal solution and can be computed offline or online.

5   Discussion

In this paper we present a novel approach to solve a stochastic shortest path Markov decision process (SSP MDP) as probabilistic inference. The SSP MDP generalizes many models, including finite and infinite MDPs. Crucially, the dynamic programming algorithms (such as policy iteration) classically used to solve an SSP MDP are valid for *indefinite horizons* (finite but of unknown length); this is not the case for active inference approaches.

The exact connections between solving an MDP using policy iteration and the presented algorithm are discussed. Afterwards, we discussed the gap between solving an MDP in active inference and the approaches in the artificial intelligence community. This included the difference between world states and temporal states, the difference between plans, probabilistic plans and policies. An interesting question now is, which approach is more appropriate? This depends on the problem at hand and whether it can be solved online or offline.

Online and Offline Planning. As discussed in Sect. 4.3, a policy is mapping from states to actions and can be used for offline and online planning. Computing a policy is somewhat computationally expensive; however, a look-up is very cheap. Thus if one operates in an environment where the transition and cost function do not change, it is best to compute an optimal policy offline then use it online (while interacting with the environment). This is the case for many planning and scheduling problems, such as a set of elevators operating in sync [6], task-level planning in robotics [19], multi-objective planning [11, 23] and playing

---

[3] The distinction between a plan and a policy when using active inference has been briefly discussed in [20]. Additionally, other methods computing plans as probabilistic inference have been proposed before active inference in [1, 33].

games [4, 25]. The challenges in these problems are often that the state-space is incredibly large and thus approximations are needed. However, the problem is fully observable and the cost and transition models are static; the rules of chess do not change half way, for instance.

If the transition or cost functions vary while interacting with the environment (e.g. [10,24]), an offline solution is not optimal. In this case, the agent can plan online by re-evaluating a policy or computing probabilistic plans (as done in active inference). Computing the latter is cheaper and requires less memory. This is because a probabilistic plan is a distribution over actions $p(a_t)$ up to a time horizon T while a (finite policy) is a conditional distribution $p(a_t|s_t)$ over all world states in S. For any time-step, the posterior over the action is related to the policy such that $p(a_t) = \sum_s p(a_t|s_t)p(s_t)$.

Consider the work in [24, 29]. In both cases a robot operates in an environment susceptible to changes. If the environment changes, the agent can easily construct a new model by varying the cost or transition function but needs to recompute a solution. In [29] the authors recompute a policy at every time-step while in [24] a probabilistic plans is computed using active inference. Since in both cases the solution is recomputed at every time-step, active inference is preferred since it requires less memory and can be computationally cheaper. On the other hand, if the environment only changes occasionally, computing a policy might remain preferable.

To conclude, if the transition and cost functions (T and C) are static, it is preferable to compute a policy offline. If T and C change occasionally, one may still compute an offline policy and recompute a policy only when a change occurs. However, if the environment is dynamic, computing a probabilistic plan (using active inference) is preferable to recomputing a policy at every time-step.

A   Appendix: Illustrations

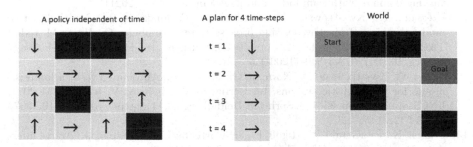

Fig. 1. An illustration of a 4×4 grid world (**right**). The initial state is blue and goal state is green. An illustration for a policy (**left**) and a plan (**middle**).

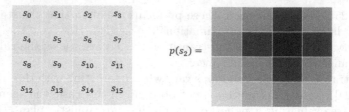

Fig. 2. Annotated world states (**left**) and a posterior over a temporal state (**right**).

References

1. Attias, H.: Planning by probabilistic inference. In: AISTATS (2003)
2. Bertsekas, D.P., Tsitsiklis, J.N.: An analysis of stochastic shortest path problems. Math. Oper. Res. **16**(3), 580–595 (1991)
3. Bertsekas, D.P., Tsitsiklis, J.N.: Neuro-dynamic programming: an overview. In: Proceedings of 1995 34th IEEE Conference on Decision and Control, vol. 1, pp. 560–564. IEEE (1995)
4. Campbell, M., Hoane, A.J., Hsu, F.: Deep blue. Artif. Intell. **134**, 57–83 (2002)
5. Cormen, T.H., Leiserson, C.E., Rivest, R.L., Stein, C.: Introduction to Algorithms. MIT Press (2009)
6. Crites, R.H., Barto, A.G., et al.: Improving elevator performance using reinforcement learning. In: Advances in Neural Information Processing Systems, pp. 1017–1023 (1996)
7. Da Costa, L., Parr, T., Sajid, N., Veselic, S., Neacsu, V., Friston, K.: Active inference on discrete state-spaces: a synthesis. arXiv preprint arXiv:2001.07203 (2020)
8. Da Costa, L., Sajid, N., Parr, T., Friston, K., Smith, R.: The relationship between dynamic programming and active inference: The discrete, finite-horizon case. arXiv preprint arXiv:2009.08111 (2020)
9. d'Epenoux, F.: A probabilistic production and inventory problem. Manage. Sci. **10**(1), 98–108 (1963)
10. Duckworth, P., Lacerda, B., Hawes, N.: Time-bounded mission planning in time-varying domains with semi-mdps and gaussian processes (2021)
11. Etessami, K., Kwiatkowska, M., Vardi, M.Y., Yannakakis, M.: Multi-objective model checking of Markov decision processes. In: Grumberg, O., Huth, M. (eds.) TACAS 2007. LNCS, vol. 4424, pp. 50–65. Springer, Heidelberg (2007). https://doi.org/10.1007/978-3-540-71209-1_6
12. Forejt, V., Kwiatkowska, M., Norman, G., Parker, D.: Automated verification techniques for probabilistic systems. In: Bernardo, M., Issarny, V. (eds.) SFM 2011. LNCS, vol. 6659, pp. 53–113. Springer, Heidelberg (2011). https://doi.org/10.1007/978-3-642-21455-4_3
13. Friston, K., FitzGerald, T., Rigoli, F., Schwartenbeck, P., Pezzulo, G.: Active inference: a process theory. Neural Comput. **29**(1), 1–49 (2017)
14. Grondman, I., Busoniu, L., Lopes, G.A., Babuska, R.: A survey of actor-critic reinforcement learning: standard and natural policy gradients. IEEE Trans. Syst. Man Cybern. Part C (Appl. Rev.) **42**(6), 1291–1307 (2012)
15. Kaplan, R., Friston, K.J.: Planning and navigation as active inference. Biol. Cybern. **112**(4), 323–343 (2018). https://doi.org/10.1007/s00422-018-0753-2

16. Koller, D., Friedman, N.: Probabilistic Graphical Models: Principles and Techniques. MIT Press, Cambridge (2009)
17. Kolobov, A.: Planning with Markov Decision Processes: An AI Perspective, vol. 6. Morgan & Claypool Publishers, San Rafael (2012)
18. Kumar, A., Zilberstein, S., Toussaint, M.: Probabilistic inference techniques for scalable multiagent decision making. J. Artif. Intell. Res. **53**, 223–270 (2015)
19. Lacerda, B., Faruq, F., Parker, D., Hawes, N.: Probabilistic planning with formal performance guarantees for mobile service robots. Int. J. Robot. Res. **38**(9), 1098–1123 (2019)
20. Millidge, B., Tschantz, A., Seth, A.K., Buckley, C.L.: On the relationship between active inference and control as inference. In: IWAI 2020. CCIS, vol. 1326, pp. 3–11. Springer, Cham (2020). https://doi.org/10.1007/978-3-030-64919-7_1
21. Murphy, K.P.: Machine Learning: A Probabilistic Perspective. MIT Press, Massachusetts (2012)
22. Nazareth, J.L., Kulkarni, R.B.: Linear programming formulations of Markov decision processes. Oper. Res. Lett. **5**(1), 13–16 (1986)
23. Painter, M., Lacerda, B., Hawes, N.: Convex hull Monte-Carlo tree-search. In: Proceedings of the International Conference on Automated Planning and Scheduling, vol. 30, pp. 217–225 (2020)
24. Pezzato, C., Hernandez, C., Wisse, M.: Active inference and behavior trees for reactive action planning and execution in robotics. arXiv preprint arXiv:2011.09756 (2020)
25. Silver, D., et al.: Mastering chess and shogi by self-play with a general reinforcement learning algorithm. arXiv preprint arXiv:1712.01815 (2017)
26. Sutton, R.S., Barto, A.G., et al.: Introduction to Reinforcement Learning, vol. 135. MIT Press, Cambridge (1998)
27. Sutton, R.S., McAllester, D.A., Singh, S.P., Mansour, Y.: Policy gradient methods for reinforcement learning with function approximation. In: Advances in Neural Information Processing Systems, pp. 1057–1063 (2000)
28. Thomas, P.S., Brunskill, E.: Policy gradient methods for reinforcement learning with function approximation and action-dependent baselines. arXiv preprint arXiv:1706.06643 (2017)
29. Tomy, M., Lacerda, B., Hawes, N., Wyatt, J.L.: Battery charge scheduling in long-life autonomous mobile robots via multi-objective decision making under uncertainty. Robot. Auton. Syst. **133**, 103629 (2020)
30. Toussaint, M., Charlin, L., Poupart, P.: Hierarchical pomdp controller optimization by likelihood maximization. In: UAI, vol. 24, pp. 562–570 (2008)
31. Toussaint, M., Harmeling, S., Storkey, A.: Probabilistic inference for solving (po) mdps. University of Edinburgh, School of Informatics Research Report EDI-INF-RR-0934 (2006)
32. Toussaint, M., Storkey, A.: Probabilistic inference for solving discrete and continuous state markov decision processes. In: Proceedings of the 23rd International Conference on Machine Learning, pp. 945–952. ACM (2006)
33. Verma, D., Rao, R.P.: Goal-based imitation as probabilistic inference over graphical models. In: Advances in Neural Information Processing Systems, pp. 1393–1400 (2006)
34. Yoon, S.W., Fern, A., Givan, R.: Ff-replan: a baseline for probabilistic planning. In: ICAPS, vol. 7, pp. 352–359 (2007)

Habitual and Reflective Control in Hierarchical Predictive Coding

Paul F. Kinghorn[1]($\boxtimes$), Beren Millidge[2], and Christopher L. Buckley[1]

[1] School of Engineering and Informatics, University of Sussex, Brighton, UK
{p.kinghorn,c.l.buckley}@sussex.ac.uk
[2] MRC Brain Networks Dynamics Unit, University of Oxford, Oxford, UK
beren@millidge.name

Abstract. In cognitive science, behaviour is often separated into two types. Reflexive control is habitual and immediate, whereas reflective is deliberative and time consuming. We examine the argument that Hierarchical Predictive Coding (HPC) can explain both types of behaviour as a continuum operating across a multi-layered network, removing the need for separate circuits in the brain. On this view, "fast" actions may be triggered using only the lower layers of the HPC schema, whereas more deliberative actions need higher layers. We demonstrate that HPC can distribute learning throughout its hierarchy, with higher layers called into use only as required.

Keywords: Hierarchical predictive coding · Decision making · Action selection

1 Introduction

In the field of cognitive science, behaviour is widely considered to be separated into two classes. Habitual (reflexive) behaviour responds rapidly and instinctively to stimuli, while reflective behaviour involves slower top-down processing and a period of deliberation. These different classes of behaviour have been labelled System 1 and System 2 by Stanovich and West in 2000 [30] and the topic was popularised in Daniel Kahneman's book "Thinking Fast and Slow" in 2011 [21]. However, it has remained unclear how the two processes are implemented in the brain.

This paper investigates how a single system operates when generating behaviours which require different amounts of deliberation. The system does not carry out planning or evaluation of prospective outcomes, but simply compares the triggering of actions which require more or less time to select the correct action for a presented situation. This distinction has some parallels with, but is separate from, the split between goal based planning (often called model-based) and habitual (often called model-free) control [11,31]. Although there is some evidence that separate brain systems underpin goal based planning and habits [11,18,33], there are also indications from fMRI and lesion studies

© Springer Nature Switzerland AG 2021
M. Kamp et al. (Eds.): ECML PKDD 2021 Workshops, CCIS 1524, pp. 830–842, 2021.
https://doi.org/10.1007/978-3-030-93736-2_59

that these processes can co-exist within the same regions of the brain [9,11,33], challenging the notion of separate systems. Since it is possible that even these extremes of behaviour are computed together in the brain, a useful contribution to the topic would be to analyse how reflexive and reflective behaviour can arise from a unified system.

Independent from this debate, there has been much progress in the last decade on the idea that perception and action are both facets of the principle of free energy minimization in the brain [14,16]. According to this approach, the brain maintains a generative model of its environment and uses variational inference to approximate Bayesian inference [20,29]. One way of implementing this (under Gaussian assumptions) is to use a hierarchical predictive coding architecture [12,13,28], which has successive layers of descending predictions and ascending prediction errors [3,5,24,32]. This theory is also often referred to as Predictive Processing (PP) [7,25]. In this paper, we investigate how PP approaches can explain both reflexive and reflective behaviour simultaneously using a single hierarchical predictive coding network architecture and inference procedure.

In it basic form, PP uses a generative model to try and correctly infer hidden causes for incoming observations. In a hierarchical predictive coding network, all layers of the hierarchy are updated to minimize prediction errors until a fixed point is reached, with the resultant top layer being the best explanation of the hidden causes of the observations at the bottom layer [3,5,7].

PP can also be used to explain action, with the network modelling how actions and sensations interact [2,4,6,8,25–27]. Actions are triggered by descending predictions which cause low level prediction errors. These errors are rectified through reflex arcs [1,17,19]. In theory, this means that motor behaviour need not wait for full end-to-end inference to be completed but, rather, action takes place once a threshold has been crossed on the reflex muscle.

This paper investigates the extent to which action selection in a predictive coding network (PCN) relies on all the layers of the PCN. To do this, we train a network to associate actions and observations with each other. We then investigate whether inference across the full network is required in order to trigger the correct action for a given observation. We show that a decision making task with a higher degree of complexity will use more of the layers and may be strongly dependent on the top layer being correctly inferred. Conversely, a decision which is a simple function of sensory observations can operate without involvement of higher layers, despite the fact that learning included those higher layers. This demonstrates that learning allows a hierarchy of action/sensation linkages to be built up in the network, with agents able to use information from lower layers to infer the correct actions without necessarily needing to engage the whole network. These findings suggest that a single PCN architecture could explain both reflexive and reflective behaviour.

In the general case of state space models, the fixed point of a PCN is often in a moving frame of reference. However, the implementation described in this paper ignores state transitions or dynamics and restricts itself to static images. It should therefore not be confused with the notion of predictive coding sometimes seen in the engineering or active inference literature which rests on a state space model for generating timeseries. Rather, our formulation follows the approach of Rao and Ballard's seminal paper [28] and ignores any temporal prediction components, whilst retaining what Friston describes as "the essence of predictive coding, namely any scheme that finds the mode of the recognition density by dynamically minimising prediction error in an input-specific fashion" [12].

The remainder of this paper is set out as follows. Section 2 outlines the HPC model which is used to implement variational inference. Section 3 describes the experiments which we use to analyse inference of labels and actions in PCNs. Section 4 presents the experimental results, demonstrating that learning to act need not rely on high level hidden states. Moreover, we show that the number of higher layers which can be ignored in decision making relates to the complexity of information needed to make that decision.

2 Hierarchical Predictive Coding (HPC)

This section presents a quick overview of how HPC can be used to approximate variational inference. For a more guided derivation, see [3, 5, 16, 24].

At the core of the free-energy principle is the concept that, in order to survive, an agent must strive to make its model of the world a good fit for incoming observations, o. If the model of observations $p(o)$ can be explained by hidden states of the world s then, in theory, a posterior estimate for s could be obtained using Bayes rule over a set of observations:

$$p(s \mid o) = \frac{p(o \mid s)\, p(s)}{p(o)} = \frac{p(o \mid s)\, p(s)}{\int p(o \mid s)\, p(s)\, ds} \tag{1}$$

but the denominator is likely to be intractable. Therefore s is approximated using variational inference. An auxiliary model (the variational distribution) is created, $q(s; \psi)$, and the divergence between q and the true posterior $p(s \mid o)$ minimized. The KL divergence is used to measure this:

$$KL[q(s; \psi) \parallel p(s \mid o)] = \int q(s; \psi) log \frac{q(s; \psi)}{p(s \mid o)} ds = \mathcal{F} + log\, p(o) \tag{2}$$

where the variational free energy $\mathcal{F}$ is defined as:

$$\mathcal{F} = \int q(s; \psi) log \frac{q(s; \psi)}{p(s, o)} ds \tag{3}$$

The value $-log\, p(o)$, is an information theoretic measure of the unexpectedness of an observation, variously called surprise, suprisal or negative of log model evidence. By adjusting s to minimize surprisal, the model becomes a better fit of the environment. Noting that KL is always positive, it can be seen from Eq. (2) that $\mathcal{F}$ is an upper bound on surprisal. Therefore, to make the model a good fit for the data, it suffices to minimize $\mathcal{F}$.

The next step is to consider how this would be implemented in the brain via HPC. In HPC, the generative model $p(s, o)$ is implemented in Markovian hierarchical layers, where the priors are simply the values of the layer above, mapped through a weight matrix and a nonlinear function. The prior at the top layer may either be a flat prior or set externally. With N layers, the top layer is labelled as layer 1, and the observation at the bottom as layer N. Thus:

$$p(s, o) = p(s_N \mid s_{N-1}) \ldots p(s_2 \mid s_1) \, p(s_1) \text{ where } s_N = o \tag{4}$$

The generative model is assumed to be Gaussian at each layer,

$$p(s_{n+1} \mid s_n) = N(s_{n+1}; f(\Theta_n \, s_n), \Sigma_{n+1}) \tag{5}$$

where s_n is a vector representing node values on layer n, Θ_n is a matrix giving the connection weights between layer n and layer $n+1$, f is a non-linear function and z_n is Gaussian noise at each layer. Note that the network also has a bias at each layer which is updated in a similar manner to the weights. This has not been included here for brevity. [Here we have shown the form where the argument of f is a weighted linear mixture of hidden states, in order to make clear how we have implemented the hierarchy. But this could equally be generalised to any non linear function f.]

Making the assumption that q is a multivariate Gaussian distribution $q(s) \sim N(s; \mu, \Sigma)$, and further assuming that the distribution of q is tightly packed around μ (to enable use of the Laplace assumption), $\mathcal{F}$ reduces to:

$$\mathcal{F} \approx -log \, p(\mu, o) \tag{6}$$

where o is a vector representing observations and μ is the mean of the brain's probability distribution for s. It is important to note that in this paper the observations are not confined to incoming senses but also include actions, in the form of proprioceptive feedback. Exteroceptive observations cause updates to model beliefs which, in turn, result in updated beliefs on proprioceptive observations. These drive motoneurons to eliminate any prediction error through reflex arcs [15, 17, 29]. Action can therefore be thought of as just a particular type of observation.

Using the distribution for a multivariate Gaussian, the estimate of $\mathcal{F}$ can be transformed into:

$$\mathcal{F} \approx -log \, p(\mu, o) = \sum_n log \, p(\mu_{n+1} \mid \mu_n) = \sum_n -\frac{1}{2}\epsilon_{n+1}^T \Sigma_{n+1}^{-1} \epsilon_{n+1} - \frac{1}{2}log(2\pi \, |\Sigma_{n+1}|) \tag{7}$$

where $\epsilon_{n+1} := \mu_{n+1} - f(\Theta_n \, \mu_n)$ is the difference between value of layer $n+1$ and the value predicted by layer n. $\mathcal{F}$ is then minimized following the Expectation-Minimization approach [10, 23], by using gradient descent to alternately update node values (μ) on a fast timescale and weight values (Θ) on a slower timescale.

The gradient for node updates in a hidden layer uses the values of ϵ_n and ϵ_{n+1}, and is given by the partial derivative:

$$\frac{\partial \mathcal{F}}{\partial \mu_n} = \epsilon_{n+1} \; \Sigma_{n+1}^{-1} \; \Theta_n^T \; f'(\Theta_n \; \mu_n) - \epsilon_n \; \Sigma_n^{-1} \tag{8}$$

but if the node values of the top layer are being updated then this is truncated to only use the difference compared to the layer below:

$$\frac{\partial \mathcal{F}}{\partial \mu_1} = \epsilon_2 \; \Sigma_2^{-1} \; \Theta_1^T \; f'(\Theta_1 \; \mu_1) \tag{9}$$

As pointed out earlier, downward predictions not only predict exteroceptive (sensory) signals, but also create a proprioceptive prediction error in the motor system (which is cancelled by movement via a reflex arc). In this paper we simply intend to monitor the signals being sent to the motor system and do not wish to include the error cancellation signal being fed back from the reflex arc. For this reason, the update of the "observation node" in the motor system is shown as only using the difference to the layer above:

$$\frac{\partial \mathcal{F}}{\partial \mu_N} = -\epsilon_N \; \Sigma_N^{-1} \tag{10}$$

After the node values have been changed, $\mathcal{F}$ is then further minimized by updating the weights using:

$$\frac{\partial \mathcal{F}}{\partial \Theta_n} = \epsilon_{n+1} \; \Sigma_{n+1}^{-1} \; \mu_n^T \; f'(\Theta_n \; \mu_n) \tag{11}$$

Since the impact of variance is not the primary focus here, our simulations assume that all Σ have fixed values of the identity matrix and therefore the gradient update for Σ has not been included.

Figure 1a summarises the flow of information in the network during gradient descent update of node values.

3   Methods

Three sets of experiments were designed, to investigate how the process of inference is distributed through hierarchical layers. The first two experiments were each run on three different tasks. The third experiment was run on a single task. The experiments are described below.

In the first set of experiments, we trained three PCNs to carry out separate inference tasks, based on selecting the correct action for a given MNIST image [22]. In all three networks, the observation layer at the bottom of the network contains 785 nodes, made up of 784 sensory nodes (representing the pixels of an MNIST image) and a single binary action node. The top layer uses a one-hot representation of each of the possible MNIST labels. There are two hidden layers of size 100 and 300. Thus, if there are 10 possible labels, there is a four-layer network of size [10,100,300,785], whose generative model produces an MNIST image and an action value from a given MNIST label. The role of each of the networks is, on presentation of an MNIST image, to infer the correct MNIST label at the top and the correct action associated with that image (Fig. 1a).

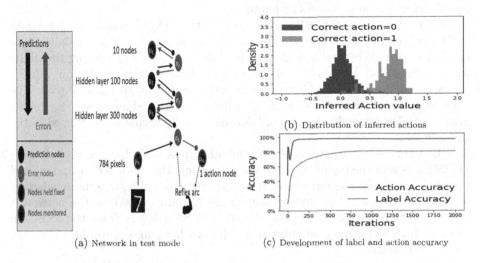

(a) Network in test mode   (c) Development of label and action accuracy

Fig. 1. Results for MNIST-digit1: correct action = 1 if label − 1, or 0 otherwise. (a) Test configuration. The observation nodes representing the MNIST picture are held fixed and both the label and the action nodes are updated using variational inference. (b) Distribution of inferred actions at end of inference period. (c) Simultaneous development of accuracy for label and action as inference progresses through iterations. Using argmax on label nodes, label accuracy = 80.8%. Using a heaviside function centred on 0.5, action accuracy = 97.4%. This indicates that the correct label is not required in order to select the correct action.

We investigated the relationship between the accuracies of action inference and label inference. Specifically, we asked: to what extent can the action be correctly triggered without correct label inference?

In the first task, MNIST-digit1, we trained the action node to output value 1 if the presented MNIST image has label 1, and value 0 for all other digits, i.e. the job of the action node is to fire when an image of the digit 1 is presented. The network is trained in a supervised manner to learn the generative model, by fixing the top and bottom layers with the training labels and observations respectively, and then, minimizing $\mathcal{F}$ in an expectation-maximization (EM) fashion [10,23], as described in Sect. 2. Once trained, the network is then tested for its ability to infer the correct label and action for a given image. This is done by presenting an MNIST image to the 784 sensory states and allowing both the labels at the top and the action at the bottom to update via the variational inference process, according to Eqs. (8)–(10). Updates to the network are applied over a large number of iterations and, at any stage of this process, the current inferred label can be read out as the argmax of the top-layer nodes while the selected action is read out according to a heaviside function applied to the action node value, centred on 0.5.

In the second task, MNIST-groups, we trained the action node to fire if the MNIST label is less than 5, and not fire otherwise. This network is trained and tested using the same process as above.

In the third task, MNIST-barred, half of the MNIST images had a white horizontal bar applied across the middle of the image. A new set of labels was created so that there were now 20 possible labels - 0 to 9 representing the digits without bars, and 10 to 19 representing the digits with bars. Action value 1 was associated with labels 10 to 19, and action value 0 with labels 0 to 9. The network for this task has size [20,100,300,785]. It is trained and tested as for the first two tasks. Appendix A gives full details of the hyperparameters used in the three PCNs.

The second set of experiments used the same three tasks but, instead of fixing MNIST labels to the top of the network in training, the top layer was populated with random noise. The purpose of these experiments was to determine whether the provision of label information in training had any impact on the network's ability to infer the correct action. We then ablated layers from the PCNs in order to investigate the contribution which each layer makes towards inferring the correct action.

The third experiment trained a network where both the MNIST image and the MNIST one hot-labels were placed at the bottom. Above this were 6 layers, all initialized with noisy values. The top layer was allowed to vary freely (see Fig. 4a). This was used to investigate how label inference performed in this scenario (rather than the traditional case of label at the top and image at the bottom), and how performance reacted to ablation of layers in test mode.

4   Results

We first investigated the relationship between accuracy of action and label inference for MNIST-digit1 (where the action node should fire if the MNIST label is 1). When run on a test set of images, the network generates values on the action node which correctly split into two groups centred near to 0 and 1, with a small overlap (Fig. 1b). As a result, the network is able to correctly infer the action for a presented image in over 97% of cases. On the other hand, the label is only correctly inferred in 81% of cases, demonstrating that action selection does not depend entirely on correct label inference. Figure 1c presents the development of label and action accuracies as iterations progress, confirming that a) action accuracy is always better than label accuracy, b) further iterations will not change this and c) action inference reaches asymptotic performance quicker than label inference.

Figure 2 compares label and action accuracy for all three tasks. In the MNIST-group task, action accuracy appears to be constrained by label accuracy. In the MNIST-barred task, the correct action is always inferred, even though the network has relatively poor label accuracy. It would therefore seem that the MNIST-group task is reliant on upper layer values in order to select the correct action, whereas the simpler tasks can reach, or approach, optimal action performance regardless of the upper layer values.

However, it is not clear from these results whether the MNIST-group task is relying on the fact that the higher layers contain information about the image labels (recall that this is how the network was trained) or whether it is simply that the existence of the higher layers is providing more compute power. To investigate this, the second set of experiments were run, where the three networks are trained with random noise at the top layer instead of the image label. In testing, label accuracy was now no better than random (as one would expect), but action accuracy was indistinguishable from the original results of Fig. 2. This demonstrates that it is

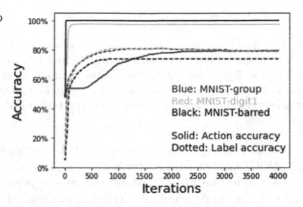

Fig. 2. Development of inference accuracy for label (at top of network) and action (at bottom of network), for 3 different tasks. Blue lines: MNIST-group. Action accuracy is constrained by label accuracy at around 80%. Red lines: MNIST-digit1. Action selection is above 97% accuracy, with label accuracy around 80%. Black lines: MNIST-barred. Accuracy of action selection quickly reaches 100% despite label accuracy being low. On a poorly trained network the results are even more striking, with action accuracy still perfect despite label accuracy of only 20% (results not shown). (Color figure online)

the existence of the layers, rather than provision of label information in training which is driving action inference.

To confirm that the three tasks make different use of the higher layers, action accuracy was measured when the top two layers were ablated in test mode (they were still present in training). Performance on the MNIST-group task (Fig. 3a) deteriorates significantly as the layers are ablated. Conversely, ablation of the top layer has no impact on the action accuracy of either the MNIST-barred (Fig. 3c) or MNIST-digit1 tasks (Fig. 3b). Both suffer slightly if the top 2 layers are ablated, although in the case of MNIST-barred the accuracy only moves from 100% to 99.9%. It can be concluded from these ablation experiments that reliance on higher layers varies with the nature of the task. Tasks which are more challenging may rely on the higher layers, while simple tasks may not suffer at all if the layers are ablated - presumably because all the information required for action selection is entirely available in the lower layers.

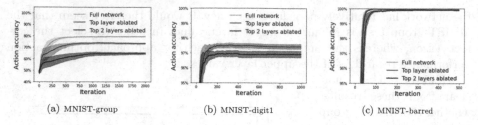

(a) MNIST-group   (b) MNIST-digit1   (c) MNIST-barred

Fig. 3. Effect of ablating layers on action accuracy. The three tasks cope differently with ablation of layers, as shown in (a), (b) and (c). Note that different y-scales are used on the figures for clarity. Each network was trained using 6 different seeds, and error bars show standard error. Results suggest that, if the lower layers are sufficient for action selection then the higher layers can be ignored.

In the third experiment, we constructed a network with both MNIST image and MNIST one hot-labels at the bottom, representing 10 different binary actions to select from (see Fig. 4a). Above this were 6 layers, all initialized with noisy values (details in Appendix A). Training was carried out as before, presenting a set of images and labels at the bottom of the network and leaving the network to learn weights throughout the hierarchy. The effect of layer ablation on the ability of the network to select the correct action (which in this experiment is the one-hot label) was then tested. When using all the layers, this network produces comparable results to the more standard PCN setup with label at the top and image at the bottom.[1] Ablation results are shown in Fig. 4b. These are consistent with the previous experiments, with accuracy reducing (but still much better than chance value of 10%) as the layers are ablated. In this case it would appear that the top 2 layers are adding nothing to the network's action selection ability. A key point to note is that the learning of the weights was not dependent on the provision of any information at the top of the network - all the learning comes about as a result of information presented at the bottom. Despite this, the network has distributed its ability through several layers, with the major part of successful inference relying on information towards the bottom of the network.

---

[1] At approximately 78%, the accuracy we achieved is significantly lower than standard non-PCN deep learning methods. This is partly because the model has not been fine-tuned (e.g. hyper-parameters, using convolutional layers, etc.). But it is also true that generative models tend to underperform discriminative models in classification tasks. This will be particularly true in our implementation which uses flat priors.

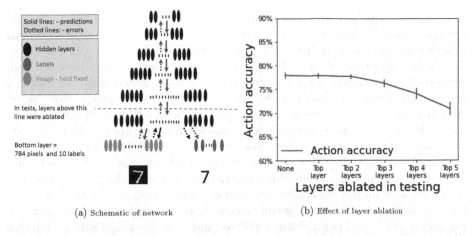

(a) Schematic of network   (b) Effect of layer ablation

Fig. 4. A 7 layer PCN where 10 binary actions are associated with MNIST images. (a) Image and one-hot labels both at the bottom. For ease of reading, the nodes shown on each layer represent both value and error nodes. Red lines show flow of information with no ablation. Black line shows flow if 5 layers are ablated. (b) Ablation of top two layers has no effect on accuracy of action selection. Ablation of the next 3 layers steadily reduces accuracy. Error bars are standard deviations across 10 differently seeded networks.

5   Discussion

We have demonstrated that, when training a PCN with senses and actions at the bottom layer, it is not necessary to provide a high level "hidden state" in training in order to learn the correct actions for an incoming sensation. Furthermore, the network appears to distribute its learning throughout the layers, with higher layers called into use only as required. In our experiments, this meant that higher layers could be ignored if the lower layers alone contained sufficient information to select the correct action. In effect, the network has learned a sensorimotor shortcut to select the correct actions. On the other hand, if the higher layers contain information which improves action selection, then ablation of those layers reduces, but doesn't destroy, performance - ablation leads to graceful degradation. This flexibility is inherent in the nature of PCNs, unlike feed forward networks, which operate end to end.

Importantly, this suggests that a PCN framework can help explain the development of fast reaction to a stimulus, even though the learning process involves all layers. For example, driving a car on an empty road might only require involvement of lower layers, whereas heavy traffic or icy conditions would require higher layers to deal with the more complex task. The fact that simple short-cuts can arise automatically during training and that the agent can dynamically select actions without involvement of higher layers could possibly also help explain why well-learned tasks can be carried out without conscious perception.

While we have provided an illustrative 'proof of principle' of this approach, much more can be done to investigate how this leads to a continuum of behaviour in active agents, which we list below in no particular order. Firstly, in our experiments inference took place with no influence from above and we have not considered the impact which exogenous priors would have. Secondly, we included no concept of a causal link between action and the subsequent sensory state. Action in real-life situations is a rolling process, with actions impacting subsequent decisions. Because our generative model did not consider time or state transitions, we cannot generalise to active inference in the sense of planning. One might argue that policy selection in active inference is a better metaphor for reflective behaviour, leading to a distinction between reflexive 'homeostatic' responses and more deliberative 'allostatic' plans. Having said this, it seems likely that the same conclusions will emerge. In other words, the same hierarchical generative model can explain reflective and reflexive behaviour at different hierarchical levels. Thirdly, the role of precisions has not been examined. Updating precisions should allow investigation of the role of attention. Finally, we have assumed the existence of a well trained network, and only touched on the performance of a partially trained network. It would be instructive to investigate how reliance on higher layers changes during the learning process.

These results support the view that a predictive coding network in the brain does not need to work from end to end, and can restrict itself to the number of lower layers required for the task at hand, possibly only in the sensorimotor system. There is the possibility of some tentative links here with more enactivist theories of the brain which posit that "representations" encode predicted action opportunities, rather than specify an abstract state of the world, but much further analysis is needed to investigate possible overlaps.

Acknowledgements. PK would like to thank Alec Tschantz for sharing the "Predictive Coding in Python" codebase https://github.com/alec-tschantz/pypc on which the experimental code was based. Thanks also to three anonymous reviewers whose comments helped improve the clarity of this paper, particularly in relation to temporal aspects of predictive coding. PK is funded by the Sussex Neuroscience 4-year PhD Programme. CLB is supported by BBRSC grant number BB/P022197/1.

A   Network parameters

Network size: 4 layer
Number of nodes on each layer: 10, 100, 300, 785 for MNIST-group and MNIST-digit1. 20, 100, 300, 785 for MNIST-barred. In the bottom layer, 784 nodes were fixed to the MNIST image, the 785th node was an action node which updates in testing. In initial set of experiments, top layer was fixed to a one-hot representation of MNIST label in training. In second set of experiments this was set to random value and allowed to update.
Non-linear function: tanh
Bias used: yes
Training set size: full MNIST training set of 60,000 images, in batches of 640

Number of training epochs: 10
Testing set size: 1280 images selected randomly from MNIST test set
Learning parameters used in weight update of EM process: Learning Rate = 1e–4, Adam
Learning parameters used in node update of EM process: Learning Rate = 0.025, SGD
Number of SGD iterations in training: 200
Number of SGD iterations in test mode: 200 * epoch number. The size is increased as epochs progress to allow for the decreasing size of the error between layers (as discussed in the text, this would normally be counteracted by increase in precision values).
Random initialisation: Except where fixed, all nodes were initialized with a random values selected from $\mathcal{N}(0.5, 0.05)$

In the experiment using a 7 layer network, the number of nodes on each layer were: 10, 25, 50, 100, 200, 300, 794. All other parameters the same as above

References

1. Adams, R.A., Shipp, S., Friston, K.J.: Predictions not commands: active inference in the motor system. Brain Struct. Funct. **218**(3), 611–643 (2013)
2. Baltieri, M., Buckley, C.L.: Generative models as parsimonious descriptions of sensorimotor loops. Behav. Brain Sci. **42**, e218–e218 (2019)
3. Bogacz, R.: A tutorial on the free-energy framework for modelling perception and learning. J. Math. Psychol. **76**, 198–211 (2017)
4. Bruineberg, J., Kiverstein, J., Rietveld, E.: The anticipating brain is not a scientist: the free-energy principle from an ecological-enactive perspective. Synthese **195**(6), 2417–2444 (2016). https://doi.org/10.1007/s11229-016-1239-1
5. Buckley, C.L., Chang, S.K., McGregor, S., Seth, A.K.: The free energy principle for action and perception: a mathematical review (2017)
6. Burr, C.: Embodied decisions and the predictive brain. In: Wiese, T.M.W. (ed.) Philosophy and predictive processing. MIND Group, Frankfurt a. M. (2016)
7. Clark, A.: Whatever next? predictive brains, situated agents, and the future of cognitive science. Behav. Brain Sci. **36**(3), 181–204 (2013)
8. Clark, A.: Predicting peace: The end of the representation wars. MIND Group, Open MIND. Frankfurt a. M. (2015)
9. Daw, N.D., Gershman, S.J., Seymour, B., Dayan, P., Dolan, R.J.: Model-based influences on humans' choices and striatal prediction errors. Neuron **69**(6), 1204–1215 (2011)
10. Dempster, A.P., Laird, N.M., Rubin, D.B.: Maximum likelihood from incomplete data via the EM algorithm. J. Roy. Stat. Soc. Ser. B (Methodol.) **39**(1), 1–22 (1977)
11. Dolan, R., Dayan, P.: Goals and habits in the brain. Neuron (Cambridge, Mass.) **80**(2), 312–325 (2013)
12. Friston, K.: Learning and inference in the brain. Neural Netw. **16**(9), 1325–1352 (2003)
13. Friston, K.: A theory of cortical responses. Philos. Trans. Roy. Soc. B Biol. Sci. **360**(1456), 815–836 (2005)

14. Friston, K.: The free-energy principle: a unified brain theory? Nat. Rev. Neurosci. **11**(2), 127–138 (2010)
15. Friston, K.: What is optimal about motor control? Neuron **72**(3), 488–498 (2011)
16. Friston, K., Kilner, J., Harrison, L.: A free energy principle for the brain. J. Physiol.-Paris **100**(1–3), 70–87 (2006)
17. Friston, K.J., Daunizeau, J., Kilner, J., Kiebel, S.J.: Action and behavior: a free-energy formulation. Biol. Cybern. **102**(3), 227–260 (2010)
18. Gläscher, J., Daw, N., Dayan, P., O'Doherty, J.P.: States versus rewards: dissociable neural prediction error signals underlying model-based and model-free reinforcement learning. Neuron **66**(4), 585–595 (2010)
19. Hipólito, I., Baltieri, M., Friston, K., Ramstead, M.J.: Embodied skillful performance: Where the action is. Synthese, pp. 1–25 (2021)
20. Hohwy, J.: The Predictive Mind. Oxford University Press, Oxford (2013)
21. Kahneman, D.: Thinking, fast and slow. Macmillan (2011)
22. LeCun, Y., Cortes, C.: MNIST handwritten digit database (2010). http://yann.lecun.com/exdb/mnist/
23. MacKay, D.J., Mac Kay, D.J.: Information Theory, Inference and Learning Algorithms. Cambridge University Press, Cambridge (2003)
24. Millidge, B.: Combining active inference and hierarchical predictive coding: a tutorial introduction and case study. PsyArXiv (2019)
25. Pezzulo, G., Donnarumma, F., Iodice, P., Maisto, D., Stoianov, I.: Model-based approaches to active perception and control. Entropy (Basel, Switzerland) **19**(6), 266 (2017)
26. Pezzulo, G., Rigoli, F., Friston, K.: Active inference, homeostatic regulation and adaptive behavioural control. Prog. Neurobiol. **134**, 17–35 (2015)
27. Ramstead, M.J., Kirchhoff, M.D., Friston, K.J.: A tale of two densities: active inference is enactive inference. Adapt. Behav. **28**(4), 225–239 (2020)
28. Rao, R.P., Ballard, D.H.: Predictive coding in the visual cortex: a functional interpretation of some extra-classical receptive-field effects. Nat. Neurosci. **2**(1), 79–87 (1999)
29. Seth, A.K.: The cybernetic bayesian brain: from interoceptive inference to sensorimotor contingencies (2015)
30. Stanovich, K.E., West, R.F.: Individual differences in reasoning: implications for the rationality debate? Behav.l Brain Sci. **23**(5), 645–665 (2000)
31. Sutton, R.S.: Reinforcement learning: an introduction (2018)
32. Whittington, J.C., Bogacz, R.: An approximation of the error backpropagation algorithm in a predictive coding network with local hebbian synaptic plasticity. Neural Comput. **29**(5), 1229–1262 (2017)
33. Wunderlich, K., Dayan, P., Dolan, R.J.: Mapping value based planning and extensively trained choice in the human brain. Nat. Neurosci. **15**(5), 786–791 (2012)

Deep Active Inference for Pixel-Based Discrete Control: Evaluation on the Car Racing Problem

N. T. A. van Hoeffelen[(✉)] and Pablo Lanillos

Department of Artificial Intelligence, Donders Institute for Brain, Cognition, and Behaviour, Radboud University, Montessorilaan 3, 6525HR Nijmegen, The Netherlands
niels.vanhoeffelen@ru.nl, p.lanillos@donders.ru.nl

Abstract. Despite the potential of active inference for visual-based control, learning the model and the preferences (priors) while interacting with the environment is challenging. Here, we study the performance of a deep active inference (dAIF) agent on OpenAI's car racing benchmark, where there is no access to the car's state. The agent learns to encode the world's state from high-dimensional input through unsupervised representation learning. State inference and control are learned end-to-end by optimizing the expected free energy. Results show that our model achieves comparable performance to deep Q-learning. However, vanilla dAIF does not reach state-of-the-art performance compared to other world model approaches. Hence, we discuss the current model implementation's limitations and potential architectures to overcome them.

Keywords: Deep active inference · Deep learning · POMDP · Visual-based control

1 Introduction

Learning from scratch which actions are relevant to succeed in a task using only high-dimensional visual input is challenging and essential for artificial agents and robotics. Reinforcement learning (RL) [26] is currently leading the advances in pixel-based control, e.g., the agent learns an action policy that maximizes the accumulated discounted rewards. Despite its dopamine biological inspiration, RL is far from capturing the physical processes happening in the brain. We argue, that prediction in any form (e.g., visual input, muscle feedback or dopamine) may be the driven motif of the general learning process of the brain [14]. Active inference [4,8], a general framework for perception, action and learning, proposes that the brain uses hierarchical generative models to predict incoming sensory data [6] and tries to minimize the difference between the predicted and observed sensory signals. This difference, mathematically described as the variational free energy (VFE), needs to be minimized to generate better predictions about the

© Springer Nature Switzerland AG 2021
M. Kamp et al. (Eds.): ECML PKDD 2021 Workshops, CCIS 1524, pp. 843–856, 2021.
https://doi.org/10.1007/978-3-030-93736-2_60

world that causes these sensory signals [8,17]. Through acting on its environment, an agent can affect sensory signals to be more in line with predicted signals, which in turn leads to a decrease of the error between observed and predicted sensory signals.

AIF models have shown great potential in low-dimensional and discrete state spaces. To work in higher-dimensional state spaces, deep active inference (dAIF) has been proposed, which uses deep neural networks for approximating probability density functions (e.g., amortised inference) [2,17,23,27,29]. Interestingly, dAIF can be classified as a generalization of the world models approach [10] and can incorporate reward-based learning, allowing for a direct comparison to RL methods. Since the first attempt of dAIF [29], developments have happened concurrently in adaptive control [16,23] and in planning, exploiting discrete-time optimization, e.g., using the expected free energy [21,28]. In [17], a dIAF agent was tested on several environments in which the state is observable (Cartpole, Acrobot, Lunar-lander). In [5], a dAIF agent using Monte-Carlo sampling was tested on the Animal-AI environment. In [3], dAIF tackled the mountain car problem and was also tested on OpenAI's car racing environment. For the car racing environment, they trained the dAIF agent on a handful of demonstration rollouts and compared it to a DQN that interacted with the environment itself. Their results showed that DQN needs a lot more interaction with the environment to start obtaining rewards compared to dAIF trained on human demonstrations of the task. Relevant for this work, in [11], a dAIF agent solved the Cartpole environment as both MDP and as POMPD instances, training the agent on just visual input.

In this paper, we study a dAIF agent[1] based on the proposed architectures in [11,17] for a more complex pixel-based control POMDP task, namely the OpenAI's Car Racing environment [13], and discuss its advantages and limitations compared to other state-of-the-art models. The performance of the dAIF agent was shown to be in line with previous works and on-par with deep Q-learning. However, it did not achieve the performance of other world model approaches [1]. Hence, we discuss the reasons for this, as well as architectures that may help to overcome these limitations.

2   Deep Active Inference Model

The dAIF architecture studied is based on [11] and described in Fig. 1. It makes use of five networks to approximate the densities of Eq. (2): observation (encoding and decoding), transition, policy, and value. The full parameter description of the networks can be found in the Appendix A (Tables 2, 3 and 4).

Variational Free Energy (VFE). AIF agents infer their actions by minimizing the VFE expressed at instant k (with explicit action 1-step ahead) as:

$$\mathcal{F}_k = -\mathbf{KL}[q(s_k, a_k) \,||\, p(o_k, s_{0:k}, a_{0:k})] \tag{1}$$

---

[1] The code can be found at https://github.com/NTAvanHoeffelen/DAIF_CarRacing.

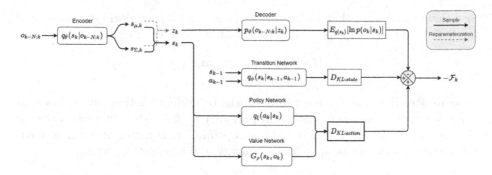

Fig. 1. Deep Active Inference architecture. Five deep artificial neural networks are used to model the observation encoding and decoding, the state transition, the policy and the EFE values. The architecture is trained end-to-end by optimizing the expected variational free energy.

where s_k, o_k, a_k are the state, the observation and action respectively, $q(s_k, a_k)$ is the recognition density, and $p(o_k, s_{0:k}, a_{0:k})$ the generative model. Under the Markov assumption and factorizing [11,17], Eq. (1) can be rewritten as:

$$\mathcal{F}_k = \underbrace{E_{q(s_k)}[\ln p(o_k|s_k)]}_{\text{sensory prediction}} - \underbrace{\mathbf{KL}[q(s_k) \parallel p(s_k|s_{k-1}, a_{k-1})]}_{\text{state prediction}} - \underbrace{\mathbf{KL}[q(a_k|s_k) \parallel p(a_k|s_k)]}_{\text{action prediction}}$$

(2)

Sensory Prediction. The observation network predicts the observations—first term in Eq. (2)—and encodes the high-dimensional input to states $q_\theta(s_k|o_{k-N:k})$ and decodes latent spaces to observations $p_\vartheta(o_{k-N:k}|z_k)$. It can be implemented with a variational autoencoder, where latent representation is described as a multivariate Gaussian distribution with mean s_μ and variance s_Σ. The latent space z_k is obtained using the reparametrisation trick. To train the network, we use the binary cross-entropy and the KL regularizer to force the latent space to be Gaussian:

$$L_{o,k} = -E_{q_\theta(s_k|o_{k-N:k})}[\ln p_\vartheta(o_{k-N:k}|z_k)]$$
$$= BCE(\hat{o}_{k-N:k}, o_{k-N:k}) - \frac{1}{2}\sum(1 + \ln s_{\Sigma,k} - s_{\mu,k}^2 - s_{\Sigma,k})$$

(3)

State Prediction. The transition network models a distribution that allows the agent to predict the state at time k given the state and action at time $k - 1$, where the input state consists of both the mean s_μ and variance s_Σ. Under the AIF approach this is the difference between the state distribution (generated by the transition network) and the actual observed state (from the encoder): $\mathbf{KL}[q(s_k) \parallel p(s_k|s_{k-1}, a_{k-1})]$. For the sake of simplicity, we define a feed-forward network that computes the maximum-a-posteriori estimate of the predicted state $\hat{s}_k = q_\phi(s_{k-1}, a_{k-1})$ and train it using the mean squared error:

$$MSE(s_{\mu,k}, q_\phi(s_{k-1}, a_{k-1})) \tag{4}$$

Action Prediction. We use two networks to evaluate action: the policy and value network. The action prediction part of Eq. (2) is the difference between the model's action distribution and the "optimal" true distribution. It is a KL divergence, which can be split into an energy and an entropy term:

$$\mathbf{KL}[q(a_k|s_k) \,||\, p(a_k|s_k)] = -\sum_a q(a_k|s_k) \ln \frac{p(a_k|s_k)}{q(a_k|s_k)}$$

$$= \underbrace{-\sum_a q(a_k|s_k) \ln p(a_k|s_k)}_{\text{energy}} - \underbrace{\sum_a q(a_k|s_k) \ln q(a_k|s_k)}_{\text{entropy}}$$

$$\tag{5}$$

The policy network models the distribution over actions at time k given the state at time k $q_\xi(a_k|s_k)$. It is implemented as a feed-forward neural network that returns a distribution over actions given a state.

The value network computes the Expected Free Energy (EFE) [11,17,21] which is used to model the true action posterior $p(a_k|s_k)$. As the true action posterior is not exactly known, we assume that prior belief makes the agent select policies that minimize the EFE. We model the distribution over actions as a precision-weighted Boltzmann distribution over the EFE [5,17,21,24]:

$$p(a_k|s_k) = \sigma(-\gamma G(s_{k:N}, o_{k:N})) \tag{6}$$

where $G(s_{k:N}, o_{k:N})$ is the EFE for a set of states and observations up to some future time N. As we are dealing with discrete time steps, it can be written as a sum over these time steps:

$$G(s_{k:N}, o_{k:N}) = \sum_k^N G(s_k, o_k) \tag{7}$$

The EFE is evaluated for every action, because of this we implicitly conditioned on every action [17]. We then define the EFE of a single time step as[2]:

$$G(s_k, o_k) = \mathbf{KL}[q(s_k) \,||\, p(s_k, o_k)]$$
$$\approx -\ln p(o_k) + \mathbf{KL}[q(s_k) \,||\, q(s_k|o_k)] \tag{8}$$
$$\approx -r(o_k) + \mathbf{KL}[q(s_k) \,||\, q(s_k|o_k)]$$

The negative log-likelihood (or surprise) of an observation $-\ln p(o_t)$, is replaced by the reward $-r(o_k)$ [7,11,17]. As AIF agents act to minimize their surprise, by replacing the surprise with the negative reward, we encode the agent

---

[2] The full derivation can be found in Appendix D.

with the prior that its goal is to maximize reward. This formulation needs the EFE computation for all of the possible states and observations up to some time N, making it computationally intractable. Tractability has been achieved through bootstrapping [11,17] and combining Monte-Carlo tree search and amortized inference [5]. Here we learn a bootstrapped estimate of the EFE. The value network is used to get an estimate of the EFE for all of the possible actions. It is modelled as a feed-forward neural network $G_\rho(s_k, o_k) = f_\rho(s_k)$.

To train the value network, we use another bootstrapped EFE estimate which uses the EFE of the current time step and a $\beta \in (0, 1]$ discounted value-net estimate of the EFE under $q(a_{k+1}|s_{k+1})$ for the next time step:

$$\hat{G}(s_k, o_k) = -r(o_k) + \mathbf{KL}[q(s_k) \,||\, q(s_k|o_k)] + \beta E_{q(a_{k+1}|s_{k+1})}[G_\rho(s_{k+1}, o_{k+1})]$$
(9)

Using gradient descent, we can optimize the parameters of the value network by computing the MSE between $G_\rho(s_k, o_k)$ and $\hat{G}(s_k, o_k)$:

$$L_{f_\rho, k} = MSE(G_\rho(s_k, o_k), \hat{G}(s_k, o_k))$$
(10)

In summary, with our implementation, the VFE loss function becomes:

$$\begin{aligned}
-\mathcal{F}_k =\ & BCE(\hat{o}_{k-N:k}, o_{k-N:k}) - \frac{1}{2}\sum(1 + \ln s_{\Sigma,k} - s_{\mu,k}^2 - s_{\Sigma,k}) \\
& + MSE(s_{\mu,k}, q_\phi(s_{k-1}, a_{k-1})) \\
& + \mathbf{KL}[q_\xi(a_k|s_k) \,||\, \sigma(-\gamma G_\rho(s_k, o_k))]
\end{aligned}$$
(11)

3 Experimental Setup

We evaluated the algorithm on OpenAI's CarRacing-v0 environment [13] (Fig. 2a). It is considered a Partial Observable Markov Decision Process (POMDP) problem as there is no access to the state of the agent/environment. The input is the top-view image (96×96 RGB) of part of the racing track centred on the car. The goal of this 2D game is to maximize the obtained reward by driving as fast and precise as possible. A reward of $+1000/N$ is received for every track tile that is visited, where N is the total number of tiles (placed on the road), and a reward of -0.1 is received for every frame that passes. Solving the game entails that an agent scores an average of more than 900 points over 100 consecutive episodes. By definition, an episode is terminated after the agent visited all track tiles, the agent strayed out of bounds of the environment, or when 1000 time steps have elapsed. Every episode, a new race track is randomly generated.

The agent/car has three continuous control variables, namely steering $[-1, 1]$ (left and right), accelerating $[0, 1]$, and braking $[0, 1]$. The action space was discretized into 11 actions, similarly to [25,30,31], described in Table 2c.

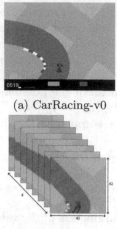

(a) CarRacing-v0

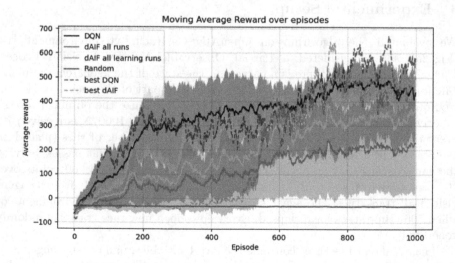

(b) Preprocessed input

action	values
do nothing	[0, 0, 0]
steer sharp left	[-1, 0, 0]
steer left	[-0.5, 0, 0]
steer sharp right	[1 ,0, 0]
steer right	[0.5, 0, 0]
accelerate 100%	[0, 1, 0]
accelerate 50%	[0, 0.5, 0]
accelerate 25%	[0, 0.25, 0]
brake 100%	[0, 0, 1]
brake 50%	[0, 0, 0.5]
brake 25%	[0, 0, 0.25]

(c) Discrete Actions

Fig. 2. (a) Screencapture of OpenAI's CarRacing-v0 environment. (b) The preprocessed model input: 8 stacked 42×42 grayscaled images. (c) The discretized action space.

4   Results

We compared our dAIF implementation with other state-of-the-art algorithms. First, Fig. 3 shows the average reward evolution while training for our dAIF architecture, Deep-Q learning (DQN) [19] (our implementation) and a random

Fig. 3. Moving average reward (MAR) comparison for OpenAI's CarRacing-v0. $MAR_e = 0.1CR_e + 0.9MAR_{e-1}$, where CR_e is the cumulative reward of the current episode. In green (solid line), the mean of the dAIF runs that were able to learn a policy, and in orange (dashed line), the best of all training runs. (Color figure online)

agent. Second, Table 1 shows the average reward over 100 consecutive episodes for the top methods in the literature. The average reward performance test and reward per episode for DQN and dAIF are provided in Appendix E (Fig. 5).

For the dAIF and the DQN implementations, observations are first preprocessed by removing the bottom part of the image. This part contains information about the accumulated rewards of the current episode, the car's true speed, four ABS sensors, steering wheel position, and gyroscope. Afterwards, the image is grey-scaled and reshaped to a size of 42×42. The input is defined as a stack of k to $k - N$ observations to provide temporal information by allowing the encoding of velocity and steering. We use experience replay [15] with a batch size of 250 and memory capacity of 300000 and 100000 transitions for DQN and dAIF respectively, and make use of target networks [19] (copy of the policy network (Table 5) for DQN and value network for dAIF) with a freeze period of 50 time steps.

The dAIF agent makes use of a pre-trained VAE (Table 6) which was frozen during training. Following a similar procedure as [3], the VAE was pre-trained on observations collected by having a human play the environment for 10000 time steps.

Table 1. Average rewards for CarRacing-v0

Method	Average reward
DQN (our implementation)	515 ± 162
dAIF (our implementation)	494 ± 241
A3C (Continuous) [18]	591 ± 45
A3C (Discrete) [12]	652 ± 10
Weight Agnostic Neural Networks [9]	893 ± 74
GA [22]	903 ± 72
World models [10]	906 ± 21

5 Discussion

The dAIF implementation described in this paper has shown to reach performance on par with Deep Q-learning. However, there are some remarks. First, it showed a slower learning curve as described in previous works [11], due to the need to learn the world model. Second, we identified some runs where the system was not able to learn—See Fig. 3 orange solid line. These runs drag down the average performance. Finally, it has failed to reach state-of-the-art performance when comparing to other world model approaches—See Table 1. Here we discuss the limitations of the current implementation and alternative architectures to overcome the challenge of learning the preferences in dAIF approaches.

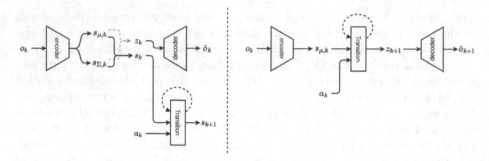

Fig. 4. Transition network outside of the observation network (left) and the transition network in between the encoder and decoder (right).

Observation and Transition Model. Our implementation does not fully exploit temporal representation learning, such as other models that use recurrent neural networks (RNN). Figure 4 describes two architectures to learn the encoding and the transitioning of the environment. Figure 4 left shows the autoencoding and transition model implemented in this work. This architecture is similar to the successful world models [10], but while we used a simple feed-forward network, they modelled the state-transition with a mixed density network RNN. Interestingly, dAIF also permits the alternative architecture described in Fig. 4 right (also proposed in [20]), in which the network learns to predict future observations. Here the transition network is part of the autoencoding. By incorporating the transition network in the structure of the observation network, we avoid the need for the dual objectives: perceptual reconstruction and dynamics learning. Preliminary testing did not show any improvements. Future work could involve more extensive testing to uncover possible performance improvements.

Dependency of Input Space. The performance of dAIF has shown a strong dependency on the learning of the observation model. Different image preprocessing methods would lead to improvements of more than 50% in the agent performance, as shown in other DQN implementations in the literature. Testing showed that without a pre-trained observation network, the model was unable to learn consistently and rarely showed performance that would suggest an indication of task comprehension. By using a pre-trained observation network learning occurred in 6 out of the 10 runs. To produce a proper action-centric representation, likelihood, transition and control should be learnt concurrently. However, parameters uncertainty may be tackled as the models are being learnt. The current implementation uses static values for the networks learning rates, future testing could investigate different variable learning rates for each network or decaying dropout temperature.

Bootstrapping of the Policy and the Value. Estimating both the policy and the value from state encoding has shown end-to-end issues when we do not pre-train the observation model. In particular, 1-step ahead action formulation in conjunction with bootstrapping might not capture a proper structure of the world, which is needed to complete the task, even if we use several consecutive input images to compute the state. N-step ahead observation optimization EFE formulations, as proposed in [5, 20, 28], may aid learning. Particularly, when substituting the negative log surprise by the rewards, the agent might loose the exploratory AIF characteristic, thus focusing only on goal-oriented behaviour. Furthermore, and very relevant, the reward implementation in CarRacing-v0 might be not the best way to provide dAIF with rewards for proper preference learning.

A   Model Parameters

Table 2. General parameters

Parameter	Value	Description
$N_{screens}$	8	Size of the observation stack
N_{colour}	1	Colour channels of the input image
N_{height}	42	Height in pixels of the input image
N_{width}	42	Width in pixels of the input image
$N_{actions}$	11	Number of actions the agent can select from
$N_{episodes}$	1000	Number of episodes the model is trained for
$N_{length\_episode}$	1000	The maximum amount of time steps in an episode
Freeze period	50	The amount of time steps the target network is frozen before copying the parameters of the policy/value network
Batch size	250	Number of items in a mini-batch

Table 3. DQN parameters

Parameter	Value	Description
Policy network		Convolutional neural network which estimates Q-values given a state see Appendix B.
Target network		Copy of the Policy network which is updated after each freeze period see Appendix B.
N_{hidden}	512	Number of hidden units in the policy and target network
λ	1e-5	Learning rate
γ	0.99	Discount factor
ϵ	0.15 → 0.05	Probability of selecting a random action. (Starts as 0.15, decreases linearly per episode with 0.00015 until a minimum of 0.05)
Memory capacity	300000	Number of transitions the replay memory can store

Table 4. dAIF parameters

Parameter	Value	Description
Observation network		VAE; see Appendix C.
Transition network		Feed-forward neural network of shape: $(2N_{latent} + 1) \times N_{hidden} \times N_{actions}$
Policy network		Feed-forward neural network of shape: $2N_{latent} \times N_{hidden} \times N_{actions}$; with a softmax function on the output
Value network		Feed-forward neural network of shape: $2N_{latent} \times N_{hidden} \times N_{actions}$
Target network		Copy of the Value network which is updated after each freeze period
N_{hidden}	512	Number of hidden units in the transition, policy, and value network.
N_{latent}	128	Size of the latent state
$\lambda_{transition}$	1e-3	Learning rate of the transition network
λ_{policy}	1e-4	Learning rate of the policy network
λ_{value}	1e-5	Learning rate of the value network
λ_{VAE}	5e-6	Learning rate of the VAE
γ	12	Precision parameter
β	0.99	Discount factor
α	18000	$\frac{1}{\alpha}$ is multiplied with the VAE loss to scale its size to that of the other term in the VFE
Memory capacity	100000	Number of transitions the replay memory can store

B   DQN: Policy Network

Table 5. Layers DQN policy network

Type	Out channels	Kernel	Stride	Input	Output
conv	64	4	2	(1, 8, 42, 42)	(1, 64, 20, 20)
batchnorm					
maxpool	–	2	2	(1, 64, 20, 20)	(1, 64, 10, 10)
relu					
conv	128	4	2	(1, 64, 10, 10)	(1, 128, 4, 4)
batchnorm					
maxpool	–	2	2	(1, 128, 4, 4)	(1, 128, 2, 2)
relu					
conv	256	2	2	(1, 128, 2, 2)	(1, 256, 1, 1)
relu					
dense	–	–	–	256	512
dense	–	–	–	512	11

C   VAE

Table 6. VAE layers

Type	out channels	kernel	stride	input	output	
conv	32	4	2	(1, 8, 42, 42)	(1, 32, 20, 20)	
batchnorm						
relu						
conv	32	4	2	(1, 32, 20, 20)	(1, 64, 9, 9)	
batchnorm						
relu						
conv	128	5	2	(1, 64, 9, 9)	(1, 128, 3, 3)	} Encoder
batchnorm						
relu						
conv	256	3	2	(1, 128, 3, 3)	(1, 256, 1, 1)	
relu						
dense	-	-	-	256	128	
dense μ	-	-	-	128	128	
dense $\log\Sigma$	-	-	-	128	128	
dense	-	-	-	128	128	
dense	-	-	-	128	256	
deconv	128	3	2	(1, 256, 1, 1)	(1, 128, 3, 3)	
batchnorm						
relu						
deconv	64	5	2	(1, 128, 3, 3)	(1, 64, 9, 9)	
batchnorm						
relu						} Decoder
deconv	32	4	2	(1, 64, 9, 9)	(1, 32, 20, 20)	
batchnorm						
relu						
deconv	8	4	2	(1, 32, 20, 20)	(1, 8, 42, 42)	
batchnorm						
relu						
sigmoid						

D   Derivations

Derivation for the EFE for a single time step:

$$G(s_k, o_k) = \mathbf{KL}[q(s_k) \| p(s_k, o_k)]$$

$$= \int q(s_k) \ln \frac{q(s_k)}{p(s_k, o_k)}$$

$$= \int q(s_k) \ln q(s_k) - \ln p(s_k, o_k)$$

$$= \int q(s_k) \ln q(s_k) - \ln p(s_k | o_k) - \ln p(o_k)$$

$$\approx \int q(s_k) \ln q(s_k) - \ln q(s_k | o_k) - \ln p(o_k)$$

$$\approx -\ln p(o_k) + \int q(s_k) \ln q(s_k) - \ln q(s_k | o_k)$$

$$\approx -\ln p(o_k) + \int q(s_k) \ln \frac{q(s_k)}{q(s_k | o_k)}$$

$$\approx -\ln p(o_k) + \mathbf{KL}[q(s_k) \| q(s_k | o_k)]$$

$$\approx -r(o_k) + \mathbf{KL}[q(s_k) \| q(s_k | o_k)]$$

E   Average Reward Over 100 Episodes

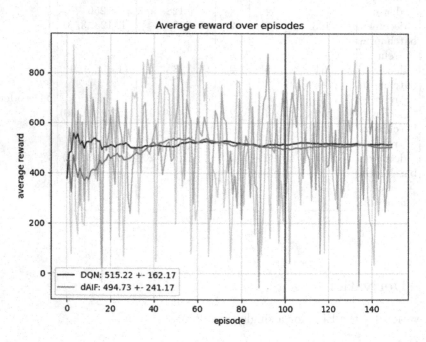

Fig. 5. Average reward test over 100 episodes for DQN and dAIF. The bright lines show the mean over episodes. The transparent lines show the reward that was obtained in a particular episode.

References

1. Openai's carracing-v0 leaderboard. https://github.com/openai/gym/wiki/Leaderboard#carracing-v0
2. Çatal, O., Nauta, J., Verbelen, T., Simoens, P., Dhoedt, B.: Bayesian policy selection using active inference. arXiv preprint arXiv:1904.08149 (2019)
3. Çatal, O., Wauthier, S., De Boom, C., Verbelen, T., Dhoedt, B.: Learning generative state space models for active inference. Front. Comput. Neurosci. **14**, 103 (2020)
4. Da Costa, L., Parr, T., Sajid, N., Veselic, S., Neacsu, V., Friston, K.: Active inference on discrete state-spaces: a synthesis. J. Math. Psychol. **99**, 102447 (2020)
5. Fountas, Z., Sajid, N., Mediano, P.A., Friston, K.: Deep active inference agents using monte-carlo methods. arXiv preprint arXiv:2006.04176 (2020)
6. Friston, K.: A theory of cortical responses. Philos. Trans. R. Soc. B Biol. Sci. **360**(1456), 815–836 (2005)
7. Friston, K., Samothrakis, S., Montague, R.: Active inference and agency: optimal control without cost functions. Biol. Cybern. **106**(8), 523–541 (2012)
8. Friston, K.J., Daunizeau, J., Kilner, J., Kiebel, S.J.: Action and behavior: a free-energy formulation. Biol. Cybern. **102**(3), 227–260 (2010)
9. Gaier, A., Ha, D.: Weight agnostic neural networks. arXiv preprint arXiv:1906.04358 (2019)
10. Ha, D., Schmidhuber, J.: World models. arXiv preprint arXiv:1803.10122 (2018)
11. van der Himst, O., Lanillos, P.: Deep active inference for partially observable MDPs. In: IWAI 2020. CCIS, vol. 1326, pp. 61–71. Springer, Cham (2020). https://doi.org/10.1007/978-3-030-64919-7_8
12. Khan, M., Elibol., O.: Car racing using reinforcement learning (2018). https://web.stanford.edu/class/cs221/2017/restricted/p-final/elibol/final.pdf
13. Klimov, O.: Carracing-v0. https://gym.openai.com/envs/CarRacing-v0/
14. Lanillos, P., van Gerven, M.: Neuroscience-inspired perception-action in robotics: applying active inference for state estimation, control and self-perception. arXiv preprint arXiv:2105.04261 (2021)
15. Lin, L.: Reinforcement learning for robots using neural networks (1992)
16. Meo, C., Lanillos, P.: Multimodal VAE active inference controller. arXiv preprint arXiv:2103.04412 (2021)
17. Millidge, B.: Deep active inference as variational policy gradients. J. Math. Psychol. **96**, 102348 (2020)
18. Min J. Jang, S., Lee, C.: Reinforcement car racing with a3c (2017). https://www.scribd.com/document/358019044/Reinforcement-Car-Racing-with-A3C
19. Mnih, V., et al.: Human-level control through deep reinforcement learning. Nature **518**(7540), 529–533 (2015)
20. Noel, A.D., van Hoof, C., Millidge, B.: Online reinforcement learning with sparse rewards through an active inference capsule. arXiv preprint arXiv:2106.02390 (2021)
21. Parr, T., Friston, K.J.: Generalised free energy and active inference. Biol. Cybern. **113**(5), 495–513 (2019)
22. Risi, S., Stanley, K.O.: Deep neuroevolution of recurrent and discrete world models. In: Proceedings of the Genetic and Evolutionary Computation Conference, pp. 456–462 (2019)

23. Sancaktar, C., van Gerven, M.A., Lanillos, P.: End-to-end pixel-based deep active inference for body perception and action. In: 2020 Joint IEEE 10th International Conference on Development and Learning and Epigenetic Robotics (ICDL-EpiRob), pp. 1–8. IEEE (2020)
24. Schwartenbeck, P., Passecker, J., Hauser, T.U., FitzGerald, T.H., Kronbichler, M., Friston, K.J.: Computational mechanisms of curiosity and goal-directed exploration. Elife **8**, e41703 (2019)
25. Slik, J.: Deep reinforcement learning for end-to-end autonomous driving (2019)
26. Sutton, R.S., Barto, A.G.: Reinforcement Learning: An Introduction. MIT Press, Cambridge (2018)
27. Tschantz, A., Baltieri, M., Seth, A.K., Buckley, C.L.: Scaling active inference. In: 2020 International Joint Conference on Neural Networks (IJCNN), pp. 1–8. IEEE (2020)
28. Tschantz, A., Millidge, B., Seth, A.K., Buckley, C.L.: Reinforcement learning through active inference. arXiv preprint arXiv:2002.12636 (2020)
29. Ueltzhöffer, K.: Deep active inference. Biol. Cybern. **112**(6), 547–573 (2018)
30. van der Wal, D., Intelligentie, B.O.K., Shang, W.: Advantage actor-critic methods for carracing (2018)
31. Zhang, Y.: Deep reinforcement learning with mixed convolutional network. arXiv preprint arXiv:2010.00717 (2020)

Robot Localization and Navigation Through Predictive Processing Using LiDAR

Daniel Burghardt[1]($\boxtimes$) and Pablo Lanillos[2]

[1] Radboud University, Houtlaan 4, 6525 XZ Nijmegen, The Netherlands
[2] Donders Institute for Brain, Cognition and Behaviour, Department of Artificial Intelligence, Radboud University, Nijmegen, The Netherlands

Abstract. Knowing the position of the robot in the world is crucial for navigation. Nowadays, Bayesian filters, such as Kalman and particle-based, are standard approaches in mobile robotics. Recently, end-to-end learning has allowed for scaling-up to high-dimensional inputs and improved generalization. However, there are still limitations to providing reliable laser navigation. Here we show a proof-of-concept of the predictive processing-inspired approach to perception applied for localization and navigation using laser sensors, without the need for odometry. We learn the generative model of the laser through self-supervised learning and perform both online state-estimation and navigation through stochastic gradient descent on the variational free-energy bound. We evaluated the algorithm on a mobile robot (TIAGo Base) with a laser sensor (SICK) in Gazebo. Results showed improved state-estimation performance when comparing to a state-of-the-art particle filter in the absence of odometry. Furthermore, conversely to standard Bayesian estimation approaches our method also enables the robot to navigate when providing the desired goal by inferring the actions that minimize the prediction error.

Keywords: Predictive processing · Robot localization · Robot navigation · Laser sensor · LiDAR

1 Introduction

Localization algorithms are part of our daily life and core for robotics. Recursive Bayesian estimation composes the current state-of-art in the field and has been essential for the development of localization, mapping, navigation and searching applications [11,20]. Bayesian filters [5], e.g., Kalman and particle filters, are able to estimate the state of a system from noisy sensor observations formalized as a hidden Markov model. These approaches are useful also in the case of non-linear modeled systems and out-of-sequence measurements [3]. The particle filter (PF) is an approximate Bayesian method that tractably computes the posterior distribution of the state of any system given the observations. The

© Springer Nature Switzerland AG 2021
M. Kamp et al. (Eds.): ECML PKDD 2021 Workshops, CCIS 1524, pp. 857–864, 2021.
https://doi.org/10.1007/978-3-030-93736-2_61

state distribution is represented by individual particles, which are evaluated and weighted recursively. Particles with higher probability get bigger weights and are re-sampled into more particles in its neighborhood, whereas particles with smaller weights get fewer new samples close to them [14].

In recent years, novel approaches based on deep neural networks, have been proposed to improve localization using high-dimensional inputs. Regression solutions, for instance, compute the absolute position of the system from only visual information [9]. However, these methods have lower accuracy than previous approaches that exploit prior information, such as geometry [19]. In particular, LiDAR-based Navigation with representation learning (e.g., using autoencoders) and reinforcement learning has shown downgraded performance in navigation tasks [8]. Alternatively to LiDAR-based approaches, however, the work of [21] has demonstrated a successful application of deep active inference in robot navigation using camera images.

We describe how the predictive processing (PP) approach to perception [4,12] can aid in localization and simple navigation tasks [13]. In this work navigation is performed having the robot move between two points in an unobstructed environment, which can be further built upon to tackle more complex environments (e.g. mazes). Under PP, the agent, following the Free Energy Principle (FEP) [6], tries to minimize the error in the predicted observations by either performing corrective actions to match the expected internal state or by updating this internal state based on what it has experienced through the senses. In this work, we present a proof-of-concept based on the Pixel-Active Inference model [18], proposed for humanoid body perception and action, to perform laser-based localization and navigation without the need for odometry. This has been successfully applied to robot manipulator control to improve adaptation [15]. Our approach combines the power of deep networks regression with variational Bayesian filtering to provide a better reliable state estimation than PFs in our proof-of-concept environment—See Fig. 1.

2   Methods

2.1   Robot

The TIAGo Base mobile robot uses the SICK TiM571 laser sensor, which has a $0.0\,\mathrm{m}$–$25\,\mathrm{m}$ range and a $270°$ aperture angle. In all experiments we limited the robot's movement to $2\,°\mathrm{C}$ of freedom, i.e., moving forward, backward and sideways.

2.2   Localization

We define the true state (position) of the robot at instant k as $\mathbf{x}_k = (x, y) \in \mathbb{R}^2$ and the position belief of the robot as $\tilde{\mathbf{x}}_k$. We further define the observation $\mathbf{o}_k$ as the laser measurements. Estimation is solved by computing the posterior distribution $p(\mathbf{x}|\mathbf{o})$ by optimizing the Variational Free Energy (VFE). The algorithm is sketched in Fig. 2a. Under the mean-field and Laplace approximation

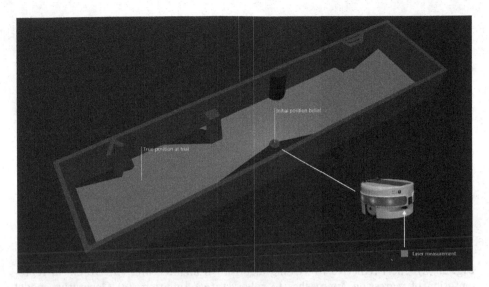

Fig. 1. Proof-of-concept environment designed for the experiments in Gazebo. In the localization experiment the robot true position is randomly set and the initial position belief is initialized to the center of the map. Laser range finder measurements are displayed as the blue shading. (Color figure online)

this is equivalent to minimizing the error between the sensory input $\mathbf{o}_k$ and the predicted sensory input $\hat{\mathbf{o}}_k$. While regression approaches (Fig. 2b) compute the pose directly from the visual input, stochastic neural filtering continuously refines the state through an error signal. We perform state estimation through perceptual inference, minimizing the VFE as follows:

$$\tilde{\mathbf{x}} = \underset{\tilde{\mathbf{x}}}{\arg\min}\, F(\tilde{\mathbf{x}}, \mathbf{o}) \rightarrow \tilde{\mathbf{x}}_{k+1} = \tilde{\mathbf{x}}_k + \alpha \partial_{\tilde{\mathbf{x}}} g(\tilde{\mathbf{x}}_k) \Sigma_{\mathbf{o}}^{-1}(\mathbf{o}_k - g(\tilde{\mathbf{x}}_k)) \tag{1}$$

Where α is the step size and $\partial_{\tilde{\mathbf{x}}}$ denotes the derivative with respect to $\tilde{\mathbf{x}}$. This is computed iteratively using gradient descent on the prediction error—sensor measurement $\mathbf{o}_k$ minus the predicted sensory input $g(\tilde{\mathbf{x}}_k)$—weighted by the variance $\Sigma_{\mathbf{o}}$. Both the predicted observations and the partial derivative of the error are computed by means of a deep neural network forward pass and its Jacobian [18], respectively.

2.3   Predicting the Observations

We compute the sensor likelihood $p(\mathbf{o}_k|\tilde{\mathbf{x}}_k)$ using a transposed convolutional neural network (Fig. 2), which augments the dimensionality of the input from $\mathbf{x}$ to the laser-sensor input size (e.g., $2 \rightarrow 622$). The input is firstly fed into two fully connected layers, and each transposed convolution layer is followed by a regular convolution layer, based on the work of [18]. At every layer, we used the ReLU activation function.

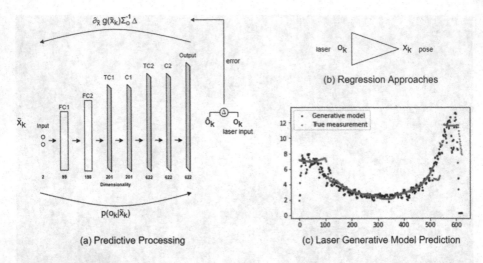

Fig. 2. (a) The Predictive Processing algorithm's architecture. (b) Graphical representation of regression approaches to robot state estimation. (c) Generative model prediction vs. true laser measurement on a test sample.

The network was trained on 13000 normalized random samples collected in the Gazebo simulation. Each sample consists of the true position of the robot and the laser-sensor measurements at that location. The training was performed with 20 batches of 500 samples, using the L1 loss and Adam optimizer [10].

2.4 Navigation

Analogously, our algorithm infers the action in the same way that state estimation is computed, namely performing *active inference*. Actions also minimize the VFE in the predicted observations.

$$\mathbf{a} = \underset{\mathbf{a}}{\operatorname{argmin}} \, F(\tilde{\mathbf{x}}, \mathbf{o}(\mathbf{a})) \tag{2}$$

We define the goal as the preference or the intention of the agent $\tilde{\mathbf{x}}_{goal}$ to arrive to a sensory state o_{goal} [17]. Estimation and control are computed as follows[1]:

$$\tilde{\mathbf{x}}_{k+1} = \tilde{\mathbf{x}}_k + \alpha \left[\partial_{\tilde{\mathbf{x}}} g(\tilde{\mathbf{x}}_k) \Sigma_{\mathbf{o}}^{-1} (\mathbf{o}_k - g(\tilde{\mathbf{x}}_k)) + \partial_{\tilde{\mathbf{x}}} g(\tilde{\mathbf{x}}_k) \Sigma_{\mathbf{x}}^{-1} \beta (\mathbf{o}_k - g(\tilde{\mathbf{x}}_{goal})) \right] \tag{3}$$

$$\mathbf{a}_{k+1} = \mathbf{a}_k + \gamma \partial_a \tilde{\mathbf{x}} \partial_{\tilde{\mathbf{x}}} g(\tilde{\mathbf{x}}_k) \Sigma_{\mathbf{o}}^{-1} (\mathbf{o}_k - g(\tilde{\mathbf{x}}_k)) \tag{4}$$

where β weights the goal attractor and γ is the action step size. Note that each term computes the weighted prediction error mapped to the latent space.

The estimated state $\tilde{\mathbf{x}}$, now biased by the desired goal, generates a new predicted observation at every new iteration that is transformed into an action $\mathbf{a}$,

---

[1] This update equation assumes that the Hessian of the goal dynamics is -1 as proposed in [18].

which minimizes the VFE. Thus, performing a movement in the direction of the goal. The pseudo-code described in Alg. 1 illustrates the process. The algorithm converges when the observation fits the predicted laser sensor measurements.

Algorithm 1: FEP localization and navigation algorithm

$\tilde{\mathbf{x}} \leftarrow$ initial belief;
$\mathbf{o}_{goal} \leftarrow g(\tilde{\mathbf{x}}_{goal})$; // `Generate goal`
while *true* **do**
    $\mathbf{o}_k \leftarrow$ Normalize(laser input);
    $\hat{\mathbf{o}}_k \leftarrow g(\tilde{\mathbf{x}})$; // `Predicted observation`
    $\tilde{\mathbf{x}} \leftarrow$ Eq. 3;
    $\mathbf{a} \leftarrow$ Eq. 4;
    PerformAction(a);
end

---

3   Results

We evaluated our laser-based Active Inference algorithm against a particle filter [14] in the Gazebo simulator, using a commercial mobile robot with a laser rangefinder sensor (TIAGo Base, pmb-2) [2], interfaced with Robot Operating System (ROS) [1]. All experiments were conducted in a designed corridor-like map described in Fig. 1. Localization and navigation results are summarized in Fig. 3.

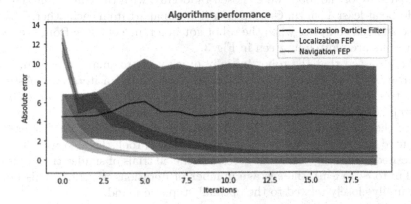

Fig. 3. Localization and navigation evaluation. Mean and standard deviation of the positioning absolute error of our model (FEP) compared against the PF algorithm. Furthermore, our model is able to perform navigation using the same Bayesian filtering framework. The green line shows the mean absolute error to the goal. (Color figure online)

3.1   Localization and Estimation

Firstly, we evaluated the localization accuracy when initializing the robot to a random position in the space when a map was given. Thus, testing absolute positing with laser measurements in static situations. For the particle filter, the particle initial probabilities were randomly spread in the environment and reset at the beginning of every trial. In our algorithm, we initialized the initial belief in the center of the environment. We computed the ground truth positional error in 100 trials for 50 iterations. Our algorithm converged much more consistently to the true state over all trials, whereas the PF struggled to deliver consistent results, as shown in Fig. 3 by the rather large standard deviation in the blue shaded area. It is important to highlight that the PF is tuned for using the robot's odometry. However, for the sake of fair comparison solely laser-sensor values were used as observations.

Secondly, we evaluated the localization performance when traversing the environment from one side to the other, by performing small teleports (to override odometry) to simulate robot movement while keeping the rotation angle constant. Results showed a more stable over time state estimation by our algorithm when compared to the PF, which seemed to suffer from the absence of odometry information.

3.2   Navigation

For the assessment of the navigation algorithm's performance, we ran an experiment consisting of 50 trials in which the robot had to navigate from a starting point to a goal position. The initial belief state was set to the robot's initial true position, to evaluate the performance of navigation without the effects of localization in the first iterations. In every trial, both the initial position and the goal state of the robot were chosen randomly, with the constraint that they should be at least 12 m (in Gazebo coordinates) apart from each other. The task was considered complete when the robot got in a range of 0.8 m from the target. The results are plotted in green in Fig. 3.

We observed that the robot initially quickly approximates the goal, with a big drop in the distance to the goal in the first couple of iterations. As it gets closer to the goal, the "velocity" of the robot (in the experiment described by the step sizes) decreases. This is a result of the diminishing gradient in every step of the algorithm, due to the stochastic gradient descent. Additionally, we computed the average number of iterations that it took the algorithm to get in the desired 0.8m range of the target. Over the 50 trials of similar travel distance ($\sim$ 11 m to $\sim$ 13.5 m), the average number of iterations was 12.5. This number is naturally closely related to the optimal step size found.

4   Conclusions

This work shows a proof-of-concept on how predictive processing, i.e. active inference agents, can perform laser-based localization and navigation tasks. The results obtained in the localization experiment, where we compared our approach against a state-of-the-art alternative (particle filter), show the potential of predictive stochastic neural filtering in robot localization, and estimation in general [7,16]. Furthermore, the navigation experiment showcased how to compute actions as a dual filtering process. Nevertheless, at its current state, the proposed algorithm suffers from a few deficiencies, most of which are related to the learning of the generative model of the world. Besides currently requiring a large dataset for training, the model is prone to mistake very similar objects in the environment, given that the estimation of the new state is independent from the previous. Additionally, because it is a supervised method trained before that the robot can do any navigation, it is unable to cope with changing environments. All in all, the environment used in our experiments is rather simplistic compared to demonstrations of current sota algorithms. Therefore, we foresee further development and experimentation in terms of integration of odometry information, the introduction of extra degrees of freedom and connection to the robot's non-linear dynamics.

References

1. Ros documentation (June 2020). http://wiki.ros.org/
2. Tiago base (October 2020). http://wiki.ros.org/Robots/TIAGo-base
3. Besada-Portas, E., Lopez-Orozco, J.A., Lanillos, P., De la Cruz, J.M.: Localization of non-linearly modeled autonomous mobile robots using out-of-sequence measurements. Sensors **12**(3), 2487–2518 (2012)
4. Clark, A.: Whatever next? predictive brains, situated agents, and the future of cognitive science. Behav. Brain Sci. **36**(3), 181–204 (2013)
5. Fox, V., Hightower, J., Liao, L., Schulz, D., Borriello, G.: Bayesian filtering for location estimation. IEEE Pervasive Comput. **2**(3), 24–33 (2003). https://doi.org/10.1109/MPRV.2003.1228524
6. Friston, K.: The free-energy principle: a unified brain theory? Nat. Rev. Neurosci. **11**(2), 127–138 (2010)
7. Friston, K.J., Trujillo-Barreto, N., Daunizeau, J.: Dem: a variational treatment of dynamic systems. Neuroimage **41**(3), 849–885 (2008)
8. Gebauer, C., Bennewitz, M.: The pitfall of more powerful autoencoders in lidar-based navigation. arXiv preprint arXiv:2102.02127 (2021)
9. Kendall, A., Grimes, M., Cipolla, R.: Posenet: a convolutional network for real-time 6-dof camera relocalization. In: Proceedings of the IEEE International Conference on Computer Vision, pp. 2938–2946 (2015)
10. Kingma, D.P., Ba, J.: Adam: A method for stochastic optimization (2017)
11. Lanillos, P.: Minimum time search of moving targets in uncertain environments. Ph.D. thesis, PhD thesis (2013)
12. Lanillos, P., Cheng, G.: Adaptive robot body learning and estimation through predictive coding. In: 2018 IEEE/RSJ International Conference on Intelligent Robots and Systems (IROS), pp. 4083–4090. IEEE (2018)
13. Lanillos, P., van Gerven, M.: Neuroscience-inspired perception-action in robotics: applying active inference for state estimation, control and self-perception. arXiv preprint arXiv:2105.04261 (2021)

14. Liu, B., Cheng, S., Shi, Y.: Particle Filter optimization: a brief Introduction. In: Tan, Y., Shi, Y., Niu, B. (eds.) Advances in Swarm Intelligence. ICSI 2016. Lecture Notes in Computer Science, vol. 9712, pp. 95–104. Springer, Cham (2016). https://doi.org/10.1007/978-3-319-41000-5_10

15. Meo, C., Lanillos, P.: Multimodal vae active inference controller. arXiv preprint arXiv:2103.04412 (2021)

16. Millidge, B., Tschantz, A., Seth, A., Buckley, C.: Neural kalman filtering. arXiv preprint arXiv:2102.10021 (2021)

17. Oliver, G., Lanillos, P., Cheng, G.: An empirical study of active inference on a humanoid robot. IEEE Transactions on Cognitive and Developmental Systems (2021)

18. Sancaktar, C., van Gerven, M.A.J., Lanillos, P.: End-to-end pixel-based deep active inference for body perception and action. In: 2020 Joint IEEE 10th International Conference on Development and Learning and Epigenetic Robotics (ICDL-EpiRob) (October 2020). https://doi.org/10.1109/icdl-epirob48136.2020.9278105, http://dx.doi.org/10.1109/ICDL-EpiRob48136.2020.9278105

19. Sattler, T., Zhou, Q., Pollefeys, M., Leal-Taixe, L.: Understanding the limitations of CNN-based absolute camera pose regression. In: Proceedings of the IEEE/CVF Conference on Computer Vision and Pattern Recognition, pp. 3302–3312 (2019)

20. Thrun, S.: Simultaneous localization and mapping. In: Jefferies, M.E., Yeap, W.K. (eds.) Robotics and Cognitive Approaches to Spatial Mapping. Springer Tracts in Advanced Robotics, vol. 38, pp. 13–41. Springer, Heidelberg (2007). https://doi.org/10.1007/978-3-540-75388-9_3

21. Çatal, O., Wauthier, S., Verbelen, T., Boom, C.D., Dhoedt, B.: Deep active inference for autonomous robot navigation (2020)

Sensorimotor Visual Perception on Embodied System Using Free Energy Principle

Kanako Esaki$^{(\boxtimes)}$ iD, Tadayuki Matsumura, Kiyoto Ito iD, and Hiroyuki Mizuno iD

Research & Development Group, Hitachi, Ltd., 1-280, Higashi-koigakubo, Kokubunji-shi, Tokyo 185-8601, Japan
`kanako.esaki.oa@hitachi.com`

Abstract. We propose an embodied system that is based on the free energy principle (FEP) for sensorimotor visual perception (SMVP). Although the FEP mathematically describes the rule that living things obey, limitation by embodiment is required to model SMVP. The proposed system is configured by a body, which partially observes the environment, and memory, which retains classified knowledge about the environment as a generative model, and executes active and perceptual inferences. Evaluation using the MNIST dataset showed that the proposed system recognizes characters by active and perceptual inferences, and the intentionality corresponding to human confirmation bias is reproduced on the system.

Keywords: Free energy · Embodiment · Sensorimotor contingency

1 Introduction

The human visual field seems to cover a wide area of the surrounding environment, but the range with high enough resolution to identify details is limited to only the central visual field of about 5° [1]. Since the human visual field has this spatial limitation, gazing-position movement is required to see the environment. Sensorimotor contingency (SMC) theory [2, 3] changes the interpretation of "seeing" [4–9] by including this movement in it. SMC theory explains that "seeing" is knowing about things to do rather than making an internal representation [10]. This means that human sensorimotor visual perception (SMVP) includes moving the gazing position using the inference of the environmental state, not the sensory input itself, to understand the environment [11]. The spatial limitation of the human visual field within this context is not a "limitation" but a "trigger" for an action that moves the gazing position [12].

Many machine-learning methods have been proposed that incorporate human characteristics. Spatial limitations, such as that described above, are treated as constraints called partial observation in the context of reinforcement learning [13–15]. These studies evaluated the performance degradation of classification and regression through partial observation. Various methods similar to SMVP have also been proposed. Auto regressive models [16–18] predict the entire image by repeating the action of obtaining a partial image. Algorithms for reinforcement learning [19, 20] generate exploring actions covering a wide range of the action space. These models and algorithms generate the

© Springer Nature Switzerland AG 2021
M. Kamp et al. (Eds.): ECML PKDD 2021 Workshops, CCIS 1524, pp. 865–877, 2021.
https://doi.org/10.1007/978-3-030-93736-2_62

actions on the basis of sensory inputs such as partial images and observations. The active vision algorithms infer hidden states from partial observations and finally understand the whole scene, but the hidden state space is constructed in terms of reconfigurabilty [21], or actions are selected on a basis of image features instead of hidden states [22–25]. None of the above studies have used the inference of environmental states to generate action, which is the essence of SMVP.

The purpose of this study is to achieve SMVP using the free energy principle (FEP) [26–30]. The FEP mathematically describes the rule that living things obey. The free energy in the FEP measures the difference between the probability distribution of environmental states that act on a biological system and an approximate posterior distribution of environmental states encoded by the configuration of that system. The biological system minimizes the free energy by changing its configuration to affect the way it samples the environment or by changing the approximate posterior distribution it encodes. These two changes correspond to "active inference" and "perceptual inference" of environmental states, respectively.

Although the FEP mathematically describes active and perceptual inferences on a biological system, limitation by embodiment of such a system to trigger action is required to model SMVP. Embodiment [31–34] provides an interaction between the biological system and environment, resulting in partial sensory inputs and actions. Their causal relationship is condensed to be stored in the embodied biological system.

We propose an embodied system that is based on the FEP to achieve SMVP. The proposed system is configured by a body, which partially observes the environment, and memory, which retains condensed knowledge about the environment. Evaluation using the MNIST dataset [35] showed that the proposed system triggers an action that moves a gazing position and repeatedly executes active and perceptual inferences by following the FEP. Moreover, the intentionality is reproduced on the proposed system, producing an equivalent of human confirmation bias. We discuss how important this bias is for taking the action in an unknown environment.

2 Sensorimotor Visual Perception

The problem settings shown in Fig. 1 are designed to list the components necessary for SMVP. Let us consider a situation in which a target object, e.g., the number 5, exists in the environment. This situation is called an environmental state x_t. The vision sensor takes on the role of the human eye and has a spatial limitation of the visual field. This limitation leads to a change in direction of the vision sensor to understand the environment. The vision sensor obtains an image of a specific region of the environment each time its direction is determined. A representative position of the region is defined as an attention position, which equals the gazing position of the human eye. The image of the region is defined as an attention image. The attention image at each time ($T = t - 2, t - 1, t$) is obtained by the previous actions $a_{t-3}, a_{t-2}, a_{t-1}$ that move the attention position. A composition image of the attention images obtained from $T = 0$ to $T = t$ is used as a sensory input (hereafter called sensory input image s_t). In our problem setting described above, SMVP is defined to infer the environment states by repeating the actions to obtain sensory input images.

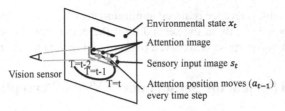

Fig. 1. Overview of problem setting to consider sensorimotor visual perception

3 Free Energy Principle

To apply the FEP to our problem setting, we make certain assumptions. The s_t is determined by only the a_{t-1} and the current x_t. The x_t does not change ($x_{t-1} = x_t = x_{t+1}$).

Under the above assumptions, the variational and expected free energy of the FEP, which are necessary components to describe the perceptual and active inferences, are expressed as follows. The variational free energy $F\left(\phi_{x_t}, a_{t-1}\right)$ is expressed as

$$F\left(\phi_{x_t}, a_{t-1}\right) = E_{q\left(x_t \mid \phi_{x_t}\right)}\left[\ln q\left(x_t \mid \phi_{x_t}\right) - \ln p_{a_{t-1}}(x_t, s_t)\right]$$
$$= D_{KL}\left[q\left(x_t \mid \phi_{x_t}\right) \| p_{a_{t-1}}(x_t \mid s_t)\right] - \ln p_{a_{t-1}}(s_t), \qquad (1)$$

where $q\left(x_t \mid \phi_{x_t}\right)$ is the approximate posterior distribution of x_t, ϕ_{x_t} is the sufficient statistics of $q(x_t)$, and $p_{a_{t-1}}(x_t, s_t)$ is a generative model that stores the causal relationship of x_t and s_t under a_{t-1}. Since a_{t-1} specifies the prior distribution of s_t, it is treated as a parameter of the generative model. Perceptual inference is aimed at minimizing $F\left(\phi_{x_t}, a_{t-1}\right)$ by changing $q\left(x_t \mid \phi_{x_t}\right)$. Since the second term of Eq. 1 is composed of a_{t-1} and s_t, which are fixed at the current time t, the purpose is achieved by changing ϕ_{x_t} to become $q\left(x_t \mid \phi_{x_t}\right) \sim p_{a_{t-1}}(x_t \mid s_t)$. The expected free energy $G\left(\phi_{x_{t+1}}, a_t\right)$, on the other hand, is expressed as

$$G\left(\phi_{x_{t+1}}, a_t\right) = E_{q\left(s_{t+1} \mid x_{t+1}, \phi_{x_{t+1}}\right)} F\left(\phi_{x_{t+1}}, a_t\right)$$
$$= E_{q\left(s_{t+1} \mid x_{t+1}, \phi_{x_{t+1}}\right)} E_{q\left(x_{t+1} \mid \phi_{x_{t+1}}\right)}\left[\ln q\left(x_{t+1} \mid \phi_{x_{t+1}}\right) - \ln p_{a_t}(x_{t+1}, s_{t+1})\right]$$
$$= E_{q\left(s_{t+1} \mid x_{t+1}, \phi_{x_{t+1}}\right)}\left[E_{q\left(x_{t+1} \mid \phi_{x_{t+1}}\right)}\left[\ln q\left(x_{t+1} \mid \phi_{x_{t+1}}\right) - \ln p(x_{t+1})\right]\right.$$
$$\left. - E_{q\left(x_{t+1} \mid \phi_{x_{t+1}}\right)} \ln p_{a_t}(s_{t+1} \mid x_{t+1})\right]$$
$$= E_{q\left(s_{t+1} \mid x_{t+1}, \phi_{x_{t+1}}\right)} D_{KL}\left[q\left(x_{t+1} \mid \phi_{x_{t+1}}\right) \| p(x_{t+1})\right]$$
$$+ E_{q\left(s_{t+1} \mid x_{t+1}, \phi_{x_{t+1}}\right)}\left[-E_{q\left(x_{t+1} \mid \phi_{x_{t+1}}\right)} \ln p_{a_t}(s_{t+1} \mid x_{t+1})\right], \qquad (2)$$

where $p_{a_t}(x_{t+1}, s_{t+1})$ is factorized into $p_{a_t}(s_{t+1} \mid x_{t+1})p(x_{t+1})$ since x_{t+1} and a_t are independent. Since s_{t+1} varies with a_t but x_{t+1} does not in our problem setting, $p_{a_t}(x_{t+1}, s_{t+1})$ is factorized into the terms with and without s_{t+1}, unlike the general active inference literature [29]. Active inference is aimed at minimizing $G(\phi_{x_{t+1}}, a_t)$ by changing action a_t. Since a_t is included only in the second term (hereafter called uncertainty) and the first term is fixed, the purpose is achieved by minimizing the uncertainty.

4 Embodied System for Sensorimotor Visual Perception

Limitation by an embodiment is key to achieving SMVP based on the FEP. The proposed embodied system is configured by a body and memory. The body has an ocular motor system for controlling the attention position. In our problem setting shown in Fig. 1, the vision sensor, which has a spatial limitation of the visual field, is the body. The body thus can only observe a partial area of the environment. The memory, on the other hand, is not a photographic memory, where x_t is observed uniformly as if it were photographed. Rather, it is a generative model that contains classified prior knowledge about the causal relationship of an s_t and x_t to be stored in a limited capacity. By limiting body and memory abilities and operating in accordance with the FEP, the proposed embodied system repeatedly executes perceptual and active inferences, as shown in Fig. 2. In perceptual inference, an s_t is generated and input to the generative model $p_{a_{t-1}}(x_t, s_t)$ to calculate an approximate posterior. In active inference, the uncertainty is calculated using the approximate posterior and used to select an attention position.

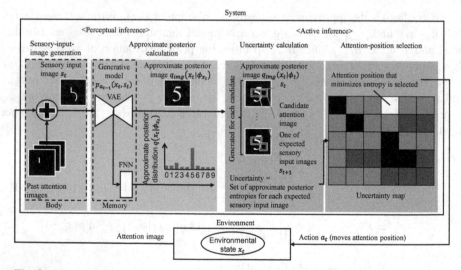

Fig. 2. Processing flow of sensorimotor visual perception with proposed embodied system

4.1 Perceptual Inference

The generation of s_t involves a current attention image obtained from the environment and the past attention images obtained from the initial time to the current time t. The attention images are composed while maintaining their relative attention positions.

The generative model $p_{a_{t-1}}(x_t, s_t)$ is implemented with a combination of a variational autoencoder (VAE) and a fully connected neural network (FNN), which is inspired by the auxiliary loss in GoogleNet [36]. The VAE contains prior knowledge about the environment, and the FNN classifies it. The combination of these is pre-trained, which is equivalent to changing ϕ_{x_t} to become $q(x_t|\phi_{x_t}) \sim p_{a_{t-1}}(x_t|s_t)$ in Eq. 1. An approximate posterior distribution $q(x_t|\phi_{x_t})$ and approximate posterior image $q_{img}(x_t|\phi_{x_t})$, which is a conversion of $q(x_t|\phi_{x_t})$ into an image format, are calculated by inputting the s_t to the combination of the VAE and FNN.

4.2 Active Inference

Calculation of uncertainty uses $q_{img}(x_t|\phi_{x_t})$. Uncertainty is the expected information amount of $p_{a_t}(s_{t+1}|x_{t+1})$, which is the probability distribution of the set of expected sensory input images s_{t+1} conditioned by x_{t+1} with action a_t as a parameter. Conditioning by $x_{t+1}(= x_t)$ is interpreted as extracting s_{t+1} from $q_{img}(x_t|\phi_{x_t})$. Each one of s_{t+1} is composed of the current s_t and candidate attention image surrounding it. All the s_{t+1} are extracted from one single $q_{img}(x_t|\phi_{x_t})$. Parameterizing a_t is interpreted as assuming the candidate attention images. Since the action space is continuous in the temporal direction under limited body, candidate images should be limited to the region surrounding s_t. Under these interpretations, the information amount of $p_{a_t}(s_{t+1}|x_{t+1})$ is calculated as that of s_{t+1}. The information amount of s_{t+1} is the entropy of the approximate posterior distribution calculated by inputting each one of s_{t+1} to $p_{a_{t-1}}(x_t, s_t)$. Since s_{t+1} is deterministically calculated from one $q_{img}(x_t|\phi_{x_t})$, the expected-value calculation is not required. Uncertainty is thus the set of the entropies for each one of s_{t+1}.

In selecting the attention position, an uncertainty map is generated where the entropy for each one of s_{t+1} corresponds to each attention position. An attention position that minimizes the entropy is selected in accordance with the uncertainty map. The a_t moves the attention position to the selected one.

5 Evaluation and Discussion

We evaluated the proposed embodied system for SMVP in a character-recognition task with the MNIST dataset. The key components of the embodiment, body and memory, were implemented as follows: the body function was achieved by setting the size of the attention image to be sufficiently small compared to the MNIST characters, and the memory (the generative model) was implemented using convolutional VAE [37, 38] that was robust against displacement. The size of both the s_t and approximate posterior image $q_{img}(x_t|\phi_{x_t})$ was 28×28 pixels, while that of the attention image was 6×6 pixels. Training of the generative model involved 60,000 images of the MNIST dataset. The stochastic variables x_t of the approximate posterior distribution $q(x_t|\phi_{x_t})$ were labels 0–9 of the MNIST dataset.

Perceptual inference was executed during the attention repetitions. Figure 3 shows the transition of $q(x_t|\phi_{x_t})$ and $q_{img}(x_t|\phi_{x_t})$ from the 1st to 20th attentions when target characters are "0" and "3". Symbols A, B, C, and D in Fig. 3 (a) are described later. Each graph plots the $q(x_t|\phi_{x_t})$ and each image is the corresponding $q_{img}(x_t|\phi_{x_t})$. The probability of the characters different from the target characters changed to the maximum. This situation is analogous to a human temporarily labeling the environmental state so that they could identify the environment on the basis of the sensory input they had collected and their past experience. For character "0", the probability of the target character was maximum after the 13th attention, and "0" appeared in the $q_{img}(x_t|\phi_{x_t})$. For character "3", however, the probability of "2" was maximum even at the 20th attention, and the $q_{img}(x_t|\phi_{x_t})$ contained a part of "3" not the entire "3".

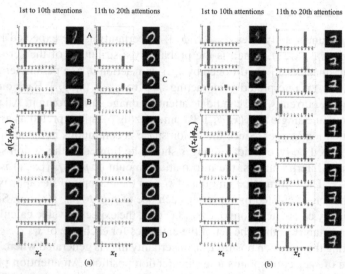

Fig. 3. Transition of approximate posterior distributions and images along with attention repetitions. Subfigures (a) and (b) correspond to making attention at characters "0" and "3", respectively.

Active inference was executed during attention repetitions. Figure 4 shows the uncertainty map at the 1st, 4th, 13th, and 20th attentions (corresponding to A, B, C, and D in Fig. 3(a)) when the target character is "0". The vertical (x) and horizontal (y) axes of the maps indicate the representative position of the candidate attention image. The uncertainty of the 1st and 20th attentions were biased toward the minimum because the change in uncertainty is similar for any next action. In the early stage of attention repetition, any next action will be useful for inference since little environmental information has been collected. In the final stage, the probability of a particular number has been high, and the entropy of $q(x_{t+1}|\phi_{x_{t+1}})$ is low with any next action. The uncertainty of the 4th and 13th attentions were dispersed because any of the next actions has the potential to specify a number.

The proposed embodied system executed SMVP in which the attention position moves on the basis of the inference of x_t. Figure 5 shows the transition of attention images, s_t, and $q_{img}(x_t|\phi_{x_t})$ from the 1st to 20th attentions. Each subset of three rows corresponds to the target character from "0" to "9". The 1st, 2nd, and 3rd rows show the attention images, s_t, and $q_{img}(x_t|\phi_{x_t})$, respectively. Before the 9th attention of target character "2", a form like "7" was inferred and the attention position moved up and down, but then a form like "2" was inferred and the attention position moved to the right. Before the 15th attention of target character "5", a form like "6" was inferred and the attention position moved around the bottom of x_t, but then a form like "5" was inferred and the attention position moved to the upper right.

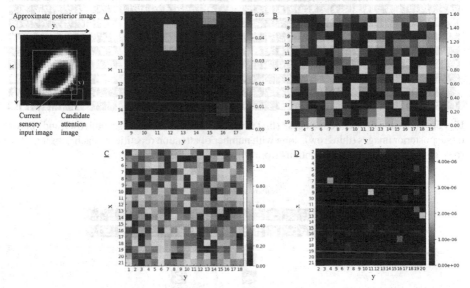

Fig. 4. Uncertainty map when making attention at "0". Subfigures A, B, C, and D correspond to 1st, 4th, 13th, and 20th attentions, respectively.

To analyze the case of character "3", the initial attention position was changed. Figure 6 shows the transition of the attention image, s_t, $q(x_t|\phi_{x_t})$, and $q_{img}(x_t|\phi_{x_t})$ from the 1st to 20th attentions when the initial attention position was different from those in Figs. 3 and 5 for character "3". Different from Fig. 3(b), the probability of "3" reached maximum, and the $q_{img}(x_t|\phi_{x_t})$ contained the entire "3" at the 20th attention. This result suggests that the proposed system has an intentionality similar to human confirmation bias that depends on what is obtained from the environment and prior knowledge about it. Although human confirmation bias has a negative impact on decision-making in various fields [39], it has the advantage of adaptability to unknown environments. In most practical cases of perception problems, incomplete models of the environment are provided. In these cases, humans take the next action on the basis of the confirmation bias. Taking the next exploring action enables the environmental information to be obtained. We believe that our results will help solve the difficult problem of triggering action in an unknown environment.

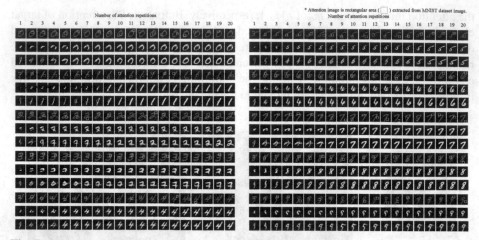

Fig. 5. Transition of attention images* (first row), sensory input images (second row), and approximate posterior images (third row) along with number of attention repetitions. Each subset of three rows corresponds to making attentions from "0" to "9".

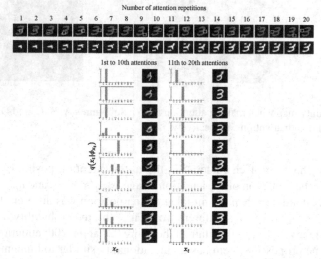

Fig. 6. Transition of attention images, sensory input images, and approximate posterior. Changes in initial attention position when making attention at "3" provides different transition

6 Conclusion

We proposed an embodied system that is based on the free energy principle (FEP) for sensorimotor visual perception (SMVP). The proposed embodied system is configured by a body and memory. By limiting body and memory abilities and operating in accordance with the FEP, the proposed system triggers an action that moves an attention position and repeatedly executes perceptual and active inferences. During the evaluation involving a character-recognition task using the MNIST dataset, as the attention was repeated, the uncertainty of the characters decreased. The probability of the correct character finally became the highest among the characters. It was thus confirmed that the proposed system greatly contributes to achieving SMVP. Moreover, changing the initial attention position provides a different final inference, suggesting that the proposed system has a confirmation bias similar to humans. We believe that these results will help solve the difficult problem of triggering action in an unknown environment.

Acknowledgements. The authors thank Dr. Qinghua Sun from Hitachi Ltd. for his constructive comments and suggestions for improving this paper.

Appendix

The generative model, described in this paper, is a combination of a variational autoencoder (VAE) and a fully connected neural network (FNN). The architecture is shown in Fig. 7. The encoder consists of four 2D convolutional layers and each layer is followed by a batch normalization and a rectified linear unit. The bottleneck consists of two linear transformation layers for calculating the average and the variance with reparameterizing function. The decoder consists of four 2D transposed convolutional layers and each layer is followed by a batch normalization and a rectified linear unit (sigmoid unit for the last layer). The classifier consists of a linear transformation layer followed by a rectified linear unit and a linear transformation layer followed by a softmax unit. The model was trained using Adam optimizer (learning rate: 0.001) with the sum of VAE loss and FNN loss.

Algorithm 1 shows the pseudo code of the processing flow. The $p_{a_{t-1}}(s_t, x_t)$ is pretrained using training data of (s_t, x_t). All the training data of s_t are pre-processed so that the center of gravity of an image is shifted to the center position. During the operation of the proposed embodied system, the process from the 2nd line to the 12th line is repeated. First, an attention image s_t' is obtained from the vision sensor. The past sensory input images are composed with the obtained s_t' while maintaining each relative attention position. The center of gravity of the composed image is calculated, and the composed image is shifted so that the center of gravity is located at the center position of the image. The shifted composed image is an s_t. Then, $q(x_t|\phi_{x_t})$ is calculated by inputting s_t to $p_{a_{t-1}}(s_t, x_t)$. After that, the sub-function starting from the 14th line is called to generate expected sensory input images s_{t+1}. In the sub-function, an $q_{img}(x_t|\phi_{x_t})$ is calculated by inputting s_t to $p_{a_{t-1}}(s_t, x_t)$. A template image is generated by detecting a bounding rectangle area of non-zero pixels in s_t and extracting the area from s_t. Template matching is carried out in the $q_{img}(x_t|\phi_{x_t})$, and the representative position of the current s_t, u_{cur}, is obtained. To calculate the next candidate attention positions u_{next}, a candidate region of u_{next} is set. The candidate region is a region obtained by adding a fixed margin pixel

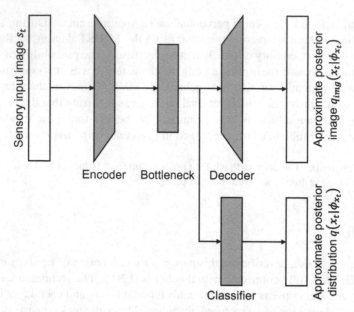

Fig. 7. Architecture of generative model

to a region of s_t in $q_{img}(x_t|\phi_{x_t})$. The region of s_t is defined by u_{cur} and the size of the template image. The u_{next} are calculated by sliding the window with the fixed stride pixel in the candidate region. The window is the size of s_t'. The representative positions of all the window positions during sliding are u_{next}. The s_{t+1} are generated by extracting the region of the s_t and the region of the next candidate attention images s_{t+1}' from $q_{img}(x_t|\phi_{x_t})$. The region of the s_t is defined by u_{cur} and the size of the template image, as mentioned above. The region of the s_{t+1}' are defined by u_{next} and the size of s_{t+1}'. The extracted images are clipped or applied with zero-padding to have the same size as s_t. Each approximate posterior distribution $q(x_{t+1}|\phi_{x_{t+1}})$ is calculated by inputting each image included in s_{t+1} to $p_{a_{t-1}}(s_t, x_t)$. Note that $q(x_t|\phi_{x_t})$ is calculated using the current s_t, while $q(x_{t+1}|\phi_{x_{t+1}})$ is calculated using s_{t+1}. The entropy of each $q(x_{t+1}|\phi_{x_{t+1}})$ is calculated and added to the uncertainty map M. Finally, the attention position having the minimum value in M is defined as the next attention position.

	Algorithm 1 Pseudo code of processing flow
1:	while system is operating do:
2:	obtain attention image s'_t
3:	compose past sensory input images with s'_t
4:	generate sensory input image s_t by shifting gravity point to center of composed image
5:	calculate approximate posterior distribution $q(x_t\|\phi_{x_t})$ using s_t and generative model $p_{a_{t-1}}(s_t, x_t)$
6:	call generate_expected_sensory_input_image
7:	for expected sensory input images s_{t+1}:
8:	calculate approximate posterior distribution $q(x_{t+1}\|\phi_{x_{t+1}})$ using $s_{t+1}[index]$ and $p_{a_{t-1}}(s_t, x_t)$
9:	calculate entropy of $q(x_{t+1}\|\phi_{x_{t+1}})$
10:	add entropy to uncertainty map M
11:	end for
12:	set attention position using M
13:	end while
14:	generate_expected_sensory_input_image
15:	generate approximate posterior image $q_{img}(x_t\|\phi_{x_t})$ using s_t and $p_{a_{t-1}}(s_t, x_t)$
16:	generate template from s_t
17:	obtain current sensory input position u_{cur} by template matching in $q_{img}(x_t\|\phi_{x_t})$
18:	calculate next candidate attention positions u_{next} using u_{cur}
19:	generate s_{t+1} using $q_{img}(x_t\|\phi_{x_t})$, u_{cur}, and u_{next}
20:	return with s_{t+1}

References

1. Mandelbaum, J., Sloan, L.L.: Peripheral visual acuity*: with special reference to scotopic illumination. Am. J. Ophthalmol. **30**(5), 581–588 (1947)
2. O'Regan, J.K., Noë, A.: A sensorimotor account of vision and visual consciousness. Behav. Brain Sci. **24**(5), 939–973 (2001)
3. Seth, A.K.: The cybernetic Bayesian brain: from interoceptive inference to sensorimotor contingencies. In: Open MIND, vol. 35 (2015)
4. Land, M.F.: Eye movements and the control of actions in everyday life. Prog. Retin. Eye Res. **25**(3), 296–324 (2006)
5. Friston, K., Kiebel, S.: Predictive coding under the free-energy principle. Philos. Trans. R. Soc. B Biol. Sci. **364**(1521), 1211–1221 (2009)
6. Seth, A.K., Suzuki, K., Critchley, H.D.: An interoceptive predictive coding model of conscious presence. Front. Psychol. **2**, 395 (2012)
7. Adams, R.A., Shipp, S., Friston, K.J.: Predictions not commands: active inference in the motor system. Brain Struct. Funct. **218**(3), 611–643 (2013)
8. Bogacz, R.: A tutorial on the free-energy framework for modelling perception and learning. J. Math. Psychol. **76**, 198–211 (2017). https://doi.org/10.1016/j.jmp.2015.11.003

9. Lotter, W., Kreiman, G., Cox, D.: Deep predictive coding networks for video prediction and unsupervised learning. In: 5th International Conference on Learning Representations, Toulon (2017)

10. O'Regan, J. K.: Experience is not something we feel but something we do: a principled way of explaining sensory phenomenology, with Change Blindness and other empirical consequences. http://nivea.psycho.univ-paris5.fr/ASSChtml/Pacherie4.html. Accessed 27 Aug 2021

11. Parr, T., Sajid, N., Da Costa, L., Mirza, M.B., Friston, K.J.: Generative models for active vision. Front. Neurorobot. **15**, 34 (2021)

12. Tang, Y., Nguyen, D., Ha, D.: Neuroevolution of self-interpretable agents. In: Proceedings of the 2020 Genetic and Evolutionary Computation Conference, Cancún, pp. 414–424. Association for Computing Machinery (2020)

13. Pineau, J., Gordon, G., Thrun, S.: Point-based value iteration: an anytime algorithm for POMDPs. In: Proceedings of the 18th International Joint Conference on Artificial Intelligence, Acapulco, pp. 1025–1030. Morgan Kaufmann Publishers Inc. (2003)

14. Ji, S., Parr, R., Li, H., Liao, X., and Carin, L.: Point-based policy iteration. In: Proceedings of the 22nd National Conference on Artificial Intelligence, Vancouver, vol. 2, pp. 1243–1249. AAAI Press (2007)

15. Silver, D., Veness, J.: Monte-Carlo planning in large POMDPs. Adv. Neural. Inf. Process. Syst. **23**, 2164–2172 (2010)

16. Gregor, K., Danihelka, I., Graves, A., Rezende, D. J., Wierstra, D.: DRAW: a recurrent neural network for image generation. In: Proceedings of the 32nd International Conference on Machine Learning, Lille, pp. 1462–1471. JMLR.org (2015)

17. Oord, A. V., Kalchbrenner, N., Kavukcuoglu, K.: Pixel recurrent neural networks. In: Proceedings of the 33rd International Conference on Machine Learning, New York, pp. 1747–1756. JMLR.org (2016)

18. Salimans, T., Karpathy, A., Chen, X., Kingma, D. P.: PixelCNN++: improving the pixelCNN with discretized logistic mixture likelihood and other modifications. In: 5th International Conference on Learning Representations, Toulon (2017)

19. Oh, J., Guo, X., Lee, H., Lewis, R., Singh, S.: Action-conditional video prediction using deep networks in Atari games. Adv. Neural. Inf. Process. Syst. **28**, 2863–2871 (2015)

20. Houthooft, R., Chen, X., Duan, Y., Schulman, J., De Turck, F., Abbeel, P.: VIME: variational information maximizing exploration. Adv. Neural. Inf. Process. Syst. **29**, 1117–1125 (2016)

21. van der Himst, O., Lanillos, P.: Deep active inference for partially observable MDPs. In: Verbelen, T., Lanillos, P., Buckley, C.L., De Boom, C. (eds.) International Workshop on Active Inference 2020. Communications in Computer and Information Science, vol. 1326, pp. 61–71. Springer, Cham (2020). https://doi.org/10.1007/978-3-030-64919-7_8

22. Daucé, E., Perrinet, L.: Visual search as active inference. In: Verbelen, T., Lanillos, P., Buckley, C.L., De Boom, C. (eds.) International Workshop on Active Inference 2020. Communications in Computer and Information Science, vol. 1326, pp. 165–178. Springer, Cham (2020). https://doi.org/10.1007/978-3-030-64919-7_17

23. Friston, K., Adams, R.A., Perrinet, L., Breakspear, M.: Perceptions as hypotheses: saccades as experiments. Front. Psychol. **3**, 151 (2012)

24. Mirza, M.B., Adams, R.A., Mathys, C.D., Friston, K.J.: Scene construction, visual foraging, and active inference. Front. Comput. Neurosci. **10**, 56 (2016)

25. Heins, R.C., Mirza, M.B., Parr, T., Friston, K., Kagan, I., Pooresmaeili, A.: Deep active inference and scene construction. Front. Artif. Intell. **3**, 81 (2020)

26. Friston, K., Kilner, J., Harrison, L.: A free energy principle for the brain. J. Physiol. Paris **100**(1–3), 70–87 (2006)

27. Friston, K.: The free-energy principle: a unified brain theory? Nat. Rev. Neurosci. **11**, 127–138 (2010)

28. McGregor, S., Baltieri, M., Buckley, C.L.: A minimal active inference agent. arXiv preprint arXiv:1503.04187 (2015)
29. Friston, K., FitzGerald, T., Rigoli, F., Schwartenbeck, P., Pezzulo, G.: Active inference: a process theory. Neural Comput. **29**(1), 1–49 (2017)
30. Buckley, C.L., Kim, C.S., McGregor, S., Seth, A.K.: The free energy principle for action and perception: a mathematical review. J. Math. Psychol. **81**, 55–79 (2017)
31. Fitzpatrick, P., Metta, G., Natale, L., Rao, S., Sandini, G.: Learning about objects through action - initial steps towards artificial cognition. In: 2003 IEEE International Conference on Robotics and Automation, Taipei, pp. 3140–3145. IEEE (2003)
32. Cheng, G., et al.: CB: a humanoid research platform for exploring neuroscience. In: 2006 6th IEEE-RAS International Conference on Humanoid Robots, Genova, pp. 182–187. IEEE (2006)
33. Friston, K.: Embodied inference: or "I think therefore I am, if I am what I think". In: Tschacher, W., Bergomi, C. (eds.) The Implications of Embodiment: Cognition and Communication, pp. 89–125. Imprint Academic (2011)
34. Gallagher, S., Allen, M.: Active inference, enactivism and the hermeneutics of social cognition. Synthese **195**(6), 2627–2648 (2016)
35. THE MNIST DATABASE of handwritten digits. http://yann.lecun.com/exdb/mnist/. Accessed 27 Aug 2021
36. Szegedy, C., et al.: Going deeper with convolutions. In: 2015 IEEE Conference on Computer Vision and Pattern Recognition, Boston, pp. 1–9. IEEE (2015)
37. Kingma, D.P., Welling, M.: Auto-encoding variational bayes. In: 2nd International Conference on Learning Representations, Banff (2014)
38. Rezende, D. J., Mohamed, S., Wierstra, D.: Stochastic backpropagation and approximate inference in deep generative models. In: Proceedings of the 31st International Conference on Machine Learning, Beijing, pp. 1278–1286. JMLR.org (2014)
39. Kappes, A., Harvey, A.H., Lohrenz, T., Montague, P.R., Sharot, T.: Confirmation bias in the utilization of others' opinion strength. Nat. Neurosci. **23**, 130–137 (2020)

Author Index

Printed in the United States
by Baker & Taylor Publisher Services